AF616376

EMS Series of Congress Reports

EMS Congress Reports publishes volumes originating from conferences or seminars focusing on any field of pure or applied mathematics. The individual volumes include an introduction into their subject and review of the contributions in this context. Articles are required to undergo a refereeing process and are accepted only if they contain a survey or significant results not published elsewhere in the literature.

Previously published:

Trends in Representation Theory of Algebras and Related Topics, *Andrzej Skowroński (ed.)*
K-Theory and Noncommutative Geometry, *Guillermo Cortiñas et al. (eds.)*
Classification of Algebraic Varieties, *Carel Faber, Gerard van der Geer and Eduard Looijenga (eds.)*
Surveys in Stochastic Processes, *Jochen Blath, Peter Imkeller and Sylvie Rœlly (eds.)*
Representations of Algebras and Related Topics, *Andrzej Skowroński and Kunio Yamagata (eds.)*
Contributions to Algebraic Geometry. Impanga Lecture Notes, *Piotr Pragacz (ed.)*
Geometry and Arithmetic, *Carel Faber, Gavril Farkas and Robin de Jong (eds.)*
Derived Categories in Algebraic Geometry. Toyko 2011, *Yujiro Kawamata (ed.)*
Advances in Representation Theory of Algebras, *David J. Benson, Henning Krause and Andrzej Skowroński (eds.)*
Valuation Theory in Interaction, *Antonio Campillo, Franz-Viktor Kuhlmann and Bernard Teissier (eds.)*
Representation Theory – Current Trends and Perspectives, *Henning Krause, Peter Littelmann, Gunter Malle, Karl-Hermann Neeb and Christoph Schweigert (eds.)*

Functional Analysis and Operator Theory for Quantum Physics

The Pavel Exner Anniversary Volume

Jaroslav Dittrich
Hynek Kovařík
Ari Laptev

Editors

Editors:

Dr. Jaroslav Dittrich
Department of Theoretical Physics
Nuclear Physics Institute
Czech Academy of Sciences
250 68 Řež
Czech Republic

Email: dittrich@ujf.cas.cz

Prof. Hynek Kovařík
DICATAM – Sezione di Matematica
Università degli Studi di Brescia
Via Branze 38
25123 Brescia
Italy

Email: hynek.kovarik@unibs.it

Prof. Ari Laptev
Department of Mathematics
Imperial College London
Huxley Building, 180 Queen's Gate
London SW7 2AZ
UK

Email: a.laptev@imperial.ac.uk

and

Institut Mittag-Leffler
Auravägen 17
182 60 Djursholm
Sweden

Email: laptev@mittag-leffler.se

2010 Mathematics Subject Classification: primary: 81Q37, 81Q35, 35P15, 35P25.

Key words: Schrödinger operators, point interactions, metric graphs, quantum waveguides, eigenvalue estimates, operator-valued functions, Cayley–Hamilton theorem, adiabatic theorem.

ISBN 978-3-03719-175-0

The Swiss National Library lists this publication in The Swiss Book, the Swiss national bibliography, and the detailed bibliographic data are available on the Internet at http://www.helveticat.ch.

Contact address:

European Mathematical Society Publishing House
Seminar for Applied Mathematics
ETH-Zentrum SEW A21
CH-8092 Zürich, Switzerland

Phone: +41 (0)44 632 34 36
Email: info@ems-ph.org
Homepage: www.ems-ph.org

Typeset by the editors using the authors' TeX files: M. Zunino, Stuttgart, Germany
Printing and binding: Beltz Bad Langensalza GmbH, Bad Langensalza, Germany
∞ Printed on acid free paper
9 8 7 6 5 4 3 2 1

Preface

Pavel Exner was born in Prague on March 30, 1946. After his studies at the Faculty of Technical and Nuclear Physics[1] of the Czech Technical University and at the Faculty of Mathematics and Physics (FMP) of the Charles University in Prague, he earned his MSc-equivalent degree in 1969 from the Charles University on the basis of his thesis on the theory of inelastic e-p scattering. In the subsequent years he continued to work at the Department of the Theoretical Physics of FMP. He was primarily interested in the quantum theory of unstable systems and, influenced by M. Havlíček, also in the representations of Lie algebras. In 1978 he left for the Joint Institute for Nuclear Research (JINR) in Dubna, where he spent 12 fruitful years.

In the 1970's he was not allowed to defend his CSc (PhD-equivalent) thesis on unstable systems at the Charles University, for the reasons which had nothing to do with science and which nowadays nobody would understand. In 1984, for the same reasons, he changed his home affiliation to the Nuclear Physics Institute of the Czechoslovak Academy of Sciences[2] at Řež near Prague where he still works. In Dubna, Pavel started to be interested in path integrals and earned his CSc degree on this subject from JINR in 1983. The results of his efforts in the study of open quantum systems and path integrals are summarized in the monograph *Open quantum systems and Feynman integrals* [1]. He was awarded several prizes, in particular, the JINR Prize in theoretical physics.

Starting from the 1980's, Pavel initiated his works on solvable models in quantum mechanics with particular attention to contact interactions supported by points, curves and surfaces. A long series of his papers in this field is still far from its end. His mathematically rigorous studies of quantum mechanical problems and his university lectures also gave rise to a monograph on the theory of linear operators, written jointly with J. Blank and M. Havlíček; first as a text book for graduate students and then as a book for active researchers in mathematical physics and applied mathematics. By now the book exists in three editions, each substantially upgraded: [2], [3], and [4].

One of the most important of Pavel's results is the discovery of the existence of bound states in curved quantum waveguides, i.e., for quantum particles confined in the two or three dimensional tube-like regions. His early papers on this subject with P. Šeba and P. Šťovíček [5] and [6], together with that of Goldstone and Jaffe [7],

[1] Presently Faculty of Nuclear Sciences and Physical Engineering.

[2] Presently Nuclear Physics Institute of Czech Academy of Sciences.

started the development of this new field in mathematical physics in which Pavel remains to be one of the leading scientists. Theory of quantum waveguides is summarized in the recent book [8].

In recent years Pavel has been working mainly on the theory of the so-called leaky quantum graphs where the particle is transversally bounded by a contact type interaction to the graph-like structure, bounded or with unlimited leads. These structures have attracted a lot of attention in the mathematical physics community over the past decade. Pavel has contributed to this rapidly developing research area by publishing numerous works on the subject on one hand, and by organizing a series of meetings and programmes for specialists in the field on the other hand.

At present, Pavel Exner is an author of more than 250 original papers with about 3300 total citations. He is also a member of several editorial boards and professional societies among which is the Academia Europaea, just to mention one of them.

A substantial part of Pavel Exner's scientific activity is dedicated to collaborations with students and young scientists. Since his return from Dubna in the early 1990s more than twenty Ph.D. students and postdocs worked under his supervision. Many of them have later continued their career in the academy and became independent researchers.

Besides his research and teaching activities, Pavel has not failed to serve the mathematical physics community also as an organizer. He founded the series of conferences "Mathematical Results in Quantum Theory" (QMath) and personally organized a number of them. The first QMath conference was held at Dubna in 1987, the QMath13 took place at Atlanta in 2016. In 2009, Pavel was the main organizer of the XVI International Congress on Mathematical Physics in Prague. He initiated the foundation, and for a number of years he has been serving as the scientific director, of the Doppler Institute for mathematical physics and applied mathematics, a group of mathematical physicists and mathematicians from a few Czech institutions collaborating and having common seminars since 1993. Pavel was the president of the International Association of Mathematical Physics in 2009–2011, vicepresident of European Research Council in 2011-2014, president of the European Mathematical Society for 2015-2018 to mention just his most important duties. Needless to say that Pavel always tries to support and push up his students and colleagues. The picture would not be complete without mentioning Pavel's family, his wife Jana with whom he had lived since marriage in 1971, three daughters, Milena, Hana, and Věra, and five grandchildren.

The present proceedings collect papers submitted to celebrate Pavel's seventies birthday. Most contributions treat subjects closely related to Pavel's scientific interests; quantum graphs, waveguides and layers, contact interactions including time-dependent ones, Schrödinger and similar operators on manifolds or on certain special domains with special potentials, product formulas for operator semigroups. Other papers deal with infinite finite-band matrices, abstract perturbation theory, nodal properties of the Laplacian eigenfunctions, non-linear equations on manifolds, stochastic and adiabatic problems, and some issues in quantum field theory. All together they provide various examples of applications of functional analysis in quantum physics and partial differential equations.

Jaroslav Dittrich
Hynek Kovařík
Ari Laptev

References

[1] P. Exner, *Open quantum systems and Feynman integrals.* Fundamental Theories of Physics. D. Reidel Publishing Co., Dordrecht, 1985. ISBN 90-277-1678-1 MR 0766559 Zbl 0638.46051

[2] J. Blank, P. Exner, and M. Havlíček, *Linear operators in quantum physics.* Karolinum, Prague, 1993. In Czech. ISBN 80-7066-586-6

[3] J. Blank, P. Exner, and M. Havlíček, *Hilbert space operators in quantum physics.* AIP Series in Computational and Applied Mathematical Physics. American Institute of Physics, New York, 1994. ISBN 1-56396-142-3 MR 1275370 Zbl 0873.46038

[4] J. Blank, P. Exner, and M. Havlíček, *Hilbert space operators in quantum physics.* Second edition. Theoretical and Mathematical Physics. Springer, Berlin etc., 2008. ISBN 978-1-4020-8869-8 MR 2458485 Zbl 1163.47060

[5] P. Exner and P. Šeba, Bound states in quantum waveguides. *J. Math. Phys.* **30** (1989), no. 11, 2574–2580. Zbl 0693.46066

[6] P. Exner, P. Šeba, and P. Šťovíček, On existence of a bound state in an L-shaped-waiguide. *Czech. J. Phys.* **B39** (1989), 1181–1191.

[7] J. Goldstone and R. L. Jaffe, Bound states in twisting tubes. *Phys. Rev.* **B45** (1992), 14100–14107.

[8] P. Exner and H. Kovařík, *Quantum waveguides.* Theoretical and Mathematical Physics. Springer, Cham, 2015. ISBN 978-3-319-18575-0 MR 3362506 Zbl 1314.81001

Pavel Exner in 2015 (photo M. Rychlík)

Contents

Relative partition function of Coulomb plus delta interaction

Sergio Albeverio, Claudio Cacciapuoti, and Mauro Spreafico

The authors are very pleased to dedicate this work to Pavel Exner, on the occasion of his 70th birthday. He has always been for us a source of inspiration, and we are very grateful to him for his support.

1 Introduction

The present paper discusses a problem related to three main areas of investigations, in mathematics and physics: the theory of quantum fields (in particular thermal fields), the study of determinants of elliptic (pseudo differential) operators, and the study of singular perturbations of linear operators. The problem providing the link between these areas originated with a theoretical investigation by H. B. G. Casimir [20] who predicted the possibility of an effect, called "Casimir effect," of attraction of parallel conducting plates in vacuum due to the presence of fluctuations in the vacuum energy of the electromagnetic quantum field.

Since the experimental confirmation of this effect by Spaarnay [65], about ten years after the work of Casimir, both theoretical and experimental studies of "Casimir like effects" have received a lot of attention. In particular the temperature corrections were first discussed by M. Fierz [33] and J. Mehra [49], we refer to the monograph [12] for more references and details on the effects of temperature. On the other hand, its dependence on the geometry of the plates and the medium (even attractiveness can become repulsion according to changing geometry) has been discussed in several publications, see, e.g., the books [12], [19], [28], [50], and [52], the survey papers [11] and [59], and, e.g., [10], [14], [15], [18], [20], [22], [24], [25], [26], [27], [29], [47], [54], [57], [60], and [62].

The physical discussion of the Casimir effect is also related to the one of the Van der Waals forces between molecules, see [52]. It has also many relations to condensed matter physics, hadronic physics, cosmology, and nanotechnology, see, e.g., the references in [11], [19], [12], [28], [50], [52], and [59].

Theoretically the Casimir effect arises when computing the difference between two infinite quantities, namely the vacuum energy of a quantum field with or without a certain "boundary condition." More generally it is a phenomenon related to the difference of two Green's functions associated with hyperbolic or elliptic operators. Such problems are also of interest in geometric analysis, particularly since the work by W. Müller [53] and M. Spreafico and S. Zerbini [70]. The latter works are related to the introduction by Ray and Singer [61] of a definition of determinants for elliptic operators on manifolds via a zeta-function renormalization (see also, e.g., [48] and [55]). By this procedure one can define $\log(\det A)^{-1/2}$, for A self-adjoint, positive, in some Hilbert space, via the analytic continuation at $s = 1/2$ of the zeta-function associated with A, defined for $\operatorname{Re} s$ sufficiently large as

$$\zeta(s; A) := \sum_{\lambda \in \sigma^+(A)} \lambda^{-s}$$

$\sigma^+(A)$ being the positive part of the spectrum of A. Setting

$$Z := (\det A)^{-1/2},$$

one has the definition of the "partition function"

$$\text{“} Z = \int_\Phi e^{-S(\varphi)} \, d\varphi \text{”},$$

$S(\varphi) := (\varphi, A\varphi)$, associated with a (Euclidean) quantum field with covariance operator given by the inverse of A (φ is the field, Φ the space of "fields configurations").

In turn, it is well known that partitions functions Z arise as normalizations in heuristic Euclidean path integrals

$$\text{“} Z^{-1} \int_\Phi e^{-S(\varphi)} f(\varphi) \, d\varphi \text{”},$$

f being complex valued functions (related to "observables"), see, e.g., [1] and [71].

On the other hand it was pointed out by Hawking [41] and, independently, Figari, Høegh-Krohn, and Nappi [34], that there is a strict relation between Euclidean vacuum states in de Sitter spaces of fixed curvature and temperature states of Euclidean states. Hawking used the Ray-Singer definition of a partition function related to A to compute physical quantities of the Euclidean model. For wide-ranging extensions of these connections see, e.g., [6], [7], [31], [32], [35], [36], [37], [51], [53], [66], and [68].

Another application of the zeta function is in the computation of the high temperature asymptotics of several thermodynamic functions such as the Helmholtz free energy, internal energy, and entropy, see, e.g., [13] and references therein.

As pointed out in [53] and [67], [68] and [69], considering the relative zeta-function of a pair of elliptic operators A, A_0, leads to define, via a relative zeta-function, a relative determinant including A and A_0, and a Casimir effect can be discussed relatively to the pair (A, A_0). In fact, the strength of the Casimir effect is expressed by the derivative of the relative zeta-function at 0. These considerations are also related to the study of relative traces of semigroups resp. resolvents associated with pairs of operators. The study of such relative traces has its origins in quantum statistical mechanics [8].

The case where A_0 is the Laplace–Beltrami operator on $S^1 \times \mathbb{R}^3$, and A is a point perturbation of A_0 has been discussed in details in [70] and [3]. For the extended study of point interactions on $\mathbb{R}^d$, $d = 1, 2, 3$, see [1], [4], and [5]. The case where $\mathbb{R}^d$ is replaced by a Riemannian manifold occurs particularly in [23] (who points out its possible relevance in number theory), see also [21], [32], and [46].

For further particular studies of point interactions in relation with the Casimir effect see [2], [4], [14], [15], [38], [40], [50], [43], [58], [63], and [64].

Particularly close to our work is the result in [3] where A_0 is the half space $x^3 > 0$ in $\mathbb{R}^3$ and A is taken to be the sum of two point interactions located at (a_1, a_2, a_3) and $(a_1, a_2, -a_3)$, $a_1, a_2 \in \mathbb{R}$, $a_3 \in \mathbb{R}_+$. The relative trace of the resolvents was computed at values of the spectral parameter λ such that $\operatorname{Im} \sqrt{\lambda} > 0$, and the spectral measure was constructed. Moreover the asymptotics for small and large values of the spectral parameter was found. Furthermore the relative zeta-function and its derivative at 0 has been computed and related to the Casimir effect [3].

The present paper extends this kind of relations to the case of the pair (A, A_0), where A_0 is the operator $-\Delta$ with a Coulomb interaction at the origin acting in $L^2(\mathbb{R}^3)$, and A is a perturbation of A_0 obtained by adding a point interaction at the origin. The construction of A_0 and A is based on [4], Chapter I.2. In order to define and study the relative partition function we use explicit formulae for the integrals of the Whittaker's functions which enter the explicit expression of the resolvent of $-\Delta$ with a Coulomb interaction.

Such explicit formulae do not exist in the situation where the point interaction is not centered at the origin. In this situation an alternative approach would be to use series expansions to compute the integrals. It turns out that this idea does not seem feasible due to the slow decay of the Coulomb interaction at infinity. On the other hand, the case of potentials with faster decay at infinity should be treatable in this way, replacing the explicit formulae by methods of regular perturbations theory.

The structure of the paper is as follows. In Section 2 we recall the general definition of the relative partition function associated to a pair of non-negative self-adjoint operators and its relation with the relative zeta function. In Section 3 we study the perturbation of the Laplacian by a Coulomb and a delta potential centered at the origin. In Section 4 we study the associated relative partition function of the Coulomb plus delta interaction.

2 Relative partition function associated to a pair of non-negative self-adjoint operators

This section presents a generalization of the method introduced in [70] to study the analytic properties of the relative zeta function associated to a pair of operators (A, A_0) as described below (see also [53]). We assume here that logarithmic terms appear in the expansion of the relative trace, and this will produce a double pole in the relative zeta function, and in turn a simple pole in the relative partition function.

2.1 Relative zeta function

We denote by $R(\lambda; A) \equiv (\lambda - A)^{-1}$ the resolvent of a linear operator A. λ is in the resolvent set, $\rho(A)$, of A, a subset of $\mathbb{C}$. The relative zeta function $\zeta(s; A, A_0)$ for a pair of non-negative self-adjoint operators (A, A_0) is defined when the relative resolvent $R(\lambda; A) - R(\lambda; A_0)$ is of trace class and some conditions on the asymptotic expansions of the trace of the relative resolvent $r(\lambda; A, A_0)$ are satisfied, as in Section 2 of [70]. These conditions imply that similar conditions on the trace of the relative heat operator $\mathrm{tr}(\mathrm{e}^{-tA} - \mathrm{e}^{-tA_0})$ are satisfied, according to Section 2 of [53]. The conditions in [70] on the asymptotic expansions ensure that the relative zeta function is regular at $s = 0$. In the present work we consider a wider class of pairs, and we admit a more general type of asymptotic expansions, as follows. Let $\mathcal{H}$ be a separable Hilbert space, and let A and A_0 be two self-adjoint non-negative linear operators in $\mathcal{H}$. Suppose that $\mathrm{SpA} = \mathrm{Sp_cA}$, is purely continuous, and assume both 0 and ∞ are accumulation points of SpA.

Then, by a standard argument (see for example the proof of the corresponding result in [70]), we prove Lemma 2.1 below.

Let us recall first the definition of asymptotic expansion. If $f(\lambda)$ is a complex valued function, we write

$$f(\lambda) \sim \sum_{n=0}^{\infty} a_n \lambda^n, \quad a_n \in \mathbb{C},\ \lambda \to 0,$$

if for any $N \in \mathbb{N}_0$ one has

$$\frac{f(\lambda) - \sum_{n=0}^{N} a_n \lambda^n}{\lambda^N} \longrightarrow 0 \quad \text{as } \lambda \to 0,$$

and we say that f *has the asymptotic expansion* $\sum_{n=0}^{\infty} a_n \lambda^n$. Then the following result holds true.

Lemma 2.1. *Let (A, A_0) be a pair of non-negative self-adjoint operators as above satisfying the following conditions*:

(B.1) *the operator $R(\lambda; A) - R(\lambda; A_0)$ is of trace class for all $\lambda \in \rho(A) \cap \rho(A_0)$*;

(B.2) *as $\lambda \to \infty$ in $\rho(A) \cap \rho(A_0)$, there exists an asymptotic expansion of the form*

$$\operatorname{tr}(R(\lambda; A) - R(\lambda; A_0)) \sim \sum_{j=0}^{\infty} \sum_{k=0}^{K_j} a_{j,k} (-\lambda)^{\alpha_j} \log^k(-\lambda),$$

where $a_{j,k} \in \mathbb{C}$, $-\infty < \cdots < \alpha_1 < \alpha_0$, $\alpha_j \to -\infty$, for large j;

(B.3) *as $\lambda \to 0$, there exists an asymptotic expansion of the form*

$$\operatorname{tr}(R(\lambda; A) - R(\lambda; A_0)) \sim \sum_{j=0}^{\infty} b_j (-\lambda)^{\beta_j};$$

where $b_j \in \mathbb{C}$, $-1 \le \beta_0 < \beta_1 < \cdots$, and $\beta_j \to +\infty$, for large j;

(C) $\alpha_0 < \beta_0$.

Then the relative zeta function is defined by

$$\zeta(s; A, A_0) = \frac{1}{\Gamma(s)} \int_0^{\infty} t^{s-1} \operatorname{tr}(\mathrm{e}^{-tA} - \mathrm{e}^{-tA_0})\, \mathrm{d}t,$$

when $\alpha_0 + 1 < \operatorname{Re}(s) < \beta_0 + 1$, and by analytic continuation elsewhere. Here Γ is the classical Gamma function and

$$\operatorname{tr}(\mathrm{e}^{-tA} - \mathrm{e}^{-tA_0}) = \frac{1}{2\pi i} \int_{\Lambda} \mathrm{e}^{-\lambda t} \operatorname{tr}(R(\lambda; A) - R(\lambda; A_0))\, \mathrm{d}\lambda,$$

where Λ is some contour of Hankel type (see, e.g., [30] and [68]). The analytic extension of $\zeta(s; A, A_0)$ is regular except for possible simple poles at $s = \beta_j$ and possible further poles at $s = \alpha_j$.

Note that the poles of the relative zeta function at $s = \alpha_j$ can be of higher orders, differently from the case investigated in [70].

Introducing the relative spectral measure, we have the following useful representation of the relative zeta function.

Lemma 2.2. *Let (A, A_0) be a pair of non-negative self-adjoint operators as above satisfying conditions* (B.1)–(B.3) *and* (C) *of Lemma* 2.1. *Then,*

$$\zeta(s; A, A_0) = \int_0^\infty v^{-2s} e(v; A, A_0)\,\mathrm{d}v,$$

where the relative spectral measure is defined by

$$e(v; A, A_0) = \frac{v}{\pi i} \lim_{\epsilon \to 0+} \left(r(v^2 \mathrm{e}^{2i\pi - i\epsilon}; A, A_0) - r(v^2 \mathrm{e}^{i\epsilon}; A, A_0) \right) \quad v \geq 0, \tag{1}$$

$$r(\lambda; A, A_0) = \operatorname{tr}(R(\lambda; A) - R(\lambda; A_0)) \qquad \lambda \in \rho(A) \cap \rho(A_0). \tag{2}$$

The integral, the limit and the trace exist.

Proof. Since (A, A_0) satisfies (B.1)–(B.3), we can write

$$\operatorname{tr}(\mathrm{e}^{-tA} - \mathrm{e}^{-tA_0}) = \frac{1}{2\pi i} \int_\Lambda \mathrm{e}^{-\lambda t} \operatorname{tr}(R(\lambda; A) - R(\lambda; A_0))\,\mathrm{d}\lambda.$$

Changing the spectral variable λ to $k = \lambda^{1/2}$, with the principal value of the square root, i.e., with $0 < \arg k < \pi$, we get

$$\operatorname{Tr}(\mathrm{e}^{-tA} - \mathrm{e}^{-tA_0}) = \frac{1}{\pi i} \int_\gamma \mathrm{e}^{-k^2 t} \operatorname{tr}(R(k^2; A) - R(k^2; A_0)) k\,\mathrm{d}k,$$

where γ is the line $k = -ic$, for some $c > 0$. Writing $k = v\mathrm{e}^{i\theta}$, $0 \leq \theta < 2\pi$, and $r(\lambda; A, A_0) = \operatorname{tr}(R(\lambda; A) - R(\lambda; A_0))$, a standard computation leads to

$$\operatorname{Tr}(\mathrm{e}^{-tA} - \mathrm{e}^{-tA_0}) = \int_0^\infty \mathrm{e}^{-v^2 t} e(v; A, A_0)\,\mathrm{d}v,$$

$$\zeta(s; A, A_0) = \int_0^\infty v^{-2s} e(v; A, A_0)\,\mathrm{d}v. \qquad \square$$

Remark 2.3. The relative spectral measure is discussed in general, e.g., in [53]. It is expressed by (2) in terms of $r(\lambda; A, A_0)$ which is the Laplace transform of $\mathrm{tr}(\mathrm{e}^{-tA} - \mathrm{e}^{-tA_0})$, which in turn is simply related to the spectral shift function (see eq. (0.6) in [53]). The derivative of the latter is essentially the density of states used, e.g., in [56] in connection with the Casimir effect, and going back to the original work by M. Š. Birman and M. G. Kreĭn [9], [44], and [45].

It is clear by construction that the analytic properties of the relative zeta function are determined by the asymptotic expansions required in conditions (B.1) and (B.2). More precisely, such conditions imply similar conditions on the expansion of the relative spectral measure, and hence on the analytic structure of the relative zeta function. This is in the next lemmas.

Lemma 2.4. *As in Lemma 2.2, let (A, A_0) be a pair of non-negative self-adjoint operators . Then the relative spectral measure $e(v; A, A_0)$ has the following asymptotic expansions. For small $v > 0$,*

$$e(v; A, A_0) \sim \sum_{j=0}^{\infty} c_j v^{2\beta_j+1},$$

where

$$c_j = -\frac{2b_j \sin \pi\beta_j}{\pi},$$

and the β_j and the b_j are the numbers appearing in condition (B3) *of Lemma* 2.1; *for large $v \geq 0$ and $j \in \mathbb{N}_0$,*

$$e(v; A, A_0) \sim \sum_{j=0}^{\infty} \sum_{h=0}^{H_j} e_{j,h} v^{2\alpha_j+1} \log^h v$$

$$\sim \sum_{j=0}^{\infty} \sum_{k=0}^{K_j} \sum_{h=0}^{k} e_{j,k,h} v^{2\alpha_j+1} \log^h v^2,$$

where

$$e_{j,k,h} = -a_{j,k}(\pi i)^{k-h-1} \binom{k}{h} (\mathrm{e}^{i\alpha_j\pi} - (-1)^{k-h}\mathrm{e}^{-i\alpha_j\pi})$$

and the $a_{j,k}$, α_j, and K_j are the numbers appearing in condition (B2) *of Lemma* 2.1. *The coefficients $e_{j,h}$ can be expressed in terms of the coefficients $e_{j,k,h}$.*

Proof. Note that the cut $(0, \infty)$ in the complex λ-plane corresponds to the cut $(-\infty, 0)$ in the complex $-\lambda$-plane. Thus $-\lambda = x\mathrm{e}^{i\theta}$, with $-\pi\theta < \pi$, and $\theta = 0$ corresponds to positive real values of $-\lambda$.

Thus, inserting the expansion (B3) for small λ in the definition of the relative spectral measure, equation (1), we obtain, for small v,

$$e(v; A, A_0) \sim -\frac{v}{i\pi} \lim_{\epsilon\to 0+} \sum_{j=0}^{\infty} b_j v^{2\beta_j} (\mathrm{e}^{(\pi i - i\epsilon)\beta_j} - \mathrm{e}^{(-\pi i + i\epsilon)\beta_j}),$$

and the first part of the statement follows. For the expansion for large v, we insert (B2) into the definition of the relative spectral measure. This gives, for large v,

$$\begin{aligned} e(v; A, A_0) \sim -\frac{v}{i\pi} \lim_{\epsilon\to 0+} \sum_{j=0}^{\infty} \sum_{k=0}^{K_j} a_{j,k} v^{2\alpha_j} (\mathrm{e}^{(\pi i - i\epsilon)\alpha_j} (\log v^2 + \pi i - i\epsilon)^k \\ - \mathrm{e}^{(-\pi i + i\epsilon)\alpha_j} (\log v^2 - \pi i + i\epsilon)^k), \end{aligned}$$

$$\begin{aligned} e(v; A, A_0) \sim -\sum_{j=0}^{\infty} v^{2\alpha_j+1} \sum_{k=0}^{K_j} \frac{a_{j,k}}{i\pi} \bigg(\mathrm{e}^{\pi i\alpha_j} \sum_{h=0}^{k} \binom{k}{h} (i\pi)^{k-h} \log^k v^2 \\ - \mathrm{e}^{-\pi i\alpha_j} \sum_{h=0}^{k} \binom{k}{h} (-i\pi)^{k-h} \log^k v^2 \bigg), \end{aligned}$$

and the thesis follows. □

Remark 2.5. We give more details on the first coefficients that are more relevant in the present work. Direct calculation gives

$$e_{j,0} = e_{j,0,0} + 2\sum_{k=1}^{K_j} e_{j,k,0} = -\sum_{k=0}^{K_j} a_{j,k} (\pi i)^{k-1} (\mathrm{e}^{i\alpha_j\pi} - (-1)^k \mathrm{e}^{-i\alpha_j\pi}),$$

$$e_{j,0,0} = -\frac{2\sin\pi\alpha_j}{\pi} a_{j,0}.$$

2.2 Relative partition function

Let W be a smooth Riemannian manifold of dimension n, and consider the product $X = S^1_{\beta/2\pi} \times W$, where S^1_r is the circle of radius r, $\beta > 0$. Let ξ be a complex line bundle over X, and L a self-adjoint non-negative linear operator on the Hilbert space $\mathcal{H}(W)$ of the L^2 sections of the restriction of ξ onto W, with respect to some fixed metric g on W. Let L be the self-adjoint non-negative operator $L = -\partial_u^2 + A$, on the Hilbert space $\mathcal{H}(X)$ of the L^2 sections of ξ, with respect to the product metric $du^2 \oplus g$ on X, and with periodic boundary conditions on the circle. Assume that there exists a second operator A_0 defined on $\mathcal{H}(W)$, such that the pair (A, A_0) satisfies the assumptions (B.1)–(B.3) of Lemma 2.1. Then, by a proof similar to the one of Lemma 2.1 of [70], it is possible to show that there exists a second operator L_0 defined in $\mathcal{H}(X)$, such that the pair (L, L_0) satisfies those assumptions too. Under these requirements, we define the regularized relative zeta partition function of the model described by the pair of operators (L, L_0) by

$$\log Z_{\mathrm{R}} = \frac{1}{2}\operatorname{Res}_{0\,s=0}\zeta'(s; L, L_0) - \frac{1}{2}\operatorname{Res}_{0\,s=0}\zeta(s; L, L_0)\log \ell^2, \tag{3}$$

where ℓ is some renormalization constant (introduced by Hawking [41], see also, e.g., [51], in connection with the scaling behavior in path integrals in curved spaces), and we have the following result, in which $\log Z_R$ is essentially expressed in terms of the relative Dedekind eta function $\eta(\beta; A, A_0)$.

Proposition 2.6. *Let A be a non-negative self-adjoint operator on W and suppose $L = -\partial_u^2 + A$, on $S^1_{\beta/(2\pi)} \times W$ as defined above. Assume there exists an operator A_0 such that the pair (A, A_0) satisfies conditions* (B.1)–(B.3) *of Lemma* 2.1. *Then, the relative zeta function $\zeta(s; L, L_0)$ (defined analogously to the one given in Lemma* 2.1) *has a simple pole at $s = 0$ with residua*

$$\begin{aligned}
\operatorname{Res}_{1\,s=0}\zeta(s; L, L_0) &= -\beta \operatorname{Res}_{2\,s=-1/2}\zeta(s; A, A_0),\\
\operatorname{Res}_{0\,s=0}\zeta(s; L, L_0) &= -\beta \operatorname{Res}_{1\,s=-1/2}\zeta(s; A, A_0)\\
&\quad - 2\beta(1-\log 2)\operatorname{Res}_{2\,s=-1/2}\zeta(s; A, A_0),\\
\operatorname{Res}_{0\,s=0}\zeta'(s; L, L_0) &= -\beta \operatorname{Res}_{0\,s=-1/2}\zeta(s; A, A_0)\\
&\quad - 2\beta(1-\log 2)\operatorname{Res}_{1\,s=-1/2}\zeta(s; A, A_0)\\
&\quad - \beta\Big(2 + \frac{\pi^2}{6} + 2(1-\log 2)^2\Big)\operatorname{Res}_{2\,s=-1/2}\zeta(s; A, A_0)\\
&\quad - 2\log\eta(\beta; A, A_0),
\end{aligned}$$

where $L_0 = -\partial_u^2 + A_0$, and the relative Dedekind eta function is defined by

$$\log\eta(\tau; A, A_0) = \int_0^\infty \log(1-\mathrm{e}^{-\tau v})e(v; A, A_0)\,\mathrm{d}v, \quad \tau > 0.$$

$\operatorname{Res}_{k\,s=s_0}\zeta(s)$ is understood as the coefficient of the term $(s-s_0)^{-k}$ in the Laurent expansion of $\zeta(s)$ around $s = s_0$.

The residua and the integral are finite.

Proof. Since (A, A_0) satisfies (B.1)–(B.3), we deduce that the (L, L_0) relative zeta function $\zeta(s; L, L_0)$ is defined by

$$\zeta(s; L, L_0) = \frac{1}{\Gamma(s)} \int_0^\infty t^{s-1} \operatorname{Tr}(e^{-tL} - e^{-tL_0})\, dt,$$

when $\alpha_0 + 1 < \operatorname{Re}(s) < \beta_0 + 1$ (with α_0 and β_0 as in Lemma 2.1). Since (see for example Lemma 2.2 of [70])

$$\operatorname{Tr}(e^{-Lt} - e^{-L_0 t}) = \sum_{n\in\mathbb{Z}} e^{-(n^2/r^2)t} \operatorname{Tr}(e^{-tA} - e^{-tA_0}),$$

where $r = \beta/(2\pi)$ and $t > 0$. Using the Jacobi summation formula and dominated convergence to exchange summation and integration we obtain

$$\begin{aligned}
\zeta(s; L, L_0) &= \frac{1}{\Gamma(s)} \int_0^\infty t^{s-1} \sum_{n\in\mathbb{Z}} e^{-(n^2/r^2)t} \operatorname{Tr}(e^{-tA} - e^{-tA_0})\, dt \\
&= \frac{\sqrt{\pi} r}{\Gamma(s)} \int_0^\infty t^{s-(1/2)-1} \operatorname{Tr}(e^{-tA} - e^{-tA_0})\, dt \\
&\quad + \frac{2\sqrt{\pi} r}{\Gamma(s)} \int_0^\infty t^{s-(1/2)-1} \sum_{n=1}^\infty e^{-\pi^2 r^2 n^2/t} \operatorname{Tr}(e^{-tA} - e^{-tA_0})\, dt \\
&= z_1(s) + z_2(s),
\end{aligned}$$

with

$$\begin{aligned}
z_1(s) &:= \frac{\sqrt{\pi} r}{\Gamma(s)} \Gamma\Big(s - \frac{1}{2}\Big) \zeta\Big(s - \frac{1}{2}; A, A_0\Big), \\
z_2(s) &:= \frac{2\sqrt{\pi} r}{\Gamma(s)} \sum_{n=1}^\infty \int_0^\infty t^{s-(1/2)-1} e^{-\pi^2 r^2 n^2/t} \operatorname{Tr}(e^{-tA} - e^{-tA_0})\, dt.
\end{aligned}$$

The first term, $z_1(s)$, can be expanded near $s = 0$, and this gives the result stated, by Lemma 2.1. By Lemma 2.2, the second term $z_2(s)$ is

$$z_2(s) = \frac{2\sqrt{\pi} r}{\Gamma(s)} \sum_{n=1}^\infty \int_0^\infty t^{s-1/2-1} e^{-\pi^2 n^2 r^2/t} \int_0^\infty e^{-v^2 t} e(v; A, A_0)\, dv\, dt,$$

and we can do the t integral using for example (3.471.9) of [39]. We obtain

$$z_2(s) = \frac{4\sqrt{\pi r}}{\Gamma(s)} \sum_{n=1}^{\infty} \int_0^{\infty} \Big(\frac{\pi n r}{v}\Big)^{s-1/2} K_{s-1/2}(2\pi n r v) e(v; A, A_0)\,\mathrm{d}v. \tag{4}$$

Since the Bessel function $K_{s-1/2}(2\pi n r v)$ is analytic in its parameter, regular at $-1/2$, and

$$K_{-1/2}(z) = \sqrt{\frac{\pi}{2z}}\mathrm{e}^{-z},$$

equation (4) gives the formula for the analytic extension of the relative zeta function $\zeta(s; H, H_0)$ near $s = 0$. We obtain

$$z_2(0) = 0, \quad z_2'(0) = -2\int_0^{\infty} \log(1 - \mathrm{e}^{-2\pi r v}) e(v; L, L_0)\,\mathrm{d}v,$$

and the integral converges by assumptions (B.2) and (B.3). □

It is clear by the previous result that all information on the relative partition function comes from the analytic structure of the spatial relative spectral function $\zeta(s; A, A_0)$ near $s = -1/2$. Such information is based on the asymptotic expansion assumed for the relative resolvent, and contained in the following lemma.

Lemma 2.7. *As in Lemma 2.2, let (A, A_0) be a pair of non-negative self-adjoint operators. Then, the relative zeta function $\zeta(s; A, A_0)$ extends analytically to the following meromorphic function in a neighborhood of $s = -1/2$:*

$$\begin{aligned}\zeta(s; A, A_0) = {}& \frac{1}{2}\sum_{j=0}^{J_0-1} \frac{c_j}{\beta_j + 1 - s} + \sum_{j=0}^{J_\infty-1}\sum_{h=0}^{H_j} \frac{(-1)^{h+1} e_{j,h}}{2^{h+1}(\alpha_j + 1 - s)^{h+1}} \\ & + \int_0^1 v^{-2s}\Big(e(s; A, A_0) - \sum_{j=0}^{J_0} c_j v^{2\beta_j+1}\Big)\,\mathrm{d}v \\ & + \int_1^{\infty} v^{-2s}\Big(e(s; A, A_0) - \sum_{j=0}^{J_\infty}\sum_{h=0}^{H_j} e_{j,h} v^{2\alpha_j+1}\log^h v\Big)\,\mathrm{d}v,\end{aligned} \tag{5}$$

where J_0 is the smallest integer such that $\beta_{J_0} > -3/2$, and J_∞ is the largest integer such that $\alpha_{J_\infty} < -3/2$ (the α_j and β_j, resp. the c_j, $e_{j,h}$, and H_j, are as in Lemma 2.1, resp. Lemma 2.2).

Proof. Set

$$\zeta_0(s; A, A_0) = \int_0^1 v^{-2s} e(v; A, A_0)\,dv$$

and

$$\zeta_\infty(s; A, A_0) = \int_1^\infty v^{-2s} e(v; A, A_0)\,dv,$$

then

$$\zeta(s; A, A_0) = \zeta_0(s; A, A_0) + \zeta_\infty(s; A, A_0).$$

Consider the expansion of $e(s; A, A_0)$ for small v given in Lemma 2.4. Let J_0 be the smallest integer such that $\beta_{J_0} > -3/2$, and write

$$\begin{aligned}\zeta_0(s; A, A_0) = {} & \int_0^1 v^{-2s}\Big(\sum_{j=0}^{J_0} c_j v^{2\beta_j+1}\Big)\,dv \\ & + \int_0^1 v^{-2s}\Big(e(s; A, A_0) - \sum_{j=0}^{J_0} c_j v^{2\beta_j+1}\Big)\,dv.\end{aligned} \tag{6}$$

The last integral in equation (6) is convergent, while the first one can be computed explicitly. This gives the statement for ζ_0, in the sense that ζ_0 has a representation like in equation (5). For ζ_∞ consider the expansion of $e(s; A, A_0)$ for large v given in Lemma 2.4. Let J_∞ be the largest integer such that $\alpha_{J_\infty} < -3/2$, and write

$$\begin{aligned}\zeta_\infty(s; A, A_0) = {} & \int_1^\infty v^{-2s}\Big(\sum_{j=0}^{J_\infty}\sum_{h=0}^{H_j} e_{j,h} v^{2\alpha_j+1}\log^h v\Big)\,dv \\ & + \int_1^\infty v^{-2s}\Big(e(s; A, A_0) - \sum_{j=0}^{J_\infty}\sum_{h=0}^{H_j} e_{j,h} v^{2\alpha_j+1}\log^h v\Big)\,dv.\end{aligned} \tag{7}$$

The last integral in equation (7) is convergent, while the first one can be computed explicitly. This gives the statement for ζ_∞, in the sense that ζ_∞ has a representation like in equation (5). Putting together the representations of ζ_0 and ζ_∞ concludes the proof. □

Corollary 2.8. *Let (A, A_0) be a pair of non-negative self-adjoint operators as in Lemma 2.2. With the notation of that lemma,*

$$\mathrm{Res}_{2\,s=-1/2}\,\zeta(s; A, A_0) = \frac{e_{a,1}}{4},$$

$$\mathrm{Res}_{1\,s=-1/2}\,\zeta(s; A, A_0) = \frac{e_{a,0}}{2} - \frac{c_b}{2},$$

$$
\begin{aligned}
\operatorname{Res}_{0\,s=-1/2}\zeta(s;A,A_0) &= \frac{1}{2}\sum_{j=0,j\neq b}^{J_0}\frac{c_j}{\beta_j+\frac{3}{2}}\\
&\quad+\sum_{j=0,j\neq a}^{J_\infty}\sum_{h=0}^{H_j}\frac{(-1)^{h+1}e_{j,h}}{2^{h+1}\left(\alpha_j+\frac{3}{2}\right)^{h+1}}\\
&\quad+\int_0^1 v\Big(e\Big(-\frac{1}{2};A,A_0\Big)-\sum_{j=0}^{J_0}c_j v^{2\beta_j+1}\Big)\,dv\\
&\quad+\int_1^\infty v\Big(e\Big(-\frac{1}{2};A,A_0\Big)-\sum_{j=0}^{J_\infty}\sum_{h=0}^{H_j}e_{j,h}v^{2\alpha_j+1}\log^h v\Big)\,dv,
\end{aligned}
$$

where in the lower limits of the sums a is the index in the sequence $\{\alpha_j\}$, such that $\alpha_a=-3/2$, and b is the index in the sequence $\{\beta_j\}$, such that $\beta_b=-3/2$, and

$$
\begin{aligned}
\operatorname{Res}_{1\,s=0}\zeta(s;L,L_0) &= -\frac{e_{a,1}}{4}\beta,\\
\operatorname{Res}_{0\,s=0}\zeta(s;L,L_0) &= -\frac{1}{2}(c_b-e_{a,0}-(1-\log 2)e_{a,1})\beta,\\
\operatorname{Res}_{0\,s=0}\zeta'(s;L,L_0) &= -\beta\Big(\frac{1}{2}\sum_{j=0,j\neq b}^{J_0}\frac{c_j}{\beta_j+\frac{3}{2}}\\
&\qquad+\sum_{j=0,j\neq a}^{J_\infty}\sum_{h=0}^{H_j}\frac{(-1)^{h+1}e_{j,h}}{2^{h+1}\left(\alpha_j+\frac{3}{2}\right)^{h+1}}\\
&\qquad+\int_0^1 v\Big(e\Big(-\frac{1}{2};A,A_0\Big)-\sum_{j=0}^{J_0}c_j v^{2\beta_j+1}\Big)\,dv\\
&\qquad+\int_1^\infty v\Big(e\Big(-\frac{1}{2};A,A_0\Big)\\
&\qquad\qquad-\sum_{j=0}^{J_\infty}\sum_{h=0}^{H_j}e_{j,h}v^{2\alpha_j+1}\log^h v\Big)\,dv\Big)\\
&\quad-\beta(1-\log 2)(e_{a,0}-c_b)\\
&\quad-\beta\Big(\frac{1}{2}+\frac{\pi^2}{24}+\frac{(1-\log 2)^2}{2}\Big)e_{a,1}\\
&\quad-2\log\eta(\beta;A,A_0).
\end{aligned}
$$

Proof. This is a simple consequence of Lemma 2.7 and Proposition 2.6. □

3 Coulomb potential plus delta interaction centered at the origin

3.1 Preliminaries

Recall that we denote by $\rho(A)$ the resolvent set of A and by $R(\lambda; A)$ the resolvent operator $(\lambda I - A)^{-1}$, for $\lambda \in \rho(A)$. If $R(\lambda; A)$ operates in $L^2(\mathbb{R}^3)$, we denote by $k(\lambda; A) = k(\lambda; A)(x, y)$ the integral kernel of $R(\lambda; A)$, $x, y \in \mathbb{R}^3$.

Let H_0 be the self-adjoint realization of the operator $-\Delta + {}^{\gamma}/_{|x|}$ in $L^2(\mathbb{R}^3)$, namely the Laplace operator plus a Coulomb potential centered at the origin in the three dimensional Euclidean space, with parameter $\gamma \in \mathbb{R}$. The kernel of the resolvent of H_0 is, see, e.g., eq. (I.2.1.16) in [4] or [16] and [17],

$$k(\lambda; H_0)(x, y) = -\frac{\Gamma(1 + {}^{\gamma}/_{2\sqrt{-\lambda}})}{4\pi|x - y|} f_\lambda(x, y),$$

where

$$\begin{aligned} f_\lambda(x, y) = {} & W_{-\gamma/2\sqrt{-\lambda}, 1/2}(\sqrt{-\lambda}\, x_+) M'_{-\gamma/(2\sqrt{-\lambda}), 1/2}(\sqrt{-\lambda}\, x_-) \\ & - W'_{-\gamma/(2\sqrt{-\lambda}), 1/2}(\sqrt{-\lambda}\, x_+) M_{-\gamma/(2\sqrt{-\lambda}), 1/2}(\sqrt{-\lambda}\, x_-), \end{aligned}$$

with $\lambda \in \rho(H_0)$, $\operatorname{Re}\sqrt{-\lambda} > 0$, $x_\pm = |x| + |y| \pm |x - y|$, and where $M_{\kappa,\mu}$ and $W_{\kappa,\mu}$ are Whittaker functions, see, e.g., [39]. In the next proposition we recall some results on the spectrum of H_0, see, e.g., [4] and [42].

Proposition 3.1. *For all $\gamma \in \mathbb{R}$ the essential spectrum of H_0 is purely absolutely continuous, moreover*

$$\sigma_{\mathrm{ess}}(H_0) = \sigma_{\mathrm{ac}}(H_0) = [0, +\infty).$$

If $\gamma \geq 0$ the point spectrum of H_0 is empty. If $\gamma < 0$ the point spectrum of H_0 is

$$\sigma_{\mathrm{pp}}(H_0) = \Big\{ -\frac{\gamma^2}{4(n+1)^2} \Big\}_{n=0}^{\infty} \qquad \gamma < 0.$$

Following [4], we introduced a perturbation of H_0, by adding a singular one center point interaction, also centered at the origin. For all $-\infty < \alpha \leq \infty$ we denote by H_α the operator formally written as $-\Delta + {}^{\gamma}/_{|x|} + \alpha\delta_0$. The concrete operator is defined in Theorem 2.1.2 of [4], and the integral kernel of the resolvent of H_α is

$$k(\lambda; H_\alpha)(x, y) = k(\lambda; H_0)(x, y) - \frac{4\pi}{4\pi\alpha - \gamma F_\gamma\Big(\frac{\gamma}{2\sqrt{-\lambda}}\Big)} g(\lambda; x) g(\lambda; y), \tag{8}$$

where $\lambda \in \rho(H_\alpha) \cap \rho(H_0)$, Re $\sqrt{-\lambda} > 0$, $x, y \in \mathbb{R}^3$, with

$$g(\lambda; x) := \frac{\Gamma\Big(1 + \frac{\gamma}{2\sqrt{-\lambda}}\Big)}{4\pi|x|} W_{-\gamma/(2\sqrt{-\lambda}),1/2}(2\sqrt{-\lambda}\,|x|) \quad x \neq 0,$$

and

$$F_\gamma(z) := \begin{cases} \psi(1+z) - \log(z) - \dfrac{1}{2z} - \psi(1) - \psi(2), & \gamma > 0, \\[2ex] \psi(1+z) - \log(-z) - \dfrac{1}{2z} - \psi(1) - \psi(2), & \gamma < 0. \end{cases}$$

Here $z \in \mathbb{C}$ and ψ is the digamma function, i.e., $\psi(z) = {}^{d}/_{dz} \log\Gamma(z)$, see (8.36) of [39]. We note that the function $F_\gamma(z)$ is indeed a function of z and of $\operatorname{sgn}(\gamma)$ only.

In the following proposition we recall some results on the spectrum of the operator H_α, see, e.g., Theorem I.2.1.3 in [4].

Proposition 3.2. *Let $-\infty < \alpha \leq \infty$. For all $\gamma \in \mathbb{R}$ the essential spectrum of H_α is purely absolutely continuous, moreover*

$$\sigma_{\mathrm{ess}}(H_\alpha) = \sigma_{\mathrm{ac}}(H_\alpha) = [0, +\infty)\,.$$

The eigenvalues of H_α associated with the s-wave ($l = 0$) are given by the solutions of the equation

$$4\pi\alpha - \gamma F_\gamma\Big(\frac{\gamma}{2\sqrt{-E}}\Big) = 0, \quad E < 0, \tag{9}$$

where we set

$$\gamma F_\gamma\Big(\frac{\gamma}{2\sqrt{-E}}\Big)\Big|_{\gamma=0} = \lim_{\gamma\to 0} \gamma F_\gamma\Big(\frac{\gamma}{2\sqrt{-E}}\Big) = -\sqrt{-E}.$$

If $\gamma \geq 0$ and $\alpha \geq -\gamma[\psi(1)+\psi(2)]/(4\pi)$, equation (9) *has no solutions, moreover the point spectrum of H_α is empty.*

If $\gamma \geq 0$ and $\alpha < -\gamma[\psi(1)+\psi(2)]/(4\pi)$, equation (9) *has precisely one solution, and the operator H_α has precisely one negative eigenvalue.*

If $\gamma < 0$ equation (9) *has infinitely many solutions. Correspondingly there are infinitely many simple eigenvalues associated with the s-wave* ($l = 0$)*, moreover for $l \geq 1$ the eigenvalues of H_α are given by the usual Coulomb levels $E_m = -\gamma^2/(4m^2)$, $m \in \mathbb{N}$, $m \geq 2$.*

Because of the results of the previous proposition, we proceed our analysis only in the case of repulsive Coulomb potential, namely for $\gamma \geq 0$.

3.2 Trace of the relative resolvent

We first note that for any $\lambda \in \rho(H_0) \cap \rho(H_\alpha)$ the difference $\mathrm{tr}(R(\lambda; H_\alpha) - R(\lambda; H_0))$ is a rank one operator (see, e.g., [4]), then the trace of the relative resolvent of the pair (H_α, H_0) is well defined by

$$r(\lambda; H_\alpha, H_0) = \mathrm{tr}(R(\lambda; H_\alpha) - R(\lambda; H_0)).$$

By equation (8) and by the definition of $g(\lambda, x)$ it follows that

$$r(\lambda; H_\alpha, H_0) = -\frac{\Gamma\Big(1 + \frac{\gamma}{2\sqrt{-\lambda}}\Big)^2}{4\pi\alpha - \gamma F_\gamma\Big(\frac{\gamma}{2\sqrt{-\lambda}}\Big)} \int_0^\infty W^2_{-\gamma/(2\sqrt{-\lambda}),1/2}(2\sqrt{-\lambda}\,|x|)\,\mathrm{d}|x|,$$

with $\mathrm{Re}\sqrt{-\lambda} > 0$.

From (7.625.4) and (9.302.1) of [39] and by using the identities $\Gamma(1+z) = z\Gamma(z)$ and $\Gamma(1-z)\Gamma(z) = \pi/\sin(\pi z)$, $z \in \mathbb{C}\backslash\mathbb{Z}$, we get for the integral in the latter equation the expression

$$\begin{aligned}
&\int_0^\infty W^2_{-\gamma/(2\sqrt{-\lambda}),1/2}(2\sqrt{-\lambda}\,|x|)\,\mathrm{d}|x| \\
&\quad = \frac{1}{2\sqrt{-\lambda}} \frac{1}{\Gamma\Big(1 + \frac{\gamma}{2\sqrt{-\lambda}}\Big)\Gamma\Big(\frac{\gamma}{2\sqrt{-\lambda}}\Big)} \\
&\qquad \frac{1}{2\pi i} \int_L \frac{(1-s)s}{\Big(s + \frac{\gamma}{2\sqrt{-\lambda}}\Big)\Big(s - 1 + \frac{\gamma}{2\sqrt{-\lambda}}\Big)} \frac{\pi^2}{\sin^2(\pi s)}\,\mathrm{d}s,
\end{aligned}$$

where L is a path in $\mathbb{C}$ from $-\infty$ to $+\infty$ such that the set $\{1, 2, 3, ...\}$ is on the right of L and the set

$$\Big\{0, -1, -2, \dots, 1 - \frac{\gamma}{2\sqrt{-\lambda}}, -\frac{\gamma}{2\sqrt{-\lambda}}, -1 - \frac{\gamma}{2\sqrt{-\lambda}}, \dots\Big\}$$

is on the left of L. We notice that for $\gamma > 0$ one can choose

$$L = \Big\{s = x_0 + iy \text{ with } 1 - \gamma\frac{\mathrm{Re}\sqrt{-\lambda}}{2|\lambda|} < x_0 < 1, -\infty < y < \infty\Big\}.$$

This gives

$$r(\lambda; H_\alpha, H_0) = -\frac{1}{4\pi\alpha - \gamma F_\gamma\Big(\frac{\gamma}{2\sqrt{-\lambda}}\Big)} \frac{1}{2\sqrt{-\lambda}} I\Big(\frac{\gamma}{2\sqrt{-\lambda}}\Big), \tag{10}$$

where

$$I(z)=\frac{z}{2\pi i}\int_L\frac{(1-s)s}{(s+z)(s-1+z)}\frac{\pi^2}{\sin^2(\pi s)}\mathrm{d}s, \tag{11}$$

and we used again the identity $\Gamma(1+z)=z\Gamma(z)$, $z\in\mathbb{C}$. In order to analyze the function $I(z)$ appearing in the formula for the relative trace of the resolvent we need the formulas in the following lemma.

Lemma 3.3. *Let L be the path*

$$L=\{z=x_0+iy|\,1-a<x_0<1\,,\ -\infty<y<\infty\},$$

with $\mathrm{Re}(a)>0$, *then*

$$\frac{1}{2\pi i}\int_L\frac{\pi^2}{\sin^2\pi z}\mathrm{d}z=1,$$

and

$$\frac{1}{2\pi i}\int_L\frac{1}{z+a}\frac{\pi^2}{\sin^2\pi z}\mathrm{d}z=\psi'(a)-\frac{1}{a^2}.$$

Proof. For the first, we just integrate

$$\begin{aligned}\frac{1}{2\pi i}\int_L\frac{\pi^2}{\sin^2\pi z}\mathrm{d}z&=-\frac{1}{2i}[\cot(\pi z)]_{x_0-iy}^{x_0+iy}\\&=\lim_{y\to\infty}-\frac{1}{2}\Big(\frac{\mathrm{e}^{ix_0-y}+\mathrm{e}^{-ix_0+y}}{\mathrm{e}^{ix_0-y}-\mathrm{e}^{-ix_0+y}}-\frac{\mathrm{e}^{ix_0+y}+\mathrm{e}^{-ix_0-iy}}{\mathrm{e}^{ix_0+y}-\mathrm{e}^{-ix_0-y}}\Big)\\&=1.\end{aligned}$$

For the second one, we first integrate twice by parts. This gives

$$\int_L\frac{1}{a+z}\frac{\pi^2}{\sin^2\pi z}dz=-2\int_L\frac{\log\sin\pi z}{(a+z)^3}\mathrm{d}z.$$

Next,we use the product representation for the sine function:

$$\int_L\frac{\log\sin\pi z}{(a+z)^3}dz=\int_L\frac{\log\pi z}{(a+z)^3}\mathrm{d}z+\int_L\frac{1}{(a+z)^3}\sum_{k=1}^{\infty}\log\Big(1-\frac{z^2}{k^2}\Big)\mathrm{d}z.$$

The first term gives no contribution, for

$$\begin{aligned}\int_L\frac{\log\pi z}{(a+z)^3}\mathrm{d}z&=-\frac{1}{2}\Big[\frac{\log\pi z}{(a+x_0+iy)^2}\Big]_{y=-\infty}^{y=+\infty}+\frac{1}{2}\int_L\frac{1}{z(a+z)^2}\mathrm{d}z\\&=0+\frac{1}{2}\Big[\frac{1}{a(a+z)}-\frac{1}{a^2}\log\Big(1+\frac{a}{z}\Big)\Big]_{y=-\infty}^{y=+\infty}\\&=0.\end{aligned}$$

In the second term, due to uniform convergence, we can twist the sum with the integration. We have

$$\int_L \frac{1}{(a+z)^3} \log\left(1 - \frac{z^2}{k^2}\right) \mathrm{d}z = 0 - \int_L \frac{z}{(a+z)^2} \frac{1}{k^2 - z^2} \mathrm{d}z.$$

Assuming $\mathrm{Re}(a) > 0$, we can deform the path L to a contour of Hankel type: starting at infinity on the upper side of the real axis, turning around the point $z = k$ and going back to infinity below the real axis. Since the integrand vanishes as z^{-3} for large $\mathrm{Re}(z)$, we can further deform the path of integration to a circle around the point $z = k$. This gives

$$\int_L \frac{z}{(a+z)^2} \frac{1}{k^2 - z^2} \mathrm{d}z = -\frac{\pi i}{(a+k)^2},$$

and hence the second formula of the lemma follows recalling the definition of the digamma function $\psi(z)$, see (8.36) of [39]. □

Now we can use the result of the latter lemma to give an explicit expression for the function $I(z)$.

Lemma 3.4. *Let $I(z)$ be the function defined in equation* (11)*, then*

$$I(z) = 1 - 2z + 2\psi'(1+z)z^2.$$

Proof. We observe that

$$\frac{(1-s)s}{(s+z)(s-1+z)} = -1 + \frac{z(1+z)}{s+z} + \frac{z(1-z)}{s-1+z},$$

from which it follows that

$$\begin{aligned}\frac{z}{2\pi i} \int_L \frac{(1-s)s}{(s+z)(s-1+z)} \frac{\pi^2}{\sin^2 \pi s} \mathrm{d}s = &-\frac{z}{2\pi i} \int_L \frac{\pi^2}{\sin^2 \pi s} \mathrm{d}s \\ &+ \frac{z^2(1+z)}{2\pi i} \int_L \frac{1}{s+z} \frac{\pi^2}{\sin^2 \pi s} \mathrm{d}s \\ &+ \frac{z^2(1-z)}{2\pi i} \int_L \frac{1}{s-1+z} \frac{\pi^2}{\sin^2 \pi s} \mathrm{d}s.\end{aligned}$$

Using Lemma 3.3, and recalling that $\psi(z+1) = \psi(z) + 1/z$, after some calculation we have the stated formula. □

Proposition 3.5. *For any $\lambda \in \rho(H_\alpha) \cap \rho(H_0)$, the trace of the relative resolvent of the pair of operators (H_α, H_0) is given by*

$$\begin{aligned} r(\lambda; H_\alpha, H_0) &= -\frac{zI(z)}{\gamma(4\pi\alpha - \gamma F_\gamma(z))}\bigg|_{z=\gamma/(2\sqrt{-\lambda})} \\ &= -\frac{z(2\psi'(1+z)z^2 - 2z + 1)}{\gamma\Big(4\pi\alpha - \gamma(\psi(1+z) - \log z - \frac{1}{2z} - \psi(1) - \psi(2))\Big)}\Bigg|_{z=\gamma/(2\sqrt{-\lambda})}, \end{aligned} \tag{12}$$

with $\operatorname{Re}\sqrt{-\lambda} > 0$. *Moreover the following asymptotic expansion holds true for small λ,*

$$r(\lambda; H_\alpha, H_0) = \sum_{k=0}^{\infty} b_k(-\lambda)^k, \tag{13}$$

with $b_k \in \mathbb{R}$. The first coefficients are given by

$$b_0 = -\frac{1}{3\gamma(\gamma - 2C\gamma + 4\pi\alpha)}, \quad b_1 = \frac{(17-24C)\gamma + 48\pi\alpha}{45\gamma^3(2C\gamma - \gamma - 4\pi\alpha)^2}.$$

For large λ, we have the following asymptotic expansion

$$r(\lambda; H_\alpha, H_0) = \sum_{j=2,k=0}^{\infty} a_{j,k}(-\lambda)^{-\frac{j}{2}} \log^k(-\lambda), \tag{14}$$

with $a_{j,k} \in \mathbb{R}$. The first coefficients are given by

$$a_{2,0} = -\frac{1}{2}, \quad a_{2,k>0} = 0,$$

$$a_{3,0} = \frac{4\pi\alpha + (2-C)\gamma + \gamma(\log\gamma - \log 2)}{2}, \quad a_{3,1} = -\frac{\gamma}{4}, \quad a_{3,k>1} = 0.$$

Proof. Formula (12) follows directly from the equation (10) and from Lemma 3.4. The asymptotic expansions of the relative trace follow easily from classical expansion of the poly Gamma function. Recalling the expansions of the digamma function (see for example (8.344) of [39]) for large $|z|$, $z \in \mathbb{C}$,

$$\psi(1+z) = \log z + \frac{1}{2}\frac{1}{z} - \sum_{k=1}^{\infty} \frac{B_{2k}}{2k}\frac{1}{z^{2k}},$$

B_{2k} being Bernoulli numbers, and for small $z \in \mathbb{C}$ (see for example (8.342) of [39]):

$$\psi(1+z) = -C + \sum_{k=2}^{\infty}(-1)^k \zeta(k) z^{k-1},$$

where C is the Euler constant, and where ζ denotes the Riemann's zeta function. Since $z = \gamma/(2\sqrt{-\lambda})$ one obtains the expansions (13) and (14). □

4 The relative partition function of the Coulomb plus delta interaction

In this section we study the relative zeta function and the relative partition function of the model described in Section 3.

It is clear, by the result in Proposition 3.5, that the conditions (B.1), (B.2) and (B.3) of Lemma 2.1, necessary to define the relative zeta function are satisfied. Also, by the same proposition, the minimum value for the index j is $j = 2$, corresponding to $\alpha_2 = -1$, then the first terms in the expansion of the relative spectral measure, according to Lemma 2.4, are: first, the term corresponding to $\alpha_2 = -1$, that gives only a term in $1/v$, since $K_2 = 0$; second, the terms corresponding to $\alpha_3 = -3/2$, that gives a term in $1/v^2$ and a term in $1/v^2 \log v^2$, since $K_3 = 1$. Applying the formula in Lemma 2.4, the coefficients are

$$e_{2,0,0} = 0;$$

$$e_{3,0,0} = \frac{4\pi\alpha + (2-C)\gamma + \gamma\log\frac{\gamma}{2}}{\pi}, \quad e_{3,1,0} = 0, \quad e_{3,1,1} = -\frac{\gamma}{2\pi}.$$

Whence, we have the following expansion of the relative spectral measure:

$$e(v; H_\alpha, H_0) = O(v^k), \quad k > 0 \tag{15}$$

for $v \to 0^+$,

$$e(v; H_\alpha, H_0) = -\frac{\gamma}{\pi}\frac{1}{v^2}\log v + \frac{4\pi\alpha + (2-C)\gamma + \gamma\log\frac{\gamma}{2}}{\pi}\frac{1}{v^2} + O(v^{-3}\log v) \tag{16}$$

for $v \to +\infty$, and

$$e_{3,0} = \frac{4\pi\alpha + (2-C)\gamma + \gamma\log\frac{\gamma}{2}}{\pi}, \quad e_{3,1} = -\frac{\gamma}{\pi}.$$

All the coefficients $e_{j,h}$ with smaller indices vanish.

We are now in the position of analyzing the relative zeta function $\zeta(s; H_\alpha, H_0)$. In fact, what we are interested in is the expansion near $s = -1/2$.

Proposition 4.1. *The relative zeta function $\zeta(s; H_\alpha, H_0)$ has an analytic expansion to a meromorphic function analytic in the strip $0 \le \mathrm{Re}(s) \le 1$, up to a double pole at $s = -1/2$. Near $s = -1/2$, the following expansion holds:*

$$\begin{aligned}\zeta(s; H_\alpha, H_0) = {} & \frac{\frac{e_{3,1}}{4}}{\left(s+\frac{1}{2}\right)^2} + \frac{\frac{e_{3,0}}{2}}{s+\frac{1}{2}} + \int_0^1 v e(v; H_\alpha, H_0)\,\mathrm{d}v \\ & + \int_1^\infty v\Big(e(v; H_\alpha, H_0) - \frac{e_{2,1}}{v^2}\log v - \frac{e_{2,0}}{v^2}\Big)\mathrm{d}v + O\Big(s+\frac{1}{2}\Big),\end{aligned}$$

where

$$e_{3,1} = -\frac{\gamma}{\pi},$$

$$e_{3,0} = \frac{8\pi\alpha - 2C\gamma + 4\gamma + \gamma\log\frac{\gamma^2}{4}}{2\pi}.$$

Proof. By the expansions in equations (15) and (16) for the relative spectral measure, we see that the indices J_0 and J_∞ defined in Lemma 2.7 are respectively: $J_0 = 0$ and $J_\infty = 4$. Hence, by the same lemma, there are no poles arising from the expansion of the spectral measure for small v, and since the minimum value for the index j of α_j is $j = 2$, there are three terms arising from the expansion for large v. The first term is with $j = 2$, and vanishes since $e_{2,0} = 0$. The other two terms are with $j = 3$, and $k = 0$ and $k = H_3 = 1$. Applying the formula in Lemma 2.7 we compute these terms. □

Corollary 4.2. *The relative zeta function of the pair of operators $(L = -\partial_u^2 + H_\alpha$, $L_0 = -\partial_u^2 + H_0)$ on $S^1_{\beta/(2\pi)} \times \mathbb{R}^3$ has a simple pole at $s = 0$ with residua*

$$\mathrm{Res}_{1\,s=0}\,\zeta(s; L, L_0) = \frac{\gamma}{4\pi}\beta,$$

$$\mathrm{Res}_{0\,s=0}\,\zeta(s; L, L_0) = -\frac{8\pi\alpha - 2C\gamma + 4\gamma + \gamma\log\frac{\gamma^2}{4}}{4\pi}\beta + \frac{(1-\log 2)\gamma}{2\pi}\beta,$$

$$
\begin{aligned}
\operatorname{Res}_{0_{s=0}} \zeta'(s; L, L_0) = &-\Big(\int_0^1 v e(v; H_\alpha, H_0)\, dv \\
&+ \int_1^\infty v\Big(e(v; H_\alpha, H_0) - \frac{e_{3,1}}{v^2}\log v - \frac{e_{3,0}}{v^2}\Big) dv \Big)\beta \\
&- \frac{(1-\log 2)\Big(8\pi\alpha - 2C\gamma + 4\gamma + \gamma\log\frac{\gamma^2}{4}\Big)}{2\pi}\beta \\
&+ \Big(2 + \frac{\pi^2}{6} + 2(1-\log 2)^2\Big)\frac{\gamma}{4\pi}\beta \\
&- 2\int_0^\infty \log(1 - e^{-\beta v}) e(v; H_\alpha, H_0)\, dv.
\end{aligned}
$$

Proof. This is a simple consequence of Proposition 4.1 and Corollary 2.8. □

Using the formula in equation (3), we obtain the following result for the relative partition function, where ℓ is some renormalization constant,

$$
\begin{aligned}
\log Z_{\mathrm{R}} = &-\frac{1}{2}\Big(\int_0^1 v e(v; H_\alpha, H_0)\, dv \\
&+ \int_1^\infty v(e(v; H_\alpha, H_0) - \frac{e_{3,1}}{v^2}\log v - \frac{e_{3,0}}{v^2}) dv \Big)\beta \\
&- \frac{(1-\log 2)\Big(8\pi\alpha - 2C\gamma + 4\gamma + \gamma\log\frac{\gamma^2}{4}\Big)}{4\pi}\beta \\
&+ \Big(2 + \frac{\pi^2}{6} + 2(1-\log 2)^2\Big)\frac{\gamma}{8\pi}\beta \\
&- \int_0^\infty \log(1 - e^{-\beta v}) e(v; H_\alpha, H_0)\, dv \\
&+ \Big(4\pi\alpha - C\gamma + 2\gamma + \gamma\log\frac{\gamma}{2} - 1 + \gamma\log 2\Big)\frac{\beta}{2\pi}\log\ell.
\end{aligned}
$$

Acknowledgements. We thank the referee for the very useful suggestions and for pointing out additional references. The authors acknowledge the hospitality and financial support of the CIRM, University of Trento, Italy. C. Cacciapuoti also gratefully acknowledges the hospitality of the Mathematical Institute, Tohoku University, Sendai, Japan, of the Hausdorff Center for Mathematics, University of Bonn, Germany, and the support of the FIR 2013 project, code RBFR13WAET.

References

[1] S. Albeverio, Wiener and Feynman–path integrals and their applications. In V. Mandrekar and P. R. Masani (eds.), *Proceedings of the Norbert Wiener Centenary Congress. 1994.* Proceedings of Symposia in Applied Mathematics, 52. American Mathematical Society, Providence, R.I., 1997, 163–194. MR 1440913 Zbl 0899.60058

[2] S. Albeverio, Z. Brzeźniak, and L. Dabrowski, Fundamental solution of the heat and Schrödinger equations with point interaction. *J. Funct. Anal.* **130** (1995), no. 1, 220–254. MR 1331982 Zbl 0822.35002

[3] S. Albeverio, G. Cognola, M. Spreafico, and S. Zerbini, Singular perturbations with boundary conditions and the Casimir effect in the half space. *J. Math. Phys.* **51** (2010), no. 6, 063502, 38 pp. MR 2676479 Zbl 1311.47030

[4] S. Albeverio, F. Gesztesy, R. Høegh-Krohn, and H. Holden, *Solvable models in quantum mechanics.* Second edition. AMS Chelsea Publishing, Providence, R.I., 2005. With an Appendix by P. Exner. ISBN 0-8218-3624-2/hbk MR 2105735 Zbl 1078.81003

[5] S. Albeverio and P. Kurasov, *Singular perturbations of differential operators.* Solvable Schrödinger type operators. London Mathematical Society Lecture Note Series, 271. Cambridge University Press, Cambridge, 2000. ISBN 0-521-77912-X MR 1752110 Zbl 0945.47015

[6] M. Asorey, A. Ibort, and G. Marmo, Global theory of quantum boundary conditions and topology change. *Internat. J. Modern Phys. A* **20** (2005), no. 5, 1001–1025. MR 2123428 Zbl 1134.58302

[7] M. F. Atiyah, V. K. Patodi, and I. M. Singer, Spectral asymmetry and Riemannian geometry. III. *Math. Proc. Cambridge Philos. Soc.* **79** (1976), no. 1, 71–99. MR 0397799 Zbl 0325.58015

[8] E. Beth and G. E. Uhlenbeck, The quantum theory of the non-ideal gas. II. Behaviour at low temperatures. Physica **4** (1937), 915–924. JFM 63.1416.02

[9] M. Š. Birman and M. G. Kreĭn, On the theory of wave operators and scattering operators. *Dokl. Akad. Nauk SSSR* **144** (1962), 475–478. In Russian. English translation, *Soviet Math. Dokl* **3** (1962), 740–744. MR 0139007 Zbl 0196.45004

[10] L. Boi, *The quantum vacuum.* A scientific and philosophical concept, from electrodynamics to string theory and the geometry of the microscopic world. Johns Hopkins University Press, Baltimore, MD, 2011. MR 2849219 Zbl 1266.81003

[11] M. Bordag (ed.), *The Casimir effect 50 years later.* Proceedings of the 4th Workshop on Quantum Field Theory under the Influence of External Conditions held at the University of Leipzig, Leipzig, September 14–18, 1998. World Scientific, River Edge, N.J., 1999. ISBN 981-02-3820-7 MR 1890606 Zbl 0923.00028

[12] M. Bordag, G. L. Klimchitskaya, U. Mohideen, and V. M. Mostepanenko, *Advances in the Casimir effect.* International Series of Monographs on Physics, 145. Oxford University Press, Oxford, 2009. ISBN 978-0-19-923874-3/hbk Zbl 1215.81002

[13] M. Bordag, V. V. Nesterenko, and I .G. Pirozhenko, High temperature asymptotics of thermodynamic functions of an electromagnetic field subjected to boundary conditions on a sphere and cylinder. *Phys. Rev. D* **65** (2002), no. 4, 045011.

[14] M. Bordag, I. G. Pirozhenko, and V. V. Nesterenko, *Spectral analysis of a flat plasma sheet model. J. Phys. A* **38** (2005), no. 50, 11027–11043. MR 2199740 Zbl 1092.78003

[15] M. Bordag and D. V. Vassilevich, Heat kernel expansion for semitransparent boundaries. *J. Phys. A* **32** (1999), no. 47, 8247–8259. MR 1733335 Zbl 0969.81044

[16] J. Brüning, V. Geyler, and K. Pankrashkin, On-diagonal singularities of the Green functions for Schrödinger operators. *J. Math. Phys.* **46** (2005), no. 11, 113508, 16 pp. MR 2186783 Zbl 1111.81055

[17] J. Brüning and R. Seeley, The resolvent expansion for second-order regular singular operators. *J. Funct. Anal.* **73** (1987), no. 2, 369–429. MR 0899656 Zbl 0625.47040

[18] G. Bressi, G. Carugno, R. Onofrio, and G. Ruoso, Measurement of the Casimir force between parallel metallic plates. *Phys. Rev. Lett.* **88** (2002), no. 4, 041804.

[19] A. A. Bytsenko, E. Elizalde, S. Odintsov, A. Romeo, and S. Zerbini, *Zeta regularization techniques with applications.* World Scientific, River Edge, N.J., 1994. ISBN 981-02-1441-3 MR 1346490 Zbl 1050.81500

[20] H. B. G. Casimir, On the attraction between two perfectly conducting plates. *Proc. Kongl. Nedel. Akad. Wetensch* **51** (1948), 793–795. Zbl 0031.19005

[21] J. Choi and J. R. Quine, Zeta regularized products and functional determinants on spheres. *Rocky Mountain J. Math.* **26** (1996), no. 2, 719–729. MR 1406503 Zbl 0864.47024

[22] G. Cognola, L. Vanzo, and S. Zerbini, Regularization dependence of vacuum energy in arbitrarily shaped cavities. *J. Math. Phys.* **33** (1992), no. 1, 222–228. MR 1141520

[23] Y. Colin de Verdière, Pseudo-Laplaciens II. *Ann. Inst. Fourier* (*Grenoble*) **33** (1983), no. 2, 87–113. MR 0699488 Zbl 0496.58016

[24] M. V. Cougo Pinto, C. Farina, M. R. Negrão, and A. Tort, Casimir effect at finite temperature of charged scalar field in an external magnetic field. Preprint 1998. arXiv:hep-th/9810033

[25] J. S. Dowker and G. Kennedy, Finite temperature and boundary effects in static space-times. *J. Phys. A* **11** (1978), no. 5, 895–920. MR 0479266

[26] J. P. Dowling, The mathematics of the Casimir effect. *Math. Mag.* **62** (1989), no. 5, 324–331. MR 1031431

[27] G. V. Dunne and K. Kirsten, Simplified vacuum energy expressions for radial backgrounds and domain walls. *J. Phys. A* **42** (2009), no. 7, 075402, 22 pp. MR 2525471 Zbl 1156.81029

[28] E. Elizalde, *Ten physical applications of spectral zeta functions.* Lecture Notes in Physics. New Series m: Monographs, 35. Springer, Berlin, 1995. ISBN 3-540-60230-5 MR 1448403 Zbl 0855.00002

[29] E. Elizalde, L. Vanzo, and S. Zerbini, Zeta-function regularization, the multiplicative anomaly and the Wodzicki Residue. *Comm. Math. Phys.* **194** (1998), no. 3, 613–630. MR 1631485 Zbl 0938.58028

[30] A. Erdélyi, W. Magnus, F. Oberhettinger, and F. G. Tricomi (eds.), *Tables of integral transforms.* Vol. II. Based, in part, on notes left by H. Bateman. Bateman Manuscript Project. California Institute of Technology. McGraw–Hill, New York etc., 1954. MR 0065685 Zbl 0058.34103

[31] D. Fermi and L. Pizzocchero, Local zeta regularization and the Casimir effect. *Prog. Theor. Phys.* **126** (2011), no. 3, 419-434. Zbl 1242.81122

[32] D. Fermi and L. Pizzocchero, Local zeta regularization and the Casimir effect. I. A general approach based on integral kernels. Preprint 2015. II. Some explicitly solvable cases. Preprint 2015. III. The case with a background harmonic potential. *Int. J. Mod. Phys. A* **30** (2015), no. 35, 1550213, 42 pp. IV. The case of a rectangular box. *Int. J. Mod. Phys. A* **31** (2016), no. 4-5, 1650003, 56 pp. arXiv:1505.00711 [math-ph] (I) arXiv:1505.01044 [math-ph] (II) Zbl 1337.81109 (III) Zbl 1333.81387 (IV)

[33] M. Fierz, On the attraction of conducting planes in vacuum. *Helv. Phys. Acta* **33** (1960), 855–858.

[34] R. Figari, R. Höegh-Krohn, and C. R. Nappi, Interacting relativistic boson fields in the de Sitter universe with two space-time dimensions. *Comm. Math. Phys.* **44** (1975), no. 3, 265–278. MR 0384010

[35] S. A. Fulling, *Aspects of quantum field theory in curved space-time.* London Mathematical Society Student Texts, 17. Cambridge University Press, Cambridge, 1989. ISBN 0-521-34400-X MR 1071177 Zbl 0677.53081

[36] G. Gibbons, Thermal zeta functions. *Phys. Letters A* **60** (1977), 385–386.

[37] P. B. Gilkey, *Invariance theory, the heat equation, and the Atiyah–Singer index theorem.* Second edition. Studies in Advanced Mathematics. CRC Press, Boca Raton, FL, 1995. ISBN 0-8493-7874-4 MR 1396308 Zbl 0856.58001

[38] P. B. Gilkey, K. Kirsten, and D. V. Vassilevich, Heat trace asymptotics defined by transfer boundary conditions. *Lett. Math. Phys.* **63** (2003), no. 1, 29–37. MR 1967534 Zbl 1044.58034

[39] I. S. Gradshteyn and I. M. Ryzhik, *Table of integrals, series, and products.* Seventh edition. Translated from the Russian. Translation edited and with a preface by A. Jeffrey and D. Zwillinger. With one CD-ROM (Windows, Macintosh and UNIX). Elsevier/Academic Press, Amsterdam, 2007. ISBN 978-0-12-373637-6, 0-12-373637-4 MR 2360010 Zbl 1208.65001

[40] N. Graham, R. L. Jaffe, V. Khemani, M. Quandt, M. Scandurra, and H. Weigel, Calculating vacuum energies in renormalizable quantum field theories: A new approach to the Casimir problem. *Nucl. Phys. B* **645** (2002), 49–84. Zbl 0999.81059

[41] S. W. Hawking, Zeta function regularization of path integral in curved spacetime. *Comm. Math. Phys.* **55** (1977), no. 2, 133–148. MR 0524257

[42] L. Hostler, Runge–Lenz vector and the Coulomb Green's functions. *J. Math. Phys.* **8** (1967), no. 3, 642–646.

[43] N. R. Khusnutdinov, Zeta-function approach to Casimir energy with singular potentials. *Phys. Rev. D* (3) **73** (2006), no. 2, 025003, 13 pp. MR 2209635

[44] M. G. Kreĭn, On the trace formula in perturbation theory. *Mat. Sbornik N.S.* **33(75)** (1953), 597–626. In Russian. MR 0060742

[45] M. G. Kreĭn, Perturbation determinants and a formula for the traces of unitary and self-adjoint operators. *Dokl. Akad. Nauk SSSR* **144** (1962), 268–271. In Russian. English translation, *Soviet Math. Dokl* **3** (1962), 707–710. MR 0139006

[46] N. Kurokawa, *Multiple sine functions and Selberg zeta functions. Proc. Japan Acad. Ser. A Math. Sci.* **67** (1991), no. 3, 61–64. MR 1105522 Zbl 0738.11041

[47] S. K. Lamoureaux, Demonstration of the Casimir force in the 0.6 to 6μm range. *Phys. Rev. Lett.* **78** (1997), 5–8.

[48] M. Lesch, Determinants of regular singular Sturm–Liouville operators. *Math. Nachr.* **194** (1998), 139–170. MR 1653090 Zbl 0924.58107

[49] J. Mehra, Temperature correction to the Casimir effect. Physica **37** (1967), no. 1, 145–152.

[50] K. A. Milton, The Casimir effect: the physical manifestation of zero-point energy. World Scientific, River Edge, N.J., 2001. ISBN 981-02-4397-9 MR 1876766 Zbl 1025.81003

[51] V. Moretti, Local ζ-function techniques vs. point-splitting procedure: a few rigorous results. *Comm. Math. Phys.* **201** (1999), no. 2, 327–363. MR 1682226 Zbl 0937.58024

[52] V. M. Mostepanenko and N. N. Trunov, *The Casimir effect and its applications.* Oxford Science Publications. Translated by R. L. Znajek. Oxford University Press, The Clarendon Press, Oxford, 1997. ISBN 0198539983

[53] W. Müller, Relative zeta functions, relative determinants and scattering theory. *Comm. Math. Phys.* **192** (1998), no. 2, 309–347. MR 1617554 Zbl 0947.58025

[54] J. M. Muñoz Castañeda and J. Mateos Guilarte, $\delta - \delta'$ generalized Robin boundary conditions and quantum vacuum fluctuations. *Phys. Rev. D* **91** (2015), 025028, 21 pp.

[55] V. V. Nesterenko, G. Lambiase, and G. Scarpetta, Calculation of the Casimir energy at zero and finite temperature: Some recent results. *Riv. Nuovo Cimento* **27** (2004), no. 6, 1–74.

[56] V. V. Nesterenko and I. G. Pirozhenko, Lifshitz formula by a spectral summation method. *Phys. Rev. A* **86** (2012), 052503.

[57] G. Ortenzi and M. Spreafico, Zeta function regularization for a scalar field in a bounded domain. *J. Phys. A* **37** (2004), 11499–11517.

[58] D. K. Park, Green's function approach to two and three-dimensional delta function potentials and application to the spin $1/2$ Aharonov–Bohm problem. *J. Math. Phys.* **36** (1995), no. 10, 5453–5464. MR 1353099 Zbl 0865.46057

[59] G. Plunien, B. Müller, and W. Greiner, The Casimir effect. *Phys. Rep.* **134** (1986), no. 2-3, 87–193. MR 0828676

[60] E. A. Power, *Introductory quantum electrodynamics.* American Elsevier Publishing Co.,New York, 1965. MR 0193904 Zbl 0124.22903

[61] D. B. Ray and I. M. Singer, R-torsion and the Laplacian on Riemannian manifolds. *Advances in Math.* **7** (1971), 145–210. MR 0295381 Zbl 0239.58014

[62] M. Scandurra, The ground state energy of a massive scalar field in the background of a semi-transparent spherical shell. *J. Phys. A* **32** (1999), no. 30, 5679–5691. MR 1720384 Zbl 1042.81564

[63] S. Scarlatti and A. Teta, Derivation of the time dependent propagator for the three-dimensional Schrödinger equation with one point interaction. *J. Phys. A* **23** (1990), no. 19, L1033–L1035. MR 1076898 Zbl 0729.35108

[64] S. N. Solodukhin, Exact solution for a quantum field with delta like interaction. *Nucl. Phys. B* **541** (1999), 461–482.

[65] M. J. Spaarnay, Measurements of attractive forces between flat plates. *Physica* **24** (1958), 751–764.

[66] M. Spreafico, Zeta function and regularized determinant on projective spaces. *Rocky Mountain J. Math.* **33** (2003), no. 4, 1499–1512. MR 2052503 Zbl 1086.11042

[67] M. Spreafico, A generalization of the Euler Gamma function. *Funktsional. Anal. i Prilozhen.* **39** (2005), no. 2, 87–91. In Russian. English translation, *Funct. Anal. Appl.* **39** (2005), no. 2, 156–159. MR 2161522 Zbl 1115.33002

[68] M. Spreafico, Zeta invariants for sequences of spectral type, special functions and the Lerch formula. *Proc. Roy. Soc. Edinburgh Sect. A* **136** (2006), no. 4, 863–887. MR 2250451 Zbl 1219.11134

[69] M. Spreafico, Zeta determinant and operator determinants. *Osaka J. Math.* **48** (2011), no. 1, 41–50. MR 2802591 Zbl 1222.58028

[70] M. Spreafico and S. Zerbini, Finite temperature quantum field theory of non compact domain and application to delta interaction. *Rep. Math. Phys.* **63** (2009), 163–177.

[71] K. Symanzik, Schrödinger representation and Casimir effect in renormalizable quantum field theory. *Nuclear Phys. B* **190** (1981), no. 1, FS 3, 1–44. MR 0623382

Inequivalence of quantum Dirac fields of different masses and the underlying general structures involved

Asao Arai

Dedicated to Professor Pavel Exner on the occasion of his 70th birthday

1 Introduction

In a previous paper [2], the author showed that there exists a general mathematical structure behind the fact (Theorem X.46 in [6]) that the time-zero Hermitian scalar fields of different masses as representations of the canonical commutation relations (CCR) over $L^2_{\mathbb{R}}(\mathbb{R}^3)$ (the real Hilbert space of square integrable functions on $\mathbb{R}^3$) are mutually inequivalent, an interesting fact which may allow one to view the boson masses as objects distinguishing elements in a family of inequivalent representations of the CCR over $L^2_{\mathbb{R}}(\mathbb{R}^3)$, giving a representation theoretic meaning to the boson masses. A point in [2] lies in understanding that the family of time-zero Hermitian scalar fields, which is indexed by the mass parameter, is a special example of a general class of representations of the CCR over an abstract Hilbert space indexed by a set of (unbounded) self-adjoint operators and the inequivalence of the time-zero scalar fields of different masses can be derived as a simple application of a theorem on inequivalence of the representations under consideration in the abstract framework. Based on this structure, a new class of representations of the CCR over $L^2_{\mathbb{R}}(\mathbb{R}^d)$ with $d \in \mathbb{N}$ arbitrary, including as a special case the time-zero Hermitian scalar fields mentioned above, was found [2].

As a next step of research, it is natural to ask if there exist similar structures in the case of Fermi fields, typically quantum Dirac fields. In this paper, we show that the answer to the question is in the affirmative.

In Section 2, we introduce a family of irreducible representations of the canonical anticommutation relations (CAR) over an abstract Hilbert space $\mathcal{H}$, where their representation space is taken to be the fermion Fock space over $\mathcal{H}$. To the author's best knowledge, this family of representations of CAR may be new. We prove a theorem on inequivalence of these representations. In Section 3, we construct a free quantum

Dirac field in the $(1 + d)$-dimensional space-time on the fermion Fock space over $L^2(\mathbb{R}^{d*}; \mathbb{C}^\nu)$, the Hilbert space of $\mathbb{C}^\nu$-valued square integrable functions on $\mathbb{R}^{d*}$ (the d-dimensional wave vector space or the d-dimensional momentum space[1]), where ν is defined by (13) below. As an application of the inequivalence theorem in Section 2, we prove that the free quantum Dirac fields of different masses as well as interacting ones are inequivalent. In Section 4, we construct a general class of inequivalent representations of the CAR over $L^2(\mathbb{R}^{d*}; \mathbb{C}^\nu)$, which is a generalization of time-zero quantum Dirac fields. In the last section, we consider quantum Dirac fields on a d-dimensional box M for comparison with those on $\mathbb{R}^d$. We prove that the quantum Dirac fields on M of different positive masses are equivalent if and only if $d = 1$.

2 A family of irreducible representations of the CAR over a Hilbert space

The inner product and the norm of a Hilbert space $\mathcal{H}$ are denoted by $\langle \cdot, \cdot \rangle_{\mathcal{H}}$ (antilinear in the left variable) and $\| \cdot \|_{\mathcal{H}}$ respectively. But we sometimes omit the subscript $\mathcal{H}$ in $\langle \cdot, \cdot \rangle_{\mathcal{H}}$ and $\| \cdot \|_{\mathcal{H}}$ if there is no danger of confusion. For a linear operator A on a Hilbert space, we denote its domain by $D(A)$. If A is densely defined, then we denote its adjoint by A^*.

We first recall the concept of representation of CAR.

Definition 2.1. Let $\mathfrak{F}$ and $\mathcal{H}$ be complex Hilbert spaces, and

$$\mathfrak{A}(\mathcal{H}) := \{\psi(f), \psi(f)^* \mid f \in \mathcal{H}\}$$

be a subset of $\mathfrak{B}(\mathfrak{F})$, the Banach space of everywhere defined bounded linear operators on $\mathfrak{F}$.

1. The pair $(\mathfrak{F}, \mathfrak{A}(\mathcal{H}))$ is called a *representation of the CAR* over $\mathcal{H}$ if the following conditions (a) and (b) hold.

   (a) *Antilineality.* For all $f, g \in \mathcal{H}$ and $\alpha, \beta \in \mathbb{C}$,

$$\psi(\alpha f + \beta g) = \alpha^* \psi(f) + \beta^* \psi(g).$$

[1] We use the physical unit system where $\hbar$ (the Planck constant divided by 2π) and the light speed c are equal to 1.

(b) *CAR*. For all $f, g \in \mathcal{H}$,

$$\{\psi(f), \psi(g)\} = 0, \quad \{\psi(f), \psi(g)^*\} = \langle f, g\rangle,$$

where $\{X, Y\} := XY + YX$.

2. The representation $(\mathfrak{F}, \mathfrak{A}(\mathcal{H}))$ is *irreducible* if there exist no proper subspaces of $\mathfrak{F}$ which remain invariant under the action of all operators $\psi(f)$ and $\psi(f)^*$ ($f \in \mathcal{H}$) in $\mathfrak{A}(\mathcal{H})$.

3. Let $(\mathfrak{F}', \mathfrak{A}'(\mathcal{H}))$, with $\mathfrak{A}'(\mathcal{H}) := \{\psi'(f), \psi'(f)^* \mid f \in \mathcal{H}\})$, be another representation of the CAR over $\mathcal{H}$. Then the two representations $(\mathfrak{F}', \mathfrak{A}'(\mathcal{H}))$ and $(\mathfrak{F}, \mathfrak{A}(\mathcal{H}))$ are *equivalent* if there exists a unitary operator $\mathbb{U}\colon \mathfrak{F} \to \mathfrak{F}'$ such that, for all $f \in \mathcal{H}$, $\psi'(f) = \mathbb{U}\psi(f)\mathbb{U}^*$.

Remark 2.2. Taking the adjoint of the first equation in (b) in Definition 2.1(1), we have $\{\psi(f)^*, \psi(g)^*\} = 0$, $f, g \in \mathcal{H}$.

As is well known, representations of the CAR over a Hilbert space can be constructed on fermion Fock spaces (see, e.g., Chapter 5 of [1], §5.2 of [3], [4], and [8]). For the reader's convenience, we first review elementary aspects of fermion Fock spaces.

Let $\mathcal{H}$ be a complex Hilbert space. Then the fermion Fock space $\mathfrak{F}(\mathcal{H})$ over $\mathcal{H}$ is defined as the infinite direct sum Hilbert space

$$\mathfrak{F}(\mathcal{H}) := \bigoplus_{n=0}^{\infty} \bigwedge^{n} \mathcal{H}$$

of the n-fold antisymmetric tensor product Hilbert space $\bigwedge^n \mathcal{H}$ of $\mathcal{H}$ with convention $\bigwedge^0 \mathcal{H} := \mathbb{C}$.

For each $f \in \mathcal{H}$, the creation operator $A^\dagger(f)$ with test vector f acting in $\mathfrak{F}(\mathcal{H})$ is defined by

$$D(A^\dagger(f)) := \Big\{\Psi = \{\Psi^{(n)}\}_{n=0}^{\infty} \in \mathfrak{F}(\mathcal{H}) \;\Big|\; \sum_{n=1}^{\infty} n\|A_n(f \otimes \Psi^{(n-1)})\|^2 < \infty\Big\},$$

$$(A^\dagger(f)\Psi)^{(0)} := 0, \quad (A^\dagger(f)\Psi)^{(n)} := \sqrt{n}A_n(f \otimes \Psi^{(n-1)}),$$

for $n \geq 1$ and $\Psi \in D(A^\dagger(f))$, where A_n denotes the antisymmetrization operator on the n-fold tensor product Hilbert space $\bigotimes^n \mathcal{H}$ of $\mathcal{H}$. Since $D(A^\dagger(f))$ includes the finite particle subspace

$$\mathfrak{F}_0 := \{\Psi \in \mathfrak{F}(\mathcal{H}) \mid \text{there exists } n_0 \text{ such that, for all } n \geq n_0,\ \Psi^{(n)} = 0\},$$

which is dense in $\mathfrak{F}(\mathcal{H})$, it follows that $D(A^\dagger(f))$ is dense in $\mathfrak{F}(\mathcal{H})$. Hence the adjoint

$$A(f) := (A^\dagger(f))^*$$

of $A^\dagger(f)$ exists and is called the *annihilation operator with test vector* f. It follows that $A(f)^* = A^\dagger(f)$. We denote by $A(f)^\#$ either $A(f)$ or $A(f)^*$.

It is proved (see, e.g., Chapter 5 of [1] and §5.2 of [3]) that $D(A(f)^\#) = \mathfrak{F}(\mathcal{H})$ and $A(f)^\#$ is bounded satisfying the following anticommutation relations

$$\{A(f), A(g)\} = 0, \quad \{A(f), A(g)^*\} = \langle f, g\rangle, \quad f, g \in \mathcal{H}. \tag{1}$$

The correspondence $\mathcal{H} \ni f \mapsto A(f)$ (resp. $A(f)^*$) is complex antilinear (resp. linear). Thus $(\mathfrak{F}(\mathcal{H}), \{A(f), A(f)^* \mid f \in \mathcal{H}\})$ is a representation of the CAR over $\mathcal{H}$, which is called the *Fock representation* of the CAR over $\mathcal{H}$. It is well known that it is irreducible (see, e.g., Theorem 10.2 in [8]).

We now fix an orthogonal decomposition

$$\mathcal{H} = \mathcal{H}_+ \oplus \mathcal{H}_-$$

of $\mathcal{H}$ with $\mathcal{H}_+$ and $\mathcal{H}_-$ being mutually orthogonal nontrivial closed subspaces ($\mathcal{H}_+ \neq \{0\}, \mathcal{H}$) and a conjugation C on $\mathcal{H}$ (C is an antilinear mapping on $\mathcal{H}$ satisfying $C^2 = I$ (identity) and $\|Cf\| = \|f\|$, $f \in \mathcal{H}$). We have

$$\langle Cf, Cg\rangle = \langle g, f\rangle, \quad f, g \in \mathcal{H}.$$

We denote by $\mathfrak{B}_\pm$ the Banach space of everywhere defined bounded linear operators from $\mathcal{H}$ to $\mathcal{H}_\pm$ and introduce a subset of the direct product space $\mathfrak{B}_+ \times \mathfrak{B}_-$:

$$\mathfrak{T}(\mathcal{H}) := \{T = (T_+, T_-) \mid T_\pm \in \mathfrak{B}_\pm,\ T_+^* T_+ + \bar{T}_-^* \bar{T}_- = I\},$$

where

$$\bar{T}_- := C T_- C.$$

It is easy to see that $\bar{T}_-$ is a bounded linear operator on $\mathcal{H}$ with operator norm $\|\bar{T}_-\| = \|T_-\|$.

Each $T \in \mathfrak{T}(\mathcal{H})$ defines an element of $\mathfrak{B}(\mathcal{H})$ by

$$Tf := (T_+ f, T_- f), \quad f \in \mathcal{H}.$$

For each $T \in \mathfrak{T}(\mathcal{H})$, we define an antilinear mapping $\psi_T : \mathcal{H} \to \mathfrak{B}(\mathfrak{F})$ by

$$\psi_T(f) := A(T_+ f, 0) + A(0, T_- C f)^*, \quad f \in \mathcal{H}.$$

It is obvious that

$$\psi_T(f)^* = A(T_+ f, 0)^* + A(0, T_- C f).$$

Let

$$\mathfrak{A}_T(\mathcal{H}) := \{\psi_T(f), \psi_T^*(f) \mid f \in \mathcal{H}\}.$$

Lemma 2.3. *For all $f, g \in \mathcal{H}$, the following anticommutation relations hold*:

$$\{\psi_T(f), \psi_T(g)\} = \{\psi_T(f)^*, \psi_T(g)^*\} = 0, \tag{2}$$

$$\{\psi_T(f), \psi_T(g)^*\} = \langle f, g\rangle. \tag{3}$$

Proof. By direct computations using (1), we have

$$\{\psi_T(f), \psi_T(g)^*\} = \langle T_+ f, T_+ g\rangle + \langle \bar{T}_- f, \bar{T}_- g\rangle.$$

Since $T_+^* T_+ + \bar{T}_-^* \bar{T}_- = I$, we obtain (3). Similarly one can prove (2). □

Lemma 2.3 shows that $(\mathfrak{F}(\mathcal{H}), \mathfrak{A}_T(\mathcal{H}))$ is a representation of the CAR over $\mathcal{H}$.

Remark 2.4. The standard choice of $T = (T_+, T_-)$ in the literature is given by $T_+ = P$ and $T_- = C(I - P)C$, where P is the orthogonal projection onto $\mathcal{H}_+$ (see, e.g., pp. 22–23 of [4] and §10.1.3 in [8]). In this case, the representation is called a *quasi-free representation.* Hence the representation $(\mathfrak{F}(\mathcal{H}), \mathfrak{A}_T(\mathcal{H}))$ gives a generalization of of the quasi-free representation.

To find more detailed properties of the representation $(\mathfrak{F}(\mathcal{H}), \mathfrak{A}_T(\mathcal{H}))$, we introduce the following additional conditions (T.1)–(T.3) for $T = (T_+, T_-) \in \mathfrak{T}(\mathcal{H})$:

(T.1) $T_+ T_+^* = I$;

(T.2) $T_- T_-^* = I$;

(T.3) $T_- \bar{T}_+^* = 0$.

We define a subset of $\mathfrak{T}(\mathcal{H})$:

$$\mathfrak{T}_*(\mathcal{H}) := \{T \in \mathfrak{T}(\mathcal{H}) \mid \text{(T.1)–(T.3) hold}\}.$$

Lemma 2.5. *Let $T \in \mathfrak{T}_*(\mathcal{H})$. Then*

$$A(f) = \psi_T(T_+^* f_+) + \psi_T(C T_-^* f_-)^*, \quad f = (f_+, f_-),\ f_\pm \in \mathcal{H}_\pm. \tag{4}$$

Proof. By (T.1) and (T.3), we have

$$\psi_T(T_+^* f_+) = A(f_+, 0), \quad f_+ \in \mathcal{H}_+.$$

Property (T.3) implies that

$$\bar{T}_+ T_-^* = 0.$$

By this fact and (T.2), we obtain

$$\psi_T(C T_-^* f_-) = A(0, f_-)^*, \quad f_- \in \mathcal{H}_-.$$

We have

$$A(f) = A(f_+, 0) + A(0, f_-).$$

Thus we obtain (4). □

Lemma 2.6. *For all $T \in \mathfrak{T}_*(\mathcal{H})$, $(\mathfrak{F}(\mathcal{H}), \mathfrak{A}_T(\mathcal{H}))$ is irreducible.*

Proof. Let $\mathfrak{D}$ be a closed subspace which is invariant under the action of all $\psi_T(f)$ and $\psi_T(f)^*$ $(f \in \mathcal{H})$. Let Q be the orthogonal projection onto $\mathfrak{D}$. Then it follows that, for all $f \in \mathcal{H}$, $Q\psi_T(f)^\# = \psi_T(f)^\# Q$. Hence, by (4), $QA(f)^\# = A(f)^\# Q$ for all $f \in \mathcal{H}$. It is well known (or easy to see) that $\{A(f), A(f)^* \mid f \in \mathcal{H}\}$ is irreducible. Hence $Q = \alpha I$ for some $\alpha \in \mathbb{C}$. But, since Q is an orthogonal projection, it follows that $\alpha = 0$ or $\alpha = 1$. This means that $\mathfrak{D} = \{0\}$ or $\mathfrak{D} = \mathfrak{F}(\mathcal{H})$. Thus $\mathfrak{A}_T(\mathcal{H})$ is irreducible. □

Lemmas 2.3 and 2.6 immediately yield the following theorem.

Theorem 2.7. *For each $T \in \mathfrak{T}_*(\mathcal{H})$, $(\mathfrak{F}(\mathcal{H}), \mathfrak{A}_T(\mathcal{H}))$ is an irreducible representation of the CAR over $\mathcal{H}$.*

Thus we have a family $\{(\mathfrak{F}(\mathcal{H}), \mathfrak{A}_T(\mathcal{H})\}_{T \in \mathfrak{T}_*(\mathcal{H})}$ of irreducible representations of the CAR over $\mathcal{H}$.

We next consider equivalence or inequivalence of two representations $(\mathfrak{F}(\mathcal{H}), \mathfrak{A}_T(\mathcal{H}))$ and $(\mathfrak{F}(\mathcal{H}), \mathfrak{A}_S(\mathcal{H}))$ with $S \neq T$ $(S, T \in \mathfrak{T}_*(\mathcal{H}))$.

For each pair $(S, T) \in \mathfrak{T}_*(\mathcal{H}) \times \mathfrak{T}_*(\mathcal{H})$, we define linear operators V and W on $\mathcal{H}$ as follows:

$$Vf := (S_+ T_+^* f_+, S_- T_-^* f_-),$$

$$Wf := (S_+ \bar{T}_-^* f_-, S_- \bar{T}_+^* f_+),$$

for $f = (f_+, f_-) \in \mathcal{H}$.

Lemma 2.8. *The following equations hold:*

$$V^*V + \overline{W}^*\overline{W} = I, \tag{5}$$

$$VV^* + WW^* = I, \tag{6}$$

$$\overline{V}^*\overline{W} + W^*V = 0, \tag{7}$$

$$\overline{V}W^* + \overline{W}V^* = 0. \tag{8}$$

Proof. The operators V and W have the operator matrix representations

$$V = \begin{pmatrix} S_+T_+^* & 0 \\ 0 & S_-T_-^* \end{pmatrix},$$

$$W = \begin{pmatrix} 0 & S_+\overline{T}_-^* \\ S_-\overline{T}_+^* & 0 \end{pmatrix}.$$

Using these representations and properties of S and T, one can easily prove (5)–(8) by direct computations. □

We define

$$B(f) := A(Vf) + A(WCf)^*, \quad f \in \mathcal{H}.$$

Then it is easy to see that $\{B(f), B(f)^* \mid f \in \mathcal{H}\}$ is a representation of the CAR over $\mathcal{H}$.

Lemma 2.9. *There exists a unitary operator* $\mathbb{U}$ *on* $\mathfrak{F}(\mathcal{H})$ *such that*

$$A(f) = \mathbb{U}B(f)\mathbb{U}^*, \quad f \in \mathcal{H}, \tag{9}$$

if and only if W *is Hilbert–Schmidt.*

Proof. Since we have (5)–(8), the lemma follows from the general theory of Bogoliubov transformations on $\mathfrak{F}(\mathcal{H})$ (see, e.g., §10.3 of [8]). □

Theorem 2.10. *Let* T *and* S *be in* $\mathfrak{T}_*(\mathcal{H})$ *with* $T \neq S$. *Then the two representations* $(\mathfrak{F}(\mathcal{H}), \mathfrak{A}_T(\mathcal{H}))$ *and* $(\mathfrak{F}(\mathcal{H}), \mathfrak{A}_S(\mathcal{H}))$ *are equivalent if and only if* $S_+\overline{T}_-^*$ *and* $S_-\overline{T}_+^*$ *are Hilbert–Schmidt.*

Proof. Suppose that $(\mathfrak{F}(\mathcal{H}), \mathfrak{A}_T(\mathcal{H}))$ and $(\mathfrak{F}(\mathcal{H}), \mathfrak{A}_S(\mathcal{H}))$ are equivalent. Then there exists a unitary operator $\mathbb{U}$ on $\mathfrak{F}(\mathcal{H})$ such that

$$\mathbb{U}\psi_S(f)\mathbb{U}^* = \psi_T(f), \quad f \in \mathcal{H}. \tag{10}$$

Then, by (T.1)–(T.3), we have

$$A(f_+, 0) = \mathbb{U}(A(S_+T_+^* f_+, 0) + A(0, S_-CT_+^* f_+)^*)\mathbb{U}^*, \quad f_+ \in \mathcal{H}_+ \tag{11}$$

and

$$A(0, f_-)^* = \mathbb{U}(A(S_+CT_-^* f_-, 0) + A(0, S_-T_-^* f_-)^*)\mathbb{U}^*, \quad f_- \in \mathcal{H}_-. \tag{12}$$

Hence, adding the first equation to the adjoint of the second one, we obtain (9). Therefore, by Lemma 2.9, W is Hilbert–Schmidt, which is equivalent to that $S_+\bar{T}_-^*$ and $S_-\bar{T}_+^*$ are Hilbert–Schmidt.

Conversely, suppose that $S_+\bar{T}_-^*$ and $S_-\bar{T}_+^*$ are Hilbert–Schmidt. Then W is Hilbert–Schmidt. Hence, by Lemma 2.9, there exists a unitary operator $\mathbb{U}$ on $\mathfrak{F}(\mathcal{H})$ such that (9) holds. Then both (11) and (12) hold. Using $T_+^*T_+ + \bar{T}_-^*\bar{T}_- = I$, we obtain (10). □

The contraposition of Theorem 2.10 gives an inequivalence theorem on the two representations $(\mathfrak{F}(\mathcal{H}), \mathfrak{A}_T(\mathcal{H}))$ and $(\mathfrak{F}(\mathcal{H}), \mathfrak{A}_S(\mathcal{H}))$:

Theorem 2.11. *Let T and S be in $\mathfrak{T}_*(\mathcal{H})$ with $T \neq S$. Then the two representations $(\mathfrak{F}(\mathcal{H}), \mathfrak{A}_T(\mathcal{H}))$ and $(\mathfrak{F}(\mathcal{H}), \mathfrak{A}_S(\mathcal{H}))$ are inequivalent if and only if $S_+\bar{T}_-^*$ or $S_-\bar{T}_+^*$ is not Hilbert–Schmidt.*

3 Inequivalence of quantum Dirac fields of different masses

In this section, we consider a free quantum Dirac field on the $(1 + d)$-dimensional space-time and an interacting one as well. We take the space dimension $d \in \mathbb{N}$ to be arbitrary, because we want to see the dependence or independence of properties of the Dirac field on d. The main purpose of this section is to show by applying Theorem 2.11 that the time-zero quantum Dirac fields of different masses, which are representations of the CAR over a Hilbert space, are mutually inequivalent. This implies that the time-t quantum Dirac fields of different masses ($t \in \mathbb{R}$) also are mutually inequivalent.

3.1 The d-dimensional free Dirac operator

For each $d \in \mathbb{N}$, we define $\nu \in \mathbb{N}$ as follows:

$$\nu := \begin{cases} 2^{(d+1)/2} & \text{if } d \text{ is odd,} \\ 2^d & \text{if } d \text{ is even.} \end{cases} \tag{13}$$

It is well known that there exist $\nu \times \nu$ Hermitian matrices $\{\alpha_j\}_{j=1}^d$ and β satisfying the following anticommutation relations (a representation of the Clifford algebra associated with the Euclidean vector space $\mathbb{R}^{1+d}$):

$$\{\alpha_j, \alpha_k\} = 2\delta_{jk} I_\nu, \quad j,k = 1,\ldots,d, \tag{14}$$

$$\{\alpha_j, \beta\} = 0, \quad j = 1,\ldots,d, \tag{15}$$

$$\beta^2 = I_\nu, \tag{16}$$

where δ_{jk} is the Kronecker delta and I_n ($n \in \mathbb{N}$) is the $n \times n$ identity matrix.

The free Dirac equation with mass $m \geq 0$ in the $(1+d)$-dimensional space-time $\mathbb{R}^{1+d} := \{(t,x) \mid t \in \mathbb{R}, x = (x_1,\ldots,x_d) \in \mathbb{R}^d\}$ is of the form:

$$i\frac{\partial \psi(t,x)}{\partial t} = H_m \psi(t,x), \tag{17}$$

where $\psi\colon \mathbb{R}^{1+d} \to \mathbb{C}^\nu$ (or a ν component distribution on $\mathbb{R}^{1+d}$), i is the imaginary unit and

$$H_m := \sum_{j=1}^d \alpha_j p_j + m\beta$$

with $p_j := -iD_j$ (D_j is the generalized partial differential operator in the variable x_j).

We use the symbol

$$\mathbb{R}^{d*} = \{k = (k_1,\ldots,k_d) \mid k_j \in \mathbb{R}, j = 1,\ldots,d\}$$

for the dual space of $\mathbb{R}^d$ (the space of d-dimensional wave vectors) and denote by $\mathcal{F}_d\colon L^2(\mathbb{R}^d) \to L^2(\mathbb{R}^{d*})$ the d-dimensional Fourier transform:

$$(\mathcal{F}_d f)(k) := \frac{1}{(2\pi)^{d/2}} \int_{\mathbb{R}^d} f(x) e^{-ikx}\,\mathrm{d}x, \quad f \in L^2(\mathbb{R}^d),\ k \in \mathbb{R}^{d*},$$

in the L^2-sense, where $kx := \sum_{j=1}^d k_j x_j$.

In what follows, we treat H_m as an operator acting in

$$\mathcal{H}_{\mathrm{D}} := L^2(\mathbb{R}^d; \mathbb{C}^\nu) = \{f = (f_r)_{r=1}^\nu \mid f_r \in L^2(\mathbb{R}^d), r = 1, \dots, \nu\},$$

the Hilbert space of $\mathbb{C}^\nu$-valued square integrable functions on $\mathbb{R}^d$. We set

$$\widehat{\mathcal{H}}_{\mathrm{D}} := \mathcal{F}_d \mathcal{H}_{\mathrm{D}} = L^2(\mathbb{R}^{d*}; \mathbb{C}^\nu).$$

For each $k \in \mathbb{R}^{d*}$, we define a $\nu \times \nu$ Hermitian matrix $h_m(k)$ by

$$h_m(k) := \alpha k + m\beta,$$

where $\alpha k := \sum_{j=1}^d \alpha_j k_j$, and denote the multiplication operator by the matrix-valued function $h_m(\cdot)$ on $\widehat{\mathcal{H}}_{\mathrm{D}}$ by $\widehat{H}_m$:

$$D(\widehat{H}_m) := \left\{ f \in \widehat{\mathcal{H}}_{\mathrm{D}} \;\middle|\; \int_{\mathbb{R}^{d*}} \|h_m(k) f(k)\|_{\mathbb{C}^\nu}^2 \, \mathrm{d}k < \infty \right\},$$

$$\widehat{H}_m f(k) := h_m(k) f(k), \quad f \in D(\widehat{H}_m), \text{ a.e. } k \in \mathbb{R}^{d*}.$$

By the theory of Fourier transform, we have $\mathcal{F}_d p_j \mathcal{F}_d^{-1} = k_j$ $(j = 1, \dots, d)$, where the right hand side denotes the multiplication operator by the variable k_j. Hence it follows that

$$\mathcal{F}_d H_m \mathcal{F}_d^{-1} = \widehat{H}_m. \tag{18}$$

Using (14), one has

$$(\alpha k)^2 = k^2, \quad k \in \mathbb{R}^{d*}. \tag{19}$$

By this fact, (15), and (16), one obtains that

$$\|h_m(k) f(k)\|_{\mathbb{C}^\nu}^2 = (k^2 + m^2) \|f(k)\|_{\mathbb{C}^\nu}^2, \quad k \in \mathbb{R}^{d*}.$$

Hence

$$D(\widehat{H}_m) = \left\{ f \in \widehat{\mathcal{H}}_{\mathrm{D}} \;\middle|\; \int_{\mathbb{R}^{d*}} k^2 \|f(k)\|_{\mathbb{C}^\nu}^2 \, \mathrm{d}k < \infty \right\}.$$

Let

$$E_m(k) := \sqrt{k^2 + m^2}, \quad k \in \mathbb{R}^{d*}$$

and

$$d_m(k) := \frac{m + E_m(k) + \beta \alpha k}{\sqrt{2E_m(k)(m + E_m(k))}} \qquad \text{(the case } m > 0\text{)},$$

$$d_0(k) := \begin{cases} \dfrac{1}{\sqrt{2}}\left(1 + \beta \dfrac{\alpha k}{|k|}\right) & \text{for } k \neq 0, \\ I_\nu & \text{for } k = 0 \end{cases} \qquad \text{(the case } m = 0\text{)}.$$

As in the case $d = 3$ (see, e.g., §1.4 of [8]), one can show that $d_m(k)$ is unitary and

$$d_m(k)h_m(k)d_m(k)^{-1} = E_m(k)\beta, \quad k \in \mathbb{R}^{d*}. \tag{20}$$

We denote by $\widehat{D}_m$ the multiplication operator by $d_m(\cdot)$. The operator

$$U_m := \widehat{D}_m \mathcal{F}_d$$

is a unitary operator from $\mathcal{H}_\mathrm{D}$ to $\widehat{\mathcal{H}}_\mathrm{D}$. By (18) and (20), we have

$$U_m H_m U_m^{-1} = E_m \beta. \tag{21}$$

Namely H_m is unitarily equivalent to $E_m\beta$. It is obvious that $E_m\beta$ is self-adjoint. Hence H_m is self-adjoint. Thus, for each $\psi_0 \in D(H_m)$, the free Dirac equation (17) with initial condition $\psi(0,\cdot) = \psi_0$ has the unique solution

$$\psi(t,\cdot) = e^{-itH_m}\psi_0, \quad t \in \mathbb{R}.$$

Relation (21) clarifies spectral properties of H_m too. For a linear operator A on a Hilbert space, we denote by $\sigma(A)$ (resp. $\sigma_\mathrm{p}(A)$) the spectrum (resp. the point spectrum) of A. By the unitary invariance of spectra, (21) implies that

$$\sigma(H_m) = \sigma(E_m\beta), \quad \sigma_\mathrm{p}(H_m) = \sigma_\mathrm{p}(E_m\beta).$$

It follows from (16) and $\beta \neq \pm I_\nu$ that $\sigma(\beta) = \sigma_\mathrm{p}(\beta) = \{\pm 1\}$. Hence

$$\sigma(E_m\beta) = \{E_m(k) \mid k \in \mathbb{R}^{d*}\} \cup \{-E_m(k) \mid k \in \mathbb{R}^{d*}\} = (-\infty,-m] \cup [m,\infty)$$

and $\sigma_\mathrm{p}(E_m\beta) = \emptyset$. Therefore

$$\sigma(H_m) = (-\infty,-m] \cup [m,\infty), \quad \sigma_\mathrm{p}(H_m) = \emptyset.$$

3.2 Eigenvectors of $h_m(k)$ and some operators

One can easily show that $\dim\ker(\beta \pm 1) = \nu/2$. Hence, by diagonalization (if necessary), we can assume without loss of generality that

$$\beta = \begin{pmatrix} I_{\nu/2} & 0 \\ 0 & -I_{\nu/2} \end{pmatrix}.$$

We denote by $\{e_r\}_{r=1}^{\nu}$ the standard basis of $\mathbb{C}^\nu$: $e_r = (\delta_{rr'})_{r'=1}^{\nu}$. For all $k \in \mathbb{R}^{d*}$ and $s = 1,\dots,\nu/2$, we define the following vectors in $\mathbb{C}^\nu$:

$$u_m(k,s) := d_m(k)^{-1}e_s \in \mathbb{C}^\nu, \quad v_m(k,s) := d_m(k)^{-1}e_{s+(\nu/2)} \in \mathbb{C}^\nu \tag{22}$$

By (20), we have

$$h_m(k)u_m(k,s) = E_m(k)u_m(k,s), \quad h_m(k)v_m(k,s) = -E_m(k)v_m(k,s).$$

Namely $u_m(k,s)$ (resp. $v_m(k,s)$) is an eigenvector of $h_m(k)$ with positive (resp. negative) energy $E_m(k)$ (resp. $-E_m(k)$). Since $d_m(k)^{-1}$ is unitary, it follows that, for each $k \in \mathbb{R}^{d*}$, the set $\{u_m(k,s), v_m(k,s) \mid s = 1,\dots,\nu/2\}$ is a complete orthonormal basis of $\mathbb{C}^\nu$. Hence

$$\langle u_m(k,s), u_m(k,s')\rangle_{\mathbb{C}^\nu} = \delta_{ss'}, \tag{23}$$

$$\langle v_m(k,s), v_m(k,s')\rangle_{\mathbb{C}^\nu} = \delta_{ss'},$$

$$\langle u_m(k,s), v_m(k,s')\rangle_{\mathbb{C}^\nu} = 0, \quad s,s' = 1,\dots,\frac{\nu}{2}. \tag{24}$$

and

$$\sum_{s=1}^{\nu/2}(u_{mr}(k,s)u_{mr'}(k,s)^* + v_{mr}(k,s)v_{mr'}(k)^*) = \delta_{rr'}, \quad r,r' = 1,\dots,\nu, \tag{25}$$

where $u_{mr}(k,s)$ (resp. $v_{mr}(k,s)$) is the rth component of the vector $u_m(k,s)$ (resp. $v_m(k,s)$) and, for a complex number $z \in \mathbb{C}$, z^* denotes the complex conjugate of z.

The Hilbert space $\widehat{\mathcal{H}}_{\mathrm{D}}$ has the orthogonal decomposition

$$\widehat{\mathcal{H}}_{\mathrm{D}} = \widehat{\mathcal{H}}_{\mathrm{D}+} \oplus \widehat{\mathcal{H}}_{\mathrm{D}-} \tag{26}$$

with

$$\widehat{\mathcal{H}}_{\mathrm{D}\pm} := L^2(\mathbb{R}^{d*}; \mathbb{C}^{\nu/2}).$$

We define linear operators $T_{m\pm}\colon \widehat{\mathcal{H}}_{\mathrm{D}} \to \widehat{\mathcal{H}}_{\mathrm{D}\pm}$ by

$$T_{m+}f := (u_m(\cdot,s)^* f)_{s=1}^{\nu/2} \in \widehat{\mathcal{H}}_{\mathrm{D}+}, \tag{27}$$

$$T_{m-}f := (\tilde{v}_m(\cdot,s)\tilde{f})_{s=1}^{\nu/2} \in \widehat{\mathcal{H}}_{\mathrm{D}-}, \quad f \in \widehat{\mathcal{H}}_{\mathrm{D}}, \tag{28}$$

where, for $w = (w_r)_{r=1}^{\nu}\colon \mathbb{R}^{d*} \to \mathbb{C}^\nu$,

$$(wf)(k) := \sum_{r=1}^{\nu} w_r(k) f_r(k), \quad f = (f_r)_{r=1}^{\nu} \in \widehat{\mathcal{H}}_{\mathrm{D}}, \ k \in \mathbb{R}^{d*}$$

and

$$\tilde{w}(k) := w(-k), \quad k \in \mathbb{R}^{d*}.$$

It follows that $T_{m\pm}$ are bounded with $\|T_{m\pm}\| \le \sqrt{\nu/2}$.

It is easy to see that

$$(T_{m+}^* f_+)_r = \sum_{s=1}^{\nu/2} u_{mr}(\cdot, s) f_{+s}, \qquad f_+ = (f_{+s})_{s=1}^{\nu/2} \in \widehat{\mathcal{H}}_{\mathrm{D}+}, \tag{29}$$

$$(T_{m-}^* f_-)_r = \sum_{s=1}^{\nu/2} v_{mr}(\cdot, s)^* \tilde{f}_{-s}, \quad f_- = (f_{-s})_{s=1}^{\nu/2} \in \widehat{\mathcal{H}}_{\mathrm{D}-},\ r = 1, \dots, \nu. \tag{30}$$

We denote by C the complex conjugation on $\widehat{\mathcal{H}}_{\mathrm{D}}$:

$$Cf := (f_r^*)_{r=1}^{\nu}.$$

For a linear operator A on $\widehat{\mathcal{H}}_{\mathrm{D}}$, we define $\bar{A}$ by

$$\bar{A} := CAC.$$

We regard $\widehat{\mathcal{H}}_{\mathrm{D}\pm}$ as subspaces of $\widehat{\mathcal{H}}_{\mathrm{D}}$ in the natural way so that C acts also on $\widehat{\mathcal{H}}_{\mathrm{D}\pm}$.

Remarkable properties of $T_{m\pm}$ are summarized in the following lemma:

Lemma 3.1. *We have*

$$T_{m+}^* T_{m+} + \bar{T}_{m-}^* \bar{T}_{m-} = I, \tag{31}$$

$$T_{m\pm} T_{m\pm}^* = I, \tag{32}$$

$$T_{m+} \bar{T}_{m-}^* = 0, \quad T_{m-} \bar{T}_{m+}^* = 0. \tag{33}$$

Proof. Throughout the proof, we write $T_\pm$ (resp. $u(k,s), v(k,s)$) for $T_{m\pm}$ (resp. $u_m(k,s), v_m(k,s)$). Let $f, g \in \widehat{\mathcal{H}}_{\mathrm{D}}$. Then

$$\begin{aligned}
&\langle T_+ f, T_+ g\rangle + \langle \bar{T}_- f, \bar{T}_- g\rangle \\
&= \sum_{s=1}^{\nu/2} \int_{\mathbb{R}^d} \{(u(\cdot,s) f^*)(k)(u(\cdot,s)^* g)(k) + (v(\cdot,s) f^*)(k)(v(\cdot,s)^* g)(k)\}\,\mathrm{d}k \\
&= \sum_{r,r'=1}^{\nu} \int_{\mathbb{R}^d} f_r(k)^* g_{r'}(k) \Big(\sum_{s=1}^{\nu/2} (u_r(k,s) u_{r'}(k,s)^* + v_r(k,s) v_{r'}(k,s)^*) \Big)\,\mathrm{d}k \\
&= \sum_{r=1}^{\nu} \int_{\mathbb{R}^d} f_r(k)^* g_r(k)\,\mathrm{d}k,
\end{aligned}$$

where we have used (25) to obtain the last equality. Hence

$$\langle T_+ f, T_+ g\rangle + \langle \bar{T}_- f, \bar{T}_- g\rangle = \langle f, g\rangle.$$

This implies (31).

We have by (29)

$$\begin{aligned}\langle T_+^* f, T_+^* g\rangle &= \sum_{r=1}^{\nu}\sum_{s,s'=1}^{\nu/2} \langle u_r(\cdot,s) f_s, u_r(\cdot,s') g_{s'}\rangle \\ &= \int_{\mathbb{R}^{d*}} \langle u(k,s), u(k,s')\rangle f_s(k)^* g_{s'}(k)\,\mathrm{d}k \\ &= \langle f, g\rangle,\end{aligned}$$

where we have used (23). Hence $T_+ T_+^* = I$. Similarly, using (30), one can show that $T_- T_-^* = I$. One can see that orthogonality (24) implies (33). □

Lemma 3.1 immediately yields the following result.

Lemma 3.2. *For all $m \geq 0$,*

$$T_m := (T_{m+}, T_{m-})$$

is an element of $\mathfrak{T}_*(\widehat{\mathcal{H}}_{\mathrm{D}})$.

It follows from (32) that $\|T_{m\pm}^*\| = 1$ and hence

$$\|T_{m\pm}\| = 1. \tag{34}$$

3.3 A free quantum Dirac field

We construct a free quantum Dirac field on the fermion Fock space $\mathfrak{F}(\widehat{\mathcal{H}}_{\mathrm{D}})$ over $\widehat{\mathcal{H}}_{\mathrm{D}}$. Note that we work with momentum representation. We denote by $a(f)$ ($f \in \widehat{\mathcal{H}}_{\mathrm{D}}$) the annihilation operator on $\mathfrak{F}(\widehat{\mathcal{H}}_{\mathrm{D}})$.

For each $f \in \widehat{\mathcal{H}}_{\mathrm{D}}$, we define

$$\widehat{\psi}_m(f) := a(T_{m+} f, 0) + a(0, T_{m-} f^*)^*, \quad f \in \widehat{\mathcal{H}}_{\mathrm{D}},$$

and set

$$\hat{\rho}_m := \{\widehat{\psi}_m(f), \widehat{\psi}_m(f)^* \mid f \in \widehat{\mathcal{H}}_{\mathrm{D}}\}.$$

Remark 3.3. A free quantum Dirac field on the $(1+d)$ (resp. 4)-dimensional space-time was considered in [5] (resp. [7]). In these papers (also in §10.1 of [8]), the projection method as mentioned in Remark 2.4 is used. But we find that the projection method is somewhat inconvenient to discuss a family of quantum Dirac fields indexed by mass m, since the orthogonal decomposition of $\mathcal{H}_D$ in the projection method depends on m. Thus we take a slightly different approach in which the orthogonal decomposition (26) is fixed independently of m.

By Lemma 3.2 and an application of Theorem 2.7, we obtain the following proposition.

Proposition 3.4. *For each* $m \geq 0$, $(\mathfrak{F}(\widehat{\mathcal{H}}_{\mathrm{D}}), \hat{\rho}_m)$ *is an irreducible representation of the CAR over* $\widehat{\mathcal{H}}_{\mathrm{D}}$.

Let

$$\psi_m(t, f) := \hat{\psi}_m(e^{it\widehat{H}_m} \hat{f}), \quad t \in \mathbb{R},\ f \in D(H_m),$$

and

$$\rho_m(t) := \{\psi_m(t, f), \psi_m(t, f)^* \mid f \in \mathcal{H}_{\mathrm{D}}\}.$$

Since $e^{it\widehat{H}_m}$ is unitary for all $t \in \mathbb{R}$, Proposition 3.4 yields the following result.

Proposition 3.5. *For each* $m \geq 0$ *and* $t \in \mathbb{R}$, $(\mathfrak{F}(\widehat{\mathcal{H}}_{\mathrm{D}}), \rho_m(t))$ *is an irreducible representation of the CAR over* $\mathcal{H}_{\mathrm{D}}$.

The following proposition shows that $\psi_m(t, f)$ is a free quantum Dirac field on the $(1 + d)$-dimensional space-time and hence $\psi_m(f) = \psi_m(0, f)$ is the time-zero field of it.

Proposition 3.6. *Let* $m \geq 0$ *and* $f \in D(H_m)$. *Then the operator-valued function:* $t \mapsto \psi_m(t, f) \in \mathfrak{B}(\mathfrak{F}(\widehat{\mathcal{H}}_{\mathrm{D}}))$ *on* $\mathbb{R}$ *is differentiable in the operator norm topology and obeys the free functional Dirac equation*

$$i \frac{d\psi_m(t, f)}{dt} = \psi_m(t, H_m f).$$

Proof. Let $\varepsilon \in \mathbb{R} \setminus \{0\}$ and

$$\phi_\varepsilon := i \frac{\psi_m(t + \varepsilon, f) - \psi_m(t, f)}{\varepsilon} - \psi_m(t, H_m f).$$

Then we have

$$\phi_\varepsilon = \hat{\psi}_m(e^{it\widehat{H}_m} g_\varepsilon),$$

where

$$g_\varepsilon := \Big(-i \frac{(e^{i\varepsilon\widehat{H}_m} - 1)}{\varepsilon} - \widehat{H}_m\Big)\hat{f}.$$

We have for all $h \in \widehat{\mathcal{H}}_{\mathrm{D}}$

$$\|\hat{\psi}_m(h)\| \leq \|T_{m+}h\| + \|T_{m-}h^*\| \leq 2\|h\|,$$

where we have used (34). Hence $\|\phi_\varepsilon\| \leq 2\|g_\varepsilon\|$. Since $\hat{f} \in D(\widehat{H}_m)$, it follows that $\lim_{\varepsilon\to 0} \|g_\varepsilon\| = 0$. Hence $\lim_{\varepsilon\to 0} \|\phi_\varepsilon\| = 0$. □

One can show that, for all $f \in \mathcal{H}_{\mathrm{D}}$,

$$\psi_m(t, f) = e^{it\mathbb{H}_{0,m}} \widehat{\psi}_m(\hat{f}) e^{-it\mathbb{H}_{0,m}}, \quad t \in \mathbb{R}, \tag{35}$$

where $\mathbb{H}_{0,m} := d\Gamma(E_m)$ is the second quantization of the multiplication operator E_m (e.g., p. 8 of [3]). The operator $\mathbb{H}_{0,m}$ is the Hamiltonian of the free quantum Dirac field with mass m in momentum representation.

We now state and prove one of the main results in this paper.

Theorem 3.7. *Let $m_1 \neq m_2$ ($m_1, m_2 \geq 0$). Then $(\mathfrak{F}(\widehat{\mathcal{H}}_{\mathrm{D}}), \rho_{m_1})$ and $(\mathfrak{F}(\widehat{\mathcal{H}}_{\mathrm{D}}), \rho_{m_2})$ are inequivalent.*

Proof. By Theorem 2.11, we need only to prove that, if $m_1 \neq m_2$, then $T_{m_1+}\bar{T}^*_{m_2-}$ or $T_{m_1-}\bar{T}^*_{m_2+}$ is not Hilbert–Schmidt. It is easy to see that

$$(T_{m_1+}\bar{T}^*_{m_2-} f)_r(k) = \sum_{s=1}^{\nu/2} K_{rs}(k) \tilde{f}_s(k), \quad f = (f_s)_{s=1}^{\nu/2} \in \widehat{\mathcal{H}}_{\mathrm{D}-},\ r = 1, \dots, \frac{\nu}{2}, \tag{36}$$

where

$$K_{rs}(k) := \langle u_{m_1}(k, r), v_{m_2}(k, s) \rangle_{\mathbb{C}^\nu}, \quad k \in \mathbb{R}^{d*}, r, s = 1, \dots, \frac{\nu}{2}. \tag{37}$$

By (22), we have $K_{rs}(k) = \langle e_r, d_{m_1}(k) d_{m_2}(k)^* e_{s+\nu/2} \rangle$. Using (15), (19), and the orthogonality $\langle e_r, e_{s+\nu/2} \rangle = 0$ ($r, s = 1, \dots, \nu/2$), we obtain

$$K_{rs}(k) = (m_2 - m_1) F_{m_1,m_2}(k) \langle e_r, \alpha k e_{s+\nu/2} \rangle, \quad r, s = 1, \dots, \frac{\nu}{2},$$

where

$$F_{m_1,m_2}(k) := \frac{1}{2\sqrt{E_{m_1}(k) E_{m_2}(k)(m_1 + E_{m_1}(k))(m_2 + E_{m_2}(k))}} \left(1 + \frac{m_1 + m_2}{E_{m_1}(k) + E_{m_2}(k)}\right).$$

It follows from (15) that $\alpha k e_{s+\nu/2} \in \ker(\beta - 1)$, $s = 1, \dots, \nu/2$. Hence

$$\sum_{r=1}^{\nu/2} |\langle e_r, \alpha k e_{s+\nu/2} \rangle|^2 = \|\alpha k e_{s+\nu/2}\|^2 = k^2,$$

where we have used (19). Therefore

$$\sum_{r=1}^{\nu/2} |K_{rs}(k)|^2 = (m_2 - m_1)^2 F_{m_1,m_2}(k)^2 k^2. \tag{38}$$

Let $m_1 \neq m_2$. Suppose that $T_{m_1+}\bar{T}^*_{m_2-}$ were Hilbert–Schmidt. Denote by $\widehat{K}_{rs}$ the multiplication operator by the function $K_{rs}(k)$ on $L^2(\mathbb{R}^{d*})$. Then, by (36), $\widehat{K}_{rs}$ is Hilbert–Schmidt. Hence

$$L := \sum_{r=1}^{\nu/2} \widehat{K}^*_{rs}\widehat{K}_{rs}$$

is Hilbert–Schmidt. By equation (38), L is the multiplication operator by the function $(m_2 - m_1)^2 F_{m_1,m_2}(k)^2 k^2$. This function is continuous on $\mathbb{R}^{d*}$ and positive for all $|k| > 0$. Hene $\sigma(L)$ includes an open interval in $[0, \infty)$. Hence L is not Hilbert–Schmidt, since the spectrum of a Hilbert–Schmidt operator is purely discrete in $\mathbb{C} \setminus \{0\}$. Therefore we have a contradiction. Thus $T_{m_1+}\bar{T}^*_{m_2-}$ is not Hilbert–Schmidt. □

3.4 An interacting quantum Dirac field

Let $\mathbb{H}_m$ be a self-adjoint operator on $\mathfrak{F}(\widehat{\mathcal{H}}_{\mathrm{D}})$ which may depend on m. Then the time-t quantum Dirac field of mass m with Hamiltonian $\mathbb{H}_m$ is defined by

$$\psi_m(t, f) := e^{it\mathbb{H}_m}\widehat{\psi}_m(\hat{f})e^{-it\mathbb{H}_m}, \quad f \in \mathcal{H}_{\mathrm{D}}, t \in \mathbb{R}.$$

This is a general form of interacting quantum Dirac fields whose time-zero field is taken to be $\widehat{\psi}_m(\hat{f})$. The time-$t$ free quantum Dirac field is given by the case where $\mathbb{H}_m = \mathbb{H}_{0,m}$ (see (35)).

Let

$$\rho_m(t) := \{\psi_m(t, f), \psi(t, f)^* \mid f \in \mathcal{H}_{\mathrm{D}}\}.$$

Then $(\mathfrak{F}(\widehat{\mathcal{H}}_{\mathrm{D}}), \rho_m(t))$ is an irreducible representation of the CAR over $\mathcal{H}_{\mathrm{D}}$. Since $e^{it\mathbb{H}_m}$ is unitary, the following corollary immediately follows from Theorem 3.7:

Corollary 3.8. *Let $m_1 \neq m_2$. Then, for all $t \in \mathbb{R}$, $(\mathfrak{F}(\widehat{\mathcal{H}}_{\mathrm{D}}), \rho_{m_1}(t))$ and $(\mathfrak{F}(\widehat{\mathcal{H}}_{\mathrm{D}}), \rho_{m_2}(t))$ are inequivalent.*

4 A generalization

In this section we briefly describe a generalization of the time-zero quantum Dirac field $\widehat{\psi}(\hat{f})$ $(f \in \mathcal{H}_{\mathrm{D}})$.

Let $\mathcal{U}$ be the set of pairs (u, v) of $\mathbb{C}^\nu$-valued Borel measurable functions u and v on $\mathbb{R}^{d*} \times \{1, \ldots, \nu/2\}$ such that, for a.e. $k \in \mathbb{R}^{d*}$, $\{u(k, s), v(k, s) \mid s = 1, \ldots, \nu/2\}$

is an orthonormal basis of $\mathbb{C}^\nu$. For each $(u, v) \in \mathfrak{U}$, we define $T_+(u): \widehat{\mathcal{H}}_{\mathrm{D}} \to \widehat{\mathcal{H}}_{\mathrm{D}+}$ and $T_-(v): \widehat{\mathcal{H}}_{\mathrm{D}} \to \widehat{\mathcal{H}}_{\mathrm{D}-}$ as follows (cf. (27) and (28)):

$$T_+(u)f := (u(\cdot, s)^* f)_{s=1}^{\nu/2} \in \widehat{\mathcal{H}}_{\mathrm{D}+},$$

$$T_-(v)f := (\tilde{v}(\cdot, s)\tilde{f})_{s=1}^{\nu/2} \in \widehat{\mathcal{H}}_{\mathrm{D}-},$$

for $f \in \widehat{\mathcal{H}}_{\mathrm{D}}$. Then, in the same way as in the proof of Lemma 3.1, one can prove the following relations:

$$T_+(u)^*T_+(u) + \bar{T}_-(v)^*\bar{T}_-(v) = I,$$

$$T_+(u)T_+(u)^* = I, \quad T_-(v)T_-(v)^* = I,$$

$$T_+(u)\overline{T_-(v)}^* = 0, \quad T_-(v)\overline{T_+(u)}^* = 0.$$

Hence $(T_+(u), T_-(v))$ is an element of $\mathfrak{T}_*(\widehat{\mathcal{H}}_{\mathrm{D}})$. Therefore, introducing the operators

$$\hat{\psi}_{u,v}(f) := a(T_+(u)f, 0) + a(0, T_-(v)f^*)^*, \quad f \in \widehat{\mathcal{H}}_{\mathrm{D}}$$

and

$$\hat{\rho}(u, v) := \{\hat{\psi}_{u,v}(f), \hat{\psi}_{u,v}(f)^* \mid f \in \widehat{\mathcal{H}}_{\mathrm{D}}\},$$

we see that $(\mathfrak{F}(\widehat{\mathcal{H}}_{\mathrm{D}}), \hat{\rho}(u, v))$ is an irreducible representation of the CAR over $\widehat{\mathcal{H}}_{\mathrm{D}}$. Clearly this class of representations includes $(\mathfrak{F}(\widehat{\mathcal{H}}_{\mathrm{D}}), \hat{\rho}_m)$, hence being a generalization of it.

As for the family $\{(\mathfrak{F}(\widehat{\mathcal{H}}_{\mathrm{D}}), \hat{\rho}(u, v)) \mid (u, v) \in \mathfrak{U}\}$ of representations of the CAR over $\widehat{\mathcal{H}}_{\mathrm{D}}$, we have the following inequivalence theorem.

Theorem 4.1. *Let (u_1, v_1) and (u_2, v_2) be in $\mathfrak{U}$. Then the two representations $(\mathfrak{F}(\widehat{\mathcal{H}}_{\mathrm{D}}), \hat{\rho}(u_1, v_1))$ and $(\mathfrak{F}(\widehat{\mathcal{H}}_{\mathrm{D}}), \hat{\rho}(u_2, v_2))$ are inequivalent if and only if, for some (r, s), there exists a non-null Borel set $B \subset \mathbb{R}^{d*}$ such that, for all $k \in B$,*

$$\langle u_1(k, r), v_2(k, s)\rangle_{\mathbb{C}^\nu} \neq 0 \quad \text{or} \quad \langle u_2(k, r), v_1(k, s)\rangle_{\mathbb{C}^\nu} \neq 0.$$

Proof. By Theorem 2.11, the two representations under consideration are inequivalent if and only if $T_+(u_1)\overline{T_-(v_2)}^*$ or $T_-(v_1)\overline{T_+(u_2)}^*$ is not Hilbert–Schmidt.

In the same way as in the case of (36), one can show that

$$(T_+(u_1)\overline{T_-(v_2)}^* f)_r(k) = \sum_{s=1}^{\nu/2} G_{rs}(k)\tilde{f}_s(k), \quad f \in \widehat{\mathcal{H}}_{\mathrm{D}},\ r = 1, \dots, \frac{\nu}{2}, \text{ a.e. } k,$$

where

$$G_{rs}(k) := \langle u_1(k, r), v_2(k, s)\rangle_{\mathbb{C}^\nu}.$$

It is easy to see that $T_+(u_1)\overline{T_-(v_2)}^*$ is not Hilbert–Schmidt if and only if, for some (r,s), the multiplication operator $\widehat{G}_{rs}$ by the function G_{rs} is not Hilbert–Schmidt. In general, the multiplication operator on $L^2(\mathbb{R}^{d*})$ by a function F is Hilbert–Schmidt if and only if $F(k) = 0$ a.e.$k \in \mathbb{R}^{d*}$. Hence $\widehat{G}_{rs}$ is not Hilbert–Schmidt if and only if there exists a non-null Borel set $B \subset \mathbb{R}^{d*}$ such that, for all $k \in B$, $G_{rs}(k) \neq 0$. Thus $T_+(u_1)\overline{T_-(v_2)}^*$ is not Hilbert–Schmidt if and only if, for some (r,s), there exists a non-null Borel set $B \subset \mathbb{R}^{d*}$ such that, for all $k \in B$, $G_{rs}(k) \neq 0$.

Note that $T_-(v_1)\overline{T_+(u_2)}^*$ is not Hilbert–Schmidt if and only if $T_+(u_2)\overline{T_-(v_1)}^*$ is not Hilbert–Schmidt. Hence one can apply the preceding result to the case (u_1, v_2) replaced by (u_2, v_1) to conclude that $T_-(v_1)\overline{T_+(u_2)}^*$ is not Hilbert–Schmidt if and only if, for some (r,s), there exists a non-null Borel set $B \subset \mathbb{R}^{d*}$ such that, $\langle u_2(k,r), v_1(k,s)\rangle_{\mathbb{C}^\nu} \neq 0$. □

5 The case of quantum Dirac fields on d-dimensional boxes

In Theorem 6.12 in [2], it is shown that the time-zero quantum scalar fields of different positive masses on a bounded region in $\mathbb{R}^d$ are equivalent if and only if $d \leq 3$, in contrast to the case of the infinite space $\mathbb{R}^d$. It is natural to ask if quantum Dirac fields on a bounded region in $\mathbb{R}^d$ have similar properties. In this section we give an answer to this question. For simplicity, we consider quantum Dirac fields on the d-dimensional box

$$M := \mathbb{I}_1 \times \cdots \times \mathbb{I}_d, \quad \mathbb{I}_j := \Big[-\frac{L_j}{2}, \frac{L_j}{2}\Big] \quad (L_j > 0,\ j = 1, \ldots, d).$$

Let

$$\Gamma_j := \Big\{\frac{2\pi}{L_j} n \ \Big|\ n \in \mathbb{Z}\Big\},$$

where $\mathbb{Z}$ is the set of all integers, and

$$\Gamma := \Gamma_1 \times \cdots \times \Gamma_d = \{k = (k_1, \ldots, k_d) \mid k_j \in \Gamma_j,\ j = 1, \ldots, d\}.$$

For each $k \in \Gamma$, we define a function ϕ_k on M by

$$\phi_k(x) := \frac{1}{\sqrt{L_1 \ldots L_d}} e^{ikx}, \quad x \in M.$$

It is well known that $\{\phi_k \mid k \in \Gamma\}$ is a complete orthonormal system (CONS) of $L^2(M)$. Hence the mapping $U_d : L^2(M) \to \ell^2(\Gamma)$ (the Hilbert space of absolutely square summable sequences on Γ) defined by

$$U_d f(k) := \langle \phi_k, f \rangle_{L^2(M)}, \quad f \in L^2(M),\ k \in \Gamma,$$

is unitary.

For each $j = 1, \ldots, d$, the multiplication operator by the jth coordinate function k_j in Γ is self-adjoint. We denote it by the same symbol k_j. Then the operator

$$p_j^{(M)} := U_d^{-1} k_j U_d$$

is a self-adjoint operator on $L^2(M)$ with $D(p_j^{(M)}) = U_d^{-1} D(k_j)$. The operator $p_j^{(M)}$ is called the *j-th momentum operator* in M with the periodic boundary condition.

As a free Dirac operator on the Hilbert space

$$\mathcal{H}_{\mathrm{D}}^{(M)} := L^2(M; \mathbb{C}^\nu)$$

with mass $m > 0$, we take the following one:

$$H_m^{(M)} := \sum_{j=1}^{d} \alpha_j p_j^{(M)} + m\beta.$$

The operator $H_m^{(M)}$ is self-adjoint and the operator equality

$$U_d H_m^{(M)} U_d^{-1} = h_m^{(M)}$$

holds with $h_m^{(M)}$ being the multiplication operator on the Hilbert space

$$\widehat{\mathcal{H}}_{\mathrm{D}}^{(M)} := U_d \mathcal{H}_{\mathrm{D}}^{(M)} = \bigoplus^{\nu} \ell^2(\Gamma).$$

by the matrix-valued function

$$h_m^{(M)}(k) := \sum_{j=1}^{d} \alpha_j k_j + m\beta, \quad k \in \Gamma.$$

As in the case of the infinite space $\mathbb{R}^d$, we have

$$\widehat{\mathcal{H}}_{\mathrm{D}}^{(M)} = \widehat{\mathcal{H}}_{\mathrm{D}+}^{(M)} \oplus \widehat{\mathcal{H}}_{\mathrm{D}-}^{(M)},$$

where $\widehat{\mathcal{H}}_{\mathrm{D}\pm}^{(M)} := \bigoplus^{\nu/2} \ell^2(\Gamma)$.

Objects in Section 3 such as E_m, d_m, u_m, v_m have counterparts in the theory on M in an obvious way. We write them with upper suffix "(M)" (for instance, $E_m^{(M)}(k) := E_m(k)$, $k \in \Gamma$). We define the operators $T_{m\pm}^{(M)}: \widehat{\mathcal{H}}_{\mathrm{D}}^{(M)} \to \widehat{\mathcal{H}}_{\mathrm{D}\pm}^{(M)}$ by $T_{m\pm}$ with u_m and v_m replaced by $u_m^{(M)}$ and $v_m^{(M)}$ respectively. Then it is easy to see that Lemma 3.1 holds with $T_{m\pm}$ replaced by $T_{m\pm}^{(M)}$.

We denote by $a_M(\cdot)$ the annihilation operator on the fermion Fock space $\mathfrak{F}(\widehat{\mathcal{H}}_{\mathrm{D}}^{(M)})$ and define a quantum Dirac field on M by

$$\psi_m^{(M)}(f) := a_M(T_{m+}^{(M)} U_d f, 0) + a_M(0, T_{m-}^{(M)}(U_d f)^*)^*, \quad f \in \mathcal{H}_{\mathrm{D}}^{(M)}.$$

Let

$$\rho_m^{(M)} := \{\psi_m^{(M)}(f), \psi_m^{(M)}(f)^* \mid f \in \mathcal{H}_{\mathrm{D}}^{(M)}\}.$$

Then, as in Proposition 3.4, we can show that $(\mathfrak{F}(\widehat{\mathcal{H}}_{\mathrm{D}}^{(M)}), \rho_m^{(M)})$ is an irreducible representation of the CAR over $\mathcal{H}_{\mathrm{D}}^{(M)}$. Now we are ready to prove the following theorem.

Theorem 5.1. *Let $m_1 \neq m_2$ $(m_1, m_2 > 0)$. Then $(\mathfrak{F}(\widehat{\mathcal{H}}_{\mathrm{D}}^{(M)}), \rho_{m_1}^{(M)})$ is equivalent to $(\mathfrak{F}(\widehat{\mathcal{H}}_{\mathrm{D}}^{(M)}), \rho_{m_2}^{(M)})$ if and only if $d = 1$.*

Proof. Throughout the proof, we write $T_{j\pm} := T_{m_j\pm}^{(M)}$, $j = 1, 2$. It is easy to see that

$$(T_{1+}\bar{T}_{2-}^* w)_r(k) = \sum_{s=1}^{\nu/2} K_{rs}(k)\tilde{w}_s(k),$$

for $w = (w_s)_{s=1}^{\nu/2} \in \widehat{\mathcal{H}}_{\mathrm{D}-}^{(M)}$, $r = 1, \dots, \nu/2$, $k \in \Gamma$, where K_{rs} is defined by (37) with domain Γ. For each $k \in \Gamma$, we define a vector $e_k \in \ell^2(\Gamma)$ by

$$e_k(k') = \delta_{kk'}, \quad k' \in \Gamma.$$

It is easy to see that $\{e_k\}_{k\in\Gamma}$ is a CONS of $\ell^2(\Gamma)$. For $s = 1, \dots, \nu/2$, we define

$$e_k^{(s)} = (0, \dots, 0, \overset{s\text{-th}}{e_k}, 0, \dots, 0) \in \widehat{\mathcal{H}}_{\mathrm{D}-}^{(M)}.$$

Then $\{e_k^{(s)} \mid k \in \Gamma,\ s = 1, \dots, \nu/2\}$ is a CONS of $\widehat{\mathcal{H}}_{\mathrm{D}-}^{(M)}$. We have

$$(T_{1+}\bar{T}_{2-}^* e_k^{(s)})_r(k') = K_{rs}(k')\delta_{k,-k'}, \quad k', k \in \Gamma.$$

Hence

$$\sum_{s=1}^{\nu/2}\sum_{k\in\Gamma} \|T_{1+}\bar{T}_{2-}^* e_k^{(s)}\|^2 = \sum_{r,s=1}^{\nu/2}\sum_{k\in\Gamma} |K_{rs}(k)|^2.$$

Hence $T_{1+}\bar{T}_{2-}^*$ is Hilbert–Schmidt if and only if $\sum_{k\in\Gamma} |K_{rs}(k)|^2 < \infty$ for $r, s = 1, \ldots, \nu/2$. In the present case too, we have (38). Hence it follows that $T_{1+}\bar{T}_{2-}^*$ is Hilbert–Schmidt if and only if $\sum_{k\in\Gamma} F_{m_1,m_2}(k)^2 k^2 < \infty$.

It is easy to see that

$$\frac{1}{2(m_1 + E_{m_1}(k))(m_2 + E_{m_2}(k))} \leq F_{m_1,m_2}(k) \leq \frac{1}{E_{m_1}(k)E_{m_2}(k)}, \quad k \in \Gamma.$$

Hence, for each $R > 0$, there exist positive constants C_1 and C_2 such that, for all $|k| \geq R$,

$$\frac{C_1}{k^2} \leq F_{m_1,m_2}(k) \leq \frac{C_2}{k^2}.$$

Therefore

$$\frac{C_1^2}{k^2} \leq F_{m_1,m_2}(k)^2 k^2 \leq \frac{C_2^2}{k^2}.$$

It is obvious that $\sum_{k\in\Gamma} 1/k^2 < \infty$ if and only if $d = 1$. Hence $T_{1+}\bar{T}_{2-}^*$ is Hilbert–Schmidt if and only if $d = 1$. Similarly, we see that $T_{1-}\bar{T}_{2+}^*$ is Hilbert–Schmidt if and only if $d = 1$. Thus, by Theorem 2.10, we obtain the desired result. □

Remark 5.2. Theorem 5.1 is interesting in that it is different from a corresponding theorem in the theory of quantum scalar fields on M (Theorem 6.12 in [2]), where, in the cases $d = 2, 3$ too, the time-zero quantum scalar fields of different masses are equivalent.

Acknowledgement. The author was supported by JSPS KAKENHI No. 15K04888.

References

[1] A. Arai, *Fock spaces and quantum fields.* In Japanese. Nippon-hyouronsha, Tokyo, 2000.

[2] A. Arai, A family of inequivalent Weyl representations of canonical commutation relations with applications to quantum field theory. *Rev. Math. Phys.* **28** (2016), no. 4, 1650007, 26 pp. MR 3513945 Zbl 1342.81624

[3] O. Bratteli and D. W. Robinson, *Operator algebras and quantum statistical mechanics.* 2. Equilibrium states. Models in quantum statistical mechanics. Second edition. Texts and Monographs in Physics. Springer, Berlin etc., 1997. ISBN 3-540-61443-5 MR 1441540 Zbl 0903.46066

[4] J. T. Ottesen, *Infinite dimensional groups and algebras in quantum physics.* Lecture Notes in Physics. New Series m: Monographs, 27. Springer, Berlin, 1995. ISBN 3-540-58914-7 MR 1345151 Zbl 0848.22026

[5] J. Palmer, Scattering automorphisms of the Dirac field. *J. Math. Anal. Appl.* **64** (1978), no. 1, 189–215. MR 0495986 Zbl 0399.46057

[6] M. Reed and B. Simon, *Methods of modern mathematical physics.* II. Fourier analysis, self-adjointness. Academic Press [Harcourt Brace Jovanovich, Publishers], New York and London, 1975. MR 0493420 Zbl 0308.47002

[7] S. N. M. Ruijsenaars, Charged particles in external fields. II. The quantized Dirac and Klein–Gordon theories. *Comm. Math. Phys.* **52** (1977), no. 3, 267–294. MR 0446207

[8] B. Thaller, *The Dirac equation.* Texts and Monographs in Physics. Springer, Berlin, 1992. ISBN 3-540-54883-1 MR 1219537 Zbl 0765.47023

On a class of Schrödinger operators exhibiting spectral transition

Diana Barseghyan and Olga Rossi

This work is dedicated to Pavel Exner on the occasion of his anniversary.

1 Introduction

The idea of Hermann Weyl to analyze spectra of quantum systems semiclassically by looking at the phase space allowed for the corresponding classical motion is one of the most seminal in modern mathematical physics. However, its validity is not universal. Various examples of systems which have purely discrete spectrum despite the fact that the respective phase space volume is infinite were constructed in the last three decades. A classical one belongs to B. Simon [12] and describes a two-dimensional Schrödinger operator with the potential $|xy|^p$. A modification of this example was studied in [8]. A related problem concerns spectral properties of Dirichlet Laplacians in regions with hyperbolic cusps: we refer to the paper [6]. Another example of this type is the so-called *Smilansky model* ([13], [14], [9], [4], and [5]) and its regular version [1].

In [3] and [2] it has been shown that a similar spectral behaviour can occur also for Schrödinger operators with potentials unbounded from below. Furthermore, a model was constructed, which exhibits a nontrivial spectral transition as the coupling constant changes. It is represented by the class of operators

$$L_p(\lambda)\colon L_p(\lambda)\psi = -\Delta\psi + (|xy|^p - \lambda(x^2+y^2)^{p/(p+2)})\psi, \quad p \geq 1, \tag{1}$$

on $L^2(\mathbb{R}^2)$ in the standard Cartesian coordinates (x, y) in $\mathbb{R}^2$; the parameter λ in the second term of the potential is non-negative.

It was established that there is a critical value of the coupling constant λ, expressed explicitly as the ground-state eigenvalue of the corresponding (an)harmonic oscillator Hamiltonian $-{}^{d^2}/{}_{dx^2} + |x|^p$ on $L^2(\mathbb{R})$, such that the spectrum of $L_p(\lambda)$ is below bounded and purely discrete for $\lambda < \lambda_{\text{crit}}$, while for $\lambda > \lambda_{\text{crit}}$ it covers the whole real axis. Moreover, in the critical case the essential spectrum of the operator $L_p(\lambda_{\text{crit}})$ covers the half line $[0, \infty)$, while the negative spectrum can be only discrete, and there is a range of values of p for which $L_p(\lambda_{\text{crit}})$ has a single negative eigenvalue.

Since the operator (1) is considered for any $p \geq 1$, it is natural to ask about the limit $p \to \infty$ which corresponds to a particle confined to a region with four hyperbolic "horns," $D = \{(x, y) \in \mathbb{R}^2 : |xy| < 1\}$, described by the Schrödinger operator $H_D(\lambda)$: $H_D(\lambda)\psi = -\Delta\psi - \lambda(x^2 + y^2)\psi$ with a non-negative parameter λ and Dirichlet condition on the boundary ∂D. This model was studied in the paper [3] for $0 \leq \lambda < 1$. The aim of the present paper is to study the properties of the operator $H_D(\lambda)$ for arbitrary $\lambda \geq 0$. This means, in particular, that we extend the results to the "critical case" $\lambda = \pi^2/4$ and "supercritical case" $\lambda > \pi^2/4$, not covered in [3].

We shall consider the operator $H_D(\lambda)$ initially defined on the set

$$\widetilde{C}_0^2(\bar{D}) = \{u \in C^2(\bar{D}) : u = 0 \text{ on } \partial D,\ \operatorname{supp}(u) \text{ is a compact set}\}.$$

We show that for $\lambda \leq \pi^2/4$ the operator $H_D(\lambda)$ is non-negative and therefore one can construct its self-adjoint extension using the Friedrichs method.

Using the fact that densily defined and symmetric operator is always closable, in case if $\lambda > \pi^2/4$ we deal with its closure $\overline{H_D}(\lambda)$. The question of its self-adjointness we postpone to a next paper.

Since the ground state eigenvalue of (an)harmonic oscillator converges to $\pi^2/4$ as $p \to \infty$, it is natural to expect that the spectrum of $H_D(\lambda)$ is discrete for $\lambda < \pi^2/4$. We show that the spectrum of (the self-adjoint extension of) $H_D(\lambda)$ is purely discrete for all $0 \leq \lambda < \pi^2/4$ and study the properties of the eigenvalues. In the critical case $\lambda = \pi^2/4$ we establish that the spectrum coincides with the half line $[0, \infty)$. In the remaining case, $\lambda > \pi^2/4$, we show that the spectrum of the operator $\overline{H_D}(\lambda)$ contains the whole real line.

2 The subcritical case

The first important observation is that spectral properties of the operator $H_D(\lambda)$ depend crucially on the value of the parameter λ. Actually, we have to distinguish two cases.

The spectral regime we are primarily interested in occurs for small values of λ. We have the following result.

Theorem 2.1. *For any $\lambda \in [0, \pi^2/4]$ the operator $H_D(\lambda)$ initially defined on $\widetilde{C}_0^2(\bar{D})$ is non-negative.*

Proof. Since the ground state eigenvalue of the one dimensional Dirichlet Laplacian $-\mathrm{d}^2/\mathrm{d}x^2$ on the interval $(-1/|y|, 1/|y|)$ with fixed non zero y is $\pi^2 y^2/4$, then, with a slight abuse of notation, for any function $u \in \mathcal{H}_0^1(D)$, that is satisfying the condition $u|_{\partial D} = 0$, we can write

$$\begin{aligned}\int_D \left|\frac{\partial u}{\partial x}\right|^2 \mathrm{d}x\,\mathrm{d}y &= \int_{-\infty}^0 \int_{1/y}^{-1/y} \left|\frac{\partial u}{\partial x}\right|^2 \mathrm{d}x\,\mathrm{d}y + \int_0^{\infty}\int_{-1/y}^{1/y} \left|\frac{\partial u}{\partial x}\right|^2 \mathrm{d}x\,\mathrm{d}y \\ &\geq \frac{\pi^2}{4}\int_{-\infty}^0 \int_{1/y}^{-1/y} y^2|u(x,y)|^2\,\mathrm{d}x\,\mathrm{d}y \\ &\quad + \frac{\pi^2}{4}\int_0^{\infty}\int_{-1/y}^{1/y} y^2|u(x,y)|^2\,\mathrm{d}x\,\mathrm{d}y \\ &= \frac{\pi^2}{4}\int_D y^2|u(x,y)|^2\,\mathrm{d}x\,\mathrm{d}y.\end{aligned} \tag{2}$$

In a similar way one establishes that

$$\int_D \left|\frac{\partial u}{\partial y}\right|^2 \mathrm{d}x\,\mathrm{d}y \geq \frac{\pi^2}{4}\int_D x^2|u(x,y)|^2\,\mathrm{d}x\,\mathrm{d}y. \tag{3}$$

Hence, in view of (2) and (3), for any function $u \in \mathcal{H}_0^1(D)$ we have

$$\begin{aligned}&\int_D (H_D(\lambda)(u))\,\bar{u}\,\mathrm{d}x\,\mathrm{d}y \\ &= \Big(1-\frac{4\lambda}{\pi^2}\Big)\int_D |\nabla u|^2\,\mathrm{d}x\,\mathrm{d}y + \frac{4\lambda}{\pi^2}\Big(\int_D |\nabla u|^2\,\mathrm{d}x\,\mathrm{d}y \\ &\qquad\qquad - \frac{\pi^2}{4}\int_D (x^2+y^2)|u|^2\,\mathrm{d}x\,\mathrm{d}y\Big) \\ &= \Big(1-\frac{4\lambda}{\pi^2}\Big)\int_D |\nabla u|^2\,\mathrm{d}x\,\mathrm{d}y + \frac{4\lambda}{\pi^2}\Big(\int_D \left|\frac{\partial u}{\partial x}\right|^2 \mathrm{d}x\,\mathrm{d}y + \int_D \left|\frac{\partial u}{\partial y}\right|^2 \mathrm{d}x\,\mathrm{d}y \\ &\qquad\qquad - \frac{\pi^2}{4}\int_D (x^2+y^2)|u|^2\,\mathrm{d}x\,\mathrm{d}y\Big) \\ &\geq \Big(1-\frac{4\lambda}{\pi^2}\Big)\int_D |\nabla u|^2\,\mathrm{d}x\,\mathrm{d}y,\end{aligned} \tag{4}$$

which establishes the theorem. □

Since the quadratic form corresponding to operator $H_D(\lambda)$ initially defined on $\widetilde{C}_0^2(\bar{D})$ is non-negative for all $\lambda \in [0, \pi^2/4]$, thus, the construction of the self-adjoint extension via Friedrichs method is possible. For the sake of brevity, we will use for it the same symbol H_D or $H_D(\lambda)$.

Theorem 2.2. *If $\lambda < \pi^2/4$ the spectrum of $H_D(\lambda)$ is purely discrete. Moreover, for the corresponding eigenvalues, denoted by $\{\beta_j(\lambda)\}_{j=1}^{\infty}$, $\lambda < \pi^2/4$, the following expression holds*

$$\beta_j(\lambda) = c_j(\lambda)\, \mu_j, \qquad j = 1, 2, \ldots, \tag{5}$$

where $(1 - 4\lambda/\pi^2) \leq c_j(\lambda) \leq 1$ and μ_j, $j = 1, 2, \ldots$, are the eigenvalues of the Dirichlet Laplacian $-\Delta_D$ arranged in the ascending order and $\mu_j \sim \pi\, j/\ln j$.

Proof. For $\lambda < \pi^2/4$ estimates (2)–(4) show that

$$\operatorname{Dom}(Q(H_D(\lambda))) = \mathcal{H}_0^1(D),$$

where $Q(H_D(\lambda))$ is the quadratic form corresponding to operator $H_D(\lambda)$. Next, again using (4) one gets

$$H_D(\lambda) \geq -\Big(1 - \frac{4\lambda}{\pi^2}\Big)\Delta_D.$$

Thus combining the minimax principle with the classical result of B. Simon [12] we find that the spectrum of $H_D(\lambda)$ for any $\lambda < \pi^2/4$ is purely discrete. Moreover, the following eigenvalue bound holds

$$\beta_j(\lambda) \geq \Big(1 - \frac{4\lambda}{\pi^2}\Big)\mu_j, \quad j = 1, 2, \ldots, \tag{6}$$

where μ_j, $j = 1, 2, \ldots$, are the eigenvalues of $-\Delta_D$ arranged in the ascending order.

Using the non-negativeness of λ and the inequality

$$\int_D (H_D(\lambda)(u))\, \bar{u}\, \mathrm{d}x\, \mathrm{d}y \leq \int_D |\nabla u|^2 \mathrm{d}x\, \mathrm{d}y,$$

we have that

$$H_D(\lambda) \leq -\Delta_D$$

and therefore

$$\beta_j(\lambda) \leq \mu_j, \ j = 1, 2, \ldots. \tag{7}$$

The asymptotic eigenvalue distribution of the Dirichlet Laplacian $-\Delta_D$ is well known (see [7]),

$$N(\Lambda) \sim \frac{1}{\pi}\Lambda \ln \Lambda.$$

By means of the known inverse asymptotic formula (see Section 9 of [11]) we get for the spectrum of $-\Delta_D$ the expression

$$\mu_j \sim \frac{\pi j}{\ln j} \tag{8}$$

as $j \to \infty$, and this together with (6) and (7) establishes (5). □

Remark 2.3. Let $\lambda < \pi^2/4$. Then for any $\epsilon > 0$ there exists a natural number $M(\epsilon)$ such that for the eigenvalue sum of operator $H_D(\lambda)$ the following lower bound holds true:

$$\sum_{j=1}^{N} \beta_j(\lambda) \geq (1-\epsilon)\frac{(\pi^2 - 4\lambda)}{4\pi}\frac{(N-2)^2}{\ln N}, \qquad N > M(\epsilon).$$

On the other hand, the following upper bound is valid

$$\sum_{j=1}^{N} \beta_j(\lambda) \leq (1+\epsilon)\pi\frac{N^2}{\ln N}, \qquad N > M(\epsilon). \tag{9}$$

Proof. The asymptotics (8) means that for any positive ϵ there exists $M(\epsilon) \in \mathbb{N}$ such that

$$(1-\epsilon)\frac{\pi j}{\ln j} \leq \mu_j \leq (1+\epsilon)\frac{\pi j}{\ln j}, \tag{10}$$

for all $j > M(\epsilon)$. Thus (5) provides

$$\begin{aligned}
\sum_{j=1}^{N} \beta_j(\lambda) &\geq \sum_{j=[N/2]}^{2[N/2]} \beta_j(\lambda) \\
&\geq \beta_{[N/2]}(\lambda)\Big[\frac{N}{2}\Big] \\
&\geq \Big(1 - \frac{4\lambda}{\pi^2}\Big)\Big[\frac{N}{2}\Big]\mu_{[N/2]} \\
&\geq (1-\epsilon)\Big(1 - \frac{4\lambda}{\pi^2}\Big)\pi\Big[\frac{N}{2}\Big]^2\frac{1}{\ln\Big[\frac{N}{2}\Big]} \\
&\geq (1-\epsilon)\frac{(\pi^2 - 4\lambda)}{4\pi}\frac{(N-2)^2}{\ln N}.
\end{aligned}$$

To establish (9) one needs to carry out a similar proof. More precisely, in view of (5) and (10),

$$\sum_{j=1}^{N} \beta_j(\lambda) \leq N\,\beta_N(\lambda) \leq N\,\mu_N \leq (1+\epsilon)\pi\frac{N^2}{\ln N}, \qquad \square$$

3 The critical case

Now we consider the critical case $\lambda = \lambda_{\text{crit}} = \pi^2/4$. The following theorem holds true.

Theorem 3.1. *The spectrum of $H_D(\lambda_{\text{crit}})$ coincides with the half line $[0, \infty)$.*

Proof. In view of Theorem 2.1 the operator $H_D(\lambda_{\text{crit}})$ is non-negative. To demonstrate that any non-negative number μ belongs to the essential spectrum of the operator $H_D(\lambda_{\text{crit}})$ we are going to use Weyl's criterion (Theorem VII.12 in [10]): we have to find a sequence $\{\psi_k\}_{k=1}^{\infty} \subset \text{Dom}(H_D(\lambda_{\text{crit}}))$ such that $\|\psi_k\| = 1$ which contains no convergent subsequence and

$$\|H_D(\lambda_{\text{crit}})\psi_k - \mu\psi_k\| \longrightarrow 0 \quad \text{as } k \to \infty.$$

We define the following sequence defined on D

$$\psi_k(x, y) := \cos\Big(\frac{\pi x y}{2}\Big) f\Big(\frac{y}{k}\Big) e^{i\sqrt{\mu} y},$$

where f is a smooth function with $\operatorname{supp} f \subset [1, 2]$ satisfying $\int_1^2 f^2(z)\,\mathrm{d}z = 1$.

We note that for a given k one can achieve that $\|\psi_k\|_{L^2(D)} \geq \frac{1}{2}$ as the following estimates show,

$$\begin{aligned}
\int_D \Big|\cos\Big(\frac{\pi x y}{2}\Big) f\Big(\frac{y}{k}\Big) e^{i\sqrt{\mu} y}\Big|^2 \mathrm{d}x\,\mathrm{d}y
&= \int_k^{2k}\int_{-1/y}^{1/y} \cos^2\Big(\frac{\pi x y}{2}\Big) f^2\Big(\frac{y}{k}\Big) \mathrm{d}x\,\mathrm{d}y \\
&= \int_k^{2k}\int_{-1/y}^{1/y} \frac{1}{2}(1 + \cos(\pi x y))\, f^2\Big(\frac{y}{k}\Big) \mathrm{d}x\,\mathrm{d}y \\
&= \int_k^{2k}\int_{-1}^{1} \frac{1}{2y}(1 + \cos(\pi t)) f^2\Big(\frac{y}{k}\Big) \mathrm{d}t\,\mathrm{d}y \\
&= \int_k^{2k} \frac{1}{y} f^2\Big(\frac{y}{k}\Big) \mathrm{d}y \\
&= \int_1^2 \frac{f^2(z)}{z}\,\mathrm{d}z \\
&\geq \frac{1}{2}\int_1^2 f^2(z)\,\mathrm{d}z \\
&= \frac{1}{2}.
\end{aligned}$$

The second derivatives of the functions ψ_k are

$$\frac{\partial^2\psi_k}{\partial x^2} = -\frac{\pi^2 y^2}{4} \cos\Big(\frac{\pi x y}{2}\Big) f\Big(\frac{y}{k}\Big) e^{i\sqrt{\mu}y}$$

and

$$\begin{aligned}\frac{\partial^2\psi_k}{\partial y^2} = \Big(&-\frac{\pi^2 x^2}{4}\cos\Big(\frac{\pi x y}{2}\Big) f\Big(\frac{y}{k}\Big) - \frac{\pi x}{k}\sin\Big(\frac{\pi x y}{2}\Big) f'\Big(\frac{y}{k}\Big)\\ &- i\pi\sqrt{\mu}\, x \sin\Big(\frac{\pi x y}{2}\Big) f\Big(\frac{y}{k}\Big) + \frac{2i\sqrt{\mu}}{k}\cos\Big(\frac{\pi x y}{2}\Big) f'\Big(\frac{y}{k}\Big)\\ &+ \frac{1}{k^2}\cos\Big(\frac{\pi x y}{2}\Big) f''\Big(\frac{y}{k}\Big) - \mu\cos\Big(\frac{\pi x y}{2}\Big) f\Big(\frac{y}{k}\Big)\Big) e^{i\sqrt{\mu}y}.\end{aligned} \tag{11}$$

Our aim is to show that choosing k sufficiently large one can make all terms except of the last one at the right-hand side of (11) as small as we wish. Changing the integration variables, we get for the first term the following estimate,

$$\begin{aligned}&\int_D \Big| e^{i\sqrt{\mu}y}\frac{\pi^2 x^2}{4}\cos\Big(\frac{\pi x y}{2}\Big) f\Big(\frac{y}{k}\Big)\Big|^2 \mathrm{d}x\,\mathrm{d}y\\ &= \frac{\pi^4}{16}\int_k^{2k}\int_{-1/y}^{1/y} x^4\cos^2\Big(\frac{\pi x y}{2}\Big) f^2\Big(\frac{y}{k}\Big)\mathrm{d}x\,\mathrm{d}y\\ &= \frac{\pi^4}{16}\int_k^{2k}\int_{-1}^{1}\frac{t^4}{y^5}\cos^2\Big(\frac{\pi t}{2}\Big) f^2\Big(\frac{y}{k}\Big)\mathrm{d}t\,\mathrm{d}y\\ &= \frac{\pi^4}{16}\int_k^{2k}\frac{1}{y^5} f^2\Big(\frac{y}{k}\Big)\mathrm{d}y\int_{-1}^{1} t^4\cos^2\Big(\frac{\pi t}{2}\Big)\mathrm{d}t\\ &\le \frac{\pi^4}{16k^4}\int_1^2 |f(z)|^2\mathrm{d}z\int_{-1}^{1} t^4\cos^2\Big(\frac{\pi t}{2}\Big)\mathrm{d}t\\ &= \frac{\pi^4}{16k^4}\int_{-1}^{1} t^4\cos^2\Big(\frac{\pi t}{2}\Big)\mathrm{d}t,\end{aligned}$$

where the right-hand side tends to zero as $k \to \infty$. In the same way we establish that for large enough k all the terms except of the last one in (11) can be made small.

Similarly one can prove that for large k the integral

$$\int_D x^4\cos^2\Big(\frac{\pi x y}{2}\Big) f^2\Big(\frac{y}{k}\Big)\mathrm{d}x\,\mathrm{d}y$$

is again as small as we wish.

Consequently, for any fixed $\varepsilon > 0$ one can choose k large enough such that

$$\begin{aligned}
&\int_D |H_D(\lambda_{\mathrm{crit}})\psi_k - \mu\psi_k|^2(x,y)\,\mathrm{d}x\,\mathrm{d}y \\
&= \int_D \left| -\frac{\partial^2\psi_k}{\partial x^2} - \frac{\partial^2\psi_k}{\partial y^2} - \frac{\pi^2}{4}(x^2+y^2)\psi_k - \mu\psi_k \right|^2 \mathrm{d}x\,\mathrm{d}y \\
&\le \int_D \left| \frac{\pi^2 y^2}{4}\cos\left(\frac{\pi x y}{2}\right) f\left(\frac{y}{k}\right) + \mu\cos\left(\frac{\pi x y}{2}\right) f\left(\frac{y}{k}\right) \right. \\
&\qquad \left. - \frac{\pi^2}{4}(x^2+y^2)\cos\left(\frac{\pi x y}{2}\right) f\left(\frac{y}{k}\right) - \mu\cos\left(\frac{\pi x y}{2}\right) f\left(\frac{y}{k}\right)\right|^2 \mathrm{d}x\,\mathrm{d}y + \varepsilon \\
&= \frac{\pi^4}{16}\int_D x^4\cos^2\left(\frac{\pi x y}{2}\right) f^2\left(\frac{y}{k}\right)\mathrm{d}x\,\mathrm{d}y + \varepsilon \\
&< 2\varepsilon.
\end{aligned} \tag{12}$$

To complete the proof we fix a sequence $\{\varepsilon_j\}_{j=1}^{\infty}$ such that $\varepsilon_j \searrow 0$ holds as $j \to \infty$ and to any j we construct a function $\psi_{k(\varepsilon_j)}$ such that the supports for different j's do not intersect each other; this can be achieved by choosing $k(\varepsilon_j) > 2k(\varepsilon_{j-1})$. The norms of $H_D(\lambda_{\mathrm{crit}})\psi_{k(\varepsilon_j)} - \mu\psi_{k(\varepsilon_j)}$ satisfy the inequality (12) with $2\varepsilon_j$ on the right-hand side, and by construction the sequence $\psi_{k(\varepsilon_j)}$ converges weakly to zero; this yields the desired Weyl sequence for any non-negative number μ. □

4 The supercritical case

Let $\lambda > \pi^2/4$ and let $\overline{H_D}(\lambda)$ denote the closure of the operator $H_D(\lambda)$ initially defined on $\tilde{C}_0^2(\bar{D})$. Our next result is the following.

Theorem 4.1. *For any $\lambda > \pi^2/4$ the spectrum of $\overline{H_D}(\lambda)$ contains the real line $\mathbb{R}$.*

Proof. We contruct a Weyl sequence $\{\psi_k\}_{k=1}^{\infty} \subset \mathrm{Dom}(\overline{H_D}(\lambda))$ (for any real number μ) which satisfies

$$\int_D |\psi_k(x,y)|^2\,\mathrm{d}x\,\mathrm{d}y \ge \frac{1}{2}, \quad k = 1,2,\ldots,$$

and

$$\int_D \left|\overline{H_D}(\lambda)\psi_k - \mu\psi_k\right|^2 \mathrm{d}x\,\mathrm{d}y \longrightarrow 0 \quad \text{as } k \to \infty.$$

This shows that number μ can not belong to the resolvent set of $\overline{H_D}(\lambda)$, Q.E.D.

We fix a positive ε and choose a natural number $k = k(\varepsilon)$ with which we associate a function $\chi_k \subset C_0^2(1,k)$ satisfying the following conditions

$$\int_1^k \frac{1}{z}\chi_k^2(z)\,\mathrm{d}z = 1 \quad \text{and} \quad \int_1^k z(\chi_k'(z))^2\,\mathrm{d}z < \varepsilon. \tag{13}$$

To give an example, consider the function constructed in [1]

$$\begin{aligned}\tilde{\chi}_k(z) = {} & \frac{8\ln^3 z}{\ln^3 k}\chi_{\{1\le z\le\sqrt{k}\}}(z) + \frac{2\ln k - 2\ln z}{\ln k}\chi_{\{\sqrt{k}+1\le z\le k-1\}}(z) \\ & + g_k(z)\chi_{\{\sqrt{k}<z<\sqrt{k}+1\}}(z) + q_k(z)\chi_{\{k-1<z\le k\}}(z),\end{aligned}$$

where g_k and q_k are interpolating functions chosen in such a way that $\tilde{\chi}_k \in C_0^2(1,k)$. The first integral in (13) is positive for $\tilde{\chi}_k$, in fact we have

$$\int_1^{\sqrt{k}} \frac{1}{z}\tilde{\chi}_k^2(z)\,\mathrm{d}z \ge \frac{1}{4},$$

hence we can define

$$\chi_k(z) = \Big(\int_1^k \frac{1}{z}\tilde{\chi}_k^2(z)\,\mathrm{d}z\Big)^{-1/2}\tilde{\chi}_k(z).$$

This function satisfies by definition the first condition of (13) and one can check that it also satisfies the second one provided k is sufficiently large; this follows from the fact that

$$\int_1^k z(\chi_k'(z))^2\,\mathrm{d}z = \mathcal{O}\Big(\frac{1}{\ln k}\Big) \quad \text{as } k \to \infty.$$

Such functions allow us to construct the Weyl sequence we seek. Given a function χ_k with the described properties, we define

$$\psi_k(x,y) := \cos\Big(\frac{\pi x y}{2}\Big)e^{i\eta_\mu(y)}\chi_k\Big(\frac{y}{n_k}\Big) + \frac{f(xy)}{y^2}e^{i\eta_\mu(y)}\chi_k\Big(\frac{y}{n_k}\Big),$$

where

$$\eta_\mu(y) = \int_{\frac{|\mu|^{1/2}}{|\alpha|}}^{y} \sqrt{\alpha^2 t^2 + \mu}\,\mathrm{d}t$$

with the real number α to be chosen later, the smooth function $f\colon[-1,1] \to \mathbb{R}$ satisfying the boundary conditions $f(\pm 1) = 0$, and $n_k \in \mathbb{N}$ also will be chosen later. First we establish that for any given k one can achieve that $\|\psi_k\|_{L^2(D)} \ge 1/2$ holds by choosing n_k large enough.

The following estimates show

$$
\begin{aligned}
\int_D \Big|\cos\Big(\frac{\pi x y}{2}\Big) e^{i\eta_\mu(y)} \chi_k\Big(\frac{y}{n_k}\Big)\Big|^2 \mathrm{d}x\, \mathrm{d}y
&= \int_{n_k}^{k n_n} \int_{-1/y}^{1/y} \Big|\cos\Big(\frac{\pi x y}{2}\Big) \chi_k\Big(\frac{y}{n_k}\Big)\Big|^2 \mathrm{d}x\, \mathrm{d}y \\
&= \int_{n_k}^{k n_n} \int_{-1}^{1} \frac{1}{y}\Big|\cos\Big(\frac{\pi t}{2}\Big) \chi_k\Big(\frac{y}{n_k}\Big)\Big|^2 \mathrm{d}t\, \mathrm{d}y \\
&= \int_{-1}^{1} \cos^2\Big(\frac{\pi t}{2}\Big) \mathrm{d}t \int_{n_k}^{k n_n} \frac{1}{y}\Big|\chi_k\Big(\frac{y}{n_k}\Big)\Big|^2 \mathrm{d}y \\
&= \int_{n_k}^{k n_n} \frac{1}{y}\Big|\chi_k\Big(\frac{y}{n_k}\Big)\Big|^2 \mathrm{d}y \\
&= \int_{1}^{k} \frac{1}{z}|\chi_k(z)|^2 \mathrm{d}z \\
&= 1
\end{aligned}
$$

and

$$
\begin{aligned}
\int_D \Big|\frac{1}{y^2} f(xy) e^{i\eta_\mu(y)} \chi_k\Big(\frac{y}{n_k}\Big)\Big|^2 \mathrm{d}x\, \mathrm{d}y &= \int_{n_k}^{k n_k} \int_{-1/y}^{1/y} \Big|\frac{1}{y^2} f(xy) \chi_k\Big(\frac{y}{n_k}\Big)\Big|^2 \mathrm{d}x\, \mathrm{d}y \\
&= \int_{n_k}^{k n_k} \int_{-1}^{1} \frac{1}{y}\Big|\frac{1}{y^2} f(t) \chi_k\Big(\frac{y}{n_k}\Big)\Big|^2 \mathrm{d}t\, \mathrm{d}y \\
&\le \frac{1}{n_k^5} \int_{n_k}^{k n_k} \int_{-1}^{1} \Big|f(t) \chi_k\Big(\frac{y}{n_k}\Big)\Big|^2 \mathrm{d}t\, \mathrm{d}y \\
&= \frac{1}{n_k^4} \int_{-1}^{1} |f(t)|^2 \mathrm{d}t \int_{1}^{k} |\chi_k(z)|^2 \mathrm{d}z,
\end{aligned}
$$

which establish our claim. Let us now prove that the inequality

$$
\|\overline{H_D}(\lambda)\psi_k - \mu\psi_k\|_{L^2(D)} < c\varepsilon
$$

with a fixed constant c holds for $k = k(\varepsilon)$.

By a straightforward calculation one gets

$$
\frac{\partial^2 \psi_k}{\partial x^2} = -\frac{\pi^2 y^2}{4} \cos\Big(\frac{\pi x y}{2}\Big) e^{i\eta_\mu(y)} \chi_k\Big(\frac{y}{n_k}\Big) + f''(xy) e^{i\eta_\mu(y)} \chi_k\Big(\frac{y}{n_k}\Big)
$$

and

$$
\begin{aligned}
\frac{\partial^2 \psi_k}{\partial y^2} = & -\frac{\pi^2 x^2}{4} \cos\left(\frac{\pi x y}{2}\right) e^{i\eta_\mu(y)} \chi_k\left(\frac{y}{n_k}\right) \\
& - i\pi x \sqrt{\alpha^2 y^2 + \mu} \sin\left(\frac{\pi x y}{2}\right) e^{i\eta_\mu(y)} \chi_k\left(\frac{y}{n_k}\right) \\
& - \frac{\pi x}{n_k} \sin\left(\frac{\pi x y}{2}\right) e^{i\eta_\mu(y)} \chi_k'\left(\frac{y}{n_k}\right) \\
& + \frac{i\alpha^2 y}{\sqrt{\alpha^2 y^2 + \mu}} \cos\left(\frac{\pi x y}{2}\right) e^{i\eta_\mu(y)} \chi_k\left(\frac{y}{n_k}\right) \\
& - (\alpha^2 y^2 + \mu) \cos\left(\frac{\pi x y}{2}\right) e^{i\eta_\mu(y)} \chi_k\left(\frac{y}{n_k}\right) \\
& + \frac{2i\sqrt{\alpha^2 y^2 + \mu}}{n_k} \cos\left(\frac{\pi x y}{2}\right) e^{i\eta_\mu(y)} \chi_k'\left(\frac{y}{n_k}\right) \\
& + \frac{1}{n_k^2} \cos\left(\frac{\pi x y}{2}\right) e^{i\eta_\mu(y)} \chi_k''\left(\frac{y}{n_k}\right) \\
& + \frac{6}{y^4} f(xy) e^{i\eta_\mu(y)} \chi_k\left(\frac{y}{n_k}\right) \\
& - \frac{4x}{y^3} f'(xy) e^{i\eta_\mu(y)} \chi_k\left(\frac{y}{n_k}\right) \\
& - \frac{4i\sqrt{\alpha^2 y^2 + \mu}}{y^3} f(xy) e^{i\eta_\mu(y)} \chi_k\left(\frac{y}{n_k}\right) \\
& + \frac{i\alpha^2}{y\sqrt{\alpha^2 y^2 + \mu}} f(xy) e^{i\eta_\mu(y)} \chi_k\left(\frac{y}{n_k}\right) \\
& - \frac{4}{n_k y^3} f(xy) e^{i\eta_\mu(y)} \chi_k'\left(\frac{y}{n_k}\right) \\
& + \frac{x^2}{y^2} f''(xy) e^{i\eta_\mu(y)} \chi_k\left(\frac{y}{n_k}\right) \\
& + \frac{2ix\sqrt{\alpha^2 y^2 + \mu}}{y^2} f'(xy) e^{i\eta_\mu(y)} \chi_k\left(\frac{y}{n_k}\right) \\
& + \frac{2x}{n_k y^2} f'(xy) e^{i\eta_\mu(y)} \chi_k'\left(\frac{y}{n_k}\right) \\
& - \frac{(\alpha^2 y^2 + \mu)}{y^2} f(xy) e^{i\eta_\mu(y)} \chi_k\left(\frac{y}{n_k}\right) \\
& + \frac{2i\sqrt{\alpha^2 y^2 + \mu}}{n_k y^2} f(xy) e^{i\eta_\mu(y)} \chi_k'\left(\frac{y}{n_k}\right) \\
& + \frac{1}{n_k^2 y^2} f(xy) e^{i\eta_\mu(y)} \chi_k''\left(\frac{y}{n_k}\right).
\end{aligned} \tag{14}
$$

We want to show that choosing n_k sufficiently large one can make some terms at the right hand side of (14) as small as we wish. Changing the integration variables, we get the following estimate,

$$\begin{aligned}\int_D \Big|\frac{\pi^2 x^2}{4}\cos\Big(\frac{\pi x y}{2}\Big)e^{i\eta_\mu(y)}\chi_k\Big(\frac{y}{n_k}\Big)\Big|^2 \mathrm{d}x\,\mathrm{d}y \\ = \frac{\pi^4}{16}\int_{n_k}^{kn_k}\int_{-1/y}^{1/y}\Big|x^2\cos\Big(\frac{\pi x y}{2}\Big)\chi_k\Big(\frac{y}{n_k}\Big)\Big|^2 \mathrm{d}x\,\mathrm{d}y \\ = \frac{\pi^4}{16}\int_{n_k}^{kn_k}\frac{1}{y^5}\Big|\chi_k\Big(\frac{y}{n_k}\Big)\Big|^2 \mathrm{d}y\int_{-1}^{1} t^4\cos^2\Big(\frac{\pi t}{2}\Big)\mathrm{d}t \\ \le \frac{\pi^4}{16 n_k^4}\int_1^k |\chi_k(z)|^2\mathrm{d}z\int_{-1}^{1} t^4\cos^2\Big(\frac{\pi t}{2}\Big)\mathrm{d}t,\end{aligned}$$

where the last integral is small. In the same way we establish the smallness of all the terms of (14) except of the following:

$$-i\pi x\sqrt{\alpha^2 y^2+\mu}\,\sin\Big(\frac{\pi x y}{2}\Big)e^{i\eta_\mu(y)}\chi_k\Big(\frac{y}{n_k}\Big),$$

$$\frac{i\alpha^2 y}{\sqrt{\alpha^2 y^2+\mu}}\cos\Big(\frac{\pi x y}{2}\Big)e^{i\eta_\mu(y)}\chi_k\Big(\frac{y}{n_k}\Big),$$

$$-(\alpha^2 y^2+\mu)\cos\Big(\frac{\pi x y}{2}\Big)e^{i\alpha y^2}\chi_k\Big(\frac{y}{n_k}\Big),$$

$$\frac{2i\sqrt{\alpha^2 y^2+\mu}}{n_k}\cos\Big(\frac{\pi x y}{2}\Big)e^{i\eta_\mu(y)}\chi_k'\Big(\frac{y}{n_k}\Big),$$

$$-\frac{\alpha^2 y^2+\mu}{y^2}f(xy)e^{i\eta_\mu(y)}\chi_k\Big(\frac{y}{n_k}\Big).$$

To continue the proof we are going to use also the smallness of the integrals

$$\int_D \Big|i\pi x(\sqrt{\alpha^2 y^2+\mu}-\alpha y)\sin\Big(\frac{\pi x y}{2}\Big)e^{i\eta_\mu(y)}\chi_k\Big(\frac{y}{n_k}\Big)\Big|^2 \mathrm{d}x\,\mathrm{d}y,$$

$$\int_D \Big|\Big(\frac{i\alpha^2 y}{\sqrt{\alpha^2 y^2+\mu}}-i\alpha\Big)\cos\Big(\frac{\pi x y}{2}\Big)e^{i\eta_\mu(y)}\chi_k\Big(\frac{y}{n_k}\Big)\Big|^2 \mathrm{d}x\,\mathrm{d}y,$$

$$\int_D \Big|\Big(\frac{2i\sqrt{\alpha^2 y^2+\mu}}{n_k}-\frac{2i\alpha y}{n_k}\Big)\cos\Big(\frac{\pi x y}{2}\Big)e^{i\eta_\mu(y)}\chi_k'\Big(\frac{y}{n_k}\Big)\Big|^2 \mathrm{d}x\,\mathrm{d}y$$

and

$$\int_D \Big|\lambda x^2 \cos\Big(\frac{\pi x y}{2}\Big) e^{i\eta_\mu(y)} \chi_k\Big(\frac{y}{n_k}\Big)\Big|^2 \mathrm{d}x\,\mathrm{d}y,$$

$$\int_D \Big|\frac{\lambda x^2}{y^2} f(xy) e^{i\eta_\mu(y)} \chi_k\Big(\frac{y}{n_k}\Big)\Big|^2 \mathrm{d}x\,\mathrm{d}y,$$

$$\int_D \Big|\frac{\mu}{y^2} f(xy) e^{i\eta_\mu(y)} \chi_k\Big(\frac{y}{n_k}\Big)\Big|^2 \mathrm{d}x\,\mathrm{d}y,$$

which can be proved in a similar way.

Consequently, for the fixed $\varepsilon > 0$ one can choose n_k large enough such that

$$\begin{aligned}
&\int_D |\overline{H_D}(\lambda)\psi_k - \mu\psi_k|^2(x,y)\,\mathrm{d}x\,\mathrm{d}y \\
&= \int_D \Big| -\frac{\partial^2\psi_k}{\partial x^2} - \frac{\partial^2\psi_k}{\partial y^2} - \lambda(x^2+y^2)\psi_k - \mu\psi_k\Big|^2 \mathrm{d}x\,\mathrm{d}y \\
&\le \int_{n_k}^{kn_k}\int_{-1/y}^{1/y} \Big|\frac{\pi^2 y^2}{4}\cos\Big(\frac{\pi x y}{2}\Big)\chi_k\Big(\frac{y}{n_k}\Big) - f''(xy)\chi_k\Big(\frac{y}{n_k}\Big) \\
&\qquad + i\alpha\pi x y \sin\Big(\frac{\pi x y}{2}\Big)\chi_k\Big(\frac{y}{n_k}\Big) - i\alpha\cos\Big(\frac{\pi x y}{2}\Big)\chi_k\Big(\frac{y}{n_k}\Big) \\
&\qquad + \alpha^2 y^2 \cos\Big(\frac{\pi x y}{2}\Big)\chi_k\Big(\frac{y}{n_k}\Big) + \mu\cos\Big(\frac{\pi x y}{2}\Big)\chi_k\Big(\frac{y}{n_k}\Big) \\
&\qquad - \frac{2i\alpha y}{n_k}\cos\Big(\frac{\pi x y}{2}\Big)\chi_k'\Big(\frac{y}{n_k}\Big) + \alpha^2 f(xy)\chi_k\Big(\frac{y}{n_k}\Big) \\
&\qquad - \lambda y^2\cos\Big(\frac{\pi x y}{2}\Big)\chi_k\Big(\frac{y}{n_k}\Big) - \lambda f(xy)\chi\Big(\frac{y}{n_k}\Big) \\
&\qquad - \mu\cos\Big(\frac{\pi x y}{2}\Big)\chi_k\Big(\frac{y}{n_k}\Big)\Big|^2 \mathrm{d}x\,\mathrm{d}y + \varepsilon \\
&\le 2\int_{n_k}^{kn_k}\int_{-1/y}^{1/y} \Big|\frac{\pi^2 y^2}{4}\cos\Big(\frac{\pi x y}{2}\Big)\chi_k\Big(\frac{y}{n_k}\Big) - f''(xy)\chi_k\Big(\frac{y}{n_k}\Big) \\
&\qquad + i\alpha\pi x y \sin\Big(\frac{\pi x y}{2}\Big)\chi_k\Big(\frac{y}{n_k}\Big) - i\alpha\cos\Big(\frac{\pi x y}{2}\Big)\chi_k\Big(\frac{y}{n_k}\Big) \\
&\qquad + \alpha^2 y^2 \cos\Big(\frac{\pi x y}{2}\Big)\chi_k\Big(\frac{y}{n_k}\Big) + \alpha^2 f(xy)\chi_k\Big(\frac{y}{n_k}\Big) \\
&\qquad - \lambda y^2\cos\Big(\frac{\pi x y}{2}\Big)\chi_k\Big(\frac{y}{n_k}\Big) \\
&\qquad - \lambda f(xy)\chi\Big(\frac{y}{n_k}\Big)\Big|^2 \mathrm{d}x\,\mathrm{d}y \\
&\quad + 2\int_{n_k}^{kn_k}\int_{-1/y}^{1/y} \Big|\frac{2i\alpha y}{n_k}\cos\Big(\frac{\pi x y}{2}\Big)\chi_k'\Big(\frac{y}{n_k}\Big)\Big|^2 \mathrm{d}x\,\mathrm{d}y + \varepsilon.
\end{aligned} \tag{15}$$

Now we choose $\alpha = \sqrt{4\lambda-\pi^2}/2$ and $f(t) = ig(t)$, where $g(t)$ is a solution of equation

$$g''(t) + \frac{\pi^2}{4}g(t) - \alpha\pi t \sin\Big(\frac{\pi t}{2}\Big) + \alpha\cos\Big(\frac{\pi t}{2}\Big) = 0 \tag{16}$$

defined on interval $[-1,1]$ with Dirichlet boundary conditions. Since the solution of the Dirichlet problem $h'' + \pi^2/4h = 0$ on interval $[-1,1]$ corresponds with one dimensional subspace generated by $\cos(\pi t/2)$ then the equation

$$\int_{-1}^{1}\Big(-\alpha\pi t\sin\Big(\frac{\pi t}{2}\Big) + \alpha\cos\Big(\frac{\pi t}{2}\Big)\Big)\cos\Big(\frac{\pi t}{2}\Big)\,\mathrm{d}t = 0$$

guarantees the existence of the solution for (16). Thus, in view of construction (13), the inequality (15) establishes that

$$\int_D |\overline{H_D}(\lambda)\psi_k - \mu\psi_k|^2\,\mathrm{d}x\,\mathrm{d}y < (8\lambda - 2\pi^2 + 1)\varepsilon. \tag{17}$$

Finally we finish the proof in a similar way as in the previous section: we fix a sequence $\{\varepsilon_j\}_{j=1}^{\infty}$ such that $\varepsilon_j \searrow 0$ holds as $j \to \infty$ and to any j we construct a function $\psi_{k(\varepsilon_j)}$ such that the supports for different j's do not intersect each other; this can be achieved by choosing each next $n_{k(\varepsilon_j)} > k(\varepsilon_{j-1})n_{k(\varepsilon_{j-1})}$. The norms of $\overline{H_D}(\lambda)\psi_{k(\varepsilon_j)} - \mu\psi_{k(\varepsilon_j)}$ satisfy the inequality (17) with $(8\lambda - 2\pi^2 + 1)\varepsilon_j$ on the right-hand side, and by construction the sequence $\psi_{k(\varepsilon_j)}$ converges weakly to zero; this yields the desired Weyl sequence for any real number μ. □

Acknowledgements. The research was supported by the Czech Science Foundation (GACR), project No. 14-02476S "Variations, geometry and physics." Diana Barseghyan is supported by the Statutory City of Ostrava (grant 0924/2016/ŠaS). We are obliged to Vladimir Lotoreichik and Vladimir Khrabustovskyi for useful discussions. We thank an anonymous referee for valuable comments.

References

[1] D. Barseghyan and P. Exner, A regular version of Smilansky model. *J. Math. Phys.* **55** (2014), no. 4, 042104, 13 pp. MR 3390580

[2] D. Barseghyan, P. Exner, A. Khrabustovskyi, and M. Tater, Spectral analysis of a class of Schrödinger operators exhibiting *J. Phys. A* **49** (2016), no. 16, 165302, 19 pp. MR 3479137 Zbl 06610631

[3] P. Exner and D. Barseghyan, Spectral estimates for a class of Schrödinger operators with infinite phase space and potential unbounded from below. *J. Phys. A* **45** (2012), no. 7, 075204, 14 pp. MR 2881073 Zbl 1241.81072

[4] W. D. Evans and M. Solomyak, Smilansky's model of irreversible quantum graphs. I. The absolutely continuous spectrum. *J. Phys. A* **38** (2005), no. 21, 4611–4627. II. The point spectrum. *J. Phys. A* **38** (2005), no. 35, 7661–7675. MR 2147079 (I) MR 2169482 (II) Zbl 1064.81037 (I) Zbl 1083.81519 (II)

[5] I. Guarneri, Irreversible behaviour and collapse of wave packets in a quantum system with point interactions. *J. Phys. A* **44** (2011), no. 48, 485304, 22 pp. MR 2860866 Zbl 1235.81014

[6] L. Geisinger and T. Weidl, Sharp spectral estimates in domains of infinite volume. *Rev. Math. Phys.* **23** (2011), no. 6, 615–641. MR 2819231 Zbl 1219.35156

[7] V. Jakšić, S. Molchanov, and B. Simon, Eigenvalue asymptotics of the Neumann Laplacian of regions and manifolds with cusps. *J. Funct. Anal.* **106** (1992), no. 1, 59–79. MR 1163464 Zbl 0783.35040

[8] D. Lundholm, Weighted supermembrane toy model. *Lett. Math. Phys.* **92** (2010), no. 2, 125–141. MR 2645822 Zbl 1190.81048

[9] S. Naboko and M. Solomyak, On the absolutely continuous spectrum in a model of an irreversible quantum graph. *Proc. London Math. Soc.* (3) **92** (2006), no. 1, 251–272. MR 2192392 Zbl 1098.81036

[10] M. Reed and B. Simon, *Methods of modern mathematical physics.* II. Fourier analysis, self-adjointness. Academic Press [Harcourt Brace Jovanovich, Publishers], New York and London, 1975. MR 0493420 Zbl 0308.47002

[11] G. Rozenblum, M. Solomyak, and M. Shubin, Spectral theory of differential operators. In R. V. Gamkrelidze (ed.), *Current problems in mathematics.* Fundamental directions, 64. Itogi Nauki i Tekhniki, Akad. Nauk SSSR, Vsesoyuz. Inst. Nauchn. i Tekhn. Inform., Moscow, 1989, 5–248. In Russian. MR 1033500 Zbl 0715.35057

[12] B. Simon, Some quantum operators with discrete spectrum but classically continuous spectrum. *Ann. Physics* **146** (1983), no. 1, 209–220. MR 0701264 Zbl 0547.35039

[13] U. Smilansky, Irreversible quantum graphs. *Waves Random Media* **14** (2004), no. 1, S143–S153. Special section on quantum graphs. MR 2042550 Zbl 1063.82025

[14] M. Solomyak, On a differential operator appearing in the theory of irreversible quantum graphs. *Waves Random Media* **14** (2004), no. 1, S173–S185. MR 2042551 Zbl 1063.35124

On the quantum mechanical three-body problem with zero-range interactions

Giulia Basti and Alessandro Teta

Dedicated to Pavel

1 Introduction

The quantum mechanical three-body problem with pairwise, local zero-range interactions is a subject of considerable interest both for physical applications and for its peculiar mathematical structure.

The model has been introduced around the middle of the last century to describe nuclear interactions at low energy. More recently, interesting applications have been developed also in the physics of cold atoms, particularly in connection with the study of the Efimov effect. This is essentially due to the experimental possibility to realize, via the so-called Feshbach resonance, situations where the interaction is well described by a zero-range force, in particular in the unitary limit. Roughly speaking, unitary limit means that the two-body interaction is characterized by a zero-energy resonance or, equivalently, by an infinite value of the scattering length.

The correct definition of the Hamiltonian, the conditions for the occurrence of the Efimov effect and the analysis of the stability problem, i.e., the existence of a finite lower bound for the Hamiltonian, have been widely studied both in the physical ([4], [5], [6], [7], [10], [14], [23], and [24]) and in the mathematical ([2], [3], [8], [9], [11], [12], [13], [16], [18], [19], [20], and [22]) literature. Let us mention that in [2] and in [22] special classes of Hamiltonians with zero-range interactions that are bounded from below are studied. More precisely, in [2] a positive n-body Hamiltonian with non local zero-range interactions is defined using the theory of Dirichlet forms, while in [22] three-body zero-range Hamiltonians with internal structure are constructed and their spectral properties characterized.

Here we shall review the state of the art concerning the construction of the Hamiltonian as a self-adjoint operator. Exploiting a quadratic form method, we also prove lower boundedness of the Hamiltonian in the case of three identical bosons when the Hilbert space is suitably restricted, i.e., excluding the "s-wave" subspace.

The formal Hamiltonian describing three quantum particles in $\mathbb{R}^d$, $d = 1, 2, 3$, interacting via a zero-range, two-body interaction can be written as

$$\mathcal{H} = -\sum_{i=1}^{3} \frac{1}{2m_i} \Delta_{\boldsymbol{x}_i} + \sum_{\substack{i,j=1 \\ i<j}}^{3} \nu_{ij}\, \delta(\boldsymbol{x}_i - \boldsymbol{x}_j), \tag{1}$$

where $\boldsymbol{x}_i \in \mathbb{R}^d$, $i = 1, 2, 3$, is the coordinate of the i-th particle, m_i is the corresponding mass, $\Delta_{\boldsymbol{x}_i}$ is the Laplacian relative to $\boldsymbol{x}_i$, and $\nu_{ij} \in \mathbb{R}$ is the strength of the interaction between particles i and j. To simplify the notation we set $\hbar = 1$.

In order to give a rigorous meaning to (1) as a self-adjoint operator in $L^2(\mathbb{R}^{3d})$, the first step is to give a mathematical definition, i.e., to establish the conditions that such Hamiltonian must satisfy. We first notice that, in any reasonable definition, the interaction term of the Hamiltonian must be non trivial only on the hyperplanes $\bigcup_{i<j}\{\boldsymbol{x}_i = \boldsymbol{x}_j\}$, where the coordinates of two particles coincide. As a starting point, it is therefore natural to consider the operator $\dot{\mathcal{H}}_0$ defined as the free Hamiltonian restricted to a domain of smooth functions vanishing in the neighbourhood of each hyperplane $\{\boldsymbol{x}_i = \boldsymbol{x}_j\}$. Such operator is symmetric but not self-adjoint and one (trivial) self-adjoint extension is obviously the free Hamiltonian. Then we define a Hamiltonian for a system of three quantum particles in $\mathbb{R}^d$ with a two-body, zero-range interaction as a non trivial self-adjoint extension of $\dot{\mathcal{H}}_0$. As a consequence of the definition, any such Hamiltonian acts as the free Hamiltonian outside the hyperplanes $\bigcup_{i<j}\{\boldsymbol{x}_i = \boldsymbol{x}_j\}$ and it is characterized by a specific boundary condition satisfied by the wave function at each hyperplane $\{\boldsymbol{x}_i = \boldsymbol{x}_j\}$.

The second and more important step is the explicit construction of the self-adjoint extensions. The two most frequently used techniques are Krein's theory of self-adjoint extensions and approximation by regularized Hamiltonians, in the sense of the limit of the resolvent or of the quadratic form. In dimension one the problem is relatively simple due to the fact that the interaction term is a small perturbation of the free Hamiltonian in the sense of quadratic forms. In dimension two a natural class of Hamiltonians with local zero-range interactions was constructed in [11] and it was also shown that such Hamiltonians are all bounded from below. In dimension three the analysis is more delicate and in the rest of the paper we shall discuss the problem in some detail.

In order to explain the difficulty, we first consider the simpler two-body case where, in the center of mass reference frame, one is reduced to study a one-body problem in the relative coordinate $\boldsymbol{x}$ with a fixed δ-interaction placed at the origin. In this case (see, e.g., [1]) the entire class of self-adjoint extensions describing Hamiltonians with point interaction can be explicitly constructed. One can show that the domain $D(h_\alpha)$ of each Hamiltonian h_α consists of functions $\psi \in L^2(\mathbb{R}^3) \cap H^2(\mathbb{R}^3 \setminus \{0\})$

such that

$$\psi(x) = \frac{q}{|x|} + r + o(1), \quad \text{with } r = \alpha q, \tag{2}$$

for $|x| \to 0$, where $q \in \mathbb{C}$ and $\alpha \in \mathbb{R}$ is a parameter proportional to the inverse of the scattering length. The relation $r = \alpha q$ in (2) should be understood as the generalized boundary condition satisfied at the origin by all the elements of the domain. Moreover, by definition h_α satisfies

$$(h_\alpha \psi)(x) = -\frac{1}{2\mu}(\Delta\psi)(x), \quad \text{for } x \neq 0, \tag{3}$$

where μ denotes the reduced mass of the two-body problem.

In the three-particle case the characterization of all possible self-adjoint extensions of $\dot{\mathcal{H}}_0$ is more involved. In order to circumvent the difficulty, a natural strategy is to construct a class of extensions based on the analogy with the two-body case. More precisely, one considers an extension of $\dot{\mathcal{H}}_0$, called Skornyakov–Ter-Martirosyan (STM) operator H_α, which, roughly speaking, is a symmetric operator acting on functions $\Psi \in L^2(\mathbb{R}^9) \cap H^2(\mathbb{R}^9 \setminus \bigcup_{i<j}\{x_i = x_j\})$ satisfying the following condition for $|x_i - x_j| \to 0$:

$$\Psi(x_1, x_2, x_3) = \frac{Q_{ij}(r_{ij}, x_k)}{|x_i - x_j|} + R_{ij}(r_{ij}, x_k) + o(1), \quad \text{with } R_{ij} = \alpha_{ij} Q_{ij}, \tag{4}$$

where

$$r_{ij} = \frac{m_i x_i + m_j x_j}{m_i + m_j}, \tag{5}$$

$k \neq i, j$, Q_{ij} is a suitable function defined on the hyperplane $\{x_i = x_j\}$ and $\{\alpha_{ij}\}$ is a collection of real parameters labelling the extension. Notice that in the above limiting procedure for $|x_i - x_j| \to 0$ we keep fixed the center of mass of the particles i, j and the position of the remaining particle. Furthermore, one has

$$(H_\alpha \Psi)(x_1, x_2, x_3) = (H_f \Psi)(x_1, x_2, x_3), \quad \text{for } x_i \neq x_j, \tag{6}$$

where H_f is the free Hamiltonian.

Noticeably, the boundary condition (4) defining the STM extension of $\dot{\mathcal{H}}_0$ is a natural generalization to the three-body case of the condition (2) that characterizes the two-body case. Unfortunately, unlike (2), (4) does not necessarily define a self-adjoint operator. Indeed, for a system of three identical bosons it was shown in [12] that the STM operator is not self-adjoint and all its self-adjoint extensions are unbounded from below owing to the presence of an infinite sequence of energy levels E_k going to $-\infty$ for $k \to \infty$. In [17] this result was generalized to the case of three distinguishable particles with different masses. This kind of instability is known in the literature as the Thomas effect. It should be stressed that the Thomas effect is strongly related to the well-known Efimov effect (see, e.g., [4]) even if, to our knowledge, a rigorous mathematical investigation of this connection is still lacking.

Here we describe an approach to the stability problem based on the theory of quadratic forms. In particular, in Section 2 we explicitly construct the quadratic form naturally associated to the STM operator in the general case of three particles with different masses.

In Sections 3 and 4 we consider two particular cases where the Hilbert space of states is suitably restricted, e.g., introducing symmetry constraints on the wave function. In such cases the quadratic form is shown to be closed and bounded from below, thus defining a self-adjoint and bounded from below Hamiltonian of the system.

In the first case we consider a system of three identical bosons and we show that instability occurs only in the "s-wave" subspace. More precisely, we restrict the Hilbert space to the wave functions which are not invariant under rotation of the coordinates of each particle and we prove that the quadratic form is closed and bounded from below on such subspace.

In the second case we discuss the antisymmetry constraint. In fact, a wave function that is antisymmetric under exchange of coordinates of two particles necessarily vanishes at the coincidence points of such two particles, thus making their mutual zero-range interaction ineffective. Therefore, it is reasonable to expect that in a system of two identical fermions plus a different particle the interaction term in the Hamiltonian is less singular, thus making the system stable. Indeed, it has been shown that this is in fact the case for suitable values of the mass ratio (see, e.g., [8], [9], [19], and [20]).

2 The energy form

We start illustrating the construction of the quadratic form in the simple case of the one-body Hamiltonian h_α, formally introduced in Section 1. The idea is to represent the generic element of $D(h_\alpha)$ in the form

$$\psi = w + qg \tag{7}$$

where w is a smooth function, $q \in \mathbb{C}$ and

$$g(\boldsymbol{x}) = \frac{1}{|\boldsymbol{x}|}.$$

The singular part qg in the decomposition (7) can be thought as the electrostatic potential produced by the point charge q placed at the origin. According to decomposition (7), the boundary condition (2) can be rewritten as

$$w(0) = \alpha q. \tag{8}$$

Taking into account (3) and (7), the expectation value of h_α can be represented as

$$\begin{aligned} F_\alpha(\psi) &= (\psi, h_\alpha \psi) \\ &= \lim_{\varepsilon \to 0} \int_{|\boldsymbol{x}|>\varepsilon} \mathrm{d}\boldsymbol{x}\, \bar{\psi}(\boldsymbol{x}) \Big(-\frac{1}{2\mu} \Delta \psi \Big)(\boldsymbol{x}) \\ &= \frac{1}{2\mu} \lim_{\varepsilon \to 0} \int_{|\boldsymbol{x}|>\varepsilon} \mathrm{d}\boldsymbol{x}\, \bar{w}(\boldsymbol{x})(-\Delta w)(\boldsymbol{x}) + \frac{\bar{q}}{2\mu} \lim_{\varepsilon \to 0} \int_{|\boldsymbol{x}|>\varepsilon} \mathrm{d}\boldsymbol{x}\, g(\boldsymbol{x})(-\Delta w)(\boldsymbol{x}). \end{aligned}$$

Integrating by parts, taking the limit $\varepsilon \to 0$ and using (8), we arrive at the following quadratic form

$$F_\alpha(\psi) = \frac{1}{2\mu} \int \mathrm{d}\boldsymbol{x}\, |\nabla w(\boldsymbol{x})|^2 + \frac{2\pi}{\mu} \alpha |q|^2 \tag{9}$$

which is defined on the natural domain

$$D(F_\alpha) = \{\psi \in L^2(\mathbb{R}^3) \mid \psi = w + qg, |\nabla w| \in L^2(\mathbb{R}^3), q \in \mathbb{C}\}. \tag{10}$$

It is a simple exercise to show that the form (9)–(10) is closed and bounded from below. Therefore it defines a self-adjoint and bounded from below operator which obviously coincides with h_α. One can also notice that, defining

$$g^\lambda(\boldsymbol{x}) = \frac{e^{-\sqrt{\lambda}|\boldsymbol{x}|}}{|\boldsymbol{x}|}, \qquad \lambda > 0,$$

the following equivalent representation of the form domain holds

$$D(F_\alpha) = \{\psi \in L^2(\mathbb{R}^3) \mid \psi = w^\lambda + qg^\lambda, w^\lambda \in H^1(\mathbb{R}^3), q \in \mathbb{C}\},$$

where $H^s(\mathbb{R}^d)$ denotes the standard Sobolev space in $\mathbb{R}^d$ of order $s \in \mathbb{R}$. Accordingly one has

$$F_\alpha(\psi) = \frac{1}{2\mu} \int \mathrm{d}\boldsymbol{x}\, (|\nabla w^\lambda(\boldsymbol{x})|^2 + \lambda |w^\lambda(\boldsymbol{x})|^2 - \lambda |\psi(\boldsymbol{x})|^2) + \frac{2\pi}{\mu}(\alpha + \sqrt{\lambda})|q|^2.$$

In the three-particle case we follow the same idea. We first introduce the notation $\boldsymbol{X} = (\boldsymbol{x}_1, \boldsymbol{x}_2, \boldsymbol{x}_3)$, $\boldsymbol{P} = (\boldsymbol{p}_1, \boldsymbol{p}_2, \boldsymbol{p}_3)$ for positions and momenta of the particles, $M = m_1 + m_2 + m_3$ for the total mass,

$$\mu_{ij} = \frac{m_i m_j}{(m_i + m_j)}$$

for the reduced masses and $\hat{f}$ for the Fourier transform of f. We set $x = |\boldsymbol{x}|$ for $\boldsymbol{x} \in \mathbb{R}^3$. Then we introduce the "potential" produced by the "charges" $Q = \{Q_{ij}\}$ distributed on the hyperplanes $\{\boldsymbol{x}_i = \boldsymbol{x}_j\}$. With an abuse of notation, we set

$$(GQ)(\boldsymbol{X}) = \sum_{i \prec j} (GQ_{ij})(\boldsymbol{X}) = \sum_{i \prec j} \frac{1}{(2\pi)^5 \mu_{ij}} \int \mathrm{d}\boldsymbol{P}\, e^{i\boldsymbol{X}\cdot\boldsymbol{P}}\, \frac{\widehat{Q}_{ij}(\boldsymbol{p}_i + \boldsymbol{p}_j, \boldsymbol{p}_k)}{H_f(\boldsymbol{P})},$$

where $k \neq i, j$, $H_f(\boldsymbol{P})$ denotes the free Hamiltonian in the momentum variables and with $\prec$ we refer to the order $1 \prec 2$, $2 \prec 3$, $3 \prec 1$. Following the line of Proposition 6.3 in [13], one shows that GQ solves in the distributional sense the equation

$$H_f(GQ)(\boldsymbol{X}) = 2\pi \sum_{i \prec j} \frac{1}{\mu_{ij}} Q_{ij}(\boldsymbol{r}_{ij}, \boldsymbol{x}_k)\, \delta(\boldsymbol{x}_i - \boldsymbol{x}_j) \tag{11}$$

where $\boldsymbol{r}_{ij}$ is defined in (5). In particular this implies

$$H_f(GQ)(\boldsymbol{x}_1, \boldsymbol{x}_2, \boldsymbol{x}_3) = 0 \quad \text{if } \boldsymbol{x}_i \neq \boldsymbol{x}_j.$$

Moreover GQ has the following behaviour when $|\boldsymbol{x}_i - \boldsymbol{x}_j| \to 0$

$$(GQ)(\boldsymbol{X}) = \frac{Q_{ij}(\boldsymbol{r}_{ij}, \boldsymbol{x}_k)}{|\boldsymbol{x}_i - \boldsymbol{x}_j|} - (\Gamma Q)_{ij}(\boldsymbol{r}_{ij}, \boldsymbol{x}_k) + o(1) \tag{12}$$

where

$$\begin{aligned}(\Gamma Q)_{ij}(\boldsymbol{r}_{ij}, \boldsymbol{x}_k) = {} & \frac{1}{(2\pi)^3} \int \mathrm{d}\boldsymbol{s}\, \mathrm{d}\boldsymbol{t}\, e^{i(\boldsymbol{r}_{ij}\cdot\boldsymbol{s} + \boldsymbol{x}_k\cdot\boldsymbol{t})} \sqrt{\frac{\mu_{ij}}{m_i + m_j} s^2 + \frac{\mu_{ij}}{m_k} t^2}\, \widehat{Q}_{ij}(\boldsymbol{s}, \boldsymbol{t}) \\ & - \frac{1}{(2\pi)^5} \int \mathrm{d}\boldsymbol{P}\, \frac{e^{i\boldsymbol{r}_{ij}(\boldsymbol{p}_i + \boldsymbol{p}_j) + i\boldsymbol{x}_k\cdot\boldsymbol{p}_k}}{H_f(\boldsymbol{P})} \Big[\frac{\widehat{Q}_{ik}(\boldsymbol{p}_i + \boldsymbol{p}_k, \boldsymbol{p}_j)}{\mu_{ik}} \\ & + \frac{\widehat{Q}_{jk}(\boldsymbol{p}_j + \boldsymbol{p}_k, \boldsymbol{p}_i)}{\mu_{jk}} \Big].\end{aligned} \tag{13}$$

Proceeding in analogy with the one-body case we decompose the generic element Ψ in $D(H_\alpha)$ as

$$\Psi = u + GQ \tag{14}$$

where u is a smooth function. Then the boundary condition (4), using (12), can be rewritten as

$$u(X)|_{x_i=x_j} = (\Gamma Q)_{ij}(r_{ij}, x_k) + \alpha_{ij} Q_{ij}(r_{ij}, x_k). \tag{15}$$

Using the decomposition (14), we obtain the explicit expression of the quadratic form $\mathcal{E}_\alpha$ associated to the operator H_α. We set

$$\mathcal{D}_\varepsilon = \{X \in \mathbb{R}^9 \mid |x_i - x_j| > \varepsilon \text{ for all } i, j\}.$$

Then taking into account (6) and (11) and the boundary condition (15) we have

$$\begin{aligned}
\mathcal{E}_\alpha(\Psi) &= (\Psi, H_\alpha \Psi) \\
&= \lim_{\varepsilon \to 0} \int_{\mathcal{D}_\varepsilon} \mathrm{d}X\, \overline{\Psi(X)}(H_f \Psi)(X) \\
&= (u, H_f u) + \lim_{\varepsilon \to 0} \int_{\mathcal{D}_\varepsilon} \mathrm{d}X\, \overline{GQ(X)}\,(H_f u)(X) \\
&= (u, H_f u) + \sum_{i<j} \frac{2\pi}{\mu_{ij}} \Big[\alpha_{ij} \| Q_{ij} \|^2 \\
&\qquad\qquad + \int \mathrm{d}r_{ij}\, \mathrm{d}x_k\, \overline{Q_{ij}(r_{ij}, x_k)} (\Gamma Q)_{ij}(r_{ij}, x_k) \Big] \\
&= (u, H_f u) + \sum_{i<j} \frac{2\pi}{\mu_{ij}} \Big[\alpha_{ij} \| Q_{ij} \|^2 \\
&\qquad\qquad + \int \mathrm{d}s\, \mathrm{d}t\, |\hat{Q}_{ij}(s,t)|^2 \sqrt{\frac{\mu_{ij}}{m_i + m_j} s^2 + \frac{\mu_{ij}}{m_k} t^2} \\
&\qquad\qquad - \frac{1}{(2\pi)^2 \mu_{jk}} 2 \operatorname{Re} \int \mathrm{d}P\, \frac{\overline{\hat{Q}_{ij}(p_i + p_j, p_k)} \hat{Q}_{jk}(p_j + p_k, p_i)}{H_f(P)} \Big],
\end{aligned}$$

where in the last equality we have used the definition of ΓQ given in (13). For later use, it is convenient to rewrite in a different form the last two integrals in the above formula. Let us introduce the change of variables

$$\begin{cases}
p = p_1 + p_2 + p_3, \\
k_1 = \dfrac{m_j + m_k}{M} p_i - \dfrac{m_i}{M} p_j - \dfrac{m_i}{M} p_k, \\
k_2 = \dfrac{m_i + m_j}{M} p_k - \dfrac{m_k}{M} p_i - \dfrac{m_k}{M} p_j.
\end{cases} \tag{16}$$

Then defining

$$\hat{\zeta}_{ij}(\boldsymbol{k},\boldsymbol{p}) = \widehat{Q}_{ij}\Big(\frac{m_i+m_j}{M}\boldsymbol{p}-\boldsymbol{k}, \frac{m_k}{M}\boldsymbol{p}+\boldsymbol{k}\Big)$$

we have

$$\int \mathrm{d}\boldsymbol{P}\, \frac{\overline{\widehat{Q}_{ij}(\boldsymbol{p}_i+\boldsymbol{p}_j,\boldsymbol{p}_k)}\widehat{Q}_{jk}(\boldsymbol{p}_j+\boldsymbol{p}_k,\boldsymbol{p}_i)}{H_f(\boldsymbol{P})}$$
$$= \int \mathrm{d}\boldsymbol{p}\,\mathrm{d}\boldsymbol{k}_1\,\mathrm{d}\boldsymbol{k}_2\, \frac{\overline{\hat{\zeta}_{ij}(\boldsymbol{k}_2,\boldsymbol{p})}\hat{\zeta}_{jk}(\boldsymbol{k}_1,\boldsymbol{p})}{\dfrac{k_1^2}{2\mu_{ij}}+\dfrac{k_2^2}{2\mu_{jk}}+\dfrac{\boldsymbol{k}_1\cdot\boldsymbol{k}_2}{m_j}+\dfrac{p^2}{2M}}.$$

Moreover, defining the variables

$$\boldsymbol{p} = \boldsymbol{t}+\boldsymbol{s}, \quad \boldsymbol{k} = \frac{m_i+m_j}{M}\boldsymbol{t} - \frac{m_k}{M}\boldsymbol{s},$$

we also have

$$\int \mathrm{d}\boldsymbol{s}\,\mathrm{d}\boldsymbol{t}\,|\widehat{Q}_{ij}(\boldsymbol{s},\boldsymbol{t})|^2 \sqrt{\frac{\mu_{ij}}{m_i+m_j}s^2+\frac{\mu_{ij}}{m_k}t^2}$$
$$= \int \mathrm{d}\boldsymbol{p}\,\mathrm{d}\boldsymbol{k}\,|\hat{\zeta}_{ij}(\boldsymbol{k},\boldsymbol{p})|^2 \sqrt{\frac{\mu_{ij}M}{m_k(m_i+m_j)}k^2+\frac{\mu_{ij}}{M}p^2}.$$

Noticing that $\|Q_{ij}\| = \|\hat{\zeta}_{ij}\|$, we obtain the following equivalent expression for $\mathcal{E}_\alpha$

$$\begin{aligned}\mathcal{E}_\alpha(\Psi) &= (u, H_f u)\\ &+ \sum_{i\prec j}\frac{2\pi}{\mu_{ij}}\Bigg[\alpha_{ij}\|\hat{\zeta}_{ij}\|^2 \\ &\qquad + \sqrt{2\mu_{ij}}\int \mathrm{d}\boldsymbol{p}\,\mathrm{d}\boldsymbol{k}\,|\hat{\zeta}_{ij}(\boldsymbol{k},\boldsymbol{p})|^2\sqrt{\frac{Mk^2}{2m_k(m_i+m_j)}+\frac{p^2}{2M}} \\ &- \frac{1}{(2\pi)^2\mu_{jk}} 2\,\mathrm{Re}\int \mathrm{d}\boldsymbol{p}\,\mathrm{d}\boldsymbol{k}_1\,\mathrm{d}\boldsymbol{k}_2\, \frac{\overline{\hat{\zeta}_{ij}(\boldsymbol{k}_2,\boldsymbol{p})}\,\hat{\zeta}_{jk}(\boldsymbol{k}_1,\boldsymbol{p})}{\dfrac{k_1^2}{2\mu_{ij}}+\dfrac{k_2^2}{2\mu_{jk}}+\dfrac{1}{m_j}\boldsymbol{k}_1\cdot\boldsymbol{k}_2+\dfrac{p^2}{2M}}\Bigg].\end{aligned} \tag{17}$$

We define the form domain as follows (see Remark 2.1 at the end of this section)

$$\begin{aligned} D(\mathcal{E}_\alpha) = \{\Psi \in L^2(\mathbb{R}^9) \mid \Psi = u + \mathcal{G}_p\zeta,\ |\nabla u| \in L^2(\mathbb{R}^9),\\ \zeta = \{\zeta_{ij}\},\ \zeta_{ij}\in H^{1/2}(\mathbb{R}^6)\}, \end{aligned} \tag{18}$$

where

$$\mathcal{G}_p \zeta = \sum_{i \prec j} \mathcal{G}_p \zeta_{ij}$$

is given by

$$(GQ_{ij})(X) = (\mathcal{G}_p \zeta_{ij})(x_i - x_j, x_k - x_j, x_{cm}), \quad x_{cm} = \frac{m_1 x_1 + m_2 x_2 + m_3 x_3}{M}.$$

In particular

$$(\widehat{\mathcal{G}_p \zeta_{ij}})(k_{ij}, k_{kj}, p) = \frac{1}{\sqrt{2\pi}\,\mu_{ij}} \frac{\hat{\zeta}_{ij}(k_{kj}, p)}{\dfrac{k_{ij}^2}{2\mu_{ij}} + \dfrac{k_{kj}^2}{2\mu_{jk}} + \dfrac{k_{ij} \cdot k_{kj}}{m_j} + \dfrac{p^2}{2M}},$$

where with k_{ij}, k_{kj} we denote the conjugate variables to $x_i - x_j$ and $x_k - x_j$ respectively.

We remark that the dependence on the variable p (the total momentum) in the last two integrals in (17) is essentially irrelevant. This fact can be seen by introducing a different decomposition for the elements of $D(\mathcal{E}_\alpha)$. More precisely, we define

$$\mathcal{G}\zeta = \sum_{i \prec j} \mathcal{G}\zeta_{ij},$$

where

$$(\widehat{\mathcal{G}\zeta_{ij}})(k_{ij}, k_{kj}, p) = \frac{1}{\sqrt{2\pi}\,\mu_{ij}} \frac{\hat{\zeta}_{ij}(k_{kj}, p)}{\dfrac{k_{ij}^2}{2\mu_{ij}} + \dfrac{k_{kj}^2}{2\mu_{jk}} + \dfrac{k_{ij} \cdot k_{kj}}{m_j}},$$

and we set

$$\Psi = u + \mathcal{G}_p \zeta = v + \mathcal{G}\zeta, \quad \Psi \in D(\mathcal{E}_\alpha).$$

By a direct computation we find

$$\begin{aligned}
\mathcal{E}_\alpha(\Psi) &= (\Psi, h_{cm}\Psi) + (v, h_f v) \\
&\quad + \sum_{i \prec j} \frac{2\pi}{\mu_{ij}} \Bigg[\alpha_{ij} \|\hat{\zeta}_{ij}\|^2 + \sqrt{2\mu_{ij}} \int \mathrm{d}p \, \mathrm{d}k \, |\hat{\zeta}_{ij}(k, p)|^2 \sqrt{\frac{Mk^2}{2m_k(m_i + m_j)}} \\
&\qquad - \frac{1}{(2\pi)^2 \mu_{jk}} 2 \operatorname{Re} \int \mathrm{d}p \, \mathrm{d}k_1 \, \mathrm{d}k_2 \, \frac{\overline{\hat{\zeta}_{ij}(k_2, p)}\, \hat{\zeta}_{jk}(k_1, p)}{\dfrac{k_1^2}{2\mu_{ij}} + \dfrac{k_2^2}{2\mu_{jk}} + \dfrac{1}{m_j} k_1 \cdot k_2} \Bigg],
\end{aligned} \tag{19}$$

where

$$h_{cm} = \frac{p^2}{2M}, \quad h_f = H_f - h_{cm}.$$

From (19) it is clear that the dependence on the variable $\boldsymbol{p}$ is only parametric and therefore irrelevant. In particular, for factorized wave function $\Psi = f \cdot \psi$, where f is a function of the center of mass coordinate and ψ is a function of the relative coordinates, we obtain

$$\mathcal{E}_\alpha(\Psi) = \|\psi\|^2 (f, h_{cm} f) + \|f\|^2 \mathcal{F}_\alpha(\psi),$$

where

$$\begin{aligned} D(\mathcal{F}_\alpha) = \{\psi \in L^2(\mathbb{R}^6) \mid \psi = w + \mathcal{G}\xi,\ |\nabla w| \in L^2(\mathbb{R}^6), \\ \xi = \{\xi_{ij}\}, \xi_{ij} \in H^{1/2}(\mathbb{R}^3)\}, \end{aligned} \tag{20}$$

$$\begin{aligned} \mathcal{F}_\alpha(\psi) = (w, h_f w) + \sum_{i \prec j} \frac{2\pi}{\mu_{ij}} \Bigg[& \alpha_{ij} \|\hat{\xi}_{ij}\|^2 \\ & + \sqrt{2\mu_{ij}} \int \mathrm{d}\boldsymbol{k}\, |\hat{\xi}_{ij}(\boldsymbol{k})|^2 \sqrt{\frac{M k^2}{2m_k(m_i + m_j)}} \\ & - \frac{1}{(2\pi)^2 \mu_{jk}} 2\,\mathrm{Re} \int \mathrm{d}\boldsymbol{k}_1\, \mathrm{d}\boldsymbol{k}_2 \frac{\overline{\hat{\xi}_{ij}(\boldsymbol{k}_2)}\, \hat{\xi}_{jk}(\boldsymbol{k}_1)}{\frac{k_1^2}{2\mu_{ij}} + \frac{k_2^2}{2\mu_{jk}} + \frac{1}{m_j} \boldsymbol{k}_1 \cdot \boldsymbol{k}_2} \Bigg]. \end{aligned} \tag{21}$$

This means that, choosing the center of mass reference frame, one can reduce the analysis to the quadratic form $\mathcal{F}_\alpha$.

We underline that the above construction procedure has the only aim to get (17) and (18) or, if one chooses the center of mass reference frame, (20) and (21). Such definitions are our starting point for the rigorous construction of the Hamiltonian of the three particle system under suitable symmetry constraints.

Remark 2.1. We note that in (18) the choice of the charges $\zeta_{ij} \in H^{1/2}(\mathbb{R}^6)$ (or in (20) the choice $\xi_{ij} \in H^{1/2}(\mathbb{R}^3)$) guarantees that all terms in the square brackets of (17) (or (21)) are finite. However, it is not a priori clear for which class of charges the *sum* of the last two terms in the square brackets is finite. Therefore our choice has some degree of arbitrariness and in fact, in some relevant cases, a larger class of charges must be considered ([9]).

3 Three bosons for non zero angular momentum

For a system of three identical bosons of unitary masses, considered in the center of mass reference frame, the Hilbert space of states is $L^2_s(\mathbb{R}^6)$, i.e., the space of square-integrable functions symmetric under the exchange of particle coordinates. In the Fourier space, we fix a pair of coordinates $\boldsymbol{k}_1,\boldsymbol{k}_2$ defined in (16) (with $\boldsymbol{p}=0$), e.g., $\boldsymbol{k}_1=\boldsymbol{p}_1, \boldsymbol{k}_2=\boldsymbol{p}_3$ and then $\boldsymbol{p}_2=-\boldsymbol{k}_1-\boldsymbol{k}_2$, so that the symmetry condition reads $\widehat{\psi}(\boldsymbol{k}_1,\boldsymbol{k}_2)=\widehat{\psi}(\boldsymbol{k}_2,\boldsymbol{k}_1)=\widehat{\psi}(\boldsymbol{k}_1,-\boldsymbol{k}_1-\boldsymbol{k}_2)$.

Moreover the symmetry condition implies that $\alpha_{ij}=\alpha$ for all $i \prec j$ and, from (4), that $Q_{12}=Q_{23}=Q_{31}$ and hence $\xi_{12}=\xi_{23}=\xi_{31}=\xi$. Then we have the following expression for the potential

$$(\widehat{\mathcal{G}\xi})(\boldsymbol{k}_1,\boldsymbol{k}_2)=\frac{2}{\sqrt{2\pi}}\frac{\hat{\xi}(\boldsymbol{k}_1)+\hat{\xi}(\boldsymbol{k}_2)+\hat{\xi}(-\boldsymbol{k}_1-\boldsymbol{k}_2)}{k_1^2+k_2^2+\boldsymbol{k}_1\cdot\boldsymbol{k}_2}.$$

With an abuse of notation we define the quadratic form associated to the STM operator in the bosonic case as

$$D(\mathcal{F}_\alpha)=\{\psi\in L^2_s(\mathbb{R}^6)\mid \psi=w+\mathcal{G}\xi,\ |\nabla w|\in L^2_s(\mathbb{R}^6),\ \xi\in H^{1/2}(\mathbb{R}^3)\}, \tag{22}$$

and

$$\mathcal{F}_\alpha(\psi)=(w,h_f w)+\frac{12}{\pi}\Phi_\alpha(\xi), \tag{23}$$

where the form Φ_α acting on the charge $\xi\in D(\Phi_\alpha)=H^{1/2}(\mathbb{R}^3)$ is given by

$$\Phi_\alpha(\xi)=\Phi^{\text{diag}}(\hat{\xi})+\Phi^{\text{off}}(\hat{\xi})+\alpha\int \mathrm{d}\boldsymbol{k}\,|\hat{\xi}(\boldsymbol{k})|^2 \tag{24}$$

and the diagonal part and the off-diagonal part are defined respectively by

$$\Phi^{\text{diag}}(f)=\frac{\sqrt{3}\pi^2}{2}\int \mathrm{d}\boldsymbol{k}\, k|f(\boldsymbol{k})|^2, \tag{25}$$

$$\Phi^{\text{off}}(f)=-\int \mathrm{d}\boldsymbol{k}_1\,\mathrm{d}\boldsymbol{k}_2\,\frac{\overline{f(\boldsymbol{k}_1)}f(\boldsymbol{k}_2)}{k_1^2+k_2^2+\boldsymbol{k}_1\cdot\boldsymbol{k}_2}. \tag{26}$$

It easy to see that if one can find an f_0 such that $\Phi^{\text{diag}}(f_0)+\Phi^{\text{off}}(f_0)<0$ then, by a scaling argument, one shows that the form (23) is unbounded from below. As a matter of fact, such f_0 can be explicitly constructed and it is rotationally invariant (for the proof one can follows the line of [13], Section 4). This fact is not surprising since it is known that the STM operator is not self-adjoint and all its self-adjoint extensions are unbounded from below, showing the occurrence of the Thomas effect (see [12]).

Following [17], we define

$$\mathcal{H}_0 = \{\psi \in L^2_s(\mathbb{R}^6) \mid \hat{\psi} = \hat{\psi}(|\boldsymbol{k}_1|, |\boldsymbol{k}_2|)\},$$

which is an invariant subspace for the STM operator, and we consider its orthogonal complement $\mathcal{H}_0^\perp$. In the next theorem we characterize our quadratic form in $\mathcal{H}_0^\perp$.

Theorem 3.1. *The quadratic form* (22)–(26) *restricted to the subspace $\mathcal{H}_0^\perp$ is bounded from below and closed for any $\alpha \in \mathbb{R}$.*

We start with some preliminaries, following the line of [8]. Given $f \in L^2(\mathbb{R}^3)$, we consider the expansion

$$f(\boldsymbol{k}) = \sum_{l=0}^{\infty} \sum_{n=-l}^{l} f_{ln}(k) Y_l^n(\theta, \varphi),$$

where Y_l^n is the the spherical harmonic of order l, n. Using the above expansion one can obtain the following decompositions for Φ^{off} and Φ^{diag} (see [8], Lemma 3.1)

$$\Phi^{\text{diag}}(f) = \sum_{l=0}^{+\infty} \sum_{n=-l}^{l} F^{\text{diag}}(f_{ln}), \tag{27}$$

$$\Phi^{\text{off}}(f) = \sum_{l=0}^{+\infty} \sum_{n=-l}^{l} F_l^{\text{off}}(f_{ln}), \tag{28}$$

with F^{diag} and F_l^{off} acting as

$$F^{\text{diag}}(g) = \frac{\sqrt{3}\pi^2}{2} \int_0^{+\infty} \mathrm{d}k\, k^3\, |g(k)|^2,$$

$$F_l^{\text{off}}(g) = -2\pi \int_0^{+\infty} \mathrm{d}k_1 \int_0^{+\infty} \mathrm{d}k_2\, k_1^2\, \overline{g(k_1)} k_2^2\, g(k_2) \int_{-1}^{1} \mathrm{d}y\, \frac{P_l(y)}{k_1^2 + k_2^2 + k_1 k_2 y},$$

where P_l denotes the Legendre polynomial of order l. Proceeding as in Lemma 3.2 of [8], one proves that

$$F_l^{\text{off}}(g) \geq 0, \quad \text{for } l \text{ odd}, \tag{29}$$

$$F_l^{\text{off}}(g) \leq 0, \quad \text{for } l \text{ even}. \tag{30}$$

Moreover F_l^{off} can be diagonalized. Setting

$$g^\sharp(k) = \frac{1}{\sqrt{2\pi}} \int \mathrm{d}x\, e^{-ikx} e^{2x} g(e^x),$$

we have (for details see [8], Lemma 3.3)

$$F^{\rm diag}(g) = \frac{\sqrt{3}\,\pi^2}{2} \int {\rm d}k\, |g^\sharp(k)|^2, \tag{31}$$

$$F_l^{\rm off}(g) = -\int {\rm d}k\; S_l(k)|g^\sharp(k)|^2, \tag{32}$$

where

$$S_l(k) = \begin{cases} \pi^2 \displaystyle\int_{-1}^{1} {\rm d}y\; P_l(y) \frac{\cosh(k \arcsin {}^y\!/\!_2)}{\cos(\arcsin {}^y\!/\!_2)\cosh(k\,{}^\pi\!/\!_2)} & \text{for } l \text{ even},\\[2ex] -\pi^2 \displaystyle\int_{-1}^{1} {\rm d}y\; P_l(y) \frac{\sinh(k \arcsin {}^y\!/\!_2)}{\cos(\arcsin {}^y\!/\!_2)\sinh(k\,{}^\pi\!/\!_2)} & \text{for } l \text{ odd}. \end{cases}$$

Therefore the comparison between $F_l^{\rm off}$ and $F^{\rm diag}$ is reduced to the study of $S_l(k)$. We first notice that $S_l(k)$ as a function of l (and for any fixed k) is decreasing for l even and increasing for l odd (see Lemma 3.5 in [8]). For the estimate, we distinguish the cases of even and odd l.

Lemma 3.2. *For l even and any $k \in \mathbb{R}$*

$$0 \le S_l(k) \le \pi^2\Big(\frac{50}{27}\pi - \frac{10}{3}\sqrt{3} + \frac{\sqrt{11}}{9} - \frac{10}{9}t_0\Big), \quad l \ne 0, \tag{33}$$

where $t_0 = \arcsin({}^1\!/\!\sqrt{12}) \simeq 0.293$ and

$$0 \le S_0(k) \le 4\pi^2. \tag{34}$$

Furthermore, for l odd and any $k \in \mathbb{R}$

$$\pi^2\Big(\frac{4}{3}\sqrt{3} - \frac{8}{\pi}\Big) \le S_l(k) \le 0.$$

Proof. Let us consider the case $l \ne 0$ and even. The positivity of $S_l(k)$ follows from (30) and (32). Since $S_l(k)$ is decreasing in l, we have $S_l(k) \le S_2(k)$, where $S_2(k)$ is an even function. An explicit integration gives

$$\begin{aligned} S_2(0) &= \pi^2 \int_{-1}^{1} {\rm d}y\; (3y^2 - 1)\frac{1}{2\cos(\arcsin {}^y\!/\!_2)}\\ &= \pi^2 \int_{-\pi/6}^{\pi/6} {\rm d}x\; (12\sin^2 x - 1)\\ &= \pi^2\Big(\frac{5}{3}\pi - 3\sqrt{3}\Big). \end{aligned}$$

Let us estimate the difference $S_2(0) - S_2(k)$ for any positive k. We have

$$\begin{aligned} S_2(0) - S_2(k) &= \pi^2 \int_{-1}^{1} \mathrm{d}y \, \frac{3y^2 - 1}{2\cos(\arcsin y/2)} \Big(1 - \frac{\cosh(k \arcsin y/2)}{\cosh(k\,\pi/2)}\Big) \\ &= 2\pi^2 I(k), \end{aligned}$$

where

$$I(k) = \int_0^{\pi/6} \mathrm{d}t \, (12\sin^2 t - 1)\Big(1 - \frac{\cosh(kt)}{\cosh(k\,\pi/2)}\Big).$$

Since $s(t) = 12\sin^2 t - 1$ is negative if $t < t_0$ and positive otherwise we can write

$$\begin{aligned} I(k) &= -\int_0^{t_0} \mathrm{d}t \, |s(t)| + \int_{t_0}^{\pi/6} \mathrm{d}t \, s(t) \\ &\quad + \frac{1}{\cosh(k\,\pi/2)} \Big[\int_0^{t_0} \mathrm{d}t \, |s(t)| \cosh(kt) - \int_{t_0}^{\pi/6} \mathrm{d}t \, s(t) \cosh(kt) \Big] \qquad (35) \\ &\geq \frac{1}{\cosh(k\,\pi/2)} \Big[(b-a)\cosh\Big(k\frac{\pi}{2}\Big) + a - b\cosh\Big(k\frac{\pi}{6}\Big) \Big], \end{aligned}$$

where

$$a = \int_0^{t_0} \mathrm{d}t \, |s(t)| \quad \text{and} \quad b = \int_{t_0}^{\pi/6} \mathrm{d}t \, s(t), \quad \text{with } b - a > 0.$$

Denoting

$$g(k) = a + \Big(\frac{10}{9}b - a\Big)\cosh\Big(k\frac{\pi}{2}\Big) - b\cosh\Big(k\frac{\pi}{6}\Big),$$

we can rewrite (35) as

$$I(k) \geq \frac{g(k)}{\cosh(k\,\pi/2)} - \frac{b}{9}.$$

Let us show that $g(k) \geq 0$. We have

$$g'(k) = \frac{\pi}{2}\Big(\frac{10}{9}b - a\Big)\sinh\Big(k\frac{\pi}{2}\Big)\Big[1 - A\,\frac{3\sinh(k\,\pi/6)}{\sinh(k\,\pi/2)}\Big] \qquad (36)$$

where

$$A = \frac{b}{10b - 9a}.$$

The term in square bracket in (36) is positive, then $g'(k) \geq 0$ which, together with $g(0) = b/9$, implies $g(k) \geq 0$. Thus we find

$$S_2(0) - S_2(k) \geq -\frac{2\pi^2}{9} b.$$

Inserting the explicit expression for b, we obtain the estimate (33).

In the case $l = 0$ the estimate (34) is straightforward.

Let us consider the case l odd. From (29) and (32) it follows that $S_l(k) \leq 0$. Noticing that $S_l(k)$ is an even function and it is increasing in l, we have $S_l(k) \geq S_1(k)$.

Since $S_1(0) = \pi^2(4/3\sqrt{3} - 8/\pi) < 0$, $\lim_{k\to\infty} S_1(k) = 0$ and $S_1'(k) \neq 0$ for $k > 0$ we obtain the thesis. □

The following estimate, which is the main tool in the proof of Theorem 3.1, is a direct consequence of the above lemma.

Proposition 3.3. *Let $f \in D(\Phi_\alpha)$ such that $f(\mathbf{k}) = \sum_{l=1}^{+\infty} \sum_{n=-l}^{l} f_{ln}(k) Y_l^n(\theta, \phi)$. Then*

$$-\Gamma\, \Phi^{\text{diag}}(f) \leq \Phi^{\text{off}}(f) \leq \Lambda\, \Phi^{\text{diag}}(f)$$

where

$$\Gamma = \frac{100}{27\sqrt{3}}\pi - \frac{20}{3} + \frac{2\sqrt{11}}{9\sqrt{3}} - \frac{20}{9\sqrt{3}} t_0 \simeq 0.101, \quad \Lambda = -\frac{8}{3} + \frac{16}{\sqrt{3}\,\pi} \simeq 0.274.$$

Proof. Using (28), (29), (32), (33), (31), and (27), we have

$$\begin{aligned}
\Phi^{\text{off}}(f) &= \sum_{l=1}^{+\infty} \sum_{n=-l}^{l} F_l^{\text{off}}(f_{ln}) \\
&\geq \sum_{\substack{l=2 \\ l \text{ even}}}^{+\infty} \sum_{n=-l}^{l} F_l^{\text{off}}(f_{ln}) \\
&= -\sum_{\substack{l=2 \\ l \text{ even}}}^{+\infty} \sum_{n=-l}^{l} \int dk\, S_l(k)\, |f_{ln}^{\sharp}(k)|^2 \\
&\geq -\Gamma \sum_{\substack{l=2 \\ l \text{ even}}}^{+\infty} \sum_{n=-l}^{l} \frac{\sqrt{3}\,\pi^2}{2} \int dk\, |f_{ln}^{\sharp}(k)|^2 \\
&\geq -\Gamma\, \Phi^{\text{diag}}(f)
\end{aligned}$$

and analogously one also proves the estimate $\Phi^{\text{off}}(f) \leq \Lambda\, \Phi^{\text{diag}}(f)$. □

Proof of Theorem 3.1. We first consider the simpler case $\alpha > 0$. From the definition (23) and Proposition 3.3 we obtain the positivity of $\mathcal{F}_\alpha$,

$$\begin{aligned}\mathcal{F}_\alpha(\psi) &= (w, h_f w) + \frac{12}{\pi}\Phi_\alpha(\xi)\\ &\geq \frac{12}{\pi}\Big[\Phi^{\mathrm{off}}(\hat{\xi}) + \Phi^{\mathrm{diag}}(\hat{\xi}) + \alpha\int \mathrm{d}\boldsymbol{k}\,|\hat{\xi}(\boldsymbol{k})|^2\Big]\\ &\geq \frac{12}{\pi}\Big[(1-\Gamma)\Phi^{\mathrm{diag}}(\hat{\xi}) + \alpha\int \mathrm{d}\boldsymbol{k}\,|\hat{\xi}(\boldsymbol{k})|^2\Big]\\ &\geq 0.\end{aligned}$$

Let us prove the closure of $\mathcal{F}_\alpha$. Let $\{\psi_n\} = \{w_n + \mathcal{G}\xi_n\}$ be a sequence in $D(\mathcal{F}_\alpha)$ such that $\psi_n \to \psi \in L^2_s(\mathbb{R}^6)$ and $\mathcal{F}_\alpha(\psi_n - \psi_m) \to 0$.

From $\mathcal{F}_\alpha(\psi_n - \psi_m) \to 0$, the positivity of h_f and the lower bound for Φ_α it follows

$$\int \mathrm{d}\boldsymbol{k}_1\,\mathrm{d}\boldsymbol{k}_2\,(k_1^2 + k_2^2)|(\widehat{w}_n - \widehat{w}_m)(\boldsymbol{k}_1,\boldsymbol{k}_2)|^2 \longrightarrow 0, \quad \|\xi_n - \xi_m\|_{H^{1/2}} \longrightarrow 0.$$

Thus there exist $v \in L^2_s(\mathbb{R}^6)$ and $\xi \in H^{1/2}(\mathbb{R}^3)$ such that

$$\int \mathrm{d}\boldsymbol{k}\,\Big|\sqrt{k_1^2 + k_2^2}\,w_n(\boldsymbol{k}_1,\boldsymbol{k}_2) - v(\boldsymbol{k}_1,\boldsymbol{k}_2)\Big|^2 \longrightarrow 0, \quad \|\xi_n - \xi\|_{H^{1/2}} \longrightarrow 0.$$

Defining

$$\widehat{w} = \frac{\hat{v}}{\sqrt{k_1^2 + k_2^2}},$$

for any $\varepsilon > 0$ we have

$$\int_{\mathbb{R}^6_\varepsilon} \mathrm{d}\boldsymbol{k}_1\,\mathrm{d}\boldsymbol{k}_2\,|(\widehat{w}_n - \widehat{w})(\boldsymbol{k}_1,\boldsymbol{k}_2)|^2 \longrightarrow 0, \tag{37}$$

$$\int_{\mathbb{R}^6_\varepsilon} |(\widehat{\mathcal{G}\xi_n} - \widehat{\mathcal{G}\xi})(\boldsymbol{k}_1,\boldsymbol{k}_2)|^2 \longrightarrow 0. \tag{38}$$

where $\mathbb{R}^d_\varepsilon = \{\boldsymbol{x} \in \mathbb{R}^d \mid x \geq \varepsilon\}$. From (37) and (38) in particular we obtain

$$\psi = w + \mathcal{G}\xi \in D(\mathcal{F}_\alpha)$$

and also $\mathcal{F}_\alpha(\psi_n - \psi) \to 0$. This concludes the proof in the case $\alpha > 0$.

In order to study the case $\alpha \leq 0$ it is convenient to consider the following decomposition for the generic ψ in the domain of $\mathcal{F}_\alpha$:

$$\psi = w^\lambda + \mathcal{G}^\lambda \xi$$

where $\lambda > 0$ and

$$\mathcal{G}^\lambda \xi(\boldsymbol{k}_1, \boldsymbol{k}_2) = \frac{2}{\sqrt{2\pi}} \frac{\hat{\xi}(\boldsymbol{k}_1) + \hat{\xi}(\boldsymbol{k}_2) + \hat{\xi}(-\boldsymbol{k}_1 - \boldsymbol{k}_2)}{k_1^2 + k_2^2 + \boldsymbol{k}_1 \cdot \boldsymbol{k}_2 + \lambda}.$$

Thus $\mathcal{G}^\lambda \xi$ belongs to $L^2_s(\mathbb{R}^6)$ and w^λ is in $H^1(\mathbb{R}^6)$. Moreover the quadratic form can be rewritten as

$$\mathcal{F}_\alpha(\psi) = (w^\lambda, h_f w^\lambda) + \lambda \|w^\lambda\|^2 - \lambda \|\psi\|^2 + \frac{12}{\pi} \Phi^\lambda_\alpha(\xi),$$

where

$$\Phi^\lambda_\alpha(\xi) = \left[\Phi^{\text{diag}}_\lambda(\hat{\xi}) + \Phi^{\text{off}}_\lambda(\hat{\xi}) + \alpha \int d\boldsymbol{k}\, |\hat{\xi}(\boldsymbol{k})|^2\right]$$

and

$$\Phi^{\text{diag}}_\lambda(f) = \pi^2 \int d\boldsymbol{k}\, |f(k)|^2 \sqrt{\frac{3}{4}k^2 + \lambda},$$

$$\Phi^{\text{off}}_\lambda(f) = -\int d\boldsymbol{k}_1\, d\boldsymbol{k}_2\, \frac{\overline{f(\boldsymbol{k}_1)} f(\boldsymbol{k}_2)}{k_1^2 + k_2^2 + \boldsymbol{k}_1 \cdot \boldsymbol{k}_2 + \lambda}.$$

Proceeding as in the case $\lambda = 0$ (see [8]), one has

$$-\Gamma\, \Phi^{\text{diag}}_\lambda \leq \Phi^{\text{off}}_\lambda \leq \Lambda\, \Phi^{\text{diag}}_\lambda$$

Therefore the quadratic form is bounded from below

$$\mathcal{F}_\alpha(\psi) \geq -\frac{\alpha^2}{\pi^4(1-\Gamma)^2} \|\psi\|^2.$$

The proof that $\mathcal{F}_\alpha$ is closed follows exactly the same line of the proof of Theorem 2.1 in [8] and it is omitted for the sake of brevity. □

We conclude observing that Theorem 3.1 implies the existence of a self-adjoint operator $H^\perp_{\alpha,0}$ in $\mathcal{H}^\perp_0$ which, at least formally, coincides with the STM operator restricted to $\mathcal{H}^\perp_0$. Such operator $H^\perp_{\alpha,0}$ is positive for $\alpha \geq 0$ and bounded from below by

$$-\frac{\alpha^2}{(\pi^4(1-\Gamma)^2)}$$

for $\alpha < 0$.

4 System of fermions

In a system of identical fermions the wave function, due to the antisymmetry under exchange of coordinates, vanishes at the coincident points of any pair of particles and therefore the zero-range interaction is ineffective. On the other hand, in physical applications it is relevant in the case of a mixture of N identical fermions of one species and M identical fermions of another species. Here the dynamics is non trivial since each fermion of one species feels the zero-range interaction with all the fermions of the other species. In particular, numerical simulations seem to suggest (see [15]) that the system is stable at least for mass ratio equal to one but, in this generality, no rigorous result is available (see [13] for a formulation of the problem in terms of quadratic forms). A significant aspect of the fermionic problem is that the stability of the system depends on the value of the mass ratio. This has been explicitly shown in the case of $N \geq 2$ identical fermions of mass one plus a different particle of mass m. More precisely one defines

$$\Lambda(m, N) = 2\pi^{-1}(N-1)(m+1)^2\Big[\frac{1}{\sqrt{m(m+2)}} - \arcsin\Big(\frac{1}{m+1}\Big)\Big].$$

For each fixed N, the function $\Lambda(\cdot, N)$ is positive, decreasing and satisfies the conditions $\lim_{m\to 0}\Lambda(m, N) = \infty$, $\lim_{m\to\infty}\Lambda(m, N) = 0$. Therefore, for each N the equation $\Lambda(m, N) = 1$ admits exactly one solution $m^*(N) > 0$, increasing with N and such that $m > m^*(N)$ if and only if $\Lambda(m, N) < 1$. Furthermore, following the strategy outlined in Section 2, we consider the STM operator for this fermionic case and construct the associated quadratic form, still denoted by $\mathcal{F}_\alpha$. In [8] it is proved the following result.

Theorem 4.1. 1. Stability. *If $m > m^*(N)$ then $\mathcal{F}_\alpha$ is closed and bounded from below. In particular $\mathcal{F}_\alpha$ is positive for $\alpha \geq 0$ and bounded from below by*

$$-\frac{\alpha^2}{4\pi^4(1-\Lambda(m, N))^2}$$

for $\alpha < 0$. Therefore the corresponding STM operator H_α is self-adjoint and bounded from below, with the same lower bound.

2. Instability. *If $m < m^*(2)$ then $\mathcal{F}_\alpha$ is unbounded from below for any $\alpha \in \mathbb{R}$.*

The above theorem provides an optimal result in the case $N = 2$, i.e., stability for $m > m^*(2)$ and instability for $m < m^*(2)$, where $m^*(2) \simeq 0.0735$, in agreement with previous heuristic results in the physical literature ([4]) and also with other mathematical results ([21] and [20]). On the other hand, in the case $N > 2$ we get only a partial result since no information is given for $m \in (m^*(2), m^*(N))$ and, in order to fill this gap, a more careful analysis of the role of the antisymmetry is required (for other results in this direction we refer to [5] and [19]).

The special case $N = 2$ in the unitary limit, i.e., for $\alpha = 0$, exhibits a further interesting behavior that we want to discuss in the rest of this section. In the center of mass reference frame we choose relative coordinates $\boldsymbol{y}_1 = \boldsymbol{x}_1 - \boldsymbol{x}_0$, $\boldsymbol{y}_2 = \boldsymbol{x}_2 - \boldsymbol{x}_0$, where $\boldsymbol{x}_1, \boldsymbol{x}_2$ are the coordinates of the fermions and $\boldsymbol{x}_0$ denotes the coordinate of the other particle. Let $L^2_a(\mathbb{R}^6)$ be the Hilbert space of states, i.e., the space of square integrable functions anisymmetric under the exchange of coordinates. Moreover we have $\xi_{12} = 0$ and the antisymmetry condition implies $\xi_{20} = -\xi_{10} := -\xi$. Then the potential in the Fourier space takes the form

$$(\widehat{\mathcal{G}\xi})(\boldsymbol{k}_1, \boldsymbol{k}_2) = \frac{2}{\sqrt{2\pi}} \frac{\hat{\xi}(\boldsymbol{k}_1) - \hat{\xi}(\boldsymbol{k}_2)}{k_1^2 + k_2^2 + \frac{2}{m+1} \boldsymbol{k}_1 \cdot \boldsymbol{k}_2}$$

and the quadratic form associated to the STM operator is

$$D(\mathcal{F}_0) = \{\psi \in L^2_a(\mathbb{R}^6) \mid \psi = w + \mathcal{G}\xi,\ |\nabla w| \in L^2_a(\mathbb{R}^6),\ \xi \in H^{1/2}(\mathbb{R}^3)\},$$

$$\mathcal{F}_0(\psi) = (w, h_f w) + \frac{2(m+1)}{\pi m} \Phi_0(\xi),$$

where

$$\Phi_0(\xi) = \Phi^{\mathrm{diag}}(\hat{\xi}) + \Phi^{\mathrm{off}}(\hat{\xi}),$$

$$\Phi^{\mathrm{diag}}(f) = \frac{2\pi^2 \sqrt{m(m+2)}}{m+1} \int \mathrm{d}\boldsymbol{k}\, k |f(\boldsymbol{k})|^2,$$

$$\Phi^{\mathrm{off}}(f) = \int \mathrm{d}\boldsymbol{k}_1\, \mathrm{d}\boldsymbol{k}_2 \frac{\overline{f(\boldsymbol{k}_1)} f(\boldsymbol{k}_2)}{k_1^2 + k_2^2 + \frac{2}{m+1} \boldsymbol{k}_1 \cdot \boldsymbol{k}_2}. \tag{39}$$

From Theorem 4.1 we know that the form is closed and bounded from below for $m > m^*$ and unbounded from below for $m < m^*$ (here we have used the shorthand notation $m^* = m^*(2)$). We also notice the main differences with respect to the form in the bosonic case, i.e., the dependence on m and, more important, the sign $+$ in front of the integral in (39). This implies that for the estimate of $\Phi^{\mathrm{off}}(f)$ one has to study the terms for l odd, and in particular the case $l = 1$, in the expansion in spherical harmonics of f.

As a matter of fact, for suitable values of the mass m the above quadratic form can be modified by enlarging the class of admissible charges and the new quadratic form turns out to be closed and bounded from below. Therefore it defines a Hamiltonian, different from H_α, describing an additional three-body interaction besides the standard two-body zero-range interaction (see [9] for details).

In order to explain the above assertion, we proceed formally. Let us define

$$\hat{\xi}_n^-(\boldsymbol{k}) = \frac{1}{k^{2-s}} Y_1^n(\theta,\phi), \quad 0 < s < 1,\ n = 0, \pm 1. \tag{40}$$

Notice that $\xi_n^- \notin L^2(\mathbb{R}^3)$ but this fact is not relevant since for $\alpha = 0$ the condition $\Phi_0(\xi) < \infty$ does not require square-integrability of ξ. The crucial point is that both $\Phi^{\text{diag}}(\hat{\xi}_n^-)$ and $\Phi^{\text{off}}(\hat{\xi}_n^-)$ diverge, due to the behavior of $\hat{\xi}_n^-(\boldsymbol{k})$ for large k and the two infinities can compensate for an appropriate value of the mass. Indeed, by a direct computation one finds

$$\begin{aligned}\Phi_0(\xi_n^-) &= 2\pi \left[\frac{\pi\sqrt{m(m+2)}}{m+1} + \int_{-1}^{1} \mathrm{d}t\, t \int_0^\infty \mathrm{d}q\, \frac{q^s}{q^2 + 1 + \frac{2}{m+1} tq} \right] \int_0^\infty \mathrm{d}k\, \frac{1}{k^{1-2s}} \\ &:= g(m,s) \int_0^\infty \mathrm{d}k\, \frac{1}{k^{1-2s}} = \infty \quad \text{unless } g(m,s) = 0.\end{aligned}$$

The problem is then reduced to the study of the equation $g(m,s) = 0$. One can show that for $s \in [0,1]$ there is a unique solution $m(s)$, monotonically increasing, with $m(0) = m^*$ and $m(1) := m^{**} \simeq 0.116$. For $m \in (m^*, m^{**})$ we can therefore define the inverse function $s(m)$, with $0 < s(m) < 1$, which satisfies $g(m, s(m)) = 0$. This means that for each $m \in (m^*, m^{**})$ the charge (40) with $s = s(m)$ can be considered to enlarge the class of admissible charges and to construct a more general quadratic form. Starting from the above argument, one can prove the following result.

Theorem 4.2. *For any $m \in (m^*, m^{**})$ and $\beta := \{\beta_n\}$, $n = 0, \pm 1$, the quadratic form in $L_a^2(\mathbb{R}^6)$*

$$D(\mathcal{F}_{0,\beta}) = \Big\{ \psi \in L_a^2(\mathbb{R}^6) \;\Big|\; \psi = w + \mathcal{G}\eta,\ |\nabla w| \in L_a^2(\mathbb{R}^6),\ \eta \in H^{-1/2}(\mathbb{R}^3),$$

$$\eta = \xi + \sum_{n=-1}^{1} q_n \xi_n^-,\ \Phi^{\text{diag}}(\hat{\xi}) < \infty,\ q_n \in \mathbb{C} \Big\},$$

$$\mathcal{F}_{0,\beta}(\psi) = (w, h_f w) + \frac{2(m+1)}{\pi m} \Phi_0(\xi) + \sum_{n=-1}^{1} \beta_n |q_n|^2$$

is closed and bounded from below. Then it uniquely defines a self-adjoint and bounded from below Hamiltonian $H_{0,\beta}$, $D(H_{0,\beta})$.

We conclude with some comments.

i) At heuristic level, the Hamiltonian $H_{0,\beta}$ has been introduced and studied in the physical literature (see, e.g., [23]). From the mathematical point of view, an analogous result has been found in [20] using an appoach based on the theory of self-adjoint extensions. Nevertheless, in [20] the analysis is done for $\alpha \neq 0$, which requires charges in L^2. Therefore the parameter s in (40) is chosen in the interval $(0, 1/2)$ and for this reason the new Hamiltonian is constructed only for a smaller range of mass, i.e., for $m \in (m^*, m^{**}_{\text{Minlos}})$, with $m^{**}_{\text{Minlos}} = m(s)|_{s=1/2} < m^{**}$.

ii) The quadratic form $\mathcal{F}_{0,\beta}$ constructed in Theorem 4.2 generalizes the previous one $\mathcal{F}_0$ in the sense that $\lim_{\beta\to\infty} \mathcal{F}_{0,\beta} = \mathcal{F}_0$.

iii) A final, and more important, comment concerns the boundary condition satisfied by an element of $D(H_{0,\beta})$. Denoting $R = \sqrt{y_1^2 + y_2^2}$ and choosing for simplicity $\xi^-_{\pm 1} = 0$, for $R \to 0$ one finds

$$\psi(y_1, y_2) = \frac{q_0}{R^{2+s(m)}} + \frac{\nu(m)\beta_0 q_0}{R^{2-s(m)}} + o(R^{s(m)-2})$$

where $\nu(m)$ is a given positive function of m. In analogy with the case of a point interaction (see (2)), such boundary condition describes an interaction supported in $y_1 = y_2 = 0$, i.e., when the positions of all the three particles coincide. Therefore the new Hamiltonian $H_{0,\beta}$ describes the two-body (resonant) zero-range interactions plus an effective three-body point interactions.

References

[1] S. Albeverio, F. Gesztesy, R. Høegh-Krohn, and H. Holden, *Solvable models in quantum mechanics.* Second edition. With an appendix by P. Exner. AMS Chelsea Publishing, Providence, R.I., 2005. ISBN 0-8218-3624-2 MR 2105735 Zbl 1078.81003

[2] S. Albeverio, R. Høegh-Krohn, and L. Streit, Energy forms, Hamiltonians, and distorted Brownian paths. *J. Mathematical Phys.* **18** (1977), no. 5, 907–917. MR 0446236 Zbl 0368.60091

[3] S. Albeverio and P. Kurasov, *Singular perturbations of differential operators.* Solvable Schrödinger type operators. London Mathematical Society Lecture Note Series, 271. Cambridge University Press, Cambridge, 2000. ISBN 0-521-77912-X MR 1752110 Zbl 0945.47015

[4] E. Braaten and H. W. Hammer, Universality in few-body systems with large scattering length. *Phys. Rep.* **428** (2006), no. 5-6, 259–390. MR 2221432

[5] Y. Castin, C. Mora, and L. Pricoupenko, Four-body Efimov effect for three fermions and a lighter particle. *Phys. Rev. Lett.* **105** (2010), 223201.

[6] Y. Castin and E. Tignone, Trimers in the resonant $(2+1)$-fermion problem on a narrow Feshbach resonance: Crossover from Efimovian to hydrogenoid spectrum. *Phys. Rev. A* **84** (2011), 062704.

[7] Y. Castin and F. Werner, The unitary gas and its symmetry properties. *Lect. Notes Phys.* **836** (2011), 127-189.

[8] M. Correggi, Dell'Antonio, D. Finco, A. Michelangeli, and A. Teta, Stability for a system of N fermions plus a different particle with zero-range interactions. *Rev. Math. Phys.* **24** (2012), no. 7, 1250017, 32 pp.
MR 2957709 Zbl 1272.81074

[9] M. Correggi, G. Dell'Antonio, D. Finco, A. Michelangeli, and A. Teta, A class of Hamiltonians for a three-particle fermionic system at unitarity. *Math. Phys. Anal. Geom.* **18** (2015), no. 1, article 32, 36 pp. MR 3437992 Zbl 1338.47002

[10] M. Correggi, D. Finco, and A. Teta, Energy lower bound for the unitary $N+1$ fermionic model. *Euro Phys. Lett.* **111** (2015), 10003.

[11] G. Dell'Antonio, R. Figari, and A. Teta, Hamiltonians for systems of N particles interacting through point interactions. *Ann. Inst. H. Poincaré Phys. Théor.* **60** (1994), no. 3, 253–290. MR 1281647 Zbl 0808.35113

[12] L. Faddeev and R. A. Minlos, On the point interaction for a three-particle system in quantum Mechanics. *Dokl. Akad. Nauk SSSR* **141** (1962), 1335–1338. In Russian. English translation, *Soviet Phys. Dokl.* **6** (1962), 1072–1074. MR 0147136

[13] D. Finco and A. Teta, Quadratic forms for the fermionic unitary gas model. *Rep. Math. Phys.* **69** (2012), no. 2, 131–159. MR 2935450 Zbl 1252.82067

[14] O. I. Kartavtsev and A. V. Malykh, Recent advances in description of few two-component fermions. *Physics of atomic nuclei* **77** (2014), 430–437.

[15] A. Michelangeli and P. Pfeiffer, Stability of the $2+2$-fermionic system with zero-range interaction. *J. Phys. A* **49** (2016), no. 10, 105301, 27 pp. MR 3462328 Zbl 1342.81756

[16] A. Michelangeli and C. Schmidbauer, Binding properties of the $(2+1)$-fermion system with zero-range interspecies interaction. *Phys. Rev. A* **87** (2013), 053601.

[17] A. M. Melnikov and R. A. Minlos, On the pointlike interaction of three different particles. In R. A. Minlos (ed.), *Many-particle Hamiltonians: spectra and scattering*. Translated from the Russian. Translation edited by A. B. Sossinsky. Advances in Soviet Mathematics, 5. American Mathematical Society, Providence, R.I., 1991, 99–112. MR 1130185 Zbl 0753.46043

[18] R. A. Minlos, On the point interaction of three particles. In P. Exner and P. Šeba (ed.), *Applications of selfadjoint extensions in quantum physics.* Proceedings of the conference held at the Laboratory of Theoretical Physics, JINR, Dubna, USSR, September 29–October 1, 1987. Lecture Notes in Physics, 324. Springer, Berlin etc., 1989, 138–145. MR 1009846 Zbl 0738.47008

[19] R. A. Minlos, On point-like interaction between n fermions and another particle. *Mosc. Math. J.* **11** (2011), no. 1, 113–127, 182. MR 2808213 Zbl 1222.81174

[20] R. A. Minlos, A system of three quantum particles with point-like interactions. *Uspekhi Mat. Nauk* **69** (2014), no. 3(417), 145–172. In Russian. English translation, *Russian Math. Surveys* **69** (2014), no. 3, 539–564. MR 3287506 Zbl 1300.81039

[21] R. A. Minlos and M. K. Shermatov, On pointlike interaction of three particles. *Vestnik Mosk. Univ. Ser. Math. Mekh.* **6** (1989), 7–14. In Russian. English translation in R. A. Minlos (ed.), *Many-particle Hamiltonians: spectra and scattering.* Advances in Soviet Mathematics, 5. American Mathematical Society, Providence, R.I., 1991, 99–112. Zbl 0753.46043

[22] B. S. Pavlov, The theory of extensions and explicitly solvable models. *Uspekhi Mat. Nauk* **42** (1987), no. 6(258), 99–131, 247. In Russian. English translation, *Russian Math. Surveys* **42**, no. 6 (1987), 127–168. MR 0933997 Zbl 0648.47010

[23] F. Werner and Y. Castin, Unitary gas in an isotropic harmonic trap: symmetry properties and applications. *Phys. Rev. A* **74** (2006), 053604.

[24] F. Werner and Y. Castin, Unitary quantum three-body problem in a harmonic trap. *Phys. Rev. Lett.* **97** (2006), 150401.

On the index of meromorphic operator-valued functions and some applications

Jussi Behrndt, Fritz Gesztesy, Helge Holden, and Roger Nichols

We dedicate this paper with great pleasure to Pavel Exner, whose tireless efforts as an ambassador for Mathematical Physics have led him to nearly every corner of this globe.

Happy Birthday, Pavel, we hope our modest contribution to operator and spectral theory will cause some joy.

1 Introduction

The purpose of this paper is fourfold:

- first, to recall recent results on factorizations of analytic operator-valued Fredholm functions following Howland [18] and more recently, [11];
- second, apply this to algebraic multiplicities of bounded, analytic operator-valued Fredholm functions;
- third, discuss the notion of an index of meromorphic operator-valued functions;
- fourth, apply this to Birman–Schwinger operators in connection with abstract perturbation theory and to operator-valued Weyl–Titchmarsh functions associated to closed extensions of dual pairs of closed operators.

In Section 2, we recall the notion of finitely-meromorphic $\mathcal{B}(\mathcal{H})$-valued functions and some of their basic properties, state the analytic Fredholm theorem, and recall in Theorems 2.5 and 2.6 a factorization of analytic operator-valued Fredholm functions originally due to Howland [18] and recently revisited under somewhat more general hypotheses in [11].

Section 3 recalls the notion of zeros of finite-type of bounded, analytic operator-valued functions $A(\cdot)$, revisits the algebraic multiplicity (8) of a zero of finite-type of $A(\cdot)$, relates the latter to the operator-valued argument principle (i.e., an operator Rouché-type theorem) and to appropriate traces of contour integrals, and finally proves equality of this notion of multiplicity with the multiplicity notion (5) originally introduced by Howland [18] in Theorem 3.3, the principal result of this section.

The topic of meromorphic operator-valued functions and the notion of their index is the principal subject of Section 4. In particular, we revisit the notion of $\mathcal{B}(\mathcal{H})$-valued finitely meromorphic functions $M(\cdot)$, introduce the notion of their index via the operator-valued argument principle and taking the trace of a contour integral as in (19), and finally recall the meromorphic Fredholm theorem.

Abstract perturbation theory and applications to Birman–Schwinger-type operators $K(\cdot)$ are treated in Section 5. This should be viewed as a refinement of recent results of this genre in Section 5 in [11]. Following Kato [19], Konno and Kuroda [23], and Howland [17], we recall a class of factorable non-self-adjoint perturbations of a given unperturbed non-self-adjoint operator H_0, giving rise to an operator H as refined in [12] (cf. Theorem 5.2), and then prove analogs of Weinstein–Aronszajn formulas, relating the difference of the algebraic multiplicity of an eigenvalue of H and H_0 to the index of the meromorphic operator-valued function $I - K(\cdot)$ in Theorem 5.5.

Finally, Section 6 focuses on closed extensions A_0, A_Θ (where Θ is an appropriate bounded operator parameter), associated to dual pairs $\{A, B\}$ of operators and their associated Weyl–Titchmarsh functions $M(\cdot)$, following work of Malamud, Mogilevskii, and Hassi [26], [27], and [28]. Our principal new result, Theorem 6.4, relates the difference of the algebraic multiplicity of a discrete eigenvalue of A_Θ and A_0 to the index of the meromorphic operator-valued function $\Theta - M(\cdot)$.

Next, we summarize the basic notation used in this paper: Let $\mathcal{H}$ and $\mathcal{K}$ be separable complex Hilbert spaces, $(\cdot\,,\cdot)_{\mathcal{H}}$ and $(\cdot\,,\cdot)_{\mathcal{K}}$ the scalar products in $\mathcal{H}$ and $\mathcal{K}$ (linear in the second factor), and $I_{\mathcal{H}}$ and $I_{\mathcal{K}}$ the identity operators in $\mathcal{H}$ and $\mathcal{K}$, respectively. Next, let T be a closed linear operator from $\operatorname{dom}(T) \subseteq \mathcal{H}$ to $\operatorname{ran}(T) \subseteq \mathcal{K}$, with $\operatorname{dom}(T)$ and $\operatorname{ran}(T)$ denoting the domain and range of T. The closure of a closable operator S is denoted by $\overline{S}$. The kernel (null space) of T is denoted by $\ker(T)$. The spectrum, point spectrum, and resolvent set of a closed linear operator in $\mathcal{H}$ will be denoted by $\sigma(\cdot)$, $\sigma_p(\cdot)$, and $\rho(\cdot)$; the discrete spectrum of T (i.e., points in $\sigma_p(T)$ which are isolated from the rest of $\sigma(T)$, and which are eigenvalues of T of finite algebraic multiplicity) is abbreviated by $\sigma_d(T)$. The *algebraic multiplicity* $m_a(z_0; T)$ of an eigenvalue $z_0 \in \sigma_d(T)$ is the dimension of the range of the corresponding *Riesz projection* $P(z_0; T)$,

$$m_a(z_0; T) = \dim(\operatorname{ran}(P(z_0; T))) = \operatorname{tr}_{\mathcal{H}}(P(z_0; T)),$$

where (with the symbol $\oint$ denoting contour integrals)

$$P(z_0; T) = \frac{-1}{2\pi i} \oint_{C(z_0;\varepsilon)} d\zeta \, (T - \zeta I_{\mathcal{H}})^{-1},$$

for $0 < \varepsilon < \varepsilon_0$ and $D(z_0; \varepsilon_0)\backslash\{z_0\} \subset \rho(T)$; here $D(z_0; r_0) \subset \mathbb{C}$ is the open disk with center z_0 and radius $r_0 > 0$, and $C(z_0; r_0) = \partial D(z_0; r_0)$ the corresponding circle. The *geometric multiplicity* $m_g(z_0; T)$ of an eigenvalue $z_0 \in \sigma_p(T)$ is defined by

$$m_g(z_0; T) = \dim(\ker((T - z_0 I_{\mathcal{H}}))).$$

The essential spectrum of T is defined by $\sigma_{\text{ess}}(T) = \sigma(T)\backslash\sigma_d(T)$.

The Banach spaces of bounded and compact linear operators in $\mathcal{H}$ are denoted by $\mathcal{B}(\mathcal{H})$ and $\mathcal{B}_\infty(\mathcal{H})$, respectively. Similarly, the Schatten–von Neumann (trace) ideals will subsequently be denoted by $\mathcal{B}_p(\mathcal{H})$, $p \in [1, \infty)$, and the subspace of all finite rank operators in $\mathcal{B}_1(\mathcal{H})$ will be abbreviated by $\mathcal{F}(\mathcal{H})$. Analogous notation $\mathcal{B}(\mathcal{H}_1, \mathcal{H}_2)$, $\mathcal{B}_\infty(\mathcal{H}_1, \mathcal{H}_2)$, etc., will be used for bounded, compact, etc., operators between two Hilbert spaces $\mathcal{H}_1$ and $\mathcal{H}_2$. In addition, $\operatorname{tr}_{\mathcal{H}}(T)$ denotes the trace of a trace class operator $T \in \mathcal{B}_1(\mathcal{H})$.

The set of bounded Fredholm operators on $\mathcal{H}$ (i.e., the set of operators $T \in \mathcal{B}(\mathcal{H})$ such that $\dim(\ker(T)) < \infty$, $\operatorname{ran}(T)$ is closed in $\mathcal{H}$, and $\dim(\ker(T^*)) < \infty$) is denoted by the symbol $\Phi(\mathcal{H})$. The corresponding (Fredholm) index of $T \in \Phi(\mathcal{H})$ is then given by $\operatorname{ind}(T) = \dim(\ker(T)) - \dim(\ker(T^*))$. For a linear operator S in $\mathcal{H}$ with closed range one defines the *defect of* S, denoted by $\operatorname{def}(S)$, by the codimension of $\operatorname{ran}(S)$ in $\mathcal{H}$, that is,

$$\operatorname{def}(S) = \dim(\operatorname{ran}(S)^\perp).$$

The symbol $\dot{+}$ denotes a direct (but not necessary orthogonal direct) decomposition in connection with subspaces of Banach spaces. Finally, we find it convenient to abbreviate $\mathbb{N}_0 = \mathbb{N} \cup \{0\}$.

2 On factorizations of analytic operator-valued functions

In this section, we recall factorizations of bounded, analytic operator-valued Fredholm functions following Howland [18] and more recently [11].

Assuming $\Omega \subseteq \mathbb{C}$ to be open and $M(\cdot)$ to be a $\mathcal{B}(\mathcal{H})$-valued meromorphic function on Ω that has the norm convergent Laurent expansion around $z_0 \in \Omega$ of the type

$$M(z) = \sum_{k=-N_0}^{\infty} (z-z_0)^k M_k(z_0), \quad M_k(z_0) \in \mathcal{B}(\mathcal{H}),\ k \in \mathbb{Z},\ k \geq -N_0, \\ 0 < |z-z_0| < \varepsilon_0,$$

for some $N_0 = N_0(z_0) \in \mathbb{N}$ and some $0 < \varepsilon_0 = \varepsilon_0(z_0)$ sufficiently small, we denote the principal part, $\mathrm{pp}_{z_0}\{M(\cdot)\}$, of $M(\cdot)$ at z_0 by

$$\mathrm{pp}_{z_0}\{M(z)\} = \sum_{k=-N_0}^{-1} (z-z_0)^k M_k(z_0), \quad M_k(z_0) \in \mathcal{B}(\mathcal{H}),\ -N_0 \leq k \leq -1, \\ 0 < |z-z_0| < \varepsilon_0. \tag{1}$$

Given the notation (1), we start with the following definition.

Definition 2.1. Let $\Omega \subseteq \mathbb{C}$ be open and connected. Suppose that $M(\cdot)$ is a $\mathcal{B}(\mathcal{H})$-valued analytic function on Ω except for isolated singularities in a neighborhood of which it is meromorphic. Then $M(\cdot)$ is called *finitely meromorphic at* $z_0 \in \Omega$ if $M(\cdot)$ is analytic on the punctured disk $D(z_0;\varepsilon_0)\backslash\{z_0\} \subset \Omega$ centered at z_0 with sufficiently small $\varepsilon_0 > 0$, and the principal part of $M(\cdot)$ at z_0 is of finite rank, that is, the principal part of $M(\cdot)$ is of the type (1), and one has

$$M_k(z_0) \in \mathcal{F}(\mathcal{H}), \quad -N_0 \leq k \leq -1.$$

In addition, $M(\cdot)$ is called *finitely meromorphic on* Ω if it is meromorphic on Ω and finitely meromorphic at each of its poles.

In this context, we mention the following useful result.

Lemma 2.2 (Lemma XI.9.3 in [13] and Proposition 4.2.2 in [15]). *Let $\Omega \subseteq \mathbb{C}$ be open and connected and $M_j(\cdot)$, $j = 1,2$, be $\mathcal{B}(\mathcal{H})$-valued finitely meromorphic functions at $z_0 \in \Omega$. Then $M_1(\cdot)M_2(\cdot)$ and $M_2(\cdot)M_1(\cdot)$ are finitely meromorphic at $z_0 \in \Omega$, and for $0 < \varepsilon < \varepsilon_0$ sufficiently small,*

$$\oint_{C(z_0;\varepsilon)} d\zeta\, M_1(\zeta)M_2(\zeta) \in \mathcal{F}(\mathcal{H})$$

and

$$\oint_{C(z_0;\varepsilon)} d\zeta\, M_2(\zeta)M_1(\zeta) \in \mathcal{F}(\mathcal{H}),$$

and the identity

$$\mathrm{tr}_{\mathcal{H}}\bigg(\oint_{C(z_0;\varepsilon)} d\zeta\, M_1(\zeta)M_2(\zeta)\bigg) = \mathrm{tr}_{\mathcal{H}}\bigg(\oint_{C(z_0;\varepsilon)} d\zeta\, M_2(\zeta)M_1(\zeta)\bigg) \tag{2}$$

holds. Moreover, for $0 < |z - z_0| < \varepsilon_0$ *one has*

$$\operatorname{tr}_{\mathcal{H}}(\operatorname{pp}_{z_0}\{M_1(z)M_2(z)\}) = \operatorname{tr}_{\mathcal{H}}(\operatorname{pp}_{z_0}\{M_2(z)M_1(z)\}).$$

For the remainder of this section we make the following assumptions:

Hypothesis 2.3. Let $\Omega \subseteq \mathbb{C}$ be open and connected, suppose that $A\colon \Omega \to \mathcal{B}(\mathcal{H})$ is analytic and that

$$A(z) \in \Phi(\mathcal{H}) \quad \text{for all } z \in \Omega.$$

One then recalls the analytic Fredholm theorem in the following form.

Theorem 2.4 (Section 4.1 of [15], [16], [17], Theorem VI.14 in [32], and [40]). *Assume that* $A\colon \Omega \to \mathcal{B}(\mathcal{H})$ *satisfies Hypothesis* 2.3. *Then either*

(i) $A(z)$ *is not boundedly invertible for any* $z \in \Omega$,

or else,

(ii) $A(\cdot)^{-1}$ *is finitely meromorphic on* Ω. *More precisely, there exists a discrete subset* $\mathcal{D}_1 \subset \Omega$ *(possibly,* $\mathcal{D}_1 = \emptyset$*) such that* $A(z)^{-1} \in \mathcal{B}(\mathcal{H})$ *(and hence lies in* $\Phi(\mathcal{H})$*) for all* $z \in \Omega\backslash\mathcal{D}_1$, $A(\cdot)^{-1}$ *is analytic on* $\Omega\backslash\mathcal{D}_1$, *meromorphic on* Ω, *and if* $z_1 \in \mathcal{D}_1$ *then*

$$A(z)^{-1} = \sum_{k=-N_0}^{\infty} (z - z_1)^k C_k(z_1), \quad 0 < |z - z_1| < \varepsilon_0,$$

for some $N_0 = N_0(z_1) \in \mathbb{N}$ *and some* $0 < \varepsilon_0 = \varepsilon_0(z_1)$ *sufficiently small, with*

$$C_k(z_1) \in \mathcal{F}(\mathcal{H}), \quad -N_0 \leq k \leq -1,$$

$$C_k(z_1) \in \mathcal{B}(\mathcal{H}), \quad k \in \mathbb{N}_0.$$

In addition,

$$C_0(z_1) \in \Phi(\mathcal{H}).$$

Finally, if $[I_{\mathcal{H}} - A(z)] \in \mathcal{B}_\infty(\mathcal{H})$ *for all* $z \in \Omega$, *then*

$$[I_{\mathcal{H}} - A(z)^{-1}] \in \mathcal{B}_\infty(\mathcal{H}), \quad z \in \Omega\backslash\mathcal{D}_1,$$

$$[I_{\mathcal{H}} - C_0(z_1)] \in \mathcal{B}_\infty(\mathcal{H}), \quad z_1 \in \mathcal{D}_1.$$

The following fundamental results are due to Howland [18] (see also [11] for more general hypotheses, replacing Howland's assumption that $[A(\cdot) - I_{\mathcal{H}}] \in \mathcal{B}_\infty(\mathcal{H})$ by the assumption that $A(\cdot)$ is Fredholm):

Theorem 2.5 ([18]). *Assume that $A\colon \Omega \to \mathcal{B}(\mathcal{H})$ satisfies Hypothesis* 2.3, *suppose that $A(z)$ is boundedly invertible for some $z \in \Omega$ (i.e., case* (ii) *in Theorem* 2.4 *applies), and let $z_0 \in \Omega$ be a pole of $A(\cdot)^{-1}$ of order $n_0 \in \mathbb{N}$. Denote by Q_1 any projection onto* $\operatorname{ran}(A(z_0))$ *and let $P_1 = I_{\mathcal{H}} - Q_1$. Then,*

$$A(z) = [Q_1 - (z - z_0)P_1]A_1(z), \quad z \in \Omega, \tag{3}$$

where

- *$A_1(\cdot)$ is analytic on Ω,*
- *$A_1(z) \in \Phi(\mathcal{H}), \quad z \in \Omega$,*
- $\operatorname{ind}(A(z)) = \operatorname{ind}(A_1(z)) = 0, \quad z \in \Omega$, *$|z - z_0|$ sufficiently small,*
- $\operatorname{def}(A_1(z_0)) \leq \operatorname{def}(A(z_0))$,
- *z_0 is a pole of $A_1(\cdot)^{-1}$ of order $n_0 - 1$.*

In particular, $A_1\colon \Omega \to \mathcal{B}(\mathcal{H})$ satisfies Hypothesis 2.3. *Finally,*

$$[I_{\mathcal{H}} - A(\cdot)] \in \mathcal{F}(\mathcal{H}) \iff [I_{\mathcal{H}} - A_1(\cdot)] \in \mathcal{F}(\mathcal{H})$$

and

$$[I_{\mathcal{H}} - A(\cdot)] \in \mathcal{B}_p(\mathcal{H}) \iff [I_{\mathcal{H}} - A_1(\cdot)] \in \mathcal{B}_p(\mathcal{H}) \quad \text{for some } 1 \leq p \leq \infty.$$

Assume that $A\colon \Omega \to \mathcal{B}(\mathcal{H})$ satisfies Hypothesis 2.3 and that $A(\cdot)^{-1}$ has a pole at $z_0 \in \Omega$. The Riesz projection $P(z)$ associated with $A(z)$ and z in a sufficiently small neighborhood $\mathcal{N}(z_0) \subset \Omega$ of z_0 is defined by

$$P(z) = \frac{-1}{2\pi i} \oint_{C(0;\varepsilon)} d\zeta \, (A(z) - \zeta I_{\mathcal{H}})^{-1}, \quad z \in \mathcal{N}(z_0),$$

where $0 < \varepsilon < \varepsilon_0$ sufficiently small (cf., e.g., Section III.6 of [20]). It follows that $P(\cdot)$ is analytic on $\mathcal{N}(z_0)$ and

$$\dim(\operatorname{ran}(P(z))) < \infty, \quad z \in \mathcal{N}(z_0).$$

In addition, introduce the projections

$$Q(z) = I_{\mathcal{H}} - P(z), \quad z \in \mathcal{N}(z_0),$$

and the transformations (cf. [42])

$$T(z) = P(z_0)P(z) + Q(z_0)Q(z), \quad z \in \mathcal{N}(z_0).$$

It follows that $T(\cdot)$ is analytic on $\mathcal{N}(z_0)$ and for $|z-z_0|$ sufficiently small, also $T(\cdot)^{-1}$ is analytic,

$$T(z) = I_{\mathcal{H}} + O(z - z_0), \quad |z - z_0| \text{ sufficiently small,}$$

and without loss of generality we may assume in the following that $T(\cdot)$ and $T(\cdot)^{-1}$ are analytic on $\mathcal{N}(z_0)$. This yields the decomposition of $\mathcal{H}$ into

$$\mathcal{H} = P(z_0)\mathcal{H} \dotplus Q(z_0)\mathcal{H}$$

and the associated 2×2 block operator decomposition of $T(z)A(z)T(z)^{-1}$ of the form

$$T(z)A(z)T(z)^{-1} = \begin{pmatrix} F(z) & 0 \\ 0 & G(z) \end{pmatrix}, \quad z \in \mathcal{N}(z_0), \tag{4}$$

where $F(\cdot)$ and $G(\cdot)$ are analytic on $\mathcal{N}(z_0)$, and, again without loss of generality, $G(\cdot)$ is boundedly invertible on $\mathcal{N}(z_0)$,

$$G(z)^{-1} \in \mathcal{B}(Q(z_0)\mathcal{H}), \quad z \in \mathcal{N}(z_0).$$

Given the block decomposition (4), we follow Howland in introducing the quantity $\nu(z_0; A(\cdot))$ by

$$\nu(z_0; A(\cdot)) = m(z_0; \det_{\operatorname{ran}(P(z_0))}(F(\cdot))). \tag{5}$$

Here $m(z; h)$ denotes the multiplicity function associated to a meromorphic function $h\colon \Omega \to \mathbb{C} \cup \{\infty\}$, which is defined by

$$m(z; h) = \begin{cases} k, & \text{if } z \text{ is a zero of } h \text{ of order } k, \\ -k, & \text{if } z \text{ is a pole of order } k, \\ 0, & \text{otherwise,} \end{cases}$$

if m does not vanish identically on Ω, and by $m(z; h) = \infty$ otherwise. In the former case,

$$m(z; h) = \frac{1}{2\pi i} \oint_{C(z;\varepsilon)} d\zeta \, \frac{h'(\zeta)}{h(\zeta)}, \quad z \in \Omega,$$

where the circle $C(z; \varepsilon)$ is chosen sufficiently small such that $C(z; \varepsilon)$ contains no other singularities or zeros of h except, possibly, z.

In the present context, since $F(\cdot)$ is analytic on $\mathcal{N}(z_0)$, so is $\det_{\operatorname{ran}(P(z_0))}(F(\cdot))$, and hence

$$\nu(z_0; A(\cdot)) \in \mathbb{N}_0 \text{ if } \det_{\operatorname{ran}(P(z_0))}(F(\cdot)) \not\equiv 0 \text{ on } \mathcal{N}(z_0).$$

Repeated applications of Theorem 2.5 then yield the following principal factorization result of [18] (again, extended to the case of Fredholm operators $A(\cdot)$):

Theorem 2.6 ([18]). *Assume that $A\colon \Omega \to \mathcal{B}(\mathcal{H})$ satisfies Hypothesis 2.3, suppose that $A(z)$ is boundedly invertible for some $z \in \Omega$ (i.e., case* (ii) *in Theorem 2.4 applies), and let $z_0 \in \Omega$ be a pole of $A(\cdot)^{-1}$ of order $n_0 \in \mathbb{N}$. Then there exist projections P_j and $Q_j = I_{\mathcal{H}} - P_j$ in $\mathcal{H}$ such that with $p_j = \dim(\operatorname{ran}(P_j))$, $1 \leq j \leq n_0$, one infers that*

$$A(z) = [Q_1-(z-z_0)P_1][Q_2-(z-z_0)P_2]\cdots[Q_{n_0}-(z-z_0)P_{n_0}]A_{n_0}(z), \quad z \in \Omega, \tag{6}$$

and

$$1 \leq p_{n_0} \leq p_{n_0-1} \leq \cdots \leq p_2 \leq p_1 < \infty,$$

where

- *$A_{n_0}(\cdot)$ is analytic on Ω,*
- *$A_{n_0}(z) \in \Phi(\mathcal{H}), \quad z \in \Omega$,*
- *$\operatorname{ind}(A(z)) = \operatorname{ind}(A_{n_0}(z)) = 0, \quad z \in \Omega$, $|z - z_0|$ sufficiently small,*
- *$A_{n_0}(z)^{-1} \in \mathcal{B}(\mathcal{H}), \quad z \in \Omega$, $|z - z_0|$ sufficiently small.*

In addition,

$$p_1 = \dim(\ker(A(z_0)) = m_g(0; A(z_0)),$$

and hence

$$\nu(z_0; A(\cdot)) = \sum_{j=1}^{n_0} p_j \geq m_g(0; A(z_0)), \quad \nu(z_0; A(\cdot)) \geq n_0,$$

and, in particular, z_0 is a simple pole of $A(\cdot)^{-1}$ if and only if

$$\nu(z_0; A(\cdot)) = m_g(0; A(z_0)).$$

Finally,

$$[I_{\mathcal{H}} - A(\cdot)] \in \mathcal{F}(\mathcal{H}) \iff [I_{\mathcal{H}} - A_{n_0}(\cdot)] \in \mathcal{F}(\mathcal{H})$$

and

$$[I_{\mathcal{H}} - A(\cdot)] \in \mathcal{B}_p(\mathcal{H}) \iff [I_{\mathcal{H}} - A_{n_0}(\cdot)] \in \mathcal{B}_p(\mathcal{H}),$$

for some $1 \leq p \leq \infty$.

We refer to [11] for analogous factorizations as in Theorems 2.5 and 2.6 but with the order of factors in (3) and (6) interchanged.

3 Algebraic multiplicities of zeros of analytic Fredholm operators

In this section we recall algebraic multiplicities of zeros of analytic Fredholm operators following [11] and relate this to Howland's notion in (5). The pertinent facts in this context can be found in [16] (see also, Sections XI.8 and XI.9 of [13], Chapter 4 of [15], and Section 11 of [29]). We follow the presentation in [11].

First the notion of zeros of finite-type is recalled.

Definition 3.1. Let $\Omega \subseteq \mathbb{C}$ be open and connected, $z_0 \in \Omega$, and suppose that $A\colon \Omega \to \mathcal{B}(\mathcal{H})$ is analytic on Ω. Then z_0 is called a *zero of finite-type of* $A(\cdot)$ if $A(z_0) \in \Phi(\mathcal{H})$ is a Fredholm operator, $\ker(A(z_0)) \neq \{0\}$, and $A(\cdot)$ is boundedly invertible on $D(z_0; \varepsilon_0)\backslash\{z_0\}$, for some sufficiently small $\varepsilon_0 > 0$.

Assume that $A\colon \Omega \to \mathcal{B}(\mathcal{H})$ is analytic on Ω and that z_0 is a zero of finite-type of $A(\cdot)$. Since $A(\cdot)$ is boundedly invertible on $D(z_0; \varepsilon_0)\backslash\{z_0\}$, for sufficiently small $\varepsilon_0 > 0$, it follows that

$$\operatorname{ind}(A(z_0)) = \dim(\ker(A(z_0))) - \dim(\ker(A(z_0)^*)) = 0,$$

and hence by [16] (or by Theorem XI.8.1 in [13]) there exists a neighborhood $\mathcal{N}(z_0) \subset \Omega$ and analytic and boundedly invertible operator-valued functions

$$E_j\colon \Omega \longrightarrow \mathcal{B}(\mathcal{H}), \quad j = 1, 2,$$

such that

$$A(z) = E_1(z)\widetilde{A}(z)E_2(z), \qquad z \in \mathcal{N}(z_0),$$

where $\widetilde{A}(\cdot)$ is of the form

$$\widetilde{A}(z) = \widetilde{P}_0 + \sum_{j=1}^{r}(z - z_0)^{n_j}\widetilde{P}_j, \quad z \in \mathcal{N}(z_0),$$

with

$$\begin{aligned} &\widetilde{P}_k (0 \leq k \leq r) \text{ mutually disjoint projections in } \mathcal{H}, \\ &[I_{\mathcal{H}} - \widetilde{P}_0] \in \mathcal{F}(\mathcal{H}), \quad \dim(\operatorname{ran}(\widetilde{P}_j)) = 1, \quad 1 \leq j \leq r, \\ &n_1 \leq n_2 \leq \cdots \leq n_r, \quad n_j \in \mathbb{N}, \ 1 \leq j \leq r. \end{aligned} \tag{7}$$

The integers n_j, $1 \leq j \leq r$, in (7) are uniquely determined by $A(\cdot)$, and the geometric multiplicity $m_g(0; A(z_0))$ of the eigenvalue 0 of $A(z_0)$ is given by

$$m_g(0; A(z_0)) = \dim(\operatorname{ran}(I_{\mathcal{H}} - \widetilde{P}_0)).$$

The following definition can be found in Section XI.9 of [13] or [16].

Definition 3.2. Let $\Omega \subseteq \mathbb{C}$ be open and connected, $z_0 \in \Omega$, suppose that the function $A\colon \Omega \to \mathcal{B}(\mathcal{H})$ is analytic on Ω, and assume that z_0 is a zero of finite-type of $A(\cdot)$. Then $m_a(z_0; A(\cdot))$, the *algebraic multiplicity of the zero of* $A(\cdot)$ *at* z_0, is defined to be

$$m_a(z_0; A(\cdot)) = \sum_{j=1}^{r} n_j, \tag{8}$$

with n_j, $1 \leq j \leq r$, introduced in (7).

Let $A\colon \Omega \to \mathcal{B}(\mathcal{H})$ be analytic on Ω and assume that z_0 is a zero of finite-type of $A(\cdot)$. As shown in Theorem XI.9.1 in [13] and in [16], one has an extension of the argument principle for scalar analytic functions to the operator-valued case in the form

$$\begin{aligned} m_a(z_0; A(\cdot)) &= \operatorname{tr}_{\mathcal{H}}\bigg(\frac{1}{2\pi i}\oint_{C(z_0;\varepsilon)} d\zeta\, A'(\zeta)A(\zeta)^{-1}\bigg) \\ &= \operatorname{tr}_{\mathcal{H}}\bigg(\frac{1}{2\pi i}\oint_{C(z_0;\varepsilon)} d\zeta\, A(\zeta)^{-1}A'(\zeta)\bigg) \end{aligned} \tag{9}$$

for $0 < \varepsilon < \varepsilon_0$ sufficiently small as in Definition 3.1. Since $A(\cdot)^{-1}$ is finitely meromorphic by Theorem 2.4, the integrals in (9) are finite rank operators (the analytic and non-finite-rank part under the integral in (9) yielding a zero contribution when integrated over $C(z_0;\varepsilon)$) and hence the trace in (9) is well-defined. Next, recalling our notation of the principal part of an operator-valued meromorphic function in (1), one also obtains

$$\begin{aligned} m_a(z_0; A(\cdot)) &= \operatorname{tr}_{\mathcal{H}}\bigg(\frac{1}{2\pi i}\oint_{C(z_0;\varepsilon)} d\zeta\, \mathrm{pp}_{z_0}\{A'(\zeta)A(\zeta)^{-1}\}\bigg) \\ &= \operatorname{tr}_{\mathcal{H}}\bigg(\frac{1}{2\pi i}\oint_{C(z_0;\varepsilon)} d\zeta\, \mathrm{pp}_{z_0}\{A(\zeta)^{-1}A'(\zeta)\}\bigg). \end{aligned}$$

Note that in the special case where $A(z) = A - zI_{\mathcal{H}}$, $z \in \Omega$, one has from (9)

$$\begin{aligned} m_a(z_0; A(\cdot)) &= \operatorname{tr}_{\mathcal{H}}\bigg(\frac{1}{2\pi i}\oint_{C(z_0;\varepsilon)} d\zeta\, A'(\zeta)A(\zeta)^{-1}\bigg) \\ &= \operatorname{tr}_{\mathcal{H}}\bigg(\frac{-1}{2\pi i}\oint_{C(z_0;\varepsilon)} d\zeta\, (A - \zeta I_{\mathcal{H}})^{-1}\bigg) \\ &= m_a(z_0; A). \end{aligned}$$

However, in general the algebraic multiplicity $m_a(z_0; A(\cdot))$ of a zero of $A(\cdot)$ at z_0 must be distinguished from the algebraic multiplicity $m_a(0; A(z_0))$ of the eigenvalue 0 of the operator $A(z_0)$.

We conclude this section with the connection between the algebraic multiplicity $m_a(z_0; A(\cdot))$ of a zero of $A(\cdot)$ at z_0 in Definition 3.2 and Howland's notion of multiplicity $\nu(z_0; A(\cdot))$ in (5). Note that if $A\colon \Omega \to \mathcal{B}(\mathcal{H})$ is analytic on Ω and z_0 is a zero of finite-type then Hypothesis 2.3 is automatically satisfied on a sufficiently small open neighborhood of z_0 and hence the quantity $\nu(z_0; A(\cdot))$ is well defined.

Theorem 3.3. *Assume that z_0 is a zero of finite-type of $A(\cdot)$. Then the algebraic multiplicity $m_a(z_0; A(\cdot))$ of the zero of $A(\cdot)$ at z_0 and the quantity $\nu(z_0; A(\cdot))$ coincide, that is,*

$$m_a(z_0; A(\cdot)) = \nu(z_0; A(\cdot)).$$

Proof. Without loss of generality we may assume that $z_0 = 0$ for the remainder of the proof of Theorem 3.3. According to (9) we then have

$$m_a(0; A(\cdot)) = \operatorname{tr}_{\mathcal{H}}\left(\frac{1}{2\pi i}\oint_{C(0;\varepsilon)} d\zeta\, A(\zeta)^{-1}A'(\zeta)\right) \tag{10}$$

for $0 < \varepsilon < \varepsilon_0$ sufficiently small. An application of Theorem 2.6 (using the notation employed in the latter) yields

$$A(z) = [Q_1 - zP_1][Q_2 - zP_2]\cdots[Q_{n_0} - zP_{n_0}]A_{n_0}(z), \quad z \in D(0;\varepsilon_0), \tag{11}$$

and

$$\nu(0; A(\cdot)) = \sum_{j=1}^{n_0} p_j, \tag{12}$$

where

$$p_j = \dim(\operatorname{ran}(P_j)) \quad \text{and} \quad Q_j = I_{\mathcal{H}} - P_j,\ 1 \le j \le n_0. \tag{13}$$

In the following we compute the trace of the integral in (10). For this one notes that by (11)

$$A(z)^{-1} = [A_{n_0}(z)]^{-1}[Q_{n_0} - zP_{n_0}]^{-1}\cdots[Q_1 - zP_1]^{-1},$$

and

$$\begin{aligned} A'(z) = {} & [-P_1][Q_2 - zP_2]\cdots[Q_{n_0} - zP_{n_0}]A_{n_0}(z) \\ & + [Q_1 - zP_1][-P_2]\cdots[Q_{n_0} - zP_{n_0}]A_{n_0}(z) \\ & \vdots \\ & + [Q_1 - zP_1][Q_2 - zP_2]\cdots[Q_{n_0-1} - zP_{n_0-1}][-P_{n_0}]A_{n_0}(z) \\ & + [Q_1 - zP_1][Q_2 - zP_2]\cdots[Q_{n_0} - zP_{n_0}]A'_{n_0}(z). \end{aligned}$$

Hence one obtains

$$
\begin{aligned}
A(z)^{-1}A'(z) = {} & [A_{n_0}(z)]^{-1}[Q_{n_0} - zP_{n_0}]^{-1}\cdots[Q_1 - zP_1]^{-1} \\
& \times \{[-P_1][Q_2 - zP_2]\cdots[Q_{n_0} - zP_{n_0}] \\
& \quad + [Q_1 - zP_1][-P_2]\cdots[Q_{n_0} - zP_{n_0}] \\
& \quad \vdots \\
& \quad + [Q_1 - zP_1]\cdots[Q_{n_0-1} - zP_{n_0-1}][-P_{n_0}]\}A_{n_0}(z) \\
& \quad + [A_{n_0}(z)]^{-1}A'_{n_0}(z),
\end{aligned} \tag{14}
$$

and since the last term on the right-hand side of (14) is analytic at $z_0 = 0$, its contour integral over $C(0;\varepsilon)$, $0 < \varepsilon < \varepsilon_0$, vanishes,

$$
\begin{aligned}
& \oint_{C(0;\varepsilon)} d\zeta\, A(\zeta)^{-1}A'(\zeta) \\
& \quad = \oint_{C(0;\varepsilon)} d\zeta [A_{n_0}(\zeta)]^{-1}[Q_{n_0} - \zeta P_{n_0}]^{-1}\cdots[Q_1 - \zeta P_1]^{-1} \\
& \qquad \times \{[-P_1][Q_2 - \zeta P_2]\cdots[Q_{n_0} - \zeta P_{n_0}] \\
& \qquad\quad + [Q_1 - \zeta P_1][-P_2]\cdots[Q_{n_0} - \zeta P_{n_0}] \\
& \qquad\quad \vdots \\
& \qquad\quad + [Q_1 - \zeta P_1]\cdots[Q_{n_0-1} - \zeta P_{n_0-1}][-P_{n_0}]\}A_{n_0}(\zeta).
\end{aligned} \tag{15}
$$

Now one obtains from (15) upon repeatedly applying cyclicity of the trace (i.e., $\mathrm{tr}_{\mathcal{H}}(CD) = \mathrm{tr}_{\mathcal{H}}(DC)$ for $C, D \in \mathcal{B}(\mathcal{H})$, with $CD, DC \in \mathcal{B}_1(\mathcal{H})$),

$$
\begin{aligned}
& \oint_{C(0;\varepsilon)} d\zeta\, \mathrm{tr}_{\mathcal{H}}(A(\zeta)^{-1}A'(\zeta)) \\
& \quad = \oint_{C(0;\varepsilon)} d\zeta\, \mathrm{tr}_{\mathcal{H}}([Q_{n_0} - \zeta P_{n_0}]^{-1}\cdots[Q_1 - \zeta P_1]^{-1} \\
& \qquad \times \{[-P_1][Q_2 - \zeta P_2]\cdots[Q_{n_0} - \zeta P_{n_0}] \\
& \qquad\quad + [Q_1 - \zeta P_1][-P_2]\cdots[Q_{n_0} - \zeta P_{n_0}] \\
& \qquad\quad \vdots \\
& \qquad\quad + [Q_1 - \zeta P_1]\cdots[Q_{n_0-1} - \zeta P_{n_0-1}][-P_{n_0}]\}) \\
& \quad = \oint_{C(0;\varepsilon)} d\zeta \sum_{j=1}^{n_0} \mathrm{tr}_{\mathcal{H}}([Q_j - \zeta P_j]^{-1}[-P_j])
\end{aligned} \tag{16}
$$

and since

$$
[Q_j - \zeta P_j]^{-1}[-P_j] = [Q_j - \zeta^{-1} P_j][-P_j] = \zeta^{-1} P_j, \tag{17}
$$

one concludes from (10), (16), (17), (13), and (12) that

$$\begin{aligned}
m_a(0; A(\cdot)) &= \operatorname{tr}_{\mathcal{H}}\bigg(\frac{1}{2\pi i}\oint_{C(0;\varepsilon)} d\zeta\, A(\zeta)^{-1}A'(\zeta)\bigg) \\
&= \frac{1}{2\pi i}\oint_{C(0;\varepsilon)} d\zeta\, \operatorname{tr}_{\mathcal{H}}(A(\zeta)^{-1}A'(\zeta)) \\
&= \frac{1}{2\pi i}\oint_{C(0;\varepsilon)} d\zeta \sum_{j=1}^{n_0} \operatorname{tr}_{\mathcal{H}}([Q_j - \zeta P_j]^{-1}[-P_j]) \\
&= \frac{1}{2\pi i}\oint_{C(0;\varepsilon)} d\zeta \Big(\sum_{j=1}^{n_0} \operatorname{tr}_{\mathcal{H}}(P_j)\Big)\zeta^{-1} \\
&= \sum_{j=1}^{n_0} p_j \\
&= \nu(0; A(\cdot)). \qquad \square
\end{aligned}$$

4 On the notion of an index of meromorphic operator-valued functions

In this section we recall the notion of the index of meromorphic operator functions and the meromorphic Fredholm theorem.

Hypothesis 4.1. Let $\Omega \subseteq \mathbb{C}$ be open and connected and assume that $M(\cdot)$ is a $\mathcal{B}(\mathcal{H})$-valued finitely meromorphic function on Ω, that is, there is a discrete set $\mathcal{D}_0 \subset \Omega$ (i.e., a set without limit points in Ω) such that $M: \Omega \backslash \mathcal{D}_0 \to \mathcal{B}(\mathcal{H})$ is analytic and for all $z_0 \in \mathcal{D}_0$ one has

$$M(z) = \sum_{k=-N_0}^{\infty} (z - z_0)^k M_k(z_0), \quad 0 < |z - z_0| < \varepsilon_0, \tag{18}$$

for some $N_0 = N_0(z_0) \in \mathbb{N}$ and some $0 < \varepsilon_0 = \varepsilon_0(z_0)$ sufficiently small, with

$$M_k(z_0) \in \mathcal{F}(\mathcal{H}), \ -N_0 \leq k \leq -1, \quad M_k(z_0) \in \mathcal{B}(\mathcal{H}), \ k \in \mathbb{N}_0.$$

One observes that if $M(\cdot)$ is finitely meromorphic on Ω, then also the function $M'(\cdot)$ is a $\mathcal{B}(\mathcal{H})$-valued finitely meromorphic function on Ω. It follows from Lemma 2.2 that the notion of the index of $M(\cdot)$ in the next definition is well-defined.

Definition 4.2. Assume Hypothesis 4.1, let $z_0 \in \Omega$ and suppose that $M(\cdot)$ is boundedly invertible on $D(z_0;\varepsilon_0)\backslash\{z_0\}$ for some $0<\varepsilon_0$ sufficiently small. Assume, in addition, that the function $M(\cdot)^{-1}$ is finitely meromorphic on $D(z_0;\varepsilon_0)$. Then the *index of $M(\cdot)$ with respect to the counterclockwise oriented circle $C(z_0;\varepsilon)$*, $\mathrm{ind}_{C(z_0;\varepsilon)}(M(\cdot))$, is defined by

$$\begin{aligned}\mathrm{ind}_{C(z_0;\varepsilon)}(M(\cdot)) &= \mathrm{tr}_{\mathcal{H}}\bigg(\frac{1}{2\pi i}\oint_{C(z_0;\varepsilon)} d\zeta\, M'(\zeta)M(\zeta)^{-1}\bigg) \\ &= \mathrm{tr}_{\mathcal{H}}\bigg(\frac{1}{2\pi i}\oint_{C(z_0;\varepsilon)} d\zeta\, M(\zeta)^{-1}M'(\zeta)\bigg), \quad 0<\varepsilon<\varepsilon_0.\end{aligned} \tag{19}$$

We note that this notion of an index is a bit more general than the one employed in Chapter 4 in [15] and [16] and hence it is not *a priori* clear if the right-hand side of (19) is an integer. However, in the special case depicted in Theorem 4.4 (ii) (see also (22)) under the additional Hypothesis 4.3, and in the applications in the following sections, the index indeed turns out to be an integer.

We also note that in the special case of an analytic function $M:\Omega\to\mathcal{B}(\mathcal{H})$ and z_0 a zero of finite-type of $M(\cdot)$, it follows from Theorem 2.4 that $M(\cdot)^{-1}$ is finitely meromorphic on $D(z_0;\varepsilon_0)$ for some $0<\varepsilon_0$ sufficiently small. Therefore, (9) implies that the index of $M(\cdot)$ in (19) coincides with the algebraic multiplicity of the zero of $M(\cdot)$ at z_0,

$$\mathrm{ind}_{C(z_0;\varepsilon)}(M(\cdot)) = m_a(z_0;M(\cdot)). \tag{20}$$

Moreover, if $M_j(\cdot)$, $j=1,2$, are $\mathcal{B}(\mathcal{H})$-valued finitely meromorphic functions that are boundedly invertible on $D(z_0;\varepsilon_0)\backslash\{z_0\}$ for some $z_0\in\Omega$ and $0<\varepsilon_0$ sufficiently small, and $M_j(\cdot)^{-1}$, $j=1,2$, are finitely meromorphic on $D(z_0;\varepsilon_0)\backslash\{z_0\}$, then employing the identity

$$\begin{aligned}&[M_1(z)M_2(z)]'[M_1(z)M_2(z)]^{-1} \\ &\quad = M_1'(z)M_1(z)^{-1} + M_1(z)[M_2'(z)M_2(z)^{-1}]M_1(z)^{-1},\end{aligned}$$

and taking the trace on either side yields the familiar formula

$$\mathrm{ind}_{C(z_0;\varepsilon)}(M_1(\cdot)M_2(\cdot)) = \mathrm{ind}_{C(z_0;\varepsilon)}(M_1(\cdot)) + \mathrm{ind}_{C(z_0;\varepsilon)}(M_2(\cdot)),$$

in particular,

$$\mathrm{ind}_{C(z_0;\varepsilon)}(M(\cdot)^{-1}) = -\mathrm{ind}_{C(z_0;\varepsilon)}(M(\cdot)).$$

For interesting applications of this circle of ideas see also [1], [3], [6], and [37].

Next we strengthen Hypothesis 4.1 as follows:

Hypothesis 4.3. Suppose $M(\cdot)$ satisfies Hypothesis 4.1 and assume that for every $z_0 \in \mathcal{D}_0$ the operator $M_0(z_0)$ in the Laurent series in (18) is a Fredholm operator, that is,

$$M_0(z_0) \in \Phi(\mathcal{H}), \quad z_0 \in \mathcal{D}_0.$$

In addition, suppose that

$$M(z) \in \Phi(\mathcal{H}), \quad z \in \Omega \backslash \mathcal{D}_0.$$

One then recalls the meromorphic Fredholm theorem in the following form:

Theorem 4.4 ([16], [17], Theorem XIII.13 in [33], [34], and [40]). *Assume that $M(\cdot)$ satisfies Hypothesis 4.3. Then either*

(i) *$M(z)$ is not boundedly invertible for any $z \in \Omega \backslash \mathcal{D}_0$,*

or else,

(ii) *$M(\cdot)^{-1}$ is finitely meromorphic on Ω. More precisely, there exists a discrete subset $\mathcal{D}_1 \subset \Omega$ (possibly, $\mathcal{D}_1 = \emptyset$) such that $M(z)^{-1} \in \mathcal{B}(\mathcal{H})$ for all $z \in \Omega \backslash \{\mathcal{D}_0 \cup \mathcal{D}_1\}$, $M(\cdot)^{-1}$ extends to an analytic function on $\Omega \backslash \mathcal{D}_1$, meromorphic on Ω such that*

$$M(z)^{-1} \in \Phi(\mathcal{H}) \quad \text{for all } z \in \Omega \backslash \mathcal{D}_1,$$

and if $z_1 \in \mathcal{D}_1$, then

$$M(z)^{-1} = \sum_{k=-N_0}^{\infty} (z - z_1)^k D_k(z_1), \quad 0 < |z - z_1| < \varepsilon_0,$$

for some $N_0 = N_0(z_1) \in \mathbb{N}$ and some $0 < \varepsilon_0 = \varepsilon_0(z_1)$ sufficiently small, with

$$D_k(z_1) \in \mathcal{F}(\mathcal{H}), \ -N_0 \le k \le -1, \quad D_k(z_1) \in \mathcal{B}(\mathcal{H}), \ k \in \mathbb{N}_0.$$

In addition,

$$D_0(z_1) \in \Phi(\mathcal{H}). \tag{21}$$

Finally, if $[I_{\mathcal{H}} - M(z)] \in \mathcal{B}_\infty(\mathcal{H})$ for all $z \in \Omega \backslash \mathcal{D}_0$, then

$$[I_{\mathcal{H}} - M(z)^{-1}] \in \mathcal{B}_\infty(\mathcal{H}), \quad z \in \Omega \backslash \mathcal{D}_1,$$

$$[I_{\mathcal{H}} - D_0(z_1)] \in \mathcal{B}_\infty(\mathcal{H}), \quad z_1 \in \mathcal{D}_1.$$

Assume Hypothesis 4.3, let $z_0 \in \Omega$ and suppose that $M(\cdot)$ is boundedly invertible on $D(z_0;\varepsilon_0)\backslash\{z_0\}$ for some $0 < \varepsilon_0$ sufficiently small (i.e., case (ii) in Theorem 4.4 applies). Then the function $M(\cdot)^{-1}$ is finitely meromorphic on $D(z_0;\varepsilon_0)$ and it follows from the operator-valued version of the argument principle proved in [16] (see also Theorem 4.4.1 in [15]) that

$$\operatorname{ind}_{C(z_0;\varepsilon)}(M(\cdot)) \in \mathbb{Z}. \tag{22}$$

5 Abstract perturbation theory and applications to Birman–Schwinger-type operators

In this section, following Kato [19], Konno and Kuroda [23], and Howland [17], we first recall a class of factorable non-self-adjoint perturbations of a given unperturbed non-self-adjoint operator. We recall the treatment in [12] (in which H_0 is explicitly permitted to be non-self-adjoint, cf. Hypothesis 5.1 (i) below) and refer to the latter for detailed proofs.

The principal result of this section then consists of the index formulas in Theorem 5.5, which are variants of Theorems 4.5 and 5.5 in [11]. We start with the following set of hypotheses.

Hypothesis 5.1. (i) Suppose that $H_0\colon \operatorname{dom}(H_0) \to \mathcal{H}$, $\operatorname{dom}(H_0) \subseteq \mathcal{H}$, is a densely defined, closed, linear operator in $\mathcal{H}$, with nonempty resolvent set,

$$\rho(H_0) \neq \emptyset,$$

$V_1\colon \operatorname{dom}(V_1) \to \mathcal{K}$, $\operatorname{dom}(V_1) \subseteq \mathcal{H}$, a densely defined, closed, linear operator from $\mathcal{H}$ to $\mathcal{K}$, and $V_2\colon \operatorname{dom}(V_2) \to \mathcal{K}$, $\operatorname{dom}(V_2) \subseteq \mathcal{H}$, a densely defined, closed, linear operator from $\mathcal{H}$ to $\mathcal{K}$ such that

$$\operatorname{dom}(V_1) \supseteq \operatorname{dom}(H_0), \quad \operatorname{dom}(V_2) \supseteq \operatorname{dom}(H_0^*).$$

In the following we denote

$$R_0(z) = (H_0 - zI_{\mathcal{H}})^{-1}, \quad z \in \rho(H_0).$$

(ii) For some (and hence for all) $z \in \rho(H_0)$, the operator $-V_1 R_0(z) V_2^*$, defined on $\operatorname{dom}(V_2^*)$, has a bounded extension in $\mathcal{K}$, denoted by $K(z)$,

$$K(z) = -\overline{V_1 R_0(z) V_2^*} \in \mathcal{B}(\mathcal{K}). \tag{23}$$

(iii) $1 \in \rho(K(\zeta_0))$ for some $\zeta_0 \in \rho(H_0)$.

Next, following Kato [19], one introduces

$$R(z) = R_0(z) - \overline{R_0(z)V_2^*}[I_{\mathcal{K}} - K(z)]^{-1} V_1 R_0(z) \tag{24}$$

for $z \in \{\zeta \in \rho(H_0) \mid 1 \in \rho(K(\zeta))\}$.

Theorem 5.2 ([19]). *Assume Hypothesis* 5.1 *and* $z \in \{\zeta \in \rho(H_0) \mid 1 \in \rho(K(\zeta))\}$. *Then,* $R(z)$ *in* (24) *defines a densely defined, closed, linear operator* H *in* $\mathcal{H}$ *by*

$$R(z) = (H - zI_{\mathcal{H}})^{-1}.$$

Moreover,

$$V_1 R(z), V_2 R(z)^* \in \mathcal{B}(\mathcal{H}, \mathcal{K})$$

and

$$\begin{aligned} R(z) &= R_0(z) - \overline{R(z)V_2^*} V_1 R_0(z) \\ &= R_0(z) - \overline{R_0(z)V_2^*} V_1 R(z). \end{aligned}$$

Finally, H *is an extension of the operator*

$$(H_0 + V_2^* V_1) \restriction \big(\mathrm{dom}(H_0) \cap \mathrm{dom}(V_2^* V_1)\big),$$

where the set $\mathrm{dom}(H_0) \cap \mathrm{dom}(V_2^* V_1)$ *may consist of* $\{0\}$ *only.*

Similarly, using the symmetry between H_0 and H inherent in Kato's formalism (cf. Sections 2 and 3 of [12]) one also derives

$$\overline{V_1 R(z) V_2^*} \in \mathcal{B}(\mathcal{K}), \quad z \in \rho(H),$$

and

$$I_{\mathcal{K}} - \overline{V_1 R(z) V_2^*} = [I_{\mathcal{K}} - K(z)]^{-1}, \quad z \in \{\zeta \in \rho(H_0) \mid 1 \in \rho(K(\zeta))\}. \tag{25}$$

For our purposes the following lemma is useful.

Lemma 5.3. *Assume Hypothesis* 5.1 *and let* $z_1, z_2 \in \rho(H_0)$. *Then*

$$K(z_1) = K(z_2) + (z_2 - z_1) V_1 R_0(z_1) \overline{R_0(z_2) V_2^*} \tag{26}$$

and if, in addition, $z_1, z_2 \in \rho(H)$ *then*

$$[I_{\mathcal{K}} - K(z_1)]^{-1} = [I_{\mathcal{K}} - K(z_2)]^{-1} + (z_2 - z_1) V_1 R(z_1) \overline{R(z_2) V_2^*}. \tag{27}$$

Proof. Formula (26) follows from (23) and the resolvent equation for $R_0(z)$, $z \in \rho(H_0)$; similarly, formula (27) is clear from (25) and the resolvent equation for $R(z)$, $z \in \rho(H)$. □

Note also that (26) yields the useful formula

$$K'(z) = -V_1 R_0(z)\overline{R_0(z)V_2^*}, \quad z \in \rho(H_0). \tag{28}$$

The next result represents an abstract version of the Birman–Schwinger principle due to Birman [4] and Schwinger [35] (cf. also [5], [10], [21], [22], [30], [31], [36], Chapter III of [38], and [39]). It is due to Konno and Kuroda [23] in the case where H_0 is self-adjoint. For the general case see [12].

Theorem 5.4 ([23]). *Assume Hypothesis* 5.1 *and let* $z_0 \in \rho(H_0)$. *Then,*

$$z_0 \in \sigma_p(H) \iff 1 \in \sigma_p(K(z_0)), \tag{29}$$

and

$$z_0 \in \rho(H) \iff 1 \in \rho(K(z_0)).$$

More precisely, if in (29) *one has* $Hf = z_0 f$ *for some* $f \in \operatorname{dom}(H)$, $f \neq 0$, *then*

$$0 \neq g = [I_{\mathcal{K}} - K(z_1)]^{-1} V_1 R_0(z_1) f = (z_0 - z_1)^{-1} V_1 f,$$

where $z_1 \in \{\zeta \in \rho(H_0) \,|\, 1 \in \rho(K(\zeta))\}$, $z_1 \neq z_0$, *satisfies* $K(z_0)g = g$, *and conversely, if in* (29) *one has* $K(z_0)g = g$ *for some* $g \in \mathcal{K}$, $g \neq 0$, *then*

$$0 \neq f = -\overline{R_0(z_0)V_2^*} g \in \operatorname{dom}(H)$$

satisfies $Hf = z_0 f$.

If, in addition to Hypothesis 5.1, it is assumed that $I_{\mathcal{K}} - K(z)$ is a Fredholm operator for all $z \in \rho(H_0)$, then by Theorem 2.7 in [11] (see also Theorem 3.2 in [12]) the geometric multiplicity of an eigenvalue z_0 of H coincides with the geometric multiplicity of the eigenvalue 1 of $K(z_0)$ and is finite,

$$\begin{aligned} m_g(z_0; H) &= \dim(\ker(H - z_0 I_{\mathcal{H}})) \\ &= \dim(\ker(I_{\mathcal{K}} - K(z_0))) \\ &= m_g(1; K(z_0)) \\ &< \infty. \end{aligned}$$

The next theorem is the main result in this section. Item (i) is a slight extension (cf. [11]) of a multiplicity result due to Latushkin and Sukhtyaev [25], and item (ii) resembles an analog of the Weinstein–Aronszajn-type formula (cf., e.g., [2], [17], Section IV.6 of [20], [24], and Section 9.3 of [41]) in the case where H and H_0 have common discrete eigenvalues.

Theorem 5.5. *Assume Hypothesis* 5.1. *Then the following assertions* (i)–(iv) *hold.*

(i) *If* $z_0 \in \rho(H_0) \cap \sigma_d(H)$, *then the index formula*

$$\operatorname{ind}_{C(z_0;\varepsilon)}(I_{\mathcal{K}} - K(\cdot)) = m_a(z_0; H) \tag{30}$$

holds for $\varepsilon > 0$ *sufficiently small. Furthermore,* z_0 *is a zero of finite-type of the function* $I_{\mathcal{K}} - K(\cdot)$, *and hence*

$$\nu(z_0; I_{\mathcal{K}} - K(\cdot)) = m_a(z_0; I_{\mathcal{K}} - K(\cdot)) = \operatorname{ind}_{C(z_0;\varepsilon)}(I_{\mathcal{K}} - K(\cdot)). \tag{31}$$

(ii) *If* $z_0 \in \sigma_d(H_0) \cap \sigma_d(H)$, *then the index formula*

$$\operatorname{ind}_{C(z_0;\varepsilon)}(I_{\mathcal{K}} - K(\cdot)) = m_a(z_0; H) - m_a(z_0; H_0) \tag{32}$$

holds for $\varepsilon > 0$ *sufficiently small.*

(iii) *Assume in addition that* $K(z) \in \mathcal{B}_\infty(\mathcal{K})$ *for all* $z \in \rho(H_0)$ *and either that* $\rho(H_0)$ *is connected, or else, that Hypothesis* 5.1 (iii), *that is,* $1 \in \rho(K(\zeta))$, *holds for some* $\zeta \in \mathbb{C}$ *lying in each of the connected components of* $\rho(H_0)$. *If* $z_0 \in \sigma_d(H_0)$, *then* $z_0 \in (\sigma_d(H) \cup \rho(H))$ *and hence the index formula* (32) *holds.*

(iv) *Assume in addition that* $K(z) \in \mathcal{B}_\infty(\mathcal{K})$ *for all* $z \in \rho(H_0)$ *and suppose that* $D(z_0; \varepsilon_0)\backslash\{z_0\} \cap \sigma(H) = \emptyset$ *for* $0 < \varepsilon_0$ *sufficiently small. If* $z_0 \in \sigma_d(H_0)$, *then* $z_0 \in (\sigma_d(H) \cup \rho(H))$ *and hence the index formula* (32) *holds.*

Proof. Observe first that by the assumptions in (i) and (ii) there exists $\varepsilon_0 > 0$ such that the punctured disc $D(z_0; \varepsilon_0)\backslash\{z_0\}$ is contained in $\rho(H) \cap \rho(H_0)$. Fix a point $z_2 \in \rho(H) \cap \rho(H_0)$ and recall from Lemma 5.3 (i) that

$$K(z) = K(z_2) + (z_2 - z)V_1(H_0 - zI_{\mathcal{H}})^{-1}\overline{R_0(z_2)V_2^*} \tag{33}$$

holds for all $z \in D(z_0; \varepsilon_0)$ if $z_0 \in \rho(H_0)$ and for all $z \in D(z_0; \varepsilon_0)\backslash\{z_0\}$ if $z_0 \in \sigma_d(H_0)$. Therefore, since $(H_0 - zI_{\mathcal{H}})^{-1}$ is analytic on $D(z_0; \varepsilon_0)$ if $z_0 \in \rho(H_0)$ and finitely meromorphic if $z_0 \in \sigma_d(H_0)$ – see, e.g., Chapter 1, §2, Theorem 2.1 and (2.3) in [14] or [20] – it follows from (33) and $V_1(H_0 - zI_{\mathcal{H}})^{-1} \in \mathcal{B}(\mathcal{H}, \mathcal{K})$

that the same is true for the functions $K(\cdot)$ and $I_{\mathcal{K}} - K(\cdot)$. The same argument using the resolvent of H and formula (27) in Lemma 5.3 shows that the function $[I_{\mathcal{K}} - K(\cdot)]^{-1}$ is analytic on the punctured disc $D(z_0; \varepsilon_0)\backslash\{z_0\}$ and finitely meromorphic on $D(z_0; \varepsilon_0)$. Hence the index of $I_{\mathcal{K}} - K(\cdot)$ with respect to the counterclockwise oriented circle $C(z_0, \varepsilon)$, $0 < \varepsilon < \varepsilon_0$, is well-defined and we compute with the help of (28), the cyclicity of the trace, and (24)

$$\begin{aligned}
\operatorname{ind}_{C(z_0;\varepsilon)}(I_{\mathcal{K}} - K(\cdot)) &= \operatorname{tr}_{\mathcal{K}}\bigg(\frac{1}{2\pi i}\oint_{C(z_0;\varepsilon)} d\zeta\, [I_{\mathcal{K}} - K(\zeta)]^{-1}(-K'(\zeta))\bigg) \\
&= \frac{1}{2\pi i}\operatorname{tr}_{\mathcal{K}}\bigg(\oint_{C(z_0;\varepsilon)} d\zeta [I_{\mathcal{K}} - K(\zeta)]^{-1} V_1 R_0(\zeta)\overline{R_0(\zeta) V_2^*}\bigg) \\
&= \frac{1}{2\pi i}\operatorname{tr}_{\mathcal{H}}\bigg(\oint_{C(z_0;\varepsilon)} d\zeta\, \overline{R_0(\zeta) V_2^*}[I_{\mathcal{K}} - K(\zeta)]^{-1} V_1 R_0(\zeta)\bigg) \\
&= \operatorname{tr}_{\mathcal{H}}\bigg(\frac{-1}{2\pi i}\oint_{C(z_0;\varepsilon)} d\zeta\, [R(\zeta) - R_0(\zeta)]\bigg) \\
&= m_a(z_0; H) - m_a(z_0; H_0),
\end{aligned}$$

where the third equality is justified in a manner analogous to (2) (cf. Proposition 4.2.2 in [15]), and the last equality holds if $z_0 \in \sigma_d(H_0)$. This proves the index formula (32). Clearly, if $z_0 \in \rho(H_0)$ then

$$\operatorname{tr}_{\mathcal{H}}\bigg(\frac{-1}{2\pi i}\oint_{C(z_0;\varepsilon)} d\zeta\, R_0(\zeta)\bigg) = 0$$

and hence the term $m_a(z_0; H_0)$ is absent in the above computation; this implies the index formula (30).

Next, we show that $z_0 \in \rho(H_0) \cap \sigma_d(H)$ is a zero of finite-type of $I_{\mathcal{K}} - K(\cdot)$; the first equality in (31) then follows from Theorem 3.3 and the second equality is clear by (20). In order to see that z_0 is a zero of finite-type recall that $I_{\mathcal{K}} - K(z)$ is boundedly invertible for all $z \in D(z_0; \varepsilon_0)\backslash\{z_0\}$ and that $z_0 \in \sigma_d(H)$ and Theorem 5.4 imply $\dim(\ker(I_{\mathcal{K}} - K(z_0))) < \infty$. Moreover, the function $[I_{\mathcal{K}} - K(\cdot)]^{-1}$ is finitely meromorphic on $D(z_0; \varepsilon_0)$ and it follows from the particular form of the Laurent series of $(H - zI_{\mathcal{H}})^{-1}$ in a neighborhood of $z_0 \in \sigma_d(H)$ (see, e.g., Chapter 1, §2, Theorem 2.1 and (2.3) in [14]) and Lemma 5.3 (ii) that the zero order coefficient of the Laurent series of $[I_{\mathcal{K}} - K(\cdot)]^{-1}$ in a neighborhood of z_0 is a Fredholm operator, that is, Hypothesis 4.3 is satisfied for $[I_{\mathcal{K}} - K(\cdot)]^{-1}$. Hence, Theorem 4.4 applies to the function $[I_{\mathcal{K}} - K(\cdot)]^{-1}$ and from (21) we obtain that $I_{\mathcal{K}} - K(z_0)$ is a Fredholm operator. Summing up we have shown that z_0 is a zero of finite-type of $I_{\mathcal{K}} - K(\cdot)$.

We turn to a discussion of item (iii). If $z_0 \in \rho(H)$, no proof is required and the index formula (5.23) takes the form

$$\operatorname{ind}_{C(z_0;\varepsilon)}(I_{\mathcal{K}} - K(\cdot)) = -m_a(z_0; H_0).$$

So we focus on $z_0 \in \sigma(H)$. From the outset it is clear that for $0 < \varepsilon_0$ sufficiently small, $\overline{R_0(z)V_2^*}$, $V_1 R_0(z)$, and $K(z)$ are analytic on $D(z_0;\varepsilon_0)\backslash\{z_0\}$, and $K(z)$ is finitely meromorphic on $D(z_0;\varepsilon_0)$. In particular, $K(z)$, $z \in D(z_0;\varepsilon_0)\backslash\{z_0\}$, is of the form,

$$K(z) = \sum_{k=-N_0}^{\infty} (z-z_0)^k K_k(z_0), \quad 0 < |z-z_0| < \varepsilon_0,$$

for some $N_0 \in \mathbb{N}$, with $K_k(z_0) \in \mathcal{F}(\mathcal{K})$, $-N_0 \le k \le -1$, $K_k(z_0) \in \mathcal{B}(\mathcal{K})$, $k \in \mathbb{N}_0$. Hence,

$$\Big[\sum_{k=0}^{\infty}(z-z_0)^k K_k(z_0)\Big] \in \mathcal{B}_\infty(\mathcal{K}), \quad 0 < |z-z_0| < \varepsilon_0,$$

implying that the norm limit,

$$K_0(z_0) = \lim_{z\to z_0}\Big[\sum_{k=0}^{\infty}(z-z_0)^k K_k(z_0)\Big] \in \mathcal{B}_\infty(\mathcal{K}),$$

exists and is compact. In particular, this implies

$$[I_{\mathcal{K}} - K_0(z_0)] \in \Phi(\mathcal{K}). \tag{34}$$

If $\rho(H_0)$ is connected then Hypothesis 5.1 (iii) and $K(z) \in \mathcal{B}_\infty(\mathcal{K})$, $z \in \rho(H_0)$, imply that $I_{\mathcal{K}} - K(z)$ is boundedly invertible for some $z \in D(z_0;\varepsilon_0)\backslash\{z_0\}$. If $\rho(H_0)$ is not connected then the assumption $1 \in \rho(K(\zeta))$ for some $\zeta \in \mathbb{C}$ in each of the connected components of $\rho(H_0)$ implies in the same way that $I_{\mathcal{K}} - K(z)$ is boundedly invertible for some $z \in D(z_0;\varepsilon_0)\backslash\{z_0\}$. Consequently, Theorems 2.4 (ii), respectively, 4.4 (ii), apply, and hence $[I_{\mathcal{K}} - K(z)]^{-1}$ is analytic on $D(z_0;\varepsilon_0)\backslash\{z_0\}$, respectively, finitely meromorphic on $D(z_0;\varepsilon_0)$ (possibly, upon further diminishing $\varepsilon_0 > 0$). By (24), then also $R(z)$ is analytic on $D(z_0;\varepsilon_0)\backslash\{z_0\}$ and finitely meromorphic on $D(z_0;\varepsilon_0)$, implying $z_0 \in \sigma_d(H)$.

Finally, we briefly turn to item (iv) again assuming $z_0 \in \sigma(H)$ without loss of generality. By (25), the condition $(D(z_0;\varepsilon_0) \setminus \{z_0\}) \cap \sigma(H) = \emptyset$ guarantees the bounded invertibility of $I_{\mathcal{K}} - K(z)$ for $z \in D(z_0;\varepsilon_0)\backslash\{z_0\}$ and one can now basically follow the proof of item (iii); we omit the details. □

Remark 5.6. In connection with Theorem 5.5 (iii), one notes that since the resolvent set $\rho(H_0) \subset \mathbb{C}$ is open, its connected components are open and at most countable (see, e.g., Theorem II.2.9 in [9]). In particular, in the important special case where $\sigma(H_0) \subseteq \mathbb{R}$, there are at most two components and in quantum mechanical applications associated with short-range potentials one frequently encounters that

$$\lim_{y \to \pm\infty} \|K(iy)\|_{\mathcal{B}(\mathcal{K})} = 0,$$

and hence the condition $1 \in \rho(K(\zeta))$ is obviously satisfied for $\zeta = iy$ with $0 < |y|$ sufficiently large.

In this context we note that condition (34), that is, $[I_{\mathcal{K}} - K_0(z_0)] \in \Phi(\mathcal{K})$, was inadvertently omitted in Theorem 5.5 in [11] and hence needs to be added to its hypotheses.

6 An index formula for the Weyl–Titchmarsh function associated to closed extensions of dual pairs

In this section we derive the index associated with the Weyl–Titchmarsh function associated to closed extensions of dual pairs of operators.

Let $\mathcal{K}$ be a separable, complex Hilbert space with scalar product $(\cdot, \cdot)_{\mathcal{K}}$, and let A and B be densely defined, closed, linear operators in $\mathcal{K}$ such that

$$(Bf, g)_{\mathcal{K}} = (f, Ag)_{\mathcal{K}}, \quad f \in \operatorname{dom}(B),\ g \in \operatorname{dom}(A). \tag{35}$$

A pair of operators $\{A, B\}$ that satisfies (35) is called a *dual pair*. It follows immediately from (35) that

$$A \subset B^* \quad \text{and} \quad B \subset A^*.$$

We recall the notion of a boundary triple for a dual pair from [26] (see also [27] and [28]).

Definition 6.1. Let $\{A, B\}$ be a dual pair of operators in $\mathcal{K}$. A triple $\{\mathcal{H}, \Gamma^B, \Gamma^A\}$, where $\mathcal{H} = \mathcal{H}_0 \oplus \mathcal{H}_1$ is a Hilbert space and

$$\Gamma^B = (\Gamma_0^B, \Gamma_1^B)^\top : \operatorname{dom}(B^*) \longrightarrow \mathcal{H}_0 \oplus \mathcal{H}_1 \tag{36}$$

and

$$\Gamma^A = (\Gamma_0^A, \Gamma_1^A)^\top : \operatorname{dom}(A^*) \longrightarrow \mathcal{H}_1 \oplus \mathcal{H}_0, \tag{37}$$

are linear mappings, is called a *boundary triple* for the dual pair $\{A, B\}$ if the following two conditions hold.

(i) For all $f \in \operatorname{dom}(B^*)$ and $g \in \operatorname{dom}(A^*)$, the following abstract Green's identity holds:

$$(B^* f, g)_{\mathcal{K}} - (f, A^* g)_{\mathcal{K}} = (\Gamma_1^B f, \Gamma_0^A g)_{\mathcal{H}_1} - (\Gamma_0^B f, \Gamma_1^A g)_{\mathcal{H}_0}. \tag{38}$$

(ii) The mappings Γ^B and Γ^A in (36) and (37) are both onto.

Next, assume that $\{A, B\}$ is a dual pair of operators in $\mathcal{K}$ and that $\{\mathcal{H}, \Gamma^B, \Gamma^A\}$, $\mathcal{H} = \mathcal{H}_0 \oplus \mathcal{H}_1$, is a boundary triple for $\{A, B\}$. Then one has

$$A = B^* \restriction \ker(\Gamma^B) \quad \text{and} \quad B = A^* \restriction \ker(\Gamma^A),$$

and the mappings in (36) and (37) are continuous when $\operatorname{dom}(B^*)$ and $\operatorname{dom}(A^*)$ are equipped with the graph norm. Moreover, the closed operators

$$A_0 = B^* \restriction \ker(\Gamma_0^B) \quad \text{and} \quad A_1 = B^* \restriction \ker(\Gamma_1^B),$$

$$B_0 = A^* \restriction \ker(\Gamma_0^A) \quad \text{and} \quad B_1 = A^* \restriction \ker(\Gamma_1^A),$$

satisfy $B_0 = A_0^*$ and $B_1 = A_1^*$, and

$$A \subset A_0,\ A_1 \subset B^* \quad \text{and} \quad B \subset B_0,\ B_1 \subset A^*.$$

More generally, with the help of a boundary triple for the dual pair $\{A, B\}$ one can describe all closed extensions A_Θ of A that are restrictions of B^*, that is,

$$A \subset A_\Theta \subset B^*$$

with the help of closed linear subspaces Θ in $\mathcal{H}_0 \times \mathcal{H}_1$. We refer the reader to [26] and [27] for more details and concentrate on the special case of extensions of A of the form

$$A_\Theta = B^* \restriction \ker(\Gamma_1^B - \Theta\Gamma_0^B), \tag{39}$$

where we assume that $\Theta \in \mathcal{B}(\mathcal{H}_0, \mathcal{H}_1)$ is a bounded operator from $\mathcal{H}_0$ into $\mathcal{H}_1$.

In order to state our main result in this context some more definitions are necessary. First, we recall the notion of *γ-field* and *Weyl–Titchmarsh function* associated to a boundary triple for a dual pair treated in [26] and [27]. Suppose that $\rho(A_0) \neq \emptyset$, $\rho(B_0) \neq \emptyset$, and observe that the direct sum decompositions

$$\operatorname{dom}(B^*) = \operatorname{dom}(A_0) \dotplus \ker(B^* - zI_{\mathcal{K}}), \quad z \in \rho(A_0), \tag{40}$$

and

$$\operatorname{dom}(A^*) = \operatorname{dom}(B_0) \dotplus \ker(A^* - z' I_{\mathcal{K}}), \quad z' \in \rho(B_0), \tag{41}$$

hold. Since

$$\operatorname{dom}(A_0) = \ker(\Gamma_0^B), \quad \operatorname{dom}(B_0) = \ker(\Gamma_0^A),$$

it follows from (40) that the mapping Γ_0^B is invertible on $\ker(B^* - zI_{\mathcal{K}})$, and it follows from (41) that the mapping Γ_0^A is invertible on $\ker(A^* - z'I_{\mathcal{K}})$.

Definition 6.2. Let $\{A, B\}$ be a dual pair of operators in $\mathcal{K}$ and let $\{\mathcal{H}, \Gamma^B, \Gamma^A\}$ be a boundary triple. The *γ-fields* $\gamma(\cdot)$ and $\gamma_*(\cdot)$ associated to $\{\mathcal{H}, \Gamma^B, \Gamma^A\}$ are defined by

$$\gamma(z) = (\Gamma_0^B \upharpoonright \ker(B^* - zI_{\mathcal{K}}))^{-1}, \qquad z \in \rho(A_0),$$

and

$$\gamma_*(z') = (\Gamma_0^A \upharpoonright \ker(A^* - z'I_{\mathcal{K}}))^{-1}, \qquad z' \in \rho(B_0),$$

respectively. The *Weyl–Titchmarsh function* $M(\cdot)$ associated to $\{\mathcal{H}, \Gamma^B, \Gamma^A\}$ is defined by

$$M(z) = \Gamma_1^B(\Gamma_0^B \upharpoonright \ker(B^* - zI_{\mathcal{K}}))^{-1}, \quad z \in \rho(A_0).$$

It is important to note that the γ-field satisfies

$$\gamma(z_1) = (I_{\mathcal{K}} + (z_1 - z_2)(A_0 - z_1 I_{\mathcal{K}})^{-1})\gamma(z_2), \quad z_j \in \rho(A_0),\ j = 1, 2. \tag{42}$$

Moreover, the values $M(z)$ of the Weyl–Titchmarsh function are bounded operators from $\mathcal{H}_0$ to $\mathcal{H}_1$,

$$M(z) \in \mathcal{B}(\mathcal{H}_0, \mathcal{H}_1), \quad z \in \rho(A_0),$$

and the Weyl–Titchmarsh function and the γ-fields are related via

$$M(z_1) - M(z_2) = (z_1 - z_2)\gamma_*(\overline{z_2})^*\gamma(z_1), \quad z_j \in \rho(A_0),\ j = 1, 2. \tag{43}$$

We shall assume from now on that $\{A, B\}$ is a dual pair and $\{\mathcal{H}, \Gamma^B, \Gamma^A\}$ is a boundary triple with the additional property $\mathcal{H}_0 = \mathcal{H}_1$, which can be viewed as a non-symmetric analog of the case of equal deficiency indices of an underlying symmetric operator. Consider a closed extension A_Θ of A as in (39) with $\Theta \in \mathcal{B}(\mathcal{H}_0)$, and assume that $z \in \rho(A_0)$. Then by Proposition 5.2 in [26] one has

$$z \in \sigma_p(A_\Theta) \iff 0 \in \sigma_p(\Theta - M(z)),$$

and

$$z \in \rho(A_\Theta) \iff 0 \in \rho(\Theta - M(z)).$$

Moreover, for all $z \in \rho(A_0) \cap \rho(A_\Theta)$, the following Krein-type resolvent formula holds,

$$(A_\Theta - zI_{\mathcal{K}})^{-1} = (A_0 - zI_{\mathcal{K}})^{-1} + \gamma(z)[\Theta - M(z)]^{-1}\gamma_*(\bar{z})^*. \tag{44}$$

The next lemma will be useful in the proof of our main result Theorem 6.4 below (cf. Corollary 4.9 in [26]). For the convenience of the reader we provide a simple direct proof in the present situation.

Lemma 6.3. *Let $\{A, B\}$ be a dual pair of operators in $\mathcal{K}$, let $\{\mathcal{H}, \Gamma^B, \Gamma^A\}$ be a boundary triple with $A_0 = B^* \restriction \ker(\Gamma_0^B)$ and Weyl–Titchmarsh function $M(\cdot)$, and assume that $\mathcal{H}_0 = \mathcal{H}_1$. Suppose that $\Theta \in \mathcal{B}(\mathcal{H}_0)$ and let A_Θ be defined as in* (39). *Then $\{\mathcal{H}, \Gamma^{B,\Theta}, \Gamma^{A,\Theta}\}$, where*

$$\Gamma^{B,\Theta} = \begin{pmatrix} \Gamma_0^{B,\Theta} \\ \Gamma_1^{B,\Theta} \end{pmatrix}, \quad \Gamma_0^{B,\Theta} = \Gamma_1^B - \Theta\Gamma_0^B, \quad \Gamma_1^{B,\Theta} = -\Gamma_0^B,$$

and

$$\Gamma^{A,\Theta} = \begin{pmatrix} \Gamma_0^{A,\Theta} \\ \Gamma_1^{A,\Theta} \end{pmatrix}, \quad \Gamma_0^{A,\Theta} = \Gamma_1^A - \Theta^*\Gamma_0^A, \quad \Gamma_1^{A,\Theta} = -\Gamma_0^A,$$

is a boundary triple for the dual pair $\{A, B\}$ with $A_\Theta = B^* \restriction \ker(\Gamma_0^{B,\Theta})$. The corresponding Weyl–Titchmarsh function $M_\Theta(\cdot)$ is given by

$$M_\Theta(z) = (\Theta - M(z))^{-1}, \quad z \in \rho(A_\Theta) \cap \rho(A_0). \tag{45}$$

Proof. Let $f \in \operatorname{dom}(B^*)$ and $g \in \operatorname{dom}(A^*)$. Then it follows with the help of the abstract Green's identity (38) for the boundary triple $\{\mathcal{H}, \Gamma^B, \Gamma^A\}$ that

$$\begin{aligned}
&(\Gamma_1^{B,\Theta} f, \Gamma_0^{A,\Theta} g)_{\mathcal{H}_0} - (\Gamma_0^{B,\Theta} f, \Gamma_1^{A,\Theta} g)_{\mathcal{H}_0} \\
&\quad = (-\Gamma_0^B f, \Gamma_1^A g - \Theta^*\Gamma_0^A g)_{\mathcal{H}_0} - (\Gamma_1^B f - \Theta\Gamma_0^B f, -\Gamma_0^A g)_{\mathcal{H}_0} \\
&\quad = (\Gamma_1^B f, \Gamma_0^A g)_{\mathcal{H}_0} - (\Gamma_0^B f, \Gamma_1^A g)_{\mathcal{H}_0} \\
&\quad = (B^* f, g)_{\mathcal{K}} - (f, A^* g)_{\mathcal{K}},
\end{aligned}$$

and hence the triple $\{\mathcal{H}, \Gamma^{B,\Theta}, \Gamma^{A,\Theta}\}$ satisfies the abstract Green's identity in Definition 6.1 (i). Moreover, as

$$\begin{pmatrix} \Gamma_0^{B,\Theta} \\ \Gamma_1^{B,\Theta} \end{pmatrix} = W_B^\Theta \begin{pmatrix} \Gamma_0^B \\ \Gamma_1^B \end{pmatrix}, \quad W_B^\Theta = \begin{pmatrix} -\Theta & I_{\mathcal{H}_0} \\ -I_{\mathcal{H}_0} & 0 \end{pmatrix}, \tag{46}$$

and
$$\begin{pmatrix}\Gamma_0^{A,\Theta}\\ \Gamma_1^{A,\Theta}\end{pmatrix} = W_A^{\Theta^*}\begin{pmatrix}\Gamma_0^{A}\\ \Gamma_1^{A}\end{pmatrix}, \quad W_A^{\Theta^*} = \begin{pmatrix}-\Theta^* & I_{\mathcal{H}_0}\\ -I_{\mathcal{H}_0} & 0\end{pmatrix}, \tag{47}$$

and the 2×2 block operator matrices W_B^{Θ} and $W_A^{\Theta^*}$ in (46) and (47) are boundedly invertible, it follows that both mappings

$$\Gamma^{B,\Theta} = (\Gamma_0^{B,\Theta}, \Gamma_1^{B,\Theta})^\top : \operatorname{dom}(B^*) \longrightarrow \mathcal{H}_0 \oplus \mathcal{H}_0$$

and

$$\Gamma^{A,\Theta} = (\Gamma_0^{A,\Theta}, \Gamma_1^{A,\Theta})^\top : \operatorname{dom}(A^*) \longrightarrow \mathcal{H}_0 \oplus \mathcal{H}_0,$$

are onto. Hence also condition (ii) in Definition 6.1 holds for $\{\mathcal{H}, \Gamma^{B,\Theta}, \Gamma^{A,\Theta}\}$, and it follows that $\{\mathcal{H}, \Gamma^{B,\Theta}, \Gamma^{A,\Theta}\}$ is a boundary triple for the dual pair $\{A, B\}$. By construction, one has (cf. (39))

$$B^* \restriction \ker(\Gamma_0^{B,\Theta}) = B^* \restriction \ker(\Gamma_1^B - \Theta\Gamma_0^B) = A_\Theta.$$

Next, it will be verified that the Weyl–Titchmarsh function $M_\Theta(\cdot)$ corresponding to the boundary triple $\{\mathcal{H}, \Gamma^{B,\Theta}, \Gamma^{A,\Theta}\}$ has the form (45). Assume that $f_z \in \ker(B^* - zI_{\mathcal{K}})$ and that $z \in \rho(A_0) \cap \rho(A_\Theta)$. Since $M(\cdot)$ is the Weyl–Titchmarsh function of the boundary triple $\{\mathcal{H}, \Gamma^B, \Gamma^A\}$, one has $M(z)\Gamma_0^B f_z = \Gamma_1^B f_z$, and hence it follows that

$$\begin{aligned}[\Theta - M(z)]\Gamma_1^{B,\Theta} f_z &= -[\Theta - M(z)]\Gamma_0^B f_z \\ &= -\Theta\Gamma_0^B f_z + \Gamma_1^B f_z \\ &= \Gamma_0^{B,\Theta} f_z.\end{aligned} \tag{48}$$

From the direct sum decomposition

$$\begin{aligned}\operatorname{dom}(B^*) &= \operatorname{dom}(A_\Theta) \dotplus \ker(B^* - zI_{\mathcal{K}}) \\ &= \ker(\Gamma_0^{B,\Theta}) \dotplus \ker(B^* - zI_{\mathcal{K}}), \quad z \in \rho(A_\Theta),\end{aligned}$$

and the fact that $\Gamma_0^{B,\Theta}$ maps onto $\mathcal{H}_0$ one then concludes together with (48) that $[\Theta - M(z)]$ maps onto $\mathcal{H}_0$. Moreover, one has

$$\ker(\Theta - M(z)) = \{0\}. \tag{49}$$

In fact, if $\Theta\varphi = M(z)\varphi$ for some $\varphi \in \mathcal{H}_0$, then by (40) there exists an element $f_z \in \ker(B^* - zI_{\mathcal{K}})$ such that $\Gamma_0^B f_z = \varphi$. This leads to

$$\Theta\Gamma_0^B f_z = \Theta\varphi = M(z)\varphi = M(z)\Gamma_0^B f_z = \Gamma_1^B f_z,$$

and hence $f_z \in \operatorname{dom}(A_\Theta) \cap \ker(B^* - zI_{\mathcal{K}})$. Therefore, $f_z \in \ker(A_\Theta - zI_{\mathcal{K}})$, and as $z \in \rho(A_\Theta)$ by assumption, we conclude $f_z = 0$ and $\varphi = \Gamma_0^B f_z = 0$. This shows (49). Now it follows from (48) that

$$[\Theta - M(z)]^{-1}\Gamma_0^{B,\Theta} f_z = \Gamma_1^{B,\Theta} f_z$$

for all $f_z \in \ker(B^* - zI_{\mathcal{K}})$ and $z \in \rho(A_0) \cap \rho(A_\Theta)$. This finally implies that the Weyl–Titchmarsh function $M_\Theta(\cdot)$ has the form (45). □

The next theorem is the main result of this section. As in Lemma 6.3 we shall assume here that the boundary triple $\{\mathcal{H}, \Gamma^B, \Gamma^A\}$ has the additional property $\mathcal{H}_0 = \mathcal{H}_1$.

Theorem 6.4. *Let $\{A, B\}$ be a dual pair of operators in $\mathcal{K}$, let $\{\mathcal{H}, \Gamma^B, \Gamma^A\}$ be a boundary triple with $A_0 = B^* \upharpoonright \ker(\Gamma_0^B)$ and Weyl–Titchmarsh function $M(\cdot)$, and assume that $\mathcal{H}_0 = \mathcal{H}_1$. Furthermore, let $\Theta \in \mathcal{B}(\mathcal{H}_0)$ be a bounded operator and consider the extension*

$$A_\Theta = B^* \upharpoonright \ker(\Gamma_1^B - \Theta\Gamma_0^B).$$

Then the following assertions (i) *and* (ii) *hold.*

(i) *If $z_0 \in \rho(A_0) \cap \sigma_d(A_\Theta)$, then the index formula*

$$\operatorname{ind}_{C(z_0;\varepsilon)}(\Theta - M(\cdot)) = m_a(z_0; A_\Theta)$$

holds for $\varepsilon > 0$ sufficiently small. Furthermore, z_0 is a zero of finite-type of the function $\Theta - M(\cdot)$, and hence

$$\nu(z_0; \Theta - M(\cdot)) = m_a(z_0; \Theta - M(\cdot)) = \operatorname{ind}_{C(z_0;\varepsilon)}(\Theta - M(\cdot)). \tag{50}$$

(ii) *If $z_0 \in \sigma_d(A_0) \cap \sigma_d(A_\Theta)$, then the index formula*

$$\operatorname{ind}_{C(z_0;\varepsilon)}(\Theta - M(\cdot)) = m_a(z_0; A_\Theta) - m_a(z_0; A_0)$$

holds for $\varepsilon > 0$ sufficiently small.

Proof. Choose $\varepsilon_0 > 0$ such that $D(z_0; \varepsilon_0)\backslash\{z_0\} \subset \rho(A_0) \cap \rho(A_\Theta)$. Then it follows from (42) and (43) that the Weyl–Titchmarsh function admits the representation

$$M(z_1) = M(z_2) + (z_1 - z_2)\gamma_*(\overline{z_2})^*(I_{\mathcal{K}} + (z_1 - z_2)(A_0 - z_1 I_{\mathcal{K}})^{-1})\gamma(z_2), \tag{51}$$

with $z_j \in \rho(A_0)$, $j = 1, 2$. If z_0 is a point in $\rho(A_0)$, then the resolvent $(A_0 - zI_{\mathcal{K}})^{-1}$ is analytic on a disc $D(z_0; \varepsilon_0)$ with $\varepsilon_0 > 0$ sufficiently small, and if z_0 is a discrete eigenvalue of A_0 the resolvent $(A_0 - zI_{\mathcal{K}})^{-1}$ is analytic on a punctured disc

$D(z_0;\varepsilon_0)\backslash\{z_0\}$ with $\varepsilon_0 > 0$ sufficiently small, and finitely meromorphic on the disc $D(z_0;\varepsilon_0)$. Hence one concludes from (51) that in the case $z_0 \in \rho(A_0)$ also the Weyl–Titchmarsh function $M(\cdot)$ is analytic on the disc $D(z_0;\varepsilon_0)$, and in the case $z_0 \in \sigma_d(A_0)$ the Weyl–Titchmarsh function $M(\cdot)$ is analytic on the punctured disc $D(z_0;\varepsilon_0)\backslash\{z_0\}$ and finitely meromorphic on $D(z_0;\varepsilon_0)$. It is clear that the same is also true for the function

$$\Theta - M(\cdot).$$

Similarly, consider the boundary triple $\{\mathcal{H}, \Gamma^{B,\Theta}, \Gamma^{A,\Theta}\}$ in Lemma 6.3 and the corresponding Weyl–Titchmarsh function

$$M_\Theta(z) = [\Theta - M(z)]^{-1}, \tag{52}$$

where the operators $M_\Theta(z) \in \mathcal{B}(\mathcal{H}_0)$ are well-defined for all $z \in \rho(A_0) \cap \rho(A_\Theta)$. If $\gamma_\Theta(\cdot)$ and $\gamma_{*\Theta}(\cdot)$ denote the corresponding γ-fields, then one has (cf. (51))

$$M_\Theta(z_1) = M_\Theta(z_2) + (z_1 - z_2)\gamma_{*\Theta}(\overline{z_2})^*[I_{\mathcal{K}} + (z_1 - z_2)(A_\Theta - z_1 I_{\mathcal{K}})^{-1}]\gamma_\Theta(z_2), \tag{53}$$

with $z_j \in \rho(A_\Theta)$, $j = 1, 2$. Since $z_0 \in \sigma_d(A_\Theta)$ by assumption, it follows as above from the properties of the resolvent $(A_\Theta - zI_{\mathcal{K}})^{-1}$ and (53) that the function $M_\Theta(\cdot)$ in (52) is finitely meromorphic on the disc $D(z_0;\varepsilon_0)$. Furthermore, one obtains from (43) that

$$\frac{d}{dz}[\Theta - M(z)] = -\frac{d}{dz}M(z) = -\gamma_*(\bar{z})^*\gamma(z), \quad z \in \rho(A_0), \tag{54}$$

and hence one computes for $0 < \varepsilon < \varepsilon_0$ with the help of (54) and (44),

$$\begin{aligned}
\operatorname{ind}_{C(z_0;\varepsilon)}(\Theta - M(\cdot)) &= \frac{1}{2\pi i}\operatorname{tr}_{\mathcal{H}_0}\bigg(\oint_{C(z_0;\varepsilon)} d\zeta\, [\Theta - M(\zeta)]^{-1}(-M'(\zeta))\bigg) \\
&= \frac{-1}{2\pi i}\operatorname{tr}_{\mathcal{H}_0}\bigg(\oint_{C(z_0;\varepsilon)} d\zeta [\Theta - M(\zeta)]^{-1}\gamma_*(\bar{\zeta})^*\gamma(\zeta)\bigg) \\
&= \frac{-1}{2\pi i}\operatorname{tr}_{\mathcal{K}}\bigg(\oint_{C(z_0;\varepsilon)} d\zeta\gamma(\zeta)\, [\Theta - M(\zeta)]^{-1}\gamma_*(\bar{\zeta})^*\bigg) \\
&= \frac{-1}{2\pi i}\operatorname{tr}_{\mathcal{K}}\bigg(\oint_{C(z_0;\varepsilon)} d\zeta\, [(A_\Theta - \zeta I)^{-1} - (A_0 - \zeta I)^{-1}]\bigg) \\
&= m_a(z_0; A_\Theta) - m_a(z_0; A_0),
\end{aligned}$$

where the third equality is justified in a manner analogous to (2) (cf. Proposition 4.2.2 in [15]). This shows the index formula in assertion (ii). Clearly, if $z_0 \in \rho(A_0)$ then the term $m_a(z_0; A_0)$ is absent and hence the index formula reduces to the one in assertion (i).

The same argument as in the proof of Theorem 5.5 shows that the point $z_0 \in \rho(A_0) \cap \sigma_d(A_\Theta)$ is a zero of finite-type of the function $[\Theta - M(\cdot)]$. Then (50) follows from Theorem 3.3 and (20). □

We conclude this section with the observation that the far simpler case of Krein-type resolvent formulas in terms of boundary data maps for one-dimensional Schrödinger and Sturm–Liouville operators discussed in [7] and [8], readily yield analogous formulas for the index of these boundary data maps in terms of (algebraic) multiplicities of eigenvalues. One can follow the computation in the proof of Theorem 6.4 line by line; we omit further details.

Acknowledgments. Fritz Gesztesy gratefully acknowledges kind invitations to the Institute for Numerical Mathematics at the Graz University of Technology, Austria, and to the Department of Mathematical Sciences of the Norwegian University of Science and Technology, Trondheim, for parts of May and June 2015. The extraordinary hospitality by Jussi Behrndt and Helge Holden at each institution, as well as the stimulating atmosphere at both places, are greatly appreciated.

Jussi Behrndt and Fritz Gesztesy gratefully acknowledge support by the Austrian Science Fund (FWF), project P 25162-N26. Fritz Gesztesy and Helge Holden were supported in part by the project Waves and Nonlinear Phenomena (WaNP) from the Research Council of Norway. Roger Nichols gratefully acknowledges support from a UTC College of Arts and Sciences RCA Grant.

References

[1] H. Ammari, H. Kang, and H. Lee, *Layer potential techniques in spectral analysis.* Mathematical Surveys and Monographs, 153. American Mathematical Society, Providence, R.I., 2009. ISBN 978-0-8218-4784-8 MR 2488135 Zbl 1167.47001

[2] N. Aronszajn and R. D. Brown, Finite-dimensional perturbations of spectral problems and variational approximation methods for eigenvalue problems. I. Finite-dimensional perturbations. *Studia Math.* **36** (1970), 1–76. MR 0271766 Zbl 0203.45202

[3] H. Bart, T. Ehrhardt, and B. Silbermann, Logarithmic residues, Rouché's theorem, and spectral regularity: The C^*-algebra case. *Indag. Math. (N.S.)* **23** (2012), no. 4, 816–847. MR 2991925 Zbl 1279.47028

[4] M. Š. Birman, The spectrum of singular boundary problems. *Mat. Sb. (N.S.)* **55 (97)** (1961), 125–174. In Russian. English translation, *Amer. Math. Soc. Transl. Ser.* 2 **53** (1966), 23–80. MR 0142896 Zbl 0174.42502

[5] M. Š. Birman and M. Z. Solomyak, Estimates for the number of negative eigenvalues of the Schrödinger operator and its generalizations. In M. Š. Birman (ed.), *Estimates and asymptotics for discrete spectra of integral and differential equations.* Papers from the Seminar on Mathematical Physics held in Leningrad, 1989–90. Translated from the Russian. Translation edited by A. B. Sossinsky. Advances in Soviet Mathematics, 7. American Mathematical Society, Providence, R.I., 1991, 1–55. MR 1306507 Zbl 0749.35026

[6] J.-F. Bony, V. Bruneau, and G. Raikov, Counting function of characteristic values and magnetic resonances. *Comm. Partial Differential Equations* **39** (2014), no. 2, 274–305. MR 3169786 Zbl 1288.35057

[7] C. Clark, F. Gesztesy, and M. Mitrea, Boundary data maps for Schrödinger operators on a compact interval. *Math. Model. Nat. Phenom.* **5** (2010), no. 4, 73–121. MR 2662451 Zbl 1208.34028

[8] S. Clark, F. Gesztesy, R. Nichols, and M. Zinchenko, Boundary data maps and Krein's resolvent formula for Sturm–Liouville operators on a finite interval. *Oper. Matrices* **8** (2014), no. 1, 1–71. MR 3202926 Zbl 1311.34037

[9] J. B. Conway, *Functions of one complex variable.* Part I. Second edition. Graduate Texts in Mathematics, 11. Springer, Berlin etc., 1978. MR 0503901 Zbl 0277.30001

[10] F. Gesztesy and H. Holden, A unified approach to eigenvalues and resonances of Schrödinger operators using Fredholm determinants. *J. Math. Anal. Appl.* **123** (1987), no. 1, 181–198. MR 0881540 Zbl 0622.35049

[11] F. Gesztesy, H. Holden, and R. Nichols, On factorizations of analytic operator-valued functions and eigenvalue multiplicity questions. *Integral Equations Operator Theory* **82** (2015), no. 1, 61–94. MR 3335509 Zbl 1316.47014

[12] F. Gesztesy, Y. Latushkin, M. Mitrea, and M. Zinchenko, Nonselfadjoint operators, infinite determinants, and some applications. *Russ. J. Math. Phys.* **12** (2005), no. 4, 443–471. MR 2201310 Zbl 1201.47028

[13] I. C. Gohberg, S. Goldberg, and M. A. Kaashoek, *Classes of linear operators.* Vol. I. Operator Theory: Advances and Applications, 49. Birkhäuser, Basel, 1990. ISBN 3-7643-2531-3 MR 1130394 Zbl 0745.47002

[14] I. C. Gohberg and M. G. Kreĭn, *Introduction to the theory of linear nonselfadjoint operators.* Translated from the Russian by A. Feinstein. Translations of Mathematical Monographs, 18. American Mathematical Society, Providence, R.I., 1969. MR 0246142 Zbl 0181.13504

[15] I. C. Gohberg and J. Leiterer, *Holomorphic operator functions of one variable and applications.* Methods from complex analysis in several variables. Operator Theory: Advances and Applications, 192. Birkhäuser, Basel, 2009. ISBN 978-3-0346-0125-2 MR 2527384 Zbl 1182.47014

[16] I. C. Gohberg and E. I. Sigal, An operator generalizations of the logarithmic residue theorem and the theorem of Rouché. *Mat. Sb.* (*N.S.*) **84 (126)** (1971), 607–629. In Russian. English translation, *Math. USSR-Sb.* **13** (1971), 603–625. MR 0313856

[17] J. S. Howland, On the Weinstein–Aronszajn formula. *Arch. Rational Mech. Anal.* **39** (1970), 323–339. MR 0273455 Zbl 0225.47013

[18] J. S. Howland, Simple poles of operator-valued functions. *J. Math. Anal. Appl.* **36** (1971), 12–21. MR 0433238 Zbl 0234.47009

[19] T. Kato, Wave operators and similarity for some non-selfadjoint operators. *Math. Ann.* **162** (1965/1966), 258–279. MR 0190801 Zbl 0139.31203

[20] T. Kato, *Perturbation theory for linear operators.* Corrected printing of the second edition. Grundlehren der Mathematischen Wissenschaften, 132. Springer, Berlin etc., 1976. MR 0407617 Zbl 0435.47001

[21] M. Klaus, Some applications of the Birman–Schwinger principle. *Helv. Phts. Acta* **55** (1982), 49–68. MR 0674865

[22] M. Klaus and B. Simon, Coupling constant thresholds in nonrelativistic quantum mechanics. I. Short-range two-body case. *Ann. Physics* **130** (1980), no. 2, 251–281. MR 0610664 Zbl 0455.35112

[23] R. Konno and S. T. Kuroda, On the finiteness of perturbed eigenvalues. *J. Fac. Sci. Univ. Tokyo Sect.* I **13** (1966), 55–63. MR 0201971 Zbl 0149.10203

[24] S. T. Kuroda, On a generalization of the Weinstein–Aronszajn formula and the infinite determinant. *Sci. Papers Coll. Gen. Ed. Univ. Tokyo* **11** (1961), 1–12. MR 0138008 Zbl 0099.10003

[25] Y. Latushkin and A. Sukhtayev, The algebraic multiplicity of eigenvalues and the Evans function revisited. *Math. Model. Nat. Phenom.* **5** (2010), no. 4, 269–292. MR 2662459 Zbl 1193.35110

[26] M. M. Malamud and V. I. Mogilevskii, Krein type formula for canonical resolvents of dual pairs of linear relations. *Methods Funct. Anal. Topology* **8** (2002), no. 4, 72–100. MR 1942823 Zbl 1074.47501

[27] M. M. Malamud and V. I. Mogilevskii, Generalized resolvents of symmetric operators. *Mat. Zametki* **73** (2003), no. 3, 460–465. In Russian. English translation, *Math. Notes* **73** (2003), no. 3-4, 429–435. MR 1992607 Zbl 1076.47002

[28] M. M. Malamud, V. I. Mogilevskii, and S. Hassi, On the unitary equivalence of the proper extensions of a Hermitian operator and the Weyl function. *Mat. Zametki* **91** (2012), no. 2, 316–320. *Math. Notes* **91** (2012), no. 1-2, 302–307. MR 3201419 Zbl 1285.47019

[29] A. S. Markus, *Introduction to the spectral theory of polynomial operator pencils.* With an appendix by M. V. Keldysh. Translated from the Russian by H. H. McFaden. Translation edited by B. Silver. Translations of Mathematical Monographs, 71. American Mathematical Society, Providence, R.I., 1988. ISBN 0-8218-4523-3 MR 0971506 Zbl 0678.47005

[30] R. G. Newton, Bounds on the number of bound states for the Schrödinger equation in one and two dimensions. *J. Operator Theory* **10** (1983), no. 1, 119–125. MR 0715561 Zbl 0527.35062

[31] J. Rauch, Perturbation theory of eigenvalues and resonances of Schrödinger Hamiltonians. *J. Funct. Anal.* **35** (1980), no. 3, 304–315. MR 0563558 Zbl 0432.35013

[32] M. Reed and B. Simon, *Methods of modern mathematical physics.* I. Functional Analysis. Revised, and enlarged edition. Academic Press [Harcourt Brace Jovanovich, Publishers], New York, 1980. ISBN 0-12-585050-6 MR 0751959 Zbl 0459.46001

[33] M. Reed and B. Simon, *Methods of modern mathematical physics.* IV. Analysis of operators. Academic Press [Harcourt Brace Jovanovich, Publishers], New York and London, 1978. ISBN 0-12-585004-2 MR 0493421 Zbl 0401.47001

[34] M. Ribaric and I. Vidav, Analytic properties of the inverse $A(z)^{-1}$ of an analytic linear operator valued function $A(z)$. *Arch. Rational Mech. Anal.* **32** (1969), 298–310. MR 0236741 Zbl 0174.18002

[35] J. Schwinger, On the bound states of a given potential. *Proc. Nat. Acad. Sci. U.S.A.* **47** (1961) 122–129. MR 0129798

[36] N. Setô, Bargmann's inequalities in spaces of arbitrary dimension. *Publ. Res. Inst. Math. Sci.* **9** (1973/74), 429–461. MR 0340846 Zbl 0276.35013

[37] H. Siedentop, On a generalization of Rouché's theorem for trace ideals with applications for resonances of Schrödinger operators. *J. Math. Anal. Appl.* **140** (1989), no. 2, 582–588. MR 1001877 Zbl 0692.47002

[38] B. Simon, *Quantum mechanics for Hamiltonians defined as quadratic forms.* Princeton Series in Physics. Princeton University Press, Princeton, N.J., 1971. MR0455975 Zbl 0232.47053

[39] B. Simon, On the absorption of eigenvalues by continuous spectrum in regular perturbation problems. *J. Functional Analysis* **25** (1977), no. 4, 338–344. MR 0445317 Zbl 0363.47014

[40] S. Steinberg, Meromorphic families of compact operators. *Arch. Rational Mech. Anal.* **31** (1968/1969), 372–379. MR 0233240 Zbl 0167.43002

[41] A. Weinstein and W. Stenger, *Methods of intermediate problems for eigenvalues.* Theory and ramifications. Mathematics in Science and Engineering, 89. Academic Press, New York and London, 1972. MR 0477971 Zbl 0291.49034

[42] F. Wolf, Analytic perturbation of operators in Banach spaces. *Math. Ann.* **124** (1952). 317–333. MR 0050167 Zbl 0046.12402

Trace formulae for Schrödinger operators with singular interactions

Jussi Behrndt, Matthias Langer,
and Vladimir Lotoreichik

Dedicated with great pleasure
to our teacher, colleague and friend Pavel Exner
on the occasion of his 70th birthday

1 Introduction

This paper is strongly inspired by the work of Pavel Exner on Schrödinger operators with singular interactions of δ and δ'-type supported on hypersurfaces in $\mathbb{R}^d$. Such operators play an important role in mathematical physics, for instance, in nuclear physics or solid state physics or in connection with photonic crystals or other nano-structures. In the case of a curve in $\mathbb{R}^2$ such models are also called "leaky quantum wires." The first rigorous investigations of such operators started in the late 1980s (see, e.g., [1], [14], and [15]), and the interest in these operators grew steadily in the last two decades. We refer the reader to the review paper [26], the monograph [30], the references therein and also to the more recent papers [6], [7], [10], [23], [24], [29], [35], and [44].

Let $\Sigma \subset \mathbb{R}^d$, where $d \geq 2$, be a C^∞-smooth closed compact hypersurface without boundary, which naturally splits the Euclidean space $\mathbb{R}^d$ into a bounded domain Ω_- and an exterior domain Ω_+. Moreover, let $\alpha, \omega \in L^\infty(\Sigma)$ be real-valued functions. The Schrödinger operator $\mathrm{H}_{\alpha,\Sigma}$ with δ-interaction of strength α and the Schrödinger operator $\mathrm{K}_{\omega,\Sigma}$ with δ'-interaction of strength ω are formally given by

$$-\Delta - \alpha\delta(x-\Sigma) \quad \text{and} \quad -\Delta - \omega\delta'(x-\Sigma). \tag{1}$$

We define these operators rigorously via quadratic forms; see Definition 1.1 below. Let us first fix some notation. Since the space $L^2(\mathbb{R}^d)$ naturally decomposes as $L^2(\mathbb{R}^d) = L^2(\Omega_+) \oplus L^2(\Omega_-)$, we can write functions $u \in L^2(\mathbb{R}^d)$ as $u = u_+ \oplus u_-$

with $u_\pm = u \upharpoonright \Omega_\pm \in L^2(\Omega_\pm)$. The L^2-based Sobolev spaces of order $s \geq 0$ over $\mathbb{R}^d$ and $\Omega_\pm$ are denoted by $H^s(\mathbb{R}^d)$ and $H^s(\Omega_\pm)$, respectively. Note that the hypersurface Σ coincides with the boundaries $\partial\Omega_\pm$ of the domains $\Omega_\pm$. Hence, for any $u \in H^1(\mathbb{R}^d)$ and $u_\pm \in H^1(\Omega_\pm)$ the *traces* $u|_\Sigma$ and $u_\pm|_\Sigma$ on Σ are well defined as functions in $L^2(\Sigma)$. Further, for a function $u \in H^1(\mathbb{R}^d \setminus \Sigma) := H^1(\Omega_+) \oplus H^1(\Omega_-)$ we define its *jump* on Σ as $[u]_\Sigma := u_+|_\Sigma - u_-|_\Sigma$.

Let us now introduce the following quadratic forms that correspond to the formal expression in (1). According to §2 of [14], §3.4 of [10] and Proposition 3.1 in [6], the symmetric quadratic forms

$$\mathfrak{h}_{\alpha,\Sigma}[u] := \|\nabla u\|^2 - (\alpha u|_\Sigma, u|_\Sigma)_\Sigma, \qquad \operatorname{dom} \mathfrak{h}_{\alpha,\Sigma} := H^1(\mathbb{R}^d),$$

$$\mathfrak{k}_{\omega,\Sigma}[u] := \|\nabla u_+\|_+^2 + \|\nabla u_-\|_-^2 - (\omega[u]_\Sigma, [u]_\Sigma)_\Sigma, \quad \operatorname{dom} \mathfrak{k}_{\omega,\Sigma} := H^1(\mathbb{R}^d \setminus \Sigma),$$

in $L^2(\mathbb{R}^d)$ are closed, densely defined and bounded from below; here $u = u_+ \oplus u_-$ with $u_\pm$ as above, and $\|\cdot\|_\pm$ denotes the norm on $L^2(\Omega_\pm; \mathbb{C}^d)$.

Definition 1.1. Let $\mathrm{H}_{\alpha,\Sigma}$ and $\mathrm{K}_{\omega,\Sigma}$ be the self-adjoint operators in $L^2(\mathbb{R}^d)$ corresponding to the forms $\mathfrak{h}_{\alpha,\Sigma}$ and $\mathfrak{k}_{\omega,\Sigma}$, respectively, via the first representation theorem (Theorem VI.2.1 in [37]). Moreover, set $\mathrm{H}_{\mathrm{free}} := \mathrm{H}_{0,\Sigma}$ ($\alpha \equiv 0$) and $\mathrm{K}_{\mathrm{N}} := \mathrm{K}_{0,\Sigma}$ ($\omega \equiv 0$).

The operator $\mathrm{H}_{\alpha,\Sigma}$ is called a Schrödinger operator with δ-interaction of strength α supported on Σ; the operator $\mathrm{K}_{\omega,\Sigma}$ is called a Schrödinger operator with δ'-interaction of strength[1] ω supported on Σ. The operator $\mathrm{H}_{\mathrm{free}}$ is the usual *free Laplacian* on $\mathbb{R}^d$, and K_{N} is the orthogonal sum of the standard *Neumann Laplacians* on Ω_+ and Ω_-. Let us mention that the operators $\mathrm{H}_{\alpha,\Sigma}$ and $\mathrm{K}_{\omega,\Sigma}$ can also be introduced via interface conditions at the hypersurface Σ; see, e.g., [10].

The aim of this paper is to derive trace formulae for $\mathrm{H}_{\alpha,\Sigma}$ and $\mathrm{K}_{\omega,\Sigma}$. According to [10] for $m \in \mathbb{N}$ the resolvent power differences

$$(\mathrm{H}_{\alpha,\Sigma} - \lambda)^{-m} - (\mathrm{H}_{\mathrm{free}} - \lambda)^{-m}, \quad m > \frac{d-2}{2}, \ \lambda \in \rho(\mathrm{H}_{\alpha,\Sigma}), \tag{2a}$$

$$(\mathrm{K}_{\omega,\Sigma} - \lambda)^{-m} - (\mathrm{H}_{\mathrm{free}} - \lambda)^{-m}, \quad m > \frac{d-1}{2}, \ \lambda \in \rho(\mathrm{K}_{\omega,\Sigma}), \tag{2b}$$

are in the trace class. Their traces as functions of λ or as functions of the interaction strengths are expected to encode a lot of information on the operators $\mathrm{H}_{\alpha,\Sigma}$ and $\mathrm{K}_{\omega,\Sigma}$ themselves and on the shape of Σ. Such non-trivial connections have been observed

[1] We point out that, in the case of invertible ω, not ω itself, but its inverse is frequently called the strength of the δ'-interaction.

in various other settings in the classical papers [18], [36], and [42] and more recently in, e.g., [5], [34], [38], [39], and [46].

The main results of the paper (see Theorems 1.2 and 1.3) are formulae that express the traces of the resolvent power differences in (2) in terms of traces of derivatives of certain operator-valued functions in the boundary space $L^2(\Sigma)$. These operator-valued functions are, in turn, expressed in terms of Neumann–to–Dirichlet maps on $\Omega_\pm$ corresponding to the differential expression $-\Delta-\lambda$ and in terms of the coupling functions α, ω. Trace formulae of this kind are useful (see, e.g., [19], [20], and [32]) in connection with the estimation of the spectral shift function.

1.1 Traces, Neumann–to–Dirichlet maps and some operator functions

We first recall some notions that are needed in order to formulate the main results of this paper. For a compact operator K in a Hilbert space $\mathcal{H}$ we define its *singular values* $s_k(K)$, $k = 1, 2, \ldots$, as the eigenvalues of the non-negative compact operator $|K| = (K^*K)^{1/2} \geq 0$ in $\mathcal{H}$ ordered in a non-increasing way and with multiplicities taken into account. If $\sum_{k=1}^{\infty} s_k(K) < \infty$, we say that K belongs to the *trace class* and define its *trace* as

$$\operatorname{Tr} K := \sum_{k=1}^{\infty} \lambda_k(K),$$

where $\lambda_k(K)$ are the eigenvalues of K repeated with their *algebraic multiplicities*. Note also that the series in the definition of the trace converges absolutely.

Let us also define some auxiliary maps associated with partial differential equations. For the sake of brevity, we introduce the spaces

$$H^{3/2}_\Delta(\Omega_\pm) := \{u_\pm \in H^{3/2}(\Omega_\pm) \colon \Delta u_\pm \in L^2(\Omega_\pm)\}. \tag{3}$$

For any $u_\pm \in H^{3/2}_\Delta(\Omega_\pm)$ its *Neumann trace* $\partial_{\nu_\pm} u_\pm|_\Sigma$ exists as a function in $L^2(\Sigma)$; see, e.g., §2.7.3 of [43]. For every $\lambda \in \mathbb{C} \setminus \mathbb{R}_+$ (where $\mathbb{R}_+ := [0, \infty)$) and every $\varphi \in L^2(\Sigma)$ the boundary value problems

$$\begin{aligned} -\Delta u_\pm &= \lambda u_\pm && \text{in } \Omega_\pm, \\ \partial_{\nu_\pm} u_\pm\big|_\Sigma &= \varphi && \text{on } \Sigma, \end{aligned}$$

have unique solutions $u_{\lambda,\pm}(\varphi) \in H^{3/2}_\Delta(\Omega_\pm)$; see, e.g., §2.7.3 of [43]. The operator-valued functions $\lambda \mapsto M_\pm(\lambda)$, $\lambda \in \mathbb{C} \setminus \mathbb{R}_+$, are then defined as

$$M_\pm(\lambda) \colon L^2(\Sigma) \longrightarrow L^2(\Sigma), \quad M_\pm(\lambda)\varphi := u_{\lambda,\pm}(\varphi)|_\Sigma.$$

For fixed $\lambda \in \mathbb{C} \setminus \mathbb{R}_+$ the operators $M_\pm(\lambda)$ are the *Neumann–to–Dirichlet maps* for the differential expression $-\Delta - \lambda$ on the domains $\Omega_\pm$. The operators $M_\pm(\lambda)$ are compact and injective and their inverses are called *Dirichlet–to–Neumann maps*. Recently, there has been a considerable growth of interest in the investigation of these maps (see, e.g., [3], [4], and [25]), in particular also with the aim to derive spectral properties of the corresponding partial differential operators (see, e.g., [2], [13], and [31]).

Further, we define the following operator-valued functions $\lambda \mapsto \widetilde{M}(\lambda), \widehat{M}(\lambda)$, $\lambda \in \mathbb{C} \setminus \mathbb{R}_+$, by

$$\widetilde{M}(\lambda) := (M_+(\lambda)^{-1} + M_-(\lambda)^{-1})^{-1}, \quad \widehat{M}(\lambda) := M_+(\lambda) + M_-(\lambda). \tag{4}$$

We should mention that for every $\lambda \in \mathbb{C} \setminus \mathbb{R}_+$ the operator $M_+(\lambda)^{-1} + M_-(\lambda)^{-1}$ is invertible and therefore $\widetilde{M}(\lambda)$ is well defined. Moreover, $\widetilde{M}(\lambda)$ and $\widehat{M}(\lambda)$ are compact operators in $L^2(\Sigma)$ for every $\lambda \in \mathbb{C} \setminus \mathbb{R}_+$; see Propositions 3.2 and 3.8 of [10]. It is worth mentioning that $\widetilde{M}(\lambda)$ and the inverse of $\widehat{M}(\lambda)$ appear naturally in the theory of boundary integral operators. They are used in the treatment of partial differential equations from both analytical [45] and computational [48] viewpoints. The operator-valued function $\widetilde{M}(\cdot)$ was successfully applied to the spectral analysis of the operator $\mathrm{H}_{\alpha,\Sigma}$ in quite a few papers; see, e.g., [27], [28], [30], [41], and the survey paper [26].

1.2 Statement of the main results

In the first main result of this note we obtain a trace formula for the resolvent power difference of the operators $\mathrm{H}_{\alpha,\Sigma}$ and $\mathrm{H}_{\mathrm{free}}$.

Theorem 1.2. *Let the self-adjoint operators* $\mathrm{H}_{\mathrm{free}}$ *and* $\mathrm{H}_{\alpha,\Sigma}$ *with* $\alpha \in L^\infty(\Sigma; \mathbb{R})$ *be as in Definition* 1.1, *and let the operator-valued function* $\widetilde{M}$ *be as in* (4). *Then for all* $m \in \mathbb{N}$ *such that* $m > {(d-2)}/{2}$ *and all* $\lambda \in \rho(\mathrm{H}_{\alpha,\Sigma})$ *the resolvent power difference*

$$\widetilde{D}_{\alpha,m}(\lambda) := (\mathrm{H}_{\alpha,\Sigma} - \lambda)^{-m} - (\mathrm{H}_{\mathrm{free}} - \lambda)^{-m}$$

belongs to the trace class, and its trace can be expressed as

$$\mathrm{Tr}(\widetilde{D}_{\alpha,m}(\lambda)) = \frac{1}{(m-1)!} \mathrm{Tr}\left(\frac{\mathrm{d}^{m-1}}{\mathrm{d}\lambda^{m-1}}((I - \alpha\widetilde{M}(\lambda))^{-1}\alpha\widetilde{M}'(\lambda))\right).$$

In the second main result of this note we obtain trace formulae for the resolvent power differences of the pairs of operators $\{\mathrm{K}_{\omega,\Sigma}, \mathrm{K}_{\mathrm{N}}\}$ and $\{\mathrm{K}_{\omega,\Sigma}, \mathrm{H}_{\mathrm{free}}\}$.

Theorem 1.3. *Let the self-adjoint operators* $\mathrm{H}_{\mathrm{free}}$, K_{N}, *and* $\mathrm{K}_{\omega,\Sigma}$ *with* $\omega \in L^\infty(\Sigma;\mathbb{R})$ *be as in Definition* 1.1, *and let the operator-valued function* $\widehat{M}$ *be as in* (4). *Then the following statements hold.*

(i) *For all* $m \in \mathbb{N}$ *such that* $m > {(d-2)}/{2}$ *and all* $\lambda \in \rho(\mathrm{K}_{\omega,\Sigma})$ *the resolvent power difference*

$$\widehat{E}_{\omega,m}(\lambda) := (\mathrm{K}_{\omega,\Sigma} - \lambda)^{-m} - (\mathrm{K}_{\mathrm{N}} - \lambda)^{-m}$$

belongs to the trace class, and its trace can be expressed as

$$\operatorname{Tr}(\widehat{E}_{\omega,m}(\lambda)) = \frac{1}{(m-1)!}\operatorname{Tr}\Big(\frac{\mathrm{d}^{m-1}}{\mathrm{d}\lambda^{m-1}}((I-\omega\widehat{M}(\lambda))^{-1}\omega\widehat{M}'(\lambda))\Big).$$

(ii) *For all* $m \in \mathbb{N}$ *such that* $m > {(d-1)}/{2}$ *and all* $\lambda \in \rho(\mathrm{K}_{\omega,\Sigma})$ *the resolvent power difference*

$$\widehat{D}_{\omega,m}(\lambda) := (\mathrm{K}_{\omega,\Sigma} - \lambda)^{-m} - (\mathrm{H}_{\mathrm{free}} - \lambda)^{-m}$$

belongs to the trace class, and its trace can be expressed as

$$\operatorname{Tr}(\widehat{D}_{\omega,m}(\lambda)) = \frac{1}{(m-1)!}\operatorname{Tr}\Big(\frac{\mathrm{d}^{m-1}}{\mathrm{d}\lambda^{m-1}}((I-\omega\widehat{M}(\lambda))^{-1}\widehat{M}(\lambda)^{-1}\widehat{M}'(\lambda))\Big).$$

We remark that it is also implicitly shown that the derivatives of the operator-valued functions appearing in the trace formulae exist in a suitable sense and that these derivatives belong to the trace class.

The main ingredients used in the proofs, which are given in Section 3, are Krein-type resolvent formulae, properties of weak Schatten–von Neumann classes, asymptotics of eigenvalues of the Laplace–Beltrami operator on Σ, and elements of elliptic regularity theory. We point out that for the proof of Theorem 1.3 (ii) an auxiliary trace formula for the resolvent power differences of $\mathrm{H}_{\mathrm{free}}$ and K_{N} is derived in Lemma 3.4. This trace formula is also of certain independent interest. We also mention that a similar strategy of proof was employed in our previous paper [12] where we proved trace formulae for generalized Robin Laplacians.

2 Preliminaries

This section consists of five subsections. In Subsection 2.1 we recall the notion of weak Schatten–von Neumann classes and their connection with the trace class, and in Subsection 2.2 we collect certain formulae that involve derivatives of holomorphic operator-valued functions. Next, in Subsection 2.3 we recall the definitions of quasi boundary triples and associated γ-fields and Weyl functions. Krein's resolvent formulae and sufficient conditions for self-adjointness of extensions are discussed in Subsection 2.4. Finally, in Subsection 2.5 we introduce specific quasi boundary triples, which are used to parameterize Schrödinger operators with singular interactions from Definition 1.1.

2.1 $\mathfrak{S}_{p,\infty}$-classes and the trace mapping

Let $\mathcal{H}$ and $\mathcal{K}$ be Hilbert spaces. Denote by $\mathfrak{S}_\infty(\mathcal{H},\mathcal{K})$ the class of all compact operators $K\colon \mathcal{H} \to \mathcal{K}$. Recall that, for $p > 0$, the *weak Schatten–von Neumann ideal* $\mathfrak{S}_{p,\infty}(\mathcal{H},\mathcal{K})$ is defined by

$$\mathfrak{S}_{p,\infty}(\mathcal{H},\mathcal{K}) := \{K \in \mathfrak{S}_\infty(\mathcal{H},\mathcal{K})\colon s_k(K) = \mathrm{O}(k^{-1/p}),\ k \to \infty\}.$$

Often we just write $\mathfrak{S}_{p,\infty}$ instead of $\mathfrak{S}_{p,\infty}(\mathcal{H},\mathcal{K})$. For $0 < p' < p$ the inclusion

$$\mathfrak{S}_{p',\infty} \subset \mathfrak{S}_{p,\infty} \tag{5}$$

holds, and for $s, t > 0$ one has

$$\mathfrak{S}_{1/s,\infty} \cdot \mathfrak{S}_{1/t,\infty} = \mathfrak{S}_{1/(s+t),\infty}, \tag{6}$$

where a product of operator ideals is defined as the set of all products of operators. We refer the reader to §III.7 and §III.14 of [33] and Chapter 2 of [47] for a detailed study of the classes $\mathfrak{S}_{p,\infty}$; see also Lemma 2.3 in [11]. If $K \in \mathfrak{S}_{p,\infty}$ with $p < 1$, then K belongs to the trace class. It is well known (see, e.g., §III.8 [33]) that, for trace class operators K_1, K_2, the operator $K_1 + K_2$ is also in the trace class, and that

$$\operatorname{Tr}(K_1 + K_2) = \operatorname{Tr} K_1 + \operatorname{Tr} K_2. \tag{7}$$

Moreover, if $K_1 \in \mathcal{B}(\mathcal{H},\mathcal{K})$ and $K_2 \in \mathcal{B}(\mathcal{K},\mathcal{H})$ are such that both products K_1K_2 and K_2K_1 are in the trace class, then

$$\operatorname{Tr}(K_1K_2) = \operatorname{Tr}(K_2K_1). \tag{8}$$

The next useful lemma is a special case of Lemma 4.7 in [11] and is based on the asymptotics of the eigenvalues of the Laplace–Beltrami operator. For a smooth compact manifold Σ we denote the usual L^2-based Sobolev spaces by $H^r(\Sigma)$, $r \geq 0$.

Lemma 2.1. *Let Σ be a $(d-1)$-dimensional compact C^∞-manifold without boundary, let $\mathcal{K}$ be a Hilbert space and let $K \in \mathcal{B}(\mathcal{K}, L^2(\Sigma))$ with* $\operatorname{ran} K \subset H^r(\Sigma)$*, where $r > 0$. Then K is compact and $K \in \mathfrak{S}_{\frac{d-1}{r},\infty}$.*

2.2 Derivatives of holomorphic operator-valued functions

In the following we shall often use product rules for holomorphic operator-valued functions. Let $\mathcal{H}_i$, $i = 1, \dots, 4$, be Hilbert spaces, $\mathcal{U}$ a domain in $\mathbb{C}$ and let

$$A: \mathcal{U} \longrightarrow \mathcal{B}(\mathcal{H}_3, \mathcal{H}_4), \quad B: \mathcal{U} \longrightarrow \mathcal{B}(\mathcal{H}_2, \mathcal{H}_3), \quad C: \mathcal{U} \longrightarrow \mathcal{B}(\mathcal{H}_1, \mathcal{H}_2)$$

be holomorphic operator-valued functions. Then for $\lambda \in \mathcal{U}$ we have

$$\frac{\mathrm{d}^m}{\mathrm{d}\lambda^m}(A(\lambda)B(\lambda)) = \sum_{\substack{p+q=m \\ p,q\geq 0}} \binom{m}{p} A^{(p)}(\lambda)B^{(q)}(\lambda), \tag{9a}$$

$$\frac{\mathrm{d}^m}{\mathrm{d}\lambda^m}(A(\lambda)B(\lambda)C(\lambda)) = \sum_{\substack{p+q+r=m \\ p,q,r\geq 0}} \frac{m!}{p!\,q!\,r!} A^{(p)}(\lambda)B^{(q)}(\lambda)C^{(r)}(\lambda). \tag{9b}$$

If $A(\lambda)^{-1}$ is invertible for every $\lambda \in \mathcal{U}$, then relation (9a) implies the following formula for the derivative of the inverse,

$$\frac{\mathrm{d}}{\mathrm{d}\lambda}(A(\lambda)^{-1}) = -A(\lambda)^{-1}A'(\lambda)A(\lambda)^{-1}. \tag{10}$$

2.3 Quasi boundary triples, Weyl functions and γ-fields

We begin this subsection by recalling the abstract concept of *quasi boundary triples* introduced in [8] as a generalization of the notion of (ordinary) *boundary triples*, see [16] and [40]. For the theory of ordinary boundary triples and associated Weyl functions the reader may consult, e.g., [17], [21], and [22]. Recent developments on quasi boundary triples and their applications to PDEs can be found in, e.g., [9], [11], [12], and [13].

Definition 2.2. Let S be a closed, densely defined, symmetric operator in a Hilbert space $(\mathcal{H}, (\cdot,\cdot)_{\mathcal{H}})$. A triple $\{\mathcal{G}, \Gamma_0, \Gamma_1\}$ is called a *quasi boundary triple* for S^* if $(\mathcal{G}, (\cdot,\cdot)_{\mathcal{G}})$ is a Hilbert space, and for some linear operator $T \subset S^*$ with $\overline{T} = S^*$ the following assumptions are satisfied:

(i) $\Gamma_0, \Gamma_1: \operatorname{dom} T \to \mathcal{G}$ are linear mappings, and the mapping $\Gamma := \binom{\Gamma_0}{\Gamma_1}$ has dense range in $\mathcal{G} \times \mathcal{G}$;

(ii) $A_0 := T \upharpoonright \ker \Gamma_0$ is a self-adjoint operator in $\mathcal{H}$;

(iii) for all $f, g \in \operatorname{dom} T$ the *abstract Green identity* holds:

$$(Tf, g)_{\mathcal{H}} - (f, Tg)_{\mathcal{H}} = (\Gamma_1 f, \Gamma_0 g)_{\mathcal{G}} - (\Gamma_0 f, \Gamma_1 g)_{\mathcal{G}}.$$

Next, we recall the definitions of the γ-field and the Weyl function associated with a quasi boundary triple $\{\mathcal{G}, \Gamma_0, \Gamma_1\}$ for S^*. Note that the decomposition

$$\operatorname{dom} T = \operatorname{dom} A_0 \dotplus \ker(T - \lambda)$$

holds for all $\lambda \in \rho(A_0)$, so that $\Gamma_0 \upharpoonright \ker(T - \lambda)$ is injective for all $\lambda \in \rho(A_0)$. The (operator-valued) functions γ and M defined by

$$\gamma(\lambda) := (\Gamma_0 \upharpoonright \ker(T - \lambda))^{-1} \quad \text{and} \quad M(\lambda) := \Gamma_1 \gamma(\lambda), \quad \lambda \in \rho(A_0),$$

are called the *γ-field* and the *Weyl function* corresponding to the quasi boundary triple $\{\mathcal{G}, \Gamma_0, \Gamma_1\}$. The adjoint of $\gamma(\bar{\lambda})$ has the following representation:

$$\gamma(\bar{\lambda})^* = \Gamma_1 (A_0 - \lambda)^{-1}, \quad \lambda \in \rho(A_0); \tag{11}$$

see Proposition 2.6 (ii) in [8] and also Proposition 6.13 in [9]. According to Proposition 2.6 in [8] the operator-valued functions $\lambda \mapsto \gamma(\lambda)$, $\lambda \mapsto \gamma(\bar{\lambda})^*$, and $\lambda \mapsto M(\lambda)$ are holomorphic on $\rho(A_0)$. Finally, we recall formulae for their derivatives: for $k \in \mathbb{N}$, $\varphi \in \operatorname{ran} \Gamma_0$ and $\lambda \in \rho(A_0)$ we have

$$\gamma^{(k)}(\lambda)\varphi = k!(A_0 - \lambda)^{-k} \gamma(\lambda)\varphi, \tag{12a}$$

$$\frac{\mathrm{d}^k}{\mathrm{d}\lambda^k} (\gamma(\bar{\lambda}))^* = k!\, \gamma(\bar{\lambda})^* (A_0 - \lambda)^{-k}, \tag{12b}$$

$$M^{(k)}(\lambda)\varphi = k!\, \gamma(\bar{\lambda})^* (A_0 - \lambda)^{-(k-1)} \gamma(\lambda)\varphi; \tag{12c}$$

see Lemma 2.4 in [12].

2.4 Self-adjoint extensions and abstract Krein's resolvent formula

In this subsection we parameterize subfamilies of self-adjoint extensions via quasi boundary triples and provide a couple of useful Krein-type formulae for resolvent differences of these extensions.

The following hypothesis will be convenient in the following.

Hypothesis 2.3. Let S be a closed, densely defined, symmetric operator in a Hilbert space $\mathcal{H}$ and let $\{\mathcal{G}, \Gamma_0, \Gamma_1\}$ be a quasi boundary triple for S^* such that $\operatorname{ran} \Gamma_0 = \mathcal{G}$. Moreover, let γ and M be the associated γ-field and Weyl function, respectively.

We remark that the quasi boundary triple $\{\mathcal{G}, \Gamma_0, \Gamma_1\}$ in Hypothesis 2.3 is also a generalized boundary triple in the sense of [22]. In this case the γ-field and the Weyl function associated with $\{\mathcal{G}, \Gamma_0, \Gamma_1\}$ are defined on the whole space $\mathcal{G}$, and the formulae (12a) and (12c) are valid for all $\varphi \in \mathcal{G}$.

Next, we state a Krein-type formula for the resolvent difference of

$$A_j := T \upharpoonright \ker \Gamma_j, \quad j = 0, 1.$$

Proposition 2.4 (Theorem 2.5 in [12]). *Assume that Hypothesis 2.3 is satisfied and that A_1 is self-adjoint in $\mathcal{H}$. Then the formula*

$$(A_0 - \lambda)^{-1} - (A_1 - \lambda)^{-1} = \gamma(\lambda) M(\lambda)^{-1} \gamma(\bar{\lambda})^*$$

holds for all $\lambda \in \rho(A_0) \cap \rho(A_1)$.

In the next proposition, we formulate a sufficient condition for self-adjointness of the extension of A defined by

$$A_{[B]} := T \upharpoonright \ker(B\Gamma_1 - \Gamma_0),$$

and provide a Krein-type formula for the resolvent difference of $A_{[B]}$ and A_0.

Proposition 2.5 (Theorem 2.6 in [12]). *Assume that Hypothesis 2.3 is satisfied, that $M(\lambda_0) \in \mathfrak{S}_\infty(\mathcal{G})$ for some $\lambda_0 \in \rho(A_0)$, and that $B \in \mathcal{B}(\mathcal{G})$ is self-adjoint in $\mathcal{G}$. Then the extension $A_{[B]}$ of A is self-adjoint in $\mathcal{H}$, and the formula*

$$(A_{[B]} - \lambda)^{-1} - (A_0 - \lambda)^{-1} = \gamma(\lambda)(I - BM(\lambda))^{-1} B \gamma(\bar{\lambda})^*$$

holds for all $\lambda \in \rho(A_{[B]}) \cap \rho(A_0)$. In this formula the middle term satisfies

$$(I - BM(\lambda))^{-1} \in \mathcal{B}(\mathcal{G})$$

for all $\lambda \in \rho(A_{[B]}) \cap \rho(A_0)$.

2.5 Quasi boundary triples for coupled problems

We recall particular quasi boundary triples, which are used to parameterize the self-adjoint operators from Definition 1.1. Furthermore, we reformulate some of the abstract statements from Subsections 2.3 and 2.4 for these quasi boundary triples.

First, we introduce the subspace $H^{3/2}_\Delta(\mathbb{R}^d \setminus \Sigma)$ of $L^2(\mathbb{R}^d)$ by

$$H^{3/2}_\Delta(\mathbb{R}^d \setminus \Sigma) := H^{3/2}_\Delta(\Omega_+) \oplus H^{3/2}_\Delta(\Omega_-),$$

where $H^{3/2}_{\Delta}(\Omega_\pm)$ are as in (3). Further, to shorten the notations, we also define the *jump of the normal derivative* by

$$[\partial_\nu u]_\Sigma := \partial_{\nu_+} u_+|_\Sigma + \partial_{\nu_-} u_-|_\Sigma,$$

for $u \in H^{3/2}_{\Delta}(\mathbb{R}^d \setminus \Sigma)$. Following the lines of Section 3 in [10], we define the operators $\widetilde{T}$ and $\widehat{T}$ in $L^2(\mathbb{R}^d)$ by

$$\widetilde{T}u := (-\Delta u_+) \oplus (-\Delta u_-), \quad \operatorname{dom}\widetilde{T} := \{u \in H^{3/2}_{\Delta}(\mathbb{R}^d \setminus \Sigma) \colon [u]_\Sigma = 0\},$$

$$\widehat{T}u := (-\Delta u_+) \oplus (-\Delta u_-), \quad \operatorname{dom}\widehat{T} := \{u \in H^{3/2}_{\Delta}(\mathbb{R}^d \setminus \Sigma) \colon [\partial_\nu u]_\Sigma = 0\},$$

and their restrictions $\widetilde{S}$ and $\widehat{S}$ by

$$\widetilde{S} := \widetilde{T} \upharpoonright \{u \in H^{3/2}_{\Delta}(\mathbb{R}^d \setminus \Sigma) \colon u_\pm|_\Sigma = 0,\ [\partial_\nu u]_\Sigma = 0\},$$

$$\widehat{S} := \widehat{T} \upharpoonright \{u \in H^{3/2}_{\Delta}(\mathbb{R}^d \setminus \Sigma) \colon \partial_{\nu_\pm} u_\pm|_\Sigma = 0,\ [u]_\Sigma = 0\}.$$

It can be verified that $\widetilde{S}$ (respectively, $\widehat{S}$) is the restriction of $\mathrm{H}_{\mathrm{free}}$ to functions, whose Dirichlet trace (respectively, Neumann trace) vanishes on Σ. In particular, as a consequence of this identification we arrive at the inclusions $\operatorname{dom}\widetilde{S}, \operatorname{dom}\widehat{S} \subset H^2(\mathbb{R}^d)$. It can also be shown that the operators $\widetilde{S}$ and $\widehat{S}$ are closed, densely defined, and symmetric in $L^2(\mathbb{R}^d)$ and that the closures of $\widetilde{T}$ and $\widehat{T}$ coincide with $\widetilde{S}^*$ and $\widehat{S}^*$, respectively. Furthermore, we define the boundary mappings by

$$\widetilde{\Gamma}_0, \widetilde{\Gamma}_1 \colon \operatorname{dom}\widetilde{T} \to L^2(\Sigma), \quad \widetilde{\Gamma}_0 u := [\partial_\nu u]_\Sigma, \qquad \widetilde{\Gamma}_1 u := u|_\Sigma,$$

$$\widehat{\Gamma}_0, \widehat{\Gamma}_1 \colon \operatorname{dom}\widehat{T} \to L^2(\Sigma), \quad \widehat{\Gamma}_0 u := \partial_{\nu_+} u_+|_\Sigma, \quad \widehat{\Gamma}_1 u := [u]_\Sigma. \tag{13}$$

The identities

$$\mathrm{H}_{\mathrm{free}} = \widetilde{T} \upharpoonright \ker\widetilde{\Gamma}_0 = \widehat{T} \upharpoonright \ker\widehat{\Gamma}_1 \quad \text{and} \quad \mathrm{K}_{\mathrm{N}} = \widehat{T} \upharpoonright \ker\widehat{\Gamma}_0$$

can be checked in a straightforward way. According to Proposition 3.2 (i) in [10] the triple $\widetilde{\Pi} := \{L^2(\Sigma), \widetilde{\Gamma}_0, \widetilde{\Gamma}_1\}$ is a quasi boundary triple for $\widetilde{S}^*$, and by Proposition 3.8 (i) in [10] the triple $\widehat{\Pi} := \{L^2(\Sigma), \widehat{\Gamma}_0, \widehat{\Gamma}_1\}$ is a quasi boundary triple for $\widehat{S}^*$.

Definition 2.6. Let $\widetilde{\gamma}$, $\widetilde{M}$ and $\widehat{\gamma}$, $\widehat{M}$ be the γ-fields and the Weyl functions of the quasi boundary triples $\widetilde{\Pi}$ and $\widehat{\Pi}$, respectively.

Remark 2.7. The definitions of the operator-valued functions $\lambda \mapsto \widetilde{M}(\lambda)$ and $\lambda \mapsto \widehat{M}(\lambda)$ as Neumann–to–Dirichlet maps in (4) and as Weyl functions of the quasi boundary triples $\widetilde{\Pi}$ and $\widehat{\Pi}$ are equivalent; see Propositions 3.2 (iii) and 3.8 (iii) in [10].

Remark 2.8. According to Propositions 3.2 (ii) and 3.8 (ii) in [10], for any $\varphi \in L^2(\Sigma)$ both *transmission boundary value problems*

$$\begin{cases} -\Delta u = \lambda u & \text{in } \mathbb{R}^d \setminus \Sigma, \\ [u]_\Sigma = 0 & \text{on } \Sigma, \\ [\partial_\nu u]_\Sigma = \varphi & \text{on } \Sigma, \end{cases} \qquad \begin{cases} -\Delta u = \lambda u & \text{in } \mathbb{R}^d \setminus \Sigma, \\ \partial_{\nu_+} u_+|_\Sigma = \varphi & \text{on } \Sigma, \\ \partial_{\nu_-} u_-|_\Sigma = -\varphi & \text{on } \Sigma, \end{cases}$$

have unique solutions $\tilde{u}_\lambda(\varphi), \hat{u}_\lambda(\varphi) \in H^{3/2}_\Delta(\mathbb{R}^d \setminus \Sigma)$. Moreover, the operator-valued functions $\tilde{\gamma}$ and $\hat{\gamma}$ satisfy $\tilde{\gamma}(\lambda)\varphi = \tilde{u}_\lambda(\varphi)$ and $\hat{\gamma}(\lambda)\varphi = \hat{u}_\lambda(\varphi)$ for $\varphi \in L^2(\Sigma)$ and $\lambda \in \mathbb{C} \setminus \mathbb{R}_+$.

Thanks to (11) the adjoints of $\tilde{\gamma}(\bar{\lambda})$ and $\hat{\gamma}(\bar{\lambda})$ can be expressed as

$$\tilde{\gamma}(\bar{\lambda})^* = \tilde{\Gamma}_1(\mathrm{H}_{\mathrm{free}} - \lambda)^{-1} \quad \text{and} \quad \hat{\gamma}(\bar{\lambda})^* = \hat{\Gamma}_1(\mathrm{K}_{\mathrm{N}} - \lambda)^{-1} \tag{14}$$

for $\lambda \in \mathbb{C} \setminus \mathbb{R}_+$. We also remark that, by Propositions 3.2 (iii) and 3.8 (iii) in [10], we have

$$\operatorname{ran} \widetilde{M}(\lambda) = \operatorname{ran} \widehat{M}(\lambda) = H^1(\Sigma), \quad \lambda \in \mathbb{C} \setminus \mathbb{R}_+. \tag{15}$$

According to Proposition 2.4, the formula

$$(\mathrm{K}_{\mathrm{N}} - \lambda)^{-1} - (\mathrm{H}_{\mathrm{free}} - \lambda)^{-1} = \hat{\gamma}(\lambda)\widehat{M}(\lambda)^{-1}\hat{\gamma}(\bar{\lambda})^* \tag{16}$$

holds for all $\lambda \in \mathbb{C} \setminus \mathbb{R}_+$. Since the operators of multiplication with α and ω are bounded and self-adjoint in $L^2(\Sigma)$, by Proposition 2.5 the extensions

$$\widetilde{T} \upharpoonright \ker(\alpha\tilde{\Gamma}_1 - \tilde{\Gamma}_0) \quad \text{and} \quad \widehat{T} \upharpoonright \ker(\omega\hat{\Gamma}_1 - \hat{\Gamma}_0)$$

are self-adjoint in $L^2(\mathbb{R}^d)$. In a way similar to [10], one can check that these restrictions coincide with $\mathrm{H}_{\alpha,\Sigma}$ and $\mathrm{K}_{\omega,\Sigma}$, respectively. Moreover, by Proposition 2.5, the formulae

$$(\mathrm{H}_{\alpha,\Sigma} - \lambda)^{-1} - (\mathrm{H}_{\mathrm{free}} - \lambda)^{-1} = \tilde{\gamma}(\lambda)(I - \alpha\widetilde{M}(\lambda))^{-1}\alpha\tilde{\gamma}(\bar{\lambda})^*, \tag{17a}$$

$$(\mathrm{K}_{\omega,\Sigma} - \lambda)^{-1} - (\mathrm{K}_{\mathrm{N}} - \lambda)^{-1} = \hat{\gamma}(\lambda)(I - \omega\widehat{M}(\lambda))^{-1}\omega\hat{\gamma}(\bar{\lambda})^*, \tag{17b}$$

hold for all $\lambda \in \rho(\mathrm{H}_{\alpha,\Sigma})$ and all $\lambda \in \rho(\mathrm{K}_{\omega,\Sigma})$, respectively. In these formulae the middle terms on the right-hand sides satisfy

$$(I - \alpha\widetilde{M}(\lambda))^{-1}, \ (I - \omega\widehat{M}(\lambda))^{-1} \in \mathcal{B}(L^2(\Sigma)) \tag{18}$$

for λ in the respective resolvent sets.

3 Proofs of the main results

In this section we prove the main results of the paper: the trace formulae for the Schrödinger operators with singular interactions. Theorems 1.2 and 1.3 are proved in Subsections 3.1 and 3.2, respectively. Throughout this section we use the notations

$$R(\lambda) := (\mathrm{H}_{\mathrm{free}} - \lambda)^{-1} \quad \text{and} \quad R_{\mathrm{N}}(\lambda) := (\mathrm{K}_{\mathrm{N}} - \lambda)^{-1}.$$

3.1 Proof of Theorem 1.2

To prove Theorem 1.2 we need an auxiliary lemma.

Lemma 3.1. *Let the γ-field $\tilde{\gamma}$ and the Weyl function $\widetilde{M}$ be as in Definition* 2.6. *Then for every $\lambda \in \mathbb{C} \setminus \mathbb{R}_+$ and every $k \in \mathbb{N}_0$ the following relations hold*:

(i) $\tilde{\gamma}^{(k)}(\lambda),\ \dfrac{\mathrm{d}^k}{\mathrm{d}\lambda^k}\tilde{\gamma}(\bar{\lambda})^* \in \mathfrak{S}_{(d-1)/(2k+3/2),\infty}$;

(ii) $\widetilde{M}^{(k)}(\lambda) \in \mathfrak{S}_{(d-1)/(2k+1),\infty}$.

Proof. (i) Let $\lambda \in \mathbb{C} \setminus \mathbb{R}_+$ and $k \in \mathbb{N}_0$. We observe that $\operatorname{ran}(R(\lambda)^k) \subset H^{2k}(\mathbb{R}^d)$. By the trace theorem we have $u|_\Sigma \in H^{s-1/2}(\Sigma)$ for every $u \in H^s(\mathbb{R}^d)$ with $s > 1/2$. Hence, we obtain from (14) that

$$\operatorname{ran}(\tilde{\gamma}(\bar{\lambda})^* R(\lambda)^k) \subset H^{2k+3/2}(\Sigma).$$

Thus Lemma 2.1 with $\mathcal{K} = L^2(\mathbb{R}^d)$ and $r = 2k + 3/2$ implies that

$$\tilde{\gamma}(\bar{\lambda})^* R(\lambda)^k \in \mathfrak{S}_{(d-1)/(2k+3/2),\infty}. \tag{19}$$

By taking the adjoint in (19) and replacing λ by $\bar{\lambda}$ we obtain

$$R(\lambda)^k \tilde{\gamma}(\lambda) \in \mathfrak{S}_{(d-1)/(2k+3/2),\infty}. \tag{20}$$

From (12a), (12b), (19), and (20) we obtain (i).

(ii) For $k = 0$ we observe that by (15) we have $\operatorname{ran} \widetilde{M}(\lambda) = H^1(\Sigma)$. Therefore, Lemma 2.1 with $\mathcal{K} = L^2(\Sigma)$ and $r = 1$ implies that $\widetilde{M}(\lambda) \in \mathfrak{S}_{d-1,\infty}$. For $k \geq 1$ we derive from (12c) that

$$\begin{aligned}\widetilde{M}^{(k)}(\lambda) = k!\,\tilde{\gamma}(\bar{\lambda})^* R(\lambda)^{k-1}\tilde{\gamma}(\lambda) &\in \mathfrak{S}_{(d-1)/(2(k-1)+3/2),\infty} \cdot \mathfrak{S}_{(d-1)/(3/2),\infty} \\ &= \mathfrak{S}_{(d-1)/(2k+1),\infty},\end{aligned}$$

where we applied (19), (20), and (6). □

Proof of Theorem 1.2. In order to shorten notation and to avoid the distinction of several cases, we set

$$\mathfrak{A}_r := \begin{cases} \mathfrak{S}_{(d-1)/r,\infty}(L^2(\Sigma)) & \text{if } r > 0, \\ \mathcal{B}(L^2(\Sigma)) & \text{if } r = 0. \end{cases}$$

It follows from (6) and the fact that $\mathfrak{S}_{p,\infty}(L^2(\Sigma))$ is an ideal in $\mathcal{B}(L^2(\Sigma))$ for $p > 0$ that

$$\mathfrak{A}_{r_1} \cdot \mathfrak{A}_{r_2} = \mathfrak{A}_{r_1+r_2}, \quad r_1, r_2 \geq 0. \tag{21}$$

The remainder of the proof is divided into two steps.

STEP 1. Let $\alpha \in L^\infty(\Sigma; \mathbb{R})$ and set

$$\widetilde{T}(\lambda) := (I - \alpha \widetilde{M}(\lambda))^{-1}, \quad \lambda \in \rho(\mathrm{H}_{\alpha,\Sigma}),$$

where $\widetilde{T}(\lambda) \in \mathcal{B}(L^2(\Sigma))$ by (18). Next, we show that

$$\widetilde{T}^{(k)}(\lambda) \in \mathfrak{A}_{2k+1}, \quad k \in \mathbb{N}, \tag{22}$$

by induction. Relation (10) implies that

$$\widetilde{T}'(\lambda) = \widetilde{T}(\lambda)\, \alpha \widetilde{M}'(\lambda)\, \widetilde{T}(\lambda), \tag{23}$$

which is in $\mathfrak{A}_3$ by Lemma 3.1 (ii). Let $m \in \mathbb{N}$ and assume that (22) is true for every $k = 1, \dots, m$, which implies, in particular, that

$$\widetilde{T}^{(k)}(\lambda) \in \mathfrak{A}_{2k}, \quad k = 0, \dots, m. \tag{24}$$

Then

$$\begin{aligned} \widetilde{T}^{(m+1)}(\lambda) &= \frac{\mathrm{d}^m}{\mathrm{d}\lambda^m}(\widetilde{T}(\lambda)\, \alpha \widetilde{M}'(\lambda)\, \widetilde{T}(\lambda)) \\ &= \sum_{\substack{p+q+r=m \\ p,q,r \geq 0}} \frac{m!}{p!\,q!\,r!} \widetilde{T}^{(p)}(\lambda)\, \alpha \widetilde{M}^{(q+1)}(\lambda)\, \widetilde{T}^{(r)}(\lambda) \end{aligned}$$

by (23) and (9b). Relation (24), the boundedness of α, Lemma 3.1 (ii) and (21) imply that

$$\widetilde{T}^{(p)}(\lambda)\, \alpha \widetilde{M}^{(q+1)}(\lambda)\, \widetilde{T}^{(r)}(\lambda) \in \mathfrak{A}_{2p} \cdot \mathfrak{A}_{2(q+1)+1} \cdot \mathfrak{A}_{2r} = \mathfrak{A}_{2(m+1)+1},$$

since $p + q + r = m$. This shows (22) for $k = m + 1$ and hence, by induction, for all $k \in \mathbb{N}$. Since $\widetilde{T}(\lambda) \in \mathcal{B}(L^2(\Sigma))$, we have, in particular,

$$\widetilde{T}^{(k)}(\lambda) \in \mathfrak{A}_{2k}, \quad k \in \mathbb{N}_0,\ \lambda \in \rho(\mathrm{H}_{\alpha,\Sigma}). \tag{25}$$

STEP 2. By taking derivatives we obtain from (17a) that, for $m \in \mathbb{N}$,

$$
\begin{aligned}
(m-1)!\, \widetilde{D}_{\alpha,m}(\lambda) &= \frac{\mathrm{d}^{m-1}}{\mathrm{d}\lambda^{m-1}}(\widetilde{D}_{\alpha,1}(\lambda)) \\
&= \frac{\mathrm{d}^{m-1}}{\mathrm{d}\lambda^{m-1}}(\tilde{\gamma}(\lambda)\, \widetilde{T}(\lambda)\, \alpha \tilde{\gamma}(\bar{\lambda})^*) \\
&= \sum_{\substack{p+q+r=m-1 \\ p,q,r \geq 0}} \frac{(m-1)!}{p!\,q!\,r!}\, \tilde{\gamma}^{(p)}(\lambda) \widetilde{T}^{(q)}(\lambda)\, \alpha \frac{\mathrm{d}^r}{\mathrm{d}\lambda^r} \tilde{\gamma}(\bar{\lambda})^*.
\end{aligned}
\tag{26}
$$

By Lemma 3.1 (i) and (25), each term in the sum satisfies

$$
\begin{aligned}
\tilde{\gamma}^{(p)}(\lambda)\, \widetilde{T}^{(q)}(\lambda)\, \alpha \frac{\mathrm{d}^r}{\mathrm{d}\lambda^r} \tilde{\gamma}(\bar{\lambda})^* \in \mathfrak{A}_{2p+3/2} \cdot \mathfrak{A}_{2q} \cdot \mathfrak{A}_{2r+3/2} &= \mathfrak{A}_{2m+1} \\
&= \mathfrak{S}_{(d-1)/(2m+1),\infty}.
\end{aligned}
\tag{27}
$$

If $m \in \mathbb{N}$ is such that $m > (d-2)/2$, then $(d-1)/(2m+1) < 1$ and, by (5) and (27), all terms in the sum in (26) are trace class operators, and the same is true if we change the order in the product in (27). Hence, we can apply the trace to the expression in (26) and use (7), (8), and (12c) to obtain

$$
\begin{aligned}
&(m-1)!\ \mathrm{Tr}(\widetilde{D}_{\alpha,m}(\lambda)) \\
&= \mathrm{Tr}\Big(\sum_{\substack{p+q+r=m-1 \\ p,q,r \geq 0}} \frac{(m-1)!}{p!\,q!\,r!}\, \tilde{\gamma}^{(p)}(\lambda)\, \widetilde{T}^{(q)}(\lambda)\, \alpha \frac{\mathrm{d}^r}{\mathrm{d}\lambda^r} \tilde{\gamma}(\bar{\lambda})^* \Big) \\
&= \sum_{\substack{p+q+r=m-1 \\ p,q,r \geq 0}} \frac{(m-1)!}{p!\,q!\,r!}\ \mathrm{Tr}\Big(\tilde{\gamma}^{(p)}(\lambda)\, \widetilde{T}^{(q)}(\lambda)\, \alpha \frac{\mathrm{d}^r}{\mathrm{d}\lambda^r} \tilde{\gamma}(\bar{\lambda})^* \Big) \\
&= \sum_{\substack{p+q+r=m-1 \\ p,q,r \geq 0}} \frac{(m-1)!}{p!\,q!\,r!}\ \mathrm{Tr}\Big(\widetilde{T}^{(q)}(\lambda)\, \alpha \Big(\frac{\mathrm{d}^r}{\mathrm{d}\lambda^r} \tilde{\gamma}(\bar{\lambda})^* \Big)\, \tilde{\gamma}^{(p)}(\lambda) \Big) \\
&= \mathrm{Tr}\Big(\sum_{\substack{p+q+r=m-1 \\ p,q,r \geq 0}} \frac{(m-1)!}{p!\,q!\,r!}\, \widetilde{T}^{(q)}(\lambda)\, \alpha \Big(\frac{\mathrm{d}^r}{\mathrm{d}\lambda^r} \tilde{\gamma}(\bar{\lambda})^* \Big)\, \tilde{\gamma}^{(p)}(\lambda) \Big) \\
&= \mathrm{Tr}\Big(\frac{\mathrm{d}^{m-1}}{\mathrm{d}\lambda^{m-1}} (\widetilde{T}(\lambda)\, \alpha \tilde{\gamma}(\bar{\lambda})^*\, \tilde{\gamma}(\lambda)) \Big) \\
&= \mathrm{Tr}\Big(\frac{\mathrm{d}^{m-1}}{\mathrm{d}\lambda^{m-1}} (\widetilde{T}(\lambda)\, \alpha \widetilde{M}'(\lambda)) \Big),
\end{aligned}
$$

which finishes the proof. □

3.2 Proof of Theorem 1.3

First, we need three preparatory lemmas. The proof of the first of them is completely analogous to the proof of Lemma 3.1 and is therefore omitted.

Lemma 3.2. *Let the γ-field $\hat{\gamma}$ and the Weyl function $\widehat{M}$ be as in Definition* 2.6. *Then for every $\lambda \in \mathbb{C} \setminus \mathbb{R}_+$ and every $k \in \mathbb{N}_0$ the following relations hold*:

(i) $\hat{\gamma}^{(k)}(\lambda), \dfrac{\mathrm{d}^k}{\mathrm{d}\lambda^k}\hat{\gamma}(\bar{\lambda})^* \in \mathfrak{S}_{(d-1)/(2k+3/2),\infty}$;

(ii) $\widehat{M}^{(k)}(\lambda) \in \mathfrak{S}_{(d-1)/(2k+1),\infty}$.

Lemma 3.3. *Let the γ-field $\hat{\gamma}$ and the Weyl function $\widehat{M}$ be as in Definition* 2.6. *Then for all $s \geq 0$, and all $\lambda \in \mathbb{C} \setminus \mathbb{R}_+$ the following statements hold*:

(i) $\operatorname{ran}(\hat{\gamma}(\bar{\lambda})^* \upharpoonright H^s(\mathbb{R}^d)) \subset H^{s+3/2}(\Sigma)$;

(ii) $\operatorname{ran}(\widehat{M}(\lambda) \upharpoonright H^s(\Sigma)) = H^{s+1}(\Sigma)$.

Proof. (i) According to (14) we have

$$\hat{\gamma}(\bar{\lambda})^* = \widehat{\Gamma}_1 R_{\mathrm{N}}(\lambda).$$

Employing the regularity shift property (Theorem 4.20 in [45]) and the trace theorem (Theorem 3.37 in [45]) we conclude that

$$\operatorname{ran}(\hat{\gamma}(\bar{\lambda})^* \upharpoonright H^s(\mathbb{R}^d)) \subset H^{s+3/2}(\Sigma)$$

holds for all $s \geq 0$.

(ii) Define the space $H^s(\mathbb{R}^d \setminus \Sigma) := H^s(\Omega_+) \oplus H^s(\Omega_-)$. It follows from the decomposition

$$\operatorname{dom} \widehat{T} = \operatorname{dom} \mathrm{K}_{\mathrm{N}} \dot{+} \ker(\widehat{T} - \lambda), \quad \lambda \in \mathbb{C} \setminus \mathbb{R}_+,$$

and the properties of the Neumann trace (§2.7.3 in [43]) that the restriction of the mapping $\widehat{\Gamma}_0$ to

$$\ker(\widehat{T} - \lambda) \cap H^{s+3/2}(\mathbb{R}^d \setminus \Sigma)$$

is a bijection onto $H^s(\Sigma)$ for $s \geq 0$. This, together with the definition of the γ-field, implies that

$$\operatorname{ran}(\hat{\gamma}(\lambda) \upharpoonright H^s(\Sigma)) = \ker(\widehat{T} - \lambda) \cap H^{s+3/2}(\mathbb{R}^d \setminus \Sigma) \subset H^{s+3/2}(\mathbb{R}^d \setminus \Sigma).$$

Hence, it follows from the definition of $\widehat{M}(\lambda)$, the definition of $\widehat{\Gamma}_1$ in (13) and the trace theorem that

$$\operatorname{ran}(\widehat{M}(\lambda) \upharpoonright H^s(\Sigma)) \subset H^{s+1}(\Sigma).$$

To verify the opposite inclusion, let $\psi \in H^{s+1}(\Sigma)$. The decomposition

$$\operatorname{dom} \widehat{T} = \operatorname{dom} \mathrm{H}_{\mathrm{free}} \dotplus \ker(\widehat{T} - \lambda), \quad \lambda \in \mathbb{C} \setminus \mathbb{R}_+,$$

implies that there exists a function $f_\lambda \in \ker(\widehat{T} - \lambda) \cap H^{s+3/2}(\mathbb{R}^d \setminus \Sigma)$ such that $\widehat{\Gamma}_1 f_\lambda = \psi$. Thus,

$$\widehat{\Gamma}_0 f_\lambda = \varphi \in H^s(\Sigma) \quad \text{and} \quad \widehat{M}(\lambda)\varphi = \psi,$$

that is, $H^{s+1}(\Sigma) \subset \operatorname{ran}(\widehat{M}(\lambda) \upharpoonright H^s(\Sigma))$, and the assertion is shown. □

Lemma 3.4. *Let the self-adjoint operators* $\mathrm{H}_{\mathrm{free}}$ *and* K_{N} *be as in Definition* 1.1, *and let the operator-valued function* $\widehat{M}$ *be as in* (4). *Then for all* $m \in \mathbb{N}$ *such that* $m > (d-1)/2$ *and all* $\lambda \in \mathbb{C} \setminus \mathbb{R}_+$ *the resolvent power difference*

$$\widehat{D}_m(\lambda) := (\mathrm{K}_{\mathrm{N}} - \lambda)^{-m} - (\mathrm{H}_{\mathrm{free}} - \lambda)^{-m}$$

belongs to the trace class, and its trace can be expressed as

$$\operatorname{Tr}(\widehat{D}_m(\lambda)) = \frac{1}{(m-1)!} \operatorname{Tr}\Big(\frac{\mathrm{d}^{m-1}}{\mathrm{d}\lambda^{m-1}}(\widehat{M}(\lambda)^{-1}\widehat{M}'(\lambda))\Big).$$

Proof. The proof is divided into three steps.

STEP 1. Let us introduce the operator-valued function

$$S(\lambda) := \widehat{M}(\lambda)^{-1}\hat{\gamma}(\bar{\lambda})^*, \quad \lambda \in \mathbb{C} \setminus \mathbb{R}_+.$$

Note that the product is well defined since by Lemma 3.3 (i)

$$\operatorname{ran}(\hat{\gamma}(\bar{\lambda})^*) \subset H^1(\Sigma) = \operatorname{dom}(\widehat{M}(\lambda)^{-1}).$$

The closed graph theorem implies that $S(\lambda) \in \mathcal{B}(L^2(\mathbb{R}^d), L^2(\Sigma))$ for all $\lambda \in \mathbb{C} \setminus \mathbb{R}_+$. Next we prove the following smoothing property for the derivatives of S:

$$\operatorname{ran}(S^{(k)}(\lambda) \upharpoonright H^s(\mathbb{R}^d)) \subset H^{s+2k+1/2}(\Sigma), \quad s \geq 0,\ k \in \mathbb{N}_0, \tag{28}$$

by induction. Since, by Lemma 3.3 (i), $\hat{\gamma}(\bar{\lambda})^*$ maps $H^s(\mathbb{R}^d)$ into $H^{s+3/2}(\Sigma)$ for all $s \geq 0$ and $\widehat{M}(\lambda)^{-1}$ maps $H^{s+3/2}(\Sigma)$ into $H^{s+1/2}(\Sigma)$ by Lemma 3.3 (ii), relation (28) is true for $k = 0$. Now let $l \in \mathbb{N}_0$ and assume that (28) is true for every $k = 0, 1, \dots, l$. It follows from (9a), (10), (12b), (12c), and (16) that for $\lambda \in \mathbb{C} \setminus \mathbb{R}_+$,

$$\begin{aligned} S'(\lambda) &= \frac{\mathrm{d}}{\mathrm{d}\lambda}(\widehat{M}(\lambda)^{-1})\hat{\gamma}(\bar{\lambda})^* + \widehat{M}(\lambda)^{-1}\frac{\mathrm{d}}{\mathrm{d}\lambda}\hat{\gamma}(\bar{\lambda})^* \\ &= -\widehat{M}(\lambda)^{-1}\widehat{M}'(\lambda)\widehat{M}(\lambda)^{-1}\hat{\gamma}(\bar{\lambda})^* + \widehat{M}(\lambda)^{-1}\hat{\gamma}(\bar{\lambda})^* R_{\mathrm{N}}(\lambda) \\ &= -\widehat{M}(\lambda)^{-1}\hat{\gamma}(\bar{\lambda})^*\hat{\gamma}(\lambda)\widehat{M}(\lambda)^{-1}\hat{\gamma}(\bar{\lambda})^* + \widehat{M}(\lambda)^{-1}\hat{\gamma}(\bar{\lambda})^* R_{\mathrm{N}}(\lambda) \\ &= S(\lambda)[R_{\mathrm{N}}(\lambda) - \hat{\gamma}(\lambda)\widehat{M}(\lambda)^{-1}\hat{\gamma}(\bar{\lambda})^*] \\ &= S(\lambda)R(\lambda). \end{aligned}$$

Hence, with the help of (9a) we obtain

$$\begin{aligned} S^{(l+1)}(\lambda) &= \frac{\mathrm{d}^l}{\mathrm{d}\lambda^l}(S(\lambda)R(\lambda)) \\ &= \sum_{\substack{p+q=l \\ p,q\geq 0}} \binom{l}{p} S^{(p)}(\lambda)R^{(q)}(\lambda) \\ &= \sum_{\substack{p+q=l \\ p,q\geq 0}} \frac{l!}{p!} S^{(p)}(\lambda)R(\lambda)^{q+1}. \end{aligned} \tag{29}$$

Using the induction hypothesis, formula (29) and smoothing properties of $R(\lambda)$, we deduce that, for $p, q \geq 0$, $p + q = l$,

$$\begin{aligned} \operatorname{ran}(S^{(p)}(\lambda)R(\lambda)^{q+1} \restriction H^s(\mathbb{R}^d)) &\subset \operatorname{ran}(S^{(p)}(\lambda) \restriction H^{s+2(q+1)}(\mathbb{R}^d)) \\ &\subset H^{s+2(p+q+1)+1/2}(\Sigma) \\ &= H^{s+2(l+1)+1/2}(\Sigma), \end{aligned}$$

which shows (28) for $k = l + 1$ and hence, by induction, for all $k \in \mathbb{N}_0$. Therefore, an application of Lemma 2.1 with $\mathcal{K} = L^2(\Sigma)$ and $r = 2k + 1/2$ yields that

$$S^{(k)}(\lambda) \in \mathfrak{S}_{(d-1)/(2k+1/2),\infty}, \quad k \in \mathbb{N}_0,\ \lambda \in \mathbb{C} \setminus \mathbb{R}_+. \tag{30}$$

STEP 2. Using Krein's formula in (16) and (9a) we obtain that, for $m \in \mathbb{N}$ and $\lambda \in \mathbb{C} \setminus \mathbb{R}_+$,

$$\widehat{D}_m(\lambda) = \frac{1}{(m-1)!} \cdot \frac{\mathrm{d}^{m-1}}{\mathrm{d}\lambda^{m-1}}(\widehat{D}_1(\lambda)) \tag{31}$$

$$= \frac{1}{(m-1)!} \cdot \frac{\mathrm{d}^{m-1}}{\mathrm{d}\lambda^{m-1}}(\hat{\gamma}(\lambda)S(\lambda))$$

$$= \frac{1}{(m-1)!} \sum_{\substack{p+q=m-1\\ p,q\geq 0}} \binom{m-1}{p} \hat{\gamma}^{(p)}(\lambda)S^{(q)}(\lambda). \tag{32}$$

By Lemma 3.2 (i), (30), and (6) we have

$$\begin{aligned}\hat{\gamma}^{(p)}(\lambda)S^{(q)}(\lambda) \in \mathfrak{S}_{(d-1)/(2p+3/2),\infty} \cdot \mathfrak{S}_{(d-1)/(2q+1/2),\infty} &= \mathfrak{S}_{(d-1)/(2(p+q)+2),\infty} \\ &= \mathfrak{S}_{(d-1)/(2m),\infty}\end{aligned} \tag{33}$$

for p, q with $p + q = m - 1$.

STEP 3. If $m > (d-1)/2$, then $(d-1)/(2m) < 1$ and, by (33), each term in the sum in (32) is a trace class operator and, by a similar argument, also $S^{(q)}(\lambda)\hat{\gamma}^{(p)}(\lambda)$. Hence, the resolvent power difference $\widehat{D}_m(\lambda)$ is a trace class operator, and we can apply the trace to (32) and use (7), (8), and (12c) to obtain

$$\begin{aligned}(m-1)!\,\mathrm{Tr}(\widehat{D}_m(\lambda)) &= \mathrm{Tr}\Bigg(\sum_{\substack{p+q=m-1\\ p,q\geq 0}} \binom{m-1}{p} \hat{\gamma}^{(p)}(\lambda)\, S^{(q)}(\lambda)\Bigg) \\ &= \sum_{\substack{p+q=m-1\\ p,q\geq 0}} \binom{m-1}{p} \mathrm{Tr}(\hat{\gamma}^{(p)}(\lambda)\, S^{(q)}(\lambda)) \\ &= \sum_{\substack{p+q=m-1\\ p,q\geq 0}} \binom{m-1}{p} \mathrm{Tr}(S^{(q)}(\lambda)\, \hat{\gamma}^{(p)}(\lambda)) \\ &= \mathrm{Tr}\Bigg(\sum_{\substack{p+q=m-1\\ p,q\geq 0}} \binom{m-1}{p} S^{(q)}(\lambda)\, \hat{\gamma}^{(p)}(\lambda)\Bigg) \\ &= \mathrm{Tr}\Big(\frac{\mathrm{d}^{m-1}}{\mathrm{d}\lambda^{m-1}}(S(\lambda)\, \hat{\gamma}(\lambda))\Big)\end{aligned}$$

$$= \operatorname{Tr}\Big(\frac{\mathrm{d}^{m-1}}{\mathrm{d}\lambda^{m-1}}(\widehat{M}(\lambda)^{-1}\,\hat{\gamma}(\bar{\lambda})^*\,\hat{\gamma}(\lambda))\Big)$$

$$= \operatorname{Tr}\Big(\frac{\mathrm{d}^{m-1}}{\mathrm{d}\lambda^{m-1}}(\widehat{M}(\lambda)^{-1}\,\widehat{M}'(\lambda))\Big),$$

which finishes the proof. □

Proof of Theorem 1.3. (i) The proof of this statement is fully analogous to the proof of Theorem 1.2. One has to replace in the argument $\mathrm{H}_{\mathrm{free}}$, α, $\mathrm{H}_{\alpha,\Sigma}$, $\widetilde{M}$, $\tilde{\gamma}$, by K_{N}, ω, $\mathrm{K}_{\omega,\Sigma}$, $\widehat{M}$, $\hat{\gamma}$, respectively. Moreover, Krein's resolvent formula in (17b) is used instead of Krein's formula in (17a) and Lemma 3.2 instead of Lemma 3.1.

(ii) By item (i) of this theorem and by Lemma 3.4, for every $m \in \mathbb{N}$ such that $m > (d-1)/2$ and every $\lambda \in \rho(\mathrm{K}_{\omega,\Sigma})$ both operators $\widehat{D}_m(\lambda)$ and $\widehat{D}_{\omega,m}(\lambda)$ belong to the trace class. In view of the identity $\widehat{E}_{\omega,m}(\lambda) = \widehat{D}_m(\lambda) + \widehat{D}_{\omega,m}(\lambda)$, we infer that $\widehat{E}_{\omega,m}(\lambda)$ is also in the trace class. Using the formula (7) we have

$$\operatorname{Tr}(\widehat{E}_{\omega,m}(\lambda)) = \operatorname{Tr}(\widehat{D}_{\omega,m}(\lambda)) + \operatorname{Tr}(\widehat{D}_m(\lambda)).$$

Combining the trace formula in (i) of this theorem and the trace formula in Lemma 3.4 we obtain

$$\begin{aligned}\operatorname{Tr}(\widehat{E}_{\omega,m}(\lambda)) &= \frac{1}{(m-1)!}\operatorname{Tr}\Big(\frac{\mathrm{d}^{m-1}}{\mathrm{d}\lambda^{m-1}}((I-\omega\widehat{M}(\lambda))^{-1}\omega\widehat{M}'(\lambda) + \widehat{M}(\lambda)^{-1}\widehat{M}'(\lambda))\Big)\\ &= \frac{1}{(m-1)!}\operatorname{Tr}\Big(\frac{\mathrm{d}^{m-1}}{\mathrm{d}\lambda^{m-1}}((I-\omega\widehat{M}(\lambda))^{-1}\widehat{M}(\lambda)^{-1}\widehat{M}'(\lambda))\Big),\end{aligned}$$

which finishes the proof. □

Acknowledgements. Jussi Behrndt gratefully acknowledges financial support by the Austrian Science Fund (FWF): Project P 25162-N26. Vladimir Lotoreichik gratefully acknowledges financial support by the Czech Science Foundation (GAČR): Project 14-06818S.

References

[1] J.-P. Antoine, F. Gesztesy, and J. Shabani, Exactly solvable models of sphere interactions in quantum mechanics. *J. Phys. A* **20** (1987), no. 12, 3687–3712. MR 0913638

[2] W. Arendt and R. Mazzeo, Friedlander's eigenvalue inequalities and the Dirichlet–to–Neumann semigroup. *Commun. Pure Appl. Anal.* **11** (2012), no. 6, 2201–2212. MR 2912743 Zbl 1267.35139

[3] W. Arendt and A. F. M. ter Elst, The Dirichlet–to–Neumann operator on rough domains. *J. Differential Equations* **251** (2011), no. 8, 2100–2124. MR 2823661 Zbl 1241.47036

[4] W. Arendt and A. F. M. ter Elst, The Dirichlet–to–Neumann operator on exterior domains. *Potential Anal.* **43** (2015), no. 2, 313–340. MR 3374114 Zbl 1331.46022

[5] C. Bardos, J.-C. Guillot, and J. Ralston, La rélation de Poisson pour l'équation des ondes dans un ouvert non borne. Application à la théorie de la diffusion. *Comm. Partial Differential Equations* **7** (1982), no. 8, 905–958. MR 0668585 Zbl 0496.35067

[6] J. Behrndt, P. Exner, and V. Lotoreichik, Schrödinger operators with δ- and δ'-interactions on Lipschitz surfaces and chromatic numbers of associated partitions. *Rev. Math. Phys.* **26** (2014), no. 8, 1450015, 43 pp. MR 3256859 Zbl 1326.47050

[7] J. Behrndt, G. Grubb, M. Langer, and V. Lotoreichik, Spectral asymptotics for resolvent differences of elliptic operators with δ and δ'-interactions on hypersurfaces. *J. Spectr. Theory* **5** (2015), no. 4, 697–729. MR 3433285 Zbl 06534463

[8] J. Behrndt and M. Langer, Boundary value problems for elliptic partial differential operators on bounded domains. *J. Funct. Anal.* **243** (2007), no. 2, 536–565. MR 2289696 Zbl 1132.47038

[9] J. Behrndt and M. Langer, Elliptic operators, Dirichlet–to–Neumann maps and quasi boundary triples. In S. Hassi, H. S. V. de Snoo and F. H. Szafraniec (eds.), *Operator methods for boundary value problems.* London Mathematical Society Lecture Note Series, 404. Cambridge University Press, Cambridge, 2012, 121–160. MR 3050306 Zbl 1331.47067

[10] J. Behrndt, M. Langer, and V. Lotoreichik, Schrödinger operators with δ and δ'-potentials supported on hypersurfaces. *Ann. Henri Poincaré* **14** (2013), no. 2, 385–423. MR 3028043 Zbl 1275.81027

[11] J. Behrndt, M. Langer, and V. Lotoreichik, Spectral estimates for resolvent differences of self-adjoint elliptic operators. *Integral Equations Operator Theory* **77** (2013), no. 1, 1–37. MR 3090162 Zbl 1311.47059

[12] J. Behrndt, M. Langer, and V. Lotoreichik, Trace formulae and singular values of resolvent power differences of self-adjoint elliptic operators. *J. Lond. Math. Soc.* (2) **88** (2013), no. 2, 319–337. MR 3106724 Zbl 1296.35097

[13] J. Behrndt and J. Rohleder, Spectral analysis of selfadjoint elliptic differential operators, Dirichlet–to–Neumann maps, and abstract Weyl functions. *Adv. Math.* **285** (2015), 1301–1338. MR 3406527 Zbl 1344.47018

[14] J. F. Brasche, P. Exner, Y. A. Kuperin, and P. Šeba, Schrödinger operators with singular interactions. *J. Math. Anal. Appl.* **184** (1994), no. 1, 112–139. MR 1275948 Zbl 0820.47005

[15] J. F. Brasche and A. Teta, Spectral analysis and scattering theory for Schrödinger operators with an interaction supported by a regular curve. In S. Albeverio, J. E. Fenstad, H. Holden and T. Lindstrøm (eds.), *Ideas and methods in quantum and statistical physics.* In memory of Raphael Høegh-Krohn (1938–1988). Vol. 2. Papers from the Symposium on Ideas and Methods in Mathematics and Physics held at the University of Oslo, Oslo, September 1988. Cambridge University Press, Cambridge, 1992, 197–211. MR 1190526 Zbl 0787.35057

[16] V. M. Bruk, A certain class of boundary value problems with a spectral parameter in the boundary condition. *Mat. Sb. (N.S.)* **100(142)** (1976), no. 2, 210–216. MR 0415381

[17] J. Brüning, V. Geyler, and K. Pankrashkin, Spectra of self-adjoint extensions and applications to solvable Schrödinger operators. *Rev. Math. Phys.* **20** (2008), no. 1, 1–70. MR 2379246 Zbl 1163.81007

[18] V. S. Buslaev and L. D. Faddeev, Formulas for traces for a singular Sturm-Liouville differential operator. *Dokl. Akad. Nauk SSSR* **132**, 13–16. In Russian. English translation, *Soviet Math. Dokl.* **1** (1960), 451–454. MR 0120417 Zbl 0129.06501

[19] G. Carron, Determinant relatif et la fonction Xi. *Amer. J. Math.* **124** (2002), no. 2, 307–352. MR 1890995 Zbl 1014.34015

[20] S. Clark, F. Gesztesy, R. Nichols, and M. Zinchenko, Boundary data maps and Krein's resolvent formula for Sturm–Liouville operators on a finite interval. *Oper. Matrices* **8** (2014), no. 1, 1–71. MR 3202926 Zbl 1311.34037

[21] V. A. Derkach and M. M. Malamud, Generalized resolvents and the boundary value problems for Hermitian operators with gaps. *J. Funct. Anal.* **95** (1991), no. 1, 1–95. MR 1087947 Zbl 0748.47004

[22] V. A. Derkach and M. M. Malamud, The extension theory of Hermitian operators and the moment problem. *J. Math. Sci.* **73** (1995), no. 2, 141–242. MR 1318517 Zbl 0848.47004

[23] J. Dittrich, P. Exner, C. Kühn, and K. Pankrashkin, On eigenvalue asymptotics for strong δ-interactions supported by surfaces with boundaries. *Asymptot. Anal.* **97** (2016), no. 1–2, 1–25. MR 3475116 Zbl 06600512

[24] V. Duchêne and N. Raymond, Spectral asymptotics of a broken δ-interaction. *J. Phys. A* **47** (2014), no. 15, 155203, 19 pp. MR 3191674 Zbl 1288.81044

[25] A. F. M. ter Elst and E. M. Ouhabaz, Analysis of the heat kernel of the Dirichlet–to–Neumann operator. *J. Funct. Anal.* **267** (2014), no. 11, 4066–4109. MR 3269871 Zbl 1317.35101

[26] P. Exner, Leaky quantum graphs: a review. In P. Exner, J. P. Keating, P. Kuchment, T. Sunada, and A. Teplyaev (eds.), *Analysis on graphs and its applications.* Papers from the program held in Cambridge, January 8–June 29, 2007. Proceedings of Symposia in Pure Mathematics, 77. American Mathematical Society, Providence, R.I., 2008, 523–564. MR 2459890 Zbl 1153.81487

[27] P. Exner, E. M. Harrell, and M. Loss, Inequalities for means of chords, with application to isoperimetric problems. *Lett. Math. Phys.* **75** (2006), no. 3, 225–233. MR 2211029 Zbl 1111.35019

[28] P. Exner and T. Ichinose, Geometrically induced spectrum in curved leaky wires. *J. Phys. A* **34** (2001), no. 7, 1439–1450. MR 1819942 Zbl 1002.81024

[29] P. Exner and M. Jex, Spectral asymptotics of a strong δ' interaction on a planar loop. *J. Phys. A* **46** (2013), no. 34, 345201, 12 pp. MR 3094827 Zbl 1273.81088

[30] P. Exner and H. Kovařík, *Quantum waveguides.* Theoretical and Mathematical Physics. Springer, Cham, 2015. ISBN 978-3-319-18575-0 MR 3362506 Zbl 1314.81001

[31] L. Friedlander, Absolute continuity of the spectra of periodic waveguides. In P. Kuchment (ed.), *Waves in periodic and random media.* Proceedings of the AMS-IMS-SIAM Joint Summer Research Conference held at Mount Holyoke College, South Hadley, MA, June 22–28, 2002. Contemporary Mathematics, 339. American Mathematical Society, Providence, R.I., 2003, 37–42. MR 2042530 Zbl 1044.35035

[32] F. Gesztesy and M. Zinchenko, Symmetrized perturbation determinants and applications to boundary data maps and Krein-type resolvent formulas. *Proc. Lond. Math. Soc.* (3) **104** (2012), no. 3, 577–612. MR 2900237 Zbl 1247.34030

[33] I. C. Gohberg and M. G. Kreĭn, *Introduction to the theory of linear nonselfadjoint operators.* Translated from the Russian by A. Feinstein. Translations of Mathematical Monographs, 18. American Mathematical Society, Providence, R.I., 1969. MR 0246142 Zbl 0181.13504

[34] G. Grubb, Remarks on trace estimates for exterior boundary problems. *Comm. Partial Differential Equations* **9** (1984), no. 3, 231–270. MR 0740091 Zbl 0553.35022

[35] M. Jex and V. Lotoreichik, On absence of bound states for weakly attractive δ'-interactions supported on non-closed curves in $\mathbb{R}^2$. *J. Math. Phys.* **57** (2016), no. 2, 022101, 20 pp. MR 3449215 Zbl 1338.81201

[36] R. Jost and A. Pais, On the scattering of a particle by a static potential. *Physical Rev.* (2) **82** (1951). 840–851. MR 0044404 Zbl 0042.45206

[37] T. Kato, *Perturbation theory for linear operators.* Reprint of the 1980 edition. Classics in Mathematics, Springer, Berlin, 1995. ISBN 3-540-58661-X MR 1335452 Zbl 0836.47009

[38] R. Killip and B. Simon, Sum rules for Jacobi matrices and their applications to spectral theory. *Ann. of Math.* (2) **158** (2003), no. 1, 253–321. MR 1999923 Zbl 1050.47025

[39] R. Killip and B. Simon, Sum rules and spectral measures of Schrödinger operators with L^2-potentials. *Ann. of Math.* (2) **170** (2009), no. 2, 739–782. MR 2552106 Zbl 1185.34131

[40] A. N. Kočubeĭ, Extensions of symmetric operators and of symmetric binary relations. *Mat. Zametki* **17** (1975), 41–48. In Russian. English translation, *Math. Notes* **17** (1975), no. 1, 25–28. MR 0365218 Zbl 0322.47006

[41] S. Kondej and V. Lotoreichik, Weakly coupled bound state of 2-D Schrödinger operator with potential-measure. *J. Math. Anal. Appl.* **420** (2014), no. 2, 1416–1438. MR 3240086 Zbl 1298.35128

[42] P. Lax and R. Phillips, Decaying modes for the wave equation in the exterior of an obstacle. *Comm. Pure Appl. Math.* **22** (1969), 737–787. MR 0254432 Zbl 0181.38201

[43] J.-L. Lions and E. Magenes, *Non-homogeneous boundary value problems and applications.* Vol. I. Translated from the French by P. Kenneth. Die Grundlehren der mathematischen Wissenschaften, Band 181. Springer, Berlin etc., 1972. MR 0350177 Zbl 0223.35039

[44] V. Lotoreichik and T. Ourmières-Bonafos, On the bound states of Schrödinger operators with δ-interactions on conical surfaces. *Comm. Partial Differential Equations* **41** (2016), no. 6, 999–1028. MR 3509521 Zbl 06617406

[45] W. McLean, *Strongly elliptic systems and boundary integral equations.* Cambridge University Press, Cambridge, 2000. ISBN 0-521-66332-6; 0-521-66375-X MR 1742312 Zbl 0948.35001

[46] R. Melrose, Weyl asymptotics for the phase in obstacle scattering. *Comm. Partial Differential Equations* **13** (1988), no. 11, 1431–1439. MR 0956828 Zbl 0686.35089

[47] B. Simon, *Trace ideals and their applications.* Second edition. Mathematical Surveys and Monographs, 120. American Mathematical Society, Providence, RI, 2005. MR 2154153 Zbl 1074.47001

[48] O. Steinbach, *Numerical approximation methods for elliptic boundary value problems.* Translated from the 2003 German original. Springer, New York, 2008. ISBN 978-0-387-31312-2 MR 2361676 Zbl 1153.65302

An improved bound for the non-existence of radial solutions of the Brezis–Nirenberg problem in $\mathbb{H}^n$

Rafael D. Benguria and Soledad Benguria

To Pavel Exner on the occasion of his 70th birthday

1 Introduction

For a long time virial theorems have played a key role in the localization of linear and nonlinear eigenvalues. In the spectral theory of Schrödinger Operators, the virial theorem has been widely used to prove the absence of positive eigenvalues for various multiparticle quantum systems (see, e.g., [12], [10], and [1]). In 1983, Brézis and Nirenberg [4] considered the existence and nonexistence of solutions of the nonlinear equation

$$-\Delta u = \lambda u + |u|^{p-1} u,$$

defined on a bounded, smooth domain of $\mathbb{R}^n$, $n > 2$, with Dirichlet boundary conditions, where $p = (n+2)/(n-2)$ is the critical Sobolev exponent. In particular, they used a virial theorem, namely the Pohozaev identity [8], to prove the nonexistence of regular solutions when the domain is star–shaped, for any $\lambda \le 0$, for any $n > 2$. After the classical paper [4] of Brézis and Nirenberg, many people have considered extensions of this problem in different settings. In particular, the Brézis–Nirenberg (BN) problem has been studied on bounded, smooth, domains of the hyperbolic space $\mathbb{H}^n$ (see, e.g., [11], [2], [6], and [3]), where one replaces the Laplacian by the Laplace–Beltrami operator in $\mathbb{H}^n$. Stapelkamp [11] proved the analog of the above mentioned nonexistence result of Brézis–Nirenberg in $\mathbb{H}^n$. Namely she proved that there are no regular solutions of the BN problem for bounded, smooth, star–shaped domains in $\mathbb{H}^n$ ($n > 2$), if $\lambda \le n(n-2)/4$. The purpose of this manuscript is to give an improved bound on λ for the nonexistence of (not necessarily positive) radial, regular solutions of the BN problem on geodesic balls of $\mathbb{H}^n$ for $2 < n < 4$ (see Theorem 2.1 below). Notice that for the case of radial solutions of the BN problem on a geodesic ball one can consider noninteger values of n, which can be considered just as a parameter.

Consider the Brezis–Nirenberg problem

$$-\Delta_{\mathbb{H}^n} u = \lambda u + |u|^{p-1} u, \tag{1}$$

on $\Omega \subset \mathbb{H}^n$, where Ω is smooth and bounded, with Dirichlet boundary conditions, i.e., $u = 0$ in $\partial\Omega$. After expressing the Laplace–Beltrami operator $\Delta_{\mathbb{H}^n}$ in terms of the conformal Laplacian, Stapelkamp [11] proved that (1) does not admit any regular solution for star-shaped domains Ω provided

$$\lambda \leq \frac{n(n-2)}{4}. \tag{2}$$

Here, we consider the BN problem (1) for radial solutions on geodesic balls of $\mathbb{H}^n$. We can prove a different bound, namely the problem for radial solutions on a geodesic ball Ω^* does not admit a solution if

$$\lambda \leq \frac{n^2(n-1)}{4(n+2)} \tag{3}$$

for $n > 2$. Our bound is better than (2) in the radial case, if $2 < n < 4$. Both bounds coincide when $n = 4$. In the rest of this manuscript we give the proof of (3).

2 Nonexistence of solutions of the BN problem on geodesic balls in $\mathbb{H}^n$, for $2 < n < 4$.

In the sequel we consider (not necessarily positive) radial solutions of the BN problem (1) on geodesic balls of $\mathbb{H}^n$. In *radial coordinates*, (1) can be written as

$$-u''(x) - (n-1)\coth(x)u'(x) = \lambda u(x) + |u|^{p-1}u(x), \tag{4}$$

with $u'(0) = u(R) = 0$, where R is the *radius* of the geodesic ball. Here, as before, $p = (n+2)/(n-2)$. Notice that (4) makes sense also if n is not an integer. For that reason henceforth we consider $n \in \mathbb{R}$, with $2 < n < 4$. Our main result is the following

Theorem 2.1. *The Boundary Value problem* (1)*, with* $u'(0) = u(R) = 0$*, has no regular solutions if*

$$\lambda \leq \frac{n^2(n-1)}{4(n+2)},$$

for $2 < n < 4$.

Remark 2.2. Notice that our bound $n^2(n-1)/(4(n+2))$ is strictly bigger than $n(n-2)/4$ for $n < 4$. Notice, on the other hand, that Stapelkamp's bound holds for all regular solutions, while our improved bound only holds for regular radial solutions. We do not know whether our bound is optimal, i.e., we do not know if there are solutions for $\lambda > n^2(n-1)/(4(n+2))$. In view of [3], there can be no positive solutions if $\lambda < \mu(n)$ (see [3] for the definition of $\mu(n)$). It is important to notice that for $n = 4$, we have that,

$$\frac{n(n-2)}{4} = \frac{n^2(n-1)}{4(n+2)} = \mu(4),$$

so at least our result is optimal as $n \to 4$.

Proof. We use a Rellich–Pohozaev argument ([9] and [8]). Multiplying equation (4) by $u(x)\sinh^{n-1}(x)$ and integrating, we obtain

$$\begin{aligned}&-\int_0^R u''(x)(u(x)\sinh^{n-1}(x))\,\mathrm{d}x - (n-1)\int_0^R u(x)u'(x)\cosh(x)\sinh^{n-2}(x)\,\mathrm{d}x\\&= \lambda\int_0^R u^2\sinh^{n-1}(x)\,\mathrm{d}x + \int_0^R |u(x)|^{p+1}\sinh^{n-1}(x)\,\mathrm{d}x.\end{aligned}$$

Integrating the first term by parts, we can write this equation as

$$\int_0^R u'^2\sinh^{n-1}(x)\,\mathrm{d}x = \lambda\int_0^R u^2\sinh^{n-1}(x)\,\mathrm{d}x + \int_0^R |u|^{p+1}\sinh^{n-1}(x)\,\mathrm{d}x. \quad (5)$$

Now let

$$G(x) = \int_0^x \sinh^{n-1}(s)\,\mathrm{d}s.$$

Multiplying equation (4) by $u'G$ and integrating, we obtain

$$\begin{aligned}&-\int_0^R \Big(\frac{u'^2}{2}\Big)' G\,\mathrm{d}x - (n-1)\int_0^R \coth(x)u'^2 G\,\mathrm{d}x\\&= \lambda\int_0^R \Big(\frac{u^2}{2}\Big)' G\,\mathrm{d}x + \int_0^R \Big(\frac{|u|^{p+1}}{p+1}\Big)' G\,\mathrm{d}x.\end{aligned}$$

After integrating by parts, and since $G(0) = 0$, we obtain

$$\begin{aligned}&\frac{u'^2(R)G(R)}{2} + \int_0^R u'^2\Big((n-1)G\coth(x) - \frac{G'}{2}\Big)\,\mathrm{d}x\\&= \lambda\int_0^R \frac{u^2}{2}G'\,\mathrm{d}x + \frac{1}{p+1}\int_0^R |u|^{p+1}G'\,\mathrm{d}x.\end{aligned} \quad (6)$$

Substituting equation (5) into equation (6), and since $1/2 - 1/(p+1) = 1/n$, it follows that

$$\int_0^R u'^2\Big((n-1)G\coth(x) - \frac{G'}{2} - \frac{G'}{p+1}\Big)\,\mathrm{d}x + \frac{u'^2(R)G(R)}{2}$$
$$= \frac{\lambda}{n}\int_0^R u^2 \sinh^{n-1}(x)\,\mathrm{d}x.$$

Notice that in equation (7) we have written $\sinh^{n-1}(x)$ as $G'(x)$. Thus, since the boundary term is positive, and since $1/2 + 1/(p+1) = (n-1)/n$, we have

$$\lambda \geq \frac{n(n-1)\displaystyle\int_0^R u'^2\Big(G\coth(x) - \frac{G'}{n}\Big)\mathrm{d}x}{\displaystyle\int_0^R u^2 G'\,\mathrm{d}x}. \tag{7}$$

Now let

$$L(x) = G\coth(x) - \frac{G'}{n}.$$

Then $L \geq 0$. In fact, we can write

$$L(x) = \frac{m(x)}{\sinh(x)},$$

where

$$m(x) = G\cosh(x) - \frac{\sinh^n(x)}{n}.$$

Then, since $G(0)$, we have $m(0) = 0$. Also,

$$m'(x) = G\sinh(x) + G'\cosh(x) - \sinh^{n-1}(x)\cosh(x) = G\sinh(x).$$

It follows that $m' \geq 0$, and therefore that $L \geq 0$.

We now use a Hardy type inequality to write the denominator integral in terms of u'^2. For a review on Hardy's inequalities see, e.g., [7] and [5]. Integrating by parts, we can write

$$\int_0^R u^2 G'\,\mathrm{d}x = -2\int_0^R (u\sinh^{(n-1)/2}(x))\Big(\frac{Gu'}{\sinh^{(n-1)/2}(x)}\Big)\mathrm{d}x.$$

Then, using Cauchy–Schwarz, it follows that

$$\Big(\int_0^R u^2 G'\,\mathrm{d}x\Big)^2 \leq 4\int_0^R u^2 G'\,\mathrm{d}x \int_0^R \frac{G^2 u'^2}{G'}\,\mathrm{d}x.$$

That is,

$$\int_0^R u^2 G' \,\mathrm{d}x < 4 \int_0^R \frac{u'^2 G^2}{G'} \,\mathrm{d}x. \tag{8}$$

Using inequality (8) in the quotient (7), we conclude that

$$\lambda > \frac{n(n-1) \displaystyle\int_0^R u'^2 \Big(G \coth(x) - \frac{G'}{n} \Big) \mathrm{d}x}{4 \displaystyle\int_0^R \frac{u'^2 G^2}{G'} \,\mathrm{d}x}.$$

In the Lemma 2.3 below, we show that

$$L(x) \geq c \frac{G^2}{G'},$$

where

$$c = \frac{n}{n+2}.$$

With this, we conclude that

$$\lambda > \frac{n^2(n-1)}{4(n+2)}.$$

□

Lemma 2.3. *Let $x \geq 0$ and let*

$$L(x) = G \coth(x) - \frac{G'}{n}.$$

Then

$$L(x) \geq \frac{n}{(n+2)} \frac{G^2}{G'}.$$

Here, as above,

$$G(x) = \int_0^x \sinh^{n-1}(s) \,\mathrm{d}s.$$

Proof. Let

$$f(x) = L(x) G'(x) - c G^2(x),$$

where

$$c = \frac{n}{n+2}.$$

It suffices to show that $f \geq 0$.

As before, we write

$$L(x) = \frac{m(x)}{\sinh(x)},$$

where

$$m(x) = G\cosh(x) - \frac{\sinh^n(x)}{n}$$

and

$$m'(x) = G\sinh(x).$$

Then,

$$f(x) = \sinh^{n-2}(x)m(x) - c\,G^2(x).$$

Notice that since $\sinh(0) = G(0) = 0$, one has that $f(0) = 0$, so it suffices to show that $f' \geq 0$. We have that

$$f'(x) = \sinh^{n-3}(x)((n-2)\cosh(x)m(x) + G\sinh^2(x)(1-2c)).$$

Let $\sigma = 2c - 1 = 1/p$, where $p = (n+2)/(n-2)$ is the critical Sobolev exponent, and let $g(x) = (n-2)\cosh(x)m(x) - \sigma G\sinh^2(x)$. It suffices to show that $g \geq 0$. Since $m(0) = 0$, then $g(0) = 0$.

Also,

$$g'(x) = 2\sinh(x)\cosh(x)G(x)(n-2-\sigma) - \sinh^{n+1}(x)\Big(\frac{(n-2)}{n} + \sigma\Big)$$

and in particular $g'(0) = 0$. Since

$$n - 2 - \sigma = \frac{(n-2)(n+1)}{(n+2)}$$

and

$$\frac{(n-2)}{n} + \sigma = \frac{2(n+1)(n-2)}{n(n+2)}$$

we can write

$$g'(x) = \frac{2(n+1)(n-2)}{n(n+2)}\sinh(x)\left[nG\cosh(x) - \sinh^n(x)\right].$$

Finally, let $h(x) = nG\cosh(x) - \sinh^n(x)$. If we show $h(x) \geq 0$, then we will have $g' \geq 0$, which will imply $g \geq 0$, and thus, that $f \geq 0$, as desired. Notice that $h(0) = 0$. Also, since $G'(x) = \sinh^{n-1}(x)$, we have

$$h'(x) = nG\sinh(x).$$

That is, $h' \geq 0$, which concludes the proof of Lemma 2.3. □

Remark 2.4. In the proof of Lemma 2.3, the constant $\sigma = 1/p$ plays a crucial role, where p is the critical Sobolev exponent. It is worth noting that for small x and g as in the proof above,

$$g(x) = x^{n+2}\Big(\frac{1}{np} - \frac{\sigma}{n}\Big) + \mathcal{O}(x^{n+4}).$$

It follows that if $\sigma \leq 1/p$, then g is positive in a neighborhood of the origin. It was this observation that led us to realize that $\sigma = 1/p$ would yield the optimal estimate.

Acknowledgements. The work of Rafael D. Benguria has been supported by Fondecyt (Chile) Projects # 112–0836 and # 114-1155 and by the Núcleo Milenio en "Física Matemática", RC–12–0002 (ICM, Chile). Soledad Benguria would like to thank the E. Schrödinger Institute in Vienna for their hospitality while part of this work was being done.

References

[1] E. Balslev, Absence of positive eigenvalues of Schrödinger operators. *Arch. Rational Mech. Anal.* **59** (1975), no. 4, 343–357. MR 0489507 Zbl 0325.35028

[2] C. Bandle and Y. Kabeya, On the positive, "radial" solutions of a semilinear elliptic equation in $\mathbb{H}^N$. *Adv. Nonlinear Anal.* **1** (2012), no. 1, 1–25. MR 3033174 Zbl 1277.35147

[3] S. Benguria, The solution gap of the Brezis–Nirenberg problem on the hyperbolic space. *Monatsh. Math.* **181** (2016), no. 3, 537–559.

[4] H. Brézis and L. Nirenberg, Positive solutions of nonlinear elliptic equations involving critical Sobolev exponents. *Comm. Pure Appl. Math.* **36** (1983), no. 4, 437–477. MR 0709644 Zbl 0541.35029

[5] E. B. Davies, A review of Hardy inequalities. In J. Rossmann, P. Takáč, and G. Wildenhain (eds.), *The Maz'ya anniversary collection.* Vol. 2. Rostock Conference on Functional Analysis, Partial Differential Equations and Applications. Papers from the conference held at the University of Rostock, Rostock, August 31–September 4, 1998. Operator Theory: Advances and Applications, 110. Birkhäuser Verlag, Basel, 1999, 55–67. MR 1747888 Zbl 0936.35121

[6] D. Ganguly and K. Sandeep, Sign changing solutions of the Brezis–Nirenberg problem in the hyperbolic space. *Calc. Var. Partial Differential Equations* **50** (2014), no. 1-2, 69–91. MR 3194676 Zbl 1295.35203

[7] B. Opic and A. Kufner, *Hardy-type inequalities.* Pitman Research Notes in Mathematics Series, 219. Longman Scientific & Technical, Harlow, 1990. ISBN 0-582-05198-3 MR 1069756 Zbl 0698.26007

[8] S. I. Pohožaev, On the eigenfunctions of the equation $\Delta u + \lambda f(u) = 0$. *Dokl. Akad. Nauk SSSR* **165** (1965), 36–39. In Russian. English translation, *Soviet Math. Dokl.* **6** (1965), 1408–1411. MR 0192184

[9] F. Rellich, Darstellung der Eigenwerte von $\Delta u + \lambda u = 0$ durch ein Randintegral. *Math. Z.* **46** (1940), 635–636. MR 0002456 JFM 66.0460.01 Zbl 0023.04204

[10] B. Simon, Absence of positive eigenvalues in a class of multiparticle quantum systems. *Math. Ann.* **207** (1974), 133–138. MR 0329485 Zbl 0253.35025

[11] S. Stapelkamp, The Brézis–Nirenberg problem on $\mathbb{H}^n$. Existence and uniqueness of solutions. In J. Bemelmans, B. Brighi, A. Brillard, M. Chipot, F. Conrad, I. Shafrir, V. Valente, and G. Vergara-Caffarelli (eds.), *Elliptic and parabolic problems.* Proceedings of the 4th European Conference held in Rolduc, June 18–22 and Gaeta, September 24–28, 2001. World Scientific, River Edge, N.J., 2002, 283–290. MR 1937548 Zbl 1109.35343

[12] J. Weidmann, The virial theorem and its application to the spectral theory of Schrödinger operators. *Bull. Amer. Math. Soc.* **73** (1967), 452–456. MR 0208197 Zbl 0156.23304

Twisted waveguide with a Neumann window

Philippe Briet and Hiba Hammedi

Dedicated to Pavel Exner on the occasion of his 70th birthday

1 Introduction

In this work, we would like to study the influence of a geometric twisting on trapped modes which occur in certain waveguides. Here the waveguide consists in a straight tubular domain $\Omega_0 := \mathbb{R} \times \omega$ having a Neumann window on its boundary $\partial\Omega_0$.

The cross section ω is supposed to be an open bounded connected subset of $\mathbb{R}^2$ of diameter $d > 0$ which is not rotationally invariant. Moreover ω is supposed to have smooth boundary $\partial\omega$.

It can be shown that the Laplace operator associated to such a straight tube has bound states [8].

Let us introduce some notations. Denote by $\mathcal{N}$ the Neumann window. It is an open bounded subset of the boundary $\partial\Omega_0$. Let $\mathcal{D}$ be its complement set in $\partial\Omega_0$. When $\mathcal{N}$ is an annulus of size $l > 0$ we will denote it by,

$$\mathcal{A}_a(l) := I_a(l) \times \partial\omega, I_a(l) := (a, l + a), \quad a \in \mathbb{R}.$$

Consider first the self-adjoint operator $H_0^{\mathcal{N}}$ associated to the following quadratic form. Let $D(Q^{\mathcal{N}}) = \{\psi \in \mathcal{H}^1(\Omega_0) \mid \psi\lceil_{\mathcal{D}} = 0\}$ and for $\psi \in D(Q^{\mathcal{N}})$,

$$Q^{\mathcal{N}}(\psi) = \int_{\Omega_0} |\nabla\psi|^2 \, \mathrm{d}x,$$

i.e., the Laplace operator defined on Ω_0 with Neumann boundary conditions (NBC) on $\mathcal{N}$ and Dirichlet boundary conditions (DBC) on $\mathcal{D}$, see [5] and [11].

It is actually shown in Section 2 of this paper that if $\mathcal{N}$ contains an annulus of size l large enough then $H_0^{\mathcal{N}}$ has at least one discrete eigenvalue. In fact it is proved in [8] that this holds true if $\mathcal{N}$ contains an annulus of any size $l > 0$.

The question we are interested in is the following: is it possible that the discrete spectrum of $H_0^{\mathcal{N}}$ disappears when we apply a geometric twisting on the guide? This question is motivated by the results of [6] and [10], where it is shown that this phenomenon occurs in some bent tubes when they are subjected to a twisting defined from an angle function θ having a derivative $\dot{\theta}$ with a compact support. In this paper we consider the situation described above which is very different from the one of [6] and [10].

Let us now define the twisting (see [4] and [7]). Choose $\theta \in C_c^1(\mathbb{R})$ and introduce the diffeomorphism

$$\begin{aligned} \mathcal{L}\colon \Omega_0 &\longrightarrow \mathbb{R}^3, \\ (s, t_2, t_3) &\longmapsto (s, t_2 \cos\theta(s) - t_3 \sin\theta(s), t_2 \sin\theta(s) + t_3 \cos\theta(s)). \end{aligned}$$

The twisted tube is given by $\Omega_\theta := \mathcal{L}(\Omega_0)$. Let

$$D(Q_\theta^{\mathcal{N}}) = \{\psi \in \mathcal{H}^1(\Omega_\theta) \mid \psi\lceil_{\mathcal{L}(\mathcal{D})} = 0\}$$

and consider the following quadratic form

$$Q_\theta^{\mathcal{N}}(\psi) := \int_{\Omega_\theta} |\nabla\psi|^2 \, dx, \quad \psi \in D(Q_\theta^{\mathcal{N}}).$$

Through unitary equivalence, we then have to consider

$$q_\theta^{\mathcal{N}}(\psi) := Q_\theta^{\mathcal{N}}(\psi o \mathcal{L}^{-1}) = \|\nabla'\psi\|^2 + \|\partial_s\psi + \dot{\theta}\partial_\tau\psi\|^2, \tag{1}$$

for $\psi \in D(q_\theta^{\mathcal{N}}) := \{\psi \in \mathcal{H}^1(\Omega_0) \mid \psi\lceil_{\mathcal{D}} = 0\}$ and where

$$\nabla' := {}^t(\partial_{t_2}, \partial_{t_3}), \quad \partial_\tau := t_2\partial_{t_3} - t_3\partial_{t_2}.$$

Denote by $H_\theta^{\mathcal{N}}$ the associated self-adjoint operator. It is defined as follows (see [5] and [11]). Let

$$D(H_\theta^{\mathcal{N}}) = \Big\{\psi \in D(q_\theta^{\mathcal{N}}) \;\Big|\; H_\theta^{\mathcal{N}}\psi \in L^2(\Omega_0), \frac{\partial\psi}{\partial n}\lceil_{\mathcal{N}} = 0\Big\},$$

with

$$H_\theta^{\mathcal{N}}\psi = (-\Delta_\omega - (\dot{\theta}\partial_\tau + \partial_s)^2)\psi, \tag{2}$$

where the transverse Laplacian $\Delta_\omega := \partial_{t_2}^2 + \partial_{t_3}^2$. If $\mathcal{N} = \mathcal{A}_a(l), l > 0$, we will denote these forms respectively as Q_θ^l, q_θ^l and the corresponding operator as H_θ^l and if $\mathcal{N} = \emptyset$ we denote the associated operator by H_θ.

Then the main result of this paper is the following one.

Theorem 1.1. i) *Under conditions stated above on* ω *and* θ, *there exists*

$$l_{\min} := l_{\min}(\omega, d) > 0$$

such as if for some $a \in \mathbb{R}$ *and* $l > l_{\min}$, $\mathcal{N} \supset \mathcal{A}_a(l)$ *then*

$$\sigma_d(H_\theta^{\mathcal{N}}) \neq \emptyset.$$

ii) *Suppose* θ *is a non zero function satisfying the same conditions as in* i) *and has a bounded second derivative. Then there exists*

$$d_{\max} := d_{\max}(\theta, \omega) > 0$$

such that for all $0 < d \leq d_{\max}$ *there exists*

$$l_{\max} := l_{\max}(\omega, d, \theta)$$

such as for all $0 < l \leq l_{\max}$, *if* $\mathcal{N} \subset \mathcal{A}_a(l)$ *and* $\operatorname{supp}(\dot{\theta}) \cap I_a(l) = \emptyset$ *for some* $a \in \mathbb{R}$ *then*

$$\sigma_d(H_\theta^{\mathcal{N}}) = \emptyset.$$

Roughly speaking this result implies that for d small enough, the discrete spectrum disappears when the width of the Neumann window decreases.

Let us describe briefly the content of the paper. In Section 2 we give the proof of Theorem 1.1 i). Section 3 is devoted to the proof of the second part of Theorem 1.1; this proof needs several steps. In particular we first establish a local Hardy inequality. This allows us to reduce the problem to the analysis of a one dimensional Schrödinger operator from which Theorem 1.1 ii) follows. Finally in the Appendix of the paper we give partial results we use in previous sections.

2 Existence of bound states

First we prove the following. Denote by $E_1, E_2, \ldots$ the eigenvalues (transverse modes) of the Laplacian $-\Delta_\omega$ defined on $\mathrm{L}^2(\omega)$ with DBC on $\partial\omega$. Let $\chi_1, \chi_2, \ldots$ be the associated eigenfunctions. Then we have

Proposition 2.1. $\sigma_{\mathrm{ess}}(H_\theta^{\mathcal{N}}) = [E_1, \infty)$.

Proof. We know that $\sigma(H_\theta) = [E_1, \infty)$ see, e.g., [2]. But by usual arguments [12], $H_\theta^{\mathcal{N}} \leq H_\theta$, then

$$[E_1, \infty) \subset \sigma_{\mathrm{ess}}(H_\theta^{\mathcal{N}}).$$

Let $a' \in \mathbb{R}$ and $l' > 0$ large enough such that $\mathcal{N} \subset \mathcal{A}_{a'}(l') = I_{a'}(l') \times \partial\omega$ and $\mathrm{supp}(\dot{\theta}) \subset I_{a'}(l')$. Let $\widetilde{H}_\theta^{l'}$ be the operator defined as in (2) but with additional Neumann boundary conditions on $\{a'\} \times \omega \cup \{a' + l'\} \times \omega$. So $H_\theta^{\mathcal{N}} \geq \widetilde{H}_\theta^{l'}$ and then $\sigma_{\mathrm{ess}}(H_\theta^{\mathcal{N}}) \subset \sigma_{\mathrm{ess}}(\widetilde{H}_\theta^{l'})$, see [12].

But $\widetilde{H}_\theta^{l'} = \widetilde{H}_i \oplus \widetilde{H}_e$. The interior operator $\widetilde{H}_i$ is the corresponding operator defined on $L^2(I_{a'}(l') \times \omega)$ with NBC on $\{a'\} \times \omega \cup \{a' + l'\} \times \omega \cup \mathcal{N}$ and DBC elsewhere on $\mathcal{A}_{a'}(l')$. By general arguments of [12] it has only discrete spectrum consequently $\sigma_{\mathrm{ess}}(\widetilde{H}_\theta^{l'}) = \sigma_{\mathrm{ess}}(\widetilde{H}_e)$.

Now the exterior operator $\widetilde{H}_e$ is defined on $L^2((-\infty, a') \times \omega \cup (a' + l', \infty) \times \omega)$ with DBC on $(-\infty, a') \times \partial\omega \cup (a' + l', \infty) \times \partial\omega$ and NBC on $\{a'\} \times \omega \cup \{a' + l'\} \times \omega$. Since $\theta = 0$ for $x < a'$ and $x > a' + l'$, it is easy to see that

$$\widetilde{H}_e = \bigoplus_{n \geq 1} (-\partial^2 + E_n)(\chi_n, \cdot)\chi_n.$$

Hence $\sigma(\widetilde{H}_e) = \sigma_{\mathrm{ess}}(\widetilde{H}_e) = [E_1, +\infty)$. □

Theorem 1.1 i) follows from

Proposition 2.2. *Under the conditions of Theorem* 1.1 i), *there exists*

$$l_{\min} := l_{\min}(\omega, d) > 0$$

such as for all $l > l_{\min}$ we have

$$\sigma_d(H_\theta^l) \neq \emptyset.$$

Proof. Let $\varphi_{l,a}$ be the function

$$\varphi_{l,a}(s) := \begin{cases} \frac{10}{l}(s - a) & \text{on } [a, a + l/10), \\ 1 & \text{on } [a + l/10, a + 9l/10), \\ -\frac{10}{l}(s - l - a) & \text{on } [a + 9l/10, a + l), \\ 0 & \text{elsewhere.} \end{cases}$$

It is easy to see that $\varphi_{l,a} \in D(q_\theta^l)$ and $\|\varphi_{l,a}\|^2 = 13l/15\,|\omega|$. Let us calculate

$$q_\theta^l(\varphi_{l,a}) - E_1\|\varphi_{l,a}\|^2 = \|\nabla'\varphi_{l,a}\|^2 + \|\dot{\theta}\partial_\tau\varphi_{l,a} + \partial_s\varphi_{l,a}\|^2 - E_1\|\varphi_{l,a}\|^2. \tag{3}$$

Evidently the first term on the right hand side of (3) is zero. For the second term on the right hand side of (3) we get,

$$\|\dot{\theta}\partial_\tau\varphi_{l,a} + \partial_s\varphi_{l,a}\|^2 = \|\partial_s\varphi_{l,a}\|^2 = \frac{20}{l}|\omega|.$$

Then

$$q^l_\theta(\varphi_{l,a}) - E_1\|\varphi_{l,a}\|^2 = |\omega|\Big(\frac{20}{l} - \frac{13l}{15}E_1\Big)$$

and thus if $l \geq l_{\min} := \sqrt{300/(13E_1)}$ we have $q^l_\theta(\varphi_{l,a}) - E_1\|\varphi_{l,a}\|^2 \leq 0$ □

Proof of Theorem 1.1 i). Using the same notation as in Theorem 1.1 i), $H^{\mathcal{N}}_\theta \leq H^l_\theta$. Moreover these operators have the same essential spectrum, then by the min–max principle the assertion follows. □

3 Absence of bound state

In this section we want to prove the second part of Theorem 1.1. Define

$$\theta_m = \inf(\operatorname{supp}(\dot{\theta})),$$

$$\theta_M = \sup(\operatorname{supp}(\dot{\theta}))$$

and

$$L = \theta_M - \theta_m.$$

Here $L > 0$. We first consider the case where the Neumann window is an annulus,

$$\mathcal{A}_a(l) = I_a(l) \times \omega.$$

Proposition 3.1. *Suppose $\mathcal{A}_a(l)$ is such that $a \geq \theta_M$. Assume that the conditions of Theorem* 1.1 ii) *hold. Then there exists $d_{\max} := d_{\max}(\omega, \theta) > 0$, such that for all $0 < d \leq d_{\max}$ there exists $l_{\max}(d, \theta, \omega) > 0$ such as for all $0 < l \leq l_{\max}$ we have*

$$\sigma_d(H^l_\theta) = \emptyset.$$

Remark 3.2. The case where $l + a \leq \theta_m$ follows from the same arguments developed below.

This proof is based on the fact that under conditions of Proposition 3.1, for every $\psi \in D(q^l_\theta)$ it holds that

$$Q(\psi) := q^l_\theta(\psi) - E_1\|\psi\|^2 \geq 0. \tag{4}$$

The proof of (4) involves several steps.

3.1 A local Hardy inequality

The aim of this paragraph is to show a Hardy type inequality needed for the proof of Proposition 3.1. It is the first step of the proof of (4). Let g be the function

$$g(s) := \begin{cases} 0 & \text{on } I_a(l), \\ E_1 & \text{elsewhere.} \end{cases} \tag{5}$$

Choose $p \in (\theta_m, \theta_M)$ such that $\dot{\theta}(p) \neq 0$ and let

$$\rho(s) := \begin{cases} \dfrac{1}{1+(s-p)^2} & \text{on } (-\infty, p], \\ 0 & \text{elsewhere.} \end{cases} \tag{6}$$

Proposition 3.3. *Under the same conditions of Proposition* 3.1, *then there exists a constant $C > 0$ depending on p and ω and $\dot{\theta}$ such that for any $\psi \in D(q_\theta^l)$,*

$$\|\nabla'\psi\|^2 + \|\dot{\theta}\partial_\tau\psi + \partial_s\psi\|^2 - \int_{\Omega_0} g(s)|\psi|^2 \,\mathrm{d}s\,\mathrm{d}t \geq C\int_{\Omega_0} \rho(s)|\psi|^2 \,\mathrm{d}s\,\mathrm{d}t. \tag{7}$$

We first show the following lemma. Define $\Omega_p := (-\infty, p) \times \omega$.

Lemma 3.4. *Under same conditions of the Proposition* 3.3. *Then for any $\psi \in D(q_\theta^l)$ we have*

$$\int_{\Omega_p} |\nabla'\psi|^2 + |\dot{\theta}\partial_\tau\psi + \partial_s\psi|^2 - E_1|\psi|^2 \,\mathrm{d}s\,\mathrm{d}t \geq C\int_{\Omega_p} \rho(s)|\psi|^2 \,\mathrm{d}s\,\mathrm{d}t.$$

In the following we will use notations suggested in [6]. For $A \subset \mathbb{R}$ denote by χ_A the characteristic function of $A \times \omega$. Let $\psi \in D(q_\theta^l)$ and define,

$$q_1^A(\psi) := \|\chi_A\nabla'\psi\|^2 - E_1\|\chi_A\psi\|^2, \quad q_2^A(\psi) := \|\chi_A\partial_s\psi\|^2,$$
$$q_3^A(\psi) := \|\chi_A\dot{\theta}\partial_\tau\psi\|^2, \quad q_{2,3}^A(\psi) := 2\,\mathrm{Re}(\partial_s\psi, \chi_A\dot{\theta}\partial_\tau\psi),$$

and

$$Q^A(\psi) = q_1^A(\psi) + q_2^A(\psi) + q_3^A(\psi) + q_{2,3}^A(\psi).$$

Here and hereafter we often use the fact that for any $\psi \in D(q_\theta^l)$,

$$q_1^A(\psi) \geq 0, \tag{8}$$

for every $A \subset \mathbb{R}$ such that $A \cap I_a(l) = \emptyset$.

Proof. Choose $r > 0$ such that $\dot{\theta}(s) \neq 0$ for any $s \in [p-r, p]$. Let f be the following function:

$$f(s) := \begin{cases} 0 & \text{on } (p, \infty), \\ \dfrac{p-s}{r} & \text{on } (p-r, p], \\ 1 & \text{elsewhere.} \end{cases}$$

For any $\psi \in D(q_\theta^l)$, simple estimates lead to

$$\begin{aligned} \int_{\Omega_p} \frac{|\psi(s,t)|^2}{1+(s-p)^2} \, \mathrm{d}s \, \mathrm{d}t &= \int_{\Omega_p} \frac{|\psi(s,t) f(s) + (1-f(s))\psi(s,t)|^2}{1+(s-p)^2} \, \mathrm{d}s \, \mathrm{d}t \\ &\leq 2\Big(\int_{\Omega_p} \frac{|f(s)\psi(s,t)|^2}{(s-p)^2} \, \mathrm{d}s \, \mathrm{d}t + \|\chi_{(p-r,p)}\psi\|^2 \Big). \end{aligned}$$

Since $f(p)\psi(p,\cdot) = 0$, we can use the usual Hardy inequality (see, e.g., [9]), then we get

$$\int_{\Omega_p} \frac{|\psi(s,t)|^2}{1+(s-p)^2} \, \mathrm{d}s \, \mathrm{d}t \leq 8 q_2^{(-\infty,p)}(f\psi) + 2\|\chi_{(p-r,p)}\psi\|^2. \tag{9}$$

Note that with our choice $[p-r, p] \cap [a, a+l] = \emptyset$. Hence to estimate the second term on the right hand side of (9) we use Theorem 6.5 of [10], then there exists $\lambda_0 = \lambda_0(\dot{\theta}, p, r) > 0$ such that for any $\psi \in D(q_\theta^l)$ we have

$$\|\chi_{(p-r,p)}\psi\|^2 \leq \frac{1}{\lambda_0} Q^{(p-r,p)}(\psi) \leq \frac{1}{\lambda_0} Q^{(-\infty,p)}(\psi). \tag{10}$$

We now want to estimate the first term on the right hand side of (9). We have

$$q_2^{(-\infty,p)}(f\psi) = \int_{\Omega_p} |\partial_s(f\psi)|^2 \mathrm{d}s \, \mathrm{d}t = q_2^{(-\infty,\theta_m)}(f\psi) + q_2^{(\theta_m,p)}(f\psi).$$

Evidently since $\dot{\theta} = 0$ and $f = 1$ in $(-\infty, \theta_m)$, from (8), we have

$$q_2^{(-\infty,\theta_m)}(f\psi) \leq Q^{(-\infty,\theta_m)}(\psi). \tag{11}$$

In the other hand since $f(p)\psi(p,\cdot) = 0$, we can apply Lemma A.1 of the Appendix. So for any $0 < \alpha < 1$ there exists $\gamma_{\alpha,1} > 0$ such that

$$|q_{2,3}^{(\theta_m,p)}(f\psi)| \leq \gamma_{\alpha,1} q_1^{(\theta_m,p)}(f\psi) + \alpha q_2^{(\theta_m,p)}(f\psi) + q_3^{(\theta_m,p)}(f\psi).$$

Let $\gamma := \max(1, \gamma_{\alpha,1})$. Then

$$\gamma^{-1}|q_{2,3}^{(\theta_m,p)}(f\psi)| \leq q_1^{(\theta_m,p)}(f\psi) + \alpha\gamma^{-1}q_2^{(\theta_m,p)}(f\psi) + \gamma^{-1}q_3^{(\theta_m,p)}(f\psi). \quad (12)$$

Hence with the decomposition, $q_{2,3}^{(\theta_m,p)} = \gamma^{-1}q_{2,3}^{(\theta_m,p)} + (1-\gamma^{-1})q_{2,3}^{(\theta_m,p)}$ and (12) we have

$$\begin{aligned} Q^{(\theta_m,p)}(f\psi) \geq (1-\gamma^{-1})(q_2^{(\theta_m,p)}(f\psi) + q_{2,3}^{(\theta_m,p)}(f\psi) + q_3^{(\theta_m,p)}(f\psi)) \\ + \gamma^{-1}(1-\alpha)q_2^{(\theta_m,p)}(f\psi) \end{aligned}$$

and since $q_3^{(\theta_m,p)} + q_{2,3}^{(\theta_m,p)} + q_2^{(\theta_m,p)} \geq 0$, we arrive at

$$q_2^{(\theta_m,p)}(f\psi) \leq \frac{\gamma}{(1-\alpha)} Q^{(\theta_m,p)}(f\psi).$$

Now by $q_1^{(\theta_m,p)}(f\psi) \leq q_1^{(\theta_m,p)}(\psi)$ and

$$\|\chi_{(\theta_m,p)}(\partial_s + \dot{\theta}\partial_\tau)(f\psi)\|^2 \leq 2\Big(\|\chi_{(\theta_m,p)}(\partial_s + \dot{\theta}\partial_\tau)\psi\|^2 + \frac{1}{r^2}\|\chi_{(p-r,p)}\psi\|^2\Big),$$

and by (10) we get

$$\begin{aligned} q_2^{(\theta_m,p)}(f\psi) &\leq \frac{2\gamma}{(1-\alpha)}\Big(Q^{(\theta_m,p)}(\psi) + \frac{1}{\lambda_0 r^2} Q^{(p-r,p)}(\psi)\Big) \\ &\leq c' Q^{(\theta_m,p)}(\psi), \end{aligned} \quad (13)$$

with

$$c' = \frac{2\gamma}{(1-\alpha)}\Big(1 + \frac{1}{\lambda_0 r^2}\Big).$$

Then (11) and (13) imply

$$q_2^{(-\infty,p)}(f\psi) \leq (1+c')Q^{(-\infty,p)}(\psi). \quad (14)$$

Hence, with

$$C^{-1} = 8(1+c') + \frac{2}{\lambda_0}, \quad (15)$$

inequalities (14) and (10) prove the lemma. □

Proof of Proposition 3.3. To prove the proposition we note that for any $\psi \in D(q_\theta^l)$ and for $p' \in \mathbb{R}$ we have

$$\int_\omega \int_{p'}^\infty |\nabla'\psi|^2 + |\dot{\theta}\partial_\tau\psi + \partial_s\psi|^2 \,\mathrm{d}s\,\mathrm{d}t \geq \int_\omega \int_{p'}^\infty g(s)|\psi|^2 \,\mathrm{d}s\,\mathrm{d}t. \quad (16)$$

Then (16) with $p' = p$ and Lemma 3.4 imply (7). □

3.2 Reduction to a one dimensional problem

We now want to prove the following result.

Proposition 3.5. *Under conditions of Proposition* 3.1, *a sufficient condition in order to get* (4) *is given by*

$$\int_{\mathbb{R}} |\psi'(s)|^2 + 2C\rho(s)|\psi(s)|^2 \,\mathrm{d}s - 4E_1 \int_a^{a+l} |\psi(s)|^2 \,\mathrm{d}s \geq 0,$$

for any $\psi \in \mathcal{H}^1(\mathbb{R})$ *where the constant* C *is defined in* (15).

Remark 3.6. This proposition means that the positivity needed here is given by the positivity of the effective one dimensional Schrödinger operator on $\mathrm{L}^2(\mathbb{R})$,

$$-\frac{d^2}{ds^2} + 2C\rho(s) - 4E_1 \mathbf{1}_{I_a(l)}. \tag{17}$$

where $\mathbf{1}_{I_a(l)}$ is the characteristic function of $I_a(l)$.

Proof. Evidently we have

$$Q(\psi) = \frac{1}{2}\Big(Q(\psi) - \int_{\Omega_0} (E_1 - g(s))|\psi|^2 \,\mathrm{d}s\,\mathrm{d}t + q_\theta^l(\psi) - \int_{\Omega_0} g(s)|\psi|^2 \,\mathrm{d}s\,\mathrm{d}t\Big),$$

where g is defined in (5). By using (7), then

$$Q(\psi) \geq \frac{1}{2}\Big(q_\theta^l(\psi) - E_1\|\psi\|^2 + C\int_{\Omega_0} \rho(s)|\psi|^2 \,\mathrm{d}s\,\mathrm{d}t - E_1\|\chi_{(a,a+l)}\psi\|^2\Big) \tag{18}$$

Rewrite the expression of q_θ^l given by (1) as follows:

$$q_\theta^l(\psi) = \|\nabla'\psi\|^2 + \|\partial_s\psi\|^2 + \|\dot\theta\partial_\tau\psi\|^2 + 2\,\mathrm{Re}(\partial_s\psi, \dot\theta\partial_\tau\psi). \tag{19}$$

We estimate the last term of the right hand side of (19). By using the formula (23) of the Appendix,

$$\begin{aligned} |q_{2,3}(\psi)| &= |q_{2,3}^{(\theta_m,\theta_M)}(\psi)| \\ &\leq \gamma_{1/2,1/2}\, q_1^{(\theta_m,\theta_M)}(\psi) + \frac{1}{2} q_2^{(\theta_m,\theta_M)}(\psi) + \frac{1}{2} q_3^{(\theta_m,\theta_M)}(\psi) \end{aligned} \tag{20}$$

where

$$\gamma_{1/2,1/2} := \tilde\gamma_{1/2,1/2} + 4d^2\|\dot\theta\|_\infty^2$$

with

$$\tilde{\gamma}_{1/2,1/2} := \max\Big\{\frac{d\|\dot{\theta}\|_\infty\|\ddot{\theta}\|_\infty\sqrt{f(L)}}{\dot{\theta}_0\sqrt{\lambda}}, \frac{d^2\|\ddot{\theta}\|_\infty^2 f(L)}{\lambda\dot{\theta}_0{}^2}, 2d^2\|\ddot{\theta}\|_\infty^2 f(L)\Big\}$$

for some constant $\lambda > 0$ depending only on the section ω and

$$f(L) := \max\Big\{2 + \frac{16L^2}{r^2}, 4L^2\Big\}.$$

Hence (19) together with (20) give

$$q_\theta^l(\psi) \geq \|\nabla'\psi\|^2 + \frac{1}{2}\|\partial_s\psi\|^2 + \frac{1}{2}\|\dot{\theta}\partial_\tau\psi\|^2 - \gamma_{1/2,1/2}\, q_1^{(\theta_m,\theta_M)}(\psi). \tag{21}$$

In view of (8) we have

$$\begin{aligned}\|\nabla'\psi\|^2 - E_1\|\psi\|^2 &\geq q_1^{(\theta_m,\theta_M)}(\psi) + q_1^{I_a(l)}(\psi)\\ &\geq q_1^{(\theta_m,\theta_M)}(\psi) - E_1\|\chi_{(a,a+l)}\psi\|^2.\end{aligned}$$

Thus this last inequality together with (21) in (18) gives

$$\begin{aligned}Q(\psi) \geq \frac{1}{2}\Big(&\frac{1}{2}\|\partial_s\psi\|^2 + \frac{1}{2}\|\dot{\theta}\partial_\tau\psi\|^2 + C\int_{\Omega_0}\rho(s)|\psi|^2\,\mathrm{d}s\,\mathrm{d}t\\ &- 2E_1\|\chi_{(a,l+a)}\psi\|^2 + (1-\gamma_{1/2,1/2})q_1^{(\theta_m\theta_M)}(\psi)\Big).\end{aligned}$$

Now if $0 < d \leq d_{\max}$ then $\gamma_{1/2,1/2} \leq 1$ so Proposition 3.5 follows.

3.3 The one dimensional Schrödinger operator

We want to show, under our conditions, that the one dimensional Schrödinger operator (17) is a positive operator. In view of Proposition 3.5 this will imply the Proposition 3.1. Here we follow a similar strategy as in [1].

Proposition 3.7. *for all* $\varphi \in \mathcal{H}^1(\mathbb{R})$, *then there exists* $l_{\max} > 0$ *such that for any* $0 < l \leq l_{\max}$ *we have*

$$\int_{\mathbb{R}} |\varphi'(s)|^2 + 2C\rho(s)|\varphi(s)|^2\,\mathrm{d}s \geq 4E_1\int_{I_a(l)}|\varphi(s)|^2\,\mathrm{d}s.$$

Proof. Introduce the function

$$\Phi(s) := \begin{cases}\Big(\dfrac{\pi}{2} + \arctan(s-p)\Big) & \text{if } s < p,\\ \dfrac{\pi}{2} & \text{if } s \geq p.\end{cases}$$

where p is the same real number as in (6). So clearly $\Phi' = \rho$. For any $t \in I_a(l)$ and $\varphi \in \mathcal{H}^1(\mathbb{R})$, we have

$$\frac{\pi}{2}\varphi(t) = \Phi(t)\varphi(t) = \int_{-\infty}^{t} (\Phi(s)\varphi(s))' \, \mathrm{d}s$$
$$= \int_{-\infty}^{t} \rho(s)\varphi(s)\, \mathrm{d}s + \int_{-\infty}^{t} \Phi(s)\varphi'(s)\, \mathrm{d}s$$

and since $\rho(s) = 0$ for any $s \in (p, \infty)$, we get,

$$\frac{\pi}{2}\varphi(t) = \int_{-\infty}^{p} \rho(s)\varphi(s)\, \mathrm{d}s + \int_{-\infty}^{t} \Phi(s)\varphi'(s)\, \mathrm{d}s.$$

Then some straightforward estimates lead to,

$$\frac{\pi^2}{4}\varphi^2(t) \leq 2\Big(\Big(\int_{-\infty}^{p} \rho(s)\varphi(s)\, \mathrm{d}s\Big)^2 + \Big(\int_{-\infty}^{t} \Phi(s)\varphi'(s)\, \mathrm{d}s\Big)^2\Big)$$
$$\leq 2\Big(\int_{-\infty}^{p} \rho(s)\, \mathrm{d}s \int_{-\infty}^{p} \rho(s)\varphi^2(s)\, \mathrm{d}s + \int_{-\infty}^{t} \Phi^2(s)\, \mathrm{d}s \int_{-\infty}^{t} \varphi'^2(s)\, \mathrm{d}s\Big).$$

By direct calculation

$$\int_{-\infty}^{p} \rho(s)\, \mathrm{d}s = \frac{\pi}{2}$$

and

$$\int_{-\infty}^{p} \Phi^2(s)\, \mathrm{d}s + \int_{p}^{t} \Phi^2(s)\, \mathrm{d}s = \pi \ln 2 + \frac{\pi^2}{4}(t-p).$$

Hence we get

$$|\varphi(t)|^2 \leq \frac{4}{\pi}\int_{\mathbb{R}} \rho(s)\varphi^2(s)\, \mathrm{d}s + \Big(\frac{8\ln 2}{\pi} + 2(t-p)\Big)\int_{\mathbb{R}} |\varphi'(s)|^2\, \mathrm{d}s. \tag{22}$$

We integrate both sides of (22) over $I_a(l)$, then

$$\int_{I_a(l)} |\varphi(t)|^2 \mathrm{d}t$$
$$\leq \frac{4l}{\pi}\int_{\mathbb{R}} \rho(s)\varphi^2(s)\, \mathrm{d}s + \Big(\Big(\frac{8\ln 2}{\pi} + 2(a-p)\Big)l + l^2\Big)\int_{\mathbb{R}} |\varphi'(s)|^2\, \mathrm{d}s$$
$$\leq c''\int_{\mathbb{R}} 2C\rho(s)\varphi^2(s) + |\varphi'(s)|^2\, \mathrm{d}s$$

where

$$c'' = 2l\Big(\frac{1}{\pi C} + \frac{4\ln 2}{\pi} + a - p\Big) + l^2.$$

Finally we get

$$4E_1\int_a^{l+a}|\varphi(t)|^2\,\mathrm{d}t \le 4E_1c''\int_{\mathbb{R}} 2\,C\rho(s)|\varphi(s)|^2+|\varphi'(s)|^2\,\mathrm{d}s.$$

So choose $0<l\le l_{\max}$ with

$$l_{\max} := -\Big(\frac{1}{\pi C}+\frac{4\ln 2}{\pi}+a-p\Big)+\sqrt{\Big(\frac{1}{\pi C}+\frac{4\ln 2}{\pi}+a-p\Big)^2+(4E_1)^{-1}}$$

then $4E_1c''\le 1$ and Proposition 3.7 follows. □

Proof of Theorem 1.1 ii). Under assumptions of Theorem 1.1 ii), $H_\theta^{\mathcal{N}}\ge H_\theta^l$, these two operators have the same essential spectrum so Theorem 1.1 ii) is proved by applying Proposition 3.1 and the min–max principle. □

Appendix

In this appendix we give a slight extension of Lemma 3 of [6] which states that under our conditions, for all $\psi\in D(q_\theta^l)$ we have for any $\alpha,\beta>0$ that there exists $\gamma_{\alpha,\beta}>0$ such that

$$|q_{2,3}(\psi)|\le\gamma_{\alpha,\beta}q_1(\psi)+\alpha q_2(\psi)+\beta q_3(\psi). \tag{23}$$

Then we have

Lemma A.1. *Let $p\in(\theta_m,\theta_M)$. For all $\psi\in D(q_\theta^l)$ such that $\psi(p,\cdot)=0$, then for any $\alpha,\beta>0$ there exists $\gamma_{\alpha,\beta}>0$ such that*

$$|q_{2,3}^{(\theta_m,p)}(\psi)|\le\gamma_{\alpha,\beta}q_1^{(\theta_m,p)}(\psi)+\alpha q_2^{(\theta_m,p)}(\psi)+\beta q_3^{(\theta_m,p)}(\psi). \tag{24}$$

Proof. Let $\psi\in D(q_\theta^l)$ such that $\psi(p,\cdot)=0$. Then $\psi\in\mathcal{H}_0^1(\Omega_p)$. We know that we may first consider vectors $\psi(s,t)=\chi_1(t)\phi(s,t)$, where $\phi\in C_0^\infty(\Omega_p)$. For such a vector ψ we have

$$q_1^{(\theta_m,p)}(\psi)=\|\chi_{(\theta_m,p)}\chi_1\nabla'\phi\|^2,$$

$$q_2^{(\theta_m,p)}(\psi)=\|\chi_{(\theta_m,p)}\chi_1\partial_s\phi\|^2,$$

$$q_3^{(\theta_m,p)}(\psi)=\|\chi_{(\theta_m,p)}\dot\theta(\chi_1\partial_\tau\phi+\phi\partial_\tau\chi_1)\|^2,$$

and

$$q_{2,3}^{(\theta_m,p)}(\psi) = 2(\dot\theta\chi_{(\theta_m,p)}\chi_1\partial_\tau\phi, \chi_1\partial_s\phi) + 2(\dot\theta\chi_{(\theta_m,p)}\phi\partial_\tau\chi_1, \chi_1\partial_s\phi). \tag{25}$$

By using simple estimates the first term on the right hand side of (25) is estimated as

$$|2(\dot\theta\chi_{(\theta_m,p)}\chi_1\partial_\tau\phi, \chi_1\partial_s\phi)| \le 2\|\dot\theta\|_\infty\|\chi_{(\theta_m,p)}\chi_1\nabla'\phi\|\|\chi_{(\theta_m,p)}\chi_1\partial_s\phi\|,$$

then

$$|2(\dot\theta\chi_{(\theta_m,p)}\chi_1\partial_\tau\phi, \chi_1\partial_s\phi)| \le c_1 q_1^{(\theta_m,p)}(\psi) + \frac{\alpha}{2}q_2^{(\theta_m,p)}(\psi), \tag{26}$$

where

$$c_1 := \frac{2}{\alpha}\, d^2\|\dot\theta\|_\infty^2$$

and $\alpha > 0$.

Integrating by parts twice and using the fact that $\dot\theta(\theta_m) = \phi(p,\cdot) = 0$, the second term of the right hand side of (25) is written as

$$2(\dot\theta\chi_{(\theta_m,p)}\phi\partial_\tau\chi_1, \chi_1\partial_s\phi) = (\chi_{(\theta_m,p)}\ddot\theta\phi\chi_1, \chi_1\partial_\tau\phi). \tag{27}$$

Then the Cauchy–Schwartz inequality implies,

$$|(\chi_{(\theta_m,p)}\ddot\theta\phi\chi_1, \chi_1\partial_\tau\phi)|^2 \le d^2\|\ddot\theta\|_\infty^2 q_1^{(\theta_m,p)}\|\chi_{(\theta_m,p)}\chi_1\phi\|^2.$$

Let $p' \in \mathbb{R}$ and $r' > 0$ such that $(p'-r,p') \subset (\theta_m,p)$ and for $s \in (p'-r,p')$, $|\dot\theta(s)| \ge \dot\theta_0$ for some $\dot\theta_0 > 0$. As in the proof of Lemma 3 of [6] we have

$$\|\chi_{(\theta_m,p)}\chi_1\phi\|^2 \le c_2(q_2^{(\theta_m,p)}(\psi) + \dot\theta_0^{-2}\|\chi_{(p'-r,p')}\dot\theta\chi_1\phi\|^2) \tag{28}$$

where

$$c_2 := \max\Big\{2 + 16\frac{(p-\theta_m)^2}{r^2},\, 4(p-\theta_m)^2\Big\}.$$

Moreover, for any $s \in \mathbb{R}$, $\dot\theta(s)\chi_1\phi(s,\cdot) \in \mathcal{H}_0^1(\Omega_p)$, then by using Lemma 1 of [6] there exists $\lambda > 0$ depending on ω such that

$$\|\chi_{(p'-r,p')}\dot\theta\chi_1\phi\|^2 \le \|\chi_{(\theta_m,p)}\dot\theta\chi_1\phi\|^2 \le \lambda^{-1}(q_3^{(\theta_m,p)}(\psi) + \|\dot\theta\|_\infty^2 q_1^{(\theta_m,p)}(\psi)). \tag{29}$$

Hence (28), (29), and (27) give

$$|(\chi_{(\theta_m,p)}\ddot\theta\phi\chi_1, \chi_1\partial_\tau\phi)|^2 \le (c_3 q_1^{(\theta_m,p)}(\psi) + \frac{\alpha}{2}q_2^{(\theta_m,p)}(\psi) + \beta q_3^{(\theta_m,p)}(\psi))^2, \tag{30}$$

where

$$c_3 := \max\Big\{\frac{d\|\ddot\theta\|_\infty\|\dot\theta\|_\infty\sqrt{c_2}}{\dot\theta_0\sqrt{\lambda}}, \frac{d^2\|\ddot\theta\|_\infty^2 c_2}{\alpha}, \frac{d^2\|\ddot\theta\|_\infty^2 c_2}{2\beta\dot\theta_0^2\lambda}\Big\}.$$

Then (26) and (30) imply (24) with $\gamma_{\alpha,\beta} := c_1 + c_3$.

Note that we can choose $\chi_1 > 0$ on ω. So that (24) holds for every $\psi \in C_0^\infty(\Omega_p)$ and by a density argument this is even true for $\psi \in \mathcal{H}_0^1(\Omega_p)$.

References

[1] D. Borisov, T. Ekholm, and H. Kovařík, Spectrum of the magnetic Schrödinger operator in a waveguide with combined boundary conditions. *Ann. Henri Poincaré* **6** (2005), no. 2, 327–342. MR 2136194 Zbl 1077.81038

[2] P. Briet, Spectral analysis for twisted waveguides. In R. Rebolledo and M. Orszag (eds.), *Quantum probability and related topics.* Proceedings of the 30th Conference held in Santiago, November 23–28, 2009. QP–PQ: Quantum Probability and White Noise Analysis, 27. World Scientific, Hackensack, N.J., 2011, 125–130. MR 2799119 Zbl 1241.81055

[3] P. Briet, H. Hammedi, and D. Krejčiřík, Hardy inequalities in globally twisted waveguides. *Lett. Math. Phys.* **105** (2015), no. 7, 939–958. MR 3357986 Zbl 1327.35458

[4] P. Briet, H. Kovařík, G. Raikov, and E. Soccorsi, Eigenvalue asymptotics in a twisted waveguide. *Comm. Partial Differential Equations* **34** (2009), no. 7-9, 818–836. MR 2560302 Zbl 1195.35119

[5] J. Dittrich and J. Kříž, Straight quantum waveguides with combined boundary conditions. In R. Weder, P. Exner, and B. Grébert (eds.), *Mathematical results in quantum mechanics.* Papers from the conference (QMATH-8) held at the Universidad Nacional Autónoma de México, Taxco, December 10–14, 2001. American Mathematical Society, Providence, R.I., 2002, 107–112. MR 1946021 Zbl 1041.81027

[6] T. Ekholm, H. Kovařík, and D. Krejčiřík, A Hardy inequality in twisted waveguides. *Arch. Ration. Mech. Anal.* **188** (2008), no. 2, 245–264. MR 2385742 Zbl 1167.35026

[7] P. Exner and H. Kovařík, Spectrum of the Schrödinger operator in a perturbed periodically twisted tube. *Lett. Math. Phys.* **73** (2005), no. 3, 183–192. MR 2188292 Zbl 1111.35014

[8] H. Hammedi, *Analyse spectrale des guides d'ondes "twistés."* Thèse de doctorat de Mathématique, Université de Toulon, Toulon, 2016.

[9] G. H. Hardy, Note on a theorem of Hilbert. *Math. Z.* **6** (1920), no. 3-4, 314–317. MR 1544414 JFM 47.0207.01

[10] D. Krejčiřík, Twisting versus bending in quantum waveguides. In P. Exner, J. P. Keating, P. Kuchment, T. Sunada, and A. Teplyaev (eds.), *Analysis on graphs and its applications.* Papers from the program held in Cambridge, January 8–June 29, 2007. Proceedings of Symposia in Pure Mathematics, 77. American Mathematical Society, Providence, R.I., 2008, 617–636. MR 2459893 Zbl 1165.81021

[11] J. Kříž, Spectral properties of planar quantum waveguides with combined boundary conditions. Ph.D. Thesis. Charles University, Prague, 2003.

[12] M. Reed and B. Simon, *Methods of modern mathematical physics.* IV. Analysis of operators. Academic Press [Harcourt Brace Jovanovich, Publishers], New York and London, 1978. MR 0493421 Zbl 0405.47007

Example of a periodic Neumann waveguide with a gap in its spectrum

Giuseppe Cardone and Andrii Khrabustovskyi

This paper is dedicated to Pavel Exner on the occasion of his jubilee.

1 Introduction

It is a well-known fact (see, e.g., [13]) that the spectrum of periodic elliptic self-adjoint differential operators has band structure, i.e., it is a locally finite union of compact intervals called *bands*. In general the bands may overlap, otherwise we have a *gap* in the spectrum - a bounded open interval having an empty intersection with the spectrum, but with ends belonging to it.

The presence of gaps in the spectrum is not guaranteed. For instance, the spectrum of the Laplace operator in $L^2(\mathbb{R}^n)$ has no gaps: $\sigma(-\Delta_{\mathbb{R}^n}) = [0,\infty)$. Therefore an interesting question arises here: to construct examples of periodic operators with non-void spectral gaps. This question is motivated by various applications, since the presence of gaps is important for the description of wave processes which are governed by differential operators under consideration: if the wave frequency belongs to a gap, then the corresponding wave cannot propagate in the medium. This feature is a main requirement for so-called *photonic crystals,* which are materials with periodic dielectric structure extensively investigating in recent years.

The problem of existence of spectral gaps for various periodic operators has been actively studied since mid 1990s. We refer to the overview [11], where one can find a lot of examples and references around this topic.

In recent years there have appeared many works in which the problem of opening of spectral gaps for operators posed in unbounded domains with a waveguide geometry (strips, tubes, graph-like domains, etc.) is studied, see, e.g., [1], [2], [4], [5], [8], [14], [15], [16], and [17]. The studies of physical processes (e.g., quantum particle motion) in such domains are of a great physical and mathematical interest because of the extensive progress in microelectronics during the last decade. We refer to the recent monograph [7] concerning spectral properties of quantum waveguides.

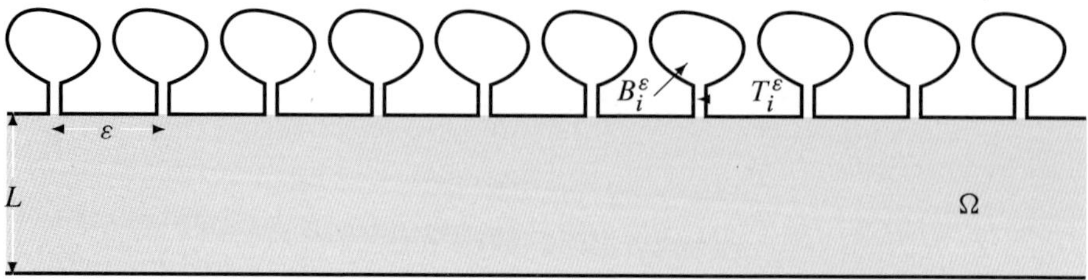

Figure 1. The waveguide Ω^ε

The simplest way to open up a gap is either to perturb a straight cylinder by a periodic nucleation of small voids (or making other "small" perturbation), see [1] and [15], or to consider a waveguide consisting of an array of identical compact domains connected by narrow "bridges," see [14] and [16]. In the first case one has small gaps separating large bands, in the second case one gets large gaps and small bands.

In the current paper we present another example of Neumann waveguide with a gap in the spectrum; the geometry of this waveguide essentially differs from previously studied examples. We are motivated by our recent work [3], where the spectrum of some Neumann problem was studied in a *bounded* domain perturbed by a lot of identical protuberances each of them consisting of two subsets - "room" and "passage" (in the simplest case, "room" is a small square and "passage" is a narrow rectangle connecting the "room" with the main domain). Peculiar spectral properties of so perturbed domains were observed for the first time by R. Courant and D. Hilbert [6]. Domains with "room-and-passage"-like geometry are widely used in order to construct examples illustrating various phenomena in Sobolev spaces theory and in spectral theory (see, for example, [9] and [10]).

Our goal is to show that perturbing a straight strip by a periodic array of "room-and-passage" protuberances one may open a spectral gap. Namely, we consider a strip of a width $L > 0$ and perturb it by a family of small identical protuberances, ε-periodically distributed along the strip axis. Here $\varepsilon > 0$ is a small parameter. Each protuberance has "room-and-passage" geometry. We denote the obtained domain by Ω^ε (see Figure 1). In Ω^ε we consider the operator $\mathcal{A}^\varepsilon = -\rho^\varepsilon \Delta_{\Omega^\varepsilon}$, where $\Delta_{\Omega^\varepsilon}$ is the Neumann Laplacian in $L^2(\Omega^\varepsilon)$. The weight ρ^ε is equal to 1 everywhere except the union of the "rooms", where it is equal to the constant $\varrho^\varepsilon > 0$.

The main result: we will prove that under suitable assumptions on L, ϱ^ε and sizes of "rooms" and "passages" the spectrum of $\mathcal{A}^\varepsilon$ converges to the spectrum of a certain spectral problem on the initial strip containing the spectral parameter in boundary conditions. Its spectrum has the form $[0,\infty) \setminus (\alpha,\beta)$, where (α,β) is a non-empty bounded interval. This, in particular, implies at least one gap in the spectrum of $\mathcal{A}^\varepsilon$ for small enough ε.

2 Setting of the problem and main result

In what follows by x and $\boldsymbol{x} = (x_1, x_2)$ we denote the Cartesian coordinates in $\mathbb{R}$ and $\mathbb{R}^2$, correspondingly.

By ε we denote a small parameter. To simplify the proof of the main theorem we suppose that it takes values from the discrete set $\mathcal{E} = \{\varepsilon\colon \varepsilon^{-1} \in \mathbb{N}\}$. The general case needs slight modifications.

We consider the unbounded strip $\Omega \subset \mathbb{R}^2$ of the width $L > 0$:

$$\Omega = \{\boldsymbol{x} \in \mathbb{R}^2\colon -L < x_2 < 0\}.$$

By Γ we denote its upper boundary: $\Gamma = \{\boldsymbol{x} \in \mathbb{R}^2\colon x_2 = 0\}$.

Let b^ε, d^ε, h^ε be positive constants, B be an open bounded domain in $\mathbb{R}^2$ having Lipschitz boundary and satisfying

$$B \subset \Big\{\mathbf{x} \in \mathbb{R}^2\colon x_1 \in \Big(-\frac{1}{2}, \frac{1}{2}\Big),\ x_2 > 0\Big\}, \tag{1}$$

$$\text{there exists } R \in (0,1) \text{ such that } \{\mathbf{x} \in \mathbb{R}^2\colon x_1 \in (-R/2, R/2),\ x_2 = 0\} \subset \partial B, \tag{2}$$

$$R^{-1} d^\varepsilon \le b^\varepsilon \le \varepsilon, \tag{3}$$

$$h^\varepsilon \longrightarrow 0 \quad \text{as } \varepsilon \to 0.$$

For $i \in \mathbb{Z}$ we set:

$$B_i^\varepsilon = \Big\{\mathbf{x} \in \mathbb{R}^2\colon \frac{1}{b^\varepsilon}(\mathbf{x} - \tilde{\mathbf{x}}^{i,\varepsilon}) \in B,\ \text{where } \tilde{\mathbf{x}}^{i,\varepsilon} = \Big(i\varepsilon + \frac{\varepsilon}{2}, h^\varepsilon\Big)\Big\}, \qquad \text{(“room”)}$$

$$T_i^\varepsilon = \Big\{\mathbf{x} \in \mathbb{R}^2\colon \Big|x_1 - i\varepsilon - \frac{\varepsilon}{2}\Big| < \frac{d^\varepsilon}{2},\ 0 \le x_2 \le h^\varepsilon\Big\}. \qquad \text{(“passage”)}$$

Conditions (1)–(3) imply that the “rooms” are pairwise disjoint and guarantee correct gluing of the i-th “room” and the i-th “passage” (the upper face of T_i^ε is contained in ∂B_i^ε). Moreover, the distance between the neighbouring “passages” is not too small, namely for $i \ne j$ one has $\operatorname{dist}(T_i^\varepsilon, T_j^\varepsilon) \ge \varepsilon - d^\varepsilon \ge \varepsilon(1 - R)$.

Attaching the “rooms” and “passages” to Ω we obtain the perturbed domain

$$\Omega^\varepsilon = \Omega \cup \Big(\bigcup_{i \in \mathbb{Z}} (T_i^\varepsilon \cup B_i^\varepsilon)\Big).$$

Let us define accurately the operator $\mathcal{A}^\varepsilon$. We denote by H^ε the Hilbert space of functions from $L^2(\Omega^\varepsilon)$ endowed with a scalar product

$$(u, v)_{H^\varepsilon} = \int_{\Omega^\varepsilon} u(\mathbf{x})\overline{v(\mathbf{x})}(\rho^\varepsilon(\mathbf{x}))^{-1}\,d\mathbf{x}, \tag{4}$$

where the function $\rho^\varepsilon(\mathbf{x})$ is defined as follows:

$$\rho^\varepsilon(\mathbf{x}) = \begin{cases} 1, & \mathbf{x} \in \Omega \cup \Big(\bigcup\limits_{i\in\mathbb{Z}} T_i^\varepsilon \Big), \\ \varrho^\varepsilon, & \mathbf{x} \in \bigcup\limits_{i\in\mathbb{Z}} B_i^\varepsilon, \end{cases} \qquad \varrho^\varepsilon > 0 \text{ is a constant.}$$

By $\mathfrak{a}^\varepsilon$ we denote the sesquilinear form in H^ε defined by

$$\mathfrak{a}^\varepsilon[u,v] = \int_{\Omega^\varepsilon} \nabla u \cdot \overline{\nabla v}\, d\mathbf{x}, \quad \operatorname{dom}(\mathfrak{a}^\varepsilon) = H^1(\Omega^\varepsilon). \tag{5}$$

The form $\mathfrak{a}^\varepsilon$ is densely defined, closed, positive and symmetric. We denote by $\mathcal{A}^\varepsilon$ the operator associated with this form, i.e.,

$$(\mathcal{A}^\varepsilon u, v)_{H^\varepsilon} = \mathfrak{a}^\varepsilon[u,v], \quad \text{for all } u \in \operatorname{dom}(\mathcal{A}^\varepsilon),\ v \in \operatorname{dom}(\mathfrak{a}^\varepsilon).$$

In other words, the operator $\mathcal{A}^\varepsilon$ is defined by the operation $-\rho^\varepsilon \Delta$ in Ω^ε and the Neumann boundary conditions on $\partial\Omega^\varepsilon$.

The goal of this work is to describe the behaviour of the spectrum $\sigma(\mathcal{A}^\varepsilon)$ as $\varepsilon \to 0$ under the assumption that the following limits exist and are positive:

$$\alpha := \lim_{\varepsilon\to 0} \frac{d^\varepsilon \varrho^\varepsilon}{h^\varepsilon (b^\varepsilon)^2 |B|}, \quad r := \lim_{\varepsilon\to 0} \frac{(b^\varepsilon)^2 |B|}{\varepsilon \varrho^\varepsilon}, \quad \alpha > 0,\ r > 0. \tag{6}$$

Also it is supposed that d^ε tends to zero not very fast, namely $\lim_{\varepsilon\to 0} \varepsilon \ln d^\varepsilon = 0$. The meaning of this condition and the meaning of α and r are explained in [3].

Now, we introduce the limit operator. By H we denote the Hilbert space of functions from $L^2(\Omega) \oplus L^2(\Gamma)$ endowed with the scalar product

$$(U,V)_H = \int_\Omega u_1(\mathbf{x})\overline{v_1(\mathbf{x})}\, d\mathbf{x} + \int_\Gamma u_2(x)\overline{v_2(x)} r\, dx, \quad U = (u_1,u_2), V = (v_1,v_2). \tag{7}$$

We introduce the sesquilinear form $\mathfrak{a}$ in H by

$$\mathfrak{a}[U,V] = \int_\Omega \nabla u_1 \cdot \overline{\nabla v_1}\, d\mathbf{x} + \int_\Gamma \alpha r (u_1|_\Gamma - u_2)\overline{(v_1|_\Gamma - v_2)}\, dx \tag{8}$$

with $\operatorname{dom}(\mathfrak{a}) = H^1(\Omega) \oplus L^2(\Gamma)$. Here by $u|_\Gamma$ we denote the trace of u on Γ. We denote by $\mathcal{A}$ the self-adjoint operator associated with this form.

Formally, the eigenvalue equation $\mathcal{A}U = \lambda U$ can be written as follows:

$$\begin{cases} -\Delta u_1 = \lambda u_1 & \text{in } \Omega, \\ \dfrac{\partial u_1}{\partial n} = \alpha r(u_2 - u_1) & \text{on } \Gamma, \\ \alpha(u_2 - u_1) = \lambda u_2 & \text{on } \Gamma, \\ \dfrac{\partial u_1}{\partial n} = 0 & \text{on } \partial\Omega \setminus \Gamma. \end{cases}$$

Here n is the outward-pointing unit normal.

Remark 2.1. Spectral properties of so-defined operators $\mathcal{A}$ were investigated in [3] and [12]. In [3] one considered the case of a bounded domain Ω, Γ is a flat subset of $\partial\Omega$. In this case the discrete spectrum of $\mathcal{A}$ consists of two sequences; one sequence accumulates at ∞, while the other one converges to α, which is the only point of the essential spectrum.

In [12] one considered, in particular, the case, when Ω is a straight unbounded strip, the line Γ is parallel to its axis and divides Ω on two unbounded strips (in this case the condition $\partial u_1/\partial u_n = \alpha(u_2 - u_1)$ becomes $\left[\partial u_1/\partial u_n\right] = \alpha(u_2 - u_1)$, where $[\cdot]$ denotes the jump of the enclosed quantity across Γ). In this case the spectrum of $\mathcal{A}$ turns out to be a union of the interval $[0, \alpha]$ and the ray $[\beta, \infty)$, where $\beta > \alpha$ provided $\alpha < (\pi/(2(L-L_\Gamma)))^2$. Here L is strip width and $L_\Gamma \in (0, L)$ is a distance from Γ to $\partial\Omega$. Let us note that, in fact, in [12] one deals with the Dirichlet conditions on $\partial\Omega$, but the Neumann case can be treated similarly – see Remark 3.2 in [12].

Using the same arguments as in [12] we arrive at the following formula for the spectrum of *our* operator $\mathcal{A}$:

$$\sigma(\mathcal{A}) = \begin{cases} [0, \alpha] \cup [\beta, \infty) & \text{if } \alpha < (\pi/(2L))^2, \\ [0, \infty) & \text{otherwise}, \end{cases} \tag{9}$$

where the number β is defined as follows. We denote by $\beta(\mu)$ (here $\mu \in \mathbb{R}$) the smallest eigenvalue of the problem

$$-u'' = \lambda u \quad \text{in } (-L, 0),$$

$$u'(0) = \mu u(0),$$

$$u'(-L) = 0.$$

It is straightforward to show that the function $\mu \mapsto \beta(\mu)$ is continuous, monotonically decreasing and moreover

$$\beta(\mu) \xrightarrow[\mu \to -\infty]{} \left(\frac{\pi}{2L}\right)^2 \quad \text{and} \quad \beta(\mu) \xrightarrow[\mu \to +\infty]{} -\infty.$$

Whence, in particular, one can conclude that there exists one and only one point β satisfying

$$\text{there exists } \mu < -\alpha r \text{ such that } \beta = \beta(\mu) = \frac{\alpha\mu}{\alpha r + \mu},$$

provided $\alpha < (\pi/(2L))^2$.

Now, we are in position to formulate the main results.

Theorem 2.2. (i) *Let the family $\{\lambda^\varepsilon \in \sigma(\mathcal{A}^\varepsilon)\}_{\varepsilon\in\mathcal{E}}$ have a convergent subsequence, i.e., $\lambda^\varepsilon \to \lambda$ as $\varepsilon = \varepsilon' \to 0$. Then $\lambda \in \sigma(\mathcal{A})$.*

(ii) *Let $\lambda \in \sigma(\mathcal{A})$. Then there exists a family $\{\lambda^\varepsilon \in \sigma(\mathcal{A}^\varepsilon)\}_{\varepsilon\in\mathcal{E}}$ such that*

$$\lim_{\varepsilon\to 0} \lambda^\varepsilon = \lambda.$$

From (9) and Theorem 2.2 we immediately obtain the following result.

Corollary 2.3. *Let $\alpha < (\pi/(2L))^2$. Let $\delta > 0$ be an arbitrary number satisfying $2\delta < \beta - \alpha$. Then there exists $\varepsilon_\delta > 0$ such that*

$$\sigma(\mathcal{A}^\varepsilon) \cap (\alpha + \delta, \beta - \delta) = \varnothing, \quad \sigma(\mathcal{A}^\varepsilon) \cap (\alpha - \delta, \beta + \delta) \neq \varnothing,$$

provided $\varepsilon < \varepsilon_\delta$.

3 Proof of Theorem 2.2

We present only the sketch of the proof since the main ideas are similar to the case of bounded domains Ω presented in [3].

Let $\{\lambda^\varepsilon \in \sigma(\mathcal{A}^\varepsilon)\}_{\varepsilon\in\mathcal{E}}$ and $\lambda^\varepsilon \to \lambda$ as $\varepsilon = \varepsilon' \to 0$. One has to show that $\lambda \in \sigma(\mathcal{A})$. In what follows we will use the index ε keeping in mind ε'.

We denote

$$\widetilde{\Omega} = (0,1) \times (-L,0), \quad \widetilde{\Omega}^\varepsilon = \Omega^\varepsilon \cap ((0,1)\times\mathbb{R}), \quad \widetilde{\Gamma} = (0,1)\times\{0\}.$$

Recall that $\varepsilon^{-1} \in \mathbb{N}$, whence $\Omega^\varepsilon + e_1 = \Omega^\varepsilon$, where $e_1 = (1,0)$, and thus $\mathcal{A}^\varepsilon$ is a periodic operator with respect to the period cell $\widetilde{\Omega}^\varepsilon$.

Using Floquet-Bloch theory (see, e.g, [13]) one can represent the spectrum of $\mathcal{A}^\varepsilon$ as a union of spectra of certain operators on $\widetilde{\Omega}^\varepsilon$. We denote by $\widetilde{H}^\varepsilon$ the space of functions from $L^2(\widetilde{\Omega}^\varepsilon)$ and the scalar product defined by (4) with $\widetilde{\Omega}^\varepsilon$ instead of Ω^ε.

Let us fix $\varphi \in [0, 2\pi)$. In $\widetilde{H}^\varepsilon$ we consider the sesquilinear form $\tilde{\mathfrak{a}}^{\varphi,\varepsilon}$ defined by (5) with $\widetilde{\Omega}^\varepsilon$ instead of Ω^ε and the definitional domain

$$\operatorname{dom}(\tilde{\mathfrak{a}}^{\varphi,\varepsilon}) = \{u \in H^1(\widetilde{\Omega}^\varepsilon) \colon u(1,\cdot) = e^{i\varphi} u(0,\cdot)\}.$$

By $\widetilde{\mathcal{A}}^{\varphi,\varepsilon}$ we denote the operator associated with this form. The spectrum of $\widetilde{\mathcal{A}}^{\varphi,\varepsilon}$ is purely discrete. We denote by $\{\tilde{\lambda}_k^{\varphi,\varepsilon}\}_{k=1}^\infty$ the sequence of eigenvalues of $\widetilde{\mathcal{A}}^{\varphi,\varepsilon}$ arranged in ascending order and with account of their multiplicity. Then one has

$$\sigma(\widetilde{\mathcal{A}}^\varepsilon) = \bigcup_{k=1}^{\infty} I_k^\varepsilon, \quad \text{where } I_k^\varepsilon = \bigcup_{\varphi \in [0,2\pi)} \{\tilde{\lambda}_k^{\varphi,\varepsilon}\} \text{ are compact intervals.} \tag{10}$$

We also introduce the operator $\widetilde{\mathcal{A}}^\varphi$ as the operator acting in

$$\widetilde{H} = \{U \in L^2(\widetilde{\Omega}) \oplus L^2(\widetilde{\Gamma}), \text{ the scalar product is defined by (7) with } \widetilde{\Omega}, \widetilde{\Gamma} \text{ instead of } \Omega, \Gamma\}$$

and generated by the sesquilinear form $\tilde{\mathfrak{a}}^\varphi$ which is defined by (8) (with $\widetilde{\Omega}, \widetilde{\Gamma}$ instead of Ω, Γ) and definitional domain $\operatorname{dom}(\tilde{\mathfrak{a}}^\varphi) = \operatorname{dom}(\tilde{\mathfrak{a}}^{\varphi,\varepsilon}) \oplus L^2(\widetilde{\Gamma})$.

Lemma 3.1. *The spectrum of $\widetilde{\mathcal{A}}^\varphi$ has the form*

$$\sigma(\widetilde{\mathcal{A}}^\varphi) = \{\alpha\} \cup \{\tilde{\lambda}_k^{\varphi,-}, k = 1,2,3\ldots\} \cup \{\tilde{\lambda}_k^{\varphi,+}, k = 1,2,3\ldots\}.$$

The points $\tilde{\lambda}_k^{\varphi,\pm}, k = 1,2,3,\ldots$ belong to the discrete spectrum, α is a point of the essential spectrum and they are distributed as follows:

$$0 \le \tilde{\lambda}_1^{\varphi,-} \le \tilde{\lambda}_2^{\varphi,-} \le \cdots \le \tilde{\lambda}_k^{\varphi,-} \le \cdots \xrightarrow[k\to\infty]{} \alpha < \tilde{\lambda}_1^{\varphi,+} \le \tilde{\lambda}_2^{\varphi,+} \le \cdots \le \tilde{\lambda}_k^{\varphi,+} \le \cdots \xrightarrow[k\to\infty]{} \infty.$$

Moreover if $\alpha < (\pi/(2d))^2$ then $\beta < \tilde{\lambda}_1^{\varphi,+}$.

This lemma was proved in [3] for the case of Neumann boundary conditions on the lateral parts of $\partial\widetilde{\Omega}^\varepsilon$. For the case of φ-periodic conditions the proof is similar.

Now, in view of (10) there exists $\varphi^\varepsilon \in [0, 2\pi)$ such that $\lambda^\varepsilon \in \sigma(\widetilde{\mathcal{A}}^{\varphi^\varepsilon,\varepsilon})$. We extract a convergent subsequence (for convenience still indexed by ε):

$$\varphi^\varepsilon \longrightarrow \varphi \in [0, 2\pi] \quad \text{as } \varepsilon \to 0. \tag{11}$$

Let u^ε be an eigenfunction of $\widetilde{\mathcal{A}}^{\varphi^\varepsilon,\varepsilon}$ corresponding to λ^ε with $\|u^\varepsilon\|_{\widetilde{\mathcal{H}}^\varepsilon} = 1$,

We introduce the operator

$$\Pi^\varepsilon : L^2\Big(\bigcup_{i=1}^{N(\varepsilon)} B_i^\varepsilon\Big) \longrightarrow L^2(\Gamma)$$

defined as follows:

$$\Pi^\varepsilon u(x) = \sum_{i=1}^{N(\varepsilon)} \Big(|B_i^\varepsilon|^{-1} \int_{B_i^\varepsilon} u(\mathbf{x})\,\mathrm{d}\mathbf{x} \Big) \chi_i^\varepsilon(x),$$

where χ_i^ε is the characteristic function of the interval $[i\varepsilon - \varepsilon, i\varepsilon]$. Using the Cauchy inequality and (6) one can easily obtain the estimate

$$\|\Pi^\varepsilon u\|^2_{L^2(\Gamma)} \le \sum_{i=1}^{N(\varepsilon)} \varrho^\varepsilon \varepsilon |B_i^\varepsilon|^{-1} \int_{B_i^\varepsilon} |u(\mathbf{x})|^2 (\varrho^\varepsilon)^{-1}\,\mathrm{d}\mathbf{x} \le C\|u\|^2_{\widetilde{\mathcal{H}}^\varepsilon}. \tag{12}$$

From (12) and $\|\nabla^\varepsilon u^\varepsilon\|^2_{L^2(\Omega^\varepsilon)} = \lambda^\varepsilon \le C$, we conclude that $\{u^\varepsilon\}_{\varepsilon\in\mathcal{E}}$ and $\{\Pi^\varepsilon u^\varepsilon\}_{\varepsilon\in\mathcal{E}}$ are bounded in $H^1(\widetilde{\Omega})$ and $L^2(\widetilde{\Gamma})$, correspondingly. Then there is a subsequence (still indexed by ε) and $u_1 \in H^1(\widetilde{\Omega})$, $u_2 \in L^2(\widetilde{\Gamma})$ such that

$$u^\varepsilon \rightharpoonup u_1 \text{ in } H^1(\widetilde{\Omega}), \quad \Pi^\varepsilon u^\varepsilon \rightharpoonup u_2 \text{ in } L^2(\widetilde{\Gamma}).$$

Also in view of the trace theorem and (11), we have $u^\varepsilon|_{\partial\widetilde{\Omega}} \to u_1$ in $L^2(\partial\widetilde{\Omega})$, whence $u_1(1,\cdot){=}e^{i\varphi}u_1(0,\cdot)$, i.e., $U = (u_1,u_2) \in \mathrm{dom}(\tilde{\mathfrak{a}}^\varphi)$.

If $u_1 = 0$ then $\lambda = \alpha$, the proof is completely similar to the proof of this fact in Theorem 2.1 in [3]. Then, in view of (9), $\lambda \in \sigma(\mathcal{A})$.

Now, let $u_1 \ne 0$. For an arbitrary $w \in \mathrm{dom}(\tilde{\mathfrak{a}}^{\varphi^\varepsilon,\varepsilon})$ we have

$$\int_{\widetilde{\Omega}^\varepsilon} \nabla u^\varepsilon(\mathbf{x}) \cdot \overline{\nabla w(\mathbf{x})}\,\mathrm{d}\mathbf{x} = \lambda^\varepsilon \int_{\widetilde{\Omega}^\varepsilon} (\rho^\varepsilon(\mathbf{x}))^{-1} u^\varepsilon(\mathbf{x})\overline{w(\mathbf{x})}\,\mathrm{d}\mathbf{x}. \tag{13}$$

Let $w_1 \in C^\infty(\bar{\widetilde{\Omega}})$, $w_2 \in C^\infty(\bar{\widetilde{\Gamma}})$, moreover $w_1(1,\cdot){=}e^{i\varphi}w_1(0,\cdot)$. We set

$$w_1^\varepsilon(\mathbf{x}) = w_1(\mathbf{x})((e^{i(\varphi^\varepsilon-\varphi)} - 1)x_1 + 1), \quad \mathbf{x} = (x_1,x_2).$$

It is easy to see that $w_1^\varepsilon(\mathbf{x})$ satisfies $w_1(1,\cdot){=}e^{i\varphi^\varepsilon}w_1(0,\cdot)$ and

$$w_1^\varepsilon \longrightarrow w_1 \text{ in } C^1(\bar{\widetilde{\Omega}}) \quad \text{as } \varepsilon \to 0. \tag{14}$$

Using these functions we construct the test-function $w(x)$ by the formula

$$w(\mathbf{x}) = \begin{cases} w_1^\varepsilon(x) + \sum\limits_{i\in\mathcal{I}^\varepsilon} (w_1^\varepsilon(\mathbf{x}^{i,\varepsilon}) - w_1^\varepsilon(\mathbf{x}))\varphi(\varepsilon^{-1}|\mathbf{x}-\mathbf{x}^{i,\varepsilon}|), & \mathbf{x} \in \widetilde{\Omega}, \\ (h^\varepsilon)^{-1}(w_2(\mathbf{x}^{i,\varepsilon}) - w_1^\varepsilon(\mathbf{x}^{i,\varepsilon}))x_2 + w_1^\varepsilon(\mathbf{x}^{i,\varepsilon}), & \mathbf{x} = (x_1,x_2) \in T_i^\varepsilon, \\ w_2(\mathbf{x}^{i,\varepsilon}), & \mathbf{x} \in B_i^\varepsilon. \end{cases}$$

Here $\mathbf{x}^{i,\varepsilon} := (i\varepsilon, 0)$, $\varphi \in C^\infty(\mathbb{R})$ satisfies $\varphi(t) = 1$ as $t \le R/2$ and $\varphi(t) = 0$ as $t \ge 1/2$, the constant $R \in (0,1)$ comes from (2) and (3). It is clear that $w^\varepsilon \in \operatorname{dom}(\mathfrak{a}^{\varphi^\varepsilon,\varepsilon})$.

We plug $w(\mathbf{x})$ into (13) and pass to $\varepsilon \to 0$. Using the same arguments as in the proof of Theorem 2.1 from [3] (but with account of (14)) we obtain:

$$\begin{aligned}\int_\Omega \nabla u_1 \cdot \overline{\nabla w_1}\, d\mathbf{x} + \int_\Gamma \alpha r (u_1|_\Gamma - u_2)\overline{(w_1|_\Gamma - w_2)}\, dx \\ = \lambda \int_{\widetilde{\Omega}} u_1 \overline{w_1}\, d\mathbf{x} + \lambda r \int_{\widetilde{\Gamma}} u_2 \overline{w_2}\, dx.\end{aligned}$$

By the density arguments this equality holds for an arbitrary $(w_1, w_2) \in \operatorname{dom}(\tilde{\mathfrak{a}}^\varphi)$ which implies $\widetilde{\mathcal{A}}^\varphi U = \lambda U, \quad U = (u_1, u_2)$. Since $u_1 \neq 0$ then $\lambda \in \sigma(\widetilde{\mathcal{A}}^\varphi)$. But in view of (9) and Lemma 3.1 for each φ one has $\sigma(\widetilde{\mathcal{A}}^\varphi) \subset \sigma(\mathcal{A})$, therefore $\lambda \in \sigma(\mathcal{A})$. Property (i) is proved.

The proof of property (ii) repeats word-by-word the proof for bounded domains Ω presented in [3].

Acknowledgements. Giuseppe Cardone is a member of GNAMPA (INdAM). Andrii Khrabustovskyi is supported by the Deutsche Forschungsgemeinschaft (DFG) through CRC 1173.

References

[1] F. L. Bakharev, S. A. Nazarov, and K. M. Ruotsalainen, A gap in the spectrum of the Neumann–Laplacian on a periodic waveguide. *Appl. Anal.* **92** (2013), no. 9, 1889–1915. MR 3169138 Zbl 1302.35266

[2] D. Borisov and K. Pankrashkin, Quantum waveguides with small periodic perturbations: gaps and edges of Brillouin zones. *J. Phys. A* **46** (2013), no. 23, 235203, 18 pp. MR 3064364 Zbl 1269.81050

[3] G. Cardone and A. Khrabustovskyi, Neumann spectral problem in a domain with very corrugated boundary. *J. Differential Equations* **259** (2015), no. 6, 2333–2367. MR 3353647 Zbl 06444028

[4] G. Cardone, V. Minutolo, and S. Nazarov, Gaps in the essential spectrum of periodic elastic waveguides. *ZAMM Z. Angew. Math. Mech.* **89** (2009), no. 9, 729–741. MR 2567306 Zbl 1189.35201

[5] G. Cardone, S. Nazarov and C. Perugia, A gap in the essential spectrum of a cylindrical waveguide with a periodic perturbation of the surface. *Math. Nachr.* **283** (2010), no. 9, 1222–1244. MR 2730490 Zbl 1213.35327

[6] R. Courant and D. Hilbert, *Methods of mathematical physics.* Vol. I. Translated and revised from the German original. Interscience Publishers, New York, N.Y., 1953. MR 0065391 Zbl 0051.28802 Zbl 0053.02805

[7] P. Exner and H. Kovařík, *Quantum waveguides.* Theoretical and Mathematical Physics. Springer, Cham, 2015. ISBN 978-3-319-18575-0 MR 3362506 Zbl 1314.81001

[8] P. Exner and O. Post, Convergence of spectra of graph-like thin manifolds. *J. Geom. Phys.* **54** (2005), no. 1, 77–115. MR 2135966 Zbl 1095.58007

[9] L. E. Fraenkel, On regularity of the boundary in the theory of Sobolev spaces. *Proc. London Math. Soc.* (3) **39** (1979), no. 3, 385–427. MR 0550077 Zbl 0406.46026

[10] R. Hempel, L. Seco, and B. Simon, The essential spectrum of Neumann Laplacians on some bounded singular domains. *J. Funct. Anal.* **102** (1991), no. 2, 448–483. MR 1140635 Zbl 0741.35043

[11] R. Hempel and O. Post, Spectral gaps for periodic elliptic operators with high contrast: an overview. In H. G. W. Begehr, R. P. Gilbert, and M. W. Wong (eds.), *Progress in analysis.* Vol. I. World Scientific, Inc., River Edge, N.J., 2003, 577–587. MR 2032728 Zbl 1061.35060

[12] A. Khrabustovskyi and M. Plum, Spectral properties of elliptic operator with double-contrast coefficients near a hyperplane. *Asymptot. Anal.* **98** (2016), no. 1-2, 91–130. Zbl 1341.47058

[13] P. Kuchment, *Floquet theory for partial differential equations.* Operator Theory: Advances and Applications, 60. Birkhäuser, Basel, 1993. ISBN 3-7643-2901-7 MR 1232660 Zbl 0789.35002

[14] S. A. Nazarov, K. Ruotsalainen, and J. Taskinen, Essential spectrum of a periodic elastic waveguide may contain arbitrarily many gaps. *Appl. Anal.* **89** (2010), no. 1, 109–124. MR 2604282 Zbl 1186.35123

[15] S. A. Nazarov, The asymptotic analysis of gaps in the spectrum of a waveguide perturbed with a periodic family of small voids. *J. Math. Sci.* (*N.Y.*) **186** (2012), no. 2, 247–301. MR 3098306 Zbl 1307.35100

[16] K. Pankrashkin, On the spectrum of a waveguide with periodic cracks. *J. Phys. A* **43** (2010), no. 47, 474030, 7 pp. MR 2738125 Zbl 1204.81069

[17] K. Yoshitomi, Band gap of the spectrum in periodically curved quantum waveguides. *J. Differential Equations* **142** (1998), no. 1, 123–166. MR 1492880 Zbl 0901.35066

Two-dimensional time-dependent point interactions

Raffaele Carlone, Michele Correggi, and Rodolfo Figari

Dedicated to Pavel Exner

1 Introduction and main results

Point interactions have been an important theoretical tool to investigate non-trivial qualitative features of the evolution of quantum systems. Since the early days of quantum mechanics they have been extensively used to provide solvable models in various fields of applied quantum physics such as solid state physics of perfect and disordered crystals, spectral nuclear structure, low energy neutron-nuclei scattering and many others. In the last edition of the reference book in the field [6], one can find a detailed and updated reading list on the subject.

The main feature of point interaction Hamiltonians is that they are characterized by a minimal set of physical parameters. All the information about the spatial geometry of the interaction potential acting on the quantum particle is included in the set of positions of the scattering centers, whereas the dynamical parameters consist in a set of real numbers characterizing boundary conditions that any function in the Hamiltonian domain has to satisfy at the scattering centers. Moreover, in one and three dimensions, it was shown that the behavior at the scattering centers at each time is sufficient to determine uniquely the solution of the Schrödinger equation at any time and at any point in space.

Later, it was recognized that a function of time (referred to as *charge* in the following) contains all the information about the behavior of the state function around one interaction point and that the charges are solutions of a system of Volterra integral equations. This extreme simplification of the Cauchy problem was used to generalize the theory to time-dependent and nonlinear point interactions. Such models were employed to investigate ionization issues and problems of quantum evolution in presence of concentrated nonlinearities in one and three dimensions, see [3], [4], [14], [9], [10], [13], [7], [8], and [16].

The two dimensional problem turned out to be decisively thornier. As an aside, let us mention that in dimension two the Laplacian and the formal delta potential scale in the same way under spacial dilation

$$-\Delta_{ax} + \lambda\delta(a\boldsymbol{x}) = \frac{1}{a^2}[-\Delta_{\boldsymbol{x}} + \lambda\delta(\boldsymbol{x})].$$

A relevant consequence of such a scale invariance is that, if E is an eigenvalue of the formal Hamiltonian, the same must be true for aE, for all $a > 0$, making the Hamiltonian either trivial or non-self-adjoint. In fact, one of the possible way to define a zero-range potential in dimension two is by "dimensional regularization," a modern quantum field renormalization scheme, which turned out to be very fruitful as a renormalization tool in Yang–Mills theory (see, e.g., [19] for further details). The extension of the model to real (or complex) dimensions $2-\epsilon$ supplies a non-scale invariant theory and a dimensional coupling constant. A (coupling constant dependent) limit $\epsilon \to 0^+$ then provides the same family of Hamiltonians obtained using one of the procedures now available to characterize all the self-adjoint extensions of the Laplacian restricted to functions supported outside the set of positions of the interaction centers [6].

From the technical point of view the presence of a logarithmic singularity in the fundamental solution of the Laplace equation $-\Delta K + \lambda K = \delta$ makes the charge equations in dimension two more difficult to deal with. In particular the techniques of fractional integration and differentiation used in dimension three to regularize the equations are no longer available. This is the main reason why the two-dimensional problem was still open, in spite of the progress in the one- and three-dimensional cases.

In the following we investigate the evolution problem generated by a time dependent point interaction Hamiltonian in dimension two. We first review notation, definitions and we state our main results. The last section is devoted to the proofs.

1.1 The model

In this paper we want to focus on the study of the time-evolution generated by a time-dependent Schrödinger operator with two-dimensional point interactions. More precisely the formal expression we start with is the following

$$\widetilde{H} = -\Delta + \sum_{j=1}^{N} \mu_j(t)\delta(\boldsymbol{x} - \boldsymbol{y}_j), \tag{1}$$

where $\boldsymbol{x}, \boldsymbol{y}_j \in \mathbb{R}^2$, $\delta(\boldsymbol{x} - \boldsymbol{y}_j)$ is the Dirac delta distribution supported at point $\boldsymbol{y}_j$ and $\mu_j(t) \in \mathbb{R}$ is its strength. The expression above is just formal because in two or more dimensions one can not give a rigorous meaning to the Dirac delta potential, not even in the sense of quadratic forms: the difficulty comes from the fact that $H^1(\mathbb{R}^2)$, the form domain of $-\Delta$, contains functions whose value at a given point $\boldsymbol{y}_j$ might not be defined. To circumvent this problem one can follow different procedures, e.g., introduce a symmetric operator [6] which coincides with (1) on a suitable subset of $H^2(\mathbb{R}^2)$ and study its self-adjoint extensions. Alternatively but equivalently it is possible to introduce (see below) a quadratic form associated with (1) and study its closedness, i.e., for any

$$\alpha_j(t) \in C^1(\mathbb{R})$$

with $\boldsymbol{\alpha}(t) = (\alpha_1(t), \dots, \alpha_N(t))$,

$$\begin{aligned} \mathcal{F}_{\boldsymbol{\alpha}(t)}[\psi] = &\int_{\mathbb{R}^2} \mathrm{d}\boldsymbol{r} \{|\nabla\phi_\lambda|^2 + \lambda|\phi_\lambda|^2 - \lambda|\psi|^2\} \\ &+ \sum_{j=1}^{N} \Big(\alpha_j(t) + \frac{1}{2\pi}\log\frac{\sqrt{\lambda}}{2} - \frac{\gamma}{2\pi}\Big)|q_j|^2 \\ &+ \frac{1}{2\pi}\sum_{j\neq k} q_j^* q_k K_0(\sqrt{\lambda}|\boldsymbol{y}_j - \boldsymbol{y}_k|) \end{aligned} \tag{2}$$

defined on the domain (notice that the domain $\mathcal{D}[\mathcal{F}]$ is actually time-independent)

$$\begin{aligned} \mathcal{D}[\mathcal{F}] = \Big\{ \psi \in L^2(\mathbb{R}^2) \;\Big|\; \psi = \phi_\lambda + \frac{1}{2\pi}\sum_{j=1}^{N} q_j K_0(\sqrt{\lambda}|\boldsymbol{x} - \boldsymbol{y}_j|), \\ \phi_\lambda \in H^1(\mathbb{R}^2), q_j \in \mathbb{C} \Big\}. \end{aligned} \tag{3}$$

Here $K_0(\sqrt{\lambda}|\boldsymbol{x}|)$ denote the anti-Fourier transform of $(|\boldsymbol{k}|^2 + \lambda)^{-1}$ for any $\lambda > 0$. $K_0(\boldsymbol{x})$ is the modified Bessel function of second kind of order 0 (also known as a Macdonald function, see Section 9.6 of [1]). It belongs to $L^2(\mathbb{R}^2)$, it is exponentially decreasing for large $|\boldsymbol{x}|$ and its asymptotic behavior for small $|\boldsymbol{x}|$ reads (see (9.6.13) in [1])

$$K_0(\sqrt{\lambda}|\boldsymbol{x}|) \underset{|\boldsymbol{x}|\to 0}{=} -\log\frac{\sqrt{\lambda}|\boldsymbol{x}|}{2} - \gamma + o(1), \tag{4}$$

with γ the Euler number. Functions in the domain of $\mathcal{F}_{\boldsymbol{\alpha}(t)}$ are thus composed by a *regular part* ϕ and a *singular part* containing a local singularity proportional to $-\log|\boldsymbol{x} - \boldsymbol{y}_j|$, whose coefficient q_j is the so-called *charge* already mentioned above.

It is easily checked that $\mathcal{D}[\mathcal{F}]$ is independent of λ: a simple way to make this apparent is to observe that for any $\lambda_1, \lambda_2 > 0$, $K_0(\sqrt{\lambda_1}|\boldsymbol{x}|) - K_0(\sqrt{\lambda_2}|\boldsymbol{x}|) \in H^1(\mathbb{R}^2)$, as one can easily verify by considering the Fourier transforms. Moreover the quadratic form (2) defined on (3) is closed and bounded from below as a consequence of the completeness of $H^1(\mathbb{R}^2)$ and $\mathbb{C}^N$. In a much more general setting the proof can be found in [15].

Therefore it defines for any $t \in \mathbb{R}$ a unique self-adjoint operator $H_{\boldsymbol{\alpha}(t)}$, whose domain is

$$\mathcal{D}(H_{\boldsymbol{\alpha}(t)}) = \Big\{\psi \in L^2(\mathbb{R}^2) \;\Big|\; \psi = \phi_\lambda + \frac{1}{2\pi}\sum_{j=1}^{N} q_j(t) K_0(\sqrt{\lambda}|\boldsymbol{x} - \boldsymbol{y}_j|), \quad \phi_\lambda \in H^2(\mathbb{R}^2), \lim_{\boldsymbol{x}\to\boldsymbol{y}_j} \phi_\lambda(\boldsymbol{x}) = (\Gamma_\lambda \boldsymbol{q})_j(t)\Big\}, \tag{5}$$

with

$$(\Gamma_\lambda)_{jk} = \begin{cases} \alpha_j(t) + \dfrac{1}{2\pi}\log\dfrac{\sqrt{\lambda}}{2} + \dfrac{\gamma}{2\pi}, & \text{if } j = k, \\ -\dfrac{1}{2\pi} K_0(\sqrt{\lambda}|\boldsymbol{y}_j - \boldsymbol{y}_k|), & \text{if } j \neq k. \end{cases}$$

In the simplest case of a single point interaction at the origin the boundary condition thus reads

$$\lim_{\boldsymbol{x}\to 0} \phi_\lambda(\boldsymbol{x}) = \Big(\alpha(t) + \frac{1}{2\pi}\log\frac{\sqrt{\lambda}}{2} + \frac{\gamma}{2\pi}\Big) q(t).$$

The action of $H_{\boldsymbol{\alpha}(t)}$ on functions of $\mathcal{D}(H_{\boldsymbol{\alpha}(t)})$ characterized in the form (5) is

$$(H_{\boldsymbol{\alpha}(t)} + \lambda)\psi = (-\Delta + \lambda)\phi_\lambda$$

and all the information on the interaction is encoded in the boundary conditions. Unlike the case of the quadratic form, the operator domain does depend on time: a generic $\psi \in \mathcal{D}(H_{\boldsymbol{\alpha}(t)})$ depends on $t \in \mathbb{R}$ through the regular part ϕ and the charge $\boldsymbol{q}(t)$. The closedness of the form clearly implies the self-adjointness of $H_{\boldsymbol{\alpha}(t)}$, provided $H_{\boldsymbol{\alpha}(t)}$ is indeed the operator associated with $\mathcal{F}_{\boldsymbol{\alpha}(t)}$, as we are going to show next. A very crucial property of functions in $\mathcal{D}(H_{\boldsymbol{\alpha}(t)})$ is that in a neighborhood of any point $\boldsymbol{y}_j$ the following asymptotic behavior holds true

$$\psi(|\boldsymbol{x}|) \underset{|\boldsymbol{x}-\boldsymbol{y}_j|\to 0}{=} \frac{1}{2\pi} q_j \log\frac{1}{|\boldsymbol{x} - \boldsymbol{y}_j|} + \alpha_j(t) q_j + o(1), \tag{6}$$

which is indeed the typical way point interactions are defined in the physics literature (see, e.g., [12] and references therein).

Notice that, unlike the three-dimensional case, the expression of the form or operator domain for $\lambda = 0$ can in principle be obtained by taking the limit $\lambda \to 0$ of (3) or (5), but, due to the singular large-$|\boldsymbol{x}|$ behavior of the Green function at $\lambda = 0$, i.e., $\log|\boldsymbol{x}|$, such a procedure does not define a well-posed domain decomposition.

For convenience of the reader we recall here how one can heuristically derive the expression of the quadratic form $\mathcal{F}_{\boldsymbol{\alpha}(t)}$ from the formal expression (1) via a sort of renormalization: pick any function ψ satisfying the required singular behavior (6) at any point $\boldsymbol{y}_j$, then it can be decomposed as

$$\psi = \phi_j + \frac{1}{2\pi} q_j \log \frac{1}{|\boldsymbol{x} - \boldsymbol{y}_i|},$$

where ϕ_j remains bounded as $\boldsymbol{x} \to \boldsymbol{y}_j$ and

$$\begin{aligned}\phi_\lambda(\boldsymbol{x}) \underset{|\boldsymbol{x}-\boldsymbol{y}_j|\to 0}{=} \; & \phi_j(\boldsymbol{y}_j) + \frac{1}{2\pi}\log\frac{\sqrt{\lambda}}{2} + \frac{1}{2\pi}\gamma \\ & - \frac{1}{2\pi}\sum_{k\neq j} q_k K_0(\sqrt{\lambda}|\boldsymbol{y}_j - \boldsymbol{y}_k|) + o(1),\end{aligned}$$

$$\phi_j(\boldsymbol{y}_j) = \alpha_j(t) q_j.$$

Now introducing an ultraviolet cut-off ε and observing that

$$(-\Delta + \lambda)\psi = (-\Delta + \lambda)\phi_\lambda,$$

if $|\boldsymbol{x} - \boldsymbol{y}_j| \geqslant \varepsilon$ for any $j = 1, \dots, N$, one has

$$\begin{aligned}
&\mathcal{F}_{\boldsymbol{\alpha}(t)}[\psi] + \lambda\|\psi\|_2^2 \\
&= \lim_{\varepsilon\to 0}\int_{\bigcup_k\{|\boldsymbol{x}-\boldsymbol{y}_k|\geqslant\varepsilon\}} \mathrm{d}\boldsymbol{x}\; \psi^*(\boldsymbol{x})[(\widetilde{H} + \lambda)\psi](\boldsymbol{x}) \\
&= \lim_{\varepsilon\to 0}\int_{\bigcup_k\{|\boldsymbol{x}-\boldsymbol{y}_k|\geqslant\varepsilon\}} \mathrm{d}\boldsymbol{x} \\
&\qquad \Big[\phi_\lambda^*(\boldsymbol{x}) + \frac{1}{2\pi}\sum_{j=1}^N q_j^* K_0(\sqrt{\lambda}|\boldsymbol{x} - \boldsymbol{y}_j|)\Big][(-\Delta + \lambda)\phi_\lambda](\boldsymbol{x}) \\
&= \|\nabla\phi_\lambda\|_2^2 + \lambda\|\phi_\lambda\|_2^2 \\
&\quad + \frac{1}{2\pi}\lim_{\varepsilon\to 0}\sum_{j=1}^N q_j^* \int_{\bigcup_k\{|\boldsymbol{x}-\boldsymbol{y}_k|\geqslant\varepsilon\}} \mathrm{d}\boldsymbol{r}\; K_0(\sqrt{\lambda}|\boldsymbol{x} - \boldsymbol{y}_j|)[(-\Delta + \lambda)\,\phi_\lambda](\boldsymbol{x}).
\end{aligned}$$

The last term can be integrated by parts twice as

$$
\begin{aligned}
\frac{1}{2\pi} \int_{\bigcup_k \{|\boldsymbol{x}-\boldsymbol{y}_k| \geq \varepsilon\}} & \mathrm{d}\boldsymbol{r} \; K_0(\sqrt{\lambda}|\boldsymbol{x}-\boldsymbol{y}_j|) \left[(-\Delta+\lambda)\,\phi_\lambda\right](\boldsymbol{x}) \\
&= -\frac{1}{2\pi} \sum_{k=1}^{N} \int_{\partial \mathcal{B}_\varepsilon(\boldsymbol{y}_k)} \mathrm{d}\sigma \; \phi_\lambda(\boldsymbol{x})\, \boldsymbol{n} \cdot \nabla K_0(\sqrt{\lambda}|\boldsymbol{x}-\boldsymbol{y}_j|) \\
&= -\frac{1}{2\pi} \int_{\partial \mathcal{B}_\varepsilon(\boldsymbol{y}_j)} \mathrm{d}\sigma \; \phi_\lambda(\boldsymbol{x})\, \boldsymbol{n} \cdot \nabla K_0(\sqrt{\lambda}|\boldsymbol{x}-\boldsymbol{y}_j|) + o(1) \\
&= \alpha_j(t) q_j + \frac{1}{2\pi} \log \frac{\sqrt{\lambda}}{2} + \frac{1}{2\pi}\gamma - \frac{1}{2\pi} \sum_{k \neq j} q_k K_0(\sqrt{\lambda}|\boldsymbol{y}_j - \boldsymbol{y}_k|) + o(1),
\end{aligned}
$$

since the asymptotics (4) implies

$$
\int_{\partial \mathcal{B}_\varepsilon(\boldsymbol{y}_j)} \mathrm{d}\sigma \; \boldsymbol{n} \cdot \nabla K_0(\sqrt{\lambda}|\boldsymbol{x}-\boldsymbol{y}_j|) = -\int_{\partial \mathcal{B}_\varepsilon(0)} \mathrm{d}\sigma \; \frac{1}{\varepsilon} = -2\pi,
$$

and the expression of the quadratic form is recovered.

It is worth mentioning that in the two-dimensional case point interactions are always attractive (see [6]), meaning that there always exists at least one bound state. For a single point interaction at $\boldsymbol{x} = 0$ with strength α, its wave function is proportional to $K_0(\sqrt{\lambda_\alpha}|\boldsymbol{x}|)$ (see [6]) and its energy E_α is

$$
E_\alpha = -\lambda_\alpha := -4 e^{2\gamma - 4\pi\alpha}.
$$

1.2 Time-evolution

Our goal is to examine the properties of the time-dependent Hamiltonians we have just defined and check under which conditions they generate a non-autonomous quantum dynamics, meaning that there exists a two parameter group of unitary operators $U(t,s)$ satisfying, in a sense which has to be specified, the Schrödinger equation

$$
i\partial_t U(t,s) = H_{\boldsymbol{\alpha}(t)} U(t,s), \tag{7}
$$

in such a way that the function

$$
\psi_t(\boldsymbol{x}) = U(t,s)\psi_s(\boldsymbol{x}).
$$

solves the Cauchy problem: for any $\psi \in \mathcal{D}(H_{\alpha(s)})$,

$$\begin{cases} i\partial_t \psi_t = H_{\boldsymbol{\alpha}(t)}\psi_t, \\ \psi_s = \psi. \end{cases}$$

In this paper the focus of our attention will be on the solution of the time-evolution problem described above. It is worth mentioning that an explicit expression of the integral kernel of the propagator $e^{-iH_\alpha t}$, when α does not depend on time, is already known [5], but its extension to the time-dependent case is not straightforward.

Our approach is based on a result, earlier proved and exploited in dimension three, stating that the solution of the Schrödinger equation is completely specified by the values of the charges $q_j(t)$, $j = 1, \dots, N$, characterizing the behavior of the wave packet around the scattering centers at each time t. The time dependent complex charges are solutions of a system of N coupled Volterra integral equations – the *charge equations* – thus reducing the complexity of the problem from the analysis of a non-autonomous flow in an infinite-dimensional Hilbert space to the search of solutions to a system of Volterra-type equation for N complex valued functions of time (see (9) below). For computational purposes as well as for possible extensions to nonlinear models such a complexity reduction is of course crucial. The procedure outlined above has been exploited in [23], for the three-dimensional analogue of the problem we are facing here, in [14], [13], [9], and [10] to investigate model-atoms ionization triggered by time dependent forces in dimension one and three and in [16] for the derivation of the time-dependent propagator in the case of three-dimensional moving point interactions. For a detailed introduction to the problem considered in this paper as well as many preliminary results we also refer to [2].

Before stating our main result we need to introduce first some notation: $U_0(t)$ will denote the free propagator, i.e.,

$$U_0(t) := e^{i\Delta t},$$

with integral kernel for $t \in \mathbb{R}$ and $\boldsymbol{x} \in \mathbb{R}^2$,

$$U_0(t; |\boldsymbol{x}|) = \frac{e^{-|\boldsymbol{x}|^2/(4it)}}{2it}.$$

The Volterra function of order -1 (see [17]) is defined as

$$\mathcal{I}(t) := \int_0^\infty \mathrm{d}\tau\, \frac{t^{\tau-1}}{\Gamma(\tau)}, \tag{8}$$

where Γ denotes the Gamma function. Some of the crucial properties of $\mathcal{I}(t)$ are listed in Section 2.1. Here we just point out that $\mathcal{I}$ is an analytic function of t with branch points at 0 and ∞.

Before stating our main result we introduce the charge equation associated to the time-evolution of the Hamiltonian $H_{\boldsymbol{\alpha}(t)}$: given any initial datum $\psi_s \in \mathcal{D}(H_{\boldsymbol{\alpha}(s)})$,

$$\boxed{\boldsymbol{q}(t) + \int_s^t \mathrm{d}\tau \, \mathcal{K}(t,\tau)\, \boldsymbol{q}(\tau) = \boldsymbol{f}(t),} \tag{9}$$

where

$$\mathcal{K}_{jk}(t,\tau) := \begin{cases} 4\pi \mathcal{I}(t-\tau)\Big(\alpha_j(\tau) - \dfrac{1}{2\pi}\log 2 + \dfrac{\gamma}{2\pi}\Big) & \text{if } j = k, \\ -2i\mathcal{I}(t-\tau) \displaystyle\int_s^\tau \mathrm{d}\sigma \; U_0(\tau-\sigma; |\boldsymbol{y}_j - \boldsymbol{y}_k|) & \text{if } j \neq k, \end{cases}$$

and

$$f_j(t) := 4\pi \int_s^t \mathrm{d}\tau \; \mathcal{I}(t-\tau)(U_0(\tau)\psi_s)(\boldsymbol{y}_j).$$

Theorem 1.1 (time-evolution). *Let $U(t,s)\colon L^2(\mathbb{R}^2) \to L^2(\mathbb{R}^2)$ be the map*

$$\boxed{(U(t,s)\psi_s)(\boldsymbol{x}) = (U_0(t-s)\psi_s)(\boldsymbol{x}) + \frac{i}{2\pi}\sum_{j=1}^N \int_s^t \mathrm{d}\tau \; U_0(t-\tau; |\boldsymbol{x}-\boldsymbol{y}_j|)\, q_j(\tau),} \tag{10}$$

where $\boldsymbol{q}(t)$ is a solution of the Volterra-type integral equation (9). *Then*

(a) *$U(t,s)$ is a two-parameter unitary group: for any $v,t,s \in \mathbb{R}$, $U(t,s)$ is unitary, $U(t,t) = \mathbb{1}$ and $U(t,s)U(s,v) = U(t,v)$;*

(b) *for any $t,s \in \mathbb{R}$, $U(t,s)$ solves the time-dependent Schrödinger equation* (7), *i.e., for any $\psi_s \in \mathcal{D}(H_{\alpha(s)})$, $\psi_t := U(t,s)\psi_s \in \mathcal{D}(H_{\boldsymbol{\alpha}(t)})$ and*

$$i\partial_t \psi_t = H_{\boldsymbol{\alpha}(t)}\psi_t. \tag{11}$$

Remark 1.2 (uniqueness of $\boldsymbol{q}(t)$). Although we did not state it explicitly the well-posedness of the time-evolution identified by $U(t,s)$ requires that the solution to (9) is unique. This is indeed the case as it is proven in Proposition 2.3.

Remark 1.3 (ansatz (10)). The statement of Theorem 1.1 says that the ansatz (10) provide the time-evolution of ψ_s whenever $\boldsymbol{q}(t)$ solves the charge equation (9). Such a statement however can be also read in the opposite direction: given ψ_t any solution to the time-dependent Schrödinger equation (11), then it can be rewritten in the form (10), with $\boldsymbol{q}(t)$ solving the charge equation (9).

2 Proofs

This Section contains the proofs of the results stated in Theorem 1.1, which are divided in several steps:

1. first we examine the integral operator defined by the Volterra kernel $\mathcal{I}$ and prove some of its relevant properties;
2. then we focus on the charge equation and prove that there exists a unique solution in the space of continuous functions;
3. such information becomes then a crucial ingredient to prove that the form domain $\mathcal{D}[\mathcal{F}]$ is invariant under the map $U(t,s)$;
4. next we show that given $\psi_s \in \mathcal{D}(H_{\boldsymbol{\alpha}(s)})$ then $U(t,s)\phi_s \in \mathcal{D}(H_{\boldsymbol{\alpha}(t)})$ and on a dense subset of the Hilbert space $U(t,s)$ defines an isometry, which coincides with the time-evolution generated by $H_{\boldsymbol{\alpha}(t)}$;
5. finally we show that U extends to a two-parameter unitary group by density.

2.1 Properties of the Volterra kernel $\mathcal{I}$

We start the discussion by recalling some useful properties of the function $\mathcal{I}[t]$ defined in (8). We refer to Section 18.3 of [17] (where $\mathcal{I}(t)$ is denoted as $\nu(t,-1)$) for further details. One striking relation involving $\mathcal{I}(t)$ is the inversion formula of the Laplace transform (see [18] and [22]): denoting by

$$(\mathcal{L}f)(p) = \int_0^\infty \mathrm{d}t\, e^{-pt} f(t),$$

the usual action of the Laplace transform, then

$$\mathcal{L}^{-1}\Big(\frac{p}{\log(p)}\Big)(t) = \nu(t,-1) = \mathcal{I}(t).$$

Since they will play some role in the following we also provide the asymptotic expansions of $\mathcal{I}(t)$ as $t \to 0$ or $t \to \infty$ (see again [17]):

$$\mathcal{I}(t) \underset{t\to 0}{=} \frac{1}{t\log^2(1/t)}[1 + \mathcal{O}(|\log t|^{-1})],$$

$$\mathcal{I}(t) \underset{t\to\infty}{=} e^t + \mathcal{O}(t^{-1}).$$

Hence, given the previous expansions, $\mathcal{I}(t) \in L^1_{\mathrm{loc}}(\mathbb{R})$.

Next we study the integral operator

$$(If)(t) := \int_0^t d\tau \, \mathcal{I}(t-\tau) f(\tau). \tag{12}$$

We also denote by J the integral operator with kernel

$$(Jf)(t) := \int_0^t d\tau \, \mathfrak{J}(t-\tau) f(\tau), \quad \mathfrak{J}(t-\tau) = -\gamma - \log(t-\tau).$$

In [11] the operator I is investigated in details and several useful properties, such as its smoothing action, are established. Here we only need a notable identity, which is stated in Lemma 2.2 and a simple estimate of the Sobolev norm of If (we refer to [11] for the proof):

Lemma 2.1. *If $f \in H^{\nu}(0,T)$ with $0 < \nu \leqslant 1$, then $If \in H^{\nu}(0,T)$, i.e., there exists $C_t < +\infty$ such that*

$$\|If\|_{H^{\nu}(0,T)} \leqslant C_T \, \|f\|_{H^{\nu}(0,T)} \, . \tag{13}$$

Moreover $C_T \to 0$ as $T \to 0$.

Lemma 2.2. *For any $t \in \mathbb{R}^+$ and $f \in L^1(0,t)$,*

$$(IJf)(t) = \int_0^t d\tau \; f(\tau).$$

Proof. We first observe that one has the identity

$$\int_0^t d\tau \, \mathcal{I}(t-\tau)(-\gamma - \log\tau) = 1. \tag{14}$$

In Lemma 32.1 in [22] it is indeed proven that (in the formula proved in the cited Lemma one should take $\alpha = 1, h = 0$)

$$\int_0^t d\tau \; (\log\tau - \psi(1)) \, \partial_t \nu(t-\tau) = -1,$$

but, using [17] (eq (12), Section 18.3), one can recognize that $\partial_t \nu(t) = \mathcal{I}(t)$.

Next we note that in the expression

$$(IJf)(t) = \int_0^t \mathrm{d}\tau \int_0^{t-\tau} \mathrm{d}\sigma\ \mathfrak{I}(\tau)\mathfrak{J}(t-\tau-\sigma) f(\sigma),$$

one can exchange the order of the integration, since

$$\begin{aligned}
&\int_0^t \mathrm{d}\tau \int_0^{t-\tau} \mathrm{d}\sigma\ \mathfrak{I}(\tau)\mathfrak{J}(t-\sigma-\tau) f(\sigma) + \int_0^t \mathrm{d}\sigma \int_0^{t-\sigma} \mathrm{d}\tau\ \mathfrak{I}(\tau)\mathfrak{J}(t-\sigma-\tau) f(\sigma)\\
&\quad = \int_0^t \mathrm{d}\sigma \int_0^t \mathrm{d}\tau\ \mathfrak{I}(\tau)\mathfrak{J}(|t-\sigma-\tau|) f(\sigma)\\
&\quad = 2\int_0^t \mathrm{d}\tau \int_0^{t-\tau} \mathrm{d}\sigma\ \mathfrak{I}(\tau)\mathfrak{J}(|t-\sigma-\tau|) f(\sigma),
\end{aligned}$$

Using (14) we conclude that

$$(IJf)(t) = \int_0^t \mathrm{d}\sigma \int_0^{t-\sigma} \mathrm{d}\iota\ \mathfrak{I}(\tau)\mathfrak{J}(t-\sigma-\tau) f(\sigma) = \int_0^t \mathrm{d}\sigma\ f(\sigma). \qquad \square$$

2.2 Derivation of the charge equation

Before starting to discuss the charge equation, we present a heuristic computation which motivates the ansatz (10). First of all we set $s = 0$ and assume that $q_j(0) = 0$. Neglecting any regularity issue, we can compute the time derivative of (10) and obtain

$$\begin{aligned}
&i\partial_t(U(t,0)\psi_0)(\boldsymbol{x})\\
&\quad(-\Delta U_0(t)\psi_0)(\boldsymbol{x}) - \frac{1}{2\pi}\sum_{j=1}^N q_j(t)\\
&\qquad + \frac{1}{2\pi}\sum_{j=1}^N \int_0^t \mathrm{d}\tau\ \partial_\tau U_0(t-\tau; |\boldsymbol{x}-\boldsymbol{y}_j|)\, q_j(\tau)\\
&\quad = (-\Delta U_0(t)\psi_0)(\boldsymbol{x}) - \frac{1}{2\pi}\sum_{j=1}^N \int_0^t \mathrm{d}\tau\ U_0(t-\tau; |\boldsymbol{x}-\boldsymbol{y}_j|)\, \dot{q}_j(\tau),
\end{aligned}$$

so that if we take the Fourier transform defined for a function $f \in L^2(\mathbb{R}^2)$ as

$$\hat{f}(\boldsymbol{p}) := \frac{1}{2\pi}\int_{\mathbb{R}^2} \mathrm{d}\boldsymbol{x}\ e^{-i\boldsymbol{p}\cdot\boldsymbol{x}} f(\boldsymbol{x}),$$

the above expression becomes (we set $k = |\boldsymbol{p}|$)

$$i\partial_t(\widehat{U(t,0)\psi_0})(\boldsymbol{p}) = p^2 e^{-ip^2 t}\widehat{\psi_0}(\boldsymbol{p}) - \frac{1}{2\pi}\sum_{j=1}^{N}\int_0^t \mathrm{d}\tau\, e^{-i\boldsymbol{p}\cdot\boldsymbol{y}_j} e^{-ip^2(t-\tau)}\,\dot{q}_j(\tau). \tag{15}$$

Similarly, recalling that (see, e.g., (6.532.4) in [20])

$$\frac{1}{2\pi}\int_{\mathbb{R}^2}\mathrm{d}\boldsymbol{p}\,\frac{e^{i\boldsymbol{p}\cdot\boldsymbol{x}}}{p^2+\lambda} = \int_0^\infty \mathrm{d}p\,\frac{pJ_0(p|\boldsymbol{x}|)}{p^2+\lambda} = K_0(\sqrt{\lambda}|\boldsymbol{x}|),$$

one gets

$$\begin{aligned}
&(\widehat{H_{\boldsymbol{\alpha}(t)}U(t,0)\psi_0})(\boldsymbol{p})\\
&= p^2\Big(\widehat{U(t,0)\psi_0}(\boldsymbol{p}) - \frac{1}{2\pi}\sum_{j=1}^{N}\frac{q_j e^{-i\boldsymbol{p}\cdot\boldsymbol{y}_j}}{p^2+\lambda}\Big) - \frac{\lambda}{2\pi}\sum_{j=1}^{N}\frac{q_j e^{-i\boldsymbol{p}\cdot\boldsymbol{y}_j}}{p^2+\lambda}\\
&= p^2 e^{-ip^2 t}\widehat{\psi_0}(\boldsymbol{p}) + \frac{1}{2\pi}\sum_{j=1}^{N}\int_0^t \mathrm{d}\tau\, e^{-i\boldsymbol{p}\cdot\boldsymbol{y}_j}\,\partial_\tau(e^{-ip^2(t-\tau)})\,q_j(\tau)\\
&\quad - \frac{1}{2\pi}\sum_{j=1}^{N} q_j e^{-i\boldsymbol{p}\cdot\boldsymbol{y}_j}\\
&= p^2 e^{-ip^2 t}\widehat{\psi_0}(\boldsymbol{p}) - \frac{1}{2\pi}\sum_{j=1}^{N}\int_0^t \mathrm{d}\tau\, e^{-i\boldsymbol{p}\cdot\boldsymbol{y}_j} e^{-ip^2(t-\tau)}\,\dot{q}_j(\tau),
\end{aligned} \tag{16}$$

which equals (15). Therefore, for any $\boldsymbol{q}(t)$ and ψ_0 for which the right hand side of (16) is defined, the assumed solution does solve the time-dependent Schrödinger equation, at least in a weak sense. To be solution of the charge equation is the condition that guarantees that for any $t \in \mathbb{R}$, $\psi_t \in \mathcal{D}(H_{\boldsymbol{\alpha}(t)})$. Indeed, if we impose the boundary condition as in (5), we get

$$\frac{1}{2\pi}\int_{\mathbb{R}^2}\mathrm{d}\boldsymbol{p}\, e^{i\boldsymbol{p}\cdot\boldsymbol{y}_j}\widehat{\phi_\lambda}(\boldsymbol{p}) = (\Gamma_\lambda\boldsymbol{q})_j\,(t), \tag{17}$$

and therefore

$$\frac{1}{2\pi}\int_{\mathbb{R}^2} \mathrm{d}\boldsymbol{p}\, e^{i\boldsymbol{p}\cdot\boldsymbol{y}_j}\Big\{e^{-ip^2t}\widehat{\psi_0}(\boldsymbol{p}) + \frac{i}{2\pi}\sum_{k=1}^{N}\int_0^t \mathrm{d}\tau\, e^{-i\boldsymbol{p}\cdot\boldsymbol{y}_k} e^{-ip^2(t-\tau)}\, q_k(\tau)$$

$$- \frac{1}{2\pi}\sum_{k=1}^{N}\frac{q_k(t)e^{-i\boldsymbol{p}\cdot\boldsymbol{y}_k}}{p^2+\lambda}\Big\}$$

$$= \Big(\alpha_j(t) + \frac{1}{2\pi}\log\frac{\sqrt{\lambda}}{2} - \frac{\gamma}{2\pi}\Big)q_j(t) - \frac{1}{2\pi}\sum_{k\neq j} q_k(t)K_0(\sqrt{\lambda}|\boldsymbol{y}_j - \boldsymbol{y}_k|).$$

The last off-diagonal term cancels exactly and thus the identity becomes

$$\frac{1}{2\pi}\int_{\mathbb{R}^2} \mathrm{d}\boldsymbol{p}\, e^{i\boldsymbol{p}\cdot\boldsymbol{y}_j}\Big\{e^{-ip^2t}\widehat{\psi_0}(\boldsymbol{p}) + \frac{i}{2\pi}\sum_{k=1}^{N}\int_0^t \mathrm{d}\tau\, e^{-i\boldsymbol{p}\cdot\boldsymbol{y}_k} e^{-ip^2(t-\tau)}\, q_k(\tau)$$

$$- \frac{1}{2\pi}\frac{q_j(t)e^{-i\boldsymbol{p}\cdot\boldsymbol{y}_j}}{p^2+\lambda}\Big\}$$

$$= \Big(\alpha_j(t) + \frac{1}{2\pi}\log\frac{\sqrt{\lambda}}{2} + \frac{\gamma}{2\pi}\Big)q_j(t).$$

Combining the last diverging term on the left hand side with the second one, via an integration by parts (here we implicitly assume that the charge belongs to a suitable Sobolev space), we get

$$\frac{1}{2\pi}\int_{\mathbb{R}^2} \mathrm{d}\boldsymbol{p}\,\Big\{e^{i\boldsymbol{p}\cdot\boldsymbol{y}_j} e^{-ip^2t}\widehat{\psi_0}(\boldsymbol{p})$$

$$- \frac{1}{2\pi(p^2+\lambda)}\int_0^t \mathrm{d}\tau\, e^{-ip^2(t-\tau)}\,[\dot{q}_j(\tau) - i\lambda q_j(\tau)]$$

$$+ \frac{i}{2\pi}\sum_{k\neq j}\int_0^t \mathrm{d}\tau\, e^{i\boldsymbol{p}\cdot(\boldsymbol{y}_j-\boldsymbol{y}_k)} e^{-ip^2(t-\tau)}\, q_k(\tau)\Big\}$$

$$= \Big(\alpha_j(t) + \frac{1}{2\pi}\log\frac{\sqrt{\lambda}}{2} + \frac{\gamma}{2\pi}\Big)q_j(t),$$

The $\boldsymbol{p}$ integral of the second term on the left hand side contains an infrared singularity for $t = \tau$ which goes as $\log(t-\tau)$: since (see (3.722.1) and (3.722.3) in [20])

$$\int_{\mathbb{R}^2} \mathrm{d}\boldsymbol{p}\,\frac{e^{-ip^2(t-\tau)}}{p^2+\lambda} = -\pi e^{i\lambda(t-\tau)}[\mathrm{Ci}(\lambda(t-\tau)) - i\,\mathrm{Si}(\lambda(t-\tau))]$$

$$= -\pi(\gamma + \log\lambda + \log(t-\tau)) + Q(\lambda; t-\tau)e^{i\lambda(t-\tau)},$$

where Si(·) and Ci(·) stand for the sine and cosine integral functions (see (5.2.1) and (5.2.2) in [1]) and (see, e.g., (5.2.16) in [1])

$$Q(\lambda; t-\tau) := -\pi(1-e^{i\lambda t})(\gamma + \log\lambda + \log(t-\tau)) \\ - \pi e^{i\lambda(t-\tau)}\Big(\sum_{n=1}^{\infty} \frac{(-(t-\tau)^2\lambda^2)^n}{2n(2n)!} - i\,\mathrm{Si}((t-\tau)\lambda)\Big).$$

Note that $Q(0; t-\tau) = 0$. Hence the charge equation can be rewritten

$$(U_0(t)\psi_0)(\boldsymbol{y}_j) + \frac{i}{2\pi}\sum_{k\neq j}\int_0^t \mathrm{d}\tau\, U_0(t-\tau; |\boldsymbol{y}_j - \boldsymbol{y}_k|)\, q_k(\tau) \\ - \Big(\alpha_j(t) + \frac{1}{2\pi}\log\frac{\sqrt{\lambda}}{2} + \frac{\gamma}{2\pi}\Big) q_j(t) \\ = -\frac{1}{4\pi}\int_0^t \mathrm{d}\tau\, \Big(\gamma + \log(t-\tau) + \log\lambda - \frac{1}{\pi}Q(\lambda; t-\tau)\Big)\partial_\tau(e^{i\lambda(t-\tau)}q_j(\tau))$$

and taking the limit $\lambda \to 0$ (notice the exact cancellation of the diverging $\log\lambda$ terms)

$$(U_0(t)\psi_0)(\boldsymbol{y}_j) + \frac{i}{2\pi}\sum_{k\neq j}\int_0^t \mathrm{d}\tau\, U_0(t-\tau; |\boldsymbol{y}_j - \boldsymbol{y}_k|)\, q_k(\tau) \\ - \Big(\alpha_j(t) - \frac{1}{2\pi}\log 2 + \frac{\gamma}{2\pi}\Big) q_j(t) \qquad (18) \\ = -\frac{1}{4\pi}\int_0^t \mathrm{d}\tau\, (\gamma + \log(t-\tau))\dot{q}_j(\tau).$$

If we now apply to both sides the integral operator I defined in (12) and exploit the property proven in Lemma 2.2, we finally recover the charge equation (9).

2.3 Charge equation

We now consider the charge equation and its solution.

Proposition 2.3 (existence and uniqueness of solutions to (9)). *Given $\psi_s \in \mathcal{D}(H_{\boldsymbol{\alpha}(s)})$, the solution $\boldsymbol{q}(t)$ of the charge equation* (9) *exists and is unique in $C(0,T)$ for any $T < \infty$. Moreover $\boldsymbol{q}(t) \in H^{\nu}(0,T)$ for any $\nu < 3/4$.*

Proof. According to the general theory of Volterra integral equations [21], specialized to the linear case there exists at least one continuous solution to (9) and it is unique if the following conditions are satisfied:

(a) $\boldsymbol{f}(t)$ is continuous on $t \in \mathbb{R}^+$;

(b) $\mathcal{K}(t,\tau)$ is measurable and $\mathcal{K}(t,\cdot) \in L^1(0,t)$ for any finite t.

Let us check that the integral kernel $\mathcal{K}_{jk}(t,\tau) \in L^1(0,t)$ In particular to see that $\mathcal{K}_{jk}(t,\tau)$ is integrable, it suffices to notice that the diagonal term is L^1 because $\mathcal{I}(t-\tau)$ is, while for $j \neq k$,

$$|\mathcal{K}_{jk}(t,\tau)| \leqslant 2|\mathcal{I}(t-\sigma)| \left| \int_s^\tau \mathrm{d}\sigma \, U_0(\tau-\sigma; |\boldsymbol{y}_j - \boldsymbol{y}_k|) \right| \leqslant C_t |\mathcal{I}(t-\sigma)|,$$

for some $C_t < \infty$, if $t < \infty$. Indeed the integral

$$\int_s^\tau \mathrm{d}\sigma \, U_0(\tau-\sigma; |\boldsymbol{y}_j - \boldsymbol{y}_k|)$$

is explicitly computable and it is finite for any $\tau \geqslant 0$, if $|\boldsymbol{y}_j - \boldsymbol{y}_k| > 0$.

Therefore to complete the proof we need to show that $\boldsymbol{f}(t)$ is a continuous function of t in any compact subset of $\mathbb{R}^+$. By hypothesis $\psi_s \in \mathcal{D}(H_{\boldsymbol{\alpha}(s)})$ and thus it can be decomposed as in (5), i.e.,

$$\begin{aligned} f_j(t) = 4\pi \int_s^t \mathrm{d}\tau \, \mathcal{I}(t-\tau) \Big\{ & (U_0(\tau)\phi_{\lambda,s})(\boldsymbol{y}_j) \\ & + 2q_j(s) \int_{\mathbb{R}^2} \mathrm{d}\boldsymbol{x}' \, U_0(\tau; |\boldsymbol{y}_j - \boldsymbol{x}'|) K_0(\boldsymbol{x}' - \boldsymbol{y}_j) \\ & + 2\sum_{k \neq j} q_k(s) \int_{\mathbb{R}^2} \mathrm{d}\boldsymbol{x}' \, U_0(\tau; |\boldsymbol{y}_j - \boldsymbol{x}'|) K_0(\boldsymbol{x}' - \boldsymbol{y}_k) \Big\} \end{aligned} \tag{19}$$

with $\phi_{\lambda,s} \in H^2(\mathbb{R}^2)$.

We first consider the term involving $\phi_{\lambda,s}$: applying the Fourier transform, we have

$$\begin{aligned} 4\pi (U_0(\tau)\phi_{\lambda,s})(\boldsymbol{y}_j) &= 2 \int_{\mathbb{R}^2} \mathrm{d}\boldsymbol{p} \, e^{i\boldsymbol{p}\cdot\boldsymbol{y}_j} e^{-ip^2\tau} \widehat{\phi_{\lambda,s}}(\boldsymbol{p}) \\ &= 2 \int_{\mathbb{R}^2} \mathrm{d}\boldsymbol{p} \, e^{-ip^2\tau} \widehat{(T_{\boldsymbol{y}_j}^{-1}\phi_{\lambda,s})}(\boldsymbol{p}) \\ &= 2\pi \int_0^\infty \mathrm{d}\varrho \, e^{-i\varrho\tau} \widehat{\langle (T_{\boldsymbol{y}_j}^{-1}\phi_{\lambda,s}) \rangle}(\sqrt{\varrho}) \\ &= (2\pi)^{3/2} \left(\mathcal{F}G(\varrho)\right)(t) \end{aligned}$$

where

$$G_1(\varrho) := \mathbb{1}_{[0,+\infty)}(\varrho) \, \widehat{\langle (T_{\boldsymbol{y}_j}^{-1}\phi_{\lambda,s}) \rangle}(\sqrt{\varrho}),$$

$\mathcal{F}$ stands for the Fourier transform in $L^2(\mathbb{R})$ and we have denoted by $\langle f \rangle$ the angular average of a function on $\mathbb{R}^2$, i.e.,

$$\langle f \rangle(p) = \frac{1}{2\pi} \int_0^{2\pi} \mathrm{d}\vartheta \ f(p, \vartheta).$$

In order to bound the norm of $(U_0(\tau)\phi_{\lambda,s})(\mathbf{y}_j)$ in $H^{\nu}(\mathbb{R})$, we estimate

$$\begin{aligned}
4\pi & \int_{\mathbb{R}} \mathrm{d}\varrho \ |\varrho|^{2\nu} |(\mathcal{F}(U_0(\,\cdot\,)\phi_{\lambda,s})(\mathbf{y}_j))(\varrho)|^2 \\
&= 32\pi^4 \int_{\mathbb{R}} \mathrm{d}\varrho \ |\varrho|^{2\nu} |(\mathcal{F}^{-1}(\mathcal{F} G_1))(-\varrho)|^2 \\
&= 64\pi^4 \int_0^{\infty} \mathrm{d}p \ p^{4\nu+1} |\langle \widehat{(T_{\mathbf{y}_j}^{-1} \phi_{\lambda,s})} \rangle (p)|^2 \\
&\leqslant C \int_{\mathbb{R}^2} \mathrm{d}\mathbf{p} \ p^{4\nu} |\widehat{(T_{\mathbf{y}_j}^{-1} \phi_{\lambda,s})}(\mathbf{p})|^2.
\end{aligned}$$

Since $\phi_{\lambda,s} \in H^2$, the last integral is bounded for any $0 \leqslant \nu \leqslant 1$ and therefore $\left(U_0(\,\cdot\,)\phi_{\lambda,s}\right)(\mathbf{y}_j) \in H^1(0, T)$. Thanks to Lemma 2.1 the Sobolev degree is conserved by the action of I and therefore the first term in (19) is in $H^1(0, T)$, which implies, via Sobolev inequality, that it is a continuous function of $t \in \mathbb{R}^+$.

Let us consider now the sum in the last term in (19): we first rewrite

$$\begin{aligned}
(U_0(\tau) T_{\mathbf{y}_k} K_0(\sqrt{\lambda}\,\cdot\,))(\mathbf{y}_j) &= \int_{\mathbb{R}^2} \mathrm{d}\mathbf{x}' \ U_0(\tau; |\mathbf{y}_j - \mathbf{x}'|) K_0(\mathbf{x}' - \mathbf{y}_k) \\
&= \frac{1}{2\pi} \int_{\mathbb{R}^2} \mathrm{d}\mathbf{p} \ \frac{e^{i\mathbf{p}\cdot(\mathbf{y}_j - \mathbf{y}_k)} e^{-ip^2\tau}}{p^2 + \lambda} \\
&= \frac{1}{2\pi} \int_0^{\infty} \mathrm{d}p \ p \int_0^{2\pi} \mathrm{d}\vartheta \ \frac{e^{ip|\mathbf{y}_j - \mathbf{y}_k| \cos\vartheta} e^{-ip^2\tau}}{p^2 + \lambda} \\
&= \int_0^{\infty} \mathrm{d}p \ p \ \frac{e^{-ip^2\tau} J_0(p|\mathbf{y}_j - \mathbf{y}_k|)}{p^2 + \lambda} \\
&= \pi \left(\mathcal{F} G_2\right)(\tau),
\end{aligned}$$

with

$$G_2(\varrho) = \mathbb{1}_{[0,+\infty)}(\varrho) \frac{J_0(\sqrt{\varrho}|\mathbf{y}_j - \mathbf{y}_k|)}{\varrho + \lambda}.$$

As before in order to bound the H^{ν}-norm of $(U_0(\tau)T_{y_k}K_0(\sqrt{\lambda}\,\cdot))(y_j)$ we estimate, using the asymptotics of Bessel functions for large argument (see (9.2.1) in [1])

$$\begin{aligned}
&\int_{\mathbb{R}} \mathrm{d}\varrho\, |\varrho|^{2\nu}\, |(\mathcal{F}(U_0(\tau)T_{y_k}K_0(\sqrt{\lambda}\,\cdot))(y_j))(\varrho)|^2 \\
&\quad = \pi^2 \int_{\mathbb{R}} \mathrm{d}\varrho\, |\varrho|^{2\nu}\, |(\mathcal{F}^{-1}\mathcal{F}G_2)(-\varrho)|^2 \\
&\quad = \pi^2 \int_0^{\infty} \mathrm{d}p\; p^{4\nu+1}\, \frac{J_0^2(p|y_j - y_k|)}{(p^2+\lambda)^2} \\
&\quad \leqslant C \int_0^{\infty} \mathrm{d}p\; p^{4\nu+1}\, \frac{1}{(p+1)\,(p^2+\lambda)^2},
\end{aligned}$$

which is finite for any $\nu < 3/4$. This implies that $(U_0(\tau)T_{y_k}K_0(\sqrt{\lambda}\cdot))(y_j)$ belongs to $H^{\nu}(0,T)$, for any $\nu < 3/4$. The action of I does not change the Sobolev degree and therefore the last term in (19) is continuous in t, thanks again to Sobolev inequality.

The last term to consider is the second one in (19): as before we have

$$\begin{aligned}
(U_0(\tau)T_{y_j}K_0(\sqrt{\lambda}\,\cdot))(y_j) &= \int_{\mathbb{R}^2} \mathrm{d}x'\, U_0(\tau; |y_j - x'|) K_0(x' - y_j) \\
&= \frac{1}{2}\int_0^{\infty} \mathrm{d}\varrho\, \frac{e^{-i\varrho\tau}}{\varrho+\lambda} \\
&= \frac{1}{2} e^{i\lambda\tau}\,(i\,\mathrm{Si}(\lambda\tau) - \mathrm{Ci}(\lambda\tau)) \\
&= -\frac{1}{2}\,(\gamma + \log\lambda + \log\tau) + \frac{1}{2\pi} Q(\lambda;\tau) e^{i\lambda\tau}.
\end{aligned}$$

Hence

$$I[(U_0(\tau)T_{y_j}K_0(\sqrt{\lambda}\,\cdot))(y_j)] = 1 + \int_s^t \mathrm{d}\tau\, \mathcal{I}(t-\tau)\Big[-\frac{1}{2}\log\lambda + \frac{1}{2\pi}Q(\lambda;\tau)\Big],$$

but since $Q(\lambda,\tau)$ is a smooth function on any compact set, the same applies to the second term in (19), which is continuous as well.

In order to prove the last statement it suffices to apply a bootstrap like argument: for sufficiently small times the charge equation can be solved since the operator $1+\mathcal{K}$ is invertible in $H^{\nu}(0,T)$, $\nu < 3/4$. This is a consequence of (13) and differentiability of $\alpha(t)$. Hence $q \in H^{\nu}(0,T)$, $\nu < 3/4$, for small enough T, but then one can repeat the argument with initial condition $q(T)$ so proving the statement. □

2.4 Time-evolution in the form and operator domains

In this Section we show that the form domain is invariant under $U(t,s)$.

Proposition 2.4 (invariance of $\mathcal{D}[\mathcal{F}]$ for initial data in $\mathcal{D}(H_{\boldsymbol{\alpha}(s)})$). *Let $\boldsymbol{q}(t)$ be the unique solution to* (9) *with initial condition $\boldsymbol{q}(t)|_{t=s} = \boldsymbol{q}(s)$ and $\psi_s \in \mathcal{D}(H_{\boldsymbol{\alpha}(s)})$, then $U(t,s)\psi_s \in \mathcal{D}[\mathcal{F}]$ for any $t \in \mathbb{R}$.*

Proof. In order to prove the statement we need to show that

$$(U(t,s)\psi_s)(\boldsymbol{x}) - \frac{1}{2\pi}\sum_{j=1}^{N} q_j(t) K_0(\sqrt{\lambda}|\boldsymbol{x}-\boldsymbol{y}_j|) \in H^1(\mathbb{R}^2),$$

whenever $\psi_s \in \mathcal{D}[\mathcal{F}]$. Setting for simplicity $s = 0$ and passing to the Fourier representation, this is equivalent to requiring that the following function of $\boldsymbol{p}$,

$$e^{-ip^2 t}\widehat{\psi_0}(\boldsymbol{p}) + \frac{i}{2\pi}\sum_{j=1}^{N}\int_0^t \mathrm{d}\tau\, e^{i\boldsymbol{p}\cdot\boldsymbol{y}_j} e^{-ip^2(t-\tau)} q_j(\tau) - \frac{1}{2\pi}\sum_{j=1}^{N}\frac{q_j(t)e^{i\boldsymbol{p}\cdot\boldsymbol{y}_j}}{p^2+\lambda},$$

belongs to $L^2(\mathbb{R}^2, (p^2+1)\mathrm{d}\boldsymbol{p})$. After an integration by parts the above expression becomes (here $\dot{q}_j$ stands for the weak derivative of q_j, which belongs at least to $H^{\nu}(0,T)$, $\nu < -1/4$, since $q_j(t) \in H^{\nu}$, $\nu < 3/4$ by Proposition 2.3)

$$\begin{aligned} & e^{-ip^2 t}\widehat{\psi_0}(\boldsymbol{p}) - \frac{1}{2\pi}\sum_{j=1}^{N}\frac{q_j(0)e^{i\boldsymbol{p}\cdot\boldsymbol{y}_j}e^{-ip^2 t}}{p^2+\lambda} \\ & \quad + \frac{1}{2\pi(p^2+\lambda)}\sum_{j=1}^{N}\int_0^t \mathrm{d}\tau\, e^{i\boldsymbol{p}\cdot\boldsymbol{y}_j} e^{-i(p^2+\lambda)(t-\tau)}\partial_\tau(e^{i\lambda\tau}q_j(\tau)). \end{aligned} \tag{20}$$

Now the first two terms represent the free evolution of the regular part $\phi_{\lambda,0}$ of the initial state ψ_0. Since by hypothesis $\psi_0 \in \mathcal{D}[\mathcal{F}]$, then $\phi_{\lambda,0} \in H^1(\mathbb{R}^2)$ and therefore the sum of those two terms (or rather their Fourier anti-transform) belongs to $H^1(\mathbb{R}^2)$ as well.

It remains then to prove that the last term in (20) is in $L^2(\mathbb{R}^2, (p^2+1)\mathrm{d}\boldsymbol{p})$: setting $z(t) = \partial_t(e^{i\lambda(t-\tau)}q_j(t))$ and calling each term of the sum $g_j(\boldsymbol{p})$ for short, we have

$$\int_{\mathbb{R}^2}\mathrm{d}\boldsymbol{p}\,(p^2+1)|g_j(\boldsymbol{p})|^2 = \frac{1}{8}\int_0^\infty \mathrm{d}\varrho\,\frac{\varrho+1}{(\varrho+\lambda)^2}\,|(\mathcal{F}z_{0,t})(\varrho)|^2$$

where we have denoted for any function $f\colon [0,T] \to \mathbb{C}$ and $0 \leqslant a < b \leqslant T$,

$$f_{a,b}(t) := f(t)\mathbb{1}_{[a,b]}(t).$$

Now the right hand side is obviously bounded by

$$C \|z_{0,t}\|^2_{H^{-1/2}(\mathbb{R})} \leqslant C_t(\|(q_j)_{0,t}\|^2_{H^{-1/2}(\mathbb{R})} + \|(\dot{q}_j)_{0,t}\|^2_{H^{-1/2}(\mathbb{R})}),$$

for some $C_t < \infty$ for finite t. In Lemma 2.1 in [8] it is proven that if $f \in H^\nu(0,T)$ for $0 < \nu < 1/2$ (see also Lemma 5 in [3]), then $f_{0,t} \in H^\nu(\mathbb{R})$. Therefore the first term on the right hand side of the above expression is always bounded, since by Proposition 2.3, $q_j \in H^\nu(0,T)$ for any $1/2 < \nu < 3/4$. For the second term one can not apply directly (see Lemma 2.1 in [8]) because the Sobolev degree of $\dot{q}_j$ is negative, but one can circumvent such a problem by modifying the extension of $\dot{q}_j$ (see the proof of Theorem 4 in [4]): let $Q_j(\tau)$, $\tau \in \mathbb{R}$, be the following function which extends $q_j(\tau)$,

$$Q_j(\tau) = \begin{cases} q_j(0) & \text{for } \tau < 0, \\ q_j(t) & \text{for } \tau > t, \\ q_j(\tau) & \text{for } 0 \leqslant \tau \leqslant t. \end{cases}$$

Then one has that $\dot{Q}_j = (\dot{q}_j)_{0,t}$ but $Q_j \in H^\nu_{\text{loc}}(\mathbb{R})$ for any $1/2 < \nu < 3/4$, which implies that $\dot{Q}_j \in H^{\nu-1}(\mathbb{R})$, since it is compactly supported. In conclusion $(\dot{q}_j)_{0,t} \in H^\nu(\mathbb{R})$ for any $-1/2 < \nu < -1/4$ and the second term is bounded as well. □

The above result in combination with the heuristic computation made at the beginning of Section 2.2 yields the following very important corollary.

Corollary 2.5. *Let $\psi_s \in \mathcal{D}[\mathcal{F}]$, then $U(t,s)\psi_s$ solves the time-dependent Schrödinger equation* (7) *in the quadratic form sense, i.e., for any $\phi \in \mathcal{D}[\mathcal{F}]$,*

$$i\,\partial_t \langle \phi \mid U(t,s)\psi_s\rangle = \mathcal{F}_{\boldsymbol{\alpha}(t)}[\phi, U(t,s)\psi_s],$$

for any $t \in \mathbb{R}$ and with $\mathcal{F}_{\boldsymbol{\alpha}(t)}[\cdot,\cdot]$ standing for the sesquilinear form associated to the quadratic form $\mathcal{F}_{\boldsymbol{\alpha}(t)}[\cdot]$.

Proof. It suffices to note that the identities proven at the beginning of Section 2.2, up to (16), are in fact rigorous once projected onto a state $\phi \in \mathcal{D}[\mathcal{F}]$. The result of Proposition 2.4 completes the argument. □

2.5 Completion of the proof

In order to complete the proof of Theorem 1.1, we have to show that

$$U(t,s)\colon \mathcal{D}(H_{\boldsymbol{\alpha}(s)}) \longrightarrow \mathcal{D}(H_{\boldsymbol{\alpha}(t)})$$

and it is an isometry in that subspace.

Lemma 2.6. *Let $\psi_s \in \mathcal{D}(H_{\boldsymbol{\alpha}(s)})$, then for any $t \in \mathbb{R}$, $U(t,s)\psi_s \in \mathcal{D}(H_{\boldsymbol{\alpha}(t)})$.*

Proof. Thanks to Proposition 2.4 at least $U(t,s)\psi_s \in \mathcal{D}[\mathcal{F}]$ so that it can be decomposed in a regular part in $H^1(\mathbb{R}^2)$ plus the singular terms given in (3). However this is the only information needed to make rigorous the heuristic derivation presented from eqs. (17) to (18). The property stated in Lemma 2.2 and the fact that $\boldsymbol{q}(t)$ solves the charge equation implies then the result. □

Lemma 2.7. *Let $\psi_s \in \mathcal{D}(H_{\boldsymbol{\alpha}(s)})$, then*

$$\|U(t,s)\psi_s\|_2 = \|\psi_s\|_2 .$$

Proof. We simply compute the time-derivative of the L^2 norm of the ansatz (10):

$$\begin{aligned}\partial_t \|U(t,s)\psi_s\|_2^2 &= 2\Re \left\langle U(t,s)\psi_s \mid \partial_t U(t,s)\psi_s \right\rangle \\ &= -2\Re (i\mathcal{F}_{\boldsymbol{\alpha}(t)}[U(t,s)\psi_s, U(t,s)\psi_s]) \\ &= 0,\end{aligned}$$

thanks to Lemma (2.6), Corollary 2.5 and the trivial observation that if $\psi_t \in \mathcal{D}(H_{\boldsymbol{\alpha}(t)})$ then $\psi_t \in \mathcal{D}[\mathcal{F}]$. □

Proof of Theorem 1.1. Lemma 2.6 in combination with Corollary 2.5 implies that given any $\psi_s \in \mathcal{D}(H_{\boldsymbol{\alpha}(s)})$, $U(t,s)\psi_s$ solves the time-dependent Schrödinger equation.

Moreover $U(t,t) = \mathbb{1}$ and, for any $\psi_0 \in \mathcal{D}(H_{\alpha(0)})$,

$$\begin{aligned}&(U(t,s)U(s,0)\psi_0)(\boldsymbol{x}) \\ &= (U_0(t)\psi_0)(\boldsymbol{x}) + \frac{i}{2\pi} U_0(t-s) \sum_{j=1}^{N} \int_0^s \mathrm{d}\tau \, U_0(s-\tau; |\boldsymbol{x}-\boldsymbol{y}_j|)\, q_j(\tau) \\ &\qquad + \frac{i}{2\pi} \sum_{j=1}^{N} \int_s^t \mathrm{d}\tau \, U_0(t-\tau; |\boldsymbol{x}-\boldsymbol{y}_j|)\, q_j(\tau) \\ &= (U_0(t)\psi_0)(\boldsymbol{x}) + \frac{i}{2\pi} \sum_{j=1}^{N} \int_0^t \mathrm{d}\tau \, U_0(t-\tau; |\boldsymbol{x}-\boldsymbol{y}_j|)\, q_j(\tau) \\ &= (U(t,0)\psi_0)(\boldsymbol{x}),\end{aligned}$$

i.e., the map $U(t,s)$ satisfies the group composition rules. Since $\mathcal{D}(H_{\boldsymbol{\alpha}(t)})$ is densely defined, one can extend the map $U(t,s)$ to the whole Hilbert space by density and, due to the properties above, such an extension is automatically unitary. □

Acknowledgements. The authors thank A. Teta for helpful discussions about the presentation of the model.

Raffaele Carlone and Michele Correggi acknowledge the support of MIUR through the FIR grant 2013 "Condensed Matter in Mathematical Physics (Cond-Math)" (code RBFR13WAET).

References

[1] M. Abramovitz and I. A. Stegun (ed.), *Handbook of mathematical functions with formulas, graphs, and mathematical tables.* Reprint of the 1972 edition. Dover Publications, New York, 1992. ISBN 0-486-61272-4 MR 1225604 MR MR0167642 (original) Zbl 0643.33001 Zbl 0171.38503 (original)

[2] R. Adami, *A class of Schrödinger equations with concentrated nonlinearity.* Ph.D. thesis. Università di Roma "La Sapienza", Roma, 2000.

[3] R. Adami and A. Teta, A class of nonlinear Schrödinger equations with concentrated nonlinearity. *J. Funct. Anal.* **180** (2001), no. 1, 148–175. MR 1814425 Zbl 0979.35130

[4] R. Adami, G. Dell'Antonio, R. Figari, and A. Teta, The Cauchy problem for the Schrödinger equation in dimension three with concentrated nonlinearity. *Ann. Inst. H. Poincaré Anal. Non Linéaire* **20** (2003), no. 3, 477–500. MR 1972871 Zbl 1028.35137

[5] S. Albeverio, Z. Brzeźniak, and L. Dąbrowski, Time-dependent propagator with point interaction. *J. Phys. A* **27** (1994), no. 14, 4933–4943. MR 1295192 Zbl 0841.34086

[6] S. Albeverio, F. Gesztesy, R. Høegh-Krohn, and H. Holden, *Solvable models in quantum mechanics.* Second edition. AMS Chelsea Publishing, Providence, R.I., 2005. With an Appendix by P. Exner. ISBN 0-8218-3624-2/hbk MR 2105735 Zbl 1078.81003

[7] C. Cacciapuoti, D. Finco, D. Noja, and A. Teta, The NLS equation in dimension one with spatially concentrated nonlinearities: the pointlike limit. *Lett. Math. Phys.* **104** (2014), no. 12, 1557–1570. MR 3275343 Zbl 1303.81072

[8] C. Cacciapuoti, D. Finco, D. Noja, and A. Teta, The point-like limit for a NLS equation with concentrated nonlinearity in dimension three. Preprint 2015. arXiv:1511.06731 [math-ph]

[9] M. Correggi and G. Dell'Antonio, Decay of a bound state under a time-periodic perturbation: a toy case. *J. Phys. A* **38** (2005), no. 22, 4769–4781. MR 2148620 Zbl 1071.81038

[10] M. Correggi, G. Dell'Antonio, R. Figari, and A. Mantile, Ionization for three dimensional time-dependent point interactions. *Comm. Math. Phys.* **257** (2005), no. 1, 169–192. MR 2163573 Zbl 1079.81014

[11] R. Carlone and A. Fiorenza, Regularization properties of Volterra integral equations with Sonine type kernels. In preparation.

[12] M. Correggi, D. Finco, and A. Teta, Energy lower bound for the unitary $N + 1$ fermionic model. *Europhys. Lett.* **111** (2015), no. 1, 10003.

[13] O Costin, J. L. Lebowitz, and A. Rokhlenko, Exact results for the ionization of a model quantum system. *J. Phys. A* **33** (2000), no. 36, 6311–6319. MR 1789555 Zbl 1064.81548

[14] O Costin, R. D. Costin, J. L. Lebowitz, and A. Rokhlenko, Evolution of a model quantum system under time periodic forcing: conditions for complete ionization. *Comm. Math. Phys.* **221** (2001), no. 1, 1–26. MR 1846899 Zbl 0996.81017

[15] G. Dell'Antonio, R. Figari, A. Teta, Hamiltonians for systems of N particles interacting through point interactions. *Ann. Inst. H. Poincaré Phys. Théor.* **60** (1994), no. 3, 253–290. MR 1281647 Zbl 0808.35113

[16] G. Dell'Antonio, R. Figari, A. Teta, The Schrödinger equation with moving point interactions in three dimensions. In F. Gesztesy, H. Holden, J. Jost, S. Paycha, M. Röckner, and S. Scarlatti (eds.), *Stochastic processes, physics and geometry*: *new interplays.* I. A volume in honor of S. Albeverio. Proceedings of the Conference on Infinite-dimensional (Stochastic) Analysis and Quantum Physics held in Leipzig, January 18–22, 1999.CMS Conference Proceedings, 28. Published by the American Mathematical Society, Providence, R.I., for the Canadian Mathematical Society, Ottawa, ON, 2000, 986–994. MR 1803381 Zbl 0974.81014

[17] A. Erdélyi, W. Magnus, F. Oberhettinger, and F. G. Tricomi, *Higher transcendental functions.* Vol. III. Based on notes left by H. Bateman. Bateman Manuscript Project, California Institute of Technology. Reprint of the 1955 original. Robert E. Krieger Publishing Co., Melbourne, Fla., 1981. ISBN 0-89874-069-X MR 0698781 MR 0066496 (original) Zbl 0542.33002 Zbl 0064.06302 (original)

[18] A. Erdélyi, W. Magnus, F. Oberhettinger, and F. G. Tricomi, *Tables of integral transforms*. Vol. I. Based, in part, on notes left by H. Bateman. Bateman Manuscript Project. California Institute of Technology. McGraw–Hill Book Company, New York etc., 1954. MR 0061695 Zbl 0055.36401

[19] L. D. Faddeev, Notes on divergences and dimensional transmutation in Yang–Mills theory. *Teoret. Mat. Fiz.* **148** (2006), no. 1, 133–142. In Russian. English translation, *Theor. Math. Phys.* **148** (2006), no. 1, 986–994. MR 2283654 Zbl 1177.81105

[20] I. S. Gradshteyn and I. M. Ryzhik, *Table of integrals, series, and products*. Seventh edition. Translated from the Russian. Translation edited and with a preface by A. Jeffrey and D. Zwillinger. With one CD-ROM (Windows, Macintosh and UNIX). Elsevier/Academic Press, Amsterdam, 2007. ISBN 978-0-12-373637-6, 0-12-373637-4 MR 2360010 Zbl 1208.65001

[21] R. K. Miller, *Nonlinear Volterra integral equations*. Mathematics Lecture Note Series. W. A. Benjamin, Menlo Park, CA, 1971. MR 0511193 Zbl 0448.45004

[22] S. G. Samko, A. A. Kilbas, and O. I. Marichev, *Fractional integrals and derivatives*. Theory and applications. Edited and with a foreword by S. M. Nikol′skiĭ. Translated from the 1987 Russian original. Revised by the authors. Gordon and Breach Science Publishers, Yverdon, 1993. ISBN 2-88124-864-0 MR 1347689 Zbl 0818.26003

[23] M. R. Sayapova and D. R. Yafaev, The evolution operator for time-dependent potentials of zero radius. *Trudy Mat. Inst. Steklov.* **159** (1983), 167–174. Boundary value problems of mathematical physics, 12. In Russian. English translation, *Proc. Steklov Inst. Math.* **159** (1984), 173–180.

On resonant spectral gaps in quantum graphs

Ngoc T. Do, Peter Kuchment, and Beng Ong

Dedicated to the 70th birthday of Pavel Exner

Introduction

Existence of gaps in the spectra of operators of mathematical physics plays important role in many areas (e.g., solid state physics [1], photonic crystal manufacturing [5] and [6], and in expander graphs construction [11], see also discussion in [7], Section 6.1). This also applies to constructions of thin branching structures (e.g., quantum wire circuits), which can be modeled by the so-called quantum graphs (see [2] and [3]). One of the standard ways to achieve the band-gap structure of the spectrum is by making the medium periodic, where the gaps may arise due to the Bragg scattering [1]. However, existence of spectral gaps in periodic media is not guaranteed and is not easy to achieve and manipulate (see, e.g., [7], Section 6.1). Thus, a different, resonant gaps technique has been explored, where identical resonators are distributed throughout the medium to create spectral gaps (the earliest reference known to the authors is [10]). This idea was implemented in the discrete situation in [12] by attaching to each vertex v of the graph Γ (which is the medium in this case) an identical *decoration* (resonator) G (Figure 1).

Here one hits a snag. The nice procedure in [12] does not work nearly that well when the common "boundary" between Γ and G consists of more than one point, as it is the case in the spider decorations (Figure 2).

When the boundary is a single point, by applying a rather standard technique used in considering the transmission problem between two media, one can rewrite the spectral problem on the decorated graph as the one on the original graph Γ with an additional energy (spectral parameter) dependent potential (*Dirichlet–to–Neumann operator* of the decoration, see [2], Section 5.1, for details). Poles of this potential arise at the spectrum of the decoration, which leads to the gap opening.

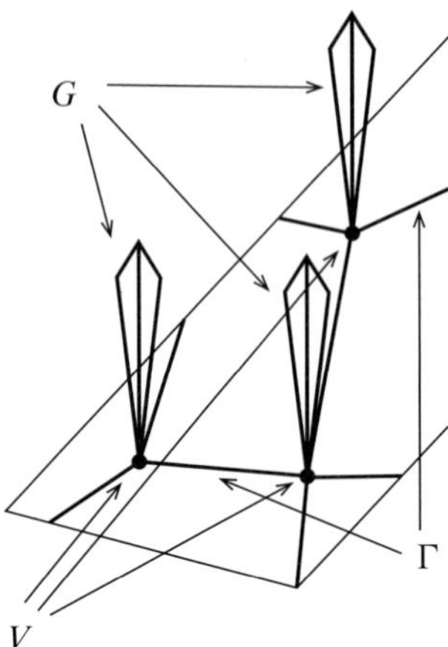

Figure 1. Decorations used by Schenker and Aizenman in [12]

This technique has been extended to the case of quantum graphs, see [2] (Section 5.1) and references therein. However, it would be more convenient in many instances (e.g., in photonic crystal theory when considering the so called inverse opal structures) to insert some internal structure into each vertex, rather than attach a decoration (resonator) to it sideways. In other words, one is looking for a *spider decoration* (Figure 2).[1]

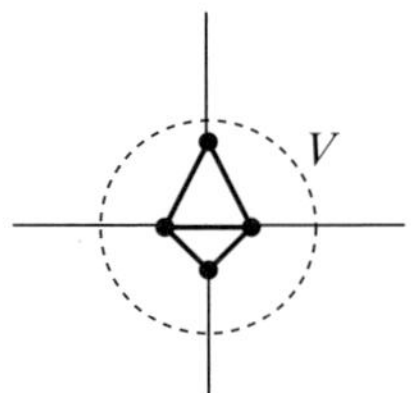

Figure 2. A "spider" decoration replacing a vertex V

With the boundary consisting of more than one point, the arising potential term is now a meromorphic matrix function, whose poles may or may not show up, depending on the vector the matrix function is applied to. Thus, it is easy to construct examples when a spider decoration does not lead to spectral gap opening (see [9], Chapter 3).

However, as it is shown in [9] (Chapter 3), there are some special decoration constructions that resolve this problem in the case of finite graphs. The goal of this paper is to extend the (unpublished) considerations of [9] (Chapter 3).

[1] Compare with the first step of the zig-zag construction of an expander [11].

We provide necessary notions and notations in Section 1, the auxiliary study of the Dirichlet–to–Neumann operator of a graph in Section 2, the main result on gap opening in Section 3, and conclusions and some remarks in Section 4.

1 Preliminaries

We consider a metric graph Γ_0, i.e., a graph such that each its edge e is equipped with a finite positive "length" l_e and a coordinate x identifying it with the segment $[0, l_e]$ (see more detailed discussion in [2], Chapter 1). We will use the standard notations $V(\Gamma_0)$ and $E(\Gamma_0)$ for the sets of vertices and edges of the graph respectively.

We will assume in this text that the following condition on the edge lengths of the graphs Γ_0 is satisfied:

$$\text{there exists } l > 0 \text{ such that } l \le l_e \le 1/l < \infty, \text{ for all } e \in E(\Gamma_0). \tag{1}$$

In particular, this is true when the graph Γ_0 is either finite (i.e., has a finitely many edges and vertices) or periodic (i.e., is equipped with the co-finite action of the group $\mathbb{Z}^p$ for some $p > 0$, see [2], Section 4.1).

Let us also assume that Γ_0 is a d-regular graph (i.e., the degree of each of its vertices is equal to d) and G is a finite metric graph with at least d vertices and a singled out subset $B \subset V(G)$ consisting of d vertices. The set B will be called the *boundary* of G. For each vertex $v \subset V(\Gamma_0)$ we establish a 1–to–1 correspondence between the edges adjacent to v and the elements of B. One can now decorate in a natural way the vertex v with the internal structure, which is a copy of G (see again Figure 2). Doing this for all vertices of Γ_0, we obtain the *decorated graph* Γ.

All graphs we consider are equipped with the self-adjoint operators,[2] H_0 in $L^2(\Gamma_0)$ and H in $L^2(\Gamma)$, as follows: on each edge they act as $-d^2/dx^2$, with the domain consisting of functions f such that

1. $f \in H^2(e)$ for each edge e;
2. f is continuous on the whole graph;
3. at each vertex, the sum of the outgoing derivatives of f along all adjacent edges is equal to zero (Kirchhoff condition);
4. the sum $\sum_e \|f\|^2_{H^2(e)}$ is finite (automatic for a finite graph).

Here $H^2(e)$ is the standard Sobolev space on the segment $e = [0, l_e]$.

[2] Usually called Kirchhoff Laplacians.

We also denote by H_G the analogous operator on G, with the exception that at the boundary vertices $v \in B$, Dirichlet conditions $f(v) = 0$ are imposed instead of Kirchhoff ones. The spectrum $\sigma(H_G)$ of this operator is discrete (see [2], Theorem 3.1.1).

We denote by Σ_D the (discrete under our conditions of finiteness or periodicity of the graph) set of Dirichlet eigenvalues of all edges of Γ_0, i.e.,

$$\Sigma_D := \{(n\pi l_e^{-1})^2\}_{n\in\mathbb{N}, e\in E(\Gamma_0)}.$$

Let us also denote by

$$N\colon \bigoplus_{e\in E(G)} H^2(e) \longrightarrow l^2(B)$$

the *Neumann* operator that for any function

$$f \in \bigoplus_{e\in E(G)} H^2(e)$$

and a vertex $v \in B$ produces the value at v equal to the sum of the outgoing derivatives of f along the edges of G adjacent to v. Here we denote by $l^2(B)$ the d-dimensional Hilbert space of functions on B.

We can now define, for any $\lambda \notin \sigma(H_G)$, the *Dirichlet–to–Neumann operator* (in fact, a $d \times d$-matrix) $\Lambda(\lambda)$ as follows: for any $\phi \in l^2(B)$ let u be the (existing and unique) solution of the following problem:

$$\begin{cases} -d^2u/dx^2 = \lambda u \text{ on each } e \in E(G), \\ u \text{ satisfies continuity and Kirchhoff condition at each vertex } v \notin B, \\ u|_B = \phi. \end{cases} \tag{2}$$

Then,

$$\Lambda(\lambda)\phi := Nu.$$

It is clear that $\Lambda(\lambda)$ is a meromorphic function with poles at $\sigma(H_D)$ only (see [2], Section 3.5, for more detailed consideration of Dirichlet–to–Neumann operator in the quantum graph case and its relation to the resolvent of H_G).

2 Auxiliary statements

Let $\lambda_0 \in \sigma(H_G)$. As it was indicated before, and as we will see clearly in the next section, it will be important for us that for any non-zero $\phi \in l^2(B)$ the vector function

$\Lambda(\lambda)\phi$ still has a pole at λ_0. It is clearly sufficient to consider vectors ϕ that belong to the unit sphere S of $l^2(G) \approx \mathbb{C}^d$.

The following auxiliary result is crucial for our goal:

Theorem 2.1 ([9]). 1. *If for a given $\phi \in S$ and $\lambda = \lambda_0$ the problem* (2) *has a solution, then $\Lambda(\lambda)\phi$ does not have singularity at λ_0.*

2. *If the problem* (2) *has no solution for $\lambda = \lambda_0$ and any $\phi \in S$, then for any ϕ, the following estimate holds in an (independent of ϕ) neighborhood of λ_0:*

$$\|\Lambda(\lambda)\phi\| \geq \frac{C}{|\lambda - \lambda_0|}\|\phi\| \tag{3}$$

with a constant C independent of ϕ.

Thus, we will be looking at graphs G with boundary B such that the problem (2) has a solution only for zero Dirichlet data ϕ. Rather than trying to describe all graphs that have this property, we will provide (for any size d of the boundary B) constructions when this does happen, which will be sufficient for our purpose of gap opening.

Theorem 2.2. *Let $l_0 > 0$ and n be an odd natural number. Suppose that the pair G, B satisfies the following conditions:*

1. *the graph G contains a cycle*[3] *Z consisting of an odd number of edges of the length l_0;*

2. *each boundary vertex $v \in B$ either belongs to Z, or is connected to a vertex of Z by a path of edges of length l_0 each.*

Then, for $\lambda_0 = (n\pi/l_0)^2$, there exist a neighborhood U of λ_0 and a constant C such that (3) *holds for any ϕ and $\lambda \in U$.*

Proof. Without loss of generality, let us assume that Z is non-self-intersecting and consider an edge $e \in Z$ (of length l_0, as all edges in the cycle). The solution of (2) for $\lambda = \lambda_0$ on this edge has the form

$$u = a\cos\left(\frac{n\pi}{l_0}x\right) + b\sin\left(\frac{n\pi}{l_0}x\right).$$

At the endpoints $x = 0, l_0$, the second term vanishes. Since n is odd, the first term changes its value from a to $-a$ at these points. Going around an odd cycle, one concludes that this is possible only if $a = 0$, and thus u vanishes at all vertices of Z. In particular, it vanishes at all boundary vertices that belong to Z. For $v \in B \setminus Z$, as

[3] which can be assumed non-self-intersecting.

there exists a path of edges of length l_0 from v to a vertex from Z, where we know that u vanishes, the same consideration shows that u vanishes at all vertices of the path. Therefore, the problem (2) does not have a solution for non-zero ϕ, and thus, due to Theorem 2.1, the inequality (3) follows. □

3 The main result

We are ready now to formulate and sketch the proof of the main result of this article:

Theorem 3.1. *Let $l_0 > 0$ and n be an odd natural number. Let also the d-regular graph Γ_0 satisfy* (1), *and finite graph G with boundary B, $|B| = d$ and the decorated graph Γ are defined as before. Suppose that the following conditions are satisfied*:

1. $\lambda_0 = (n\pi/l_0)^2 \notin \Sigma_D$, *with* $\operatorname{dist}(\lambda_0, \Sigma_D) = r > 0$;
2. *The decoration* (*resonator*) G, B *satisfies the conditions of Theorem* 2.2.

Then there exists a punctured neighborhood of λ_0, depending on G, topology of Γ_0, and r only, which does not belong to the spectrum $\sigma(H)$.

Proof. To understand the idea of the proof, let us assume first that the graph Γ_0 (and thus Γ) is finite. Then the spectrum of H is discrete. Thus, if $\lambda \in \sigma(H)$, there exists a non-zero eigenfunction u. Assume that the neighborhood of λ_0 has radius less than $r_1 < r$. Then it contains no elements of Σ_D.

Removing all internal edges and vertices from each of the decorations in Γ, one gets a disjoint union of edges of Γ_0, since each former vertex $v \in \Gamma_0$ is replaced by d vertices $v_1, \dots, v_d$, see Figure 3. We denote this new graph $\widetilde{\Gamma}$.

Denoting

$$\phi_v := (u(v_1), \dots, u(v_d))^t, \quad \phi'_v := \Big(\frac{du}{dx_{e_1}}(v_1), \dots, \frac{du}{dx_{e_d}}(v_d)\Big)^t,$$

where $e_1, \dots, e_d$ are the edges formerly adjacent to vertex v and coordinates x_{e_j} increase from the value zero at vertex v, one can rewrite the equation for u on Γ as the following one on $\widetilde{\Gamma}$:

$$\begin{cases} -u'' = \lambda u \text{ on each edge,} \\ \phi'_v = -\Lambda(\lambda)\phi_v \text{ at each vertex } v \in V(\Gamma_0). \end{cases}$$

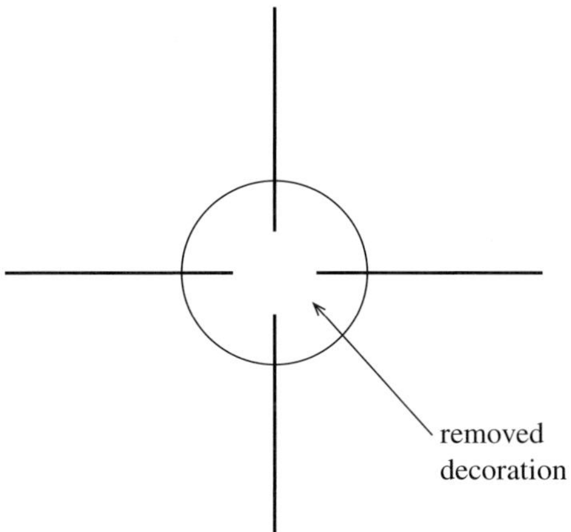

Figure 3. Decoration removed from a former vertex of degree 4

Since λ_0 is at a qualified distance from the Dirichlet spectrum Σ_D, the resolvent estimates for self-adjoint operators together with embedding theorems show that

$$\sum_e \|\phi\|^2_{H^2(e)} \leq M \sum_v \|\phi_v\|^2,$$

$$\sum_v \|\phi_v'\|^2 \leq M \sum_v \|\phi_v\|^2,$$

for some constant M depending only on the distance from the Dirichlet spectrum and topology of Γ_0. On the other hand, if λ is sufficiently close to λ_0 (how close, depends on M and the decoration G), according to Theorem 2.1, one has

$$\sum_v \|\phi_v'\|^2 \geq \frac{C^2}{|\lambda - \lambda_0|^2} \sum_v \|\phi_v\|^2 > M \sum_v \|\phi_v\|^2.$$

This contradiction proves the claim of the theorem for finite graphs.

In the case of infinite graphs, generalized eigenfunctions with control of growth need to be used (see [8] and [4]). The details will be provided elsewhere, but so far we illustrate this on the simplest example of periodic graphs,

In the case of a periodic graph, assumption that λ belongs to the spectrum of H implies existence of a quasi-periodic Bloch–Floquet generalized eigenfunction $\phi(x)$, which under the $\mathbb{Z}^p$-shifts acquires only a phase shift (see details in [2], Section 4.2). Then the above consideration for the finite graph goes smoothly, if the summation over edges and vertices of Γ_0 is replaced by the same for the compact orbit space graph $\Gamma_0/\mathbb{Z}^p$.

Another option in the periodic case is to use the Floquet–Bloch decomposition and apply the above (finite case) argument for each value of the quasi-momentum.[4] □

4 Conclusions and final remarks

1. Although it is probably not easy to understand the case of a general "spider" decoration, the main result of this article allows one to create spectral gaps rather easily at prescribed locations. Indeed, the value $(n\pi l_0^{-1})^2$ involves an arbitrary positive length l_0 and odd natural number n, which gives one a significant freedom of choosing location. As soon as this is done, one can easily produce a spider decoration G that achieves the goal. For instance, if $d = 4$, each degree 4 vertex can be replaced with the structure shown in Figure 4, where we created an odd cycle through three boundary vertices and connected the fourth one to them with a single edge.

2. It was shown in [9] on examples that for even cycles and/or even n the claim of the Theorem 3.1 is incorrect.

3. The regularity condition on the graph Γ_0 is not truly necessary, at least in the case of finite and periodic graphs. Indeed, one can manage variable degrees by choosing different decorations, adjusted to each particular vertex and such that the corresponding resonant values λ_0 agree.

4. Besides gap's existence at a given location, its size is of importance. It depends on the value of the constant C in (3). Our construction allows for a variety of decorations achieving the gap at the same location. It would be interesting to analyse the dependence of C on the decoration, to pick the most effective designs.

5. More general vertex conditions can be considered and Kirchhoff conditions are used just for simplicity.

Acknowledgments. The authors express their gratitude to the NSF for the support. They are also grateful to the referees for useful comments.

The work of Ngoc T. Do and Peter Kuchment was partially supported by the NSF DMS grant # 1517938.

[4] See [2], Chapter 4, for these notions and constructions.

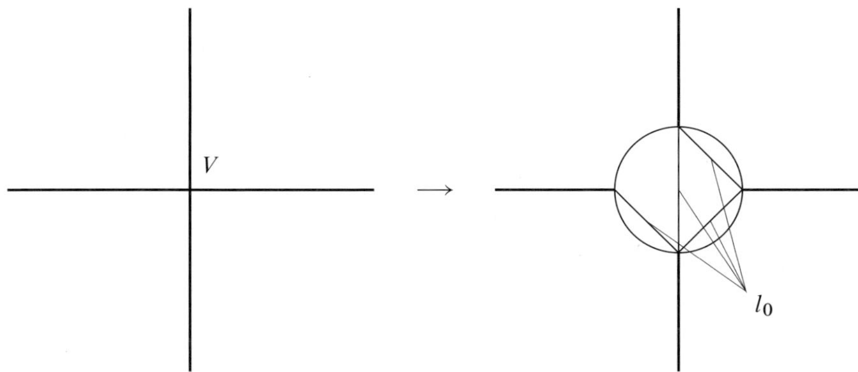

Figure 4. A degree 4 vertex replaced by a "spider" with edges of lengths l_0

References

[1] N. W. Ashcroft and N. D. Mermin, *Solid state physics.* Holt, Rinehart and Winston, New York and London, 1976.

[2] G. Berkolaiko and P. Kuchment, *Introduction to quantum graphs.* Mathematical Surveys and Monographs, 186. American Mathematical Society, Providence, R.I., 2013. ISBN 978-0-8218-9211-4 MR 3013208 Zbl 1318.81005

[3] P. Exner, J. P. Keating, P. Kuchment, T. Sunada, and A. Teplyaev (eds.), *Analysis on graphs and its applications.* Papers from the program held in Cambridge, January 8–June 29, 2007. Proceedings of Symposia in Pure Mathematics, 77. American Mathematical Society, Providence, R.I., 2008. ISBN 978-0-8218-4471-7 MR 2459860 Zbl 1143.05002

[4] P. D. Hislop and O. Post, Anderson localization for radial tree-like quantum graphs. *Waves Random Complex Media* **19** (2009), no. 2, 216–261. MR 2536455 Zbl 1267.81165

[5] J. D. Joannopoulos, S. Johnson, R. D. Meade, and J. N. Winn, *Photonic crystals: molding the flow of light.* Second edition. Princeton University Press, Princeton, N.J., 2008. ISBN 978-0-691-12456-8/hbk Zbl 1144.78303

[6] P. Kuchment, The mathematics of photonic crystals. In G. Bao, L. Cowsar, and W. Masters (eds.), *Mathematical modeling in optical science.* Frontiers in Applied Mathematics, 22. Society for Industrial and Applied Mathematics (SIAM), Philadelphia, PA, 2001, 207–272. MR 1831334 Zbl 0986.78004

[7] P. Kuchment, *An overview of periodic elliptic operators. Bull. Amer. Math. Soc. (N.S.)* **53** (2016), no. 3, 343–414. MR 3501794 Zbl 06589535

[8] D. Lenz, C. Schubert, and P. Stollman, Eigenfunction expansions for Schrödinger operators on metric graphs. *Integral Equations Operator Theory* **62** (2008), no. 4, 541–553. MR 2470123 Zbl 1170.47009

[9] B.-S. Ong, *Spectral problems of optical waveguides and quantum graphs.* Ph.D. thesis. Texas A&M University, College Station, TX, 2006. ISBN 978-0542-84115-6 MR 2709215

[10] B. Pavlov, The theory of extensions, and explicitly solvable models. *Uspekhi Mat. Nauk* **42** (1987), no. 6(258), 99–131, 247. In Russian. English translation, *Russ. Math. Surv.* **42** (1987), no. 6, 127–168 (1987). MR 0933997 Zbl 0665.47004

[11] O. Reingold, S. Vadhan, and A. Wigderson, Entropy waves, the zig-zag graph product, and new constant-degree expanders. *Ann. of Math.* (2) **155** (2002), no. 1, 157–187. MR 1888797 Zbl 1008.05101

[12] J. Schenker and M. Aizenman, The creation of spectral gaps by graph decoration. *Lett. Math. Phys.* **53** (2000), no. 3, 253–262. MR 1808253 Zbl 0990.47006

Adiabatic theorem for a class of stochastic differential equations on a Hilbert space

Martin Fraas

In honor of Pavel Exner on the occasion of his 70th birthday

1 Introduction

We study solutions of a stochastic differential equation

$$\varepsilon\,\mathrm{d}X(s) = L_1(s)X(s)\,\mathrm{d}s + \sqrt{\varepsilon}L_2(s)X(s)\,\mathrm{d}B_s, \quad s \in (0,1) \tag{1}$$

where L_1, L_2 are bounded operators on a Hilbert space $\mathcal{H}$ and B_s is a Brownian motion. The equation is expressed in the slow time $s = \varepsilon t$. The scaling of the second term reflects the Brownian scaling, $\varepsilon^{-1/2}B_{\varepsilon^{-1}s}$ is a Brownian motion in the slow time for any $\varepsilon > 0$. The adiabatic theory studies solutions of the equation in the limit $\varepsilon \to 0$.

A particular example of eq. (1) that motivates our study is a slowly driven stochastic Schrödinger equation, a classical member of the family of quantum stochastic equations derived by Hudson and Parthasarathy [10]. In their full extent quantum stochastic equations describe a system linearly coupled to a bosonic free field. When the coupling is through the position or momentum operator only the equations are equivalent to classical Îto equations with the Brownian motion representing the bath. Conditioning the dynamics on a continuous measurement on the free field gives non-linear quantum filtering equations derived by Belavkin [5]. These equations (and their time-discrete counterparts) provide basic framework for quantum closed loop feedback and control, see [20] and [7]. The goal of our line of research is to develop a feedback theory for the adiabatic quantum control. In particular we plan to develop an adiabatic theory for quantum filtering equations. The adiabatic theory for the unconditioned stochastic Schrödinger equation, derived here, is the first step in this direction.

Equation (1) has been widely studied in the deterministic case, $L_2(s) \equiv 0$, see [11], [4], [16], and references therein. The main feature of the adiabatic theory is that solutions of eq. (1) can be described algebraically as follows:

(i) the evolution generated by the equation leaves the kernel of L_1 invariant to the leading order in ε;

(ii) there is an asymptotic expansion that describes the motion inside the kernel and the tunneling out of the kernel.

Leaving aside exact assumptions, it was understood by Avron and Elgart [1] that (i) holds provided one can define the projection on the kernel in a continuous manner. On the other hand (as was long known), the expansion (ii) exists provided 0 is an eigenvalue isolated from the rest of the spectra (so-called gap condition). We will study only the case when the projection can be defined continuously irrespectively of the Brownian path. Hence a generalization of (i) might not be surprising. However, we will also derive an expansion (ii), which is somehow surprising because the gap condition cannot hold for all realizations of the Brownian motion.

The most restrictive condition of our theory is a requirement that $\ker L_1(s) \subset \ker L_2(s)$ for each instant of time. Under this assumption we derive below an asymptotic expansion for the solution of a form

$$X(s) = x_0(s) + \sqrt{\varepsilon} y_1(s) + \varepsilon x_1(s) + \cdots + \varepsilon^{N-1/2} y_N(s) + \varepsilon^N x_N(s) + O(\varepsilon^{(N+1)/2}).$$

The standard integer power terms (x's) are deterministic and given by the adiabatic expansion in the absence of the stochastic term, $L_2 = 0$. The novel half integer terms are stochastic and describe propagation of an instantaneous error to the future. They are expressed as backward Îto integrals arising from a Duhamel formula.

In the case of stochastic Schrödinger equation with a simple ground state[1] the stochastic term of order $\sqrt{\varepsilon}$ is orthogonal to the ground state and describes the tunneling out of the ground state. We derive a formula for this tunneling and describe its full statistics. This extends the work [4] where a formula for the mean tunneling was derived by studying a slowly driven Lindblad equation [15]. These two equations are closely connected, the latter is obtained from the stochastic Schrödinger equation by averaging over the randomness.

The article is organized as follows. In the remaining part of the introduction we introduce our notation and discuss basics of the stochastic calculus necessary to follow our exposition. In Section 2 we describe the stochastic calculus in more details, in particular we describe the two-sided stochastic calculus of Pardoux and Protter [14]. We also state there several technical propositions regarding the stochastic integration.

[1] Or any simple isolated eigenvalue.

The reader not interested in proofs might safely skip the section. Section 3 gives our assumptions and basic results. In the following Section 4 we apply these results to a stochastic Schrödinger equation describing dephasing and derive the full statistics of tunneling in the leading order. The last section contains the full adiabatic expansion and its proof.

Remark 1.1. In view of the application we had in mind we chose to describe the theory on a Hilbert space rather then on a Banach space. Extension to a finite dimensional Banach space is straightforward. Infinite dimensional Banach spaces introduce several technical complications (starting with the very existence of the Îto calculus) and we do not know what are natural assumptions on the geometry of the Banach space for the extension of our results.

We comment on various complications with the Banach space theory throughout the article.

We denote the scalar product on $\mathcal{H}$ by $(\cdot,\cdot)$ and the norm by $\|\cdot\|$. We suppress randomness from our notation and "$=,\le,\dots$" between random variables holds with probability 1. $\mathbb{E}[\cdot]$ stands for the expectation value with respect to the Brownian motion, and $\|\cdot\|_\infty$ is the corresponding L^∞ norm. In particular for a random variable $X\in\mathcal{H}$, $\|X\|\le\|X\|_\infty$. $O(\varepsilon^n)$ is a random variable for which $\varepsilon^{-n}\|O(\varepsilon^n)\|$, $\varepsilon\in(0,1)$ is a family of random variables with uniformly bounded moments.

We make extensive use of Îto calculus and recall that for stochastic integrals f,g and a function h it holds that

$$\mathrm{d}(fg)=\mathrm{d}f g+f\,\mathrm{d}g+\mathrm{d}f\,\mathrm{d}g,\quad d(h\circ g)=(h'\circ g)\mathrm{d}g+\frac{1}{2}(h''\circ g)(\mathrm{d}g)^2,$$

where $\mathrm{d}f\,\mathrm{d}g$ should be interpreted according to the rules $(\mathrm{d}s)^2=\mathrm{d}s\,\mathrm{d}B_s=0$, $(\mathrm{d}B_s)^2=\mathrm{d}s$. We also use the backward Îto calculus which comes with a similar set of rules given in the following section. In a nutshell backward Îto calculus integrates functions of the future, while Îto forward calculus integrates functions of the past.

To illuminate the difference between the forward/backward integrals we consider the two-parameter stochastic propagator [18], $U_\varepsilon(s,s')$, associated to Equation (1). This is a random variable that depends on the Brownian increments in the interval (s',s). As a function of s, for a fixed s', the propagator satisfies a forward Îto equation,

$$U_\varepsilon(s,s')=\mathbb{1}+\int_{s'}^{s}L_1(t)U(t,s')\,\mathrm{d}t+\int_{s'}^{s}L_2(t)U(t,s')\,\mathrm{d}B_t. \tag{2}$$

On the other hand as a function of s' it satisfies a backward Îto equation

$$U_\varepsilon(s,s')=\mathbb{1}+\int_{s'}^{s}U(s,t)L_1(t)\,\mathrm{d}t+\int_{s'}^{s}U(s,t)L_2(t)\,\mathrm{d}B_t. \tag{3}$$

We will not stress the difference between the backward and the forward integration in our notation. If the integrand refers to the past (it is non-anticipatory) it is a forward integral, if the integrand refers to the future it is a backward integral. In fact we use a shorthand differential notation,

$$\varepsilon\, \mathrm{d}_s U_\varepsilon(s,s') = \mathrm{d}L(s) U_\varepsilon(s,s'), \qquad U_\varepsilon(s',s') = \mathbb{1},$$
$$\varepsilon\, \mathrm{d}_{s'} U_\varepsilon(s,s') = -U_\varepsilon(s,s') \mathrm{d}L(s'), \quad U_\varepsilon(s,s) = \mathbb{1},$$

as an equivalent of eq. (2) resp. eq. (3), where $\mathrm{d}L(s) = L_1(s)\,\mathrm{d}s + \sqrt{\varepsilon} L_2(s)\,\mathrm{d}B_s$.

We end this short exposition with two standard relations that hold for both forward and backward integration,

$$\mathbb{E}\Big[\int_0^1 X_t\,\mathrm{d}B_t\Big] = 0, \quad \mathbb{E}\Big[\Big\|\int_0^1 X_t\,\mathrm{d}B_t\Big\|^2\Big] = \int_0^1 \mathbb{E}[\|X_t\|^2]\,\mathrm{d}t, \tag{4}$$

the latter equality can be understood using a formal relation $\mathbb{E}[\mathrm{d}B_t\,\mathrm{d}B_s] = \delta(t-s)\,\mathrm{d}t$.

2 A two-sided stochastic calculus

We are neither competent nor is it our purpose to explain the stochastic calculus in this preliminary section. Several textbooks and monographs are devoted to this topic, the author used a book of McKean [13] and the relevant chapters in a book of Simon [17]. The same applies to the two-sided integral constructed by Pardoux and Protter [14], the interested reader should consult their article for details and proofs. We merely repeat what is relevant for our exposition and we gather several lemmas that we shall need for the proofs.

We consider a one-dimensional Brownian motion $B_t, 0 \le t \le 1$ and the associated sigma algebra $\sigma(B_s, 0 \le s \le t)$. For a continuous process X_t adapted to the sigma algebra the forward Îto integral of X_t is defined as

$$\int_0^s X_t\,\mathrm{d}B_t = \lim_{N\to\infty} \sum_{k=1}^{2^N} X_{((k-1)/2^N)s}(B_{(k/2^N)s} - B_{((k-1)/2^N)s}).$$

It is an important part of the definition that the increment points to the future and hence $B_{(k/2^N)s} - B_{((k-1)/2^N)s}$ and $X_{((k-1)/2^N)s}$ are independent random variables. A consequence of this choice is that the integral, as a function of s, is a martingale and two basic formulas, cf. (4),

$$\mathbb{E}\Big[\int_0^s X_t\,\mathrm{d}B_t\Big] = 0, \quad \mathbb{E}\Big[\Big(\int_0^s X_t\,\mathrm{d}B_t\Big)^2\Big] = \int_0^s \mathbb{E}[X_t^2]\,\mathrm{d}t,$$

hold true.

Backward Îto integral is defined in an analogous manner. We consider a process Y_t adapted to a sigma algebra $\sigma(B_s - B_1, t \leq s \leq 1)$ and we define the integral of Y_t by

$$\int_s^1 Y_t \,\mathrm{d}B_t = \lim_{N\to\infty} \sum_{k=1}^{2^N} Y_{s+(1-s)k/2^N}(B_{s+(1-s)k/2^N} - B_{s+(1-s)(k-1)/2^N}).$$

Note that here the increments point to the past in order to ensure the independence with the integrand. The backward Îto integral is a backward martingale as a function of s and formulas corresponding to (4) hold true,

$$\mathbb{E}\Big[\int_s^1 Y_t \,\mathrm{d}B_t\Big] = 0, \quad \mathbb{E}\Big[\Big(\int_s^1 Y_t \,\mathrm{d}B_t\Big)^2\Big] = \int_s^1 \mathbb{E}[Y_t^2]\,\mathrm{d}t.$$

We now consider particular processes X_t, Y_t that arise as solutions of forward, resp. backward, stochastic differential equations,

$$X_t = X_0 + \int_0^s b(X_t)\,\mathrm{d}t + \int_0^s \sigma(X_t)\,\mathrm{d}B_t,$$

$$Y_t = Y_0 + \int_s^1 c(Y_t)\,\mathrm{d}t + \int_s^1 \gamma(Y_t)\,\mathrm{d}B_t,$$

for some continuous functions b, c, σ, γ. The integral in the first equation being the forward Îto integral, while the integral in the second equation being the backward Îto integral. Correspondingly the first equation has a unique solution X_t that is a non-anticipatory martingale and the second equation has a unique solution Y_t that is a backward martingale adapted to the associated sigma algebra.

We also use a differential form of these equations

$$\mathrm{d}X_t = b(X_t)\,\mathrm{d}t + \sigma(X_t)\,\mathrm{d}B_t,$$

$$\mathrm{d}Y_t = -c(Y_t)\,\mathrm{d}t - \gamma(Y_t)\,\mathrm{d}B_t.$$

Although the notation makes no distinction between the forward and the backward case one should keep in mind that these are distinct differentials.

A stochastic integral for joint functions of X_t, Y_t was constructed in [14]. Let $f(t, X_t, Y_t)$ be a continuous function of its arguments, then an integral

$$\int_{s'}^s f(t, X_t, Y_t)\,\mathrm{d}B_t$$

can be defined in such a way that if f is independent of Y_t (resp. X_t) then the integral coincides with the forward (resp. backward) Îto integral. Furthermore the integral satisfies the following chain rule:

$$\begin{aligned} f(s, X_s, Y_s) &= f(s', X_{s'}, Y_{s'}) + \int_{s'}^{s} \partial_t f(t, X_t, Y_t) \mathrm{d}t \\ &\quad + \int_{s'}^{s} \partial_X f(t, X_t, Y_t) \mathrm{d}X_t + \int_{s'}^{s} \partial_Y f(t, X_t, Y_t) \mathrm{d}Y_t \\ &\quad + \frac{1}{2} \int_{s}^{s'} \partial_{XX} f(t, X_t, Y_t) (\mathrm{d}X_t)^2 - \frac{1}{2} \int_{s}^{s'} \partial_{YY} f(t, X_t, Y_t) (\mathrm{d}Y_t)^2, \end{aligned} \tag{5}$$

where $(\mathrm{d}X_t)^2 = \sigma^2(X_t)\mathrm{d}t$ is interpreted according to the Îto rules $(\mathrm{d}t)^2 = \mathrm{d}t\,\mathrm{d}B_t = 0$, $(\mathrm{d}B_t)^2 = \mathrm{d}t$.

In the following paragraphs we employ the formula (5) in a case with no second-order derivatives to operator-valued processes X_t, Y_t. Due to the simplicity of that case the operator-valued extension is clear. The matrix-valued version is discussed in more details in [14]. We are not aware of an infinite dimensional extension of [14], the results below hence apply to that case provided the reader take such extension for granted.

To demonstrate the power of the calculus we show that equations (2, 3) define the same propagator (the value of ε is not important for the following considerations and we skip the index) and that this propagator satisfies the semigroup property

$$U(s, 0) = U(s, s')U(s', 0).$$

To this end we fix a time t and let $X_s = U(s, 0)$ be a solution of eq. (2) and $Y_s = \widetilde{U}(t, s)$ be a solution of eq. (3). Then the above chain rule implies that for any $t \leq s \leq s'$ we have

$$\widetilde{U}(t, s)U(s, 0) = \widetilde{U}(t, s')U(s', 0).$$

By choosing $s = t$ and $s' = 0$ we get the sought equivalence $\widetilde{U}(t, 0) = U(t, 0)$. Upon erasing the tilde in the above equation we then establish the semigroup property.

In the following we will need two specific results concerning stochastic differential equations. The first is a particular version of the Duhamel formula, the second is a prior bound on stochastic integrals. We formulate the bound for the forward integral, the corresponding bound holds also for the backward integral.

Lemma 2.1 (Duhamel formula). *The solution of the differential eq.* (2) *satisfies a relation*

$$U(s,s') = V(s,s') + \int_{s'}^{s} U(s,t)L_2(t)V(t,s')\,\mathrm{d}B_t,$$

where $V(s,s')$ *is the solution of a deterministic equation,*

$$\mathrm{d}V(s,s') = L_1(s)V(s,s')\,\mathrm{d}s, \quad V(s',s') = \mathbb{1}.$$

Proof. The proof is again an application of the chain rule (5). Pick $X_t = V(t,s')$, $Y_t = U(s,t)$; then for any $s \geq t \geq t' \geq s'$ the chain rule gives

$$U(s,t)V(t,s') = U(s,t')V(t',s') - \int_{t'}^{t} U(s,x)L_2(x)V(x,s')\,\mathrm{d}B_x.$$

The statement of the lemma then follows by choosing $t = s$ and $t' = s'$. □

Lemma 2.2 (prior estimates of stochastic integrals). *Let* $X_t \in \mathcal{H}$ *be a non-anticipatory stochastic process, then the following estimates hold true*:

(a) $$\mathbb{E}\Big[\Big\|\int_0^1 X(s)\,\mathrm{d}B_s\Big\|^{2n}\Big] \leq (2n^2-n)^n\,\mathbb{E}\Big[\int_0^1 \|X_s\|^{2n}\,\mathrm{d}s\Big], \quad n \geq 1,$$

(b) $$\mathrm{Prob}\Big(\Big\|\int_0^1 X_s\,\mathrm{d}B_s\Big\|^2 > \gamma\Big) \leq e^{-\gamma/(8\|X\|_\infty^2)+1/4}, \tag{6}$$

where $\|X\|_\infty := \sup_{0\leq t\leq 1}\|X_t\|_\infty$.

Proof. Denote

$$\Theta_t = \int_0^t X_s\,\mathrm{d}B_s$$

and consider a real valued stochastic process $\zeta_t = (\Theta_t, \Theta_t)$. The stochastic differentiation of this process gives $\mathrm{d}\zeta_t = ((X_t,\Theta_t) + (\Theta_t,X_t))\,\mathrm{d}B_t + (X_t,X_t)\,\mathrm{d}t$, which is equivalent to an integral relation

$$\zeta_t - \int_0^t \|X_s\|^2\,\mathrm{d}s = \int_0^t ((X_s,\Theta_s) + (\Theta_s,X_s))\,\mathrm{d}B_s. \tag{7}$$

(a) We take the expectation of $d\zeta_t^n = n\zeta_t^{n-1}\,\mathrm{d}\zeta_t + 1/2\; n(n-1)\zeta_t^{n-2}\,\mathrm{d}\zeta_t\,\mathrm{d}\zeta_t$ to get an integral relation,

$$\mathbb{E}[\zeta_t^n] = n\int_0^t \mathbb{E}[\zeta_s^{n-1}\|X_s\|^2]\,\mathrm{d}s + \frac{n(n-1)}{2}\int_0^t \mathbb{E}[\zeta_s^{n-2}((\Theta_s,\,X_s) + (X_s,\Theta_s))^2]\,\mathrm{d}s,$$

between the moments. As a first observation note that all integrands are non-negative functions and hence $\mathbb{E}[\zeta_t^n]$ is a non-decreasing function of t. Now we employ the estimate $((X_s, \Theta_s) + (\Theta_s, X_s))^2 \leq 4\|X_s\|^2 \zeta_s$ and the Hölder inequality to get

$$\begin{aligned}\mathbb{E}[\zeta_t^n] &\leq (n + 2n(n-1)) \int_0^t \mathbb{E}[\zeta_s^{n-1} \|X_s\|^2] \,\mathrm{d}s \\ &\leq (2n^2 - n) \Big(\int_0^t \mathbb{E}[\zeta_s^n] \,\mathrm{d}s \Big)^{(n-1)/n} \Big(\int_0^t \mathbb{E}[\|X_s\|^{2n}] \,\mathrm{d}s \Big)^{1/n} \\ &\leq (2n^2 - n)(\mathbb{E}[\zeta_t^n])^{n-1/n} \Big(\int_0^t \mathbb{E}[\|X_s\|^{2n}] \,\mathrm{d}s \Big)^{1/n}.\end{aligned}$$

In the last inequality we also used $0 \leq t \leq 1$. Solving for the n-th moment establishes the first inequality of the lemma.

To prove (b) we will use a well-known prior estimate on stochastic integrals [13], Chapter 2.3. Let $e_t, 0 \leq t \leq 1$ be a real non-anticipatory function and suppose that $\int_0^1 e_t^2 \,\mathrm{d}t < \infty$, then for any reals β and α the following bound holds true:

$$\mathrm{Prob}\Big[\max_{0 \leq t \leq 1} \Big(\int_0^t e_s \,\mathrm{d}B_s - \frac{\alpha}{2} \int_0^t e_s^2 \,\mathrm{d}s \Big) > \beta \Big] \leq e^{-\alpha\beta}.$$

We note that the bound is an application of Doob's martingale inequality.

Applying the bound to eq.(7) then implies

$$\mathrm{Prob}\Big[\max_{0 \leq t \leq 1} \Big(\zeta_t - \int_0^t \|X_s\|^2 \,\mathrm{d}s - \frac{\alpha}{2} \int_0^t ((X_s, \Theta_s) + (\Theta_s, X_s))^2 \,\mathrm{d}s \Big) > \beta \Big] \leq e^{-\alpha\beta}.$$

We claim that for $0 \leq t \leq 1$,

$$\begin{aligned}&\zeta_t - \int_0^t \|X_s\|^2 \,\mathrm{d}s - \frac{\alpha}{2} \int_0^t ((X_s, \Theta_s) + (\Theta_s, X_s))^2 \,\mathrm{d}s \\ &\quad \geq \zeta_t - \|X\|_\infty^2 - 2\alpha \|X\|_\infty^2 \max_{0 \leq s \leq 1} \zeta_s.\end{aligned} \tag{8}$$

In particular whenever $\max RHS > \beta$ then also $\max LHS > \beta$ and the probability of an event $\max RHS > \beta$ is smaller then the probability of an event $\max LHS > \beta$. Combining this with the probability bound above we have

$$\begin{aligned}&\mathrm{Prob}[\max_{0 \leq t \leq 1} \zeta_t - \|X\|_\infty^2 - 2\alpha \|X_\infty\|^2 \max_{0 \leq t \leq 1} \zeta_t > \beta] \\ &= \mathrm{Prob}\Big[\max_{0 \leq t \leq 1} \zeta_t > \frac{\beta + \|X\|_\infty^2}{1 - 2\alpha \|X\|_\infty^2} \Big] \\ &\leq e^{-\alpha\beta}.\end{aligned}$$

Writing

$$\gamma = \frac{\beta + \|X\|_\infty^2}{1 - 2\alpha\|X\|_\infty^2}$$

and choosing the optimal $\alpha = 1/(4\|X\|_\infty^2)$ we get Bound (6).

It remains to prove (8). The inequality follows from the inequality $\|X_s\|^2 \leq \|X\|_\infty^2$ and the inequality

$$((X_s, \Theta_s) + (\Theta_s, X_s))^2 \leq 4\|X\|_\infty^2 \|\Theta_s\|^2 \leq 4\|X\|_\infty^2 \max_{0 \leq s \leq 1} \zeta_s.$$

Note that for $0 \leq t \leq 1$ an integral $\int_0^t$ of a positive constant can be bounded by that constant. □

An important consequence of the lemma is that

$$\int_0^1 O_t(\varepsilon^n)\,\mathrm{d}B_t = O(\varepsilon^n),$$

provided the moments of $\|O_t(\varepsilon^n)\|$ are uniformly bounded with respect to t.

Generalization of Lemma 2.2 is one of the main technical obstacles of a Banach space version of the theory. For finite dimensional spaces all norms are equivalent and the above bounds hold true up to a dimension dependent constant. On the other hand we do not know if such bounds are available in the infinite dimensional Banach spaces.

The adiabatic expansion, which is the main result of our paper, has a natural formulation in terms of the backward Îto integral. On the other hand it is often easier – not principally, just thanks to a larger degree of familiarity – to perform calculations with a forward Îto integral. Due to a special structure of integrals that appear in this work we can always convert a backward integral to a forward integral.

Remark 2.3. The backward stochastic integrals of the type $\int_0^s U(s,s')f(s')\,\mathrm{d}B_{s'}$, where f is a deterministic function, can be converted into a forward integral thanks to the semigroup relation $U(s,s') = U(s,0)U(s',0)^{-1}$. The relation expresses the propagator in the future by a constant (with respect to the integration) times a propagator in the past. We still need to convert the backward to a forward integral.

To see in details how the conversion works we take a second look at the forward Îto integral that we defined by

$$I_- = \lim_{N\to\infty} \sum_{k=1}^{2^N} X_{((k-1)/2^N)s}(B_{(k/2^N)s} - B_{((k-1)/2^N)s}).$$

Alternatively one can define[2]

$$I_+ = \lim_{N\to\infty} \sum_{k=1}^{2^N} X_{(k/2^N)s}(B_{(k/2^N)s} - B_{((k-1)/2^N)s})$$

and these two definitions are related by a quadratic variation of X_t,

$$I_+ - I_- = \int_0^s \mathrm{d}X_t \,\mathrm{d}B_t = \int_0^s \sigma(X_t)\,\mathrm{d}t.$$

For the integral under consideration this now implies, we introduce back ε as this will be useful at a later point in the article,

$$\begin{aligned} \int_0^s U_\varepsilon(s,s')f(s')\,\mathrm{d}B_{s'} - U_\varepsilon(s,0)\int_0^s U_\varepsilon(s',0)^{-1}f(s')\,\mathrm{d}B_{s'} \\ = -\frac{1}{\sqrt{\varepsilon}}U_\varepsilon(s,0)\int_0^s U_\varepsilon(s',0)^{-1}L_2(s')f(s')\,\mathrm{d}s', \end{aligned} \tag{9}$$

where we have used $\varepsilon\,\mathrm{d}U_\varepsilon(s,0)^{-1} = U_\varepsilon(s,0)^{-1}(-\mathrm{d}L(s) + L_2^2(s)\,\mathrm{d}s)$. The second line seems to diverge as $\varepsilon \to 0$, but in fact it is of order 1 assuming $\|U_\varepsilon(s,0)\| = 1$. Provided all the inverse operators exist on the range of L_2 we have:

$$\int_0^s U_\varepsilon(s,s')f(s')\,\mathrm{d}B_{s'} = U_\varepsilon(s,0)\int_0^s U_\varepsilon(s',0)^{-1}\tilde{f}(s')\,\mathrm{d}B_{s'} + O(\sqrt{\varepsilon}), \tag{10}$$

with $\tilde{f}(s) = [1 + L_2(s)(L_1(s) - L_2^2(s))^{-1}L_2(s)]f(s)$. To see this we use

$$\begin{aligned} &\int_0^s U_\varepsilon(s',0)^{-1}L_2(s')f(s')\,\mathrm{d}s' \\ &\quad = \varepsilon\int_0^s \mathrm{d}U_\varepsilon(s',0)^{-1}(-L_1(s') + L_2^2(s'))^{-1}L_2(s')f(s') \\ &\qquad + \sqrt{\varepsilon}\int_0^s U_\varepsilon(s',0)^{-1}L_2(s')(-L_1(s') + L_2^2(s'))^{-1}L_2(s')f(s')\,\mathrm{d}B_{s'}. \end{aligned}$$

An integration by parts shows that the first line of the RHS of the formula is of order ε and after plugging it into eq. (9) we obtain eq. (10).

3 Assumptions and basic results

We derive a solution of eq. (1) in the adiabatic limit $\varepsilon \to 0$ under three additional assumptions.

[2] This is sometimes referred to as a backward integral, we do not use this name to avoid confusion.

Assumption 3.1. (A) For each s, $L_1(s) - 1/2 L_2^2(s)$ generates a contraction semigroup and $iL_2(s)$ is self-adjoint.

(B) $L_1(s), s \in (0,1)$ is a family of operators for which 0 remains a uniformly isolated discrete eigenvalue.

(C) $\ker L_2(s) \supseteq \ker L_1(s), s \in (0,1)$.

Condition (A) is a sufficient and necessary condition of a stochastic version of Hille–Yosida theorem; It implies that $U_\varepsilon(s,s')$ is a contraction, i.e. $\|U_\varepsilon(s,s')\| \le 1$. This prevents an exponential blow up of solutions, and it is a standard condition in the adiabatic theory, see [11], and [4].

Given assumption (A), a differentiation (see eq. (11) below) gives

$$\varepsilon \,\mathrm{d}\|x(s)\|^2 = ((2\,\mathrm{Re}\, L_1(s) - L_2^2(s))x(s), x(s))$$

for any solution $x(s) = U_\varepsilon(s,0)x(0)$. Hence assumption (A) implies that the evolution is almost surely a contraction. The opposite statement holds in the autonomous case.

Proposition 3.2 (stochastic Hille–Yosida). *Let $U(s,s')$ be a propagator associated to a stochastic differential equation*

$$\mathrm{d}U(s,s') = L_1 U(s,s')\mathrm{d}s + L_2 U(s,s')\mathrm{d}B_s, \quad U(s',s') = \mathbb{1}.$$

Then the following are equivalent:

(i) *$U(s,s')$ is a contraction, i.e. $\|U(s,s')\| \le 1$;*

(ii) *$L_1 - 1/2\, L_2^2$ is a generator of a contraction semigroup and iL_2 is self-adjoint.*

Furthermore (ii) implies that L_1 is a generator of contractions.

Proof. Without loss of generality we put $s' = 0$, and throughout the proof we denote $x(s) = U(s,0)x(0)$. The condition that $U(s,0)$ is a contraction is then equivalent to the statement that $\|x(s)\| \le \|x(0)\|$ for all initial vectors $x(0)$.

(ii) $\implies$ (i). By Îto rules we have

$$\begin{aligned} \mathrm{d}\|x(s)\|^2 &= [(L_1x(s), x(s)) + (x(s), L_1x(s)) + (L_2x(s), L_2x(s))]\mathrm{d}s \\ &\quad + [(L_2x(s), x(s)) + (x(s), L_2x(s))]\mathrm{d}B_s \\ &= ((2\mathrm{Re}L_1 - L_2^2)x(s), x(s))\mathrm{d}s, \end{aligned} \tag{11}$$

where the last line is due to the assumption $L_2^* = -L_2$. Recall that $L_1 - 1/2\, L_2^2$ is a generator of contraction on a Hilbert space if and only if it is dissipative, i.e. $\mathrm{Re}(L_1 - 1/2\, L_2^2) \le 0$. It then follows that $\mathrm{d}\|x(s)\|^2 \le 0$.

(i) $\implies$ (ii). We first prove that L_2 generates isometries, by proving that both L_2 and $-L_2$ generate contraction semigroups. Suppose to the contrary that there exists an interval $I = (I_-, I_+)$ such that for $\varphi \in I$ and some $x \in \mathcal{H}$ we have $\|e^{\varphi L_2} x\| > \|x\|$.

We consider the same decomposition of $\mathrm{d}L$ as in (i). We treat $L_1 - 1/2\ L_2^2$ as a perturbation and express $U(s) \equiv U(s,0)$ be a Duhamel formula. Since the perturbation is deterministic this is the standard version of the formula,

$$U(s) = e^{L_2 B_s} + \int_0^s e^{L_2(B_s - B_{s'})}\Big(L_1 - \frac{1}{2}L_2^2\Big)U(s')\,\mathrm{d}s'.$$

The event $E_{I,s} = \{B_s \in I \text{ and } I_- - 1 \le B_t \le I_+ + 1, 0 \le t \le s\}$ has a non-zero probability for any interval I and any s. By choosing s sufficiently small we can then achieve $\|U(s)x\| > \|x\|$, which is in contradiction with (i).

Since $L_2^* = -L_2$ we have

$$\mathrm{d}\|x(s)\|^2 = [(L_1 x(s), x(s)) + (x(s), L_1 x(s)) + (L_2 x(s), L_2 x(s))]\,\mathrm{d}s$$

and since $\mathrm{d}\|x(0)\| \le 0$ we conclude that $L_1 - 1/2\,L_2^2$ is dissipative and hence generate a contraction semigroup.

The last claim of the proposition is not related to a classification of contraction semigroups. To prove it, observe that if L_2 is anti-selfadjoint then L_2^2 is a generator of contractions. Hence $L_1 - 1/2\,L_2^2 + 1/2\,L_2^2$ is also a generator of contractions. Alternatively L_1 is a generator of the semigroup $\mathbb{E}[X_t]$. □

Remark 3.3. In a Banach space version of the proposition, the condition iL_2 is self-adjoint should be replaced by L_2 is a generator of isometries. The proof is technically more involved and requires a version of Trotter-Kato formula that does not seem to be available in the literature ([12] assumes compact state space, while [8] and [6] assume the Hilbert space structure). In particular that if $\mathrm{d}L_j$, $j = 1,2$ is the generator of a propagator $U_j(s,s')$, then the propagator $U(s,s')$ generated by a sum $\mathrm{d}L_1 + \mathrm{d}L_2$ can be expressed as

$$U(s,s') = \lim_{N\to\infty} U_1(s,s_N)U_2(s,s_N)U_1(s_N,s_{N-1})U_2\ldots U_1(s_1,s')U_2(s_1,s'),$$

where $s \ge s_N \ge \cdots \ge s_1 \ge s'$ is any partition of the interval with a mesh going to 0 as $N \to \infty$. This implies that if $\mathrm{d}L_1$ and $\mathrm{d}L_2$ generates contractions then so does $\mathrm{d}L_1 + \mathrm{d}L_2$.

The gap condition, assumption (B), is also completely standard in the adiabatic theory. Since L_1 is a generator of contraction semigroup, $\ker L_1 \cap \operatorname{ran} L_1 = 0$ (see [4]) and the gap condition implies

$$\mathcal{H} = \ker L_1(s) \oplus \operatorname{ran} L_1(s). \tag{12}$$

The rather restrictive condition (C) allows us to define the slow manifold and we cannot imagine how it can be relaxed.

Before stating our results we shortly recall concepts from the adiabatic theory, see [4] or [19] for a more thorough exposition. Let $P(s)$ be a $\mathcal{C}^1$ family of projections on $\mathcal{H}$ then the equation

$$\frac{\partial}{\partial s}T(s,s') = [\dot{P}(s), P(s)]T(s,s'), \quad T(s',s') = \mathbb{1}$$

defines parallel transport on ran $P(s)$. The name "parallel transport" is justified by two crucial properties:

(i) $T(s,s')P(s') = P(s)T(s,s')$;

(ii) a section $x(s) = T(s,s')x(s') \in \operatorname{ran} P(s)$ satisfies the equation

$$P(s)\dot{x}(s) = 0.$$

The parallel transport relevant to (1) is given by the projection $P(s)$ on $\ker L_1(s)$ in the direction of ran $L_1(s)$. This projection is well defined thanks to the decomposition eq. (12). Henceforth $T(s,s')$ shall always refer to this particular projection, unless stated otherwise.

Theorem 3.4. *Let $L_1(s), L_2(s)$ be C^3 families of operators satisfying* (A)–(C). *Then the differential equation $\varepsilon\, \mathrm{d}X(s) = \mathrm{d}L(s)X(s)$ admits solutions of the form*

$$X(s) = a_0(s) + \sqrt{\varepsilon}\int_0^s U_\varepsilon(s,s')L_2(s')b_1(s')\,\mathrm{d}B_{s'} + \varepsilon(a_1(s) + b_1(s)) + O(\varepsilon^{3/2}),$$

where

$$a_0(s) = T(s,0)a_0(0),$$

$$b_1(s) = L_1(s)^{-1}\dot{a}_0(s),$$

$$a_1(s) = \int_0^s T(s,s')P(s')\dot{b}_1(s')\,\mathrm{d}s',$$

and the initial condition $a_0(0)$ belongs to $\ker L_1(0)$.

We note that the integrand $U_\varepsilon(s,s')$ refers to the future and the integral is the backward Îto integral. The theorem is an immediate corollary of a more general Theorem 5.1 that describes the full expansion to all orders in ε. We feature it separately because we are not aware of any application of the expansion beyond the first order.

4 Stochastic Schrödinger equation

The theorem may be applied to a driven stochastic Schrödinger equation [9] (Chapter 5),

$$\varepsilon \mathrm{d}|\psi(s)\rangle = -\Big(iH(s) + \frac{1}{2}\Gamma(s)^2\Big)|\psi(s)\rangle \,\mathrm{d}s - \sqrt{\varepsilon} i\Gamma(s)|\psi(s)\rangle \,\mathrm{d}B_s,$$

where $|\psi\rangle$ is a vector in a Hilbert space and H, Γ are self-adjoint operators. The equation generates unitary evolution and the average state $\bar{\rho}(s) = \mathbb{E}[|\psi(s)\rangle\langle\psi(s)|]$ satisfies a Lindblad equation

$$\varepsilon\dot{\bar{\rho}}(s) = -i[H(s), \bar{\rho}(s)] + \Gamma(s)\bar{\rho}(s)\Gamma(s) - \frac{1}{2}(\Gamma^2(s)\bar{\rho}(s) + \bar{\rho}(s)\Gamma^2(s)). \quad (13)$$

As in the deterministic case [4] (Section 3.1), we need to subtract the dynamical phase before we can directly apply the adiabatic theorem. For an integrable function $E(s)$ and a square integrable function[3] $\sqrt{\gamma(s)}$ the transformation $H(s) \to H(s) - E(s)$, $\Gamma(s) \to \Gamma(s) - \sqrt{\gamma(s)}$ transforms the solution of the stochastic Schrödinger equation according to

$$|\psi(s)\rangle \to e^{i/\varepsilon \int_0^s E(t)\,\mathrm{d}t + i/\sqrt{\varepsilon}\int_0^s \sqrt{\gamma(t)}\,\mathrm{d}B_t}|\psi(s)\rangle.$$

For simplicity we consider a d-dimensional Hilbert space and $H(s), \Gamma(s)$ with simple eigenvalues $E_0(s) = 0, \dots, E_{d-1}(s)$, $\sqrt{\gamma_0(s)} = 0, \dots, \sqrt{\gamma_{d-1}(s)}$ corresponding to a joint normalized eigenbasis $|\psi_0(s)\rangle, \dots, |\psi_{d-1}(s)\rangle$. The eigenstate $|\psi_k(s)\rangle$ is determined only up to a phase and without loss of generality we assume that it is chosen in accordance with the parallel transport associated to the projection $|\psi_k(s)\rangle\langle\psi_k(s)|$. Primarily, we shall study solutions $|\psi_\varepsilon(s)\rangle$ of the stochastic Schrödinger equation with an initial condition $|\psi_\varepsilon(0)\rangle = |\psi_0(0)\rangle$. Likewise we can study solutions with an initial condition $|\psi_k(0)\rangle, k \in (1, d-1)$ after applying the above mentioned transformations.

Of particular interest is the tunneling out of the ground state defined as

$$\begin{aligned} T_\varepsilon(s) &= 1 - |\langle\psi_0(s) \mid \psi_\varepsilon(s)\rangle|^2 \\ &= \sum_{k=1}^{d-1} |\langle\psi_k(s) \mid \psi_\varepsilon(s)\rangle|^2. \end{aligned}$$

Theorem 4.1. *Let $H(s), \Gamma(s)$ be as above. Then the stochastic Schrödinger equation admits a solution*

$$|\psi_\varepsilon(s)\rangle = |\psi_0(s)\rangle + \sqrt{\varepsilon}\sum_{k=1}^{d-1}\Big(\int_0^s D_\varepsilon^{(k)}(s, s')t_k(s')\,\mathrm{d}B_{s'}\Big)|\psi_k(s)\rangle + O(\varepsilon),$$

[3] The artificial square root in the definition of γ was introduced in order to have the final results in the same form as in the Lindblad case.

where

$$D_\varepsilon^{(k)}(s,s') = e^{-i/\varepsilon \int_{s'}^{s} E_k(t)\,\mathrm{d}t - i/\sqrt{\varepsilon}\int_{s'}^{s}\sqrt{\gamma_k(t)}\,\mathrm{d}B_t},$$

$$t_k(s) = -i\sqrt{\gamma_k(s)}\frac{\langle \psi_k(s) \mid \dot{\psi}_0(s)\rangle}{-iE_k(s) - 1/2\,\gamma_k(s)}.$$

In particular for the tunneling we have $T_\varepsilon(s) = \varepsilon \sum_{k=1}^{d-1} T_k(s) + O(\varepsilon^{3/2})$,

$$T_k(s) = \left| \int_0^s D_\varepsilon^{(k)}(s,s') t_k(s')\,\mathrm{d}B_{s'} \right|^2.$$

In the leading order, terms $T_k(s)$ *are independent random variables, and each term has an exponential distribution with mean*

$$\mathbb{E}[T_k(s)] = \int_0^s |t_k(s')|^2\,\mathrm{d}s'. \tag{14}$$

Proof. Conditions (A)–(C) for $L_1(s) = -(iH(s) + 1/2\,\Gamma(s)^2)$ and $L_2(s) = -i\Gamma(s)$ are clearly satisfied. $U_\varepsilon(s,s')$ is a unitary propagator and $L_1(s)$ has eigenvectors $|\psi_k(s)\rangle$ corresponding to simple discrete eigenvalues $-iE_k(s) - 1/2\,\gamma_k(s)$. In view of Theorem 3.4 and the discussion above we then have in the leading order

$$U_\varepsilon(s,s')|\psi_k(s')\rangle = D_\varepsilon^{(k)}(s,s')|\psi_k(s)\rangle + O(\sqrt{\varepsilon}). \tag{15}$$

We proceed to the next order for the case with the initial condition $a_0(0) = |\psi_0(s)\rangle$. In order to do so we need to compute the coefficient $b_1(s)$. We express it in the joint eigenbasis of H and Γ,

$$b_1(s) = \sum_{k=1}^{d-1} \frac{\langle \psi_k(s) \mid \dot{\psi}_0(s)\rangle}{-iE_k(s) - 1/2\,\gamma_k(s)} |\psi_k(s)\rangle.$$

It then follows from Theorem 3.4 that

$$|\psi_\varepsilon(s)\rangle = |\psi_0(s)\rangle + \sqrt{\varepsilon}\sum_{k=1}^{d-1}\left(\int_0^s U_\varepsilon(s,s')t_k(s')|\psi_k(s')\rangle\,\mathrm{d}B_{s'}\right) + O(\varepsilon),$$

and by substituting from eq. (15) we obtain the first equation of the theorem. The expression for the tunneling is an immediate consequence. To compute the mean of the tunneling we use Formula (4). It remains to show that transitions to different excited states are independent in the leading order and that the distribution of the tunneling is exponential. This will require some effort.

We recall that exponential probability distribution with mean μ has a probability density function $p(x) = \mu^{-1}e^{-x/\mu}$ and is uniquely characterized by its moments $\int p(x)x^n = n!\mu^n$. Our strategy is to compute the moments by establishing a recurrence relation between $\mathbb{E}[T_\varepsilon^n]$ and $\mathbb{E}[T_\varepsilon^{n-1}]$.

For convenience we first express the tunneling as a forward stochastic integral. Using the computation in Remark 2.3, eq. (10), with $L_1 = (-iE_k - 1/2\,\gamma_k)$ and $L_2 = -i\sqrt{\gamma_k}$ we have

$$\int_0^s D_\varepsilon^{(k)}(s,s')t_k(s')\,\mathrm{d}B_{s'} = D_\varepsilon^{(k)}(s,0)\int_0^s D_\varepsilon^{(k)}(0,s')r_k(s')\,\mathrm{d}B_{s'} + O(\sqrt{\varepsilon}),$$

where

$$r_k(s) = -i\sqrt{\gamma_k(s)}\,\frac{\langle\psi_k(s)\mid\dot{\psi}_0(s)\rangle}{(-iE_k(s) + 1/2\,\gamma_k(s))}.$$

We hence obtain a forward expression for the tunneling in the leading order,

$$T_k(s) = \left|\int_0^s D_\varepsilon^{(k)}(0,s')r_k(s')\,\mathrm{d}B_{s'}\right|^2.$$

Note that $|t_k(s)|^2 = |r_k(s)|^2$, as it has to be for the mean to remain the same.

We start by considering a single transition $T_k(s)$. Îto rules imply

$$\mathrm{d}T_k(s) = \left(\int_0^s \bar{D}_\varepsilon^{(k)}(0,s')\bar{r}_k(s')\,\mathrm{d}B_{s'}\right)D_\varepsilon^{(k)}(0,s)r_k(s)\,\mathrm{d}B_s + \text{c.c.} + |r_k(s)|^2\,\mathrm{d}s,$$

and

$$\begin{aligned}(\mathrm{d}T_k(s))^2 = {} & \left(\int_0^s \bar{D}_\varepsilon^{(k)}(0,s')\bar{r}_k(s')\,\mathrm{d}B_{s'}\right)^2 D_\varepsilon^{(k)}(0,s)^2 r_k^2(s)\,\mathrm{d}s + \text{c.c.}\\ & + 2T_k(s)|r_k(s)|^2\,\mathrm{d}s.\end{aligned}$$

Using integral version of $\mathrm{d}T^n = nT^{n-1}\mathrm{d}T + 1/2\,n(n-1)T^{n-2}\mathrm{d}T\,\mathrm{d}T$ and taking the expectation value we have (use the first formula in eq. (4))

$$\begin{aligned}\mathbb{E}[T_k^n(s)] = {} & n\int_0^s \mathbb{E}[T_k^{n-1}(s')]|r_k(s')|^2\,\mathrm{d}s' + n(n-1)\int_0^s \mathbb{E}[T_k^{n-1}(s')]|r_k(s')|^2\,\mathrm{d}s'\\ & + \left(\frac{n(n-1)}{2}\int_0^s \mathbb{E}\Big[T^{n-2}(s')\Big(\int_0^{s'} \bar{D}_\varepsilon^{(k)}(0,s'')r_k(s'')\,\mathrm{d}B_{s''}\Big)^2\right.\\ & \qquad \left. D_\varepsilon^{(k)}(0,s')^2 r_k^2(s')\,\mathrm{d}s'\Big] + \text{c.c.}\right).\end{aligned}$$

Integrating by parts with respect to the factor $e^{-i2/\varepsilon\int_0^{s'}E_k(t)\,\mathrm{d}t}$ shows that the second line is of order $\varepsilon^{1/2}$, whence

$$\mathbb{E}[T_k^n(s)] = n^2\int_0^s \mathbb{E}[T_k^{n-1}(s')]|r_k(s')|^2\,\mathrm{d}s' + O(\varepsilon^{1/2}).$$

Using this relation recursively we arrive at

$$\begin{aligned}\mathbb{E}[T_k^n(s)] &= (n!)^2\int_{0\le s_1\le\cdots\le s_n\le s}\prod_{i=1}^n |r_k(s_i)|^2\,\mathrm{d}s_1\ldots\mathrm{d}s_n + O(\varepsilon^{1/2})\\ &= n!\Big(\int_0^s |r_k(s')|^2\,\mathrm{d}s'\Big)^n + O(\varepsilon^{1/2}),\end{aligned}$$

which is exactly the relation characterizing exponential distribution.

Now consider two terms $T_k(s)$, $T_l(s)$ for $l\neq k$. By Îto's formula we have

$$\begin{aligned}\mathbb{E}[T_k(s)T_l(s)] &= \int_0^s (\mathbb{E}[\mathrm{d}T_k(s')T_l(s')] + \mathbb{E}[T_k(s')\,\mathrm{d}T_l(s')] + \mathbb{E}[\mathrm{d}T_k(s')\,\mathrm{d}T_l(s')])\\ &= \int_0^s (|r_k(s')|^2\mathbb{E}[T_l(s')] + |r_l(s')|^2\mathbb{E}[T_k(s')])\,\mathrm{d}s' + O(\sqrt{\varepsilon})\\ &= \mathbb{E}[T_k(s)]\mathbb{E}[T_l(s)] + O(\sqrt{\varepsilon}).\end{aligned}$$

That the last term on the RHS of the first line is of order $\varepsilon^{1/2}$ can be shown by integration by parts. Hence we showed that T_k and T_l are uncorrelated and we proceed to higher powers by induction. Suppose that T_k^{n-1} and T_l^m (T_k^n and T_l^{m-1}) are uncorrelated to the leading order, then we have

$$\begin{aligned}\mathbb{E}[T_k^nT_l^m] &= \int (\mathbb{E}[\mathrm{d}(T_k^n)T_l^m] + \mathbb{E}[T_k^n\,\mathrm{d}(T_l^m)] + \mathbb{E}[\mathrm{d}(T_k^n)\,\mathrm{d}(T_l^m)])\\ &= \int (n^2|r_k|^2\mathbb{E}[T_k^{n-1}T_l^m] + m^2|r_l|^2\mathbb{E}[T_k^nT_l^{m-1}]) + O(\sqrt{\varepsilon})\\ &= \int (\mathrm{d}(\mathbb{E}[T_k^n])\mathbb{E}[T_l^n] + \mathbb{E}[T_k^n]\mathrm{d}(\mathbb{E}[T_l^n])) + O(\sqrt{\varepsilon})\\ &= \mathbb{E}[T_k^n]\mathbb{E}[T_l^n] + O(\sqrt{\varepsilon}).\end{aligned}$$

So to leading order T_k and T_l are independent, which finishes the proof. □

Remark 4.2. The main deficiency of the expansion in Theorem 3.4 is that it involves the propagator itself, albeit in a higher order. It is straightforward, although cumbersome, to recursively eliminate the propagator. We do not know of any more direct manner to derive higher order terms in the expansion.

Formula (14) for the mean tunneling has been derived in [4] using the corresponding adiabatic Lindblad equation, eq. (13), and subsequently used to study an optimal sweeping rate [2] and Landau–Zener tunneling with dephasing [3]. The mean tunneling is additive, which was interpreted as the tunneling in the dephasing case being local and unidirectional. The full statistics of the tunneling derived here, offers an unexpected twist. If the tunneling was additive it would have a Gaussian distribution, not an exponential one. It follows that only the mean tunneling is additive, while higher order cumulants exhibit non-local behavior typical for the Hamiltonian evolution.

5 Full expansion and its proof

Now we present the main theorem, that describes the expansion to all orders.

Theorem 5.1. *Let $L_1(s), L_2(s)$ be C^{N+2}-families of operators satisfying assumptions* (A)–(C).

1. *The differential equation $\varepsilon\, \mathrm{d}X = \mathrm{d}L(s)X$ admits solutions of the form*

$$X(s) = \sum_{n=0}^{N} \varepsilon^n \Big(\varepsilon^{-1/2} \int_0^s U_\varepsilon(s,s') L_2(s') b_n(s')\, \mathrm{d}B_{s'} + a_n(s) + b_n(s) \Big) + \varepsilon^N r_N(\varepsilon, s)$$

with

- $a_n(s) \in \ker L_1(s),\ b_n(s) \in \operatorname{ran} L_1(s)$,
- *initial data $x(0)$ is specified by arbitrary $a_n(0) \in \ker L_1(0)$; however, the $b_n(0)$ are determined below by the $a_n(0)$ and together define the "slow manifold."*

2. *The coefficients are determined recursively through ($n = 0, \dots, N$)*

$$\begin{aligned} b_0(s) &= 0, \\ a_n(s) &= T(s,0)a_n(0) + \int_0^s T(s,s')\dot{P}(s')b_n(s')\, \mathrm{d}s', \\ b_{n+1}(s) &= L_1(s)^{-1}(\dot{P}(s)a_n(s) + P_\perp(s)\dot{b}_n(s)). \end{aligned}$$

3. *The remainder is uniformly small in ε and is of the form*

$$r_N(\varepsilon, s) = \sqrt{\varepsilon} \int_0^s r_N^{(2)}(\varepsilon, s')\, \mathrm{d}B_{s'} + \varepsilon r_N^{(1)}(\varepsilon, s),$$

where $r_N^{(1)}(\varepsilon,s), r_N^{(2)}(\varepsilon,s)$ are uniformly bounded functions. In particular, the error term $r_N(\varepsilon,s) = O(\sqrt{\varepsilon})$.

Proof. Since $L_1(s)$ is a generator of a contraction semigroup (see the last claim in Proposition 3.2) we can use the standard deterministic adiabatic theory for an equation $\varepsilon \mathrm{d}\widetilde{X}(s) = L_1(s)\widetilde{X}(s)$. Using the expansion in Theorem 6 in [4] the equation has a solution,

$$\widetilde{X}(s) = \sum_{n=0}^{N} \varepsilon^n \left(a_n(s) + b_n(s)\right) + \varepsilon^{N+1} r_N^{(1)}(\varepsilon, s),$$

where $r_N^{(1)}(\varepsilon, s)$ is uniformly bounded. The definition of a-terms and b-terms is such that $a_n(s) \in \ker(L_1(s))$, and $b_n(s)$ belongs to the range of $L_1(s)$.

By the Duhamel formula of Lemma 2.1 and assumption (C) we then have a solution of the stochastic equation,

$$\begin{aligned} X(s) = \sum_{n=0}^{N} \varepsilon^n \Big(\varepsilon^{-1/2} \int_0^s U_\varepsilon(s, s') L_2(s') b_n(s') \, \mathrm{d}B_{s'} + a_n(s) + b_n(s) \Big) \\ + \varepsilon^{N+1} r_N^{(1)}(\varepsilon, s) + \varepsilon^{N+1/2} \int_0^s U_\varepsilon(s, s') L_2(s') r_N^{(1)}(\varepsilon, s') \, \mathrm{d}B_{s'}. \end{aligned}$$

With $r_N^{(2)}(\varepsilon, s') = U_\varepsilon(s, s') L_2(s') r_N^{(1)}(\varepsilon, s')$, this is exactly the expansion of the theorem.

That $r_n^{(2)}(\varepsilon, s)$ is uniformly bounded (with probability 1) follows from assumption (A), which implies that $\|U_\varepsilon(s, s')\| \le 1$. That the error is of the order $O(\sqrt{\varepsilon})$ follows from Lemma 2.2, or more precisely from a backward integration counterpart of the lemma. In fact, Lemma 2.2(a) is sufficient for that conclusion, Lemma 2.2(b) gives complementary estimates for the probability distribution of the error terms. □

We conclude with several remarks regarding the generality of our exposition. Including several independent noises, i.e. $L_2 \mathrm{d}B \to \sum_k L_2^{(k)} \mathrm{d}B_k$ where B_k are independent Brownian motions, is straightforward. In particular the tunneling described in Section 4 turns into a sum over the noises, each giving an independent contribution to the tunneling. Boundedness of L_1, L_2 can surely be relaxed, and an adiabatic theorem with errors of order $O(1)$ should hold without the gap condition, assumption (B). We do not plan to elaborate on any of these generalizations. On the other hand it is important to allow generators $L_1(s), L_2(s)$ to depend on the Brownian motion, B_t, for $s \ge t \ge 0$. We hope to address this question in a further work.

Acknowledgments. I thank Gian Michele Graf and Eddy Mayer-Wolf for fruitful discussions.

A part of the work was done while I visited the Isaac Newton Institute in Cambridge, UK. Support by the Swiss National Science Foundation is acknowledged.

References

[1] J. E. Avron and A. Elgart, Adiabatic theorem without a gap condition. *Comm. Math. Phys.* **203** (1999), no. 2, 445–463. MR 1697605 Zbl 0936.47047

[2] J. E. Avron, M. Fraas, G. M. Graf, and P. Grech, Optimal time-schedule for adiabatic evolution. *Phys. Rev. A* **82** (2010), 040304.

[3] J. E. Avron, M. Fraas, G. M. Graf, and P. Grech, Landau–Zener tunneling for dephasing lindblad evolutions. *Comm. Math. Phys.* **305** (2011), no. 3, 633–639. MR 2819409 Zbl 1221.81118

[4] J. E. Avron, M. Fraas, G. M. Graf, and P. Grech, Adiabatic theorems for generators of contracting evolutions. *Comm. Math. Phys.* **314** (2012), no. 1, 163–191. MR 2954513 Zbl 1256.47062

[5] V. P. Belavkin, Quantum stochastic calculus and quantum nonlinear filtering. *J. Multivariate Anal.* **42** (1992), no. 2, 171–201. MR 1183841 Zbl 0762.60059

[6] D. Dürr, G. Hinrichs, and M. Kolb, On a stochastic trotter formula with application to spontaneous localization models. *J. Stat. Phys.* **143** (2011), no. 6, 1096–1119. MR 2813787 Zbl 1226.82050

[7] J. Gough and M. R. James, Quantum feedback networks: Hamiltonian formulation. *Comm. Math. Phys.* **287** (2009), no. 3, 1109–1132. MR 2486674 Zbl 1196.81152

[8] J. Gough, O. Obrezkov, and O. G. Smolyanov, Randomized hamiltonian Feynman integrals and Schrödinger–Itô stochastic equations. *Izv. Ross. Akad. Nauk Ser. Mat.* **69** (2005), no. 6, 3–20. In Russian. English translation, *Izv. Math.* **69** (2005), no. 6, 1081–1098. MR 2190085 Zbl 1133.35109

[9] A. S. Holevo, *Statistical structure of quantum theory.* Lecture Notes in Physics. New Series m: Monographs. Physics and Astronomy - Online Library. m67. Springer, Berlin, 2001.

[10] R. L. Hudson and K. R. Parthasarathy, Quantum Ito's formula and stochastic evolutions. *Comm. Math. Phys.* **93** (1984), no. 3, 301–323. MR 0745686 Zbl 0546.60058

[11] A. Joye, General adiabatic evolution with a gap condition. *Comm. Math. Phys.* **275** (2007), no. 1, 139–162. MR 2335771 Zbl 1176.47032

[12] T. G. Kurtz, A random trotter product formula. *Proc. Amer. Math. Soc.* **35** (1972), 147–154. MR 0303347 Zbl 0255.47051

[13] H. P. McKean, jr., *Stochastic integrals.* Probability and Mathematical Statistics, 5. Academic Press, New York and London, 1969. MR 0247684 Zbl 0191.46603

[14] E. Pardoux and P. Protter, A two-sided stochastic integral and its calculus. *Probab. Theory Related Fields* **76** (1987), no. 1, 15–49. MR 0899443 Zbl 0608.60058

[15] M. S. Sarandy and D. A. Lidar, Adiabatic approximation in open quantum systems. *Phys. Rev. A* **71** (2005), 012331.

[16] J. Schmid, Adiabatic theorems with and without spectral gap condition for non-semisimple spectral values. In P. Exner, W. König, and H. Neidhardt (eds.), *Mathematical results in quantum mechanics.* Proceedings of the QMath 12 Conference held at Humboldt University, Berlin, September 10–13, 2013. World Scientific, Hackensack, N.J., 2015, 355–362. MR 3381167 Zbl 1339.81044

[17] B. Simon, *Functional integration and quantum physics.* Pure and Applied Mathematics, 86. Academic Press, New York and London, 1979. ISBN 0-12-644250-9 MR 0544188 Zbl 0434.28013

[18] A. V. Skorokhod, Operator stochastic differential equations and stochastic semigroups. *Uspekhi Mat. Nauk* **37** (1982), no. 6(228), 157–183. In Russian. English translation, *Russian Math. Surveys* **37** (1982), no. 6, 177–204. MR 0683278 Zbl 0519.60071

[19] S. Teufel, *Adiabatic perturbation theory in quantum dynamics.* Lecture Notes in Mathematics, 1821. Springer, Berlin, 2003. ISBN 3-540-40723-5 MR 2158392 Zbl 1053.81003

[20] H. M. Wiseman and G. J. Milburn, *Quantum measurement and control.* Cambridge University Press, Cambridge, 2010. ISBN 978-0-521-80442-4 MR 2761009 Zbl 05651600

Eigenvalues of Schrödinger operators with complex surface potentials

Rupert L. Frank

To Pavel Exner on the occasion of his 70th birthday

1 Introduction and main results

Recently there has been great interest in bounds on eigenvalues of Schrödinger operators with complex potentials. A conjecture of Laptev and Safronov [19] states that for a certain range of p's, all eigenvalues of a Schrödinger operator lie in a disk in the complex plane whose radius is bounded from above in terms of only the L^p norm of the potential. This conjecture was motivated by a corresponding result by Abramov, Aslanyan and Davies [1] in one dimension and with $p = 1$. In one part of the parameter regime the conjecture was proved in [9], and in the other part it was proved in [15] for radial potentials. For arbitrary potentials it is still open.

In this paper we deal with the analogue of this question for potentials supported on a hyperplane, which is a special case of what is called a 'leaky graph Hamiltonian' in [8]. More specifically, in $\mathbb{R}^d$, $d \geq 2$, we introduce coordinates $x = (x', x_d)$ with $x' \in \mathbb{R}^{d-1}$ and $x_d \in \mathbb{R}$ and consider the Schrödinger operator

$$-\Delta + \sigma(x')\delta(x_d) \quad \text{in } L^2(\mathbb{R}^d) \tag{1}$$

with a complex function σ on $\mathbb{R}^{d-1}$. If $\sigma \in L^p(\mathbb{R}^{d-1})$ for some $p > 1$ in $d = 2$ and $p \geq d - 1$ in $d \geq 3$, this formal expression can be given meaning as an m-sectorial operator in $L^2(\mathbb{R}^d)$ through the quadratic form

$$\int_{\mathbb{R}^d} |\nabla\psi(x)|^2 \, \mathrm{d}x + \int_{\mathbb{R}^{d-1}} \sigma(x')|\psi(x', 0)|^2 \, \mathrm{d}x' \tag{2}$$

with form domain $H^1(\mathbb{R}^d)$. If also $p < \infty$, it is a consequence of relative form compactness that the spectrum of this operator in $\mathbb{C} \setminus [0, \infty)$ consists of isolated eigenvalues of finite algebraic multiplicities; this is discussed below in more detail.

For *real* σ, the variational principle for the lowest eigenvalue and the Sobolev trace theorem imply that any eigenvalue E satisfies

$$E \geq -\left(C_{\gamma,d} \int_{\mathbb{R}^{d-1}} \sigma(x')_-^{2\gamma+d-1} \, \mathrm{d}x' \right)^{1/\gamma}$$

for all $\gamma > 0$ with a constant $C_{\gamma,d}$ independent of σ.

Our main result is an analogue of this bound for complex σ.

Theorem 1.1. *Let $d \geq 2$ and $0 < \gamma \leq 1/2$. There is a constant $D_{\gamma,d}$ such that for any complex $\sigma \in L^{2\gamma+d-1}(\mathbb{R}^{d-1})$ and any eigenvalue $E \in \mathbb{C}$ of $-\Delta + \sigma(x')\delta(x_d)$ in $L^2(\mathbb{R}^d)$,*

$$|E|^\gamma \leq D_{\gamma,d} \int_{\mathbb{R}^{d-1}} |\sigma(x')|^{2\gamma+d-1} \, \mathrm{d}x'.$$

When $\gamma > 1/2$ we cannot show that eigenvalues are bounded, but we can show that, if (E_j) is a sequence of eigenvalues with $\operatorname{Re} E_j \to \infty$, then $\operatorname{Im} E_j \to 0$. The following theorem gives a quantitative version of this. We use the notation

$$\delta(z) := \operatorname{dist}(z, \mathbb{C} \setminus [0,\infty)) = \begin{cases} |z| & \text{if } \operatorname{Re} z \leq 0, \\ |\operatorname{Im} z| & \text{if } \operatorname{Re} z > 0. \end{cases} \tag{3}$$

Theorem 1.2. *Let $d \geq 2$ and $\gamma > 1/2$. There is a constant $D_{\gamma,d}$ such that for any complex $\sigma \in L^{2\gamma+d-1}(\mathbb{R}^{d-1})$ and any eigenvalue $E \in \mathbb{C}$ of $-\Delta + \sigma(x')\delta(x_d)$ in $L^2(\mathbb{R}^d)$,*

$$|E|^{1/2} \delta(E)^{(2\gamma-1)/2} \leq D_{\gamma,d} \int_{\mathbb{R}^{d-1}} |\sigma(x')|^{2\gamma+d-1} \, \mathrm{d}x'.$$

In dimensions $d \geq 3$ we also obtain a criterion for the absence of eigenvalues.

Theorem 1.3. *Let $d \geq 3$. There is a constant $D_{0,d}$ such that for any complex $\sigma \in L^{d-1}(\mathbb{R}^{d-1})$, if*

$$\int_{\mathbb{R}^{d-1}} |\sigma(x')|^{d-1} \, \mathrm{d}x' < D_{0,d}^{-1},$$

then $-\Delta + \sigma(x')\delta(x_d)$ in $L^2(\mathbb{R}^d)$ has no eigenvalue.

These three theorems are the analogues of the results in [9] and [10] for Schrödinger operators with usual potentials and our proof will follow the strategy in those papers (which, in turn, was motivated by [1]).

Our final result concerns bounds on sums of powers of eigenvalues of $-\Delta + \sigma(x')\delta(x_d)$, which are analogues of the Lieb–Thirring inequalities [21]. Such bounds were shown in [11] for real σ and, using the technique from [12], extended to complex σ provided one only considers eigenvalues outside of a cone around the positive real axis. The following theorem is useful for eigenvalues close to the positive real axis.

Theorem 1.4. *Let $0 < \gamma < 1/2$ if $d = 2$ and $0 < \gamma \leq 1/2$ if $d \geq 3$. Let $\tau = 0$ if $\gamma < (d-1)/(4d-6)$ and $\tau > ((4d-6)\gamma-(d-1))/(d-1-2\gamma)$ if $\gamma \geq (d-1)/(4d-6)$. Then there is a constant $L_{\gamma,d,\tau}$ such that, for any complex $\sigma \in L^{2\gamma+d-1}(\mathbb{R}^{d-1})$, the eigenvalues (E_j) of $-\Delta + \sigma(x')\delta(x_d)$ in $L^2(\mathbb{R}^d)$, repeated according to their algebraic multiplicity, satisfy*

$$\Big(\sum_j \delta(E_j)|E_j|^{(1-\tau)/2}\Big)^{2\gamma/(1+\tau)} \leq L_{\gamma,d,\tau} \int_{\mathbb{R}^{d-1}} |\sigma(x')|^{2\gamma+d-1}\, dx'.$$

This is the analogue of a result from [13] for Schrödinger operators with usual potentials. The method from [10] can probably be used to derive bounds for $\gamma > 1/2$, but to keep the exposition brief we do not pursue this here.

Our proof of Theorem 1.4 identifies, in the spirit of [5], [3], [6], [7], [13], and [10], the eigenvalues of (1) with zeroes of an analytic function. As explained in detail in [10], a result on zeroes of analytic functions [3] plus inequalities on regularized determinants reduce the proof to resolvent bounds in trace ideals. These latter bounds are the content of Proposition 5.1 and constitute the technical main result of this paper.

In conclusion we mention that there are two further methods which yield inequalities for sums of powers of eigenvalues. One method from [6] relies on averaging the bounds from [11] with respect to the opening angle of the cone. Another method from [17] is based on an extension of an inequality of Kato; see also [10].

Remark 1.5. All the theorems reported here have an obvious analogue for the operator $-\Delta$ in $L^2(\mathbb{R}^d_+)$ with boundary condition $\partial\psi/\partial\nu = -\sigma\psi$. (Here $\mathbb{R}^d_+ = \{x \in \mathbb{R}^d : x_d > 0\}$ and $\partial/\partial\nu = -\partial/\partial x_d$.) This simply comes from the fact that the operator (1) leaves the spaces of functions which are even and odd with respect to x_d invariant and on the former subspace it is unitarily equivalent to $-\Delta$ in $L^2(\mathbb{R}^d_+)$ with boundary condition $\partial\psi/\partial\nu = -1/2\,\sigma\psi$.

2 Uniform Sobolev inequalities

In this section we prove a Sobolev inequality for functions on $\mathbb{R}^N$. (Later on, in the proof of Theorems 1.1, 1.2, and 1.3 we will choose $N = d - 1$.) The inequality involves the operator $\sqrt{-\Delta - z}$ and the crucial point is that the constant in the inequality depends only on $|z|$ but not on the argument of z. Such uniform Sobolev inequalities go back to Kenig, Ruiz and Sogge [18] for $-\Delta - z$ and, in fact, our theorem follows by modifying their proof.

Since it comes at no extra effort, we deal with the operators $(-\Delta-z)^s$ for arbitrary $0 < s \leq (N+1)/2$. This operator acts as multiplication by $(\xi^2 - z)^s$ in Fourier space. We will assume that $z \in \mathbb{C} \setminus [0,\infty)$, so $\xi^2 - z \in \mathbb{C} \setminus (-\infty, 0]$ for all $\xi \in \mathbb{R}^d$ and we can define $(\xi^2 - z)^s = \exp(s \log(\xi^2 - z))$ with the principal branch of the logarithm on $\mathbb{C} \setminus (-\infty, 0]$.

Proposition 2.1. *Let $0 < s \leq (N+1)/2$ and assume that*

$$\begin{cases} 2\dfrac{N}{N+2s} \leq p \leq 2\dfrac{N+1}{N+1+2s} & \text{if } s < N/2, \\ 1 < p \leq 2\dfrac{N+1}{N+1+2s} & \text{if } s = N/2, \\ 1 \leq p \leq 2\dfrac{N+1}{N+1+2s} & \text{if } s > N/2. \end{cases}$$

Then there is a constant $C_{N,p,s}$ such that for all $u \in W^{2s,p}(\mathbb{R}^N)$ and $z \in \mathbb{C} \setminus [0,\infty)$,

$$\|u\|_{p'} \leq C_{N,p,s}\, |z|^{-(Np+2ps-2N)/(2p)} \|(-\Delta - z)^s u\|_p. \tag{4}$$

Moreover, if $2(N+1)/(N+1+2s) < p \leq 2$, there is a constant $C_{N,p,s}$ such that for all $u \in W^{2s,p}(\mathbb{R}^N)$ and $z \in \mathbb{C} \setminus [0,\infty)$,

$$\|u\|_{p'} \leq C_{N,p,s}\, \delta(z)^{-(Np+2ps-2N-2+p)/(2p)}\, |z|^{-(2-p)/(2p)} \|(-\Delta - z)^s u\|_p. \tag{5}$$

We recall that $\delta(z)$ appearing in (5) was defined in (3). Moreover, $p' = p/(p-1)$.

Proof. Let $\zeta \in \mathbb{C}$ with $\operatorname{Re}\zeta \geq 0$ and consider the operator

$$T_\zeta(z) := e^{\zeta^2}(-\Delta - z)^{-\zeta} = e^{\zeta^2} e^{-\zeta \log(-\Delta - z)},$$

which is again defined as a multiplier in Fourier space with the same convention for the branch of the logarithm. Note that this is essentially the family of operators from [18] (proof of Theorem 2.3) with $\zeta = -\lambda$. Clearly, by bounding the multiplier in Fourier space, one finds

$$\|T_\zeta(z)\|_{L^2\to L^2} \le A \quad \text{if } \operatorname{Re}\zeta = 0 \tag{6}$$

with a constant A depending only on N. Moreover, it is shown in [18] (proof of Theorem 2.3) that

$$\|T_\zeta(z)\|_{L^1\to L^\infty} \le B_{\operatorname{Re}\zeta}|z|^{-(2\operatorname{Re}\zeta-N)/2} \quad \text{if } N/2 < \operatorname{Re}\zeta \le (N+1)/2 \tag{7}$$

with a constant $B_{\operatorname{Re}\zeta}$ depending only on N and $\operatorname{Re}\zeta$. (Strictly speaking, this bound was only shown there with a constant independent of z for $|z| \ge 1$, but the stated bound simply follows from this by scaling. Moreover, the assumption $N \ge 3$ in [18] is irrelevant for the proof of (7).)

Since $T_s(z)$ coincides, up to a multiplicative constant, with the inverse of the operator $(-\Delta - z)^s$, if $s > N/2$ and $p = 1$, we can choose $\zeta = s$ in (7) and obtain the bound in the proposition.

For $p > 1$ as in the theorem, with the extra assumption $p > 2N/(N+2s)$ if $s < N/2$, we define $t := sp/(2-p)$ and note that $N/2 < t \le (N+1)/2$ and $t > s$. Since the operators $T_\zeta(z)$ depend analytically on ζ, we can use complex interpolation with the lines $\operatorname{Re}\zeta = 0$ and $\operatorname{Re}\zeta = t$ and obtain

$$\|T_s(z)\|_{L^p\to L^{p'}} \le A^{2(p-1)/p} B_{ps/(2-p)}^{(2-p)/p} |z|^{-(Np+2ps-2N)/(2p)},$$

which again gives the claimed bounds.

For the first part of the proposition it remains to prove the bound for $p = 2N/(N+2s)$ and $s < N/2$. Again as in [18] (proof of Theorem 2.3) we consider

$$\widetilde{T}_\zeta(z) := \frac{e^{\zeta^2}}{\Gamma\Big(\frac{N-2\zeta}{2}\Big)}(-\Delta - z)^{-\zeta} = \frac{e^{\zeta^2}}{\Gamma\Big(\frac{N-2\zeta}{2}\Big)} e^{-\zeta\log(-\Delta - z)}.$$

Bound (6) remains valid for $\widetilde{T}_\zeta(z)$ and, as shown in [18] (proof of Theorem 2.3), bound (7) holds even for $\operatorname{Re}\zeta = N/2 =: t$. If $s < N/2$ one has $0 < s < t$ and therefore one can again use complex interpolation to deduce an $L^{2N/(N+2s)} \to L^{2N/(N-2s)}$ bound for $\widetilde{T}_s(z)$. This completes the proof of the proposition.

To prove the second part of the proposition we note that

$$\|u\|_2 \le \delta(z)^{-s}\|(-\Delta - z)^s u\|_2.$$

This, together with (4) for $p = 2(N+1)/(N+1+2s)$, implies (5) by standard (Riesz–Thorin) complex interpolation. □

3 Bound on the Birman–Schwinger operator

Let us give the details of the definition of the operator (1) using the method from [10] (Section 4).

We consider the operator $H_0 := -\Delta$ in the Hilbert space $\mathcal{H} := L^2(\mathbb{R}^d)$ with form domain $H^1(\mathbb{R}^d)$. Moreover, let $\mathcal{G} := L^2(\mathbb{R}^{d-1})$ and consider the operators G and G_0 from $\mathcal{H}$ to $\mathcal{G}$ with domain $H^1(\mathbb{R}^d)$ defined by

$$(G_0\psi)(x') := \sqrt{\sigma(x')}\,\psi(x',0),$$

$$(G\psi)(x') := \sqrt{|\sigma(x')|}\,\psi(x',0).$$

(Here we write $\sqrt{\sigma(x')} = \sigma(x')/\sqrt{|\sigma(x')|}$ if $\sigma(x') \neq 0$ and $\sqrt{\sigma(x')} = 0$ otherwise.) We claim that, if $\sigma \in L^p(\mathbb{R}^{d-1})$ with $1 < p < \infty$ if $d = 2$ and $d-1 \leq p < \infty$ if $d \geq 3$, then

$$G_0(H_0+1)^{-1/2} \quad \text{and} \quad G(H_0+1)^{-1/2}$$

are compact. When σ is bounded and has support in a set of finite measure, this follows from the trace version of Rellich's compactness theorem, see, e.g., Theorem 6.3 in [2]. By the trace version of Sobolev's embedding theorem (see, e.g., Theorem 4.12 in [2]) and an argument as in Lemma 4.3 in [10] we obtain the assertion in the general case.

Thus, we have verified the assumptions of Lemma B.1 in [10] and we infer that the quadratic form (2), which is the same as $\|H_0^{1/2}\psi\|^2 + (G\psi, G_0\psi)$, is closed and sectorial and generates an m-sectorial operator H. Moreover, let $z \in \mathbb{C}\setminus[0,\infty) = \rho(H_0)$ and define the Birman–Schwinger operator

$$K(z) = G_0(H_0-z)^{-1}G^* \quad \text{in } L^2(\mathbb{R}^{d-1}). \tag{8}$$

Strictly speaking, since in our case the operators G and G_0 are not closable, the operator $K(z)$ is defined as

$$K(z) = (G_0(H_0+1)^{-1/2})(H_0+1)(H_0-z)^{-1}(G(H_0+1)^{-1/2})^*.$$

The following appears as Lemma B.1 in [10] and represents a version of the Birman–Schwinger principle.

Lemma 3.1. *Let $z \in \mathbb{C}\setminus[0,\infty)$, then $1 + K(z)$ is boundedly invertible if and only if $z \in \rho(H)$.*

Remark 3.2. In passing we mention that Proposition B.2 in [10] yields that

$$[0,\infty) = \{z \in \mathbb{C} \colon \operatorname{ran}(H-z) \text{ is not closed}\} \\ \cup \{z \in \mathbb{C} \colon \dim\ker(H-z) = \operatorname{codim}\operatorname{ran}(H-z) = \infty\}$$

and

$$\sigma(H) \setminus [0,\infty) = \{z \in \mathbb{C} \colon \operatorname{ran}(H-z) \text{ is closed and} \\ 0 < \dim\ker(H-z) = \operatorname{codim}\operatorname{ran}(H-z) < \infty\}.$$

Moreover, the latter set is at most countable and consists of eigenvalues of finite algebraic multiplicities which are isolated in $\sigma(H)$. These facts, however, will not be relevant for the proof of Theorems 1.1 and 1.3.

Our next goal is to find a convenient expression for the Birman–Schwinger operator. If we denote by Γ the trace operator which restricts a function on $\mathbb{R}^d$ to $\mathbb{R}^{d-1} \times \{0\}$, then we have

$$G_0 = \sqrt{\sigma}\,\Gamma, \quad G = \sqrt{|\sigma|}\,\Gamma.$$

Moreover, let us denote the Laplacian on $\mathbb{R}^{d-1}$ by $-\Delta'$.

Lemma 3.3. *Let $z \in \mathbb{C} \setminus [0,\infty)$. Then*

$$\Gamma(-\Delta - z)^{-1}\Gamma^* = \frac{1}{2}(-\Delta' - z)^{-1/2}.$$

Technically, one can consider Γ as an unbounded operator from $L^2(\mathbb{R}^d)$ to $L^2(\mathbb{R}^{d-1})$ with domain $H^1(\mathbb{R}^d)$. The expression Γ^* is purely formal and should be interpreted in the same sense as explained after (8).

Proof. Since $(-\Delta - z)^{-1}$ has integral kernel

$$(2\pi)^{-d} \int_{\mathbb{R}^d} \frac{e^{i\xi\cdot(x-y)}}{\xi^2 - z}\,d\xi, \quad x, y \in \mathbb{R}^d,$$

the operator $\Gamma(-\Delta - z)^{-1}\Gamma^*$ has integral kernel

$$(2\pi)^{-d} \int_{\mathbb{R}^{d-1}} \int_{\mathbb{R}} \frac{e^{i\xi'\cdot(x'-y')}}{(\xi')^2 + \xi_d^2 - z}\,d\xi'\,d\xi_d, \quad x', y' \in \mathbb{R}^{d-1}.$$

The integral with respect to ξ_d can be computed using

$$\int_{\mathbb{R}} \frac{d\xi_d}{\xi_d^2 + b^2} = \frac{\pi}{b} \quad \text{if } \operatorname{Re} b > 0.$$

Thus, $\Gamma(-\Delta - z)^{-1}\Gamma^*$ has integral kernel

$$\frac{1}{2}(2\pi)^{-d+1}\int_{\mathbb{R}^{d-1}} \frac{e^{i\xi'\cdot(x'-y')}}{\sqrt{(\xi')^2 - z}}\,\mathrm{d}\xi', \quad x', y' \in \mathbb{R}^{d-1},$$

with the branch of the square root as described before Theorem 2.1. This coincides with the integral kernel of the operator $1/2\,(-\Delta' - z)^{-1/2}$. □

Remark 3.4. One can also show that ψ is an eigenfunction of (1) corresponding to an eigenvalue E if and only if $\psi(x) = (\exp(-|x_d|\sqrt{-\Delta' - E})\varphi)(x')$, where $\varphi = \Gamma\psi$ satisfies $(\sqrt{-\Delta' - E} + \sigma/2)\varphi = 0$. This is closely related to the harmonic extension and the Dirichlet–to–Neumann operator for the Laplacian on $L^2(\mathbb{R}^d_+)$; see also Remark 1.5. This observation was also crucial in [14].

Combining Lemma 3.3 with Theorem 2.1 we obtain

Corollary 3.5. *Let $0 < \gamma \leq 1/2$ if $d = 2$ and $0 \leq \gamma \leq 1/2$ if $d \geq 3$. Then there is a constant $C_{\gamma,d}$ such that for all $\alpha_1, \alpha_2 \in L^{2(2\gamma+d-1)}(\mathbb{R}^{d-1})$,*

$$\|\alpha_1\Gamma(-\Delta - z)^{-1}\Gamma^*\alpha_2\| \leq C_{\gamma,d}|z|^{-\gamma/(2\gamma+d-1)}\|\alpha_1\|_{2(2\gamma+d-1)}\|\alpha_2\|_{2(2\gamma+d-1)}.$$

Moreover, if $\gamma > 1/2$, then there is a constant $C_{\gamma,d}$ such that for all $\alpha_1, \alpha_2 \in L^{2(2\gamma+d-1)}(\mathbb{R}^{d-1})$

$$\begin{aligned}&\|\alpha_1\Gamma(-\Delta - z)^{-1}\Gamma^*\alpha_2\| \\ &\quad\leq C_{\gamma,d}\,\delta(z)^{-(2\gamma-1)/(2(2\gamma+d-1))}|z|^{-1/(2(2\gamma+d-1))}\|\alpha_1\|_{2(2\gamma+d-1)}\|\alpha_2\|_{2(2\gamma+d-1)}.\end{aligned}$$

Proof. According to Lemma 3.3,

$$\alpha_1\Gamma(-\Delta - z)^{-1}\Gamma^*\alpha_2 = \frac{1}{2}\alpha_1(-\Delta' - z)^{-1/2}\alpha_2,$$

and, for any $1 \leq p \leq \infty$,

$$\|\alpha_1(-\Delta' - z)^{-1/2}\alpha_2\| \leq \|\alpha_1\|_{L^{p'}\to L^2}\|(-\Delta' - z)^{-1/2}\|_{L^p\to L^{p'}}\|\alpha_2\|_{L^2\to L^p}.$$

By Hölder's inequality, if $1 \leq p \leq 2$,

$$\|\alpha_1\|_{L^{p'}\to L^2} = \|\alpha_1\|_{2p/(2-p)} \quad \text{and} \quad \|\alpha_2\|_{L^2\to L^p} = \|\alpha_2\|_{2p/(2-p)}.$$

We bound the norm of $(-\Delta' - z)^{-1}$ from L^p to $L^{p'}$ by Theorem 2.1 with $N = d-1$. Bound (4) holds if $1 < p \leq 4/3$ for $d = 2$ and if $2(d-1)/d \leq p \leq 2d/(d+1)$ if $d \geq 3$. These conditions correspond precisely to our assumptions on γ in the first part if we pick p such that $2p/(2-p) = 2(2\gamma+d-1)$. Similarly, bound (5) holds if $p > 2d/(d+1)$ which corresponds to $\gamma > 1/2$. □

4 Proof of Theorems 1.1, 1.2, and 1.3

Let E be an eigenvalue of the operator (1). We begin with the case $E \in \mathbb{C} \setminus [0, \infty)$, where we use the argument of [1]; see also [9]. Then, by the Birman–Schwinger principle (Lemma 3.1), $1 + K(E) = 1 + \sqrt{\sigma}\Gamma(-\Delta - E)^{-1}\Gamma^* \sqrt{|\sigma|}$ is not boundedly invertible and therefore $\|K(E)\| \geq 1$. Combining this with the upper bound on $\|K(E)\|$ in the first part of Corollary 3.5 (with $\alpha_1 = \sqrt{\sigma}$ and $\alpha_2 = \sqrt{|\sigma|}$), we obtain

$$1 \leq C_{\gamma,d} |E|^{-\gamma/(2\gamma+d-1)} \|\sigma\|_{2\gamma+d-1}.$$

This is the claimed bound on $|E|^\gamma$ for $0 < \gamma \leq 1/2$ in Theorem 1.1 and the condition for the absence of eigenvalues for $\gamma = 0$ in Theorem 1.3. Using the second part of Corollary 3.5 instead, we obtain the claimed bound on $|E|^{1/2}\delta(E)^{(2\gamma-1)/2}$ for $\gamma > 1/2$ in Theorem 1.2.

Now let $E \in [0, \infty)$ and denote by ψ a corresponding eigenfunction. We use an approximation argument similar to [15]. For $\varepsilon > 0$ let

$$\varphi_\varepsilon := G_0(-\Delta - E - i\varepsilon)^{-1}(-\Delta - E)\psi,$$

which is well-defined since $\psi \in H^1(\mathbb{R}^d)$. We claim that $\varphi_\varepsilon \to G_0\psi$ weakly in $L^2(\mathbb{R}^{d-1})$ as $\varepsilon \to 0$. (Note that $G_0\psi \in L^2(\mathbb{R}^{d-1})$ is well-defined since $\psi \in H^1(\mathbb{R}^d)$.) In fact, by dominated convergence in Fourier space we conclude that for any $f \in L^2(\mathbb{R}^{d-1})$, as $\varepsilon \to 0$,

$$\begin{aligned}(f, \varphi_\varepsilon) &= ((G_0(-\Delta + 1)^{-1/2})^* f, (-\Delta - E - i\varepsilon)^{-1}(-\Delta - E)(-\Delta + 1)^{1/2}\psi) \\ &\longrightarrow ((G_0(-\Delta + 1)^{-1/2})^* f, (-\Delta + 1)^{1/2}\psi) = (f, G_0\psi).\end{aligned}$$

On the other hand, the eigenvalue equation for ψ gives

$$\varphi_\varepsilon = (G_0(-\Delta - E - i\varepsilon)^{-1}G^*)(G_0\psi),$$

and therefore, by Corollary 3.5,

$$\|\varphi_\varepsilon\| \leq C_{\gamma,d}(E^2 + \varepsilon^2)^{-\gamma/(2(2\gamma+d-1))} \|\sigma\|_{2\gamma+d-1} \|G_0\psi\|.$$

By weak semi-continuity of the norm we conclude that

$$\|G_0\psi\| \leq \liminf_{\varepsilon\to 0} \|\varphi_\varepsilon\| \leq \liminf_{\varepsilon\to 0} C_{\gamma,d}(E^2 + \varepsilon^2)^{-\gamma/(2(2\gamma+d-1))} \|\sigma\|_{2\gamma+d-1} \|G_0\psi\|.$$

Since $G_0\psi \not\equiv 0$ (otherwise ψ would be an eigenfunction of $-\Delta$ with eigenvalue E), we finally obtain again

$$1 \leq C_{\gamma,d} |E|^{-\gamma/(2\gamma+d-1)} \|\sigma\|_{2\gamma+d-1},$$

as claimed.

5 Uniform Sobolev inequalities in trace ideals

By the argument in the proof of Corollary 3.5 we see that the uniform Sobolev inequality from Proposition 2.1 is equivalent to a bound of the operator norm of $\alpha_1(-\Delta - z)^{-s}\alpha_2$ in terms of the $L^{2p/(2-p)}(\mathbb{R}^N)$-norms of α_1 and α_2 and an inverse power of $|z|$. In this section we improve this by showing that not only the operator norm, but also a trace ideal norm can be bounded in terms of the same quantities.

Proposition 5.1. *Let $0 < s \leq (N+1)/2$ and assume that*

$$\begin{cases} 1 \leq q \leq \dfrac{N+1}{2s} & \text{if } N < 2s, \\ 1 < q \leq \dfrac{N+1}{2s} & \text{if } N = 2s, \\ \dfrac{N}{2s} \leq q \leq \dfrac{N+1}{2s} & \text{if } N > 2s. \end{cases}$$

In addition, if $N = 1$ and $s \leq 1/2$ assume that $q < 2$ and, if $N = 2$ and $s \leq 1/2$ that $q > 1/s$. Then there is a constant $C_{N,q,s}$ such that for all $\alpha_1, \alpha_2 \in L^{2q}(\mathbb{R}^N)$ and $z \in \mathbb{C} \setminus [0, \infty)$,

$$\|\alpha_1(-\Delta - z)^{-s}\alpha_2\|_r \leq C_{N,q,s}|z|^{-s+N/(2q)}\|\alpha_1\|_{2q}\|\alpha_2\|_{2q}$$

with $r = 2$ if $N = 1$ and

$$r = \max\left\{\frac{(N-1)q}{N-qs}, 2\right\} \quad \text{if } N \geq 2.$$

For $s = 1$ and $q \leq (N+1)/(2s)$ this proposition appears in [13]. There it is also shown that the trace ideal index r is smallest possible if $N \geq 3$ or if $N = 2$ and $q \geq 4/3$. We note that the technique from [10] allows one also to obtain inequalities for $q > (N+1)/(2s)$.

Proof. We distinguish the following two cases:

(A) $$\frac{N-1}{2} \leq s \leq \frac{N+1}{2}, \quad q \leq \frac{2N}{N-1+2s}, \quad q < 2;$$

(B) either

(B1) $$s < \frac{N-1}{2}$$

or

(B2) $$\frac{N-1}{2} \le s \le \frac{N+1}{2}$$

and

$$q > \frac{2N}{N-1+2s}.$$

Note that case (A) corresponds to $r = 2$ and case (B) to $r = (N-1)q/(N-qs)$.

CASE (A). We know that $(-\Delta - z)^{-s}$ is an integral operator with integral kernel

$$\int_{\mathbb{R}^N} \frac{e^{i\xi\cdot(x-y)}}{(\xi^2 - z)^s} \frac{d\xi}{(2\pi)^N} = \frac{2^{1-s}}{(2\pi)^{N/2}\Gamma(s)} \Big(\frac{\sqrt{-z}}{|x-y|}\Big)^{(N-2s)/2} K_{(N-2s)/2}(\sqrt{-z}|x-y|),$$

where we choose the branch of the square root on $\mathbb{C} \setminus (-\infty, 0]$ with positive real part; see, e.g., [16] (Section III.2.8). Bounds on Bessel functions (we give references for more precise bounds when dealing with case (B)) show that the absolute value of this kernel is bounded by

$$C_{N,\rho,s}|z|^{(N-2s)/2}(\sqrt{|z|}\,|x-y|)^{-\rho},$$

where

$$\begin{cases} 0 \le \rho \le \dfrac{N+1-2s}{2} & \text{if } N/2 < s \le (N+1)/2, \\ 0 < \rho \le \dfrac{N+1-2s}{2} & \text{if } N/2 = s, \\ N-2s \le \rho \le \dfrac{N+1-2s}{2} & \text{if } (N-1)/2 \le s < N/2. \end{cases}$$

If $0 \le 2\rho < N$ we can use the Hardy–Littlewood–Sobolev inequality to bound the Hilbert–Schmidt norm and obtain

$$\|\alpha_1(-\Delta - z)^{-s}\alpha_2\|_2 \le C'_{N,\rho,s}|z|^{(N-2s-\rho)/2}\|\alpha_1\|_{2N/(N-\rho)}\|\alpha_2\|_{2N/(N-\rho)}.$$

Substituting $q = N/(N-\rho)$, we see that the assumptions on q in case (A) correspond to the assumptions on ρ and we obtain the claimed bounds.

CASE (B). We use complex interpolation similarly as in the proof of Theorem 2.1. Since multiplication by $\alpha_j/|\alpha_j|$ is a bounded operator, we may assume that $\alpha_j \ge 0$ for $j = 1, 2$. We consider the same family $T_\zeta(z)$ of operators as in the proof of Theorem 2.1. Bound (6) implies immediately that

$$\|\alpha_1^{\zeta/s} T_\zeta(z) \alpha_2^{\zeta/s}\| \leq A \quad \text{if } \operatorname{Re}\zeta = 0$$

with a constant A depending only on N. On the other hand, the explicit form of the integral kernel of $T_\zeta(z)$ and the bounds in [18] (Proof of Theorem 2.3) – see also [4] and [13] – show that this integral kernel satisfies

$$|T_\zeta(z)(x,y)| \leq B_{\operatorname{Re}\zeta} |z|^{(N-1-2\operatorname{Re}\zeta)/4} |x-y|^{-(N+1-2\operatorname{Re}\zeta)/2} \quad \text{if } 0 < |\operatorname{Re}\zeta - N/2| \leq 1/2.$$

Thus, if we assume in addition that $\operatorname{Re}\zeta > 1/2$, we can bound the Hilbert–Schmidt norm as before by the Hardy–Littlewood–Sobolev inequality and get

$$\begin{aligned} &\|\alpha_1^{\zeta/s} T_\zeta(z) \alpha_2^{\zeta/s}\|_2 \\ &\quad \leq B'_{\operatorname{Re}\zeta} |z|^{(N-1-2\operatorname{Re}\zeta)/4} \|\alpha_1\|_{4N\operatorname{Re}\zeta/(s(N-1+2\operatorname{Re}\zeta))}^{\operatorname{Re}\zeta/s} \|\alpha_2\|_{4N\operatorname{Re}\zeta/(s(N-1+2\operatorname{Re}\zeta))}^{\operatorname{Re}\zeta/s}. \end{aligned}$$

We choose $t > s$ with $0 < |t - N/2| \leq 1/2$ and $t > 1/2$ and use complex interpolation with the lines $\operatorname{Re}\zeta = 0$ and $\operatorname{Re}\zeta = t$ to get

$$\begin{aligned} &\|\alpha_1 T_s(z) \alpha_2\|_{2t/s} \\ &\quad \leq A^{(t-s)/t} (B'_t)^{s/t} |z|^{s(N-1-2t)/(4t)} \|\alpha_1\|_{4Nt/(s(N-1+2t))} \|\alpha_2\|_{4Nt/(s(N-1+2t))}. \end{aligned}$$

Substituting $t = (N-1)qs/(2(N-qs))$, we see that the assumptions on q in case (B), plus the assumption $q \neq N^2/(2(2N-1))$, correspond to the assumptions on t.

Finally, let $q = N^2/((2N-1)s)$. In this case we use the same family $\widetilde{T}_\zeta(z)$ of operators as in the proof of Theorem 2.1 and note that

$$|\widetilde{T}_\zeta(z)(x,y)| \leq \widetilde{B}_{N/2} |z|^{-1/4} |x-y|^{-1/2} \quad \text{if } \operatorname{Re}\zeta = N/2.$$

As before, we can use the Hardy–Littlewood–Sobolev inequality to bound the Hilbert–Schmidt norm of $\alpha_1^{\zeta/s} T_\zeta(z) \alpha_2^{\zeta/s}$ for $\operatorname{Re}\zeta = N/2$, and then we can deduce the claimed bound by complex interpolation. This proves the claimed bound in case (B). □

Combining this proposition with Lemma 3.3 we get

Corollary 5.2. *Let $0 < \gamma < 1/2$ if $d = 2$, $0 < \gamma \leq 1/2$ if $d = 3$ and $0 \leq \gamma \leq 1/2$ if $d \geq 4$. Then there is a constant $C_{\gamma,d}$ such that for all $\alpha_1, \alpha_2 \in L^{2(2\gamma+d-1)}(\mathbb{R}^{d-1})$,*

$$\|\alpha_1 \Gamma(-\Delta - z)^{-1} \Gamma^* \alpha_2\|_r \leq C_{\gamma,d} |z|^{-\gamma/(2\gamma+d-1)} \|\alpha_1\|_{2(2\gamma+d-1)} \|\alpha_2\|_{2(2\gamma+d-1)},$$

where $r = 2$ if $d = 2$ and $r = 2(d-2)(2\gamma+d-1)/(d-1-2\gamma)$ if $d \geq 3$.

6 Proof of Theorem 1.4

According to Proposition 4.1 in [10] (which generalizes a result in [20]) the eigenvalues (E_j) of $-\Delta + \sigma(x')\delta(x_d)$ in $L^2(\mathbb{R}^d)$ coincide with the eigenvalues (E_j) of finite type of the analytic family $1 + K$ in $\mathbb{C} \setminus [0, \infty)$, repeated according to algebraic multiplicity. Here K denotes the Birman–Schwinger operator from (8). Corollary 5.2 combined with Theorem 3.1 in [10] (which is essentially from [13] and relies on [3]) yields

$$\sum_j \delta(z_j)|z_j|^{-1/2\,(1-((2\gamma r)/(2\gamma+d-1)-1+\varepsilon)_+)} \leq C_{\gamma,d,\varepsilon}\|\sigma\|_{2\gamma+d-1}^{(2\gamma+d-1)(1+((2\gamma r)/(2\gamma+d-1)-1+\varepsilon)_+)/(2\gamma)}$$

for any $\varepsilon > 0$ with $r = 2$ if $d = 2$ and $r = 2(d-2)(2\gamma+d-1)/(d-1-2\gamma)$ if $d \geq 3$. Setting $\tau = ((2\gamma r)/(2\gamma+d-1) - 1 + \varepsilon)_+$, we obtain the inequality in the theorem.

Acknowledgement. Support through NSF grant DMS-1363432 is acknowledged.

References

[1] A. A. Abramov, A. Aslanyan, and E. B. Davies, Bounds on complex eigenvalues and resonances. *J. Phys. A* **34** (2001), no. 1, 57–72. MR 1819914 Zbl 1123.81415

[2] R. A. Adams and J. J. F. Fournier, *Sobolev spaces.* Second edition. Pure and Applied Mathematics (Amsterdam), 140. Elsevier/Academic Press, Amsterdam, 2003. ISBN 0-12-044143-8 MR 2424078 Zbl 1098.46001

[3] A. Borichev, L. Golinskii, and S. Kupin, A Blaschke-type condition and its application to complex Jacobi matrices. *Bull. Lond. Math. Soc.* **41** (2009), no. 1, 117–123. MR 2481997 Zbl 1175.30007

[4] S. Chanillo and E. Sawyer, Unique continuation for $\Delta + v$ and the C. Fefferman–Phong class. *Trans. Amer. Math. Soc.* **318** (1990), no. 1, 275–300. MR 0958886 Zbl 0702.35034

[5] M. Demuth and G. Katriel, Eigenvalue inequalities in terms of Schatten norm bounds on differences of semigroups, and application to Schrödinger operators. *Ann. Henri Poincaré* **9** (2008), no. 4, 817–834. MR 2413204 Zbl 1156.81019

[6] M. Demuth, M. Hansmann, and G. Katriel, On the discrete spectrum of nonselfadjoint operators. *J. Funct. Anal.* **257** (2009), no. 9, 2742–2759. MR 2559715 Zbl 1183.47016

[7] M. Demuth, M. Hansmann, G. Katriel, Eigenvalues of non-selfadjoint operators: a comparison of two approaches. In M. Demuth and W. Kirsch (eds.), *Mathematical physics, spectral theory and stochastic analysis.* Advances in Partial Differential Equations (Basel). Birkhäuser/Springer Basel AG, Basel, 2013, 107–163. MR 3077277 Zbl 1280.47005

[8] P. Exner, Leaky quantum graphs: a review. In P. Exner, J. P. Keating, P. Kuchment, T. Sunada, and A. Teplyaev (eds.), *Analysis on graphs and its applications.* Papers from the program held in Cambridge, January 8–June 29, 2007. Proceedings of Symposia in Pure Mathematics, 77. American Mathematical Society, Providence, R.I., 2008, 523–564. MR 2459890 Zbl 1153.81487

[9] R. L. Frank, Eigenvalue bounds for Schrödinger operators with complex potentials. *Bull. Lond. Math. Soc.* **43** (2011), no. 4, 745–750. MR 2820160 Zbl 1228.35158

[10] R. L. Frank, Eigenvalue bounds for Schrödinger operators with complex potentials. III. To appear in *Trans. Amer. Math. Soc.*

[11] R. L. Frank and A. Laptev, Spectral inequalities for Schrödinger operators with surface potentials. In T. Suslina and D. Yafaev (eds.), *Spectral theory of differential operators.* M. Š. Birman 80th anniversary collection. American Mathematical Society Translations, Series 2, 225. Advances in the Mathematical Sciences, 62. American Mathematical Society, Providence, R.I., 2008, 91–102. MR 2509777 Zbl 1170.35486

[12] R. L. Frank, A. Laptev, E. H. Lieb, and R. Seiringer, Lieb–Thirring inequalities for Schrödinger operators with complex-valued potentials. *Lett. Math. Phys.* **77** (2006), no. 3, 309–316. MR 2260376 Zbl 1160.81382

[13] R. L. Frank and J. Sabin, Restriction theorems for orthonormal functions, Strichartz inequalities and uniform Sobolev estimates. To appear in *Amer. J. Math.*

[14] R. L. Frank, R. G. Shterenberg, On the scattering theory of the Laplacian with a periodic boundary condition. II. Additional channels of scattering. *Doc. Math.* **9** (2004), 57–77. MR 2054980 Zbl 1210.35040

[15] R. L. Frank and B. Simon, Eigenvalue bounds for Schrödinger operators with complex potentials. II. To appear in *J. Spectr. Theory.*

[16] I. M. Gel′fand and G. E. Shilov, *Generalized functions.* Vol. 1. Properties and operations. Translated from the Russian by E. Saletan. Academic Press, New York and London, 1964. MR 0435831 Zbl 0115.33101

[17] M. Hansmann, An eigenvalue estimate and its application to non-self-adjoint Jacobi and Schrödinger operators. *Lett. Math. Phys.* **98** (2011), no. 1, 79–95. MR 2836430 Zbl 1230.47008

[18] C. E. Kenig, A. Ruiz, and C. D. Sogge, Uniform Sobolev inequalities and unique continuation for second order constant coefficient differential operators. *Duke Math. J.* **55** (1987), no. 2, 329–347. MR 0894584 Zbl 0644.35012

[19] A. Laptev and O. Safronov, Eigenvalue estimates for Schrödinger operators with complex potentials. *Comm. Math. Phys.* **292** (2009), no. 1, 29–54. MR 2540070 Zbl 1185.35045

[20] Y. Latushkin and A. Sukhtayev, The algebraic multiplicity of eigenvalues and the Evans function revisited. *Math. Model. Nat. Phenom.* **5** (2010), no. 4, 269–292. MR 2662459 Zbl 1193.35110

[21] E. H. Lieb and W. Thirring, Inequalities for the moments of the eigenvalues of the Schrödinger Hamiltonian and their relation to Sobolev inequalities. In E. H. Lieb (ed.), *Studies in mathematical physics.* Essays in Honor of V. Bargmann. Princeton University Press, Princeton, 1976, pp. 269–303. Zbl 0342.35044

A lower bound to the spectral threshold in curved quantum layers

Pedro Freitas and David Krejčiřík

Dedicated to Pavel Exner on the occasion of his 70th birthday

1 Introduction

In this paper we obtain a lower bound to the lowest energy of a quantum particle confined to the space delimited by two parallel surfaces. We assume that these surfaces represent a perfect hard-wall boundary, in the sense that the particle wavefunction vanishes there, and concentrate in the case where they are unbounded. In agreement with the paper [6] where these structures were introduced, we shall use the term *quantum layers* for such systems.

This rather simple model is known to be remarkably successful in describing various aspects of electronic transport in quantum heterostructures (we refer to the monograph [20] for the physical background). One of the main questions arising within this scope is whether or not there are geometrically induced bound states. Indeed, some of the most important theoretical results in the field are a number of theorems guaranteeing the existence of such solutions under rather simple and general physical conditions [6], [3], [18], [17], [19], and [21] (see also [9], [13], [12], [14], [11], [5], [16], and [15] for other mathematical studies of quantum layers).

The main contribution of the present paper is to provide a lower bound to the ground-state energy of the bound states. However, our results are more general in the sense that this lower bound also applies to situations where the lowest energy in the spectrum does not correspond to a bound state, but rather to a scattering state; this happens, e.g., if the layer is periodically curved.

To obtain this lower bound, we follow an idea similar to that used by Pavel Exner and the present authors in [8] to derive a lower bound to the spectral threshold in *quantum tubes*, i.e. in the case of the configuration space being a d-dimensional tube about an infinite curve, with $d \geq 2$. More precisely, there it was shown that the lower bound is given by the lowest Dirichlet eigenvalue in a torus determined by the geometry of the tube. This lower bound is optimal in the sense that it is achieved by a tube (about a curve of constant curvature). However, the geometry of quantum layers is more complicated and we shall see that the optimality is one of the main features in which the present situation differs from that of quantum tubes.

In view of the above physical model, the Hamiltonian of a quantum layer can be identified with the Dirichlet Laplacian in a tubular neighbourhood of constant radius about a complete non-compact surface $\Sigma \subset \mathbb{R}^3$. In this paper, we proceed in a greater generality by considering compact surfaces, too. More precisely, we assume only that

$$\begin{aligned}&\Sigma \text{ is a connected complete orientable surface of class } C^2 \text{ embedded in } \mathbb{R}^3\\&\text{with bounded principal curvatures } k_1 \text{ and } k_2.\end{aligned} \tag{1}$$

Then, given a positive number a satisfying

$$a \max\{\|k_1\|_\infty, \|k_2\|_\infty\} < 1, \tag{2}$$

we introduce the tubular neighbourhood

$$\Omega := \{\boldsymbol{x} \in \mathbb{R}^3 \mid \operatorname{dist}(\boldsymbol{x}, \Sigma) < a\} \tag{3}$$

and denote by $-\Delta_D^\Omega$ the Dirichlet Laplacian in $L^2(\Omega)$. In addition to (2), we also assume that Ω "does not overlap itself" (cf. (7) below).

If Σ is compact, then Ω is bounded and a lower bound to the spectral threshold of the Laplacian follows by means of the Faber–Krahn inequality; i.e., $\inf \sigma(-\Delta_D^\Omega)$ is bounded from below by the lowest Dirichlet eigenvalue of the ball of volume $|\Omega|$ in this case. However, we are mainly interested in the unbounded case, where similar arguments based on the Faber–Krahn inequality may, at best, just provide a trivial bound and the location of $\inf \sigma(-\Delta_D^\Omega)$ becomes difficult, since we are actually dealing with a class of *quasi-cylindrical* domains (cf. §49 of [10] or Section X.6.1 of [7]). In this note we derive the following universal lower bound:

Theorem 1.1. *Let Ω be as above. One has*

$$\inf \sigma(-\Delta_D^\Omega) \geq \min\{\lambda_1(k_1^+, k_2^-), \lambda_1(k_1^-, k_2^+)\}, \tag{4}$$

where $k_i^\pm := \pm \sup(\pm k_i)$, $i \in \{1, 2\}$, and

$$\lambda_1(\kappa_1, \kappa_2) := \inf_{\psi \in W_0^{1,2}((-a,a))\setminus\{0\}} \frac{\displaystyle\int_{-a}^{a} |\psi'(u)|^2 \,(1-\kappa_1 u)\,(1-\kappa_2 u)\,\mathrm{d}u}{\displaystyle\int_{-a}^{a} |\psi(u)|^2 \,(1-\kappa_1 u)\,(1-\kappa_2 u)\,\mathrm{d}u} \tag{5}$$

for constants $\kappa_1, \kappa_2 \in [-1/a, 1/a]$.

In view of Theorem 1.1, the spectral threshold of the Dirichlet Laplacian in the three-dimensional tubular manifold Ω can be estimated from below by means of the one-dimensional spectral problem associated with (5). It is easy to verify that $\lambda_1(k_1, k_2)$ with constant k_1 and k_2 gives the spectral threshold of the Dirichlet Laplacian in the layer about the plane if $k_1 = k_2 = 0$, a sphere if $k_1 = k_2 > 0$ or a cylinder if $k_1 > 0$ and $k_2 = 0$. That is, Theorem 1.1 is optimal for the class of layers built about surfaces with non-negative Gauss curvature $k_1 k_2$. On the other hand, we are not aware of a geometric meaning of (5) if the Gauss curvature $k_1 k_2$ is negative and the surface is complete. In fact, since no such surface exists which satisfies hypothesis (1) and whose Gauss curvature is identically equal to a negative constant, a better lower bound than (4) is expected to hold for layers about surfaces with sign-changing or non-positive Gauss curvature.

In any case, while the right hand side of (4) diminishes as the Gauss curvature of Σ becomes more negative, it is uniformly bounded away from zero for layers about surfaces whose Gauss curvature is non-negative:

Proposition 1.2. *Let $\kappa_1, \kappa_2 \in (-1/a, 1/a)$ be such that $\kappa_1 \kappa_2 \geq 0$. Then*

$$\lambda_1(\kappa_1, \kappa_2) \geq \frac{j_{0,1}^2}{(2a)^2},$$

where $j_{0,1} \approx 2.40$ denotes the first zero of the Bessel function J_0.

The bound of Proposition 1.2 is reminiscent of the uniform lower bound obtained in [8] for strips, i.e. a two-dimensional analogy of quantum layers, by applying the Faber–Krahn inequality to a sequence of Dirichlet annuli converging to a Dirichlet disk.

If $\kappa_1 \kappa_2 < 0$, it actually turns out that it is impossible to obtain a lower bound to $\lambda_1(\kappa_1, \kappa_2)$ for all $\kappa_1, \kappa_2 \in (-1/a, 1/a)$ that would not depend on κ_1 and κ_2, as the following result shows:

Proposition 1.3. *We have*

$$\lambda_1\Big(-\frac{1}{a}, \frac{1}{a}\Big) = 0.$$

The rest of this paper consists of one section where we provide the proofs of Theorem 1.1 and Propositions 1.2 and 1.3.

2 The proofs

The central step in the proof of Theorem 1.1 is based on an idea adopted from [8]. Roughly speaking, expressing the Laplacian $-\Delta_D^\Omega$ in the natural coordinates parameterising the layer (3) by means of "longitudinal" coordinates of the reference surface Σ and a "transverse" coordinate of the normal bundle of Σ, we neglect the contribution of the former and the latter leads to a "variable" lower bound of the type (5). The constant lower bound given by the right hand side of (4) and the uniform lower bound of Proposition 1.2 then follow from an analysis of the one-dimensional spectral problem associated with (5).

We need to start with a detailed geometry of curved layers adopted from [3]. Let g be the Riemannian metric of Σ induced by the embedding. The orientation of Σ is specified by a globally defined unit normal vector field $n\colon \Sigma \to S^2$. For any point $x \in \Sigma$, we introduce the Weingarten map

$$L_x\colon T_x\Sigma \longrightarrow T_x\Sigma, \quad \xi \longmapsto -\mathrm{d}n_x(\xi).$$

The principal curvatures k_1 and k_2 at x are defined as eigenvalues of L_x with respect to $g(x)$. Although these curvatures are *a priori* defined only locally on Σ, the Gauss curvature $K := k_1 k_2$ and the mean curvature $M := 1/2(k_1+k_2)$ are globally defined continuous functions on Σ.

Let us introduce the mapping

$$\mathcal{L}\colon \Sigma \times (-a,a) \longrightarrow \mathbb{R}^3, \quad (x,u) \longmapsto x + n(x)u. \tag{6}$$

Assuming (2) and that

$$\mathcal{L} \text{ is injective,} \tag{7}$$

this mapping induces a diffeomorphism and the image $\mathcal{L}(\Sigma \times (-a,a))$ coincides with Ω defined by (3). In other words, Ω is a submanifold of $\mathbb{R}^3$ squeezed between two parallel surfaces at the distance a from Σ.

Using (6), we can identify Ω with the Riemannian manifold $\Sigma\times(-a,a)$ endowed with the metric G induced by $\mathcal{L}$. One has

$$G(x,u) = g(x) \circ (I_x - L_x u)^2 + \mathrm{d}u^2,$$

where I_x denotes the identity map on $T_x\Sigma$. By the definition of principal curvatures, it is easy to see that the measure on $\Omega \simeq (\mathbb{R} \times (-a,a), G)$ at a point (x,u) acquires the form

$$\mathrm{d}\Omega = (1 - k_1(x)\, u)(1 - k_2(x)u)\,\mathrm{d}\Sigma\,\mathrm{d}u,$$

where $\mathrm{d}\Sigma\,\mathrm{d}u$ stands for the product measure on $\Sigma \times (-a,a)$ at (x,u). Here $\mathrm{d}\Sigma = |g(x)|^{1/2}\mathrm{d}x^1\,\mathrm{d}x^2$ in a local coordinate system of Σ at x, with the usual notation $|g| := \det(g)$.

Let G^{ij} be the coefficients of the inverse of G in local coordinates (x,u) for $\Sigma \times (-a,a)$. Using the above identification, $-\Delta_D^\Omega$ is unitarily equivalent to the self-adjoint operator H associated with the quadratic form h defined in the Hilbert space $\mathcal{H} := L^2(\Sigma \times (-a,a), \mathrm{d}\Omega)$ by

$$h[\Psi] := \int_{\Sigma\times(-a,a)} \overline{(\partial_i \Psi(x,u))}\, G^{ij}(x,u)\, (\partial_j \Psi(x,u))\, \mathrm{d}\Omega,$$

$$\Psi \in \operatorname{Dom} h := W_0^{1,2}(\Sigma \times (-a,a), \mathrm{d}\Omega).$$

Here the Sobolev space $W_0^{1,2}(\Sigma\times(-a,a), \mathrm{d}\Omega)$ is defined as the completion of functions from $C_0^\infty(\Sigma\times(-a,a))$ with respect to the norm $(h[\cdot]+\|\cdot\|_{\mathcal{H}}^2)^{1/2}$. Consequently, to prove Theorem 1.1, it is equivalent to establish the lower bound (4) for the operator H.

Proof of Theorem 1.1. Let Ψ be any function defined in $C_0^\infty(\Sigma \times (-a,a))$, a dense subspace of $\operatorname{Dom} h$. Since $(G^{\mu\nu})_{\mu,\nu=1,2}$ is positive definite, one has

$$\begin{aligned} h[\Psi] &\geq \int_\Sigma \mathrm{d}\Sigma \int_{-a}^{a} \mathrm{d}u\, |\partial_u \Psi(x,u)|^2 (1-k_1(x)\,u)(1-k_2(x)u) \\ &\geq \int_\Sigma \mathrm{d}\Sigma\, \lambda_1(k_1(x),k_2(x)) \int_{-a}^{a} \mathrm{d}u\, |\Psi(x,u)|^2 (1-k_1(x)u)(1-k_2(x)u), \end{aligned}$$

where $\lambda_1(\kappa_1,\kappa_2)$ is defined by (5). It remains to show that

$$\lambda_1(k_1(x),k_2(x)) \geq \min\{\lambda_1(k_1^+,k_2^-), \lambda_1(k_1^-,k_2^+)\} \tag{8}$$

for all $x \in \Sigma$. Given constants $\kappa_1,\kappa_2 \in (-1/a, 1/a)$, the change of test function

$$\phi := \sqrt{(1-\kappa_1 u)(1-\kappa_2 u)}\, \psi$$

in (5) and an integration by parts yields

$$\lambda_1(\kappa_1,\kappa_2) = \inf_{\phi\in W_0^{1,2}((-a,a))\setminus\{0\}} \frac{\displaystyle\int_{-a}^{a} (|\phi'(u)|^2 + V(u;\kappa_1,\kappa_2)|\phi(u)|^2)\,\mathrm{d}u}{\displaystyle\int_{-a}^{a} |\phi(u)|^2\, \mathrm{d}u}, \tag{9}$$

where

$$V(u;\kappa_1,\kappa_2) := -\frac{1}{4}\, \frac{(\kappa_1-\kappa_2)^2}{(1-\kappa_1 u)^2(1-\kappa_2 u)^2}. \tag{10}$$

The constant lower bound (8) then follows by observing that

$$V(u; k_1(x), k_2(x)) \geq \min\{V(u; k_1^+, k_2^-), V(u; k_1^-, k_2^+)\}$$

for any fixed $u \in (-a, a)$ and all $x \in \Sigma$. The last inequality can be established for non-zero u's by writing

$$V(u; \kappa_1, \kappa_2) = -\frac{1}{4u^2}\Big(\frac{1}{1-\kappa_1 u} - \frac{1}{1-\kappa_2 u}\Big)^2 \tag{11}$$

and follows more easily for $u = 0$. □

Remark 2.1. Following Remark 1 in [3], since the hypothesis (2) is still enough to ensure that $(\Sigma \times (-a, a), G)$ is immersed in $\mathbb{R}^3$, we do not need to assume (7) in order to get (4) for the operator H.

Let us now derive the uniform lower bound of Proposition 1.2.

Proof of Proposition 1.2. In view of (5), without loss of generality we may assume that κ_1 and κ_2 are non-negative. By (9) with (11), we have

$$\lambda_1(\kappa_1, \kappa_2) \geq \min\{\lambda_1(\kappa_1, 0), \lambda_1(0, \kappa_2)\}.$$

However, $\lambda_1(\kappa, 0) = \lambda_1(0, \kappa)$ with $\kappa \in [0, 1/a)$ is the spectral threshold of the Dirichlet Laplacian in the strip of cross-section $(-a, a)$ built either over a circle of curvature κ if $\kappa \neq 0$ or over a straight line if $\kappa = 0$. With the help of the monotonicity properties established in Proposition 4.2 in [8] (or using again (9) with (11)), the Faber–Krahn inequality yields (cf. Proposition 4.5 in [8])

$$\lambda_1(0, \kappa) \geq \lambda_1(0, 1/a) = \frac{j_{0,1}^2}{(2a)^2}$$

for all $\kappa \in [0, 1/a)$. Notice that $\lambda_1(0, 1/a)$ is the lowest eigenvalue of the Dirichlet Laplacian in the disk of radius $2a$. □

Finally, we establish Proposition 1.3.

Proof of Proposition 1.3. For any positive number $\varepsilon < \min\{1, a\}$, let us set

$$\psi_\varepsilon(u) := \begin{cases} 1 & \text{if } |u| \leq a - \varepsilon, \\ -\dfrac{\log\left(\dfrac{a-u}{\varepsilon^2}\right)}{\log(\varepsilon)} & \text{if } a - \varepsilon \leq |u| \leq a - \varepsilon^2, \\ 0 & \text{if } a - \varepsilon^2 \leq |u|. \end{cases}$$

Then $\psi_\varepsilon \in W_0^{1,2}((-a,a))$ and using ψ_ε as a test function in the right hand side of (5) with $\kappa_1 := -1/a$ and $\kappa_2 := 1/a$, we obtain

$$\lambda_1\Big(-\frac{1}{a},\frac{1}{a}\Big) \le 2\,\frac{\displaystyle\int_0^a |\psi_\varepsilon'(u)|^2\Big(1-\frac{u}{a}\Big)\,\mathrm{d}u}{\displaystyle\int_0^a |\psi_\varepsilon(u)|^2\Big(1-\frac{u}{a}\Big)\,\mathrm{d}u},$$

where we used the bounds $1 \le 1 + u/a \le 2$. While the denominator converges to $\int_0^a (1-u/a)\,\mathrm{d}u = a/2$, an explicit computation shows that the numerator tends to zero as $\varepsilon \to 0$. □

Remark 2.2. Note that (9) yields a Hardy–Poincaré-type inequality

$$\int_{-a}^a |\phi'(u)|^2\,\mathrm{d}u \ge \lambda_1(\kappa_1,\kappa_2)\int_{-a}^a |\phi(u)|^2\,\mathrm{d}u + \int_{-a}^a |V(u;\kappa_1,\kappa_2)|\,|\phi(u)|^2\,\mathrm{d}u \quad (12)$$

for all $\phi \in W_0^{1,2}((-a,a))$ and all $\kappa_1,\kappa_2 \in [-1/a, 1/a]$, where the Hardy weight $V(\cdot\,;\kappa_1,\kappa_2)$ is given by (10) and the Poincaré constant $\lambda_1(\kappa_1,\kappa_2)$ interpolates between 0 and $\pi^2/(2a)^2$. An equivalent version of this inequality in weighted spaces follows from (5). If $\kappa_1 = \kappa_2 \ge 0$, then $V(\cdot\,;\kappa_1,\kappa_2)$ vanishes identically and $\lambda_1(\kappa_1,\kappa_2)$ equals $\pi^2/(2a)^2$, the first eigenvalue of the Dirichlet Laplacian in the interval $(-a,a)$. On the other hand, putting $\kappa_1 = 1/a$ and $\kappa_2 = -1/a$ in (12), Proposition (1.3) yields an optimal Hardy-type inequality

$$\int_{-a}^a |\phi'(u)|^2\,\mathrm{d}u \ge \int_{-a}^a \frac{a^2}{(a^2-u^2)^2}\,|\phi(u)|^2\,\mathrm{d}u$$

for all $\phi \in W_0^{1,2}((-a,a))$. We remark that this inequality is better than the well-known bound (see, e.g., [1])

$$\int_{-a}^a |\phi'(u)|^2\,\mathrm{d}u \ge \int_{-a}^a \frac{1}{4\,(a-|u|)^2}\,|\phi(u)|^2\,\mathrm{d}u$$

for all $\phi \in W_0^{1,2}((-a,a))$, which can be established by the classical Hardy inequality. Notice that the function $a - |\cdot|$ has the meaning of the distance to the boundary of the one-dimensional domain $(-a,a)$. Hardy inequalities with weights of type (10) have been recently considered for higher-dimensional domains in [4] (see also Lemma 8 in [2]).

Acknowledgements. The research of David Krejčiřík was supported by the project RVO61389005 and the GACR grant No. 14-06818S. Both authors were partially supported by FCT (Portugal) through project PTDC/MATCAL/4334/2014.

References

[1] H. Brezis and M. Marcus, *Hardy's inequalities revisited. Ann. Scuola Norm. Sup. Pisa Cl. Sci.* (4) **25** (1997), no. 1-2, 217–237. MR 1655516 Zbl 1011.46027

[2] R. Bosi, J. Dolbeault, and M. J. Esteban, Estimates for the optimal constants in multipolar Hardy inequalities for Schrödinger and Dirac operators. *Commun. Pure Appl. Anal.* **7** (2008), no. 3, 533–562. MR 2379440 Zbl 1162.35329

[3] G. Carron, P. Exner, and D Krejčiřík, Topologically nontrivial quantum layers. *J. Math. Phys.* **45** (2004), no. 2, 774–784. MR 2029097 Zbl 1070.58025

[4] C. Cazacu, New estimates for the Hardy constants of multipolar Schrödinger operators. *Commun. Contemp. Math.* **18** (2016), no. 5, 1550093, 28 pp. MR 3523190 Zbl 06617534

[5] M. Dauge, T. Ourmières-Bonafos, and N. Raymond, Spectral asymptotics of the Dirichlet Laplacian in a conical layer. *Commun. Pure Appl. Anal.* **14** (2015), no. 3, 1239–1258. MR 3320173 Zbl 06424917

[6] P. Duclos, P. Exner, and D. Krejčiřík, Bound states in curved quantum layers. *Comm. Math. Phys.* **223** (2001), no. 1, 13–28. MR 1860757 Zbl 0988.81034

[7] D. E. Edmunds and W. D. Evans, *Spectral theory and differential operators.* Oxford Mathematical Monographs. Oxford Science Publications. The Clarendon Press, Oxford University Press, New York, 1987. ISBN 0-19-853542-2 MR 0929030 Zbl 0628.47017

[8] P. Exner, P. Freitas, and D. Krejčiřík, A lower bound to the spectral threshold in curved tubes. *Proc. R. Soc. Lond. Ser. A Math. Phys. Eng. Sci.* **460** (2004), no. 2052, 3457–3467. MR 2166298 Zbl 1330.35360

[9] P. Exner and M. Tater, Spectrum of Dirichlet Laplacian in a conical layer. *J. Phys. A* **43** (2010), no. 47, 474023, 11 pp. MR 2738118 Zbl 1204.81064

[10] I. M. Glazman, *Direct methods of qualitative spectral analysis of singular differential operators.* Translated from the Russian by the IPST staff. Israel Program for Scientific Translations, Jerusalem, 1965, and Daniel Davey & Co., Inc., New York 1966. MR 0190800 Zbl 0143.36505

[11] S. Haag, J. Lampart, and S. Teufel, Generalised quantum waveguides. *Ann. Henri Poincaré* **16** (2015), no. 11, 2535–2568. MR 3411741 Zbl 1327.81213

[12] D. Krejčiřík, Spectrum of the Laplacian in narrow tubular neighbourhoods of hypersurfaces with combined Dirichlet and Neumann boundary conditions. *Math. Bohem.* **139** (2014), no. 2, 185–193. MR 3238833 Zbl 1340.35239

[13] D. Krejčiřík and Z. Lu, Location of the essential spectrum in curved quantum layers. *J. Math. Phys.* **55** (2014), no. 8, 083520, 13 pp. MR 3390760 Zbl 1301.82073

[14] D. Krejčiřík, N. Raymond, and M. Tušek, The magnetic Laplacian in shrinking tubular neighborhoods of hypersurfaces. *J. Geom. Anal.* **25** (2015), no. 4, 2546–2564. MR 3427137 Zbl 1337.35042

[15] D. Krejčiřík and M. Tušek, Nodal sets of thin curved layers. *J. Differential Equations* **258** (2015), no. 2, 281–301. MR 3274759 Zbl 1323.35109

[16] J. Lampart, Convergence of nodal sets in the adiabatic limit. *Ann. Global Anal. Geom.* **47** (2015), no. 2, 147–166. MR 3313138 Zbl 1311.58017

[17] Ch. Lin and Z. Lu, On the discrete spectrum of generalized quantum tubes. *Comm. Partial Differential Equations* **31** (2006), no. 10-12, 1529–1546. MR 2273964 Zbl 1107.53014

[18] Ch. Lin and Z. Lu, Existence of bound states for layers built over hypersurfaces in $\mathbb{R}^{n+1}$. *J. Funct. Anal.* **244** (2007), no. 1, 1–25. MR 2294473 Zbl 1111.81059

[19] Ch. Lin and Z. Lu, Quantum layers over surfaces ruled outside a compact set. *J. Math. Phys.* **48** (2007), no. 5, 053522, 14 pp. MR 2329884 Zbl 1144.81376

[20] J. T. Londergan, J. P. Carini, and D. P. Murdock, *Binding and scattering in two-dimensional systems.* Lecture Notes in Physics. New Series m: Monographs. m60. Springer, Berlin, 1999. Zbl 0997.81511

[21] Z. Lu and J. Rowlett, On the discrete spectrum of quantum layers. *J. Math. Phys.* **53** (2012), no. 7, 073519, 22 pp. MR 2985259 Zbl 1278.82081

To the spectral theory of vector-valued Sturm–Liouville operators with summable potentials and point interactions

Yaroslav Granovskyi, Mark Malamud, Hagen Neidhardt, and Andrea Posilicano

Dedicated with great pleasure to Pavel Exner on the occasion of his 70th birthday

1 Introduction

Differential operators with point interactions arise in various physical applications as exactly solvable models that describe complicated physical phenomena (numerous results as well as a comprehensive list of references may be found in [4] and its appendix [17] and in [6]). An important class of such operators is formed by the differential operators with the coefficients having singular support on a disjoint set of points. The most known example is the operator $H_{X,\alpha,q}$ associated with the formal differential expression

$$\ell_{X,\alpha,q} := -\frac{\mathrm{d}^2}{\mathrm{d}x^2} + q(x) + \sum_{x_n \in X} \alpha_n \delta(x - x_n). \tag{1}$$

This operator describes a δ-interaction on a discrete set $X = \{x_n\}_{n \in I} \subset \mathbb{R}$, and the coefficients α_n are called the *strengths* of the interaction at the point $x = x_n$. Investigation of this model was originated by Kronig and Penney [30] and Grossmann *et. al.* [24] (see also [20]). In particular, the "Kronig–Penney model" ($\ell_{X,\alpha,q}$ with $X = \mathbb{Z}$, $\alpha_n \equiv \alpha$, and $q \equiv 0$) provides a simple model for a non-relativistic electron moving in a fixed crystal lattice.

There are several ways to associate an operator with the expression $\ell_{X,\alpha,q}$. In the following we will treat Hamiltonian (1) in the framework of extension theory of symmetric operators.

The minimal symmetric operator $H_{X,\alpha,q}$ is naturally associated with (1) in $L^2(\mathbb{R}_+)$. Namely, define the operator $H^0_{X,\alpha,q}$ by the differential expression

$$\ell_q := -\frac{\mathrm{d}^2}{\mathrm{d}x^2} + q(x), \quad x \in \mathbb{R}_+ = (0,\infty), \tag{2}$$

on the domain

$$\operatorname{dom}(H^0_{X,\alpha,q}) = \left\{ \begin{array}{c} f \in W^{2,2}_{\mathrm{comp}}(\mathbb{R}_+ \setminus X), \\ n \in \mathbb{Z}_+ \end{array} \middle| \begin{array}{c} f(0) = 0,\ f(x_n+) = f(x_n-), \\ f'(x_n+) - f'(x_n-) = \alpha_n f(x_n) \end{array} \right\}. \tag{3}$$

Clearly, $H^0_{X,\alpha,q}$ is symmetric and hence admits a closure $H_{X,\alpha,q}$. In general, the operator $H_{X,\alpha,q}$ is symmetric but not automatically self-adjoint, even in the case $q \equiv 0$. Note that the Hamiltonian $H_q := H_{X;0,q}$ with $\alpha = \{\alpha_n\}_{n\in\mathbb{N}} = 0$ is identified with the Dirichlet realization of the expression (2) in $L^2(\mathbb{R}_+)$. We set

$$d_* := \inf_n d_n \quad \text{and} \quad d^* := \sup_n d_n, \qquad d_n := x_n - x_{n-1},\ x_0 := 0.$$

Numerous works are devoted to the spectral analysis of the operators $H_{X,\alpha,q}$, cf. monographs [4] and [6] and the review papers [17], [10], [14], and [27]. Spectral analysis of an operator means the characterization of continuous, absolute continuous and singular spectrum. In the following we are interested in the spectral analysis of Hamiltonians of type $H_q = H_{X,\alpha,q}$ for the vector-valued case. For the scalar case there are only a few results in this direction known in the literature. Let us recall them.

Theorem 1.1 ([5] and [26]). *Let $q(\cdot) = \overline{q(\cdot)} \in L^\infty(\mathbb{R}_+)$ and let $d^* < \infty$. Then $\sigma_{\mathrm{ac}}(H_{X,\alpha,q}) = \sigma_{\mathrm{ac}}(H_q)$ provided that*

$$\sum_{n=1}^{\infty} \frac{|\alpha_n|}{d_{n+1}} < \infty. \tag{4}$$

If in addition, $q \in L^1(\mathbb{R}_+)$, then $\sigma_{\mathrm{ac}}(H_{X,\alpha,q}) = [0,\infty)$.

For $d_* > 0$ the result was established earlier by Mikhaĭlets [35]. In this case condition (4) turns into $\sum_{n=1}^{\infty} |\alpha_n| < \infty$. The proof in [5] is based on the boundary triplet approach to the extensions. Namely, it was shown that the resolvent difference $(H_{X,\alpha,q} - i)^{-1} - (H_{X,0,q} - i)^{-1}$ is a trace class operator. Then the result is implied by the Birman– Kreĭn theorem generalizing the classical Kato–Rosenblum result (see Theorem 16.1 in [7] and Section 99 of [3]).

However Theorem 1.1 does not ensure absence of a singular part $\sigma_s(H_{X,\alpha,q})$ of the spectrum $\sigma(H_{X,\alpha,q})$. It is well known that this problem requires a special analysis which cannot be extracted from the Kato–Rosenblum theorem. Moreover, to the best of our knowledge the pure absolute continuity of $H_{X,\alpha,q}$ was established only in a few cases, for instance by Shubin Christ and Stolz [37].

Theorem 1.2 ([37]). *Let $X = \mathbb{Z}$, i.e., $X = \{n\}_{n\in\mathbb{Z}}$. Then the following holds*:

(i) *if $q(x) = 0$ and either $\sum_{n=-\infty}^{0} |\alpha_n| < \infty$ or $\sum_{n=1}^{\infty} |\alpha_n| < \infty$, then the positive part $E_{H_{\mathbb{Z},\alpha,0}}(\mathbb{R}_+)H_{\mathbb{Z},\alpha,0}$ of $H_{\mathbb{Z},\alpha,0}$ is purely absolutely continuous, i.e.,*

$$\sigma(H_{\mathbb{Z},\alpha,0}) \cap [0,+\infty) = \sigma_{\mathrm{ac}}(H_{\mathbb{Z},\alpha,0}) = [0,+\infty)$$

and

$$\sigma_s(H_{\mathbb{Z},\alpha,0}) \cap \mathbb{R}_+ = \emptyset;$$

(ii) *if $q(x) = x$ and $\sum_{n=-\infty}^{-1} |\alpha_n|/\sqrt{|n|} < \infty$, then*

$$\sigma(H_{\mathbb{Z},\alpha,x}) = \sigma_{\mathrm{ac}}(H_{\mathbb{Z},\alpha,x}) = \mathbb{R} \quad \textit{and} \quad \sigma_s(H_{\mathbb{Z},\alpha,x}) = \emptyset.$$

To prove this result the authors generalized the method of subordinacy originated and developed by D. Gilbert and D. Pearson [21] and [22] (see also [38]).

The main object of our paper is the vector-valued Sturm–Liouville differential expression with a summable matrix potential $Q(\cdot) = Q^*(\cdot) \in L^1(\mathbb{R}_+, \mathbb{C}^{m\times m})$ and a finite number of point interactions

$$\mathcal{L}_{X,\alpha,Q} := -\frac{\mathrm{d}^2}{\mathrm{d}x^2} + Q(x) + \sum_{x_n\in X} \alpha_n\delta(x - x_n), \quad x \in \mathbb{R}_+ = (0,\infty). \tag{5}$$

Here $X = \{x_n\}_{n=1}^{p} \subset \mathbb{R}_+$ is a finite strictly increasing sequence, $x_{n+1} > x_n$, $n \in \{1,\dots,p\}$, $p < \infty$, and $\alpha = \{\alpha_n\}_1^p \subset \mathbb{C}^{m\times m}$, $\alpha_n = \alpha_n^*$.

The Hamiltonian $H_{X,\alpha,Q}$ associated in $L^2(\mathbb{R}_+,\mathbb{C}^m)$ with the formal expression (5) is given by formula (3) with q replaced by Q and $\alpha = \{\alpha_n\}_1^p$ being a sequence of self-adjoint $m \times m$ matrices. Note that the Hamiltonian $H_{X,0,Q}$ with $\alpha_n = 0, n \in \mathbb{N}$, is identified with the Dirichlet realization H_Q of the expression

$$\mathcal{L}_Q := -\frac{d^2}{dx^2} + Q(x), \quad x \in \mathbb{R}_+,$$

considered in $L^2(\mathbb{R}_+, \mathbb{C}^m)$. The main result of the paper consists in a complete spectral analysis of the Hamiltonian $H_{X,\alpha,Q}$ and reads as follows.

Theorem 1.3. *Let $Q = Q^* \in L^1(\mathbb{R}_+, \mathbb{C}^{m\times m})$ and let $H_{X,\alpha,Q}$ be the Hamiltonian associated with* (5)*. Then the positive part $E_{H_{X,\alpha,Q}}(\mathbb{R}_+)H_{X,\alpha,Q}$ of $H_{X,\alpha,Q}$ is unitarily equivalent to the positive part $E_{H_Q}(\mathbb{R}_+)H_Q$ of the Dirichlet realization $H_Q = H_{X,0,Q}$. In particular, the spectrum of $E_{H_{X,\alpha,Q}}(\mathbb{R}_+)H_{X,\alpha,Q}$ is purely absolutely continuous and of constant spectral multiplicity m.*

Moreover, the Hamiltonian $H_{X,\alpha,Q}$ is semibounded below and its negative spectrum is either finite or forms a sequence tending to zero.

This result seems to be new even in the scalar case ($m = 1$) and supplements Theorems 1.1 and Theorem 1.2(i) in the case of finitely many point interactions.

At first we establish this result for the Sturm–Liouville operator H_Q without point interactions, i.e., assuming that $\alpha_n = 0$ for $n \in \{1, \dots, p\}$ in (5) (see Theorem 3.8). In this case Theorem 1.3 ensures the absolute continuity of the positive part of the Dirichlet realization $H_Q := H_{X,0,Q}$ and discreteness of its negative part. Moreover, we show that similar statements are valid for *any realization* of the expression H_Q in $L^2(\mathbb{R}_+, \mathbb{C}^m)$ (see Corollary 3.9).

The paper is organized as follows. In Section 2 we present necessary information on boundary triplets and the corresponding Weyl functions. We also present necessary facts from [9], [32], and [33] on description of absolutely continuous and singular spectra of extensions by means of the limit behavior of the Weyl function near the real axis.

In Section 3 we prove Theorem 3.8 and Corollary 3.9. This result generalizes the classical Titchmarsh's result (see Chapter 5 in [39]) to the case of Sturm–Liouville operator with a matrix-valued summable potential and coincides with it in the scalar case ($m = 1$). Emphasize that the main ingredient of these results is the absolute continuity of the positive part of *any realization* of $\mathcal{L}_Q = -\mathrm{d}^2/\mathrm{d}x^2 + Q(x)$ (any extension of the minimal operator associated with the expression $\mathcal{L}_Q$).

In Section 4 we prove Theorem 1.3 for the Hamiltonian $H_{X,\alpha,Q}$. Our proof substantially relies on two main ingredients:

(i) the form of the Weyl function and its limit values at the real axis obtained in the proof of Theorem 3.8(i);

(ii) on the Weyl function technique elaborated in [9], [32], and [33] and presented in Section 2.

Besides using the Weyl function technique we show that *any self-adjoint extension $\widetilde{H}$ of the minimal operator $H^{\min}_{X,\alpha,Q}$ associated with expression* (5) (*see definition* (63)) *has no singular continuous spectrum*, $\sigma_{\mathrm{sc}}(\widetilde{H}) = \emptyset$. Notice that statements of such type and even stronger ones like

$$\sigma_{\mathrm{sc}}(\widetilde{H}) \cap \mathbb{R}_+ = \sigma_p(\widetilde{H}) \cap \mathbb{R}_+ = \emptyset \tag{6}$$

for *any self-adjoint extension* $\widetilde{H}$ of a certain minimal symmetric differential operators with infinite deficiency indices has already been discovered earlier. For instance, the property (6), i.e., the fact that the positive part remains purely absolutely continuous for any extension, was discovered for Sturm–Liouville operators $-\mathrm{d}^2/\mathrm{d}x^2 + T$ with bounded or unbounded operator potentials T (see [33]) as well as for Schrödinger operators in $\mathbb{R}^3$ with infinitely many point interactions $X = \{x_n\}$, which form a sparse sequence [34].

However as distinct from (6) there exist extensions $\widetilde{H}$ of $H^{\min}_{X,\alpha,Q}$ having positive eigenvalues embedded in $\sigma_{\mathrm{ac}}(\widetilde{H}) = [0, \infty)$ (see Remark 4.7).

In connection with our investigation we mention the papers [18] and [19] by Exner and Fraas devoted to investigation of the ac-spectrum of Schrödinger operators in $L^2(\mathbb{R}^n)$, $n \geq 2$, with a singular interaction supported by an infinite family of concentric shells,

$$H_{R,\alpha} = -\Delta + \sum_{k=1}^{\infty} \alpha_k \delta(|x| - r_k), \quad \alpha = \{\alpha_k\}_{k=1}^{\infty} \subset \mathbb{R}.$$

Motivated by the paper of Hempel, Hinz, and Kalf [25], Exner and Fraas in [18] and [19] obtained a complete characterization of the spectrum of the Hamiltonian $H_{R,\alpha}$ with radially periodic interactions: $\alpha_k \equiv \alpha$ and $r_k = r_0 + Tk$.

It is our pleasure to dedicate the paper to our friend Pavel Exner on the occasion of his 70th birthday. He realizes very early the importance of point interactions in quantum mechanics and has contributed a lot to the spectral theory of Schrödinger operators with such interactions.

2 Preliminaries

2.1 Boundary triplets and Weyl functions

Let us recall some basic facts of the theory of abstract boundary triplets and the corresponding Weyl functions, cf. [15], [16], and [23].

The set $\widetilde{\mathcal{C}}(\mathcal{H})$ of closed linear relations in $\mathcal{H}$ is the set of closed linear subspaces of $\mathcal{H} \oplus \mathcal{H}$. Recall that $\mathrm{dom}(\Theta) = \{f \mid \{f, f'\} \in \Theta\}$, $\mathrm{ran}(\Theta) = \{f' \mid \{f, f'\} \in \Theta\}$, and $\mathrm{mul}(\Theta) = \{f' \mid \{0, f'\} \in \Theta\}$ are the domain, the range, and the multivalued part of Θ. A closed linear operator A in $\mathcal{H}$ is identified with its graph $\mathrm{gr}(A)$, so that the set $\mathcal{C}(\mathcal{H})$ of closed linear operators in $\mathcal{H}$ is viewed as a subset of $\widetilde{\mathcal{C}}(\mathcal{H})$. In particular, a linear relation Θ is an operator if and only if $\mathrm{mul}(\Theta)$ is trivial.

We recall that the adjoint relation $\Theta^* \in \widetilde{\mathcal{C}}(\mathcal{H})$ of $\Theta \in \widetilde{\mathcal{C}}(\mathcal{H})$ is defined by

$$\Theta^* = \left\{ \begin{pmatrix} h \\ h' \end{pmatrix} \,\middle|\, (f', h)_{\mathcal{H}} = (f, h')_{\mathcal{H}} \text{ for all } \begin{pmatrix} f \\ f' \end{pmatrix} \in \Theta \right\}.$$

A linear relation Θ is said to be *symmetric* if $\Theta \subset \Theta^*$ and self-adjoint if $\Theta = \Theta^*$.

For a symmetric linear relation $\Theta \subseteq \Theta^*$ in $\mathcal{H}$ the multivalued part $\mathrm{mul}(\Theta)$ is the orthogonal complement of $\mathrm{dom}(\Theta)$ in $\mathcal{H}$. Therefore setting

$$\mathcal{H}_{\mathrm{op}} := \overline{\mathrm{dom}(\Theta)}$$

and $\mathcal{H}_\infty = \mathrm{mul}(\Theta)$, one arrives at the orthogonal decomposition $\Theta = \Theta_{\mathrm{op}} \oplus \Theta_\infty$ where Θ_{op} is a symmetric operator in $\mathcal{H}_{\mathrm{op}}$, the operator part of Θ, and $\Theta_\infty = \{\left(\begin{smallmatrix} 0 \\ f' \end{smallmatrix}\right) \mid f' \in \mathrm{mul}(\Theta)\}$, a "pure" linear relation in $\mathcal{H}_\infty$.

Let A be a densely defined closed symmetric operator in a separable Hilbert space $\mathfrak{H}$ with equal deficiency indices $\mathrm{n}_\pm(A) = \dim(\mathfrak{N}_{\pm\mathrm{i}}) \leq \infty$, where $\mathfrak{N}_z := \ker(A^* - z)$ is the defect subspace.

Definition 2.1 ([23]). A triplet $\Pi = \{\mathcal{H}, \Gamma_0, \Gamma_1\}$ is called a *boundary triplet* for the adjoint operator A^* if $\mathcal{H}$ is an auxiliary Hilbert space and $\Gamma_0, \Gamma_1 \colon \mathrm{dom}(A^*) \to \mathcal{H}$ are linear mappings such that the abstract Green identity

$$(A^* f, g)_{\mathfrak{H}} - (f, A^* g)_{\mathfrak{H}} = (\Gamma_1 f, \Gamma_0 g)_{\mathcal{H}} - (\Gamma_0 f, \Gamma_1 g)_{\mathcal{H}}, \quad f, g \in \mathrm{dom}(A^*),$$

holds and the mapping

$$\Gamma := \begin{pmatrix} \Gamma_0 \\ \Gamma_1 \end{pmatrix} : \mathrm{dom}(A^*) \longrightarrow \mathcal{H} \oplus \mathcal{H}$$

is surjective.

First, note that a boundary triplet for A^* exists whenever the deficiency indices of A are equal, $\mathrm{n}_+(A) = \mathrm{n}_-(A)$. Moreover, $\mathrm{n}_\pm(A) = \dim(\mathcal{H})$ and

$$\ker(\Gamma) = \ker(\Gamma_0) \cap \ker(\Gamma_1) = \mathrm{dom}(A).$$

Note also that Γ is a bounded mapping from $\mathfrak{H}_+ = \mathrm{dom}(A^*)$ equipped with the graph norm to $\mathcal{H} \oplus \mathcal{H}$.

A boundary triplet for A^* is not unique. Moreover, for any self-adjoint extension $\widetilde{A} := \widetilde{A}^*$ of A there exists a boundary triplet $\Pi = \{\mathcal{H}, \Gamma_0, \Gamma_1\}$ such that $\ker(\Gamma_0) = \mathrm{dom}(\widetilde{A})$.

Definition 2.2. (i) A closed extension A' of A is called a *proper extension* if

$$A \subset A' \subset A^*.$$

The set of all proper extensions of A completed by the (non-proper) extensions A and A^* is denoted by Ext_A.

(ii) Two proper extensions A', A'', of A are called *disjoint* if

$$\mathrm{dom}(A') \cap \mathrm{dom}(A'') = \mathrm{dom}(A)$$

and *transversal* if in addition

$$\mathrm{dom}(A') + \mathrm{dom}(A'') = \mathrm{dom}(A^*).$$

Any self-adjoint extension $\widetilde{A}$ of A is proper, i.e., $\widetilde{A} \in \mathrm{Ext}_A$. By fixing a boundary triplet Π one can parametrize the set Ext_A in the following way.

Proposition 2.3 ([16]). *Let A be as above and let $\Pi = \{\mathcal{H}, \Gamma_0, \Gamma_1\}$ be a boundary triplet for A^*. Then the mapping*

$$\mathrm{Ext}_A \ni \widetilde{A} \longrightarrow \Gamma\,\mathrm{dom}(\widetilde{A}) = \{\{\Gamma_0 f, \Gamma_1 f\} \mid f \in \mathrm{dom}(\widetilde{A})\} =: \Theta \in \widetilde{\mathcal{C}}(\mathcal{H}) \tag{7}$$

establishes a bijective correspondence between the sets Ext_A *and* $\widetilde{\mathcal{C}}(\mathcal{H})$. *We put $A_\Theta := \widetilde{A}$ where Θ is defined by (7), i.e.,*

$$A_\Theta := A^* \upharpoonright \Gamma^{-1}\Theta = A^* \upharpoonright \{f \in \mathrm{dom}(A^*) \mid \{\Gamma_0 f, \Gamma_1 f\} \in \Theta\}.$$

(i) *A_Θ is symmetric if and only if Θ is symmetric, i.e., $\Theta \subseteq \Theta^*$. In particular, A_Θ is self-adjoint if and only if Θ is self-adjoint, i.e., $\Theta = \Theta^*$. Moreover,* $\mathrm{n}_\pm(A_\Theta) = \mathrm{n}_\pm(\Theta)$.

(ii) *The extensions A_Θ and A_0 are disjoint (transversal) if and only if Θ is an operator. In this case A_Θ is given by*

$$A_\Theta = A^* \upharpoonright \ker(\Gamma_1 - \Theta\Gamma_0). \tag{8}$$

Moreover, the extensions A_Θ and A_0 are transversal if and only if $\Theta \in \mathcal{B}(\mathcal{H})$.

The linear relation Θ (the operator B) in the correspondence (7) (resp. (8)) is called *the boundary relation* (*the boundary operator*). We emphasize that for differential operators in contrast to the von Neumann extension theory the parametrization (7)–(8) describes the set of proper extensions directly in terms of boundary conditions.

It follows immediately from Proposition 2.3 that the extensions

$$A_0 := A^* \upharpoonright \ker(\Gamma_0) \quad \text{and} \quad A_1 := A^* \upharpoonright \ker(\Gamma_1)$$

are self-adjoint. Clearly, $A_j = A_{\Theta_j}$, $j \in \{0, 1\}$, where the subspaces $\Theta_0 := \{0\} \times \mathcal{H}$ and $\Theta_1 := \mathcal{H} \times \{0\}$ are self-adjoint relations in $\mathcal{H}$. Note that Θ_0 is a "pure" linear relation.

In [15] and [16] the concept of the classical Weyl–Titchmarsh m-function from the theory of Sturm–Liouville operators was generalized to the case of symmetric operators with equal deficiency indices. The role of abstract Weyl functions in the extension theory is similar to that of the classical Weyl–Titchmarsh m-function in the spectral theory of singular Sturm–Liouville operators.

Definition 2.4 ([15]). Let A be a densely defined closed symmetric operator in $\mathfrak{H}$ with equal deficiency indices and let $\Pi = \{\mathcal{H}, \Gamma_0, \Gamma_1\}$ be a boundary triplet for A^*. The operator valued functions $\gamma(\cdot)\colon \rho(A_0) \to \mathcal{B}(\mathcal{H}, \mathfrak{H})$ and $M(\cdot)\colon \rho(A_0) \to \mathcal{B}(\mathcal{H})$ defined by

$$\gamma(z) := (\Gamma_0 \upharpoonright \mathfrak{N}_z)^{-1} \quad \text{and} \quad M(z) := \Gamma_1 \gamma(z), \quad z \in \rho(A_0), \tag{9}$$

are called the *γ-field* and the *Weyl function,* respectively, corresponding to the boundary triplet Π.

The γ-field $\gamma(\cdot)$ and the Weyl function $M(\cdot)$ in (9) are well defined. Moreover, both $\gamma(\cdot)$ and $M(\cdot)$ are holomorphic on $\rho(A_0)$ and the following relations hold (see [15])

$$\gamma(z) = (I + (z - \zeta)(A_0 - z)^{-1})\gamma(\zeta) \quad z, \zeta \in \rho(A_0), \tag{10}$$

$$M(z) - M(\zeta)^* = (z - \bar{\zeta})\gamma(\zeta)^*\gamma(z), \quad z, \zeta \in \rho(A_0). \tag{11}$$

Identities (10) and (11) mean that $\gamma(\cdot)$ and $M(\cdot)$ are the γ-field and the Q-function of the operator A_0, respectively, in the sense of M. Kreĭn (see [28]). It follows from (11) that $M(\cdot)$ is an $R[\mathcal{H}]$-function (or *Nevanlinna function*), i.e., $M(\cdot)$ is an ($\mathcal{B}(\mathcal{H})$-valued) holomorphic function on $\mathbb{C} \setminus \mathbb{R}$ satisfying

$$\operatorname{Im} z \cdot \operatorname{Im} M(z) \geq 0, \quad M(z)^* = M(\bar{z}), \quad z \in \mathbb{C} \setminus \mathbb{R}.$$

Moreover, due to (11), $M(\cdot) \in R^u[\mathcal{H}]$, i.e., it satisfies $0 \in \rho(\operatorname{Im} M(i))$.

It is well known that $M(\cdot)$ admits an integral representation (see, for instance, [2] and [3])

$$M(z) = C_0 + \int_{\mathbb{R}} \Big(\frac{1}{t - z} - \frac{t}{1 + t^2}\Big) \mathrm{d}\Sigma_M(t), \quad z \in \rho(A_0), \tag{12}$$

where $\Sigma_M(\cdot)$ is an operator-valued Borel measure on $\mathbb{R}$ satisfying the conditions $\int_{\mathbb{R}} 1/(1+t^2)\, d\Sigma_M(t) \in \mathcal{B}(\mathcal{H})$ and $C_0 = C_0^* \in \mathcal{B}(\mathcal{H})$. The integral in (12) is understood in the strong sense. Note that a linear term $C_1 z$ is missing in (12) since A is densely defined (see [15]).

Proposition 2.5. *Let $\Pi = \{\mathcal{H}, \Gamma_0, \Gamma_1\}$ and $\widetilde{\Pi} = \{\mathcal{H}, \widetilde{\Gamma}_0, \widetilde{\Gamma}_1\}$ be two boundary triplets for the operator A^*, let $M(\cdot)$ and $\widetilde{M}(\cdot)$ be the corresponding Weyl functions, and let $\widetilde{A}_0 := A^* \upharpoonright \ker(\widetilde{\Gamma}_0)$ and*

$$J := i \begin{pmatrix} 0 & -I_{\mathcal{H}} \\ I_{\mathcal{H}} & 0 \end{pmatrix}.$$

Then the following holds:

(i) *there is $J_{\mathcal{H}}$-unitary operator $X = (X_{ij})_{i,j=1}^2 \in \mathcal{B}(\mathcal{H} \oplus \mathcal{H})$, i.e., $X^* J X = J$, such that*

$$\begin{pmatrix} X_{11} & X_{12} \\ X_{21} & X_{22} \end{pmatrix} \begin{pmatrix} \Gamma_1 \\ \Gamma_0 \end{pmatrix} = \begin{pmatrix} \widetilde{\Gamma}_1 \\ \widetilde{\Gamma}_0 \end{pmatrix}; \tag{13}$$

(ii) *$0 \in \rho(X_{21} M(z) + X_{22})$ for $z \in \rho(\widetilde{A}_0)$. The Weyl functions $\widetilde{M}(\cdot)$ and $M(\cdot)$ are related by means of the linear fractional transformation*

$$\widetilde{M}(z) = X(M(z)) := (X_{11} M(z) + X_{12})(X_{21} M(z) + X_{22})^{-1}, \quad z \in \rho(\widetilde{A}_0).$$

Recall that a symmetric operator A in $\mathfrak{H}$ is said to be *simple* if there is no nontrivial subspace which reduces it to a self-adjoint operator. In other words, A is simple if it does not admit an (orthogonal) decomposition $A = A' \oplus S$ where A' is a symmetric operator and S is a self-adjoint operator acting on a nontrivial Hilbert space. It is easily seen (and well-known) that A is simple if and only if the closed linear span of $\{\mathfrak{N}_z(A) \mid z \in \mathbb{C} \setminus \mathbb{R}\}$ coincides with $\mathfrak{H}$.

If A is simple, then the Weyl function $M(\cdot)$ determines the boundary triplet Π uniquely up to unitary equivalence (see [15]). In particular, $M(\cdot)$ contains the full information about the spectral properties of A_0. Moreover, the spectrum of a proper (not necessarily self-adjoint) extension $A_\Theta \in \mathrm{Ext}_A$ can be described by means of $M(\cdot)$ and the boundary relation Θ.

Proposition 2.6 (Theorem 2.2 in [15]). *Let $\Pi = \{\mathcal{H}, \Gamma_0, \Gamma_1\}$ be a boundary triplet for A^* and let $M(\cdot)$ and $\gamma(\cdot)$ be the corresponding Weyl function and the γ-field. Then for any $\widetilde{A} = A_\Theta \in \mathrm{Ext}_A$ with $\rho(A_\Theta) \neq \emptyset$ the following Kreĭn type formula holds:*

$$(A_\Theta - z)^{-1} - (A_0 - z)^{-1} = \gamma(z)(\Theta - M(z))^{-1} \gamma^*(\bar{z}), \quad z \in \rho(A_0) \cap \rho(A_\Theta). \tag{14}$$

Moreover, if A is simple, then for any $z \in \rho(A_0)$

$$z \in \sigma_j(A_\Theta) \iff 0 \in \sigma_j(\Theta - M(z)), \quad j \in \{\mathrm{pp}, \mathrm{c}\}.$$

Formula (14) is a generalization of the known Kreĭn formula for canonical resolvents (cf. [3] and [28]). It establishes a one–to–one correspondence between the set of proper extensions $\widetilde{A} = A_\Theta$ with non-empty resolvent set and the set of linear relations Θ in $\mathcal{H}$. Note also that all quantities which enter into (14) are expressed in terms of the boundary triplet Π (see formulas (8) and (9)) (cf. [15] and [16]).

2.2 Weyl function and spectrum

In the following we are going to characterize the spectrum of the extension A_0 in terms for the Weyl function. To this end let $\Phi(\cdot)$ be a scalar Nevanlinna function. Let $\Phi(z)$ a holomorphic function in $\mathbb{C}_+$. In the following by $\lim_{z \to\!\!\!> x} \Phi(z)$ we mean that the limit $\lim_{r \downarrow 0} \Phi(x + re^{i\theta})$, $x \in \mathbb{R}$, exist uniformly in $\theta \in [\varepsilon, \pi - \varepsilon]$ for each $\varepsilon \in (0, \pi/2)$. Let us introduce the sets

$$
\begin{aligned}
\Omega_{\mathrm{s}}(\Phi) &:= \{x \in \mathbb{R} \mid |\Phi(z)| \to +\infty \text{ as } z \to\!\!\!> x\},\\
\Omega_{\mathrm{pp}}(\Phi) &:= \{x \in \mathbb{R} \mid \lim_{z \to\!\!\!> x} (z - x)\Phi(z) \neq 0\},\\
\Omega_{\mathrm{sc}}(\Phi) &:= \{x \in \mathbb{R} \mid |\Phi(z)| \to +\infty \text{ and } (z - x)\Phi(z) \to 0 \text{ as } z \to\!\!\!> x\}\\
\Omega_{\mathrm{ac}}(\Phi) &:= \{x \in \mathbb{R} \mid 0 < \operatorname{Im} \Phi(x + i0) < +\infty\}, \quad \Phi(x + i0) = \lim_{y \downarrow 0} \Phi(x + iy).
\end{aligned}
$$

Any scalar R-function $\Phi(\cdot)$ admits the representation

$$
\Phi(z) = C_0 + C_1 z + \int_{\mathbb{R}} \Big(\frac{1}{t - z} - \frac{t}{1 + t^2}\Big)\, \mathrm{d}\mu(t), \quad z \in \mathbb{C}_+, \tag{15}
$$

where $C_0, C_1 \in \mathbb{R}$, $C_1 \geq 0$ and the Borel measure $\mu(\cdot)$ obeys

$$
\int_{\mathbb{R}} \frac{\mathrm{d}\mu(t)}{1 + t^2} < \infty.
$$

The sets $\Omega_{\mathrm{s}}(\Phi)$, $\Omega_{\mathrm{pp}}(\Phi)$, $\Omega_{\mathrm{sc}}(\Phi)$ and $\Omega_{\mathrm{ac}}(\Phi)$ are mutually disjoint and, moreover, are singular, pure point, singular continuous and absolutely continuous supports of $\mu(\cdot)$, i.e.,

$$
\mu(\mathcal{X} \cap \Omega_\tau(\Phi)) = \mu_\tau(\mathcal{X}), \quad \tau = \mathrm{s}, \mathrm{pp}, \mathrm{sc}, \mathrm{ac},
$$

for any Borel set $\mathcal{X} \subseteq \mathbb{R}$. Let $\mathcal{X} \subseteq \mathbb{R}$ be a Borel set. The set

$$
\mathrm{cl}_{\mathrm{ac}}(\mathcal{X}) = \{x \in \mathbb{R} \mid \mathrm{mes}((x - \epsilon, x + \epsilon) \cap \mathcal{X}) > 0 \text{ for all } \epsilon > 0\}
$$

is called the *absolutely continuous closure* of the set $\mathcal{X}$. Obviously, the set $\mathrm{cl}_{\mathrm{ac}}(\mathcal{X})$ is always closed and one has $\mathrm{cl}_{\mathrm{ac}}(\mathcal{X}) \subseteq \overline{\mathcal{X}}$. For the Borel measure μ we consider the Lebesgue–Jordan decomposition $\mu = \mu_{\mathrm{s}} + \mu_{\mathrm{ac}}$, where μ_{s} and μ_{ac} are the corresponding singular and absolutely continuous measures, respectively. The supports of the measures μ_{s} and μ_{ac} are denoted by $S_{\mathrm{s}}(\mu)$ and $S_{\mathrm{ac}}(\mu)$, respectively.

Lemma 2.7 (Lemma 4.1 in [9]). *Let $\Phi(\cdot)$ be a scalar R-function which has the representation* (15). *Then* $S_{\mathrm{ac}}(\mu) = \mathrm{cl}_{\mathrm{ac}}(\Omega_{\mathrm{ac}}(\Phi))$.

Let $\Pi = \{\mathcal{H}, \Gamma_0, \Gamma_1\}$ be a boundary triple of A^* with Weyl function $M(\cdot)$. We set

$$M_h(z) := (M(z)h, h), \quad z \in \mathbb{C}_\pm, \quad h \in \mathcal{H}, \quad h \neq 0.$$

Further, let $\mathcal{T} = \{h_k\}_{k=1}^N$, $1 \le N \le \infty$, be a total set in $\mathcal{H}$. We set

$$\begin{aligned}
\Omega_{\mathrm{s}}(M;\mathcal{T}) &:= \textstyle\bigcup_{k=1}^N \Omega_{\mathrm{s}}(M_{h_k}),\\
\Omega_{\mathrm{pp}}(M;\mathcal{T}) &:= \textstyle\bigcup_{k=1}^N \Omega_{\mathrm{pp}}(M_{h_k}),\\
\Omega_{\mathrm{sc}}(M;\mathcal{T}) &:= \textstyle\bigcup_{k=1}^N \Omega_{\mathrm{s}}(M_{h_k}) \setminus \Omega_{\mathrm{pp}}(M;\mathcal{T}),\\
\Omega_{\mathrm{ac}}(M;\mathcal{T}) &:= \textstyle\bigcup_{k=1}^N \Omega_{\mathrm{ac}}(M_{h_k}) \setminus \Omega_{\mathrm{s}}(M;\mathcal{T}).
\end{aligned}$$

Obviously, the sets $\Omega_{\mathrm{s}}(M;\mathcal{T})$ and $\Omega_{\mathrm{ac}}(M;\mathcal{T})$ as well as the sets $\Omega_{\mathrm{pp}}(M;\mathcal{T})$, $\Omega_{\mathrm{sc}}(M;\mathcal{T})$ and $\Omega_{\mathrm{ac}}(M;\mathcal{T})$ are mutually disjoint. We note that the sets $\Omega_\tau(M;\mathcal{T})$, $\tau =$s, pp, sc, have Lebesgue measure zero, i.e., $\mathrm{mes}(\Omega_\tau(M;\mathcal{T})) = 0$, $\tau =$s, pp, sc.

Theorem 2.8 (Theorem 3.8 in [9]). *Let A be a simple densely defined closed symmetric operator on a separable Hilbert space $\mathfrak{H}$ with $n_+(A) = n_-(A)$. Further, let $\Pi = \{\mathcal{H}, \Gamma_0, \Gamma_1\}$ be a boundary triplet of A^* with Weyl function $M(\cdot)$ and let $E_{A_0}(\cdot)$ be the spectral measure of the self-adjoint extension A_0 of A.*

If $\mathcal{T} = \{h_k\}_{k=1}^N$, $1 \le N \le +\infty$, is a total set in $\mathcal{H}$, then the sets $\Omega_{\mathrm{s}}(M;\mathcal{T})$, $\Omega_{\mathrm{pp}}(M;\mathcal{T})$, $\Omega_{\mathrm{sc}}(M;\mathcal{T})$ and $\Omega_{\mathrm{ac}}(M;\mathcal{T})$ are singular, pure point, singular continuous and absolutely continuous supports of $E_{A_0}(\cdot)$, respectively, i.e, we have

$$E_{A_0}(\mathcal{X} \cap \Omega_\tau(M;\mathcal{T})) = E^\tau_{A_0}(\mathcal{X}), \quad \tau = \text{s, pp, sc, ac},$$

*for each Borel set $\mathcal{X} \subseteq \mathbb{R}$. In particular, it holds $\sigma_{\mathrm{pp}}(A_0) = \Omega_{\mathrm{pp}}(M;\mathcal{T})$ and $\sigma_\tau(A_0) \subseteq \overline{\Omega_\tau(M;\mathcal{T})} \subseteq \sigma(A_0)$ for $\tau =$*s, sc, ac.

Proposition 2.9 (Proposition 4.2 in [9]). *Let A be a simple densely defined closed symmetric operator on a separable Hilbert space $\mathfrak{H}$ with $n_+(A) = n_-(A)$. Further, let $\Pi = \{\mathcal{H}, \Gamma_0, \Gamma_1\}$ be a boundary triplet of A^* with Weyl function $M(\cdot)$.*

If $\mathcal{T} = \{h_k\}_{k=1}^N$, $1 \le N \le +\infty$, is a total set in $\mathcal{H}$, then the absolutely continuous spectrum of the self-adjoint extension $A_0 := A^* \upharpoonright \ker(\Gamma_0)$ of A is given by

$$\sigma_{\mathrm{ac}}(A_0) = \overline{\textstyle\bigcup_{k=1}^N \mathrm{cl}_{\mathrm{ac}}(\Omega_{\mathrm{ac}}(M_{h_k}))}.$$

Theorem 2.10 (Theorem 4.3 in [9]). *Let A be a simple densely defined closed symmetric operator on a separable Hilbert space $\mathfrak{H}$ with $n_+(A) = n_-(A)$. Further, let $\Pi = \{\mathcal{H}, \Gamma_0, \Gamma_1\}$ be a boundary triple of A^* with Weyl function $M(\cdot)$.*

If $\mathcal{T} = \{h_k\}_{k=1}^N$, $1 \le N \le +\infty$, is a total set in $\mathcal{H}$, then for the self-adjoint extension A_0 of A the following conclusions hold.

(i) *The self-adjoint extension A_0 of A has no point spectrum within the interval (a,b), i.e., $\sigma_{\mathrm{pp}}(A_0) \cap (a,b) = \emptyset$, if and only if for each $k = 1, 2, \dots, N$ one has*

$$\lim_{y \downarrow 0} y M_{h_k}(x + iy) = 0$$

for all $x \in (a,b)$. In this case the following relation holds

$$\sigma(A_0) \cap (a,b) = \sigma_c(A_0) \cap (a,b) = \Big(\overline{\bigcup_{k=1}^N \Omega_{\mathrm{sc}}(M_{h_k})} \cup \overline{\bigcup_{k=1}^N \Omega_{\mathrm{ac}}(M_{h_k})}\Big) \cap (a,b).$$

(ii) *The self-adjoint extension A_0 of A has no singular continuous spectrum within the interval (a,b), i.e., $\sigma_{\mathrm{sc}}(A_0) \cap (a,b) = \emptyset$, if for each $k = 1, 2, \dots, N$ the set $\Omega_{\mathrm{sc}}(M_{h_k}) \cap (a,b)$ is countable, in particular, if $(a,b) \setminus \Omega_{\mathrm{ac}}(M_{h_k})$ is countable.*

(iii) *The self-adjoint extension A_0 of A has no absolutely continuous spectrum within the interval (a,b), i.e., $\sigma_{\mathrm{ac}}(A_0) \cap (a,b) = \emptyset$, if and only if for each $k = 1, 2, \dots, N$ the condition*

$$\mathrm{Im}(M_{h_k}(x + i0)) = 0$$

holds for a.e. $x \in (a,b)$. In his case we have

$$\sigma(A_0) \cap (a,b) = \sigma_{\mathrm{s}}(A_0) \cap (a,b) = \overline{\Omega_{\mathrm{s}}(M; \mathcal{T})} \cap (a,b).$$

2.3 Weyl function and spectral multiplicity

Let $E(\cdot)$ be an orthogonal operator-valued measure defined on the Borel sets $\mathcal{B}(\mathbb{R})$ of $\mathbb{R}$. With the measure $E(\cdot)$ one associates a multiplicity function $N_E(\cdot)$ which is defined on $\mathbb{R}$, Borel measurable and takes values $\mathbb{N}_0 = \{0, 1, \dots\}$, cf. Section 7.4 of [8]. The multiplicity function is important since together with the spectral type $[E]$, cf. Section 7.3 of [8]. it characterizes the measure $E(\cdot)$ up to unitary equivalence. Every measure $E(\cdot)$ admits a unique orthogonal decomposition $E(\cdot) = E^{\mathrm{s}}(\cdot) \oplus E^{\mathrm{ac}}(\cdot)$ into a singular orthogonal operator measure $E^{\mathrm{s}}(\cdot)$ and an absolutely continuous orthogonal operator measure $E^{\mathrm{ac}}(\cdot)$. An orthogonal operator measure $E(\cdot)$ is called *singular* if there is a Borel set δ_0 of Lebesgue measure zero such that $E(\delta) = E(\delta \cap \delta_0)$

for any Borel set δ. A measure $E(\cdot)$ is *absolutely continuous* if for any Borel set δ of Lebesgue measure zero one has $E(\delta) = 0$. Obviously, the measures $E^{s}(\cdot)$ and $E^{ac}(\cdot)$ admit also multiplicity functions which are denoted by $N_{E^s}(\cdot)$ and $N_{E^{ac}}(\cdot)$. For a self-adjoint operator H we define the multiplicity function $N_H(\cdot)$ by $N_H(\lambda) := N_{E_H}(\lambda)$, $\lambda \in \mathbb{R}$, where $E_H(\cdot)$ is the orthogonal spectral measure which corresponds to the self-adjoint operator H. The unique decomposition $E_H(\cdot) = E_H^s(\cdot) \oplus E_H^{ac}(\cdot)$ yields a decomposition of H into a singular part H^s and an absolutely continuous part H^{ac} such that $H = H^s \oplus H^{ac}$. We set

$$N_{H^s}(\lambda) := N_{E_H^s}(\lambda) \quad \text{and} \quad N_{H^{ac}}(\lambda) := N_{E^{ac}}(\lambda), \quad \lambda \in \mathbb{R}.$$

Obviously we have $N_{H^s}(\lambda) = N_{E_{H^s}}(\lambda)$ and $N_{H^{ac}}(\lambda) = N_{E_{H^{ac}}}(\lambda)$ for $\lambda \in \mathbb{R}$.

In the following we are interested in the multiplicity function $N_{A_0^{ac}}(\cdot)$ of the self-adjoint extension $A_0 := A^* \ker(\Gamma_0)$. It turns out that $N_{A_0^{ac}}(\cdot)$ can be computed by using the Weyl function as follows: choosing $D \in \mathfrak{S}_2(\mathcal{H})$ such that

$$\ker(D) = \ker(D^*) = \{0\}$$

we introduce the sandwiched Weyl function $M^D(\cdot)$,

$$(M^D)(z) := D^* M(z) D, \quad z \in \mathbb{C}_+.$$

It turns out that the limit

$$(M^D)(t) := \text{s-}\lim_{y \to +0} M^D(t + iy)$$

exists for a.e. $t \in \mathbb{R}$. We set

$$d_{M^D}(t) := \dim(\operatorname{ran}(\operatorname{Im}(M^D(t)))),$$

which is well-defined for a.e. $t \in \mathbb{R}$.

Proposition 2.11 (Proposition 3.2 in [32]). *Let A be be a simple densely defined closed symmetric operator, let $\Pi = \{\mathcal{H}, \Gamma_0, \Gamma_1\}$ be a boundary triplet for A^* and let $M(\cdot)$ be the corresponding Weyl function. Further, let $E_{A_0}(\cdot)$ be the spectral measure of $A_0 = A^* \restriction \ker(\Gamma_0)$. If $D \in \mathfrak{S}_2(\mathcal{H})$ and satisfies $\ker(D) = \ker(D^*) = \{0\}$, then $N_{A_0^{ac}}(t) = d_{M^D}(t)$ for a.e. $t \in \mathbb{R}$ and $\sigma_{ac}(A_0) = \operatorname{cl}_{ac}(\operatorname{supp}(d_{M^D}))$ where $\operatorname{supp}(d_{M^D}) := \{\lambda \in \mathbb{R} \mid d_{M^D}(\lambda) > 0\}$.*

If, in addition, the limit $M(t) := \text{s-}\lim_{y\to+0} M(t + iy)$ exists for a.e. $t \in \mathbb{R}$, then $N_{A_0^{ac}}(t) = d_M(t)$ for a.e. $t \in \mathbb{R}$ and $\sigma_{ac}(A_0) = \operatorname{cl}_{ac}(\operatorname{supp}(d_M))$.

3 Weyl function for matrix Sturm–Liouville operator with integrable potential matrix

Throughout this section we assume that $Q(\cdot) = Q(\cdot)^* \in L^1(\mathbb{R}_+, \mathbb{C}^{m\times m})$, $\mathbb{R}_+ = (0,\infty)$. Note that self-adjointness of a potential matrix $Q(\cdot)$ means that $Q(x) = Q(x)^*$ for a.e. $x \in \mathbb{R}_+$. Let us consider the Sturm–Liouville differential expression

$$\mathcal{L}_Q(f(x)) := -\frac{d^2 f(x)}{dx^2} + Q(x)f(x), \quad f = (f_1,\dots,f_m)^T, \, x \in \mathbb{R}_+.$$

In this section we investigate the structure of the Weyl function of the Dirichlet realization H_Q of $\mathcal{L}_Q$ as well as its limit representation. In particular, we show that the positive part of each realization $\widetilde{H}_Q$ of $\mathcal{L}_Q$ is purely absolutely continuous. This fact is well known in the scalar case (see Chapter 5.3 in [39]). Moreover, it is known in the matrix case whenever $|Q|$ has a finite first moment (see Chapter 4.1 in [1]). In the latter case each realization of $\mathcal{L}_Q$ including the Dirichlet realization H_Q has at most finitely many negative eigenvalues.

Note also that several papers are devoted to the spectral theory of vector-valued Sturm–Liouville operators. For instance, the high-energy asymptotics for Weyl–Titchmarsh matrices associated with general matrix-valued Schrödinger operators on a half-line was obtained in [12]. Several papers are also devoted to inverse problems. For instance, an extension of Borg's classical result from the class of periodic scalar potentials to the class of reflectionless matrix-valued potentials was obtained in [13] (see also references in [13]).

3.1 Asymptotic representations of solutions and special identities

We are interested in matrix-valued solutions $Y(x,z)$ of the equation

$$\mathcal{L}_Q(Y(x,z)) = zY(x,z), \quad x \in \mathbb{R}_+, \, z \in \mathbb{C}. \tag{16}$$

Let $C(x,z)$ and $S(x,z)$ be the matrix-valued solutions of the equation (16) satisfying the initial conditions

$$C(0,z) = S'(0,z) = I_m \qquad z \in \mathbb{C},$$
$$S(0,z) = C'(0,z) = \mathbb{O}_m, \quad z \in \mathbb{C}.$$

Lemma 3.1. *Let $Q(\cdot) \in L^1(\mathbb{R}_+, \mathbb{C}^{m\times m})$ such that $Q(x) = Q(x)^*$ for a.e. $x \in \mathbb{R}_+$. For any $z = \lambda \in \mathbb{R}_+$ the solutions $S(x,\lambda)$ and $C(x,\lambda)$ admit the representation*

$$S(x,\lambda) = a_1(\lambda)\cos(x\sqrt{\lambda}) + a_2(\lambda)\frac{\sin(x\sqrt{\lambda})}{\sqrt{\lambda}} + \frac{1}{\sqrt{\lambda}} o_m(1) \quad \text{as } x \to \infty, \tag{17}$$

$$C(x,\lambda) = b_1(\lambda)\cos(x\sqrt{\lambda}) + b_2(\lambda)\frac{\sin(x\sqrt{\lambda})}{\sqrt{\lambda}} + \frac{1}{\sqrt{\lambda}} o_m(1) \quad \text{as } x \to \infty, \tag{18}$$

where

$$a_1(\lambda) = -\frac{1}{\sqrt{\lambda}}\int_0^\infty \sin(t\sqrt{\lambda})Q(t)S(t,\lambda)\,dt, \tag{19a}$$

$$a_2(\lambda) = I_m + \int_0^\infty \cos(t\sqrt{\lambda})Q(t)S(t,\lambda)\,dt, \tag{19b}$$

and

$$b_1(\lambda) = I_m - \frac{1}{\sqrt{\lambda}}\int_0^\infty \sin(t\sqrt{\lambda})Q(t)C(t,\lambda)\,dt, \tag{20a}$$

$$b_2(\lambda) = \int_0^\infty \cos(t\sqrt{\lambda})Q(t)C(t,\lambda)\,dt, \tag{20b}$$

and $o_m(1)$ denotes an $m \times m$ matrix with entries $o(1)$.

Proof. Let $\widehat{\mathbb{C}}$ be the complex plane with a cut along $[0,\infty)$. For any $z \in \widehat{\mathbb{C}}_0 := \widehat{\mathbb{C}}\setminus\{0\}$ the solution $S(x,z)$ of equation (16) satisfies the integral equation

$$S(x,z) = \frac{\sin(x\sqrt{z})}{\sqrt{z}} I_m + \frac{1}{\sqrt{z}}\int_0^x \sin\{(x-t)\sqrt{z}\}Q(t)S(t,z)\,dt, \quad z \in \widehat{\mathbb{C}}_0. \tag{21}$$

Let $\sqrt{z} = \rho + i\beta$ and $\beta \geq 0$ and let $S_1(x,z) := e^{-\beta x}S(x,z)$. Inserting this expression in (21) it yields

$$S_1(x,z) = e^{-\beta x}\frac{\sin(x\sqrt{z})}{\sqrt{z}} I_m + \frac{1}{\sqrt{z}}\int_0^x e^{-\beta(x-t)} \sin\{(x-t)\sqrt{z}\}Q(t)S_1(t,z)\,dt.$$

Since $|\cos x\sqrt{z}| \leq e^{\beta x}$ and $|\sin x\sqrt{z}| \leq e^{\beta x}$, one has

$$|S_1(x,z)| \leq \frac{1}{|\sqrt{z}|} + \frac{1}{|\sqrt{z}|}\int_0^x |Q(t)|\cdot|S_1(t,z)|\,dt, \quad z \in \widehat{\mathbb{C}}_0,\ x \in \mathbb{R}_+,$$

where $|S_1(t,z)| = |S_1(t,z)|_m$ denotes matrix $m \times m$ norm. Therefore by the Gronwall lemma one gets

$$|S_1(x,z)| \leq \frac{1}{|\sqrt{z}|}\exp\left\{\frac{1}{|\sqrt{z}|}\int_0^x |Q(t)|\,dt\right\}, \quad z \in \widehat{\mathbb{C}}_0,\ x \in \mathbb{R}_+. \tag{22}$$

Since $Q \in L^1(\mathbb{R}_+, \mathbb{C}^{m\times m})$ the matrix-function $S_1(x,z)$ is bounded for all $x \in \mathbb{R}_+$ and $z \in \widehat{\mathbb{C}}_0$.

Further let $\lambda \in \mathbb{R}_+$. Then the matrix-function $S(x,\lambda)$ is bounded. Thus we obtain from (21) that

$$\begin{aligned} S(x,\lambda) &= \frac{\sin(x\sqrt{\lambda})}{\sqrt{\lambda}} I_m + \frac{1}{\sqrt{\lambda}} \int_0^\infty \sin\{(x-t)\sqrt{\lambda}\} Q(t) S(t,\lambda)\, dt \\ &\quad + O_m\Big(\frac{1}{\sqrt{\lambda}} \int_x^\infty |Q(t)|\, dt\Big) \\ &= a_1(\lambda) \cos x(\sqrt{\lambda}) + a_2(\lambda) \frac{\sin(x\sqrt{\lambda})}{\sqrt{\lambda}} + \frac{o_m(1)}{\sqrt{\lambda}}, \quad \lambda \in \mathbb{R}_+, \end{aligned}$$

as $x \to \infty$, where $O_m(\alpha)$ is an $m\times m$ matrix with entries $O(\alpha)$, and $a_1(\cdot)$ and $a_2(\cdot)$ are given by (19). Since the integrals converge uniformly in $\lambda \geq \varepsilon > 0$ the matrix functions $a_1(\cdot)$ and $a_2(\cdot)$ are continuous in $\lambda \in \mathbb{R}_+$.

To prove (18) we note that the solution $C(\cdot, z)$ of equation (16) satisfies the integral equation

$$C(x,z) = \cos(x\sqrt{z}) I_m + \frac{1}{\sqrt{z}} \int_0^x \sin\{(x-t)\sqrt{z}\} Q(t) C(t,z)\, dt, \quad z \in \widehat{\mathbb{C}}_0. \tag{23}$$

Setting $C_1(x,z) := e^{-\beta x} C(x,z)$, inserting this expression in (23) and applying the Gronwall lemma yields

$$|C_1(x,z)| \leq \frac{1}{|\sqrt{z}|} \exp\Big\{\frac{1}{|\sqrt{z}|} \int_0^x |Q(t)|\, dt\Big\}, \quad z \in \widehat{\mathbb{C}}_0,\ x \in \mathbb{R}_+. \tag{24}$$

Inserting this inequality in (23) implies (18). □

Lemma 3.2. *Let $Q(\cdot) = Q(\cdot)^* \in L^1(\mathbb{R}_+, \mathbb{C}^{m\times m})$. Let $a_j(\cdot)$, $b_j(\cdot)$, $j \in \{1,2\}$, be the matrix functions given by* (19) *and* (20). *Then the following relations hold*:

$$a_1(\lambda) b_1(\lambda)^* = b_1(\lambda) a_1(\lambda)^*, \quad \lambda \in \mathbb{R}_+, \tag{25}$$

$$a_2(\lambda) b_2(\lambda)^* = b_2(\lambda) a_2(\lambda)^*, \quad \lambda \in \mathbb{R}_+, \tag{26}$$

$$a_2(\lambda) b_1(\lambda)^* - b_2(\lambda) a_1(\lambda)^* = b_1(\lambda) a_2(\lambda)^* - a_1(\lambda) b_2(\lambda)^*, \quad \lambda \in \mathbb{R}_+. \tag{27}$$

Proof. Differentiating (21) with respect to x one gets

$$S'(x,z) = \cos(x\sqrt{z})I_m + \int_0^x \cos\{(x-t)\sqrt{z}\}Q(t)S(t,z)\,\mathrm{d}t, \quad z \in \widehat{\mathbb{C}} \setminus \{0\}.$$

Following arguments of Lemma 3.1 we obtain

$$S'(x,\lambda) = -a_1(\lambda)\sqrt{\lambda}\sin(x\sqrt{\lambda}) + a_2(\lambda)\cos(x\sqrt{\lambda}) + o_m(1), \quad x \to \infty, \tag{28}$$

$\lambda \in \mathbb{R}_+$. Similarly one gets

$$C'(x,\lambda) = -b_1(\lambda)\sqrt{z}\sin(x\sqrt{\lambda}) + b_2(\lambda)\cos(x\sqrt{\lambda}) + o_m(1), \quad x \to \infty, \tag{29}$$

$\lambda \in \mathbb{R}_+$. Further, let us introduce the fundamental matrix solution $W(x,z)$ of equation (16) and the fundamental symmetry J by setting

$$W(x,z) := \begin{pmatrix} C(x,z) & S(x,z) \\ C'(x,z) & S'(x,z) \end{pmatrix} \quad \text{and} \quad J := \begin{pmatrix} 0 & I_m \\ -I_m & 0 \end{pmatrix}.$$

It is well known that the following relations hold

$$W(z)JW(\bar{z})^* = W(\bar{z})^*JW(z) = J, \quad z \in \widehat{\mathbb{C}} \setminus \{0\}, \tag{30}$$

cf. [31]. The first of these matrix identities splits into three "scalar" identities

$$\begin{aligned} C(x,z)S(x,\bar{z})^* &= S(x,z)C(x,\bar{z})^*, \\ C'(x,z)S'(x,\bar{z})^* &= S'(x,z)C'(x,\bar{z})^*, \end{aligned} \tag{31}$$

and

$$I_m = C(x,z)S'(x,\bar{z})^* - S(x,z)C'(x,\bar{z})^*. \tag{32}$$

Let $z = \lambda \in \mathbb{R}_+$. Inserting expressions (17) and (18) for $S(x,z)$ and $C(x,z)$ into (31) we derive

$$\begin{aligned} &\Big(b_1(\lambda)\cos(x\sqrt{\lambda}) + b_2(\lambda)\frac{\sin(x\sqrt{\lambda})}{\sqrt{\lambda}} + \frac{o_m(1)}{\sqrt{\lambda}}\Big) \\ &\quad \Big(a_1(\lambda)^*\cos(x\sqrt{\lambda}) + a_2(\lambda)^*\frac{\sin(x\sqrt{\lambda})}{\sqrt{\lambda}} + \frac{o_m(1)}{\sqrt{\lambda}}\Big) \\ &\quad = \Big(a_1(\lambda)\cos(x\sqrt{\lambda}) + a_2(\lambda)\frac{\sin(x\sqrt{\lambda})}{\sqrt{\lambda}} + \frac{o_m(1)}{\sqrt{\lambda}}\Big) \\ &\qquad \Big(b_1(\lambda)^*\cos(x\sqrt{\lambda}) + b_2(\lambda)^*\frac{\sin(x\sqrt{\lambda})}{\sqrt{\lambda}} + \frac{o_m(1)}{\sqrt{\lambda}}\Big). \end{aligned}$$

This identity is easily transformed into

$$\begin{aligned}
&\cos^2(x\sqrt{\lambda})(b_1(\lambda)a_1(\lambda)^* - a_1(\lambda)b_1(\lambda)^*)\\
&\quad + \frac{\sin^2(x\sqrt{\lambda})}{\lambda}(b_2(\lambda)a_2(\lambda)^* - a_2(\lambda)b_2(\lambda)^*)\\
&\quad + \frac{\cos(x\sqrt{\lambda})\sin(x\sqrt{\lambda})}{\sqrt{\lambda}}(b_1(\lambda)a_2(\lambda)^* - a_1(\lambda)b_2(\lambda)^*\\
&\qquad\qquad + b_2(\lambda)a_1(\lambda)^* - a_2(\lambda)b_1(\lambda)^*)\\
&\quad + \frac{o_m(1)}{\sqrt{\lambda}}\cos(x\sqrt{\lambda})(b_1(\lambda) - a_1(\lambda))\\
&\quad + \frac{o_m(1)}{\sqrt{\lambda}}\frac{\sin(x\sqrt{\lambda})}{\sqrt{\lambda}}(b_2(\lambda) - a_2(\lambda))\\
&\quad + \frac{o_m(1)}{\sqrt{\lambda}}\cos(x\sqrt{\lambda})(a_1(\lambda)^* - b_1(\lambda)^*)\\
&\quad + \frac{o_m(1)}{\sqrt{\lambda}}\frac{\sin(x\sqrt{\lambda})}{\sqrt{\lambda}}(a_2(\lambda)^* - b_2(\lambda)^*) = 0.
\end{aligned} \tag{33}$$

Fixing $\lambda \in \mathbb{R}_+$ and setting in (33) $x_k = 2\pi k/\sqrt{\lambda}$ we obtain

$$\begin{aligned}
&b_1(\lambda)a_1(\lambda)^* - a_1(\lambda)b_1(\lambda)^*\\
&\quad + \frac{o_m(1)}{\sqrt{\lambda}}(b_1(\lambda) - a_1(\lambda)) + \frac{o_m(1)}{\sqrt{\lambda}}(a_1(\lambda)^* - b_1(\lambda)^*) = 0.
\end{aligned}$$

Passing in this identity to the limit as $k \to +\infty$ we arrive at formula (25).

Next setting in (33) $x_k = (2\pi k + \pi/2)/\sqrt{\lambda}$ one derives

$$\begin{aligned}
&b_2(\lambda)a_2(\lambda)^* - a_2(\lambda)b_2(\lambda)^*\\
&\quad + \frac{o_m(1)}{\lambda}(b_2(\lambda) - a_2(\lambda)) + \frac{o_m(1)}{\lambda}(a_2(\lambda)^* - b_2(\lambda)^*) = 0.
\end{aligned}$$

Passing here to the limit as $k \to +\infty$ one arrives at formula (26).

Finally, setting in (33) $x_k = (\pi k + \pi/4)/\sqrt{\lambda}$, and tending k to $+\infty$ we obtain (27). □

Lemma 3.3. *Let $Q(\cdot) = Q(\cdot)^* \in L^1(\mathbb{R}_+, \mathbb{C}^{m\times m})$. Let $a_j(\cdot)$ and $b_j(\cdot)$, $j \in \{1,2\}$, be given by* (19) *and* (20)*, respectively. Then the following holds:*

$$a_2(\lambda)b_1(\lambda)^* - b_2(\lambda)a_1(\lambda)^* = b_1(\lambda)a_2(\lambda)^* - a_1(\lambda)b_2(\lambda)^* = I_m, \quad \lambda \in \mathbb{R}_+. \tag{34}$$

Proof. Inserting expressions (17) and (18) for $S(x,z)$ and $C(x,z)$ into (32) with $z = \lambda \in \mathbb{R}_+$ we obtain

$$\begin{aligned}
&\Big(b_1(\lambda)\cos(x\sqrt{\lambda}) + b_2(\lambda)\frac{\sin(x\sqrt{\lambda})}{\sqrt{\lambda}} + \frac{o_m(1)}{\sqrt{\lambda}}\Big)\\
&\qquad\Big(-a_1(\lambda)^*\sqrt{\lambda}\sin(x\sqrt{\lambda}) + a_2(\lambda)^*\cos(x\sqrt{\lambda}) + \frac{o_m(1)}{\sqrt{\lambda}}\Big)\\
&\quad - \Big(a_1(\lambda)\cos(x\sqrt{\lambda}) + a_2(\lambda)\frac{\sin(x\sqrt{\lambda})}{\sqrt{\lambda}} + \frac{o_m(1)}{\sqrt{\lambda}}\Big)\\
&\qquad\qquad\Big(-b_1(\lambda)^*\sqrt{\lambda}\sin(x\sqrt{\lambda}) + b_2(\lambda)^*\cos(x\sqrt{\lambda}) + \frac{o_m(1)}{\sqrt{\lambda}}\Big) = I_m.
\end{aligned}$$

This formula is obviously equivalent to

$$\begin{aligned}
&a_2(\lambda)b_1(\lambda)^* - b_2(\lambda)a_1(\lambda)^*\\
&\quad + \sqrt{\lambda}\cos(x\sqrt{\lambda})\sin(x\sqrt{\lambda})(a_1(\lambda)b_1(\lambda)^* - b_1(\lambda)a_1(\lambda)^*)\\
&\quad + \frac{1}{\sqrt{\lambda}}\cos(x\sqrt{\lambda})\sin(x\sqrt{\lambda})(b_2(\lambda)a_2(\lambda)^* - a_2(\lambda)b_2(\lambda)^*)\\
&\quad + \frac{o_m(1)}{\sqrt{\lambda}}\cos(x\sqrt{\lambda})(b_1(\lambda) + a_2(\lambda)^* - a_1(\lambda) - b_2(\lambda))\\
&\quad + \frac{o_m(1)}{\sqrt{\lambda}}\sqrt{\lambda}\sin(x\sqrt{\lambda})(b_1(\lambda)^* - a_1(\lambda)^*)\\
&\quad + \frac{o_m(1)}{\sqrt{\lambda}}\frac{\sin(x\sqrt{\lambda})}{\sqrt{\lambda}}(b_2(\lambda) - a_2(\lambda)) = I_m.
\end{aligned}$$

Taking into account identities (25) and (26) one gets

$$\begin{aligned}
&a_2(\lambda)b_1(\lambda)^* - b_2(\lambda)a_1(\lambda)^*\\
&\quad + \frac{o_m(1)}{\sqrt{\lambda}}\cos(x\sqrt{\lambda})(b_1(\lambda) + a_2(\lambda)^* - a_1(\lambda) - b_2(\lambda)^*)\\
&\quad + o_m(1)\sin(x\sqrt{\lambda})(b_1(\lambda)^* - a_1(\lambda)^*)\\
&\quad + \frac{o_m(1)}{\lambda}\sin(x\sqrt{\lambda})(b_2(\lambda) - a_2(\lambda)) = I_m.
\end{aligned} \tag{35}$$

Fix $\lambda \in \mathbb{R}_+$ and set $x_k = {}^{(2\pi k + \pi/2)}/\sqrt{\lambda}$, $k \in \mathbb{N}$. Passing in (35) to the limit as $k \to +\infty$ we obtain

$$a_2(\lambda)b_1(\lambda)^* - b_2(\lambda)a_1(\lambda)^* = I_m, \quad \lambda \in \mathbb{R}_+,$$

i.e., the first identity in (34). The second identity in (34) is implied by combining the first one with (27). □

In what follows we need also the second group of identities.

Lemma 3.4. *Let $Q(\cdot) = Q(\cdot)^* \in L^1(\mathbb{R}_+, \mathbb{C}^{m\times m})$. Let $a_j(\cdot)$ and $b_j(\cdot)$, $j \in \{1,2\}$, be the matrix functions given by* (19) *and* (20)*, Then the following relations hold*:

$$b_1(\lambda)^* a_2(\lambda) - b_2(\lambda)^* a_1(\lambda) = I_m, \quad \lambda \in \mathbb{R}_+, \tag{36}$$

$$b_1(\lambda)^* b_2(\lambda) = b_2(\lambda)^* b_1(\lambda), \quad \lambda \in \mathbb{R}_+, \tag{37}$$

$$a_1(\lambda)^* a_2(\lambda) = a_2(\lambda)^* a_1(\lambda), \quad \lambda \in \mathbb{R}_+. \tag{38}$$

Proof. The second identity of (30) yields

$$C(x,\bar{z})^* S'(x,z) - C'(x,\bar{z})^* S(x,z) = I_m, \quad z \in \widehat{\mathbb{C}}, \tag{39a}$$

and

$$C(x,\bar{z})^* C'(x,z) = C'(x,\bar{z})^* C(x,z), \quad z \in \widehat{\mathbb{C}}, \tag{39b}$$

$$S(x,z)^* S'(x,z) = S'(x,z)^* S(x,z), \quad z \in \widehat{\mathbb{C}}. \tag{39c}$$

Let us prove identity (36). Inserting asymptotic expressions (17), (18) and (28), (29) for $S(x,z)$ and $C(x,z)$ and their derivatives into (39a) we obtain

$$\begin{aligned}
&\Big(b_1(\lambda)^* \cos(x\sqrt{\lambda}) + b_2(\lambda)^* \frac{\sin(x\sqrt{\lambda})}{\sqrt{\lambda}} + \frac{o_m(1)}{\sqrt{\lambda}}\Big) \\
&\qquad \Big(-a_1(\lambda)\sqrt{\lambda}\sin(x\sqrt{\lambda}) + a_2(\lambda)\cos(x\sqrt{\lambda}) + \frac{o_m(1)}{\sqrt{\lambda}}\Big) \\
&\quad - \Big(-b_1(\lambda)^*\sqrt{\lambda}\sin(x\sqrt{\lambda}) + b_2(\lambda)^*\cos(x\sqrt{\lambda}) + \frac{o_m(1)}{\sqrt{\lambda}}\Big) \\
&\qquad\quad \Big(a_1(\lambda)\cos(x\sqrt{\lambda}) + a_2(\lambda)\frac{\sin(x\sqrt{\lambda})}{\sqrt{\lambda}} + \frac{o_m(1)}{\sqrt{\lambda}}\Big) = I_m, \quad \lambda \in \mathbb{R}_+.
\end{aligned}$$

This identity is easily transformed into

$$\begin{aligned}
&b_1(\lambda)^* a_2(\lambda) - b_2(\lambda)^* a_1(\lambda) \\
&\quad + \frac{o_m(1)}{\sqrt{\lambda}} \cos(x\sqrt{\lambda})(b_1(\lambda)^* - b_2(\lambda)^* + a_2(\lambda) - a_1(\lambda)) \\
&\quad + o_m(1)\sin(x\sqrt{\lambda})(b_1(\lambda)^* - a_1(\lambda)) \\
&\quad - \frac{o_m(1)}{\lambda}\sin(x\sqrt{\lambda})(a_1(\lambda) + a_2(\lambda)) = I_m.
\end{aligned}$$

Passing here to the limit as $x \to +\infty$ we arrive at identity (36).

Let us prove identity (37). Inserting asymptotic expressions (17), (18) and (28), (29) for $S(x,\lambda)$ and $C(x,\lambda)$ and their derivatives into (39b) we obtain

$$\begin{aligned}&\Big(b_1(\lambda)^* \cos(x\sqrt{\lambda}) + b_2(\lambda)^* \frac{\sin(x\sqrt{\lambda})}{\sqrt{\lambda}} + o_m(1)\Big)\\ &\qquad\Big(-b_1(\lambda)\sqrt{\lambda}\sin(x\sqrt{\lambda}) + b_2(\lambda)\cos(x\sqrt{\lambda}) + o_m(1)\Big)\\ &= \Big(-b_1(\lambda)^*\sqrt{\lambda}\sin(x\sqrt{\lambda}) + b_2(\lambda)^*\cos(x\sqrt{\lambda}) + o_m(1)\Big)\\ &\qquad\Big(b_1(\lambda)\cos(x\sqrt{\lambda}) + b_2(\lambda)\frac{\sin(x\sqrt{\lambda})}{\sqrt{\lambda}} + o_m(1)\Big).\end{aligned}$$

Setting here $x_k = 2\pi k/\sqrt{\lambda}$ and tending k to $+\infty$ we arrive at identity (37).

Identity (38) is obtained from (39c) in a similar manner. □

3.2 Investigating of the Weyl function

Let $Q(\cdot) = Q(\cdot)^* \in L^1(\mathbb{R}_+, \mathbb{C}^{m\times m})$. Denote by $AC_{\mathrm{loc}}(\mathbb{R}_+)$ the set of locally absolutely continuous functions on $\mathbb{R}_+$, i.e., $f \in AC_{\mathrm{loc}}(\mathbb{R}_+)$ if $f \in AC[0,b]$ for any $b \in \mathbb{R}_+$. We set

$$\begin{aligned}Af &:= \mathcal{L}_Q(f), \quad x \in \mathbb{R}_+,\ f \in \mathrm{dom}(A),\\ \mathrm{dom}(A) &:= \left\{ f \in L^2(\mathbb{R}_+, \mathbb{C}^m) \;\middle|\; \begin{array}{c} f, f' \in AC_{\mathrm{loc}}(\mathbb{R}_+, \mathbb{C}^m),\\ \mathcal{L}_Q(f) \in L^2(\mathbb{R}_+, \mathbb{C}^m), \quad f(0) = f'(0) = 0 \end{array}\right\}\end{aligned} \tag{40}$$

and note (see [3] and [36]) that the operator A coincides with the minimal operator $H_Q^{\min}$ associated with expression $\mathcal{L}_Q$. The adjoint operator A^* is given by

$$\begin{aligned}A^*f &:= \mathcal{L}_Q(f), \quad x \in \mathbb{R}_+,\ f \in \mathrm{dom}(A^*),\\ \mathrm{dom}(A^*) &:= \left\{ f \in L^2(\mathbb{R}_+, \mathbb{C}^m) \;\middle|\; \begin{array}{c} f, f' \in AC_{\mathrm{loc}}(\mathbb{R}_+, \mathbb{C}^m),\\ \mathcal{L}_Q(f) \in L^2(\mathbb{R}_+, \mathbb{C}^m) \end{array}\right\}\end{aligned}$$

and coincides with the maximal operator $H_Q^{\max}$ associated with $\mathcal{L}_Q$ (see [3] and [36]). It is important to note that the operator A is simple.

Lemma 3.5. *Let $Q(\cdot) = Q(\cdot)^* \in L^1(\mathbb{R}_+, \mathbb{C}^{m\times m})$. Then a triplet $\Pi = \{\mathcal{H}, \Gamma_0, \Gamma_1\}$ with*

$$\mathcal{H} = \mathbb{C}^m, \quad \Gamma_0 f = f(0), \quad \Gamma_1 f = f'(0), \quad f = (f_1, \dots, f_m)^T \in \mathrm{dom}(H_Q^{\max}), \tag{41}$$

is a boundary triplet for the operator $A^* = H_Q^{\max}$.

We leave the proof to the reader.

Proposition 3.6. *Let $Q(\cdot) = Q(\cdot)^* \in L^1(\mathbb{R}_+, \mathbb{C}^{m\times m})$ and let $M(\cdot)$ be the Weyl function corresponding to the boundary triplet* (41). *Then the following holds*:

(i) *the matrix-valued functions $N_1(z)$,*

$$N_1(z) = \frac{I_m}{2i\sqrt{z}} + \frac{1}{2i\sqrt{z}}\int_0^\infty e^{it\sqrt{z}}Q(t)S(t,z)\,\mathrm{d}t, \tag{42}$$

and

$$N_2(z) = \frac{I_m}{2} - \frac{1}{2i\sqrt{z}}\int_0^\infty e^{it\sqrt{z}}Q(t)C(t,z)\,\mathrm{d}t, \tag{43}$$

are both well defined for $z \in \widehat{\mathbb{C}}_0$ and continuous as well as holomorphic in $\mathbb{C}\setminus[0,\infty)$;

(ii) *the following relation holds*:

$$N_1(z)M(z) = N_2(z), \quad z \in \mathbb{C}_+. \tag{44}$$

Proof. (i) The properties of the functions $N_1(\cdot)$ and $N_2(\cdot)$ follow immediately from the estimates (22) and (24).

(ii) Fix $z \in \mathbb{C}_+$ and let $\sqrt{z} = \rho + i\beta$, $\beta > 0$. Then it follows from (21) that

$$\begin{aligned} S(x,z) = {} & -\frac{e^{-ix\sqrt{z}}}{2i\sqrt{z}}I_m + O_m(e^{-\beta x}) \\ & - \frac{1}{2i\sqrt{z}}\int_0^\infty e^{-i(x-t)\sqrt{z}}Q(t)S(t,z)\,\mathrm{d}t \\ & + \frac{1}{2i\sqrt{z}}\int_x^\infty e^{-i(x-t)\sqrt{z}}Q(t)S(t,z)\,\mathrm{d}t \\ & + O_m\Big(\int_0^x e^{-\beta(x-t)}|Q(t)S(t,z)|\,\mathrm{d}t\Big) \end{aligned}$$

as $x \to \infty$. Since $Q \in L^1(\mathbb{R}_+, \mathbb{C}^{m\times m})$ it follows from (22) that for each $z \in \mathbb{C}_+$ one has the estimate

$$|S(x,z)| = e^{\beta x}|S_1(x,z)| \le \frac{1}{|\sqrt{z}|}e^{\beta x}\exp\Big(\frac{1}{\sqrt{z}}\int_0^x |Q(t)|\,\mathrm{d}t\Big) = O_m(e^{\beta x}), \tag{45}$$

as $x \to \infty$. It follows from the obvious identity $|e^{-i(x-t)\sqrt{z}}| = e^{(x-t)\beta}$ that

$$\begin{aligned} & O_m\Big(\int_0^x e^{-\beta(x-t)}|Q(t)S(t,z)|\,\mathrm{d}t\Big) \\ & = O_m\Big(\int_0^x e^{\beta(2t-x)}|Q(t)|\,\mathrm{d}t\Big) \\ & = O_m\Big(e^{\beta(x-2\delta)}\int_0^{x-\delta}|Q(t)|\,\mathrm{d}t\Big) + O_m\Big(e^{\beta x}\int_{x-\delta}^x |Q(t)|\,\mathrm{d}t\Big) = o_m(e^{\beta x}) \end{aligned}$$

as $x \to \infty$, $z \in \mathbb{C}_+$. Moreover, the inequality (45) yields that

$$\int_x^\infty e^{-i(x-t)\sqrt{z}} Q(t) S(t,z)\,\mathrm{d}t = O_m\Big(e^{\beta x} \int_x^\infty |Q(t)|\,\mathrm{d}t\Big) = o_m(e^{\beta x})$$

as $x \to \infty$, $z \in \mathbb{C}_+$. Combining all relations we arrive at the desired asymptotic estimate

$$S(x,z) = e^{-ix\sqrt{z}}\{-N_1(z) + o_m(1)\}, \quad x \to \infty,\ z \in \mathbb{C}_+, \tag{46}$$

with $N_1(\cdot)$ given by (42).

Similarly we obtain from (23) that

$$\begin{aligned} C(x,z) = {} & \frac{e^{-ix\sqrt{z}}}{2z} I_m + O_m(e^{-\beta x}) \\ & - \frac{1}{2i\sqrt{z}} \int_0^\infty e^{-i(x-t)\sqrt{z}} Q(t) C(t,z)\,\mathrm{d}t \\ & + \frac{1}{2i\sqrt{z}} \int_x^\infty e^{-i(x-t)\sqrt{z}} Q(t) C(t,z)\,\mathrm{d}t \\ & + O_m\Big(\int_0^x e^{-\beta(x-t)} |Q(t) C(t,z)|\,\mathrm{d}t\Big), \end{aligned}$$

as $x \to \infty$, $z \in \mathbb{C}_+$. Moreover, inequality (24) yields $|C(x,z)| = O_m(e^{\beta x})$ as $x \to \infty$, $z \in \mathbb{C}_+$. Using this relation and repeating the above reasoning we derive an asymptotic formula for $C(\cdot,z)$ similar to (46):

$$C(x,z) = e^{-ix\sqrt{z}}\{N_2(z) + o_m(1)\}, \quad x \to \infty,\ z \in \mathbb{C}_+, \tag{47}$$

where $N_2(\cdot)$ is given by (43).

We emphasize that in contrast to the functions a_j, b_j, $j \in \{1,2\}$, defined on $\mathbb{R}_+$, the matrix functions $N_j(\cdot)$, $j \in \{1,2\}$, are well defined and holomorphic in $\mathbb{C}_+$.

It is easily seen that the Weyl solution of equation (16) is given by

$$Y(x,z) = C(x,z) + S(x,z)M(z) \in L^2(\mathbb{R}_+, \mathbb{C}^{m\times m}),$$

where $M(\cdot)$ is the Weyl function corresponding to the triplet (41). Combining this representation with asymptotic relations (46) and (47) one gets

$$e^{-ix\sqrt{z}}(N_2(z) - N_1(z)M(z) + o_m(1)) \in L^2(\mathbb{R}_+, \mathbb{C}^{m\times m}).$$

Since $e^{-ix\sqrt{z}} \notin L^2(0,\infty)$ for $z \in \mathbb{C}_+$, the later relation implies (44). □

In the following we need some properties of the functions $N_1(z)$ and $N_2(z)$.

Lemma 3.7. *Let $Q(\cdot) = Q(\cdot)^* \in L^1(\mathbb{R}_+, \mathbb{C}^{m\times m})$. Let also $N_1(\cdot)$ and $N_2(\cdot)$ be the functions given by* (42) *and* (43)*, respectively. In addition, the following holds.*

(i) *The non-tangential limits*

$$N_1(\lambda) = \lim_{z\to\!\!\succ\lambda} N_1(z) = -\frac{1}{2}a_1(\lambda) - \frac{i a_2(\lambda)}{2\sqrt{\lambda}}, \tag{48a}$$

$$N_2(\lambda) = \lim_{z\to\!\!\succ\lambda} N_2(z) = \frac{1}{2}b_1(\lambda) + \frac{i b_2(\lambda)}{2\sqrt{\lambda}}, \tag{48b}$$

exist and are invertible for any $\lambda \in \mathbb{R}_+$.

(ii) *For each bounded interval $[a,b] \subseteq \mathbb{R}_+$ there is a rectangle $\mathcal{R}(a,b,\varepsilon) := [a,b]\times[0,\varepsilon]$, $\varepsilon > 0$ such that $N_1(\cdot)^{-1}$ and $N_2(\cdot)^{-1}$ exist and are continuous in $\mathcal{R}(a,b,\varepsilon)$. In particular, it holds that*

$$\lim_{z\to\!\!\succ\lambda} N_1(z)^{-1} = N_1(\lambda)^{-1}, \quad \lim_{z\to\!\!\succ\lambda} N_2(z)^{-1} = N_2(\lambda)^{-1} \tag{49}$$

for $\lambda \in \mathbb{R}_+$.

(iii) *For each $\lambda \in \mathbb{R}_- := (-\infty, 0)$ one has*

$$N_1(\lambda) = \lim_{z\to\!\!\succ\lambda} N_1(z) = -\frac{I_m}{2\sqrt{|\lambda|}} - \frac{1}{2\sqrt{|\lambda|}}\int_0^\infty e^{-\sqrt{|\lambda|}t} Q(t)S(t,\lambda)\,dt, \tag{50a}$$

$$N_2(\lambda) = \lim_{z\to\!\!\succ\lambda} N_2(z) = \frac{I_m}{2} + \frac{1}{2\sqrt{|\lambda|}}\int_0^\infty e^{-\sqrt{|\lambda|}t} Q(t)C(t,\lambda)\,dt. \tag{50b}$$

Proof. (i) Using (19) and (20) we verify the right-hand sides of (48).

Let us show that $\ker(N_1(\lambda)) = \{0\}$ for every $\lambda \in \mathbb{R}_+$. It follows from (48) with account of identity (38) that

$$\begin{aligned} N_1(\lambda)^* N_1(\lambda) &= \frac{1}{4}\Big(a_1(\lambda)^* - i\frac{a_2(\lambda)^*}{\sqrt{\lambda}}\Big)\Big(a_1(\lambda) + i\frac{a_2(\lambda)}{\sqrt{\lambda}}\Big) \\ &= \frac{1}{4\lambda}(\lambda a_1(\lambda)^* a_1(\lambda) + a_2(\lambda)^* a_2(\lambda)) \\ &\quad + \frac{i}{4\sqrt{\lambda}}(a_1(\lambda)^* a_2(\lambda) - a_2(\lambda)^* a_1(\lambda)) \\ &= \frac{1}{4\lambda}[\lambda\, a_1(\lambda)^* a_1(\lambda) + a_2(\lambda)^* a_2(\lambda)] \geq 0, \quad \lambda \in \mathbb{R}_+. \end{aligned} \tag{51}$$

On the other hand, (36) implies $\ker(a_1(\lambda)) \cap \ker(a_2(\lambda)) = \{0\}$. Combining this relation with (51) one gets $\ker(N_1(\lambda)) = \{0\}$ for $\lambda \in \mathbb{R}_+$. Hence $N_1(\lambda)^{-1}$ exists for each $\lambda \in \mathbb{R}_+$. Since $N_1(z)$ is continuous in $\widehat{\mathbb{C}}_0$ one gets that for any closed interval $[a, b] \subseteq \mathbb{R}_+$ there is sufficiently small $\varepsilon > 0$ such that the matrix-valued function $N_1(z)$ is invertible and continuous in the rectangle $\mathcal{R}(a, b, \varepsilon)$.

Let us show that $\ker(N_2(\lambda)) = \{0\}$ for every $\lambda \in \mathbb{R}_+$. From the second identity of (48) we derive similarly as above that

$$\begin{aligned} N_2(\lambda)^* N_2(\lambda) &= \frac{1}{4}\Big(b_1(\lambda)^* - i\frac{b_2(\lambda)^*}{\sqrt{\lambda}}\Big)\Big(b_1(\lambda) + i\frac{b_2(\lambda)}{\sqrt{\lambda}}\Big) \\ &= \frac{1}{4\lambda}(\lambda\, b_1(\lambda)^* b_1(\lambda) + b_2(\lambda)^* b_2(\lambda)) + \frac{i}{4\sqrt{\lambda}}(b_1(\lambda)^* b_2(\lambda) - b_2(\lambda)^* b_1(\lambda)) \\ &= \frac{1}{4\lambda}[\lambda\, b_1(\lambda)^* b_1(\lambda) + b_2(\lambda)^* b_2(\lambda)] \geq 0, \quad \lambda \in \mathbb{R}_+. \end{aligned} \tag{52}$$

At the same time, taking adjoints in (36) one derives that $\ker b_1(\lambda) \cap \ker b_2(\lambda) = \{0\}$ for each $\lambda \in \mathbb{R}_+$. Combining this relation with (52) one gets $\ker N_2(\lambda) = \{0\}$ for $\lambda \in \mathbb{R}_+$.

(ii) The existence of the rectangle $\mathcal{R}(a, b, \varepsilon)$ can be proven as above. The relations (49) immediately follow.

(iii) The relations (50) follow from the definitions (42) and (43). □

Theorem 3.8. *Let $Q(\cdot) = Q(\cdot)^* \in L^1(\mathbb{R}_+, \mathbb{C}^{m\times m})$. Further, let H_Q be the Dirichlet realization of* (16) *and let $M(\cdot)$ be the Weyl function corresponding to the boundary triplet* (41)*. Then the following holds.*

(i) *The non-tangential boundary values $M(\lambda + i0) := \lim_{z \to\!\!\!\to \lambda} M(z)$ exist **for each** $\lambda \in \mathbb{R}_+$ and*

$$\begin{aligned} M(\lambda + i0) &= N_1(\lambda)^{-1} N_2(\lambda) \\ &= -(\sqrt{\lambda} a_1(\lambda) + i a_2(\lambda))^{-1}(\sqrt{\lambda} b_1(\lambda) + i b_2(\lambda)), \quad \lambda \in \mathbb{R}_+. \end{aligned} \tag{53}$$

In particular, one has

$$\begin{aligned} \operatorname{Im}(M(\lambda + i0)) &= \frac{1}{4\sqrt{\lambda}}(N_1(\lambda)^* N_1(\lambda))^{-1} \\ &= \sqrt{\lambda}(\lambda a_1(\lambda)^* a_1(\lambda) + a_2(\lambda)^* a_2(\lambda))^{-1}, \quad \lambda \in \mathbb{R}_+. \end{aligned} \tag{54}$$

(ii) *The determinant* $d_1(z) = \det(N_1(z))$ *is holomorphic in* $\mathbb{C} \setminus [0, \infty)$ *and the set of its zeros* Λ_1 *is discrete. The Weyl function admits the representation*

$$M(z) = N_1(z)^{-1} N_2(z), \quad z \in \mathbb{C} \setminus \Lambda_1. \tag{55}$$

(iii) *The corresponding spectral measure* $\Sigma_M(\cdot)$ *(see* (12)*) on* $\mathbb{R}_+$ *is given by*

$$\begin{aligned} \Sigma_M(t) &= \frac{1}{\pi} \int_0^t \sqrt{\lambda} (\lambda\, a_1(\lambda)^* a_1(\lambda) + a_2(\lambda)^* a_2(\lambda))^{-1} \mathrm{d}\lambda \\ &= \frac{1}{4\pi} \int_0^t \frac{1}{\sqrt{\lambda}} (N_1(\lambda)^* N_1(\lambda))^{-1} \mathrm{d}\lambda. \end{aligned} \tag{56}$$

In particular, the spectral measure $\Sigma_M(\cdot)$ *is absolutely continuous with* **continuous** *density* ${}^{\mathrm{d}\Sigma_M(\lambda)}/_{\mathrm{d}\lambda}$, $\lambda \in \mathbb{R}_+$, *of maximal rank.*

(iv) *The operator* H_Q *is semi-bounded from below and its negative spectrum is either finite or discrete with the only accumulation point at zero. Its non-negative part is purely absolutely continuous,* $H_Q^{\mathrm{ac}} = H_Q E_{H_Q}(\mathbb{R}_+)$ *and* $N_{H_Q^{\mathrm{ac}}}(\lambda) = m$, $\lambda \in \mathbb{R}_+$. *In particular,* $\sigma_{\mathrm{ac}}(H_Q) = [0, \infty)$, $\sigma_{\mathrm{sc}}(H_Q) = \sigma_{\mathrm{pp}}(H_Q) \cap \mathbb{R}_+ = \emptyset$, *and* $\sigma_{\mathrm{pp}}(H_Q) \cap \mathbb{R}_- \subseteq \Lambda_1 \cap \mathbb{R}_-$.

Proof. (i) For any closed interval $[a, b] \subseteq \mathbb{R}_+$ there is an $\varepsilon > 0$ such that $N_1(z)^{-1}$ exist and is continuous. From (44) we obtain the representation

$$M(z) = N_1(z)^{-1} N_2(z), \quad z \in \mathcal{R}(a, b, \varepsilon).$$

Using (49) we find the existence of the limit $M(\lambda + i0) = \lim_{z \to\!\!\!\!\to \lambda} M(z)$ and

$$M(\lambda + i0) = \lim_{z \to\!\!\!\!\to \lambda} M(z) = N_1(\lambda)^{-1} N_2(\lambda) \tag{57}$$

for any $\lambda \in \mathbb{R}_+$. Inserting (48) into (57) we prove (53).

First let us compute the imaginary part

$$M_I(\lambda + i0) = \mathrm{Im}(M(\lambda + i0))$$

of $M(\lambda + i0)$. It follows from (53) that

$$M_I(\lambda + i0) = \frac{1}{2i} N_1(\lambda)^{-1} (N_2(\lambda) N_1(\lambda)^* - N_1(\lambda) N_2(\lambda)^*)(N_1(\lambda)^*)^{-1}.$$

Notice that

$$\begin{aligned}
&N_2(\lambda)N_1(\lambda)^* - N_1(\lambda)N_2(\lambda)^* \\
&\quad = \frac{1}{4}\Big[\Big(b_1(\lambda) + i\frac{b_2(\lambda)}{\sqrt{\lambda}}\Big)\Big(-a_1(\lambda)^* + i\frac{a_2(\lambda)^*}{\sqrt{\lambda}}\Big) \\
&\qquad\quad + \Big(a_1(\lambda) + i\frac{a_2(\lambda)}{\sqrt{\lambda}}\Big)\Big(b_1(\lambda)^* - i\frac{b_2(\lambda)^*}{\sqrt{\lambda}}\Big)\Big] \\
&\quad = \frac{1}{4}\Big[(a_1(\lambda)b_1(\lambda)^* - b_1(\lambda)a_1(\lambda)^*) + \frac{a_2(\lambda)b_2(\lambda)^* - b_2(\lambda)a_2(\lambda)^*}{\lambda} \\
&\qquad\quad + \frac{i}{\sqrt{\lambda}}(b_1(\lambda)a_2(\lambda)^* - a_1(\lambda)b_2(\lambda)^* + a_2(\lambda)b_1(\lambda)^* - b_2(\lambda)a_1(\lambda)^*)\Big].
\end{aligned}$$

Using relations (25), (26), and (34) we obtain

$$N_2(\lambda)N_1(\lambda)^* - N_1(\lambda)N_2(\lambda)^* = \frac{2i}{4\sqrt{\lambda}} I_m$$

which yields

$$M_I(\lambda + i0) = \frac{1}{4\sqrt{\lambda}} N_1(\lambda)^{-1}(N_1(\lambda)^*)^{-1} = \frac{1}{4\sqrt{\lambda}}(N_1(\lambda)^* N_1(\lambda))^{-1}, \quad \lambda \in \mathbb{R}_+$$

This proves the first statement in (54). Inserting (51) into this expression we arrive at the second statement in (54).

Note that the functions $a_j(\cdot)$ are continuous on $\mathbb{R}_+$. Moreover, as it follows from (19), $a_1(0) = 0$ and $a_2(0) = I_m$.

(ii) Since the function $N_1(\cdot)$ is holomorphic in $\mathbb{C} \setminus [0, \infty)$, the determinants $d_1(\cdot)$ is holomorphic in $\mathbb{C} \setminus [0, \infty)$. Hence the set of its zeros Λ_1 is discrete. Taking into account (44) we arrive at (55).

(iii) Since the normal limit $M(\lambda + i0)$ exists for every $\lambda \in \mathbb{R}_+$, formula (56) follows either from the Stieltjes inversion formula or from the Fatou theorem.

(iv) Since the limit $M(\lambda + i0)$ for any $\lambda \in \mathbb{R}_+$ is invertible we get from Proposition 2.9 that $[0, \infty) \subseteq \sigma_{\mathrm{ac}}(H_Q)$. It follows from Theorem 2.10 (i) that $\sigma_{\mathrm{pp}}(H_Q) \cap \mathbb{R}_+ = \emptyset$. Finally, Theorem 2.10 (ii) yields that $\sigma_{\mathrm{sc}}(H_Q) \cap \mathbb{R}_+ = \emptyset$. Hence, H_Q has no positive singular spectrum and the part $H_Q E_{H_Q}(\mathbb{R}_+)$ is purely absolutely continuous. Finally, Proposition 2.11 implies $N_{A_0^{\mathrm{ac}}}(\lambda) = N_{H_Q}(\lambda) = m$ for $\lambda \in \mathbb{R}_+$. Hence, the part $H_Q E_{H_Q}(\mathbb{R}_+)$ is purely absolutely continuous of constant multiplicity m.

Since the representation (55) is valid for all points of $\mathbb{C} \setminus [0,\infty)$ outside of Λ_1 one easily gets that $M(\lambda + i0) = \lim_{z \to\!\!\to \lambda} M(z) = N_1(\lambda)^{-1} N_2(\lambda)$ holds all $\lambda \in \mathbb{R}_- \setminus \Lambda_1$, where $\mathbb{R}_- := (-\infty, 0)$. $N_1(\lambda)$ and $N_2(\lambda)$, $\lambda \in \mathbb{R}_-$ are given by (50). Hence $M(\cdot)$ is holomorphic in $\mathbb{C} \setminus \{(\Lambda_1 \cap \mathbb{R}_-) \cup [0,\infty)\}$. This immediately yields that $\mathrm{Im}(M(\lambda + i0)) = 0$ for $\lambda \in \mathbb{R}_- \setminus \Lambda_1$. By Theorem 2.10 (ii) and (iii) we find that $\sigma_{\mathrm{sc}}(H_Q) \cap \mathbb{R}_- = \emptyset$ and $\sigma_{\mathrm{ac}}(H_Q) \cap \mathbb{R}_- = \emptyset$. Hence $\sigma_{\mathrm{sc}}(H_Q) = \emptyset$, $\sigma_{\mathrm{ac}}(H_Q) = [0,\infty)$ and $\sigma_{\mathrm{pp}}(H_Q) \cap \mathbb{R}_+ = \emptyset$. However $\sigma_{\mathrm{pp}}(H_Q) \subseteq \Lambda_1 \cap \mathbb{R}_-$.

It remains to show that H_Q is semi-bounded. To this end we have to check that the set Λ_1 is semi-bounded form below. From (50) and estimate (22) we get

$$\lim_{\lambda \to -\infty} 2\sqrt{|\lambda|} N_1(\lambda) = -I_m. \tag{58}$$

Assuming now Λ_1 is not bounded from below. Then there is a sequence $\{\lambda_k\}_{k \in \mathbb{N}}$, $\lambda_k \in \Lambda_1 \cap \mathbb{R}_-$, tending to $-\infty$ as $k \to \infty$. Hence $\lim_{k\to\infty} 2\sqrt{|\lambda_k|} N_1(\lambda_k) = 0$ which contradicts (58). □

Corollary 3.9. *Let $Q(\cdot) = Q(\cdot)^* \in L^1(\mathbb{R}_+, \mathbb{C}^{m\times m})$ and let $\widetilde{A} = \widetilde{A}^*$ be any self-adjoint extension of A. Let also $\widetilde{\Pi} = \{\mathcal{H}, \widetilde{\Gamma}_0, \widetilde{\Gamma}_1\}$ be a boundary triplet for A^* such that $\widetilde{A} = A^* \restriction \ker(\widetilde{\Gamma}_0)$ and let $\widetilde{M}(\cdot)$ be the corresponding Weyl function.*

(i) *The limit $\widetilde{M}(\lambda + i0) := \lim_{z \to\!\!\to \lambda} \widetilde{M}(z)$ exists for any $\lambda \in \mathbb{R}_+$ and $\widetilde{M}_I(\lambda + i0)$ is positively definite. Moreover, the corresponding spectral measure $\Sigma_{\widetilde{M}}$ admits the representation*

$$\Sigma_{\widetilde{M}}(t) = \frac{1}{\pi} \int_0^t (M(\lambda)^* X_{21}^* + X_{22}^*)^{-1} M_I(\lambda) (X_{21} M(\lambda) + X_{22})^{-1} \, d\lambda, \quad t > 0. \tag{59}$$

(ii) *The operators $E_{\widetilde{A}}(\mathbb{R}_+)\widetilde{A}$ and $E_{H_Q}(\mathbb{R}_+) H_Q$ are unitarily equivalent and, hence $\widetilde{A}^{\mathrm{ac}} = \widetilde{A} E_{\widetilde{A}}(\mathbb{R}_+)$ and $N_{\widetilde{A}^{\mathrm{ac}}}(\lambda) = m$, $\lambda \in \mathbb{R}_+$. Moreover, $\sigma_{\mathrm{ac}}(\widetilde{A}) = [0,\infty)$ and $\sigma_{\mathrm{sc}}(\widetilde{A}) = \sigma_{\mathrm{pp}}(\widetilde{A}) \cap \mathbb{R}_+ = \emptyset$.*

The operator $\widetilde{A}$ is semi-bounded from below and its negative spectrum is either finite or discrete with the only accumulation point at zero.

Proof. (i) There is a boundary triplet $\widetilde{\Pi} = \{\mathcal{H}, \widetilde{\Gamma}_0, \widetilde{\Gamma}_1\}$ for A^* such that

$$\widetilde{A} = A^* \restriction \ker(\widetilde{\Gamma}_0).$$

By Proposition 2.5, there exists a J-unitary matrix $X = (X_{ij})_{i,j=1}^2$ such that (13) holds. If $\widetilde{M}(\cdot)$ is the Weyl function of $\widetilde{\Pi}$, then $0 \in \rho(X_{21} M(z) + X_{22})$,

$$\widetilde{M}(z) = X(M(z)) := (X_{11} M(z) + X_{12})(X_{21} M(z) + X_{22})^{-1}, \quad z \in \rho(\widetilde{A}). \tag{60}$$

Denote by $\mathbb{C}_+^{m\times m}$ the set of strictly dissipative $m \times m$ matrices, i.e.,

$$\mathbb{C}_+^{m\times m} := \{T \in \mathbb{C}^{m\times m} \mid T_I = (2i)^{-1}(T - T^*) \geq \varepsilon = \varepsilon(T) > 0\}.$$

By Theorem 3.8 $M(\lambda) := \lim_{z \to\!\!\!\!\!\to \lambda} M(z)$ exists for each $\lambda \in \mathbb{R}_+$ and $M(\lambda) \in \mathbb{C}_+^{m\times m}$. Combining this fact with the Kreĭn– Šmul′jan theorem (see Theorem 4.1 in [29]) one gets

$$0 \in \rho(X_{21}M(\lambda) + X_{22}) \quad \text{and} \quad X(M(\lambda)) \in \mathbb{C}_+^{m\times m}, \quad \lambda \in \mathbb{R}_+.$$

Hence

$$\widetilde{M}(\lambda) := \lim_{z \to\!\!\!\!\!\to \lambda} \widetilde{M}(z) = (X_{11}M(\lambda) + X_{12})(X_{21}M(\lambda) + X_{22})^{-1}, \quad \lambda \in \mathbb{R}_+. \tag{61}$$

Using the J-unitarity of $X = (X_{ij})_{i,j=1}^2$ it follows from (61) that

$$\begin{aligned}\widetilde{M}_I(\lambda) &:= (2i)^{-1}\widetilde{M}(\lambda) \\ &= (M(\lambda)^* X_{21}^* + X_{22}^*)^{-1} M_I(\lambda)(X_{21}M(\lambda) + X_{22})^{-1}, \quad \lambda \in \mathbb{R}_+.\end{aligned}$$

One obtains (59) by combining this relation with the Stieltjes inversion formula.

(ii) These conclusions can be proved similarly to those of Theorem 3.8(iv). We leave the proof to the reader. □

Remark 3.10. Theorem 3.8 generalizes the classical Titchmarsh's result (see Chapter 5 in [39]) to the case of Sturm–Liouville operator with a matrix-valued summable potential and coincides with it in the scalar case ($m = 1$).

Remark 3.11. If a potential matrix Q has a finite first moment, that is if we have $\int_{\mathbb{R}_+} x|Q(x)|\,\mathrm{d}x < \infty$, then the positive part $E_{H_Q}(\mathbb{R}_+)H_Q$ of H_Q is purely absolutely continuous of constant multiplicity m. Moreover $\sigma_p(H_Q) = \sigma_p(H_Q) \cap \mathbb{R}_-$ is finite (see Theorem 2.1.1 in [1]).

4 Sturm–Liouville operator with point interactions

Let $Q \in L^1(\mathbb{R}_+, \mathbb{C}^{m\times m})$ such that $Q(x) = Q(x)^*$ for a.e. $x \in \mathbb{R}_+$. Further, let $X = \{x_n\}_{n=1}^p \subset \mathbb{R}_+$, $p < \infty$ be a strictly increasing sequence of positive numbers, $x_{n+1} > x_n$. Denote $d_n := x_n - x_{n-1}$, $x_0 := 0$, and assume that $Q \in L^1(\mathbb{R}_+, \mathbb{C}^{m\times m})$. Let also $\{\alpha_n\}_{n=1}^p$ be the sequence of self-adjoint $m \times m$–matrices.

In this section we consider the matrix-valued Schrödinger operator $H_{X,\alpha,Q}$ in $L^2(\mathbb{R}_+, \mathbb{C}^{m\times m})$ associated with the formal differential expression

$$\mathcal{L}_{X,\alpha,Q} := -\frac{\mathrm{d}^2}{\mathrm{d}x^2} + Q(x) + \sum_{x_n\in X} \alpha_n \delta(x - x_n), \tag{62}$$

where δ denotes the Dirac delta. This operator describes δ–interactions on a set $X = \{x_n\}_{n=1}^p \subset \mathbb{R}_+$, $p < \infty$, and the coefficient $\alpha_n \in \mathbb{C}^{m\times m}$ is called the *strength* of the interaction at the point $x = x_n$.

In $L^2(\mathbb{R}_+)$ we define the closed symmetric operator A associated with (62) by

$$(Af)(x) := \mathcal{L}_Q(f)(x), \quad x \in \mathbb{R}_+ \setminus X, \quad f \in \mathrm{dom}(A),$$

$$\mathrm{dom}(A) = \left\{ f \in L^2(\mathbb{R}_+ \setminus X, \mathbb{C}^m) \;\middle|\; \begin{array}{c} f, f' \in AC_{\mathrm{loc}}(\mathbb{R}_+ \setminus X, \mathbb{C}^m), \\ \mathcal{L}_Q(f) \in L^2(\mathbb{R}_+, \mathbb{C}^m), \\ f(0) = 0, \quad f(x_n\pm) = 0, \quad n \in \{1, \dots, p\}, \\ f'(0) = 0, \quad f'(x_n\pm) = 0, \quad n \in \{1, \dots, p\} \end{array} \right\}. \tag{63}$$

Notice that A is a minimal operator associated with the differential expression (62). In what follows the minimal operator is denoted by $H^{\min}_{X,\alpha,Q}$. One has

$$\mathrm{dom}(H^{\min}_{X,\alpha,Q}) = \{f \in \mathrm{dom}(H^{\max}_Q) \mid f(x_n) = f'(x_n) = 0,\ n \in \{0, 1, \dots, p\}\}.$$

The adjoint operator A^* is given by

$$(A^* f)(x) := \mathcal{L}_Q(f)(x), \quad x \in \mathbb{R}_+ \setminus X,\ f \in \mathrm{dom}(A^*),$$

$$\mathrm{dom}(A^*) = \left\{ f \in L^2(\mathbb{R}_+ \setminus X, \mathbb{C}^m) \;\middle|\; \begin{array}{c} f, f' \in AC_{\mathrm{loc}}(\mathbb{R}_+ \setminus X, \mathbb{C}^m), \\ \mathcal{L}_Q(f) \in L^2(\mathbb{R}_+, \mathbb{C}^m) \end{array} \right\}.$$

The operator A^* is called the *maximal operator* associated with (62) and is denoted by $H^{\max}_{X,\alpha,Q}$. We consider the self-adjoint extension $H_{X,\alpha,Q}$ defined by

$$H_{X,\alpha,Q} = A^* \restriction \mathrm{dom}(H_{X,\alpha,Q}).$$

We have

$$\mathrm{dom}(H_{X,\alpha,Q}) = \left\{ f \in \mathrm{dom}(H^{\max}_{X,\alpha,Q}) \;\middle|\; \begin{array}{c} f(0) = 0, \quad f(x_n+) = f(x_n-), \\ f'(x_n+) - f'(x_n-) = \alpha_n f(x_n), \\ n \in \{1, \dots, p\} \end{array} \right\}, \tag{64}$$

where α_n are self-adjoint matrices. Clearly, if $\alpha_n = 0$ for all $n \in \{1,\dots,p\}$, then $H_{X,\alpha,Q}$ coincides with the Dirichlet realization H_Q of the previous section, i.e., $H_{X,0,Q} = H_Q$.

Let $H_Q^{\max}$ be the maximal operator associated with the differential expression $\mathcal{L}_Q$.

Proposition 4.1. *Let $Q \in L^1(\mathbb{R}_+, \mathbb{C}^{m\times m})$ such that $Q(x) = Q(x)^*$ for a.e. $x \in \mathbb{R}_+$. If the maximal operator $H_Q^{\max}$ has no positive eigenvalues, then the operator $H_{X,\alpha,Q}$ also has no positive eigenvalues, i.e., $\sigma_p(H_{X,\alpha,Q}) \cap \mathbb{R}_+ = \emptyset$.*

Proof. 1. Consider equation (16) in $L^2((x_p,+\infty),\mathbb{C}^m)$. Let us show that any solution of this equation with $\lambda = \lambda_0 > 0$ does not belong to $L^2((x_p,\infty),\mathbb{C}^m)$. Assume the contrary, that is there exists $f \in L^2((x_p,\infty),\mathbb{C}^m)$ which satisfies the equation

$$-\frac{d^2 f(x)}{dx^2} + \{Q(x) - \lambda_0 I_m\} f(x) = 0, \quad x \in (x_p,+\infty). \tag{65}$$

Let $C_0 = f(x_p+)$, $C_1 = f'(x_p+) \in \mathbb{C}^m$ and $|C_0| + |C_1| \neq 0$. Denote by $\tilde{f}$ the solution of equation (65) on $(0,x_p)$ satisfying the initial conditions

$$\tilde{f}(x_p-) = C_0, \quad \tilde{f}'(x_p-) = C_1.$$

Clearly, $\tilde{f} \in L^2((0,x_p),\mathbb{C}^m)$, so that the function

$$\hat{f}(x) := \begin{cases} f(x) & \text{if } x \in (x_p,+\infty), \\ \tilde{f}(x) & \text{if } x \in (0,x_p), \end{cases}$$

is well defined on $\mathbb{R}_+$ and belongs to $\operatorname{dom}(H_Q^{\max})$. Thus, $\hat{f} \neq 0$, satisfies equation (65) and $\hat{f} \in L^2(\mathbb{R}_+,\mathbb{C}^m)$. Hence $\lambda_0 \in \sigma(H_Q^{\max})$. This contradicts the assumption of the proposition, and hence $C_0 = C_1 = 0$.

2. Let us show that $\lambda_0 \notin \sigma_p(H_{X,\alpha,Q})$. Assume the contrary. Then there exists a non-trivial $f \in \operatorname{dom}(H_{X,\alpha,Q})$ such that

$$H_{X,\alpha,Q} f = \lambda_0 f.$$

On the interval $(x_p,+\infty)$ this equation turns into (65). According to the first step $f(x) = 0$ for $x \in (x_p,+\infty)$. Hence $f(x_p+) = f'(x_p+) = 0$.

On the other hand, due to (64) the inclusion $f \in \operatorname{dom}(H_{X,\alpha,Q})$ yields

$$f(x_p-) = f(x_p+) = 0, \quad f'(x_p-) = f'(x_p+) - \alpha_p f(x_p) = f'(x_p+) = 0.$$

Therefore equation (65) restricted to the interval (x_{p-1},x_p) leads to the Cauchy problem

$$-f''(x) + Q(x) f(x) = \lambda_0 f(x), \quad x \in (x_{p-1},x_p), \quad f(x_p-) = f'(x_p-) = 0.$$

By the Cauchy uniqueness theorem this problem has a trivial solution on (x_{p-1}, x_p), i.e., $f(x) = 0$ for $x \in (x_{p-1}, x_p)$. Repeating this procedure $p-1$ times we prove by induction that $f \equiv 0$ on $\mathbb{R}_+$. Thus, $\lambda_0 \notin \sigma_p(H_{X,\alpha,Q}) \cap \mathbb{R}_+$. □

Corollary 4.2. *Let $Q \in L^1(\mathbb{R}_+, \mathbb{C}^{m\times m})$ such that $Q(x) = Q(x)^*$ for a.e. $x \in \mathbb{R}_+$. Then $\sigma_p(H_{X,\alpha,Q}) \cap \mathbb{R}_+ = \emptyset$.*

Proof. It is well known that under the assumption $Q \in L^1(\mathbb{R}_+, \mathbb{C}^{m\times m})$ the maximal operator $H_Q^{\max}$ has no positive eigenvalues. In particular, it follows from Corollary 3.9. Proposition 4.1 completes the proof. □

Let us recall the following definition of local interactions from [11].

Definition 4.3 ([11]). Let A be the minimal operator associated with expression (62) and given by (63) where X is either finite or infinite. A symmetric extension $\widetilde{A}$ with $A \subset \widetilde{A} \subset A^*$ is called a *one-dimensional Schrödinger operator with local point interactions in X* if the Lagrange brackets

$$[f, g]_x := f'(x+)\overline{g(x+)} - f(x+)\overline{g'(x+)}$$

are continuous in $(0, \infty)$ for every pair $f, g \in \operatorname{dom} \widetilde{A}$.

Define

$$f_{+}^{-}(x) := \begin{pmatrix} f(x-) - if'(x-) \\ f(x+) + if'(x+) \end{pmatrix} \quad \text{and} \quad f_{-}^{+}(x) := \begin{pmatrix} f(x+) - if'(x+) \\ f(x-) + if'(x-) \end{pmatrix}.$$

Following [11] we define the class of extensions of A with local point interactions:

$$\operatorname{dom} A_{\{U_{x_j}\}} = \{f \in \operatorname{dom} A^* \mid f_{-}^{+}(x_j) = U_{x_j} f_{+}^{-}(x_j)\},\ j \in \{1, \ldots, p\}\}. \tag{66}$$

where $\{U_{x_j}\}$ is a family of unitary 2×2–matrices.

The following characterization of the operators with local point interactions was obtained in [11].

Theorem 4.4 ([11]). *An extension $\widetilde{A} \supset A$ is a one-dimensional Schrödinger operator with local point interactions in X if and only if $\widetilde{A}$ is a selfadjoint extension of A for some choice of unitary 2×2–matrices $\{U_{x_j}; x_j \in X\}$.*

For instance, the unitary matrix

$$U_{x_j} = e^{-i\alpha_j} \begin{pmatrix} \cos\alpha_j & -i\sin\alpha_j \\ -i\sin\alpha_j & \cos\alpha_j \end{pmatrix} \quad \text{for } \alpha_j \in \left(-\frac{\pi}{2}, \frac{\pi}{2}\right)$$

defines a δ–interaction at x_j of strength $2\tan\alpha_j$, and

$$U_{x_j} = e^{i\alpha_j}\begin{pmatrix}\cos\alpha_j & -i\sin\alpha_j\\ -i\sin\alpha_j & \cos\alpha_j\end{pmatrix} \quad \text{for } \alpha_j \in \Big(-\frac{\pi}{2},\frac{\pi}{2}\Big)$$

defines a δ'–interaction at x_j of strength $2\tan\alpha_j$.

Next we extend Corollary 4.1 to the case of operators with local point interactions in X.

Proposition 4.5. *Let A be the minimal operator associated with* (62). *Let*

$$U_{x_j} = \begin{pmatrix}u_{1,j} & u_{2,j}\\ u_{3,j} & u_{4,j}\end{pmatrix}, \quad j \in \{1,\dots,p\},$$

and let $\widetilde{H} = A_{\{U_{x_j}\}}$ be an extension of A given by (66). *Then $\sigma_p(\widetilde{H}) \cap \mathbb{R}_+ = \emptyset$ whenever $u_{1,j} \neq 0$.*

Proof. We mimic the proof of Proposition 4.1. Let $\widetilde{H}f = \lambda f$ for some $\lambda > 0$. Under the assumption $Q \in L^1(\mathbb{R}_+, \mathbb{C}^{m\times m})$, the maximal operator $H_Q^{\max}$ has no positive eigenvalues, as mentioned in the proof of Corollary 4.1. This implies $f(x) = 0$ for $x \in (x_p,\infty)$. To prove that $f(x) = 0$ for $x \in (x_{p-1}, x_p)$ it suffices to prove the implication

$$f(x_p+) = f'(x_p+) = 0 \implies f(x_p-) = f'(x_p-) = 0 \tag{67}$$

whenever $u_{1,p} \neq 0$. Inserting equalities $f(x_p+) = f'(x_p+) = 0$ in the pth relation in (66) yields

$$\begin{pmatrix}0\\ f(x_p-) + if'(x_p-)\end{pmatrix} = \begin{pmatrix}u_{1,j} & u_{2,j}\\ u_{3,j} & u_{4,j}\end{pmatrix}\begin{pmatrix}f(x_p-) - if'(x_p-)\\ 0\end{pmatrix}.$$

This identity is easily transformed into the following one:

$$\begin{pmatrix}u_{1,p}(f(x_p-) - if'(x_p-))\\ u_{3,p}(f(x_p-) - if'(x_p-))\end{pmatrix} = \begin{pmatrix}0\\ f(x_p-) + if'(x_p-)\end{pmatrix}.$$

Since $u_{1,p} \neq 0$, this identity splits into the following two ones:

$$f(x_p-) - if'(x_p-) = 0 \quad \text{and} \quad f(x_p-) + if'(x_p-) = 0.$$

In turn, the latter yields the required implication (67). The rest of the proof is similar to that of Proposition 4.1. □

Theorem 4.6. *Let $Q \in L^1(\mathbb{R}_+, \mathbb{C}^{m\times m})$ such that $Q(x) = Q(x)^*$ for a.e. $x \in \mathbb{R}_+$. Then the singular continuous spectrum of any self-adjoint extension $\widetilde{H}$ of the operator A given by* (63) *is empty, i.e., $\sigma_{\mathrm{sc}}(\widetilde{H}) \cap \mathbb{R} = \emptyset$.*

Proof. 1. Let A_j be the minimal operator associated with the differential expression $\mathcal{L}_Q$ on $L^2(\Delta_j, \mathbb{C}^m)$, $\Delta_j := (x_{j-1}, x_j)$, $j \in \{1,\dots,p\}$. It is easily seen that $n_\pm(A_j) = 2m$, $j \in \{1,\dots,p\}$. It is well known and easily checked that $\Pi_j = \{\mathcal{H}_j, \Gamma_0^{(j)}, \Gamma_1^{(j)}\}$ with

$$\mathcal{H}_j = \mathbb{C}^{2m}, \quad \Gamma_0^{(j)} f = \begin{pmatrix} f(x_{j-1}+) \\ f(x_j-) \end{pmatrix}, \quad \Gamma_1^{(j)} f = \begin{pmatrix} f'(x_{j-1}+) \\ -f'(x_j-) \end{pmatrix}, \tag{68}$$

$j \in \{1,\dots,p\}$, is a boundary triplet for the operator A_j^*. The corresponding Weyl function is given by

$$M_j(z) = \begin{pmatrix} -S_j(x_j-,z)^{-1} C_j(x_j-,z) & S_j(x_j-,z)^{-1} \\ (S_j(x_j-,\bar{z})^*)^{-1} & -S_j'(x_j-,z) S_j(x_j-,z)^{-1} \end{pmatrix},$$

$j \in \{1,\dots,p\}$, $z \in \mathbb{C}_\pm$, where $S_j(\cdot,z)$ and $C_j(\cdot,z)$ are solutions of

$$\mathcal{L}_Q(S_j(x,z)) = zS_j(x,z) \quad \text{and} \quad \mathcal{L}_Q(C_j(x,z)) = zC_j(x,z), \tag{69}$$

$x \in \Delta_j$, which satisfy the boundary conditions

$$C_j(x_{j-1}+,z) = S_j'(x_{j-1}+,z) = I_m, \tag{70a}$$

$$S_j(x_{j-1}+,z) = C_j'(x_{j-1}+,z) = \mathbb{O}_m, \tag{70b}$$

for $z \in \mathbb{C}_\pm$. Let A_{p+1} be the minimal operator associated with differential expression $\mathcal{L}_Q$ on $L^2(\Delta_{p+1}, \mathbb{C}^m)$, $\Delta_{p+1} := (x_p, \infty)$. Notice that $n_\pm(A_{p+1}) = m$. One easily checks that $\Pi_{p+1} = \{\mathcal{H}_{p+1}, \Gamma_0^{(p+1)}, \Gamma_1^{(p+1)}\}$,

$$\mathcal{H}_{p+1} = \mathbb{C}^m, \quad \Gamma_0^{(p+1)} f = f(x_p+), \quad \Gamma_1^{(p+1)} f = f'(x_p+), \tag{71}$$

is a boundary triplet for A_{p+1}^*. One easily checks that the operator A_{p+1} is unitarily equivalent to the operator A defined in $L^2(\mathbb{R}_+, \mathbb{C}^m)$ by (40), however, with potential $Q_{p+1}(x) := Q(x + x_p)$, $x \in \mathbb{R}_+$, instead of Q. We denote this operator by $A_{Q_{p+1}}$. Obviously the boundary triplet $\Pi = \{\mathbb{C}^m, \Gamma_0, \Gamma_1\}$ given by (41) is also a boundary triplet for $A_{Q_{p+1}}^*$. Let $M_{p+1}(\cdot)$ be the Weyl function of the boundary triplet Π_{p+1} with respect to A_{p+1} and let $M_{Q_{p+1}}(\cdot)$ the Weyl function of the boundary triplet Π with respect to $A_{Q_{p+1}}$. One checks that $M_{p+1}(z) = M_{Q_{p+1}}(z)$, $z \in \mathbb{C}_\pm$.

By $S_{p+1}(\cdot,z)$ and $C_{p+1}(\cdot,z)$ we denote solutions of the differential equations

$$\mathcal{L}_{Q_{p+1}}(S_{p+1}(x,z)) = zS_{p+1}(x,z), \quad \mathcal{L}_{Q_{p+1}}(C_{p+1}(x,z))(x) = zC_{p+1}(x,z),$$

considered on $\mathbb{R}_+$ and satisfying the initial conditions

$$S'_{p+1}(0,z) = C_{p+1}(0,z) = I_m \quad \text{and} \quad S_{p+1}(0,z) = C'_{p+1}(0,z) = \mathbb{O}_m, \quad z \in \mathbb{C}.$$

Similar to (69) and (70) we can introduce the solutions $S_{p+1}(x,z)$ and $C_{p+1}(x,z)$ for $x \in \Delta_{p+1}$. Notice that

$$S_{p+1}(x + x_p, z) = S_{Q_{p+1}}(x,z) \quad \text{and} \quad C_{p+1}(x + x_p, z) = C_{Q_{p+1}}(x,z),$$

$x \in \Delta_{j+1}$. By Proposition 3.6 the Weyl function $M_{Q_{p+1}}(\cdot)$ satisfies the relation

$$N_1(z)M_{Q_{p+1}}(z) = N_2(z), \quad z \in \mathbb{C}_+, \tag{72}$$

where

$$N_1(z) = \frac{I_m}{2i\sqrt{z}} + \frac{1}{2i\sqrt{z}}\int_0^\infty e^{i\sqrt{z}t} Q(t + x_p) S_{Q_{p+1}}(t,z)\,dt, \tag{73a}$$

$$N_2(z) = \frac{I_m}{2} - \frac{1}{2i\sqrt{z}}\int_0^\infty e^{i\sqrt{z}t} Q(t + x_p) C_{Q_{p+1}}(t,z)\,dt, \tag{73b}$$

$z \in \mathbb{C}_+$. Using $M_{Q_{p+1}}(z) = M_{p+1}(z)$, $z \in \mathbb{C}_\pm$, we find

$$N_1(z)M_{p+1}(z) = N_2(z), \quad z \in \mathbb{C}_+,$$

Notice that $M_j(\cdot)$, $j \in \{1,\dots,p\}$, are meromorphic functions while $M_{p+1}(\cdot)$ is not meromorphic.

2. It is easily seen that A given by (63) admits a decomposition $A = \bigoplus_{j=1}^{p+1} A_j$. Hence A is a symmetric with the deficiency indices $n_\pm(A) = \sum_{j=1}^{p+1} n_\pm(A_j) \leq (2p+1)m$. Clearly, the direct sum $\Pi = \bigoplus_{j=1}^{p+1} \Pi_j = \{\mathcal{H}, \Gamma_0, \Gamma_1\}$ of boundary triplets Π_j with

$$\mathcal{H} := \bigoplus_{j=1}^{p+1} \mathcal{H}_j, \quad \Gamma_0 := \bigoplus_{j=1}^{p+1} \Gamma_0^{(j)}, \quad \Gamma_1 := \bigoplus_{j=1}^{p+1} \Gamma_1^{(j)},$$

where $\Gamma_k^{(j)}$, $k \in \{0,1\}$, are given by (68) and (71), respectively, forms a boundary triplet for A^*. The corresponding Weyl function is

$$M(z) = \bigoplus_{j=1}^{p+1} M_j(z) := \widetilde{M}_p(z) \oplus M_{p+1}(z), \quad \text{where } \widetilde{M}_p(z) := \bigoplus_{j=1}^{p} M_j(z).$$

Clearly, the Weyl function $\widetilde{M}_p(\cdot)$ is a meromorphic matrix function in $\mathbb{C}$. Hence its singularities constitute at most countable set of real-valued poles $\Omega_p := \{\omega_j\}_{j=1}^{\infty}$.

Assume for simplicity that $\widetilde{H} = \widetilde{H}^*$ is disjoint with $A_0 = A^* \upharpoonright \ker \Gamma_0$. Then in accordance with Proposition 2.3(iii) there is a bounded operator B such that

$$\widetilde{H} = A^* \upharpoonright \ker(\Gamma_1 - B\Gamma_0) = A_B, \quad B \in \mathcal{B}(\mathcal{H}).$$

Further, let

$$B = \begin{pmatrix} B_{11} & B_{12} \\ B_{12}^* & B_{22} \end{pmatrix} = B^*, \quad B_{11} \in \mathbb{C}^{2mp \times 2mp}, \; B_{22} \in \mathbb{C}^{m \times m},$$

be the block-matrix representation of B with respect to the decomposition

$$\mathcal{H} = \widetilde{\mathcal{H}}_p \oplus \mathcal{H}_{p+1} \quad \text{where} \quad \widetilde{\mathcal{H}}_p = \bigoplus_{j=1}^{p} \mathcal{H}_j = \mathbb{C}^{2mp}, \; \mathcal{H}_{p+1} = \mathbb{C}^m.$$

Then

$$B - M(z) = \begin{pmatrix} B_{11} - \widetilde{M}_p(z) & B_{12} \\ B_{12}^* & B_{22} - M_{p+1}(z) \end{pmatrix}.$$

Since $M(\cdot)$ is the Weyl function, its imaginary part $M_I(\cdot)$ is positive definite in $\mathbb{C}_+$, i.e., $M_I(z) \geq \varepsilon(z) > 0$ and $0 \in \rho(M_I(z))$ for $z \in \mathbb{C}_+$. Hence both matrices $\operatorname{Im}(M(z) - B) = \operatorname{Im} M(z) = M_I(z)$ and $\operatorname{Im}(\widetilde{M}_p(z) - B_{11}) = \operatorname{Im} \widetilde{M}_p(z)$ are also positive definite for $z \in \mathbb{C}_+$. It follows that both matrices $M(z) - B$ and $\widetilde{M}_p(z) - B_{11}$ are invertible for $z \in \mathbb{C}_+$. Therefore the inverse matrix $(B - M(z))^{-1}$ exists for each $z \in \mathbb{C}_+$ and can be computed by the Frobenious formula:

$$(B - M(z))^{-1} = \begin{pmatrix} K_{11}(z) & K_{12}(z) \\ K_{21}(z) & M_B^{-1}(z) \end{pmatrix}, \quad z \in \mathbb{C}_+,$$

where

$$M_B(z) := B_{22} - M_{p+1}(z) - B_{12}^*(B_{11} - \widetilde{M}_p(z))^{-1} B_{12}, \tag{74}$$

$$\begin{aligned} K_{11}(z) &= (B_{11} - \widetilde{M}_p(z))^{-1} \\ &\quad + (B_{11} - \widetilde{M}_p(z))^{-1} B_{12} M_B^{-1}(z) B_{12}^* (B_{11} - \widetilde{M}_p(z))^{-1}, \\ K_{12}(z) &= -(B_{11} - \widetilde{M}_p(z))^{-1} B_{12} M_B^{-1}(z), \\ K_{21}(z) &= -M_B^{-1}(z) B_{12}^* (B_{11} - \widetilde{M}_p(z))^{-1}. \end{aligned} \tag{75}$$

3. We are going to show that $\sigma_{sc}(\widetilde{H}) \cap \mathbb{R}_+ = \emptyset$. Since the matrix function $B_{11} - \widetilde{M}_p(\cdot)$ is meromorphic in $\mathbb{C}$ and invertible in $\mathbb{C}_+$, its determinant

$$d_1(z) := \det(B_{11} - \widetilde{M}_p(z))$$

has at most a countable set of isolated zeros in $\mathbb{R} \setminus \Omega_p$. Denoting this null set by Ω_n, we get

$$\lim_{z \to\!\!\succ \lambda} (B_{11} - \widetilde{M}_p(z))^{-1} = (B_{11} - \widetilde{M}_p(\lambda))^{-1}, \quad \lambda \in \mathbb{R}_+ \setminus (\Omega_p \cup \Omega_n). \tag{76}$$

It follows from (74), (76) and Theorem 3.8 (i) that the limit

$$M_B(\lambda + i0) := \lim_{z \to\!\!\succ \lambda} M_B(z)$$

exists for each $\lambda \in \mathbb{R}_+ \setminus (\Omega_p \cup \Omega_n)$ and

$$M_B(\lambda + i0) = B_{22} - M_{p+1}(\lambda + i0) - B_{12}^*(B_{11} - \widetilde{M}_p(\lambda))^{-1} B_{12}, \tag{77}$$

$\lambda \in \mathbb{R}_+ \setminus (\Omega_p \cup \Omega_n)$. Since $B_{22} = B_{22}^*$, $B_{11} = B_{11}^*$, and $\widetilde{M}_p(\lambda) = \widetilde{M}_p(\lambda)^*$ for every $\lambda \in \mathbb{R}_+ \setminus (\Omega_p \cup \Omega_n)$, the later identity yields

$$\operatorname{Im}(-M_B(\lambda + i0)) = \operatorname{Im}(M_{p+1}(\lambda + i0)), \quad \lambda \in \mathbb{R}_+ \setminus (\Omega_p \cup \Omega_n).$$

Since $\operatorname{Im}(M_{p+1}(\lambda + i0))$ is invertible for $\lambda \in \mathbb{R}_+ \setminus (\Omega_p \cup \Omega_n)$ one gets that the limit $M_B(\lambda + i0)$ is invertible for each $\lambda \in \mathbb{R}_+ \setminus (\Omega_p \cup \Omega_n)$ and

$$M_B(\lambda + i0)^{-1} = (B_{22} - M_{p+1}(\lambda + i0) - B_{12}^*(B_{11} - \widetilde{M}_p(\lambda))^{-1} B_{12})^{-1}, \tag{78}$$

$\lambda \in \mathbb{R}_+ \setminus (\Omega_p \cup \Omega_n)$. Since the limit $M_B(\lambda + i0)$ is invertible we get from (76),(78) and formulas (74)-(75) that the limit $B - M(\lambda + i0) := \lim_{z \to\!\!\succ \lambda}(B - M(z))$

$$B - M(\lambda + i0) = \begin{pmatrix} B_{11} - \widetilde{M}_p(\lambda) & B_{12} \\ B_{12}^* & B_{22} - M_{p+1}(\lambda + i0) \end{pmatrix} \tag{79}$$

exist for $\mathbb{R}_+ \setminus (\Omega_p \cup \Omega_n)$ and is invertible. Moreover, we have

$$(B - M(\lambda + i0))^{-1} = \lim_{z \to\!\!\succ \lambda} (B - M(z))^{-1}$$

for every $\lambda \in \mathbb{R}_+ \setminus (\Omega_p \cup \Omega_n)$. One easily checks that $\Pi_B = \{\mathcal{H}, \Gamma_0^B, \Gamma_1^B\}$,

$$\Gamma_0^B := \Gamma_1 - B\Gamma_0 \quad \text{and} \quad \Gamma_1^B := -\Gamma_0$$

is a boundary triplet for A^* and $A_0 = A^* \restriction \ker(\Gamma_0^B) = \widetilde{H}$. The corresponding Weyl function $M_B(\cdot)$ is given by $M_B(z) = (B - M(z))^{-1}$, $z \in \mathbb{C}_\pm$. Since the set $\Omega_p \cup \Omega_n$ is at most countable, Theorem 2.10(ii) applies and ensures that $\sigma_{sc}(A_B) \cap \mathbb{R}_+ = \emptyset$.

4. Let us show that $\sigma_{sc}(\widetilde{H}) \cap \mathbb{R}_- = \emptyset$. We note that the resolvent difference $(\widetilde{H} - i)^{-1} - (H_Q - i)^{-1}$ is a finite dimensional operator. Further, in accordance with Theorem 3.8 (iv) we have $\sigma_{sc}(H_Q) \cap \mathbb{R}_- = \emptyset$ where H_Q is the Dirichlet realization of the differential expression $\mathcal{L}_Q$. To prove $\sigma_{sc}(\widetilde{H}) \cap \mathbb{R}_- = \emptyset$ it remains to apply the Weyl theorem on the stability of the continuous spectrum under compact perturbations.

5. It remains to consider the case of $\widetilde{H}$ not disjoint with A_0. In this case we use Lemma 2.12 of [32]. We leave the details to the reader. □

Remark 4.7. Note that a similar result is also valid for Hamiltonians with finitely many δ'-interactions since Corollary 4.2 remains true for such extension of the operator A.

In connection with Theorem 4.6 we note that similar result for point spectrum is false, i.e., $\sigma_p(\widetilde{H}) \cap \mathbb{R}_+ \neq \emptyset$ in general. For instance, one obtains a counterexample by setting $\widetilde{H} = \widetilde{H}_1 \oplus \widetilde{H}_2$, where $\widetilde{H}_1 \in \mathrm{Ext}_{A_1}$ and $\widetilde{H}_2 \in \mathrm{Ext}_{S_1}$, where $S_1 := \bigoplus_{j=2}^{p+1} A_j$.

In other words, as distinct from the Hamiltonians $H_{X,\alpha,Q}$ other extensions $\widetilde{H} \in \mathrm{Ext}_A$ might have positive eigenvalues.

Theorem 4.8. *Let $Q = Q^* \in L^1(\mathbb{R}_+, \mathbb{C}^{m \times m})$ and let $H_{X,\alpha,Q}$ be the Hamiltonian associated with* (64). *Then the positive part $E_{H_{X,\alpha,Q}}(\mathbb{R}_+)H_{X,\alpha,Q}$ of $H_{X,\alpha,Q}$ is unitarily equivalent to the positive part $E_{H_Q}(\mathbb{R}_+)H_Q$ of the Dirichlet realization $H_Q = H_{X,0,Q}$. In particular, the spectrum of $E_{H_{X,\alpha,Q}}(\mathbb{R}_+)H_{X,\alpha,Q}$ is purely absolutely continuous and of constant spectral multiplicity m.*

Moreover, the Hamiltonian $H_{X,\alpha,Q}$ is semi-bounded below and its negative spectrum is either finite or forms a sequence tending to zero.

Proof. By Theorem 4.6 the Hamiltonian $H_{X,\alpha,Q}$ has no singular continuous non-negative spectrum, $\sigma_{sc}(H_{X,\alpha,Q}) \cap \mathbb{R}_+ = \emptyset$. On the other hand, Corollary 4.2 implies absence of (embedded) positive eigenvalues, i.e., $\sigma_p(H_{X,\alpha,Q}) \cap \mathbb{R}_+ = \emptyset$. Thus, the non-negative spectrum $\sigma(H_{X,\alpha,Q}) \cap \mathbb{R}_+$ of $H_{X,\alpha,Q}$ is purely absolutely continuous, i.e., $E_{H_{X,\alpha,Q}}(\mathbb{R}_+)H_{X,\alpha,Q} = E_{H_{X,\alpha,Q}}(\mathbb{R}_+)H^{ac}_{X,\alpha,Q}$. Since the resolvent difference $(\widetilde{H} - i)^{-1} - (H_Q - i)^{-1}$ is finite-dimensional, the unitarily equivalence of the non-negative parts $E_{H_{X,\alpha,Q}}(\mathbb{R}_+)H_{X,\alpha,Q}$ and $E_{H_Q}(\mathbb{R}_+)H_Q$ follows from the Kato–Rosenblum theorem. □

Remark 4.9. Assume that $Q(\cdot)$ decays exponentially, i.e., $|Q(x)| < Ce^{-ax}$ with some $C > 0$ and $a > 0$. Then the matrix functions $N_1(z^2)$ and $N_2(z^2)$ given by (73) admit holomorphic continuation from $\mathbb{C}_+$ into $\mathbb{C}^a_- \setminus \{0\}$ where

$$\mathbb{C}^a_- := \{z = x + iy \in \mathbb{C}, -a < y \leq 0\}.$$

Therefore it follows from representation (72)-(73) that $M_{p+1}(z^2) = M_{Q_{p+1}}(z^2)$ admits a meromorphic continuation into $\mathbb{C}_-^a \setminus \{0\}$. In turn, the latter with account of (77) and (79) ensures a meromorphic continuation of $B - M(z^2)$ into $\mathbb{C}_-^a \setminus \{0\}$. Hence, $(B - M(z^2))^{-1}$ is holomorphic on the non-negative semi-axes $\mathbb{R}_+$ with a possible exception of a countable number of real poles $\Omega_- := \{\omega_j^-\}_{j=1}^\infty$. This observation essentially simplifies the proof of Theorem 4.6.

Acknowledgement. Hagen Neidhardt was supported by the European Research Council via ERC-2010-AdG no 267802 (“Analysis of Multiscale Systems Driven by Functionals”).

References

[1] Z. S. Agranovich and V. A. Marchenko, *The inverse problem of scattering theory.* Translated from the Russian by B. D. Seckler. Gordon and Breach Science Publishers, New York and London, 1963. MR 0162497 Zbl 0117.06003

[2] N. I. Akhiezer, *The classical moment problem and some related questions in analysis.* Translated from the Russian by N. Kemmer. University Mathematical Monographs. Oliver & Boyd, Edinburgh and London, 1965. MR 0184042 Zbl 0135.33803

[3] N. I. Akhiezer and I. M. Glazman, *Theory of linear operators in Hilbert space.* Vishcha Shkola, Kharkov, 1977. In Russian. English traslation by E. R. Dawson. Translation edited by W. N. Everitt. Monographs and Studies in Mathematics, 9 (Vol. I) and 10 (Vol. II). Pitman (Advanced Publishing Program), Boston, MA, and London, 1981. ISBN 0-273-08495-X (Vol. I, translation) ISBN 0-273-08496-8 (Vol. II, translation) MR 0486990 (Vol. I, Russian) MR 0615736 (Vol. I, translation) MR 0509335 (Vol. II, Russian) MR 0615737 (Vol. II, translation) Zbl 0467.47001

[4] S. Albeverio, F. Gesztesy, R. Høegh-Krohn, and H. Holden, *Solvable models in quantum mechanics.* Second edition. AMS Chelsea Publishing, Providence, R.I., 2005. With an Appendix by P. Exner. ISBN 0-8218-3624-2/hbk MR 2105735 Zbl 1078.81003

[5] S. Albeverio, A. Kostenko, M. Malamud, and H. Neidhardt, Spherical Schrödinger operators with δ-type interactions. *J. Math. Phys.* **54** (2013), no. 5, 052103, 24 pp. MR 3098915 Zbl 1282.81067

[6] S. Albeverio and P. Kurasov, *Singular perturbations of differential operators.* Solvable Schrödinger type operators. London Mathematical Society Lecture Note Series, 271. Cambridge University Press, Cambridge, 2000. ISBN 0-521-77912-X MR 1752110 Zbl 0945.47015

[7] H. Baumgärtel and M. Wollenberg, *Mathematical scattering theory.* Mathematische Lehrbücher und Monographien, II. Abteilung: Mathematische Monographien. Akademie-Verlag, Berlin, 1983. Licensed edition, Operator Theory: Advances and Applications, 9. Birkhäuser Verlag, Basel etc., 1983. ISBN 3-7643-1519-9 (licensed edition) MR 0699113 MR 0781534 (licensed edition) Zbl 0536.47007 Zbl 0536.47008 (licensed edition)

[8] M. Š. Birman and M. Z. Solomjak, *Spectral theory of self-adjoint operators in Hilbert space.* Translated from the 1980 Russian original by S. Khrushchëv and V. Peller. Mathematics and its Applications (Soviet Series). D. Reidel Publishing Co., Dordrecht, 1987. ISBN 90-277-2179-3 MR 1192782 Zbl 0744.47017

[9] J. F. Brasche, M. M. Malamud, and H. Neidhardt, Weyl function and spectral properties of self-adjoint extensions. *Integral Equations Operator Theory* **43** (2002), no. 3, 264–289. MR 1902949 Zbl 1008.47028

[10] J. Bruening, V. Geyler, and K. Pankrashkin, Spectra of self-adjoint extensions and applications to solvable Schrödinger operators. *Rev. Math. Phys.* **20** (2008), no. 1, 1–70. MR 2379246 Zbl 1163.81007

[11] D. Buschmann, G. Stolz, and J. Weidmann, One-dimensional Schrödinger operators with local point interactions. *J. Reine Angew. Math.* **467** (1995), 169–186. MR 1355927 Zbl 0833.34082

[12] S. Clark and F. Gesztesy, Weyl–Titchmarsh M–function asymptotics for matrix-valued Schrödinger operators. *Proc. London Math. Soc.* (3) **82** (2001), no. 3, 701–724. MR 1816694 Zbl 1025.34021

[13] S. Clark, F. Gesztesy, H. Holden, and B. M. Levitan, Borg-type theorems for matrix-valued Schrödinger operators. *J. Differential Equations* **167** (2000), no. 1, 181–210. MR 1785118 Zbl 0970.34073

[14] V. Derkach, S. Hassi, M. Malamud, and H. S. V. de Snoo, Boundary triplets and Weyl functions. Recent developments. In S. Hassi, H. S. V. de Snoo, and F. H. Szafraniec (eds.), *Operator methods for boundary value problems.* London Mathematical Society Lecture Note Series, 404. Cambridge University Press, Cambridge, 2012, 161–220. MR 3050307 Zbl 1320.47024

[15] V. A. Derkach and M. M. Malamud, Generalized resolvents and the boundary value problems for Hermitian operators with gaps. *J. Funct. Anal.* **95** (1991), no. 1, 1–95. MR 1087947 Zbl 0748.47004

[16] V. A. Derkach and M. M. Malamud, The extension theory of Hermitian operators and the moment problem. *J. Math. Sci.* **73** (1995), no. 2, 141–242. MR 1318517 Zbl 0848.47004

[17] P. Exner, Appendix in [4].

[18] P. Exner and M. Fraas, On the dense point and absolutely continuous spectrum for Hamiltonians with concentric δ-shells. *Lett. Math. Phys.* **82** (2007), no. 1, 25–37. MR 2367872 Zbl 1136.35019

[19] P. Exner and M. Fraas, Interlaced dense point and absolutely continuous spectra for Hamiltonians with concentric-shell singular interactions. In I. Beltita, Gh. Nenciu and R. Purice (eds.), *Mathematical results in quantum mechanics.* Selected papers from the Quantum Mathematics International Conference (QMATH 10) held in Moieciu, September 10 15, 2007. World Scientific, Hackensack, N.J., 2008, 48–65. MR 2466678 Zbl 1156.81383

[20] F. Gesztesy and H. Holden, A new class of solvable models in quantum mechanics describing point interactions on the line. *J. Phys. A* **20** (1987), no. 15, 5157–5177. MR 0914699 Zbl 0627.34030

[21] D. J. Gilbert, On subordinacy and analysis of the spectrum of Schrödinger operators with two singular endpoints. *Proc. Roy. Soc. Edinburgh Sect. A* **112** (1989), no. 3-4, 213–229. MR 1014651 Zbl 0678.34024

[22] D. J. Gilbert, D. B. Pearson, On subordinacy and analysis of the spectrum of one-dimensional Schrödinger operators. *J. Math. Anal. Appl.* **128** (1987), no. 1, 30–56. MR 0915965 Zbl 0666.34023

[23] V. I. Gorbachuk and M. L. Gorbachuk, *Boundary value problems for operator differential equations.* Translated and revised from the 1984 Russian original. Mathematics and its Applications (Soviet Series), 48. Kluwer Academic Publishers Group, Dordrecht, 1991. ISBN 0-7923-0381-4 MR 1154792 Zbl 0751.47025

[24] A. Grossmann, R. Høegh-Krohn, and M. Mebkhout, A class of explicitly soluble, local, many-center Hamiltonians for one-particle quantum mechanics in two and three dimensions. I. *J. Math. Phys.* **21** (1980), no. 9, 2376–2385. MR 0585589

[25] R. Hempel, A. M. Hinz, and H. Kalf, On the essential spectrum of Schrödinger operators with spherically symmetric potentials. *Math. Ann.* **277** (1987), no. 2, 197–211. With a comment by J. Weidmann. MR 0886419 Zbl 0629.35028 Zbl 0629.35029 (comment)

[26] A. S. Kostenko and M. M. Malamud, 1-D Schrödinger operators with local point interactions on a discrete set. *J. Differential Equations* **249** (2010), no. 2, 253–304. MR 2644117 Zbl 1195.47031

[27] A. Kostenko and M. Malamud, 1–D Schrödinger operators with local point interactions: a review. In H. Holden, B. Simon and G. Teschl (eds.), *Spectral analysis, differential equations and mathematical physics: a festschrift in honor of Fritz Gesztesy's 60th birthday.* Proceedings of Symposia in Pure Mathematics, 87. American Mathematical Society, Providence, R.I., 2013, 235–262. MR 3087909 Zbl 3087325

[28] M. G. Kreĭn and H. Langer, On defect subspaces and generalized resolvents of a Hermitian operator in a space $\Pi_{\varkappa}$. *Funkcional. Anal. i Priložen* **5** (1971), no. 2, 59–71; ibid. **5** (1971) no. 3, 54–69. In Russian. English translation, *Funct. Anal. Appl.* **5** (1971), 136–146; ibid. **6** (1972), 217–228. MR 0282238 MR 0282239 Zbl 0236.47034 Zbl 0236.47035

[29] M. G. Kreĭn and Ju. L. Šmul′jan, On linear-fractional transformations with operator coefficients. *Mat. Issled* **2** (1967), no. 3, 64–96. In Russian. English translation, *Amer. Math. Soc. Transl. Ser.* 2 **103** (1974), 125-152. MR 0239456

[30] R. de L. Kronig and W. G. Penney, Quantum mechanics of electrons in crystal lattices. *Proc. Roy. Soc. London Ser. A* **130** (1931), 499–513. JFM 57.1587.02 Zbl 0001.10601

[31] M. M. Malamud, Uniqueness of the matrix Sturm–Liouville equation given a part of the monodromy matrix, and Borg type results. In W. O. Amrein, A. M. Hinz, and D. B. Pearson (eds.), *Sturm-Liouville theory.* Past and present. Including papers from the International Colloquium held at the University of Geneva, Geneva, September 15–19, 2003. Birkhäuser Verlag, Basel, 2005, 237–270. MR 2145085 Zbl 1094.34010

[32] M. M. Malamud and H. Neidhardt, On the unitary equivalence of absolutely continuous parts of self-adjoint extensions. *J. Funct. Anal.* **260** (2011), no. 3, 613–638. MR 2737392 Zbl 1241.47011

[33] M. M. Malamud and H. Neidhardt, Sturm–Liouville boundary value problems with operator potentials and unitary equivalence. *J. Differential Equations* **252** (2012), no. 11, 5875–5922. MR 2911417 Zbl 1257.47053

[34] M. M. Malamud and K. Schmüdgen, Spectral theory of Schrödinger operators with infinitely many point interactions and radial positive definite functions. *J. Funct. Anal.* **263** (2012), no. 10, 3144–3194. MR 2973337 Zbl 1256.35030

[35] V. A. Mikhaĭlets, One-dimensional Schrödinger operator with point interactions. *Dokl. Akad. Nauk* **335** (1994), no. 4, 421–423. In Russian. English translation, *Russian Acad. Sci. Dokl. Math.* **49** (1994), no. 2, 345–349. MR 1281688 Zbl 0831.34088

[36] M. A. Naĭmark, *Linear differential operators.* Part II: Linear differential operators in Hilbert space. With additional material by the author, and a supplement by V. È. Ljance. Translated from the Russian by E. R. Dawson. English translation edited by W. N. Everitt. Frederick Ungar Publishing Co., New York 1968. MR 0262880 Zbl 0227.34020

[37] C. Shubin Christ and G. Stolz, Spectral theory of one-dimensional Schrödinger operators with point interactions. *J. Math. Anal. Appl.* **184** (1994), no. 3, 491–516. MR 1281525 Zbl 0805.34077

[38] G. Stolz, Bounded solutions and absolute continuity of Sturm-Liouville operators. *J. Math. Anal. Appl.* **169** (1992), no. 1, 210–228. MR 1180682 Zbl 0785.34052

[39] E. C. Titchmarsh, *Eigenfunction expansions associated with second-order differential equations.* Clarendon Press, Oxford, 1946. MR 0019765 Zbl 0061.13505

Spectral asymptotics for the Dirichlet Laplacian with a Neumann window via a Birman–Schwinger analysis of the Dirichlet-to-Neumann operator

André Hänel and Timo Weidl

Dedicated to Pavel Exner on the occasion of his 70th birthday

1 Introduction

In what follows we consider an infinite quantum waveguide subject to a perturbation of the boundary conditions. In spectral theory this type of perturbation is of particular interest, since it is non-additive and may not be treated with standard methods, such as the Birman–Schwinger principle. The simplest case arises by considering the Dirichlet Laplacian on an infinite strip having a so-called Neumann window. Let $\Omega = \mathbb{R} \times (0, \alpha)$. We consider $-\Delta$ on Ω with Dirichlet boundary on all of $\partial\Omega$ except for some small part of the boundary, where we impose Neumann boundary conditions. We are interested in the behaviour of the discrete eigenvalues below the essential spectrum $[\pi^2/\alpha^2, \infty)$ depending on the length of the window, cf. Figure 1.

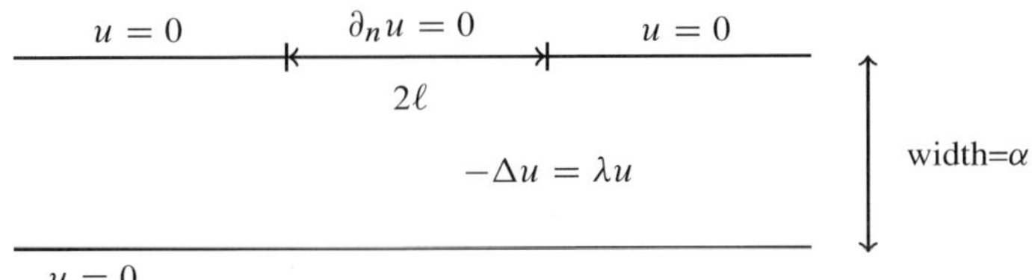

Figure 1. The Dirichlet Laplacian with a Neumann window.

This case was first investigated in [11], where the existence of an eigenvalue was proved by a variational argument. Moreover, a numerical computation given by these authors suggested that for small windows of size 2ℓ the distance of the eigenvalue to the spectral threshold π^2/α^2 is of order ℓ^4. The first analytic proof concerning this fact was given by Exner and Vugalter in [12]. They proved a two-sided asymptotic estimate, i.e., for small $\ell > 0$ there exist a unique eigenvalue $\lambda(\ell)$ below the essential spectrum $[\pi^2/\alpha^2, \infty)$ and constants $c_1, c_2 > 0$ such that

$$c_1\ell^4 \leq \frac{\pi^2}{\alpha^2} - \lambda(\ell) \leq c_2\ell^4 \quad \text{as } \ell \to 0.$$

In fact in [11] and [12] the authors considered the more general case of two quantum waveguides which are coupled through a small window, cf. Figure 2.

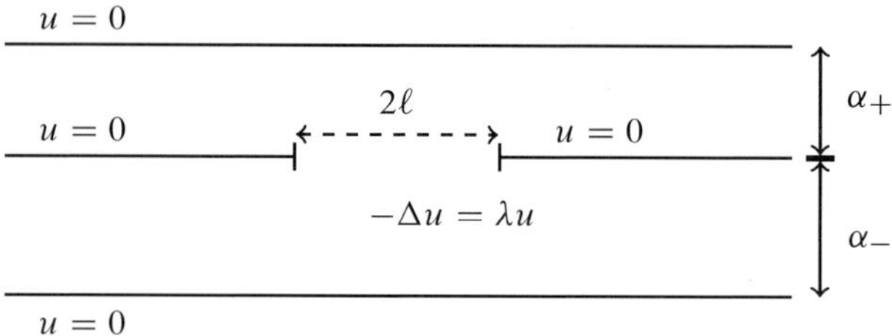

Figure 2. Two laterally coupled quantum waveguides of width α_+ and α_-.

If both waveguides have the same width $\alpha_+ = \alpha_- = \alpha$, then we may use the symmetry with respect to the horizontal direction. In this case the eigenvalue problem is equivalent to the mixed problem in Figure 1. The proof of the two-sided asymptotic estimate in [12] is based on a variational argument. The upper bound may easily be obtained using a suitable test-function and the min-max principle for self-adjoint operators. However, the more delicate part consists in finding a uniform lower bound for the variational coefficient. In order to prove such an estimate Exner and Vugalter decomposed an arbitrary test-function, using an expansion in the vertical direction.

Popov [27] refined the two-sided estimate and proved that the ground state eigenvalue $\lambda(\ell)$ satisfies the following asymptotic behaviour:

$$\frac{\pi^2}{\alpha^2} - \lambda(\ell) = \begin{cases} \left(\frac{\pi^3}{4\alpha^3}\right)^2 \ell^4 + o(\ell^4), & \alpha_+ \neq \alpha_-, \\ \left(\frac{\pi^3}{2\alpha^3}\right)^2 \ell^4 + o(\ell^4), & \alpha_+ = \alpha_-, \end{cases} \qquad \text{as } \ell \to 0.$$

His proof is based on a scheme which matches the asymptotic expansions for the eigenfunctions, cf. [23] and [17]. Popov uses different expansions for the eigenfunctions near the window and distant from the window. Using the explicit formulae for the Green's functions in the upper and lower waveguide he computes the asymptotic behaviour of the eigenvalue. Further terms in this expansion have been calculated in [15]. In [30] the approach was generalised to three-dimensional layers, cf. also [13] for a two-sided asymptotic estimate. Further extensions include e.g., higher dimensional cylinders [28] and [18], a finite or an infinite number of windows [28], [29], [5], [6], [7], and [26], the case of three coupled waveguides [16] or magnetic operators [8]. The case of two retracting distant windows has been investigated in [9] and [10]. For an overview concerning spectral problems in quantum waveguides we refer to [14].

We provide a new approach for the symmetric case which uses the explicit representation of the Dirichlet–to–Neumann operator. This allows us to reformulate the singular perturbation of the original operator into an additive perturbation of the Dirichlet–to–Neumann operator, or merely its truncated part. We replace the matching scheme for the eigenfunctions in [27] by an asymptotic expansion of the Dirichlet–to–Neumann operator and a subsequent use of the Birman–Schwinger principle. As a particular consequence we will observe that only the principal symbol has an influence on the first term of the asymptotic formula. In a similar way we treat the case of two coupled quantum waveguides. An application of the method to elastic waveguides may be found in [22].

Structure of the article

We start by treating the two-dimensional case. We consider the Laplacian $-\Delta$ on $\Omega = \mathbb{R} \times (0, \alpha)$ with Dirichlet boundary conditions except for some small set $\Sigma_\ell \times \{0\} \subseteq \partial\Omega$, where we impose Robin boundary conditions. Here $\Sigma_\ell := \ell \cdot \Sigma \subseteq \mathbb{R}$ and $\Sigma \subseteq \mathbb{R}$ is a finite union of bounded open intervals. Section 2 starts with the definition of the self-adjoint realisation of the corresponding Laplacian and the introduction of the Dirichlet–to–Neumann operator and of the Dirichlet–to–Robin operator. The asymptotic formula for the eigenvalue of the corresponding Laplacian is stated and proven in Section 2, see Theorems 2.9 and 2.10. Additionally, in Theorems 2.9 and 2.10 we prove the uniqueness of the eigenvalue for small window sizes and in Theorem 2.16 we treat the case of two quantum waveguides coupled through a small window.

Section 3 is devoted to three-dimensional layers of the form $\Omega = \mathbb{R}^2 \times (0, \alpha)$. In this case the Robin window is given by $\Sigma_\ell \times \{0\} \subseteq \partial\Omega$, where $\Sigma_\ell := \ell \cdot \Sigma \subseteq \mathbb{R}^2$ is a bounded open set with Lipschitz boundary, cf. Figure 3. We follow the same scheme as in the two-dimensional case and prove in Theorem 3.1 an asymptotic formula for the ground state eigenvalue as the window length decreases.

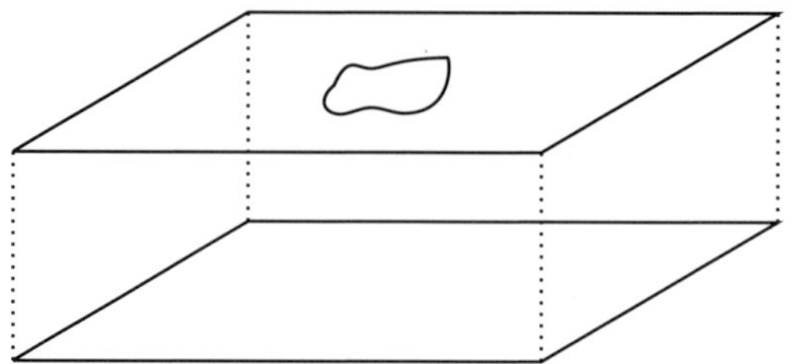

Figure 3. An infinite layer with a small window.

2 The two-dimensional case

The construction of the Dirichlet–to–Robin operator

Let $\alpha > 0$ and put $\Omega := \mathbb{R} \times (0, \alpha)$ with coordinates $(x, y) \in \Omega$. Let $\Sigma \subseteq \mathbb{R}$ be the Robin window and assume that Σ is a bounded open set, which is a finite union of open intervals. We denote the scaled window by $\Sigma_\ell := \ell \cdot \Sigma$. The Laplacian on Ω with Robin boundary conditions on $\Sigma_\ell \times \{0\}$ and Dirichlet boundary conditions on the remaining part of the boundary is defined by the quadratic form

$$a_{\ell,b}[u] := \int_\Omega |\nabla u(x, y)|^2 \,\mathrm{d}x\,\mathrm{d}y + \int_{\Sigma_\ell} b(x) \cdot |u(x, 0)|^2 \,\mathrm{d}x,$$

with the form domain

$$D[a_{\ell,b}] := \{u \in H^1(\Omega) : u|_{\mathbb{R}\times\{\alpha\}} = 0 \text{ and } \operatorname{supp}(u|_{\mathbb{R}\times\{0\}}) \subseteq \overline{\Sigma_\ell}\}.$$

Here $b \in L_\infty(\mathbb{R})$ is a real-valued function and $u|_{\mathbb{R}\times\{0\}}, u|_{\mathbb{R}\times\{\alpha\}} \in H^{1/2}(\mathbb{R})$ are the boundary traces of the function $u \in H^1(\Omega)$. Then $a_{\ell,b}$ defines a closed semi-bounded form on $L_2(\Omega)$ and gives rise to a self-adjoint operator, which we denote by $A_{\ell,b}$. The essential spectrum of the operator $A_{\ell,b}$ is given by $\sigma_{\text{ess}}(A_{\ell,b}) = [\pi^2/\alpha^2, \infty)$. This well-known fact is due to Birman [3], where he gives a proof in the case of compact boundary $\partial\Omega$.

As a first step we consider the Dirichlet–to–Neumann operator acting on the lower part of the boundary $\mathbb{R} \times \{0\}$. For $s \in \mathbb{R}$ let $H^s(\mathbb{R})$ be the standard Sobolev space on $\mathbb{R}$ with the usual norm defined via Fourier transform. Let $\omega \in \mathbb{C}$ and $g \in H^{1/2}(\mathbb{R})$. We consider a weak solution $u \in H^1(\Omega)$ of the Poisson problem

$$(-\Delta - \omega)u = 0 \text{ in } \Omega, \quad u|_{\mathbb{R}\times\{0\}} = g, \quad u|_{\mathbb{R}\times\{\alpha\}} = 0. \tag{1}$$

Applying the Fourier transform in the horizontal direction, it follows from (1) that $\hat{u}(\xi, y)$ solves

$$(-\partial_y^2 + \xi^2 - \omega)\hat{u}(\xi, y) = 0, \quad \hat{u}(\xi, 0) = \hat{g}(\xi), \quad \hat{u}(\xi, \alpha) = 0 \tag{2}$$

with $(\xi, y) \in \mathbb{R} \times (0, \alpha)$. Conversely, if $u \in H^1(\Omega)$ is given such that its Fourier transform $\hat{u}(\xi, y)$ solves the family (2) of Sturm Liouville problems, then u is a weak solution of the Poisson problem (1). For fixed $\xi \in \mathbb{R}$ with $\xi^2 \neq \omega$, the solution of (2) is given by

$$\hat{u}(\xi, y) = \frac{\hat{g}(\xi)}{\sinh(\alpha\sqrt{\xi^2 - \omega})} \cdot \sinh((\alpha - y)\sqrt{\xi^2 - \omega}). \tag{3}$$

Here and subsequently we choose the branch of the square root function such that $z \mapsto \sqrt{z}$ is holomorphic in $\mathbb{C} \setminus (-\infty, 0]$ and such that $\sqrt{z} > 0$ for $z > 0$. Moreover, we extend the definition to $z \in (-\infty, 0]$ and assume that $\operatorname{Im}(\sqrt{z}) \geq 0$ for $z \leq 0$. Actually, the expression for $\hat{u}$ is independent of the value of the square root function as long as one uses the same in the two terms.

Lemma 2.1. *Let $\omega \in \mathbb{C} \setminus [\pi^2/\alpha^2, \infty)$. For every $g \in H^{1/2}(\mathbb{R})$ there exists a unique $u \in H^1(\Omega)$ which solves* (1), *and moreover $\|u\|_{H^1(\Omega)} \leq c\|g\|_{H^{1/2}(\mathbb{R})}$ with $c = c(\omega, \alpha) > 0$ independent of g.*

For the proof of Lemma 2.1 one has to verify that the function u given by (3) belongs indeed to $H^1(\Omega)$ if $g \in H^{1/2}(\mathbb{R})$. We want to omit this simple calculation.

Remark 2.2. If $\omega \geq \pi^2/\alpha^2$, then in general $\hat{u}$ will have a singularity and the above mapping property does not hold true. This is to be expected, as in this case ω will be located in the essential spectrum $[\pi^2/\alpha^2, \infty)$.

Here and subsequently we assume that $\omega \in \mathbb{C} \setminus [\pi^2/\alpha^2, \infty)$. Let $g \in H^{1/2}(\mathbb{R})$ and let u be the solution of the Poisson problem (1). Its normal derivative $\partial_n u$ satisfies

$$\widehat{\partial_n u}(\xi, 0) = m_\omega(\xi)\hat{g}(\xi),$$

where

$$m_\omega(\xi) := \sqrt{\xi^2 - \omega} \cdot \coth(\alpha\sqrt{\xi^2 - \omega}).$$

The Dirichlet–to–Neumann operator is defined by $D_\omega \colon H^{1/2}(\mathbb{R}) \to H^{-1/2}(\mathbb{R})$,

$$\widehat{D_\omega g}(\xi) := \hat{g}(\xi) \cdot \sqrt{\xi^2 - \omega} \cdot \coth(\alpha\sqrt{\xi^2 - \omega}).$$

We note that D_ω is a classical pseudo-differential operator of order 1 with x-independent symbol m_ω. Since the operator $A_{\ell,b}$ is defined by its quadratic form we give a variational characterisation of the Dirichlet–to–Neumann operator D_ω. We also refer to Chapter 4 in McLean [25] for mixed problems formulated in their variational form.

Lemma 2.3. *Let $\omega \in \mathbb{C}\setminus[\pi^2/\alpha^2, \infty)$ and $g \in H^{1/2}(\mathbb{R})$. We denote by $u \in H^1(\Omega)$ the solution of the Poisson problem* (1). *Then for $h \in H^{-1/2}(\mathbb{R})$ the following two assertions are equivalent*:

(1) $h = D_\omega g$;

(2) *for all $v \in H^1(\Omega)$ with $v|_{\mathbb{R}\times\{\alpha\}} = 0$ we have*

$$\langle \nabla u, \nabla v\rangle_\Omega = \omega\langle u, v\rangle_\Omega + \langle h, v|_{\mathbb{R}\times\{0\}}\rangle_{\mathbb{R}}. \tag{4}$$

Here $\langle\cdot,\cdot\rangle_\Omega$ and $\langle\cdot,\cdot\rangle_{\mathbb{R}}$ denote the dual pairings with respect to the scalar product in $L_2(\Omega)$ and $L_2(\mathbb{R})$ identified with $L_2(\mathbb{R}\times\{0\})$.

Proof. Let $g \in H^{1/2}(\mathbb{R})$ and $u \in H^1(\Omega)$ be chosen as above. From (3) and integration by parts we obtain

$$\begin{aligned}\langle \nabla u, \nabla v\rangle_\Omega - \omega\langle u, v\rangle_\Omega &= \int_{\mathbb{R}} \partial_y \hat{u}(\xi, 0)\overline{\hat{v}(\xi, 0)}\,\mathrm{d}\xi \\ &= \int_{\mathbb{R}} \sqrt{\xi^2 - \omega}\cdot \coth(\alpha\sqrt{\xi^2-\omega})\hat{g}(\xi)\overline{\hat{v}(\xi,0)}\,\mathrm{d}\xi \\ &= \langle D_\omega g, v|_{\mathbb{R}\times\{0\}}\rangle_{\mathbb{R}}.\end{aligned}$$

This proves one direction of the equivalence. The converse follows as the trace operator $u \mapsto u|_{\mathbb{R}\times\{0\}}\colon H^1(\Omega) \to H^{1/2}(\mathbb{R})$ has a continuous right inverse, cf. Lemma 3.36 in [25]. In particular, for every $f \in H^{1/2}(\mathbb{R})$ there exists $v \in H^1(\Omega)$ such that $v|_{\mathbb{R}\times\{\alpha\}} = 0$ and $v|_{\mathbb{R}\times\{0\}} = f$, and thus, $D_\omega g$ is uniquely defined by (4). □

In order to treat the mixed boundary value problem we introduce for $s \in \mathbb{R}$ the following function spaces:

$$\widetilde{H}_0^s(\Sigma_\ell) := \{g \in H^s(\mathbb{R})\colon \operatorname{supp}(g) \subseteq \overline{\Sigma_\ell}\}, \tag{5}$$

$$H^s(\Sigma_\ell) := \{g \in (C_c^\infty(\Sigma_\ell))'\colon \text{there exists } G \in H^s(\mathbb{R}) \text{ with } g = G|_{\Sigma_\ell}\}. \tag{6}$$

Here $C_c^\infty(\Sigma_\ell)$ is the space of smooth functions with compact support in Σ_ℓ; we denote by $(C_c^\infty(\Sigma_\ell))'$ the space of distributions on Σ_ℓ. We note that $\widetilde{H}_0^s(\Sigma_\ell)$ is a closed subspace of distributions in $\mathbb{R}$ whereas $H^s(\Sigma_\ell)$ is a subspace of distributions in Σ_ℓ. The latter space may be identified with the quotient space

$$H^s(\mathbb{R})/\widetilde{H}_0^s(\mathbb{R}\setminus\overline{\Sigma_\ell}),$$

where $\widetilde{H}_0^s(\mathbb{R}\setminus\overline{\Sigma_\ell})$ contains, by definition, those distributions in $H^s(\mathbb{R})$ which have support in $\mathbb{R}\setminus\Sigma_\ell$. We endow the spaces in (5) and (6) with their natural topology, i.e., $\widetilde{H}_0^s(\Sigma_\ell)$ carries the subspace topology of $H^s(\mathbb{R})$ and $H^s(\Sigma_\ell)$ has the quotient topology. For $s \geq 0$ we may identify $\widetilde{H}_0^s(\Sigma_\ell)$ with the subspace of $L_2(\Sigma_\ell)$ which consists of those functions whose extension by 0 yields an element of $H^s(\mathbb{R})$. Furthermore, the space $\widetilde{H}_0^s(\Sigma_\ell)$ is an isometric realisation of the (anti-)dual of $H^{-s}(\Sigma_\ell)$ and *vice versa*. The dual pairing is given by the expression

$$\langle g, h\rangle_{\Sigma_\ell} := \langle G, h\rangle_{\mathbb{R}}, \quad g \in H^{-s}(\Sigma_\ell),\ h \in \widetilde{H}_0^s(\Sigma_\ell), \tag{7}$$

where $G \in H^{-s}(\mathbb{R})$ denotes any extension of g, cf. Theorem 3.14 in [25]. Note that $C_c^\infty(\Sigma_\ell)$ is a dense subset of $\widetilde{H}_0^{1/2}(\Sigma_\ell)$, cf. Theorem 3.29 in [25]. In particular the above expression is independent of the chosen extension G. Thus, the domain of the quadratic form $a_{\ell,b}$ may be rewritten as

$$D[a_{\ell,b}] = \{u \in H^1(\Omega) : u|_{\mathbb{R}\times\{\alpha\}} = 0 \text{ and } u|_{\mathbb{R}\times\{0\}} \in \widetilde{H}_0^{1/2}(\Sigma_\ell)\}.$$

We define the truncated Dirichlet–to–Neumann operator

$$D_{\ell,\omega} : \widetilde{H}_0^{1/2}(\Sigma_\ell) \longrightarrow H^{-1/2}(\Sigma_\ell), \quad D_{\ell,\omega} := r_\ell D_\omega e_\ell,$$

where

$$r_\ell : H^{-1/2}(\mathbb{R}) \longrightarrow H^{-1/2}(\Sigma_\ell)$$

is the restriction operator and

$$e_\ell : \widetilde{H}_0^{1/2}(\Sigma_\ell) \longrightarrow H^{1/2}(\mathbb{R})$$

is the embedding. Identifying $\widetilde{H}_0^{1/2}(\Sigma_\ell)$ with a subspace of $L_2(\Sigma_\ell)$, the operator e_ℓ is simply extension by 0. Considering the corresponding topologies one easily observes that $D_{\ell,\omega}$ is a bounded linear operator. Recalling that $b \in L_\infty(\mathbb{R})$, we define the truncated Dirichlet–to–Robin operator by

$$D_{\ell,\omega} + b : \widetilde{H}_0^{1/2}(\Sigma_\ell) \longrightarrow H^{-1/2}(\Sigma_\ell), \quad D_{\ell,\omega} + b := r_\ell(D_\omega + b)e_\ell,$$

where we identify b with the corresponding multiplication operator. The next lemma gives a characterisation of the eigenvalues of $A_{\ell,b}$ in terms of the truncated Dirichlet–to–Robin operator.

Lemma 2.4. *Let* $\omega \in \mathbb{C} \setminus [\pi^2/\alpha^2, \infty)$ *and* $\ell > 0$. *Then*

$$\dim \ker(A_{\ell,b} - \omega) = \dim \ker(D_{\ell,\omega} + b).$$

Proof. The assertion follows if we prove that the trace mapping is an isomorphism of $\ker(A_{\ell,b} - \omega)$ onto $\ker(D_{\ell,\omega} + b)$. Let us first prove that it indeed maps $\ker(A_{\ell,b} - \omega)$ into $\ker(D_{\ell,\omega} + b)$. Let $u \in \ker(A_{\ell,b} - \omega)$ and denote $g := u|_{\mathbb{R}\times\{0\}} \in \widetilde{H}_0^{1/2}(\Sigma_\ell)$ its boundary trace. Let $h \in \widetilde{H}_0^{1/2}(\Sigma_\ell)$ be an arbitrary test function and choose $v \in D[a_{\ell,b}]$, such that $v|_{\mathbb{R}\times\{0\}} = h$. The dual pairing (7) and Lemma 2.3 imply

$$\begin{aligned} \langle (D_{\ell,\omega} + b)g, h \rangle_{\Sigma_\ell} &= \langle D_\omega g + bg, h \rangle_{\mathbb{R}} \\ &= \langle \nabla u, \nabla v \rangle_\Omega + \langle bg, h \rangle_{\mathbb{R}} - \omega \langle u, v \rangle_\Omega \\ &= a_{\ell,b}[u, v] - \omega \langle u, v \rangle_\Omega \\ &= 0, \end{aligned}$$

as $D_\omega g \in H^{1/2}(\mathbb{R})$ is obviously an extension of $D_{\ell,\omega} g \in H^{1/2}(\Sigma_\ell)$ and u is an eigenfunction for the eigenvalue ω. Hence, $g \in \ker(D_{\ell,\omega} + b)$, which proves that the mapping

$$\ker(A_{\ell,b} - \omega) \ni u \longmapsto u|_{\mathbb{R}\times\{0\}} \in \ker(D_{\ell,\omega} + b)$$

is well defined. Moreover, Lemma 2.1 implies that this mapping is injective. It remains to prove surjectivity. Let $g \in \ker(D_{\ell,\omega} + b)$ and denote by $u \in H^1(\Omega)$ the unique solution of the Poisson problem (1). Then $u \in D[a_{\ell,b}]$ and for arbitrary $v \in D[a_{\ell,b}]$ we have

$$\begin{aligned} a_{\ell,b}[u, v] &= \langle \nabla u, \nabla v \rangle_\Omega + \langle bg, v|_{\mathbb{R}\times\{0\}} \rangle_{\mathbb{R}} \\ &= \omega \langle u, v \rangle_\Omega + \langle D_\omega g + bg, v|_{\mathbb{R}\times\{0\}} \rangle_{\mathbb{R}} \\ &= \omega \langle u, v \rangle_\Omega + \langle (D_{\ell,\omega} + b)g, v|_{\mathbb{R}\times\{0\}} \rangle_{\Sigma_\ell} \\ &= \omega \langle u, v \rangle_\Omega. \end{aligned}$$

Thus, $u \in D(A_{\ell,b})$ and $(A_{\ell,b} - \omega)u = 0$. This proves the assertion. □

A particular consequence of Lemma 2.4 is the observation that the Dirichlet–to–Robin operator $D_{\ell,\omega} + b$ has non-trivial kernel if and only if ω is an eigenvalue of $A_{\ell,b}$. Put $V := \widetilde{H}_0^{1/2}(\Sigma_\ell)$ and consider the Gelfand triple

$$V \longrightarrow L_2(\Sigma_\ell) \longrightarrow V^*.$$

We identify V with a subspace of $L_2(\Sigma_\ell)$ and $V^* = H^{-1/2}(\Sigma_\ell)$ is the space of antilinear functionals on $\widetilde{H}_0^{1/2}(\Sigma_\ell)$, cf. the dual pairing (7). The truncated Dirichlet–to–Robin operator $D_{\ell,\omega}$ maps

$$D_{\ell,\omega} + b\colon V \longrightarrow V^*,$$

and thus, it is completely described by the sesquilinear form

$$(d_{\ell,\omega} + b)[g,h] := \langle (D_{\ell,\omega} + b)g, h\rangle_{V^*,V} = \langle D_{\ell,\omega} g, h\rangle_{\Sigma_\ell} + \langle bg, h\rangle_{\Sigma_\ell}$$

where $g, h \in D[d_{\ell,\omega}] := \widetilde{H}_0^{1/2}(\Sigma_\ell)$. Using the dual pairing (7) we obtain

$$d_{\ell,\omega}[g,h] = \langle D_\omega g, h\rangle_{\mathbb{R}} = \int_{\mathbb{R}} \sqrt{\xi^2 - \omega}\coth(\alpha\sqrt{\xi^2 - \omega}) \cdot \hat{g}(\xi)\,\overline{\hat{h}(\xi)}\,\mathrm{d}\xi \tag{8}$$

and

$$b[g,h] = \int_{\Sigma_\ell} b(x) \cdot g(x)\,\overline{h(x)}\,\mathrm{d}x. \tag{9}$$

We note that (8) is independent of ℓ and the dependence of ℓ in (9) is manifested in the domain of integration. In particular we may consider the dependence on ℓ as a constraint on the support of the functions g and h.

Lemma 2.5. *Let $\omega \in \mathbb{C}\setminus[\pi^2/\alpha^2, \infty)$. Then $d_{\ell,\omega}$ defines a closed sectorial form in $L_2(\Sigma_\ell)$. The associated m-sectorial operator is the restriction of $D_{\ell,\omega} + b$ to the operator domain*

$$X_{\ell,\omega} := \{g \in \widetilde{H}_0^{1/2}(\Sigma_\ell)\colon D_{\ell,\omega} g \in L_2(\Sigma_\ell)\}.$$

Proof. Combining formulae (8) and (9) we have for $g, h \in \widetilde{H}_0^{1/2}(\Sigma_\ell)$

$$(d_{\ell,\omega} + b)[g,h] = \int_{\mathbb{R}} m_\omega(\xi) \cdot \hat{g}(\xi)\,\overline{\hat{h}(\xi)}\,\mathrm{d}\xi + \int_{\Sigma_\ell} b(x) \cdot g(x)\,\overline{h(x)}\,\mathrm{d}x$$

with $m_\omega(\xi) = \sqrt{\xi^2 - \omega}\coth(\alpha\sqrt{\xi^2 - \omega})$. Note that

$$m_\omega(\xi) = \sqrt{\xi^2 - \omega}\coth(\alpha\sqrt{\xi^2 - \omega}) = |\xi| + O(1) \quad \text{as } \xi \to \pm\infty.$$

As $\widetilde{H}_0^{1/2}(\Sigma_\ell)$ carries the subspace topology induced by $H^{1/2}(\mathbb{R})$ this implies

$$c_1^{-1}\|g\|^2_{\widetilde{H}_0^{1/2}(\Sigma_\ell)} \le \operatorname{Re}(d_{\ell,\omega} + b)[g] + c_2\|g\|^2_{L_2(\Sigma_\ell)} \le c_1\|g\|^2_{\widetilde{H}_0^{1/2}(\Sigma_\ell)}, \tag{10}$$

for constants $c_i = c_i(\omega, \alpha, \Sigma_\ell) \in \mathbb{R}$, $i = 1, 2$. Thus, the form $d_{\ell,\omega}$ is bounded from below and closed. Moreover, it easily follows that $d_{\ell,\omega} + b$ is sectorial. To prove the second assertion let $g \in X_{\ell,\omega}$ such that $D_{\ell,\omega} g = \tilde{f} \in L_2(\Sigma_\ell)$. Then,

$$(d_{\ell,\omega} + b)[g, h] = \langle D_{\ell,\omega} g, h\rangle_{\Sigma_\ell} + \langle bg, h\rangle_{\Sigma_\ell} = \langle \tilde{f} + bg, h\rangle_{\Sigma_\ell}$$

for all $h \in C_c^\infty(\Sigma_\ell)$. As $C_c^\infty(\Sigma_\ell)$ is dense in $\widetilde{H}_0^{1/2}(\Sigma_\ell)$ the above equality holds true for every $h \in \widetilde{H}_0^{1/2}(\Sigma_\ell)$. Thus, g lies in the operator domain of the m-sectorial realisation which acts as $D_{\ell,\omega} + b$. In the same way it follows that the domain of the m-sectorial realisation is contained in $X_{\ell,\omega}$. □

Since

$$\ker(D_{\ell,\omega} + b) \subseteq X_{\ell,\omega},$$

we can apply Hilbert space methods to determine whether zero is an eigenvalue of $D_{\ell,\omega} + b$ or not. The spectrum of the m-sectorial realisation consists of a discrete set of eigenvalues only accumulating at infinity, since $D[d_{\ell,\omega} + b] = \widetilde{H}_0^{1/2}(\Sigma_\ell)$ is compactly embedded into $L_2(\Sigma_\ell)$, cf. Theorem 3.27 in [25]. Moreover, for real $\omega \in \mathbb{R}\backslash[\pi^2/\alpha^2, \infty)$ the quadratic form $d_{\ell,\omega} + b$ is symmetric, and thus, the associated operator is self-adjoint.

Remark 2.6. The close relation between self-adjoint extensions of differential operators and self-adjoint operators acting on the boundary has been pointed out in the case of bounded domains by G. Grubb, in particular with regard to resolvent formulae, cf. [21] and the references therein.

Another consequence of Lemma 2.5 or merely its proof is the following lemma.

Lemma 2.7. *The (original) operator $D_{\ell,\omega} + b\colon \widetilde{H}_0^{1/2}(\Sigma_\ell) \to H^{-1/2}(\Sigma_\ell)$ is an Fredholm operator with zero index.*

The proof follows by combining formula (10) and Theorems 2.34 and 3.27 in [25].

As the m-sectorial realisation of the Dirichlet–to–Robin operator is simply the restriction of the original operator we do not want to introduce a separate notation for it. In fact, we will mainly work with a quadratic form, which arises after scaling the Robin window. Recall that $\Sigma_\ell := \ell \cdot \Sigma$. We define the unitary scaling operator

$$T_\ell\colon L_2(\Sigma) \longrightarrow L_2(\Sigma_\ell), \quad (T_\ell g)(x) = \ell^{-1/2} g\Big(\frac{x}{\ell}\Big).$$

Note that the operator T_ℓ bijectively maps $\widetilde{H}_0^{1/2}(\Sigma)$ into $\widetilde{H}_0^{1/2}(\Sigma_\ell)$. Set

$$\mathcal{Q}_b(\ell, \omega)\colon \widetilde{H}_0^{1/2}(\Sigma) \longrightarrow H^{-1/2}(\Sigma), \quad \mathcal{Q}_b(\ell, \omega) := T_\ell^*(D_{\ell,\omega} + b)T_\ell$$

and let

$$q_b(\ell,\omega)[g,h] := (d_{\ell,\omega} + b)[T_\ell g, T_\ell h]$$

be the associated sesquilinear form with $D[q_b(\ell,\omega)] := \widetilde{H}_0^{1/2}(\Sigma)$. Then

$$\dim\ker(A_{\ell,b} - \omega) = \dim\ker(D_{\ell,\omega} + b) = \dim\ \ker(\mathcal{Q}_b(\ell,\omega)).$$

Next we prove an asymptotic expansion of the operator $\mathcal{Q}_b(\ell,\omega)$ as $\ell \to 0$ and $\omega \to \pi^2/\alpha^2$. This expansion represents the principal tool of the proof of the main result. Here and subsequently we denote by $\mathcal{Q}_0\colon \widetilde{H}_0^{1/2}(\Sigma) \to H^{-1/2}(\Sigma)$,

$$\langle \mathcal{Q}_0 g, h\rangle_\Sigma := q_0[g,h] := \int_{\mathbb{R}} |\xi| \cdot \hat{g}(\xi)\, \overline{\hat{h}(\xi)}\, \mathrm{d}\xi$$

the Dirichlet–to–Neumann operator for the mixed problem on the upper half-space or equivalently on the lower half-space corresponding to the spectral parameter $\omega = 0$. Note that $\mathcal{Q}_0\colon \widetilde{H}_0^{1/2}(\Sigma) \to H^{-1/2}(\Sigma)$ is also a Fredholm operator with Fredholm index 0, which follows from Theorem 2.34 in [25]. The identity $\mathcal{Q}_0 g = 0$ implies

$$0 = \langle \mathcal{Q}_0 g, g\rangle_\Sigma = \int_{\mathbb{R}} |\xi| \cdot |\hat{g}(\xi)|^2\, \mathrm{d}\xi,$$

and thus, $g = 0$. Hence $\mathcal{Q}_0$ has trivial kernel and it is invertible.

In what follows we denote by P_{ct} the projection in $L_2(\Sigma)$ onto the subspace of constant functions and let $K_{\ln|x|}\colon L_2(\Sigma) \to L_2(\Sigma)$,

$$(K_{\ln|x|} f)(z) = \int_\Sigma \ln|z-x| \cdot f(x)\, \mathrm{d}x, \quad x \in \Sigma.$$

Theorem 2.8. *Let $b = 0$. There exist $\ell_0 = \ell_0(\alpha,\Sigma) > 0$ and $\varepsilon = \varepsilon(\alpha,\Sigma) > 0$ such that for $\ell \in (0,\ell_0)$ and $|\omega - \pi^2/\alpha^2| < \varepsilon$ the asymptotic expansion holds true*

$$\begin{aligned}\mathcal{Q}_0(\ell,\omega) &= \frac{1}{\ell}\mathcal{Q}_0 - \ell \cdot \Big(\frac{|\Sigma|\cdot\pi^2}{\alpha^3}\Big)\Big(\sqrt{\frac{\pi^2}{\alpha^2} - \omega}\Big)^{-1} P_{\mathrm{ct}} \\ &\quad + \sum_{k_1=1}^{\infty}\sum_{k_2=0}^{\infty} \ell^{2k_1-1}\Big(\sqrt{\frac{\pi^2}{\alpha^2} - \omega}\Big)^{k_2} (B^{(0)}_{k_1,k_2} + B^{(1)}_{k_1,k_2} \cdot \ln\ell).\end{aligned}$$

Here $B^{(i)}_{k_1,k_2} \in \mathcal{L}(L_2(\Sigma))$ for $i = 1,2$. The series converges absolutely in the operator norm of $\mathcal{L}(L_2(\Sigma))$. For the first terms we obtain

$$B^{(0)}_{1,0} = \frac{|\Sigma|\cdot\rho(\alpha)}{2\pi} P_{\mathrm{ct}} + \frac{\pi}{2\alpha^2} K_{\ln|x|}, \quad B^{(1)}_{1,0} = \frac{|\Sigma|\cdot\pi}{2\alpha^2} P_{\mathrm{ct}}, \quad B^{(0)}_{1,1} = B^{(1)}_{1,1} = 0,$$

where the constant $\rho(\alpha) \in \mathbb{R}$ is given by formula (15) *and $|\Sigma|$ is the Lebesgue measure of Σ.*

The next section is devoted to the proof of Theorem 2.8.

The proof of Theorem 2.8

For $g, h \in \widetilde{H}_0^{1/2}(\Sigma)$ we have

$$\langle \mathfrak{Q}_0(\ell,\omega)g, h\rangle_\Sigma = q_0(\ell,\omega)[g,h] = \ell \int_{\mathbb{R}} m_\omega(\xi)\cdot \hat{g}(\ell\xi)\,\overline{h(\ell\xi)}\,\mathrm{d}\xi,$$

where $m_\omega(\xi) = \sqrt{\xi^2-\omega}\cdot\coth(\alpha\sqrt{\xi^2-\omega})$. The main idea of the proof is to use the asymptotic expansion of the function $m_\omega(\xi)$ for $\xi = 0$ and $\xi \to \pm\infty$ while letting the parameter $\omega \to \pi^2/\alpha^2$.

As a first step of the proof we show that m_ω has a meromorphic extension to the complex plane and calculate explicitly its singularities and residues. To this end we use the partial fraction decomposition of the hyperbolic cotangent function, i.e., we have

$$\coth(z) = \frac{1}{z} + \sum_{k=1}^{\infty} \frac{2z}{z^2 + k^2\pi^2}, \qquad z \in \mathbb{C}\backslash\{ik\pi : k \in \mathbb{Z}\},$$

cf., e.g., [24] (Chapter V, §1.71). Hence,

$$m_\omega(\xi) = \frac{1}{\alpha} + \sum_{k=1}^{\infty} \frac{2\alpha(\xi^2-\omega)}{\alpha^2(\xi^2-\omega) + k^2\pi^2}, \qquad \xi \in \mathbb{R}, \tag{11}$$

and thus, the mermorphic extension of m_ω is given by the above series. For $\omega \in \mathbb{C}\backslash[\pi^2/\alpha^2, \infty)$ the singularities of m_ω are all simple poles which are located at

$$\pm\mathrm{i}\sqrt{\frac{k^2\pi^2}{\alpha^2} - \omega}, \qquad k \in \mathbb{N}.$$

In particular they do not lie on the real axis. As $\omega \to \pi^2/\alpha^2$ the two poles nearest the real axis converge to $0 \in \mathbb{C}$ and they give rise to a pole of order two in the limit case. Here and subsequently we fix $\beta = \beta(\alpha) \in (\pi/\alpha, \sqrt{3}\,\pi/\alpha)$. Then there exists $\varepsilon = \varepsilon(\alpha) > 0$ such that for $\omega \in \mathbb{C}\backslash[\pi^2/\alpha^2, \infty)$, $|\omega - \pi^2/\alpha^2| < \varepsilon$ the function m_ω has exactly two poles inside the strip $\mathbb{R} + \mathrm{i}[-\beta, \beta]$. The residues of the function m_ω at these points are given by

$$\operatorname{Res}_{\xi = \pm\mathrm{i}\sqrt{\pi^2/\alpha^2 - \omega}}\, m_\omega(\xi) = \pm\frac{\pi^2}{\alpha^3}\cdot\frac{\mathrm{i}}{\sqrt{\frac{\pi^2}{\alpha^2} - \omega}} \tag{12}$$

as can easily be seen from the expansion (11).

Let $g, h \in \widetilde{H}_0^{1/2}(\Sigma)$. Since g, h are compactly supported their Fourier transforms $\hat{g}, \hat{h}$ admit holomorphic extensions on the whole complex plane. Note that the function $\hat{h}^*$,

$$\hat{h}^*(\xi) := \overline{\hat{h}(\bar{\xi})}, \qquad \xi \in \mathbb{C},$$

is also an entire function on $\mathbb{C}$. We decompose the form $q_0(\ell, \omega)$ as follows

$$q_0(\ell,\omega)[g,h] = \ell\Big(\int_{[-1,1]} + \int_{[-1,1]^c}\Big) m_\omega(\xi)\cdot \hat{g}(\ell\xi)\,\hat{h}^*(\ell\xi)\,\mathrm{d}\xi, \qquad (13)$$

where we put $[-1,1]^c = \mathbb{R}\backslash[-1,1]$. Using the Taylor expansion of the function $\hat{g}\cdot\hat{h}^*$ at $0 \in \mathbb{C}$, we obtain for the first integral

$$\int_{[-1,1]} m_\omega(\xi)\cdot \hat{g}(\ell\xi)\,\hat{h}^*(\ell\xi)\,\mathrm{d}\xi = \sum_{k=0}^{\infty} \ell^k e_k[g,h] \int_{[-1,1]} \xi^k m_\omega(\xi)\,\mathrm{d}\xi$$

with

$$e_k[g,h] = \frac{1}{k!}\cdot\frac{\mathrm{d}^k}{\mathrm{d}\xi^k}(\hat{g}(\xi)\hat{h}^*(\xi))\Big|_{\xi=0} = \frac{1}{k!}\cdot\sum_{j=0}^{k}\binom{k}{j}\hat{g}^{(j)}(0)\cdot\overline{\hat{h}^{(k-j)}(0)}.$$

We note that m_ω is an even function, and thus, in the expansion the terms of odd order vanish. Let E_k be the operator associated with the form e_k. Then

$$\int_{[-1,1]} m_\omega(\xi)\cdot \hat{g}(\ell\xi)\,\hat{h}^*(\ell\xi)\,\mathrm{d}\xi = \sum_{k=0}^{\infty} \ell^{2k}\langle E_{2k}g, h\rangle_\Sigma \int_{[-1,1]} \xi^{2k} m_\omega(\xi)\,\mathrm{d}\xi.$$

Note that

$$|\hat{g}^{(j)}(0)| \le \frac{1}{\sqrt{2\pi}}\Big(\int_\Sigma |x|^{2j}\Big)^{1/2}\|g\|_{L_2(\Sigma)} \le C^j\|g\|_{L_2(\Sigma)}$$

for sufficiently large $C = C(\Sigma) > 0$, which implies

$$\|E_k\|_{\mathcal{L}(L_2(\Sigma))} \le \frac{(2C)^k}{k!}.$$

To estimate the integral $\int_{[-1,1]} \xi^{2k} m_\omega(\xi)\,\mathrm{d}\xi$ we denote by γ the path depicted in Figure 4 connecting the points -1 and 1. Its image $\mathrm{im}(\gamma)$ coincides with the boundary of the following rectangle except for the line segment $[-1,1]$.

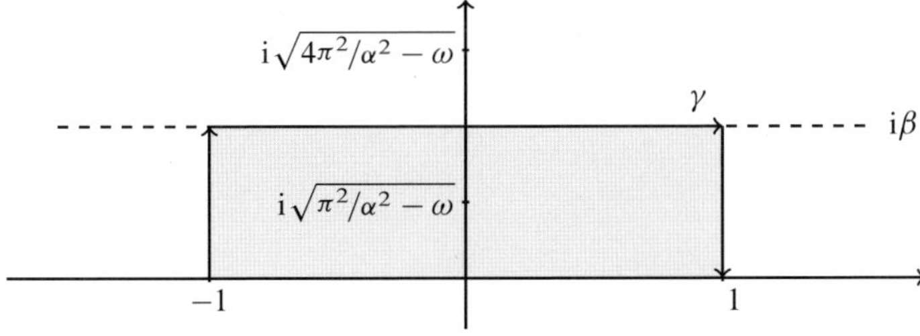

Figure 4. The path γ.

Note that $\operatorname{Im}(\mathrm{i}\sqrt{\pi^2/\alpha^2-\omega})>0$ if $\omega\in\mathbb{C}\setminus[\pi^2/\alpha^2,\infty)$. Using formula (12) the residue theorem implies

$$\int_{[-1,1]}\xi^{2k}m_\omega(\xi)\,\mathrm{d}\xi=-\frac{(-1)^k\cdot 2\pi^3}{\alpha^3}\Big(\frac{\pi^2}{\alpha^2}-\omega\Big)^{(k-1)/2}+\int_\gamma\xi^{2k}m_\omega(\xi)\,\mathrm{d}\xi.$$

Next we use for fixed $\xi\in\operatorname{im}(\gamma)$ the Taylor expansion of $m_\omega(\xi)$ at $\omega=\pi^2/\alpha^2$. Thus, there exists $\varepsilon>0$ such that for $|\omega-\pi^2/\alpha^2|<\varepsilon$ we have

$$m_\omega(\xi)=\sum_{j=0}^\infty\frac{(-1)^j\Big(\frac{\pi^2}{\alpha^2}-\omega\Big)^j}{j!}\frac{\mathrm{d}^j}{\mathrm{d}\omega^j}m_\omega(\xi)\Big|_{\omega=\pi^2/\alpha^2}.$$

This expression may be considered as a power series in ω with values in $L_1(\operatorname{im}(\gamma))$, and we obtain

$$\int_\gamma\xi^{2k}m_\omega(\xi)\,\mathrm{d}\xi=\sum_{j=0}^\infty\frac{(-1)^j\Big(\frac{\pi^2}{\alpha^2}-\omega\Big)^j}{j!}\Big[\int_\gamma\xi^{2k}\frac{\mathrm{d}^j}{\mathrm{d}\omega^j}m_\omega(\xi)\,\mathrm{d}\xi\Big]_{\omega=\pi^2/\alpha^2}.$$

Finally,

$$\begin{aligned}
&\ell\int_{[-1,1]}m_\omega(\xi)\cdot\hat{g}(\ell\xi)\,\hat{h}^*(\ell\xi)\,\mathrm{d}\xi\\
&=-\ell\Big(\frac{2\pi^3}{\alpha^3}\Big)\Big(\sqrt{\frac{\pi^2}{\alpha^2}-\omega}\Big)^{-1}\langle E_0g,h\rangle_\Sigma\\
&\quad-\Big(\frac{2\pi^3}{\alpha^3}\Big)\cdot\sum_{k=1}^\infty\ell^{2k+1}\cdot(-1)^k\cdot\Big(\frac{\pi^2}{\alpha^2}-\omega\Big)^{(k-1)/2}\cdot\langle E_{2k}g,h\rangle_\Sigma\\
&\quad+\sum_{k=0}^\infty\sum_{j=0}^\infty\ell^{2k+1}\langle E_{2k}g,h\rangle_\Sigma\frac{(-1)^j\Big(\frac{\pi^2}{\alpha^2}-\omega\Big)^j}{j!}\Big[\int_\gamma\xi^{2k}\frac{\mathrm{d}^j}{\mathrm{d}\omega^j}m_\omega(\xi)\,\mathrm{d}\xi\Big]_{\omega=\pi^2/\alpha^2}.
\end{aligned}$$

We note that the two series

$$\sum_{k=1}^\infty\ell^{2k+1}E_{2k}\cdot(-1)^k\cdot\Big(\frac{\pi^2}{\alpha^2}-\omega\Big)^{(k-1)/2}$$

and

$$\sum_{k=0}^\infty\sum_{j=0}^\infty\ell^{2k+1}E_{2k}\frac{(-1)^j\Big(\frac{\pi^2}{\alpha^2}-\omega\Big)^j}{j!}\Big[\int_\gamma\xi^{2k}\frac{\mathrm{d}^j}{\mathrm{d}\omega^j}m_\omega(\xi)\,\mathrm{d}\xi\Big]_{\omega=\pi^2/\alpha^2}$$

converge absolutely in the operator norm in $\mathcal{L}(L_2(\Sigma))$, uniformly in $\ell \in [0,1]$ and $|\omega - \pi^2/\alpha^2| < \varepsilon$. For the first series this is obvious. Considering the second series leads us to the estimate

$$\sum_{k=0}^{\infty} \ell^{2k+1} \|E_{2k}\|_{\mathcal{L}(L_2(\Sigma))} \sum_{j=0}^{\infty} \frac{\left|\frac{\pi^2}{\alpha^2} - \omega\right|^j}{j!} \left| \int_\gamma \xi^{2k} \frac{\mathrm{d}^j}{\mathrm{d}\omega^j} m_{\pi^2/\alpha^2}(\xi)\, \mathrm{d}\xi \right|$$

$$\leq \Big(\sum_{k=0}^{\infty} \ell^{2k+1} \cdot c_1^{2k} \cdot \frac{(2C)^{2k}}{(2k)!} \Big) \Big(\sum_{j=0}^{\infty} \frac{\varepsilon^j}{j!} \Big\| \frac{\mathrm{d}^j}{\mathrm{d}\omega^j} m_{\pi^2/\alpha^2} \Big\|_{L_1(\mathrm{im}(\gamma))} \Big) < \infty$$

for $\ell > 0$ and $|\omega - \pi^2/\alpha^2| < \varepsilon$. Here $c_1 := \sup\{|\xi| : \xi \in \mathrm{im}(\gamma)\}$. Calculating the first terms of the expansion we obtain

$$\ell \int_{[-1,1]} m_\omega(\xi) \cdot \hat{g}(\ell\xi)\, \hat{h}^*(\ell\xi)\, \mathrm{d}\xi$$

$$= -\ell \cdot \frac{2\pi^3 (\hat{g}\hat{h}^*)(0)}{\alpha^3 \sqrt{\frac{\pi^2}{\alpha^2} - \omega}} + \ell \cdot (\hat{g}\hat{h}^*)(0) \int_\gamma m_{\pi^2/\alpha^2}(\xi)\, \mathrm{d}\xi + \mathcal{O}\Big(\ell^3 + \frac{\pi^2}{\alpha^2} - \omega\Big).$$

Let P_{ct} be the projection in $L_2(\Sigma)$ onto the subspace of constant functions. Then

$$(\hat{g}\hat{h}^*)(0) = \frac{1}{2\pi} \Big(\int_\Sigma g(x)\, \mathrm{d}x \Big) \Big(\int_\Sigma \overline{h(x)}\, \mathrm{d}x \Big) = \frac{|\Sigma|}{2\pi} \langle P_{\mathrm{ct}} g, h \rangle_\Sigma,$$

and thus,

$$\ell \int_{[-1,1]} m_\omega(\xi) \cdot \hat{g}(\ell\xi)\, \hat{h}^*(\ell\xi)\, \mathrm{d}\xi$$

$$= \ell \Big(- \frac{|\Sigma| \cdot \pi^2}{\alpha^3 \sqrt{\frac{\pi^2}{\alpha^2} - \omega}} + \frac{|\Sigma| \cdot \rho_{0,1}(\alpha)}{2\pi} \Big) \langle P_{\mathrm{ct}} g, h \rangle_\Sigma + \mathcal{O}\Big(\ell^3 + \frac{\pi^2}{\alpha^2} - \omega\Big),$$

with $\rho_{0,1}(\alpha) = \int_\delta m_{\pi^2/\alpha^2}(\xi)\, \mathrm{d}\xi$. In order to treat the second integral in (13) we use the asymptotic expansion of $m_\omega(\xi)$ for $\xi \to \pm\infty$. For ease of notation we suppose that $\alpha > \pi$, so that $[-1,1] \subseteq (-\pi/\alpha, \pi/\alpha)$. We have

$$\ell \int_{[-1,1]^c} m_\omega(\xi) \cdot \hat{g}(\ell\xi)\, \hat{h}^*(\ell\xi)\, \mathrm{d}\xi$$

$$= \ell \int_{\mathbb{R}} |\xi| \cdot \hat{g}(\ell\xi)\, \hat{h}^*(\ell\xi)\, \mathrm{d}\xi + \ell \sum_{i=1}^{3} \int_{\mathbb{R}} m_{\omega,i}(\xi) \cdot \hat{g}(\ell\xi)\, \hat{h}^*(\ell\xi)\, \mathrm{d}\xi,$$

where

$$m_{\omega,1}(\xi) := -\mathbb{1}_{[-1,1]}(\xi) \cdot |\xi|,$$

$$m_{\omega,2}(\xi) := \mathbb{1}_{[-1,1]^c}(\xi) \cdot (\sqrt{\xi^2 - \omega} - |\xi|),$$

$$m_{\omega,3}(\xi) := \mathbb{1}_{[-1,1]^c}(\xi) \sqrt{\xi^2 - \omega}(\coth(\alpha \sqrt{\xi^2 - \omega}) - 1).$$

Here $\mathbb{1}_E$ denotes the indicator function of a Borel set $E \subseteq \mathbb{R}$. We note that the first term $\int_{\mathbb{R}} m_{\omega,1}(\xi) \cdot \hat{g}(\ell\xi)\ \hat{h}^*(\ell\xi)\ \mathrm{d}\xi$ is independent of ω and may be expanded as before into a power series with respect to the parameter ℓ; we have

$$\begin{aligned} \ell \int_{\mathbb{R}} m_{\omega,1}(\xi) \cdot \hat{g}(\ell\xi)\ \hat{h}^*(\ell\xi)\ \mathrm{d}\xi &= \ell \int_{[-1,1]} |\xi| \cdot \hat{g}(\ell\xi)\ \hat{h}^*(\ell\xi)\ \mathrm{d}\xi \\ &= \frac{\ell|\Sigma|}{2\pi} \langle P_{\mathrm{ct}} g, h\rangle_{\Sigma} \cdot \int_{[-1,1]} |\xi|\ \mathrm{d}\xi + \mathcal{O}(\ell^3) \\ &= \frac{\ell|\Sigma|}{2\pi} \langle P_{\mathrm{ct}} g, h\rangle + \mathcal{O}(\ell^3). \end{aligned}$$

To treat the second integral we use for $\xi \in \mathbb{R}$ and $|\omega - \pi^2/\alpha^2| < \varepsilon$ the following expansion

$$\begin{aligned} m_{\omega,2}(\xi) &= \mathbb{1}_{[-1,1]^c}(\xi)(\sqrt{\xi^2 - \omega} - |\xi|) = \mathbb{1}_{[-1,1]^c}(\xi) \cdot \left(|\xi| \cdot \sqrt{1 - \frac{\omega}{\xi^2}} - |\xi| \right) \\ &= \sum_{k=1}^{\infty} \binom{1/2}{k} (-\omega)^k \cdot |\xi|^{-2k+1} \mathbb{1}_{[-1,1]^c}(\xi). \end{aligned} \tag{14}$$

This series may be considered as power series in ω with values in $L_\infty(\mathbb{R}_\xi)$. Let $Y_k, Z_k \in C^\infty(\mathbb{R}\backslash\{0\}) \cap L_{1,\mathrm{loc}}(\mathbb{R})$ such that

$$\widehat{Y}_k(\xi) = \frac{1}{\sqrt{2\pi}} \cdot |\xi|^{-2k+1} \mathbb{1}_{[-1,1]^c}(\xi) \quad \text{and} \quad \widehat{Z}_k(\xi) = \frac{1}{\sqrt{2\pi}} \text{ f.p.}(|\xi|^{-2k+1}).$$

Here

$$\text{f.p.}(|\xi|^{-2k+1}) = \text{f.p.}(\xi_+^{-2k+1}) + \text{f.p.}(\xi_-^{-2k+1})$$

designates the distribution which is defined by the finite part of the singular function $|\xi|^{-2k+1}$, cf. Chapter 5 in [25]. We note that $X_k := Y_k - Z_k$ is analytic, since its Fourier transform $\widehat{X}_k = \widehat{Y}_k - \widehat{Z}_k$ has compact support. This allows us to determine the order of the singularity of Y_k at $0 \in \mathbb{R}$. Using Lemma 5.10 in [25] we have

$$Z_k(x) = \frac{1}{\pi} \cdot \frac{(-1)^k x^{2k-2}}{(2k-2)!} (\ln|x| + \gamma_0 - H_{2k-2}),$$

where γ_0 is the Euler–Mascheroni constant and $H_{2k-2} = \sum_{j=1}^{2k-2} 1/j$. We have

$$\begin{aligned}
&\ell \int_{\mathbb{R}} m_{\omega,2}(\xi) \cdot \hat{g}(\ell\xi)\, \hat{h}^*(\ell\xi)\, \mathrm{d}\xi \\
&\quad = \sum_{k=1}^{\infty} \binom{1/2}{k} (-\omega)^k \cdot \ell \cdot \int_{\mathbb{R}} \hat{Y}_k(\xi) \cdot \hat{g}(\ell\xi)\, \hat{h}^*(\ell\xi)\, \mathrm{d}\xi \\
&\quad = \sum_{k=1}^{\infty} \binom{1/2}{k} (-\omega)^k \cdot \langle Y_k * T_\ell g, T_\ell h\rangle_{\Sigma_\ell},
\end{aligned}$$

and

$$\langle Y_k * T_\ell g, T_\ell h\rangle_{\Sigma_\ell} = \ell \int_{\Sigma\times\Sigma} (X_k + Z_k)(\ell|z-x|)\, g(x)\, \overline{h(z)}\, \mathrm{d}x\, \mathrm{d}z.$$

Defining the operators $K_{|x|^{2k-2}}, K_{|x|^{2k-2}\ln|x|} \colon L_2(\Sigma) \to L_2(\Sigma)$,

$$\begin{aligned}
(K_{|x|^{2k-2}} f)(z) &= \int_{\Sigma} |z-x|^{2k-2} f(x)\, \mathrm{d}x, \quad z \in \Sigma \\
(K_{|x|^{2k-2}\ln|x|} f)(z) &= \int_{\Sigma} |z-x|^{2k-2} \ln|z-x|\, f(x)\, \mathrm{d}x, \quad z \in \Sigma
\end{aligned}$$

we obtain

$$\begin{aligned}
&\ell \int_{\Sigma\times\Sigma} Z_k(\ell|z-x|)\, g(x)\, \overline{h(z)}\, \mathrm{d}x\, \mathrm{d}z \\
&\quad = \frac{(-1)^k\, \ell^{2k-1}}{\pi \cdot (2k-2)!} \int_{\Sigma\times\Sigma} |z-x|^{2k-2} \ln(\ell|z-x|)\, g(x)\, \overline{h(z)}\, \mathrm{d}x\, \mathrm{d}z \\
&\qquad + \frac{(-1)^k\, \ell^{2k-1}}{\pi \cdot (2k-2)!} \cdot (\gamma_0 - H_{2k-2}) \int_{\Sigma\times\Sigma} |z-x|^{2k-2}\, g(x)\, \overline{h(z)}\, \mathrm{d}x\, \mathrm{d}z \\
&\quad = \frac{(-1)^k\, \ell^{2k-1}}{\pi \cdot (2k-2)!} (\ln\ell + \gamma_0 - H_{2k-2}) \langle K_{|x|^{2k-2}} g, h\rangle_{\Sigma} \\
&\qquad + \frac{(-1)^k\, \ell^{2k-1}}{\pi \cdot (2k-2)!} \langle K_{|x|^{2k-2}\ln|x|} g, h\rangle_{\Sigma}.
\end{aligned}$$

Note that

$$\ell \int_{\Sigma\times\Sigma} X_k(\ell|z-x|)\, g(x)\, \overline{h(z)}\, \mathrm{d}x\, \mathrm{d}z = \sum_{j=0}^{\infty} \frac{X_k^{(2j)}(0) \cdot \ell^{2j+1}}{(2j)!} \langle K_{|x|^{2j}} g, h\rangle_{\Sigma},$$

which follows by expanding the even function X_k into a power series. For the coefficients $X_k^{(2j)}(0)$ we obtain from the definition of the finite part

$$\begin{aligned}2\pi X_k^{(2j)}(0) &= 2\int_0^\infty (\mathrm{i}\xi)^{2j}(\hat{Y}_k - \hat{Z}_k)(\xi)\,\mathrm{d}\xi \\ &= -2(-1)^j \mathrm{f.p.}_{\varepsilon\to 0}\int_\varepsilon^1 \xi^{2j}\cdot\xi^{-2k+1}\,\mathrm{d}\xi \\ &= (-1)^{j+1}\cdot\begin{cases}\dfrac{1}{j-k+1} & \text{if } j-k+1\neq 0,\\ 0 & \text{if } j-k+1=0.\end{cases}\end{aligned}$$

Finally, we have

$$\begin{aligned}&\ell\int_{\mathbb{R}} m_{\omega,2}(\xi)\cdot\hat{g}(\ell\xi)\,\hat{h}^*(\ell\xi)\,\mathrm{d}\xi \\ &= \sum_{k=1}^\infty \binom{1/2}{k}\frac{\omega^k\cdot\ell^{2k-1}}{\pi(2k-2)!}\Big[(\ln\ell+\gamma_0-H_{2k-2})\langle K_{|x|^{2k-2}}g,h\rangle_\Sigma \\ &\qquad\qquad + \langle K_{|x|^{2k-2}\ln|x|}g,h\rangle_\Sigma\Big] \\ &\quad+\sum_{k=1}^\infty\sum_{j=0}^\infty\binom{1/2}{k}\cdot(-\omega)^k\cdot\frac{X_k^{(2j)}(0)\cdot\ell^{2j+1}}{(2j)!}\langle K_{|x|^{2j}}g,h\rangle_\Sigma.\end{aligned}$$

Note that both series,

$$\frac{1}{\pi}\sum_{k=1}^\infty\binom{1/2}{k}\cdot\omega^k\cdot\frac{\ell^{2k-1}}{(2k-2)!}[(\ln\ell+\gamma_0-H_{2k-2})K_{|x|^{2k-2}}+K_{|x|^{2k-2}\ln|x|}]$$

and

$$\sum_{k=1}^\infty\sum_{j=0}^\infty\binom{1/2}{k}(-\omega)^k\cdot\frac{X_k^{(2j)}(0)\cdot\ell^{2j+1}}{(2j)!}K_{|x|^{2j}},$$

converge uniformly in the operator norm for $\ell\in(0,\ell_0)$ and $|\omega-\pi^2/\alpha^2|<\varepsilon$. This follows from the estimates on the coefficients $X_k^{2j}(0)$ and from

$$\|K_{|x|^{2k-2}}\|_{\mathcal{L}(L_2(\Sigma))}\le C^{2k-2},\quad \|K_{|x|^{2k-2}\ln|x|}\|_{\mathcal{L}(L_2(\Sigma))}\le C^{2k-2}$$

for sufficiently large $C=C(\Sigma)>0$. Note that $\omega<1$ since we assumed that $\pi^2/\alpha^2<1$. Changing the centre of the power series in ω and calculating the first terms give us the following asymptotic estimate

$$\ell \int_{\mathbb{R}} m_{\omega,2}(\xi) \cdot \hat{g}(\ell\xi)\, \hat{h}^*(\ell\xi)\, \mathrm{d}\xi$$
$$= \frac{\ell \ln \ell \cdot |\Sigma| \cdot \pi}{2\alpha^2} \langle P_{\mathrm{ct}} g, h\rangle_\Sigma + \ell \cdot |\Sigma| \cdot \Big(\frac{\pi \cdot \gamma_0}{2\alpha^2} + \frac{\rho_{0,2}(\alpha)}{2\pi}\Big) \langle P_{\mathrm{ct}} g, h\rangle_\Sigma$$
$$+ \ell \cdot \frac{\pi}{2\alpha^2} \langle K_{\ln |x|} g, h\rangle_\Sigma + \mathcal{O}\Big(\ell^3 \ln \ell + \frac{\pi^2}{\alpha^2} - \omega\Big),$$

where

$$\rho_{0,2}(\alpha) := \pi \sum_{k=1}^{\infty} \binom{1/2}{k} \Big(-\frac{\pi^2}{\alpha^2}\Big)^k X_k(0) = \frac{1}{2} \sum_{k=2}^{\infty} \binom{1/2}{k} \Big(-\frac{\pi^2}{\alpha^2}\Big)^k \frac{1}{k-1}.$$

We note that $K_{|x|^0} = |\Sigma| \cdot P_{\mathrm{ct}}$. Thus, the only point remaining is the expansion of the integral $\ell^2 \int_{\mathbb{R}} m_{\omega,3}(\xi) \cdot \hat{g}(\ell\xi)\, \hat{h}^*(\ell\xi)\, \mathrm{d}\xi$. We have

$$m_{\omega,3}(\xi) = \mathbb{1}_{[-1,1]^c}(\xi)\, \sqrt{\xi^2 - \omega}(\coth(\alpha \sqrt{\xi^2 - \omega}) - 1).$$

It easily follows that the function

$$\omega \mapsto \mathrm{e}^{\delta|\xi|} m_{\omega,3}(\xi) \in L_\infty(\mathbb{R}_\xi)$$

is a vector-valued holomorphic function for $|\omega - \pi^2/\alpha^2| < \varepsilon$ and some $\delta > 0$. In particular, we obtain that

$$m_{\omega,3}(\xi) = \sum_{k=0}^{\infty} \frac{(-1)^k \Big(\frac{\pi^2}{\alpha^2} - \omega\Big)^k}{k!} \frac{\mathrm{d}^k}{\mathrm{d}\omega^k} m_{\omega,3}(\xi)\Big|_{\omega=\pi^2/\alpha^2}$$

and the series converges absolutely as a power series in ω with values in some exponentially weighted L_∞-space. Choose $\widetilde{X}_k \in C^\infty(\mathbb{R})$ such that

$$\widehat{\widetilde{X}_k}(\xi) = \frac{1}{\sqrt{2\pi}} \cdot \frac{\mathrm{d}^k}{\mathrm{d}\omega^k} m_{\omega,3}(\xi)\Big|_{\omega=\pi^2/\alpha^2}.$$

Then $\widetilde{X}$ is an even function and analytic in some neighbourhood of 0,

$$\widetilde{X}_k(x) = \sum_{j=0}^{\infty} \frac{\widetilde{X}_k^{(2j)}(0)}{(2j)!} x^{2j},$$

where

$$\widetilde{X}_k^{(2j)}(0) = \frac{1}{2\pi} \int_{\mathbb{R}} (\mathrm{i}\xi)^{2j} \frac{\mathrm{d}^k}{\mathrm{d}\omega^k} m_{\pi^2/\alpha^2,3}(\xi)\, \mathrm{d}\xi.$$

Note that

$$|\widetilde{X}_k^{(2j)}(0)| \le \frac{1}{2\pi} \left\| e^{\delta|\xi|} \frac{d^k}{d\omega^k} m_{\pi^2/\alpha^2,3}(\xi) \right\|_{L_\infty(\mathbb{R}_\xi)} 2 \int_0^\infty \xi^{2j} e^{-\delta\xi} \, d\xi$$
$$= \frac{\delta^{-1-2j}(2j)!}{\pi} \left\| e^{\delta|\xi|} \frac{d^k}{d\omega^k} m_{\pi^2/\alpha^2,3}(\xi) \right\|_{L_\infty(\mathbb{R}_\xi)},$$

since

$$\int_0^\infty \xi^{2j} e^{-\delta\xi} \, d\xi = \delta^{-1-2j}(2j)!,$$

cf. (2.321) in [20]. In particular, we obtain

$$\ell \int_{\mathbb{R}} m_{\omega,3}(\xi) \cdot \hat{g}(\ell\xi) \, \hat{h}^*(\ell\xi) \, d\xi$$
$$= \sum_{k=0}^{\infty} \frac{(-1)^k \left(\frac{\pi^2}{\alpha^2} - \omega\right)^k}{k!} \langle \widetilde{X}_k * T_\ell g, T_\ell h \rangle_{\Sigma_\ell}$$
$$= \sum_{k=0}^{\infty} \sum_{j=0}^{\infty} \frac{(-1)^k \left(\frac{\pi^2}{\alpha^2} - \omega\right)^k}{k!} \frac{\widetilde{X}_k^{(2j)}(0)}{(2j)!} \cdot \ell^{2j+1} \cdot \langle K_{|x|^{2j}} g, h \rangle_\Sigma.$$

Note that the estimates on the coefficients $\widetilde{X}_k^{(2j)}(0)$ imply that the series

$$\sum_{k=0}^{\infty} \sum_{j=0}^{\infty} \frac{(-1)^k \left(\frac{\pi^2}{\alpha^2} - \omega\right)^k}{k!} \frac{\widetilde{X}_k^{(2j)}(0)}{(2j)!} \cdot \ell^{2j+1} \cdot K_{|x|^{2j}},$$

converges in $\mathcal{L}(L_2(\Sigma))$, uniformly in $\ell \in [0, \ell_0]$ and in $|\omega - \pi^2/\alpha^2| < \varepsilon$. In particular, we obtain

$$\ell \int_{\mathbb{R}} m_{\omega,3}(\xi) \cdot \hat{g}(\ell\xi) \, \hat{h}^*(\ell\xi) \, d\xi$$
$$= \frac{\ell \cdot |\Sigma| \cdot \langle P_{\mathrm{ct}} g, h \rangle_\Sigma}{2\pi} \int_{\mathbb{R}} m_{\pi^2/\alpha^2,3}(\xi) \, d\xi + \mathcal{O}\Big(\ell^3 + \frac{\pi^2}{\alpha^2} - \omega\Big)$$
$$= \frac{\ell \cdot |\Sigma| \cdot \rho_{0,3}(\alpha)}{2\pi} \langle P_{\mathrm{ct}} g, h \rangle_\Sigma + \mathcal{O}\Big(\ell^3 + \frac{\pi^2}{\alpha^2} - \omega\Big),$$

where $\rho_{0,3}(\alpha) = \int_{\mathbb{R}} m_{\pi^2/\alpha^2,3}(\xi) \, d\xi$. Putting

$$\rho_0(\alpha) = \rho_{0,1}(\alpha) + \rho_{0,2}(\alpha) + \rho_{0,3}(\alpha) + 1 + \frac{\gamma_0 \pi^2}{\alpha^2}, \tag{15}$$

with

$$\begin{aligned}
\rho_{0,1}(\alpha) &= \int_{\mathcal{S}} m_{\pi^2/\alpha^2}(\xi)\, \mathrm{d}\xi, \\
\rho_{0,2}(\alpha) &= \frac{1}{2} \sum_{k=2}^{\infty} \binom{1/2}{k} \Big(-\frac{\pi^2}{\alpha^2} \Big)^k \frac{1}{k-1}, \\
\rho_{0,3}(\alpha) &= \int_{\mathbb{R}} m_{\pi^2/\alpha^2,3}(\xi)\, \mathrm{d}\xi.
\end{aligned}$$

proves Theorem 2.8.

The asymptotic behaviour of the ground state eigenvalue of $A_{\ell,b}$

Recall that $b \in L_\infty(\mathbb{R})$ and $\Sigma_\ell = \ell \cdot \Sigma \subseteq \mathbb{R}$, where $\Sigma \subseteq \mathbb{R}$ is a finite union of bounded intervals. The following theorems provide the asymptotic behaviour of the ground state eigenvalue as the window length decreases.

Theorem 2.9. *There exists $\ell_0 = \ell_0(\alpha, b, \Sigma) > 0$ such that for all $\ell \in (0, \ell_0)$ the operator $A_{\ell,b}$ has a unique eigenvalue $\lambda(\ell)$ below its essential spectrum. It satisfies*

$$\sqrt{\pi^2/\alpha^2 - \lambda(\ell)} = \ell^2 \Big(\frac{\pi^2}{\alpha^3} \Big) \cdot \tau_0(\Sigma) + \mathcal{O}(\ell^3) \quad \text{as } \ell \to 0.$$

The constant $\tau_0(\Sigma) > 0$ is given by (19). *If b is continuously differentiable in some neighbourhood of* 0*, then the next term of the asymptotic formula is given by*

$$\ell^3 \Big(\frac{b(0) \cdot \tau_1(\Sigma) \cdot \pi^2}{\alpha^3} \Big)$$

up to an error of order $\mathcal{O}(\ell^4 \cdot \ln \ell)$. The constant $\tau_1(\Sigma) > 0$ is given by (20).

For the special case $\Sigma = (-1, 1)$ we obtain:

Theorem 2.10. *Let $\Sigma_\ell = (-\ell, \ell)$ and let b be twice differentiable in some neighbourhood of* 0*. Then the eigenvalue $\lambda(\ell)$ satisfies*

$$\begin{aligned}
\sqrt{\pi^2/\alpha^2 - \lambda(\ell)} &= \ell^2 \Big(\frac{\pi^3}{2\alpha^3} \Big) + \ell^3 \Big(\frac{4b(0)\pi^2}{3\alpha^3} \Big) - \ell^4 \ln \ell \Big(\frac{\pi^5}{8\alpha^5} \Big) \\
&\quad + \ell^4 \Big(\frac{\rho_0(\alpha)\pi^3}{8\alpha^3} + \frac{\pi^5}{32\alpha^5}(1 + \ln 16) - b(0)^2 \cdot \frac{\rho_1 \cdot \pi^2}{\alpha^3} \Big) \\
&\quad + \mathcal{O}(\ell^5 \ln \ell)
\end{aligned}$$

as $\ell \to 0$. The constant $\rho_1 > 0$ is given by (21).

First we prove the existence and the uniqueness of the eigenvalue of the operator $A_{\ell,b}$ for small $\ell > 0$. To this end we use the asymptotic expansion in Theorem 2.8 in its weaker form

$$\mathcal{Q}_b(\ell,\omega) = \frac{1}{\ell}\mathcal{Q}_0 - \ell \cdot \Big(\frac{|\Sigma| \cdot \pi^2}{\alpha^3}\Big)\Big(\sqrt{\frac{\pi^2}{\alpha^2} - \omega}\Big)^{-1} P_{\mathrm{ct}} + R_b(\ell,\omega) \qquad (16)$$

with the following estimate on the remainder:

$$\sup\{\|R_b(\ell,\omega)\|_{\mathcal{L}(L_2(\Sigma))} \colon \ell \in (0,\ell_0) \text{ and } |\omega - \pi^2/\alpha^2| < \varepsilon\} < \infty. \qquad (17)$$

We note that $|\langle bT_\ell g, T_\ell h\rangle_{\Sigma_\ell}| \le \ell \|b\|_{L_\infty(\Sigma)} \|g\|_{L_2(\Sigma)} \|h\|_{L_2(\Sigma)}$.

Remark 2.11. Using a similar argumentation as in Theorem 2.8 it follows that for every compact set $K \subseteq \mathbb{C}\setminus[\pi^2/\alpha^2,\infty)$ there exists $\ell_0 = \ell_0(\alpha, b, \Sigma, K)$ such that

$$\mathcal{Q}_b(\ell,\omega) = \frac{1}{\ell}\mathcal{Q}_0 + \widetilde{R}_b(\ell,\omega),$$

and the remainder satisfies

$$\sup\{\|\widetilde{R}_b(\ell,\omega)\|_{\mathcal{L}(L_2(\Sigma))} \colon \omega \in K \text{ and } \ell \in (0,\ell_0)\} < \infty.$$

Recalling that the operator $\mathcal{Q}_0$ is a invertible, we obtain

$$\ell\mathcal{Q}_b(\ell,\omega) = \mathcal{Q}_0(I + \ell\mathcal{Q}_0^{-1}\widetilde{R}_b(\ell,\omega)).$$

Choosing $\ell > 0$ sufficiently small implies that $\mathcal{Q}_b(\ell,\omega)$ is invertible for all $\omega \in K$ and $\ell \in (0,\ell_0)$. In particular, 0 cannot be an eigenvalue of $\mathcal{Q}_b(\ell,\omega)$. As a consequence the discrete eigenvalues of the operator $A_{\ell,b}$ converge to π^2/α^2 as $\ell \to 0$.

In what follows we consider for real ω not only the kernel of the operator $\mathcal{Q}_b(\ell,\omega)$, but more generally the discrete eigenvalues of the self-adjoint realisation of $\mathcal{Q}_b(\ell,\omega)$. For $\ell > 0$ and $\omega \in \mathbb{R}\setminus[\pi^2/\alpha^2,\infty)$ we denote

$$\mu_1(\ell,\omega) \le \mu_2(\ell,\omega) \le \cdots$$

these eigenvalues counted with multiplicities.

Lemma 2.12. *Let $\ell_0 > 0$ and $\varepsilon > 0$ be chosen as in Theorem 2.8. Then the following assertions hold true*:

(1) *for fixed $\ell > 0$ the function $\mu_1(\ell, \cdot)$ is strictly decreasing in $(-\infty, \pi^2/\alpha^2)$;*

(2) *for fixed $\ell \in (0, \ell_0)$ we have $\mu_1(\ell, \omega) \to -\infty$ as $\omega \to \pi^2/\alpha^2$;*

(3) *for fixed $\omega \in (\pi^2/\alpha^2 - \varepsilon, \pi^2/\alpha^2)$ we have $\mu_1(\ell, \omega) \to \infty$ as $\ell \to 0$;*

(4) *there exists $\tilde{\ell}_0 = \tilde{\ell}_0(\alpha, \Sigma)$ such that for all $\tilde{\ell} \in (0, \ell_0)$ and for all $|\omega - \pi^2/\alpha^2| < \varepsilon$ we have $\mu_2(\ell, \omega) > 0$.*

Proof. We note that for fixed $\xi \in \mathbb{R}$ the function $m_\omega(\xi)$ is strictly decreasing in ω as can easily be seen from

$$
\begin{aligned}
m_\omega(\xi) &= \frac{1}{\alpha} + \sum_{k=1}^{\infty} \frac{2\alpha(\xi^2 - \omega)}{\alpha^2(\xi^2 - \omega) + k^2\pi^2} \\
&= \frac{1}{\alpha} + \sum_{k=1}^{\infty} \frac{2\alpha}{\alpha^2 + \dfrac{k^2\pi^2}{\xi^2 - \omega}}.
\end{aligned}
$$

Thus, for $-\infty < \omega_1 < \omega_2 < \pi^2/\alpha^2$ and $g \in \widetilde{H}_0^{1/2}(\Sigma)\setminus\{0\}$ we have

$$
q_b(\ell, \omega_1)[g] > q_b(\ell, \omega_2)[g].
$$

Then assertion (1) follows by applying the min-max principle for self-adjoint operators. Let us now prove assertion (2). Decomposition (16) and the min-max principle for self-adjoint operators yield for $\ell \in (0, \ell_0)$ and $|\omega - \pi^2/\alpha^2| < \varepsilon$ that $\mu_1(\ell, \omega) \le q_b(\ell, \omega)[g_0]$ for any $g_0 \in \widetilde{H}_0^{1/2}(\Sigma)$ with $\|g_0\|_{L_2(\Sigma)} = 1$. Choosing g_0 such that $\langle P_{\mathrm{ct}} g_0, g_0\rangle_\Sigma \neq 0$ we obtain

$$
\mu_1(\ell, \omega) \le \frac{1}{\ell}\langle \mathcal{Q}_0 g_0, g_0\rangle_\Sigma - \ell \cdot \Big(\frac{|\Sigma| \cdot \pi^2}{\alpha^3}\Big)\Big(\sqrt{\frac{\pi^2}{\alpha^2} - \omega}\Big)^{-1} \cdot \langle P_{\mathrm{ct}} g_0, g_0\rangle_\Sigma + C_1,
$$

which tends to $-\infty$ as $\omega \to \pi^2/\alpha^2$. Here $C_1 := \sup\{\|R_b(\ell, \omega)\|_{\mathcal{L}(L_2(\Sigma))} : \ell \in (0, \ell_0)$ and $|\omega - \pi^2/\alpha^2| < \varepsilon\}$. This proves (2). To deduce (3) we recall that $\mathcal{Q}_0$ is invertible and we have $q_0[g] = \langle \mathcal{Q}_0 g, g\rangle_\Sigma \ge 0$ for all $g \in \widetilde{H}_0^{1/2}(\Sigma)$. Thus, there exists $\mu_* > 0$ such that

$$
\langle \mathcal{Q}_0 g, g\rangle_\Sigma = q_0[g] \ge \mu_* \|g\|^2_{L_2(\Sigma)}, \quad g \in \widetilde{H}_0^{1/2}(\Sigma).
$$

We note that the spectrum of the self-adjoint realisation of $\mathcal{Q}_0$ is discrete since the form domain of q_0 is compactly embedded in $L_2(\Sigma)$. Thus, for $\omega \in \mathbb{R}\setminus[\pi^2/\alpha^2, \infty)$ with $|\omega - \pi^2/\alpha^2| < \varepsilon$ we have

$$\mu_1(\ell,\omega) = \inf\{q_b(\ell,\omega)[g] \colon g \in \widetilde{H}_0^{1/2}(\Sigma) \text{ and } \|g\|_{L_2(\Sigma)} = 1\} \geq \frac{\mu_*}{\ell} - C_1 \longrightarrow \infty$$

as $\ell \to 0$. This proves (3). Assertion (4) follows if we prove that the form $q_b(\ell,\omega)$ is positive on a subset of codimension 1. Choose $g \in \widetilde{H}_0^{1/2}(\Sigma)$, $\|g\|_{L_2(\Sigma)} = 1$, orthogonal to the constant functions. Then

$$q_b(\ell,\omega)[g] = \frac{1}{\ell} q_0[g] + \langle R_b(\ell,\omega)g, g\rangle_\Sigma \geq \frac{\mu_*}{\ell} - C_1 > 0$$

for $0 < \ell < \tilde{\ell}_0 := \min\{1, \mu_*/C_1\}$ and $|\omega - \pi^2/\alpha^2| < \varepsilon$. This concludes the proof of Lemma 2.12. □

Lemma 2.13. *There exists $\ell_0 = \ell_0(\alpha, b, \Sigma) > 0$ such that for all $\ell \in (0, \ell_0)$ the operator $A_{\ell,b}$ has a unique eigenvalue $\lambda(\ell)$ below its essential spectrum.*

Proof. We start by proving the uniqueness of the eigenvalue. Let $\varepsilon > 0$ be chosen as in Theorem 2.8 and Lemma 2.12. Using the remark before Lemma 2.12 we may choose $\ell_0 > 0$ such that $\inf \sigma(A_{\ell,b}) \geq \pi^2/\alpha^2 - \varepsilon$ for all $\ell \in (0, \ell_0)$. Moreover, we assume that $\mu_2(\ell,\omega) > 0$ for all $\ell \in (0,\ell_0)$ and $\omega \in (\pi^2/\alpha^2 - \varepsilon, \pi^2/\alpha^2)$. Fix $\ell \in (0,\ell_0)$ and assume that $\omega \in \sigma_d(A_{\ell,b})$. Then $\mu_1(\ell,\omega) = 0$. Lemma 2.12 (1) implies, for $\omega_1 < \omega < \omega_2 < \pi^2/\alpha^2$,

$$\mu_1(\ell,\omega_1) < \mu_1(\ell,\omega) = 0 < \mu_1(\ell,\omega_2).$$

In particular we have $\ker \mathcal{Q}_b(\ell,\omega_1) = \ker \mathcal{Q}_b(\ell,\omega_2) = \{0\}$, which proves the uniqueness of the eigenvalue of $A_{\ell,b}$.

Next we prove the existence of the eigenvalue. Using Lemma 2.12 (3) we may assume that $\mu_1(\ell, \pi^2/\alpha^2 - \varepsilon/2) > 0$ for all $\ell \in (0,\ell_0)$. Fix $\ell \in (0,\ell_0)$. Since $\mu_1(\ell,\omega) \to -\infty$ as $\omega \to \pi^2/\alpha^2$ and $\mu_1(\ell,\omega)$ depends continuously on ω it follows that there exists $\widetilde{\omega} \in (\pi^2/\alpha^2 - \varepsilon/2, \pi^2/\alpha)$ such that $\mu_1(\ell,\widetilde{\omega}) = 0$. □

Remark 2.14. Another method of proof for Lemma 2.13 may be based on a variant of operator-valued Rouché's theorem, cf., e.g., [19] or the monograph [1].

Next, using the Birman–Schwinger principle, we prove the asymptotic formula for the eigenvalue of $A_{\ell,b}$. To this end we choose $\ell_0 > 0$ such that the operator $\mathcal{Q}_0 + \ell R_b(\ell,\omega)$ is invertible for all $\ell \in (0,\ell_0)$ and $\omega \in (\pi^2/\alpha^2 - \varepsilon, \pi^2/\alpha^2)$. The existence of such an ℓ_0 follows from the estimate (17).

Lemma 2.15 (Birman–Schwinger principle). *Let $\ell \in (0, \ell_0)$ and $\omega \in (\pi^2/\alpha^2 - \varepsilon, \pi^2/\alpha^2)$. Then* 0 *is an eigenvalue of the operator*

$$\ell \mathcal{Q}_b(\ell, \omega) = \mathcal{Q}_0 - \ell^2 \Big(\frac{|\Sigma| \cdot \pi^2}{\alpha^3}\Big)\Big(\sqrt{\frac{\pi^2}{\alpha^2} - \omega}\Big)^{-1} P_{\mathrm{ct}} + \ell R_b(\ell, \omega)$$

if and only if 1 *is an eigenvalue of the Birman–Schwinger operator*

$$\ell^2 \Big(\frac{|\Sigma| \cdot \pi^2}{\alpha^3}\Big)\Big(\sqrt{\frac{\pi^2}{\alpha^2} - \omega}\Big)^{-1} \cdot P_{\mathrm{ct}}^{1/2} (\mathcal{Q}_0 + \ell R_b(\ell, \omega))^{-1} P_{\mathrm{ct}}^{1/2}.$$

A proof may be found, e.g., in [4].

Since the projection P_{ct} is a rank-one operator with $P_{\mathrm{ct}}^2 = P_{\mathrm{ct}} = P_{\mathrm{ct}}^{1/2}$, the Birman–Schwinger principle implies that ω is an eigenvalue of the operator $A_{\ell,b}$ if and only if the trace of the Birman–Schwinger operator is equal to one, i.e.,

$$\ell^2 \Big(\frac{\pi^2}{\alpha^3}\Big) \frac{1}{\sqrt{\frac{\pi^2}{\alpha^2} - \omega}} \langle (\mathcal{Q}_0 + \ell R_b(\ell, \omega))^{-1} \phi_0, \phi_0 \rangle_\Sigma = 1,$$

where $\phi_0(x) = 1$ is the (non-normalised) constant function on $L_2(\Sigma)$. For the choice $\omega = \lambda(\ell)$ we obtain

$$\sqrt{\frac{\pi^2}{\alpha^2} - \lambda(\ell)} = \ell^2 \Big(\frac{\pi^2}{\alpha^3}\Big) \langle (\mathcal{Q}_0 + \ell R_b(\ell, \lambda(\ell)))^{-1} \phi_0, \phi_0 \rangle_\Sigma.$$

Next we use an asymptotic expansion for the resolvent term. We have

$$\begin{aligned} (\mathcal{Q}_0 + \ell R_b(\ell, \omega))^{-1} &= (I + \ell \mathcal{Q}_0^{-1} R_b(\ell, \omega))^{-1} \mathcal{Q}_0^{-1}, \\ &= \sum_{k=0}^{\infty} \ell^k (-\mathcal{Q}_0^{-1} R_b(\ell, \omega))^k \mathcal{Q}_0^{-1} = \mathcal{Q}_0^{-1} + \mathcal{O}(\ell), \end{aligned} \tag{18}$$

uniformly in $\omega \in (\pi^2/\alpha^2, -\varepsilon, \pi^2/\alpha^2)$. Note that for sufficiently small ℓ the sum converges absolutely in $\mathcal{L}(L_2(\Sigma))$. Hence,

$$\langle (\mathcal{Q}_0 + \ell R_b(\ell, \lambda(\ell)))^{-1} \phi_0, \phi_0 \rangle_\Sigma = \langle \mathcal{Q}_0^{-1} \phi_0, \phi_0 \rangle_\Sigma + \mathcal{O}(\ell)$$

as $\ell \to 0$, and thus,

$$\begin{aligned} \sqrt{\pi^2/\alpha^2 - \lambda(\ell)} &= \ell^2 \Big(\frac{\pi^2}{\alpha^3}\Big) \langle (\mathcal{Q}_0 + \ell R_b(\ell, \lambda(\ell)))^{-1} \phi_0, \phi_0 \rangle_\Sigma \\ &= \Big(\frac{\pi^2}{\alpha^3}\Big) \cdot \langle \mathcal{Q}_0^{-1} \phi_0, \phi_0 \rangle_\Sigma \cdot \ell^2 + \mathcal{O}(\ell^3). \end{aligned}$$

This proves the first term of the asymptotics in Theorem 2.9 with

$$\tau_0(\Sigma) := \langle \mathcal{Q}_0^{-1}\phi_0, \phi_0\rangle_\Sigma = \langle \mathcal{Q}_0^{-1/2}\phi_0, \mathcal{Q}_0^{-1/2}\phi_0\rangle_\Sigma > 0. \tag{19}$$

Until now no additional assumptions on $b \in L_\infty(\mathbb{R})$ were necessary. Now let b be differentiable in a neighbourhood of 0. Then

$$\langle bT_\ell g, T_\ell h\rangle_{\Sigma_\ell} = \int_\Sigma b(\ell x)\cdot g(x)\,\overline{h(x)}\,\mathrm{d}x = b(0)\langle g, h\rangle_\Sigma + \mathcal{O}(\ell).$$

Note that the remainder may be estimated uniformly in the operator norm. Thus, together with Theorem 2.8 we obtain

$$R_b(\ell,\omega) = b(0)I + \mathcal{O}(\ell\ln\ell).$$

The estimate holds uniformly in $\ell \in (0,\ell_0)$ and $|\omega - \pi^2/\alpha^2| < \varepsilon$. Using formula (18) we obtain

$$\begin{aligned}(\mathcal{Q}_0 + \ell R_b(\ell,\lambda(\ell)))^{-1} &= \mathcal{Q}_0^{-1} + \ell\mathcal{Q}_0^{-1}R_b(\ell,\lambda(\ell))\mathcal{Q}_0^{-1} + \mathcal{O}(\ell^2)\\ &= \mathcal{Q}_0^{-1} + \ell\cdot b(0)\cdot I + \mathcal{O}(\ell^2\cdot\ln\ell).\end{aligned}$$

Finally,

$$\begin{aligned}\sqrt{\pi^2/\alpha^2 - \lambda(\ell)} &= \ell^2\Big(\frac{\pi^2}{\alpha^3}\Big)\langle(\mathcal{Q}_0 + \ell R_b(\ell,\omega))^{-1}\phi_0, \phi_0\rangle_\Sigma\\ &= \Big(\frac{\pi^2}{\alpha^3}\Big)\cdot\tau_0(\Sigma)\cdot\ell^2 + \Big(\frac{b(0)\pi^2}{\alpha^3}\Big)\cdot\tau_1(\Sigma)\cdot\ell^3 + \mathcal{O}(\ell^4\ln\ell)\end{aligned}$$

with

$$\tau_1(\Sigma) := \langle \mathcal{Q}_0^{-2}\phi_0, \phi_0\rangle_\Sigma = \langle \mathcal{Q}_0^{-1}\phi_0, \mathcal{Q}_0^{-1}\phi_0\rangle_\Sigma > 0. \tag{20}$$

This proves Theorem 2.9.

Now let $\Sigma = (-1,1)$. Then the operator $\mathcal{Q}_0$ becomes the composition of the standard finite Hilbert transform and the derivative. Using (4.8) in [2] or Section 5.2 of [1] we obtain

$$(\mathcal{Q}_0^{-1}\phi_0)(x) = \sqrt{1-x^2},$$

which implies

$$\langle \mathcal{Q}_0^{-1}\phi_0, \phi_0\rangle_\Sigma = \int_{-1}^{1}\sqrt{1-x^2}\,\mathrm{d}x = \frac{\pi}{2}.$$

For the sake of simplicity we assume that $b \in C^2(\mathbb{R})$. Then

$$\begin{aligned}\langle bT_\ell g, T_\ell h\rangle_{\Sigma_\ell} &= \int_\Sigma b(\ell x)\cdot g(x)\,\overline{h(x)}\,\mathrm{d}x\\ &= b(0)\langle g, h\rangle_\Sigma + \ell\cdot b'(0)\langle M_x g, h\rangle_\Sigma + \mathcal{O}(\ell^2),\end{aligned}$$

where $M_x\colon L_2(\Sigma) \to L_2(\Sigma)$ is the multiplication operator

$$(M_x f)(x) = x f(x).$$

Theorem 2.8 implies

$$R_b(\ell,\omega) = b(0)\cdot I + \ell \ln \ell \cdot \frac{\pi}{\alpha^2} P_{\mathrm{ct}} + \ell\Big(\frac{\rho_0(\alpha)}{\pi} P_{\mathrm{ct}} + \frac{\pi}{2\alpha^2} K_{\ln|x|} + b'(0) M_x\Big) + R_b^{(1)}(\ell,\omega),$$

with

$$\|R_b^{(1)}(\ell,\lambda(\ell))\|_{\mathcal{L}(L_2(\Sigma))} \le C(\ell^3 \ln \ell + \pi^2/\alpha^2 - \lambda(\ell)) = \mathcal{O}(\ell^3 \ln \ell).$$

To calculate the asymptotic behaviour of the eigenvalue we use the expansion

$$\begin{aligned}(\mathcal{Q}_0 + \ell R_b(\ell,\lambda(\ell)))^{-1} &= \mathcal{Q}_0^{-1} - \ell\cdot b(0)\mathcal{Q}_0^{-2} - \ell^2 \ln \ell \cdot \frac{\pi}{\alpha^2}\mathcal{Q}_0^{-1} P_{\mathrm{ct}} \mathcal{Q}_0^{-1} \\ &\quad - \ell^2\cdot \mathcal{Q}_0^{-1}\Big(\frac{\rho_0(\alpha)}{\pi} P_{\mathrm{ct}} + \frac{\pi}{2\alpha^2} K_{\ln|x|} + b'(0) M_x\Big)\mathcal{Q}_0^{-1} \\ &\quad + \ell^2 \cdot b(0)^2 \mathcal{Q}_0^{-3} + \mathcal{O}(\ell^3 \ln \ell).\end{aligned}$$

Then

$$\begin{aligned}&\ell^{-2}\Big(\frac{\pi^2}{\alpha^3}\Big)^{-1}\cdot\sqrt{\frac{\pi^2}{\alpha^2} - \lambda(\ell)} \\ &= \frac{\pi}{2} - \ell\cdot b(0)\langle \mathcal{Q}_0^{-2}\phi_0, \phi_0\rangle_{(-1,1)} - \ell^2 \ln \ell \cdot \frac{\pi}{\alpha^2}\langle \mathcal{Q}_0^{-1} P_{\mathrm{ct}} \mathcal{Q}_0^{-1}\phi_0,\phi_0\rangle_{(-1,1)} \\ &\quad - \ell^2 \cdot \frac{\rho_0(\alpha)}{\pi}\langle \mathcal{Q}_0^{-1} P_{\mathrm{ct}} \mathcal{Q}_0^{-1}\phi_0,\phi_0\rangle_{(-1,1)} - \ell^2\cdot\frac{\pi}{\alpha^2}\langle \mathcal{Q}_0^{-1} K_{\ln|x|} \mathcal{Q}_0^{-1}\phi_0,\phi_0\rangle_{(-1,1)} \\ &\quad - \ell^2\cdot b'(0)\langle \mathcal{Q}_0^{-1} M_x \mathcal{Q}_0^{-1}\phi_0,\phi_0\rangle_{(-1,1)} + \ell^2\cdot b(0)^2\langle \mathcal{Q}_0^{-3}\phi_0,\phi_0\rangle_{(-1,1)} \\ &\quad + \mathcal{O}(\ell^3 \ln \ell).\end{aligned}$$

In order to calculate the asymptotic behaviour of $\langle(\mathcal{Q}_0^{-1} + \ell R_{\ell,\lambda(\ell)})^{-1}\phi_0,\phi_0\rangle_{(-1,1)}$ we shall need the following identities:

$$\langle \mathcal{Q}_0^{-2}\phi_0,\phi_0\rangle_{(-1,1)} = \langle \mathcal{Q}_0^{-1}\phi_0, \mathcal{Q}_0^{-1}\phi_0\rangle_{(-1,1)} = \int_{-1}^{1}(1-x^2)\,\mathrm{d}x = \frac{4}{3},$$

$$\langle \mathcal{Q}_0^{-1} P_{\mathrm{ct}} \mathcal{Q}_0^{-1}\phi_0,\phi_0\rangle_{(-1,1)} = \langle P_{\mathrm{ct}}\mathcal{Q}_0^{-1}\phi_0, \mathcal{Q}_0^{-1}\phi_0\rangle_{(-1,1)} = \frac{\pi^2}{8},$$

$$\langle \mathcal{Q}_0^{-1} M_x \mathcal{Q}_0^{-1}\phi_0,\phi_0\rangle_{(-1,1)} = \langle M_x\mathcal{Q}_0^{-1}\phi_0, \mathcal{Q}_0^{-1}\phi_0\rangle_{(-1,1)} = \int_{-1}^{1}x(1-x^2)\,\mathrm{d}x = 0.$$

Next we calculate $\langle K_{\ln|x|}\mathcal{Q}_0^{-1}\phi_0, \mathcal{Q}_0^{-1}\phi_0\rangle_{(-1,1)}$. Recall that $(\mathcal{Q}_0^{-1}\phi_0)(x) = \sqrt{1-x^2}$. Using (5.6)–(5.9) in [1] for $\psi := K_{\ln|x|}\mathcal{Q}_0^{-1}\phi_0$ we obtain

$$\psi'(x) = \text{p. v.} \int_{-1}^{1} \frac{\phi_0(x)}{x-y}\, dy = \text{p. v.} \int_{-1}^{1} \frac{\sqrt{1-x^2}}{x-y}\, dy = \pi x.$$

Thus, $\psi(x) = \frac{\pi x^2}{2} + \psi(0)$, where

$$\psi(0) = \int_{-1}^{1} \ln|x|\sqrt{1-x^2}\, dx = 2\int_{0}^{1} \ln|x|\sqrt{1-x^2}\, dx = -\frac{\pi}{2}\Big(\frac{1}{2} + \ln 2\Big),$$

cf. Section 4.241 in [20]. Hence,

$$\begin{aligned}&\langle K_{\ln|x|}\mathcal{Q}_0^{-1}\phi_0, \mathcal{Q}_0^{-1}\phi_0\rangle_{(-1,1)}\\ &\quad= \frac{\pi}{2}\int_{-1}^{1} x^2\sqrt{1-x^2}\, dx - \frac{\pi}{2}\Big(\frac{1}{2} + \ln 2\Big)\int_{-1}^{1}\sqrt{1-x^2}\, dx = \frac{\pi^2}{16}(-1-\ln 16).\end{aligned}$$

Setting

$$\begin{aligned}\rho_1 := \langle \mathcal{Q}_0^{-3}\phi_0, \phi_0\rangle_{(-1,1)} &= \langle \mathcal{Q}_0^{-2}\phi_0, \mathcal{Q}_0^{-1}\phi_0\rangle_{(-1,1)}\\ &= \langle \mathcal{Q}_0^{-3/2}\phi_0, \mathcal{Q}_0^{-3/2}\phi_0\rangle_{(-1,1)} > 0 \end{aligned}\tag{21}$$

we obtain

$$\begin{aligned}\sqrt{\frac{\pi^2}{\alpha^2} - \lambda(\ell)} = {}& \ell^2\Big(\frac{\pi^3}{2\alpha^3}\Big) - \ell^3\Big(\frac{4b(0)\pi^2}{3\alpha^3}\Big) - \ell^4\ln\ell\Big(\frac{\pi^5}{8\alpha^5}\Big)\\ &- \ell^4\Big(\frac{\rho_0(\alpha)\pi^3}{8\alpha^3} - \frac{\pi^5}{32\alpha^5}(1+\ln 16) - b(0)^2\rho_1\cdot\frac{\pi^2}{\alpha^3}\Big)\\ &+ \mathcal{O}(\ell^5\ln\ell).\end{aligned}$$

This proves Theorem 2.10.

Concluding the two-dimensional case we briefly want to sketch what happens in the case of two waveguides of width α_+ and α_-, which are coupled through a window $\Sigma_\ell := \ell\cdot\Sigma$. We use the same ansatz and introduce the corresponding Dirichlet–to–Neumann operators $D^+_{\ell,\omega}$ and $D^-_{\ell,\omega}$ on the upper and on the lower waveguide. Comparing the normal derivatives along the window we observe that ω is an eigenvalue of the corresponding Dirichlet-Laplacian if and only if 0 is an eigenvalue of

$$D_{\ell,\omega} := D^+_{\ell,\omega} + D^-_{\ell,\omega}.$$

Using the same scaling operator T_ℓ as above leads to the analysis of the operator

$$\mathcal{Q}(\ell,\omega) = \mathcal{Q}^+(\ell,\omega) + \mathcal{Q}^-(\ell,\omega) = T_\ell^* D^+_{\ell,\omega} T_\ell + T_\ell^* D^-_{\ell,\omega} T_\ell.$$

In what follows we assume that $\alpha_+ > \alpha_-$ so that the essential spectrum of the corresponding operator A_ℓ is given by the interval $[\pi^2/\alpha_+^2, \infty)$. Using the asymptotic expansions of $\mathcal{Q}^\pm(\ell, \omega)$ as $\ell \to 0$ and $\omega \to \pi^2/\alpha^2{}_+$ we obtain

$$\mathcal{Q}(\ell, \omega) = \frac{2}{\ell}\mathcal{Q}_0 - \ell\Big(\sqrt{\frac{\pi^2}{\alpha^2} - \omega}\Big)^{-1}\Big(\frac{|\Sigma| \cdot \pi^2}{\alpha_+^3}\Big) P_{\mathrm{ct}} + \mathcal{O}(1).$$

The same approach yields now the following result.

Theorem 2.16. *In the case of two coupled waveguides the ground state eigenvalue $\lambda(\ell)$ satisfies*

$$\sqrt{\frac{\pi^2}{\alpha^2} - \lambda(\ell)} = \Big(\frac{\pi^2}{2\alpha_+^3}\Big) \cdot \tau_0(\Sigma) \cdot \ell^2 + \mathcal{O}(\ell^3) \quad \text{as } \ell \to 0.$$

Here $\tau_0(\Sigma) > 0$ is again given by (19).

3 Infinite layers

We consider the mixed problem for an infinite layer $\Omega := \mathbb{R}^2 \times (0, \alpha)$ with coordinates $(x, y) = (x_1, x_2, y) \in \mathbb{R}^2 \times (0, \alpha)$. Let $\Sigma \times \{0\} \subseteq \partial\Omega$ be the Robin window, where $\Sigma \subseteq \mathbb{R}^2$ is a bounded open subset with Lipschitz boundary. For $\ell > 0$ we denote by $\Sigma_\ell := \ell \cdot \Sigma \subseteq \mathbb{R}^2$ the scaled window. Let $b \in L_\infty(\mathbb{R}^2)$ be a real-valued function and consider the quadratic form

$$a_{\ell,b}[u] := \int_\Omega |u(x, y)|^2 \, dx \, dy + \int_{\Sigma_\ell} b(x) \cdot |u(x, 0)|^2 \, dx$$

with the form domain

$$D[a_{\ell,b}] := \{u \in H^1(\Omega) \colon u|_{\mathbb{R}^2 \times \{\alpha\}} = 0 \text{ and } \operatorname{supp}(u|_{\mathbb{R}^2 \times \{0\}}) \subseteq \overline{\Sigma_\ell}\}.$$

As in the two-dimensional case we observe that $a_{\ell,b}$ is a closed semi-bounded form in $L_2(\Omega)$, and thus, it induces a self-adjoint operator $A_{\ell,b}$. The essential spectrum of $A_{\ell,b}$ is independent of b and ℓ and given by $\sigma_{\mathrm{ess}}(A_{\ell,b}) = [\pi^2/\alpha^2, \infty)$. We prove the following theorem.

Theorem 3.1. *There exists $\ell_0 = \ell_0(\alpha, b, \Sigma) > 0$ such that the operator $A_{\ell,b}$ has a unique eigenvalue $\lambda(\ell)$ below the essential spectrum $[\pi^2/\alpha^2, \infty)$. If b is C^1 in some neighbourhood of $0 \in \mathbb{R}^2$ then the eigenvalue satisfies the asymptotic estimate*

$$\ln(\pi^2/\alpha^2 - \lambda(\ell)) = -\ell^{-3} \frac{4\alpha^3}{\tau_0(\Sigma) + \tau_1(\Sigma) b(0) \ell + \mathcal{O}(\ell^2)} \quad \text{as } \ell \to 0,$$

with constants $\tau_0(\Sigma) > 0$ *given by* (23) *and* $\tau_1(\Sigma) > 0$ *given by* (24).

Since we shall only slightly modify our approach we will merely sketch the major steps of the proof. Actually, most of the results proven in the two-dimensional case may be reused. Let $\omega \in \mathbb{C}$ and $g \in H^{1/2}(\mathbb{R}^2)$. We consider for $u \in H^1(\Omega)$ the Poisson problem

$$(-\Delta - \omega)u = 0 \text{ in } \Omega, \quad u(\cdot, 0) = g, \quad u(\cdot, \alpha) = 0. \tag{22}$$

Applying the Fourier transform with respect to the first two variables leads for every $\xi = (\xi_1, \xi_2) \in \mathbb{R}^2$ to the following Sturm–Liouville problem

$$(-\partial_y^2 + |\xi|^2 - \omega)\hat{u}(\xi, \cdot) = 0 \text{ in } (0, \alpha), \quad \hat{u}(\xi, 0) = \hat{g}(\xi), \quad \hat{u}(\xi, \alpha) = 0,$$

where $\xi \in \mathbb{R}^2$. The solution of (22) is given by

$$\hat{u}(\xi, y) := \hat{g}(\xi) \cdot \frac{\sinh((\alpha - y)\sqrt{|\xi|^2 - \omega})}{\sinh(\alpha\sqrt{|\xi|^2 - \omega})},$$

which is similar to formula (3). In the same way we obtain that the Poisson problem (22) is uniquely solvable for all $g \in H^{1/2}(\mathbb{R}^2)$ if and only if $\omega \in \mathbb{C} \setminus [\pi^2/\alpha^2, \infty)$. Moreover, there exists $c = c(\alpha, \omega) > 0$ such that $\|u\|_{H^1(\Omega)} \leq c\|g\|_{H^{1/2}(\mathbb{R}^2)}$. In what follows let $\omega \in \mathbb{C} \setminus [\pi^2/\alpha^2, \infty)$. Then the normal derivative of u satisfies

$$\widehat{\partial_n u}(\xi, 0) = m_\omega(|\xi|) \cdot \hat{g}(\xi), \quad \xi \in \mathbb{R}^2,$$

where the function m_ω is defined as in the two-dimensional case, i.e.,

$$m_\omega(|\xi|) = \sqrt{|\xi|^2 - \omega} \cdot \coth(\alpha\sqrt{|\xi|^2 - \omega}).$$

The Dirichlet–to–Neumann operator for the infinite layer is given by the Fourier integral operator

$$D_\omega \colon H^{1/2}(\mathbb{R}^2) \longrightarrow H^{-1/2}(\mathbb{R}^2), \quad \widehat{D_\omega g}(\xi) := m_\omega(|\xi|) \cdot \hat{g}(\xi).$$

The next step is to define the truncated operator on the boundary. The corresponding spaces $\widetilde{H}_0^{1/2}(\Sigma_\ell)$ and $H^{-1/2}(\Sigma_\ell)$ are defined as in (5) and (6). As both Σ and Σ_ℓ have Lipschitz boundary, the dual pairing (7) still holds true, cf. Theorems 3.14 and 3.30 in [25]. Put

$$D_{\ell,\omega} + b \colon \widetilde{H}_0^{1/2}(\Sigma_\ell) \longrightarrow H^{-1/2}(\Sigma_\ell), \quad D_{\ell,\omega} + b := r_\ell(D_\omega + b)e_\ell,$$

where

$$r_\ell \colon H^{-1/2}(\mathbb{R}^2) \longrightarrow H^{-1/2}(\Sigma_\ell)$$

denotes the restriction operator and

$$e_\ell \colon \widetilde{H}_0^{1/2}(\Sigma_\ell) \longrightarrow H^{1/2}(\mathbb{R}^2)$$

the embedding operator. As in Lemma 2.4 we obtain

$$\dim\ker(A_{\ell,b} - \omega) = \dim\ker(D_{\ell,\omega} + b).$$

Let

$$T_\ell \colon L_2(\Sigma) \longrightarrow L_2(\Sigma_\ell), \quad (T_\ell g)(x) := \ell^{-1} g(x/\ell)$$

be the unitary scaling operator. In what follows we consider the scaled operator

$$\mathcal{Q}_b(\ell,\omega) = T_\ell^*(D_{\ell,\omega} + b)T_\ell$$

together with its associated sesquilinear form

$$\begin{aligned} q_b(\ell,\omega) &:= \langle \mathcal{Q}_b(\ell,\omega)g, h\rangle_\Sigma \\ &= \ell^2 \int_{\mathbb{R}^2} m_\omega(|\xi|)\cdot \hat{g}(\ell\xi)\,\overline{\hat{h}(\ell\xi)}\,\mathrm{d}\xi + \int_\Sigma b(\ell x)\cdot g(x)\,\overline{h(x)}\,\mathrm{d}x, \end{aligned}$$

where $g, h \in D[q_b(\ell,\omega)] := \widetilde{H}_0^{1/2}(\Sigma)$. As before we define

$$\mathcal{Q}_0 \colon \widetilde{H}_0^{1/2}(\Sigma) \longrightarrow H^{-1/2}(\Sigma), \quad \langle \mathcal{Q}_0 g, h\rangle_\Sigma := q_0[g,h] := \int_{\mathbb{R}^2} |\xi|\cdot \hat{g}(\xi)\cdot \overline{\hat{h}(\xi)}\,\mathrm{d}\xi,$$

and let P_{ct} denote the projection onto the space of constant functions in $L_2(\Sigma)$. Moreover, we denote by $K_{1/|x|} \colon L_2(\Sigma) \to L_2(\Sigma)$ the convolution operator

$$(K_{1/|x|} f)(z) := \int_\Sigma \frac{f(z)}{|x - z|}\,\mathrm{d}x, \quad z \in \Sigma.$$

Theorem 3.2. *Let $b = 0$. There exists $\ell_0 > 0$ and $\varepsilon > 0$ such that for $\ell \in (0, \ell_0)$ and $|\omega - \pi^2/\alpha^2| < \varepsilon$ the following expansion holds true:*

$$\mathcal{Q}_0(\ell,\omega) = \frac{1}{\ell}\mathcal{Q}_0 + \ell^2 \cdot \frac{|\Sigma|}{4\alpha^3} \ln\Big(\frac{\pi^2}{\alpha^2} - \omega\Big) P_{\mathrm{ct}} - \ell \frac{\pi}{4\alpha^2} K_{1/|x|} + R(\ell,\omega).$$

Here $|\Sigma|$ denotes the volume of Σ and the remainder satisfies

$$\|R(\ell,\omega)\|_{\mathcal{L}(L_2(\Sigma))} \le C\Big(\ell^2 + \frac{\pi^2}{\alpha^2} - \omega\Big)$$

for some constant $C = C(\alpha,\Sigma) > 0$ which is independent of ℓ, ω.

Proof. We use the same decomposition for $q_0(\ell,\omega)$ as in the two-dimensional case and put

$$\begin{aligned} q_0(\ell,\omega)[g,h] &= \ell^2\Big(\int_{\{|\xi|\le 1\}} + \int_{\{|\xi|>1\}}\Big) m_\omega(|\xi|)\cdot \hat{g}(\ell\xi)\,\overline{h(\ell\xi)}\,\mathrm{d}\xi \\ &=: q_0^{(1)}(\ell,\omega)[g,h] + q_0^{(2)}(\ell,\omega)[g,h]. \end{aligned}$$

Recall that

$$\begin{aligned} m_\omega(\xi) &= \frac{1}{\alpha} + \sum_{k=1}^{\infty} \frac{2\alpha(\xi^2-\omega)}{\alpha^2(\xi^2-\omega)+k^2\pi^2} \\ &= -\Big(\frac{2\pi^2}{\alpha^3}\Big)\frac{1}{\xi^2-\omega+\pi^2\alpha^{-2}} + \mathcal{O}(1) \end{aligned}$$

for $|\xi| \le 1$ and $|\omega - \pi^2/\alpha^{-2}| < \varepsilon$. Thus

$$\begin{aligned} q_0^{(1)}(\ell,\omega)[g,h] = {} & -\ell^2\cdot\Big(\frac{2\pi^2}{\alpha^3}\Big)\int_{\{|\xi|\le 1\}} \frac{1}{|\xi|^2-\omega+\pi^2\alpha^{-2}}\cdot \hat{g}(\ell\xi)\,\overline{h(\ell\xi)}\,\mathrm{d}\xi \\ & + \mathcal{O}(\ell^2). \end{aligned}$$

We note that the the first expression coincides almost with the free resolvent of the Laplacian in $\mathbb{R}^2$, which with respect to the spectral parameter ω has a logarithmic singularity. Using the Taylor expansion of $\hat{g}\cdot\bar{\hat{h}}$ at 0 we have

$$\begin{aligned} & -\ell^2\cdot\Big(\frac{2\pi^2}{\alpha^3}\Big)\int_{\{|\xi|\le 1\}} \frac{1}{|\xi|^2-\omega+\pi^2\alpha^{-2}}\cdot \hat{g}(\ell\xi)\,\overline{h(\ell\xi)}\,\mathrm{d}\xi \\ &= -\ell^2\cdot\Big(\frac{2\pi^2}{\alpha^3}\Big)\cdot\sum_{\beta\in\mathbb{N}_0^2} \ell^{|\beta|}\frac{1}{\beta!}\frac{\partial^\beta}{\partial\xi^\beta}\big(\hat{g}(\xi)\bar{\hat{h}}(\xi)\big)\Big|_{\xi=0}\cdot\int_{\{|\xi|\le 1\}} \frac{\xi^{\beta+1}}{|\xi|^2+\pi^2\alpha^{-2}-\omega}\,\mathrm{d}\xi \\ &= -\ell^2\cdot\Big(\frac{2\pi^2}{\alpha^3}\Big)\hat{g}(0)\bar{\hat{h}}(0)\cdot\int_{\{|\xi|\le 1\}} \frac{1}{|\xi|^2+\pi^2\alpha^{-2}-\omega}\,\mathrm{d}\xi + \mathcal{O}(\ell^3), \end{aligned}$$

since for $|\beta| \geq 1$ we have

$$\left| \int_{\{|\xi|\leq 1\}} \frac{\xi^\beta}{|\xi|^2 + \pi^2\alpha^{-2} - \omega} \, \mathrm{d}\xi \right| \leq \int_0^1 \frac{r^2}{r^2 + \pi^2\alpha^{-2} - \omega} \, \mathrm{d}r \leq C$$

and C may be chosen independently of ω. Moreover,

$$\int_{\{|\xi|\leq 1\}} \frac{1}{|\xi|^2 + \pi^2\alpha^{-2} - \omega} \, \mathrm{d}\xi = -\frac{1}{2} \ln\Big(\frac{\pi^2}{\alpha^2} - \omega\Big) + \mathcal{O}(1),$$

and thus,

$$\begin{aligned} q_0^{(1)}(\ell,\omega)[g,h] &= \ell^2 \cdot \Big(\frac{\pi^2}{\alpha^3}\Big) \hat{g}(0) \cdot \bar{\hat{h}}(0) \cdot \ln\Big(\frac{\pi^2}{\alpha^2} - \omega\Big) + \mathcal{O}(\ell^3) \\ &= \ell^2 \cdot \Big(\frac{|\Sigma|}{4\alpha^3}\Big) \ln\Big(\frac{\pi^2}{\alpha^2} - \omega\Big) \langle P_{\mathrm{ct}} g, h\rangle_\Sigma + \mathcal{O}(\ell^3). \end{aligned}$$

Next we consider the form $q_0^{(2)}(\ell,\omega)$. The expansion (14) of m_ω for large $|\xi|$ implies

$$\begin{aligned} &q_0^{(2)}(\ell,\omega)[g,h] \\ &= \ell^2 \int_{\{|\xi|>1\}} m_\omega(|\xi|) \cdot \hat{g}(\ell\xi) \, \overline{\hat{h}(\ell\xi)} \, \mathrm{d}\xi \\ &= \frac{1}{\ell} q_0[g,h] - \ell^2 \cdot \frac{\omega}{2} \int_{\mathbb{R}^2} \frac{\hat{g}(\ell\xi) \, \overline{\hat{h}(\ell\xi)}}{|\xi|} \, \mathrm{d}\xi + \ell^2 \int_{\mathbb{R}^2} m_{\omega,\mathrm{res}}(|\xi|) \cdot \hat{g}(\ell\xi) \, \overline{\hat{h}(\ell\xi)} \, \mathrm{d}\xi, \end{aligned}$$

with

$$m_{\omega,\mathrm{res}}(|\xi|) = -\mathbb{1}_{\{|\xi|\leq 1\}}(\xi) \cdot |\xi| + \mathbb{1}_{\{|\xi|>1\}}(\xi) \cdot (m_\omega(|\xi|) - |\xi|) + \frac{\omega}{2|\xi|}.$$

We choose the functions $X_\omega, X_{\omega,\mathrm{res}} \in C^\infty(\mathbb{R}^2 \setminus \{0\})$ such that

$$\widehat{X}_\omega(\xi) = \frac{\omega}{4\pi|\xi|} \quad \text{and} \quad \widehat{X}_{\omega,\mathrm{res}}(\xi) = \frac{1}{2\pi} m_{\omega,\mathrm{res}}(\xi).$$

Calculating X_ω for $(s,\varphi) \in \mathbb{R}_+ \times (0, 2\pi)$, $x = (s\cos\varphi, s\sin\varphi)$ we have

$$\begin{aligned} X_\omega(x) &= \frac{\omega}{8\pi^2} \int_{\mathbb{R}^2} \frac{\mathrm{e}^{\mathrm{i}x\xi}}{|\xi|} \, \mathrm{d}\xi \\ &= \frac{\omega}{8\pi^2} \int_0^\infty \int_{-\pi}^{\pi} \mathrm{e}^{\mathrm{i}st(\cos\varphi \;\; \sin\varphi)\cdot(\cos u \;\; \sin u)^T} \, \mathrm{d}u \, \mathrm{d}t \\ &= \frac{\omega}{4\pi} \int_0^\infty J_0(ts) \, \mathrm{d}t. \end{aligned}$$

Here J_0 is the Bessel function of the first kind of order 0. Moreover, all integrals should be interpreted as oscillatory integrals or improper Riemann integrals. Using Section 6.511 in [20] we obtain

$$X_\omega(x) = \frac{\omega}{4\pi|x|}\int_0^\infty J_0(r)\mathrm{d}r = \frac{\omega}{4\pi|x|},$$

and thus,

$$\begin{aligned}\ell^2\cdot\frac{\omega}{2}\int_{\mathbb{R}^2}\frac{\hat{g}(\ell\xi)\,\overline{\hat{h}(\ell\xi)}}{|\xi|}\,\mathrm{d}\xi &= \frac{\omega}{4\pi}\langle K_{1/|x|}T_\ell g, T_\ell h\rangle_{\Sigma_\ell}\\ &= \frac{\omega\cdot\ell^2}{4\pi}\int_{\Sigma\times\Sigma}\frac{g(x)\,\overline{h(z)}}{\ell|z-x|}\,\mathrm{d}x\,\mathrm{d}z\\ &= \ell\cdot\frac{\pi}{4\alpha^2}\langle K_{1/|x|}g, h\rangle_\Sigma + \mathcal{O}\Big(\frac{\pi^2}{\alpha^2}-\omega\Big).\end{aligned}$$

Note that $m_{\omega,\mathrm{res}}(\xi) = \mathcal{O}(|\xi|^{-3})$ as $|\xi|\to\infty$, uniformly in $\omega\in(0,\pi^2/\alpha^2)$, and thus,

$$\sup_{\omega\in(0,\pi^2/\alpha^2)}\|X_{\omega,\mathrm{res}}\|_{L_\infty(\mathbb{R}^2)} < \infty,$$

which implies

$$\begin{aligned}\ell^2\int_{\mathbb{R}^2} m_{\omega,\mathrm{res}}(|\xi|)\cdot\hat{g}(\ell\xi)\,\overline{\hat{h}(\ell\xi)}\,\mathrm{d}\xi &= \ell^2\int_{\Sigma\times\Sigma}X_{\omega,\mathrm{res}}(\ell(z-x))\cdot g(x)\,\overline{h(z)}\,\mathrm{d}x\,\mathrm{d}z\\ &= \mathcal{O}(\ell^2).\end{aligned}$$

This concludes the proof of the theorem. □

Let us now prove the asymptotics of the ground state eigenvalue of the operator $A_{\ell,b}$ as $\ell\to 0$. We shall omit the proof of the uniqueness or the existence of the eigenvalue for small $\ell>0$ as this follows in much the same way as in Lemma 2.13. We note that the operator $\mathcal{Q}_0$ is again invertible and a Fredholm operator since $\widetilde{H}_0^{1/2}(\Sigma)$ is compactly embedded into $L_2(\Sigma)$, cf. the arguments from the previous section. Then for arbitrary $b\in L_\infty(\mathbb{R}^2)$ we have

$$\ell\mathcal{Q}_b(\ell,\omega) = \mathcal{Q}_0 + \ell^3\cdot\frac{|\Sigma|}{4\alpha^3}\ln\Big(\frac{\pi^2}{\alpha^2}-\omega\Big)\cdot P_{\mathrm{ct}} + \ell R_b(\ell,\omega)$$

with

$$\sup\Big\{\|R_b(\ell,\omega)\| : \ell\in(0,\ell_0)\text{ and }\Big|\omega-\frac{\pi^2}{\alpha^2}\Big|<\varepsilon\Big\}<\infty.$$

Applying the Birman–Schwinger principle, we obtain the following identity for the eigenvalue $\lambda(\ell)$

$$-\frac{\ell^3}{4\alpha^3}\ln\Big(\frac{\pi^2}{\alpha^2}-\lambda(\ell)\Big)\cdot\langle(\mathcal{Q}_0+\ell R_b(\ell,\lambda(\ell)))^{-1}\phi_0,\phi_0\rangle_\Sigma = 1$$

or equivalently

$$\ln\Big(\frac{\pi^2}{\alpha^2}-\lambda(\ell)\Big) = -\frac{4\alpha^3}{\ell^3\cdot\langle(\mathcal{Q}_0+\ell R_b(\ell,\lambda(\ell)))^{-1}\phi_0,\phi_0\rangle_\Sigma}.$$

Here $\phi_0(x)=1$ is again the non-normalised constant function in $L_2(\Sigma)$. As before we obtain

$$-\ln\Big(\frac{\pi^2}{\alpha^2}-\lambda(\ell)\Big) = \frac{4\alpha^3}{\ell^3\cdot\tau_1(\Sigma)+\mathcal{O}(\ell^4)}$$

as $\ell\to 0$. Here

$$\tau_1(\Sigma) := \langle\mathcal{Q}_0^{-1}\phi_0,\phi_0\rangle_\Sigma = \langle\mathcal{Q}_0^{-1/2}\phi_0,\mathcal{Q}_0^{-1/2}\phi_0\rangle_\Sigma > 0. \tag{23}$$

This proves the first term of the asymptotic formula. Higher terms of the expansion may be calculated as above; assuming smoothness of b we obtain

$$\ln\Big(\frac{\pi^2}{\alpha^2}-\lambda(\ell)\Big) = -\ell^{-3}\frac{4\alpha^3}{\tau_0(\Sigma)-\ell\cdot\tau_1(\Sigma)\cdot b(0)+\mathcal{O}(\ell^3)} \quad \text{as } \ell\to 0,$$

where

$$\tau_1(\Sigma) := \langle\mathcal{Q}_0^{-1}\phi_0,\mathcal{Q}_0^{-1}\phi_0\rangle_\Sigma > 0. \tag{24}$$

This concludes the proof of Theorem 3.1.

Acknowledgement. The research and in particular André Hänel were supported by DFG grant WE-1964/4-1.

References

[1] H. Ammari, H. Kang, and H. Lee, *Layer potential techniques in spectral analysis.* Mathematical Surveys and Monographs, 153. American Mathematical Society, Providence, RI, 2009. ISBN 978-0-8218-4784-8 MR 2488135 Zbl 1167.47001

[2] H. Ammari, H. Kang, H. Lee, and J. Lim, Boundary perturbations due to the presence of small linear cracks in an elastic body. *J. Elasticity* **113** (2013), no. 1, 75–91. MR 3092955 Zbl 1277.35037

[3] M. Š. Birman, Perturbations of the continuous spectrum of a singular elliptic operator by varying the boundary and the boundary conditions. *Vestnik Leningrad. Univ.* **17** (1962), no. 1, 22–55. In Russian. MR 0138874 Zbl 0149.31903

[4] M. Š. Birman and M. Z. Solomyak, Schrödinger operator. Estimates for number of bound states as function-theoretical problem. In S. G. Gindikin (ed.), *Spectral theory of operators.* Papers from the Fourteenth School on Operators in Functional Spaces held at Novgorod State Pedagogical Institute, Novgorod, 1989. Translated from the Russian. Translation edited by S. Ivanov. American Mathematical Society Translations, Series 2, 150. Soviet Regional Conferences. American Mathematical Society, Providence, R.I., 1992, 1–54. MR 1157648 Zbl 1157647

[5] D. Borisov, R. Bunoiu and G. Cardone, On a waveguide with frequently alternating boundary conditions: homogenized Neumann condition. *Ann. Henri Poincaré* **11** (2010), no. 8, 1591–1627. MR 2769705 Zbl 1210.82077

[6] D. Borisov, R. Bunoiu and G. Cardone, On a waveguide with an infinite number of small windows. *C. R. Math. Acad. Sci. Paris* **349** (2011), no. 1-2, 53–56. MR 2755696 Zbl 1211.35098

[7] D. Borisov, R. Bunoiu and G. Cardone, Waveguide with non-periodically alternating Dirichlet and Robin conditions: homogenization and asymptotics. *Z. Angew. Math. Phys.* **64** (2013), no. 3, 439–472. MR 3068832 Zbl 1282.35033

[8] D. Borisov, T. Ekholm, and H. Kovařík, Spectrum of the magnetic Schrödinger operator in a waveguide with combined boundary conditions. *Ann. Henri Poincaré* **6** (2005), no. 2, 327–342. MR 2136194 Zbl 1077.81038

[9] D. Borisov and P. Exner, Exponential splitting of bound states in a waveguide with a pair of distant windows. *J. Phys. A* **37** (2004), no. 10, 3411–3428. MR 2039856 Zbl 1050.81008

[10] D. Borisov and P. Exner, Distant perturbation asymptotics in window-coupled waveguides. I. The nonthreshold case. *J. Math. Phys.* **47** (2006), no. 11, 113502, 24 pp. MR 2278663 Zbl 1112.81043

[11] P. Exner and P. Šeba, M. Tater, and D. Vaněk, Bound states and scattering in quantum waveguides coupled laterally through a boundary window. *J. Math. Phys.* **37** (1996), no. 10, 4867–4887. MR 1411612 Zbl 0883.35085

[12] P. Exner and S. A. Vugalter, Asymptotic estimates for bound states in quantum waveguides coupled laterally through a narrow window. *Ann. Inst. H. Poincaré Phys. Théor.* **65** (1996), no. 1, 109–123. MR 1407168 Zbl 0858.35093

[13] P. Exner and S. A. Vugalter, Bound-state asymptotic estimates for window-coupled Dirichlet strips and layers. *J. Phys. A* **30** (1997), no. 22, 7863–7878. MR 1616939 Zbl 0930.58016

[14] P. Exner and H. Kovařík, *Quantum waveguides.* Theoretical and Mathematical Physics. Springer, Cham, 2015. ISBN 978-3-319-18575-0 MR 3362506 Zbl 1314.81001

[15] S. V. Frolov and I. Yu. Popov, Resonances for laterally coupled quantum waveguides. *J. Math. Phys.* **41** (2000), no. 7, 4391–4405. MR 1765577 Zbl 0973.81018

[16] S. V. Frolov and I. Yu. Popov, Three laterally coupled quantum waveguides: breaking of symmetry and resonance asymptotics. *J. Phys. A* **36** (2003), no. 6, 1655–1670. MR 1959957 Zbl 1070.81063

[17] R. R. Gadyl′shin, Surface potentials and the method of matching asymptotic expansions in the Helmholtz resonator problem. *Algebra i Analiz* **4** (1992), no. 2, 88–115. In Russian. English translation, *St. Petersburg Math. J.* **4** (1993), no. 2, 273–296. MR 1182395 Zbl 0792.35031

[18] R. R. Gadyl′shin, On regular and singular perturbations of acoustic and quantum waveguides. *C. R., Méc., Acad. Sci. Paris* **332** (2004), no. 8, 647–652. Zbl 1234.35086

[19] I. C. Gohberg and E. I. Sigal, An operator generalization of the logarithmic residue theorem and Rouché's theorem. *Mat. Sb.* (*N.S.*) **84(126)** (1971), 607–629. In Russian. English translation, *Math. USSR-Sb.* **13** (1971), 603–625. MR 0313856 Zbl 0254.47046

[20] I. S. Gradshteyn and I. M. Ryzhik, *Table of integrals, series, and products.* Fourth edition prepared by Ju. V. Geronimus and M. Ju. Ceĭtlin. Translation edited by A. Jeffrey. Academic Press, New York and London, 1965. MR 0197789

[21] G. Grubb, The mixed boundary value problem, Krein resolvent formulas and spectral asymptotic estimates. *J. Math. Anal. Appl.* **382** (2011), no. 1, 339–363. MR 2805518 Zbl 1223.47050

[22] A. Hänel and T. Weidl, Eigenvalue asymptotics for an elastic strip and an elastic plate with a crack. *Quart. J. Mech. Appl. Math.* **69** (2016), no. 4, 319–352.

[23] A. M. Il′in, *Matching of asymptotic expansions of solutions of boundary value problems.* Translated from the Russian by V. V. Minachin. Translations of Mathematical Monographs, 102. American Mathematical Society, Providence, R.I., 1992. ISBN 0-8218-4561-6 MR 1182791 Zbl 0754.34002

[24] M. A. Lawrentjew and B. W. Schabat, *Methoden der komplexen Funktionentheorie.* Translation and scientific editing by U. Pirl, R. Kühnau and L. Wolfersdorf. Mathematik für Naturwissenschaft und Technik, 13. VEB Deutscher Verlag der Wissenschaften, Berlin, 1967. MR 0209447 Zbl 0153.09601

[25] W. McLean, *Strongly elliptic systems and boundary integral equations.* Cambridge University Press, Cambridge, 2000. ISBN 0-521-66332-6; 0-521-66375-X MR 1742312 Zbl 0948.35001

[26] S. A. Nazarov, Asymptotics of an eigenvalue on the continuous spectrum of two quantum waveguides coupled through narrow windows. *Mat. Zametki* **93** (2013), no. 2, 227–245. In Russian. English translation, *Math. Notes* **93** (2013), no. 1-2, 266–281. MR 3205969 Zbl 06158184

[27] I. Yu. Popov, Asymptotics of bound state for laterally coupled waveguides. *Rep. Math. Phys.* **43** (1999), no. 3, 427–437. MR 1713399 Zbl 1040.78510

[28] I. Yu. Popov, Asymptotics of bound states and bands for laterally coupled three-dimensional waveguides. *Rep. Math. Phys.* **48** (2001), no. 3, 277–288. MR 1873969 Zbl 0989.78003

[29] I. Yu. Popov, Asymptotics of bound states and bands for waveguides coupled through small windows. *Appl. Math. Lett.* **14** (2001), no. 1, 109–113. MR 1793712 Zbl 1049.78012

[30] I. Yu. Popov, Asymptotics of bound states and bands for laterally coupled waveguides and layers. *J. Math. Phys.* **43** (2002), no. 1, 215–234. MR 1872495 Zbl 1059.78025

Dirichlet eigenfunctions in the cube, sharpening the Courant nodal inequality

Bernard Helffer and Rola Kiwan

To Pavel Exner on the occasion of his 70th birthday

1 Introduction and main result

Consider the Dirichlet eigenvalues of the Laplacian in a bounded domain Ω with piecewise C^1-boundary.

$$\begin{cases} -\Delta u = \lambda u & \text{in } \Omega\,, \\ u = 0 & \text{on } \partial\Omega\,. \end{cases} \tag{1}$$

We denote by

$$\{\lambda_k\}_{k\geq 1} = \{\lambda_k(\Omega)\}_{k\geq 1}$$

the sequence of eigenvalues (counted with their multiplicity):

$$\lambda_1 < \lambda_2 \leq \lambda_3 \leq \cdots \leq \lambda_k \leq \cdots.$$

It is well known that the first eigenvalue is simple and that the eigenfunction u_1 has a constant sign in Ω. All the higher order eigenfunctions must change sign inside Ω and, consequently, must vanish inside Ω.

We call nodal set of an eigenfunction u_k associated with λ_k the closure of the zero set of u_k,

$$\mathcal{N}(u_k) = \overline{\{x \in \Omega;\ u_k(x) = 0\}}.$$

This nodal set cuts the domain $\Omega \setminus \mathcal{N}(u_k)$ into $\mu_k = \mu(u_k)$ connected components called "nodal domains."

The famous Courant nodal theorem [7] of 1923 states that

$$\mu(u_k) \leq k.$$

We will say that an eigenvalue λ is Courant sharp if $\lambda = \lambda_k$ and if there exists an associate eigenfunction with k nodal domains. While it is always true in the case of dimension 1 by the Sturm–Liouville theory, Pleijel's theorem [27] asserts in 1956 that equality can only occur for a finite set of k's, when the dimension is at least two.

Since we know that the first eigenfunction does not vanish and that the second eigenfunction has exactly two nodal domains, λ_1 and λ_2 are Courant sharp ($\mu_1 = 1$ and $\mu_2 = 2$). We are now interested in checking if other eigenvalues are Courant sharp.

Starting from the founding paper by Å. Pleijel [27], many other works have investigated in which cases this inequality is sharp: Helffer, Hoffmann-Ostenhof, and Terracini [13] and [14], Helffer and Hoffmann-Ostenhof [11] and [12], Bérard and Helffer [3], [4], and [5], Helffer and Persson-Sundqvist [16], Léna [22], and Leydold [23], [24], and [25]. All these results were devoted to (2D)-cases in open sets in $\mathbb{R}^2$: squares, rectangles, equilateral triangles, disks, ..., or in surfaces like $\mathbb{S}^2$ or $\mathbb{T}^2$.

The aim of the current paper is to start the analysis of analogous results for domains in $\mathbb{R}^3$. Looking at explicitly solvable models and avoiding easy cases where the multiplicity of each eigenvalue is one (making the analysis of the nodal set easy), it is natural to first consider the case of the cube. More precisely, we will prove:

Theorem 1.1. *In the case of the cube $\mathcal{C} := (0, \pi)^3$ the only eigenvalues of the Dirichlet Laplacian which are Courant sharp are the two first eigenvalues: $\lambda_1 = 3$ and $\lambda_2 = 6$.*

2 Coming back to Pleijel's paper

Outside the proof of Pleijel's theorem in 2D, Pleijel [27] (see also [3] for a more detailed analysis) considers as an example the case of the square which reads

Theorem 2.1. *In the case of the square the only eigenvalues which are Courant sharp for the Dirichlet Laplacian are the two first eigenvalues and the fourth one.*

The proof was based on a first reduction to the analysis of the eigenvalues less than 68 (the argument will be extended to the (3D)-case below and this is a quantitative version of the proof of Pleijel's theorem), then all the other eigenvalues were eliminated using this time a more direct consequence of Faber–Krahn's inequality, except three remaining cases for which Pleijel was rather sketchy and which have to be treated by hand.

At the end of his celebrated paper Å. Pleijel wrote:

> *"In order to treat, for instance the case of the free three-dimensional membrane* $[0,\pi]^3$*, it would be necessary to use, in a special case, the theorem quoted in* [8], p. 394.[a] This theorem which generalizes part of the Liouville–Rayleigh theorem for the string asserts that a linear combination, with constant coefficients, of the n first eigenfunctions can have at most n nodal domains. However, as far as I have been able to find there is no proof of this assertion in the literature."
>
> [a] In the German version, this is p. 454 in the English version.

Pleijel was indeed speaking of a result presented in [8] as being proved in the thesis defended in 1932 at the University of Göttingen by Horst Herrmann (with Richard Courant as advisor). This so called Courant-Herrmann conjecture was asserting that, for a given $k \in \mathbb{N}$, Courant's theorem holds also for linear combinations of eigenfunctions associated with eigenvalues λ_j with $j \leq k$. It is said in [10] that the authors can not find any mention of the result in the thesis itself. We learned in a recent article by N. Kuznetsov [20] that V. Arnold [1] has discussed this question with R. Courant in the seventies and that he can prove that this theorem cannot be true in general.

Pleijel is not explicitly saying why he needs this result but one could think that he is interested, because he speaks about the "free problem" (i.e., the Neumann problem), in counting the number of components of the restriction of an eigenfunction to a face of the cube $(0,\pi)^3$. Looking for example to the zeroset of

$$(x,y,z) \longmapsto a\cos x\cos y\cos nz + b\cos y\cos z\cos nx + c\cos z\cos x\cos ny,$$

one gets for fixed $z = 0$, a linear combination of the eigenfunctions of the square $\cos x\cos y$, $\cos y\cos nx$ and $\cos x\cos ny$ corresponding to two different eigenspaces for the Neumann Laplacian in the square $(0,\pi)\times(0,\pi)$. We will not go further in this paper on the Neumann problem but similar questions could also occur in the Dirichlet problem and we typically meet below the eigenfunction

$$(x,y,z) \longmapsto a\sin x\sin y\sin nz + b\sin y\sin z\sin nx + c\sin z\sin x\sin ny,$$

and will be interested for example in the intersection of its zero set with the hyperplace $\{z = \pi/2\}$ inside the cube (in the case $n = 3$).

3 Reminder on Pleijel's theorem in $3D$

Let us first prove that there are only a finite number of eigenvalues that satisfy $\mu_k := \mu(u_k) = k$. This proof was given in dimension n by Bérard and Meyer [6].

Proposition 3.1. *If λ_k is an eigenvalue of* (1) *such that $\lambda_{k-1} < \lambda_k$, and u_k is an associated eigenfunction, then*

$$\lambda_k^{3/2}|\Omega| \geq \mu(u_k)\frac{4}{3}\pi^4, \tag{2}$$

where $|\Omega|$ denotes the volume of Ω.

Proof. Assume that the nodal set cuts the domain Ω in μ_k connected components and let us denote them by Ω_i, $1 \leq i \leq \mu_k$. Since u_k does not vanish inside Ω_i, it is equal to its first eigenvalue and now using the (3D)-Faber–Krahn inequality on each component (see for example Bérard and Meyer [6]):

$$\lambda_k^{3/2}|\Omega_i| \geq \frac{4}{3}\pi^4, \quad \text{for } 1 \leq i \leq \mu_k.$$

Adding together all the equations we get (2). □

Theorem 3.2. *One has*

$$\lim \sup_{k\to+\infty} \frac{\mu_k}{k} \leq \frac{9}{2\pi^2} < 1. \tag{3}$$

In particular, there exists only a finite number of eigenvalues satisfying $\mu_k = k$.

Proof. We start from the Weyl asymptotics for the counting function

$$N(\lambda) := \#\{k,\, \lambda_k < \lambda\}, \tag{4}$$

which reads

$$N(\lambda) \sim \frac{1}{6\pi^2}|\Omega|\lambda^{3/2}.$$

For an eigenvalue λ_k such that $\lambda_{k-1} < \lambda_k$, we have $N(\lambda_k) = k - 1$. Then from

$$\lambda_k^{3/2} \sim \frac{6\pi^2}{|\Omega|}k$$

together with (2), we get (3). □

Remark 3.3. It is clear from (3) that we cannot have an infinite number of eigenvalues satisfying $\mu_k = k$.

4 The case of the cube

Let us consider the cube $(0,\pi)^3$ for which an orthogonal basis of eigenfunctions for the Dirichlet problem is given by

$$\begin{cases} u_{\ell,m,n}(x,y,z) = \sin(\ell x)\cdot\sin(my)\cdot\sin(nz), \\ \lambda_{\ell,m,n} = \ell^2+m^2+n^2, \end{cases}$$

for $\ell,m,n \geq 1$.

Applying Proposition 3.1 for this domain, we get

Proposition 4.1. *If u_k is an eigenfunction associated with λ_k such that u_k has k nodal domains and if $\lambda_{k-1} < \lambda_k$ we have*

$$\frac{{\lambda_k}^{3/2}}{k} \geq \frac{4}{3}\pi. \tag{5}$$

Here we will try to find a lower bound for the number $N(\lambda)$, since we know the λ's are equal to $\ell^2+m^2+n^2$ where ℓ,m,n are integers, so we need to count the number of the lattice points of $\mathbb{R}^3$ inside the sphere of radius $\sqrt{\lambda}$.

Lemma 4.2. *If $\lambda \geq 3$, then*

$$N(\lambda) > \frac{\pi}{6}\lambda^{3/2} - \frac{3\pi}{4}\lambda + 3\sqrt{\lambda-2} - 1. \tag{6}$$

The proof is given in the appendix.

Lemma 4.3. *If u_k is an eigenfunction associated with λ_k such that u_k has k nodal domains we have*

$$\Big(\frac{3}{4\pi} - \frac{\pi}{6}\Big)\lambda^{3/2} + \frac{3\pi}{4}\lambda - 3\sqrt{\lambda} + 3 > 0.$$

Proof. First by the Courant theorem, we have necessarily $\lambda_{k-1} < \lambda_k$.

Applying (6), we have

$$k-1 = N(\lambda) > \frac{\pi}{6}\lambda^{3/2} - \frac{3\pi}{4}\lambda + 3\sqrt{\lambda-2} - 1,$$

i.e.,

$$k > \frac{\pi}{6}\lambda^{3/2} - \frac{3\pi}{4}\lambda + 3\sqrt{\lambda-2}.$$

Together with (5), this implies

$$\Big(\frac{3}{4\pi}-\frac{\pi}{6}\Big)\lambda^{3/2}+\frac{3\pi}{4}\lambda > 3\sqrt{\lambda-2}.$$

One immediately sees that, for $\lambda \geq 3$,

$$\sqrt{\lambda-2}-\sqrt{\lambda} = -\frac{2}{\sqrt{\lambda}+\sqrt{\lambda-2}} \geq -\frac{2}{1+\sqrt{3}} > -1. \qquad \square$$

Now setting $\mu = \sqrt{\lambda}$ we get the third order inequality

$$\Big(\frac{3}{4\pi}-\frac{\pi}{6}\Big)\mu^3+\frac{3\pi}{4}\mu^2-3\mu+3>0. \tag{7}$$

Now, consider the cubic function

$$h(x)=\Big(\frac{3}{4\pi}-\frac{\pi}{6}\Big)x^3+\frac{3\pi}{4}x^2-3x+3 \quad \text{on } (0,+\infty).$$

The coefficient of x^3 is strictly negative. One can check easily that $h(7)<0$, $h'(7)<0$ and that $h''(x)<0$ for all $x>7$. This implies that $h(x)$ is negative for $x>7$. Hence Inequality (7) for a positive μ implies $\mu \in (0,7)$ and coming back to λ: $\lambda < 49$. So we have finally proved:

Proposition 4.4. *If u_k is an eigenfunction associated with λ_k such that u_k has k nodal domains and if $\lambda_{k-1} < \lambda_k$, we have*

$$\lambda_k \leq 48.$$

5 The list

In this section, we establish the list of the eigenvalues which are less than 48 and determine which of these eigenvalues satisfy the necessary condition (5) for being Courant sharp.

Coming back to the consequences of Faber–Krahn's inequality, one can check that among all the values on Table 1, the only eigenvalues that satisfy inequality (5) and $\lambda_{k-1} < \lambda_k$ are λ_1, λ_2, λ_5, λ_8 and λ_{12}.

Proposition 5.1. *The only eigenvalues which can be "Courant sharp" are the eigenvalues λ_k with $k = 1, 2, 5, 8$ and 12.*

As λ_1 and λ_2 are Courant sharp, the only remaining cases to analyze correspond to $k = 5, 8, 12$.

In the next section we will by a finer analysis involving symmetries eliminate other cases.

Table 1. List of all eigenvalues less than 49

k	(ℓ, m, n)	λ_k
λ_1	(1,1,1)	3
$\lambda_2 = \lambda_3 = \lambda_4$	(1,1,2)	6
$\lambda_5 = \lambda_6 = \lambda_7$	(1, 2, 2)	9
$\lambda_8 = \lambda_9 = \lambda_{10}$	(1,1,3)	11
λ_{11}	(2,2,2)	12
$\lambda_{12} = \lambda_{13} = \lambda_{14} = \lambda_{15} = \lambda_{16} = \lambda_{17}$	(1,2,3)	14
$\lambda_{18} = \lambda_{19} = \lambda_{20}$	(2,2,3)	17
$\lambda_{21} = \lambda_{22} = \lambda_{23}$	(1,1,4)	18
$\lambda_{24} = \lambda_{25} = \lambda_{26}$	(1,3,3)	19
$\lambda_{27} = \lambda_{28} = \lambda_{29} = \lambda_{30} = \lambda_{31} = \lambda_{32}$	(1,2,4)	21
$\lambda_{33} = \lambda_{34} = \lambda_{35}$	(2,3,3)	22
$\lambda_{36} = \lambda_{37} = \lambda_{38}$	(2, 2, 4)	24
$\lambda_{39} = \lambda_{40} = \lambda_{41} = \lambda_{42} = \lambda_{43} = \lambda_{44}$	(1, 3, 4)	26
$\lambda_{45} = \lambda_{46} = \lambda_{47} = \lambda_{48}$	(3, 3, 3) & (1, 1, 5)	27
$\lambda_{49} = \lambda_{50} = \lambda_{51} = \lambda_{52} = \lambda_{53} = \lambda_{54}$	(2, 3, 4)	29
$\lambda_{55} = \lambda_{56} = \lambda_{57} = \lambda_{58} = \lambda_{59} = \lambda_{60}$	(1, 2, 5)	30
$\lambda_{61} = \lambda_{62} = \lambda_{63} = \lambda_{64} = \lambda_{65} = \lambda_{66}$	(1, 4, 4) & (2, 2, 5)	33
$\lambda_{67} = \lambda_{68} = \lambda_{69}$	(3, 3, 4)	34
$\lambda_{70} = \lambda_{71} = \lambda_{72} = \lambda_{73} = \lambda_{74} = \lambda_{75}$	(1, 3, 5)	35
$\lambda_{76} = \lambda_{77} = \lambda_{78}$	(2, 4, 4)	36
$\lambda_{79} = \lambda_{80} = \lambda_{81} = \lambda_{82} = \lambda_{83} = \lambda_{84} = \lambda_{85} = \lambda_{86} = \lambda_{87}$	(1, 1, 6) & (2, 3, 5)	38
$\lambda_{88} = \lambda_{89} = \lambda_{90} = \lambda_{91} = \lambda_{92} = \lambda_{93} = \lambda_{94} = \lambda_{95} = \lambda_{96}$	(1, 2, 6) & (3, 4, 4)	41
$\lambda_{97} = \lambda_{98} = \lambda_{99} = \lambda_{100} = \lambda_{101} = \lambda_{102}$	(1, 4, 5)	42
$\lambda_{103} = \lambda_{104} = \lambda_{105}$	(3, 3, 5)	43
$\lambda_{106} = \lambda_{107} = \lambda_{108}$	(2, 2, 6)	44
$\lambda_{109} = \lambda_{110} = \lambda_{111} = \lambda_{112} = \lambda_{113} = \lambda_{114}$	(2, 4, 5)	45
$\lambda_{115} = \lambda_{116} = \lambda_{117} = \lambda_{118} = \lambda_{119} = \lambda_{120}$	(1, 3, 6)	46
λ_{121}	(4, 4, 4)	48

6 Courant theorem with symmetry

We first recall some generalities which come back to Leydold [23], and were used in various contexts, see [24], [25], [16], and [14]. Suppose that there exists an isometry g such that $g(\Omega) = \Omega$ and $g^2 = \mathrm{Id}$. Then g acts naturally on $L^2(\Omega)$ by $gu(\boldsymbol{x}) = u(g^{-1}\boldsymbol{x})$, for all $\boldsymbol{x} \in \Omega$, and one can naturally define an orthogonal decomposition of $L^2(\Omega)$,

$$L^2(\Omega) = L^2_{\mathrm{odd}} \oplus L^2_{\mathrm{even}},$$

where by definition

$$L^2_{\mathrm{odd}} = \{u \in L^2,\ gu = -u\}$$

and

$$L^2_{\text{even}} = \{u \in L^2,\ gu = u\}.$$

These two spaces are left invariant by the Laplacian and one can consider separately the spectrum of the two restrictions. Let us explain for the "odd case" what could be a Courant theorem with symmetry. If u is an eigenfunction in L^2_{odd} associated with λ, we see immediately that the nodal domains appear by pairs (exchanged by g) and following the proof of the standard Courant theorem we see that if $\lambda = \lambda_j^{\text{odd}}$ for some j (that is the j-th eigenvalue in the odd space), then the number $\mu(u)$ of nodal domains of u satisfies $\mu(u) \leq j$.

We get a similar result for the "even" case (but in this case a nodal domain D is either g-invariant or $g(D)$ is a distinct nodal domain).

These remarks may lead to improvements when each eigenspace has a specific symmetry. As we shall see, this will be the case for the cube with the map

$$(x, y, z) \longmapsto (\pi - x, \pi - y, \pi - z).$$

We observe indeed that

$$u_{\ell,m,n}(\pi - x, \pi - y, \pi - z) = (-1)^{\ell+m+n+1} u_{\ell,m,n}(x, y),$$

and that

$$\ell^2 + m^2 + n^2 \equiv \ell + m + n \pmod 2.$$

Hence, for a given eigenvalue the whole eigenspace is even if $\ell + m + n$ is odd and odd if $\ell + m + n$ is even. Equivalently, the whole eigenspace is even if the eigenvalue is odd and even if the eigenvalue is odd.

Application

λ_5 is not Courant sharp. The eigenspace associated with $\lambda_5 = 9$ is even. This is the second one (in this even space). Hence it should have less than four nodal domains by Courant's theorem with symmetry and has labelling 5.

λ_{12} is not Courant sharp. $\lambda_{12} = 14$ is the fifth eigenvalue in the odd space with respect to σ. It should has less than 10 nodal domains and has labelling 12.

7 The remaining value: $k = 8$

7.1 Main result

The proof of our main theorem relies now on the analysis of the last case which is the object of the next proposition.

Proposition 7.1. *In the eigenspace associated with λ_8 the eigenfunctions have either 2, 3, or 4 nodal domains. In particular λ_8 cannot be Courant sharp.*

7.2 Preliminaries

For the value $\lambda_8 = 11$ we have to analyze the zeroset of

$$\Phi_{a,b,c}(x, y, z) := a \sin x \sin y \sin 3z + b \sin y \sin z \sin 3x + c \sin z \sin x \sin 3y,$$

for $(a, b, c) \neq (0, 0, 0)$.

This looks nice because we can divide by $\sin x \sin y \sin z$ and by making the change of coordinates $u = \cos x$, $v = \cos y$, $w = \cos z$, we get for the zero set of $\Phi_{a,b,c}$ in the new coordinates a quadric surface $\mathcal{Q}_{a,b,c}$ to analyze in the cube $\mathcal{C} = (-1, 1)^3$, whose equation is

$$(\mathcal{Q}_{a,b,c}): \quad 4\,(au^2 + bv^2 + cw^2) - (a + b + c) = 0,$$

for $(a, b, c) \neq (0, 0, 0)$.

When $a+b+c \neq 0$, we immediately see that there are no critical points inside the cube, so the nodal set is simply an hypersurface (cylinder, ellipsoid or hyperboloid with one or two sheets). In this case, this is the analysis at the six faces of the cube which will be decisive for analyzing possible changes in the number of connected components. In the case when $a + b + c = 0$, we have a double cone with a unique critical point at $(0, 0, 0)$.

In the next subsections, we discuss the different cases.

7.3 Cylinder

This corresponds to the case $abc = 0$. We can use the (2D)- analysis as done in [3]. It is known that the number of nodal domains can only be 2, 3, or 4 (See Section 3.1 and Figure 2.1 there). See Figure 1.[1]

[1] All the pictures in this paper were produced using MATLAB.

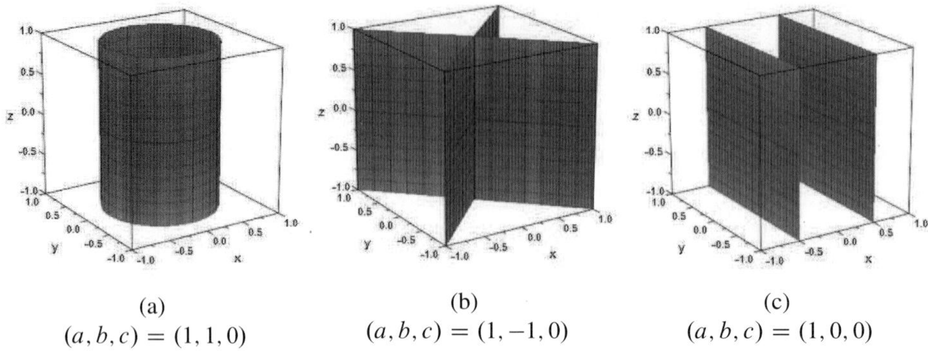

(a) $(a,b,c) = (1,1,0)$ (b) $(a,b,c) = (1,-1,0)$ (c) $(a,b,c) = (1,0,0)$

Figure 1. Cylinders

7.4 Double cone

This corresponds to $abc \neq 0$ and $a+b+c = 0$. The equation of $\mathcal{Q}_{a,b,c}$ is

$$au^2 + bv^2 = -cw^2.$$

One can verify that the intersection of this cone with each horizontal side $w = \pm 1$ is exactly at the vertices of the cube $u^2 = v^2 = 1$, and that the intersection with each vertical face is a hyperbola. Therefore there are three connected components of $\mathcal{C} \setminus \mathcal{Q}_{a,b,c}$. See Figure 2.

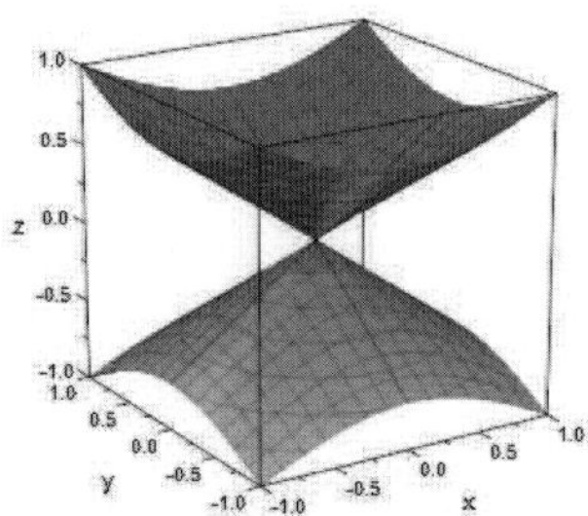

Figure 2. Double cones. $(a,b,c) = (0.2, 0.2, -0.4)$

7.5 Ellipsoid

This corresponds to $abc \neq 0$, with a, b, c of the same sign. Without loss of generality, we can assume that $0 < a \leq b \leq c$ and that $a+b+c = 1$. We note that this implies

$3/2\,(a+b) \le a + 2b \le 1$ and

$$au^2 + bv^2 + (1-a-b)w^2 = \frac{1}{4}.$$

We denote by $\Omega_{a,b,c}$ the open full ellipsoid delimited by $\mathcal{Q}_{a,b,c}$.

Let us look at the intersection of $\mathcal{Q}_{a,b,c}$ with the horizontal faces. We have

$$au^2 + bv^2 \le -\frac{3}{4} + \frac{2}{3} < 0.$$

We deduce that in this case there are no possible intersections with the horizontal faces, and therefore two subcases can occur depending on the intersection of $\mathcal{Q}_{a,b,c}$ with the vertical edges. This set is determined by

$$(1-a-b)w^2 = \frac{1}{4} - (a+b), \quad w \in (-1,+1).$$

See Figure 3.

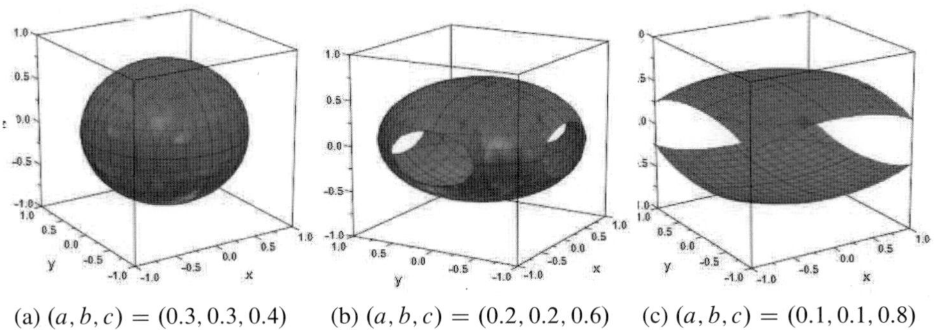

(a) $(a,b,c) = (0.3, 0.3, 0.4)$ (b) $(a,b,c) = (0.2, 0.2, 0.6)$ (c) $(a,b,c) = (0.1, 0.1, 0.8)$

Figure 3. Ellipsoids

Subcase $(a+b) > 1/4$. The ellipsoid $\mathcal{Q}_{a,b,c}$ does not touch the vertical edges and in this case $\mathcal{C} \cap \overline{\Omega_{a,b,c}}^{\,c}$ is connected and $\mathcal{C} \setminus \mathcal{Q}_{a,b,c}$ has exactly two connected components.

Subcase $(a+b) \le 1/4$. $\mathcal{Q}_{a,b,c}$ cuts each vertical edge along a segment $[-w_0, +w_0]$ with

$$w_0 = \sqrt{\frac{\frac{1}{4} - (a+b)}{1-a-b}}.$$

The intersection of $\mathcal{Q}_{a,b,c}$ with each vertical face of the cube is the union of two arcs of an ellipse. In this case it is clear that $\mathcal{C} \setminus \mathcal{Q}_{a,b,c}$ has three connected components.

7.6 One sheet hyperboloid

This corresponds to $abc \neq 0$, a, b, c not of the same sign and $(abc)(a+b+c) < 0$. Without loss of generality, we can assume that $b \geq a > 0$, $c < 0$ and $a+b+c = 1$. We note that this implies that $\mathcal{Q}_{a,b,c} \cap \{w = 0\}$ is an ellipse contained in the cube.

The equation of $\mathcal{Q}_{a,b,c}$ can be written as

$$au^2 + bv^2 = \frac{1}{4} - cw^2.$$

$\mathcal{Q}_{a,b,c}$ cuts $\mathbb{R}^3$ into two components $\Omega^+_{a,b,c}$ and $\Omega^-_{a,b,c}$ where $\Omega^+_{a,b,c}$ contains $(0,0,0)$. But we have to look inside the cube.

We first observe that $\mathcal{Q}_{a,b,c}$ has empty intersection at the vertical edges. We have indeed

$$a + b = 1 - c > \frac{1}{4} - c \geq \frac{1}{4} - cw^2.$$

We now look at the intersection with $w = 0$. We get an ellipse

$$\mathcal{E}^0_{a,b,c} := \mathcal{Q}_{a,b,c} \cap \{w = 0\},$$

whose equation is

$$au^2 + bv^2 = \frac{1}{4}.$$

We observe that this ellipse could be included in the cube, if $a > 1/4$ or not if $a \leq 1/4$.

We also look at the intersection with the upper horizontal face. We note that the ellipse

$$\mathcal{E}^1_{a,b,c} := \mathcal{Q}_{a,b,c} \cap \{w = 1\}$$

has always a non empty intersection with this face.

Four subcases appear (See Figure 4).

Subcase $a \leq 1/4$. Under this condition $\mathcal{Q}_{a,b,c} \cap \{v = 0\} \cap \mathcal{C}$ is empty. Hence $\{v = 0\} \cap \mathcal{C}$ is contained in one nodal domain which is invariant by $v \mapsto -v$. The other nodal domains are exchanged by this symmetry. This gives an odd number of nodal domains and this can not be Courant sharp because the labelling is 8. More precisely the two curves in $\mathcal{C}$ defined by

$$v = \pm\sqrt{\frac{\frac{1}{4} - au^2 - cw^2}{b}}$$

cut the cube in three components.

The three last subcases are under the condition that $a > 1/4$. We note that this condition implies that $\mathcal{E}^0_{a,b,c}$ is strictly included in the square $(-1,+1) \times (-1,+1)$ and the discussion continues according to the position of $\mathcal{E}^1_{a,b,c}$ in the horizontal face.

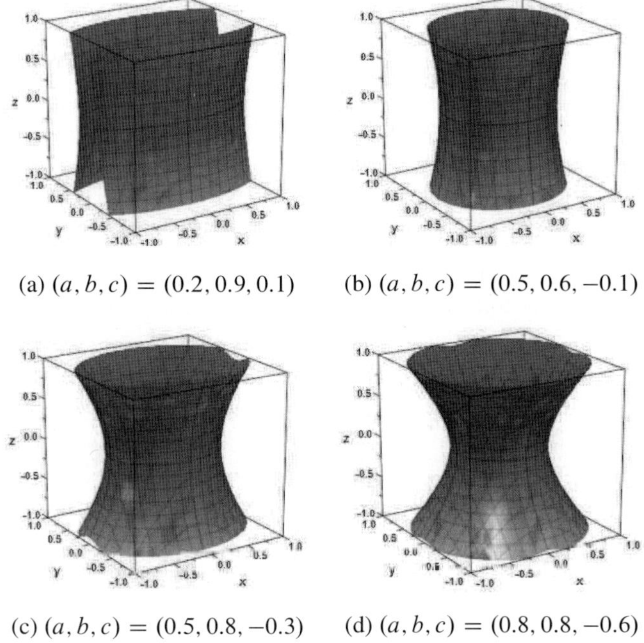

(a) $(a, b, c) = (0.2, 0.9, 0.1)$ (b) $(a, b, c) = (0.5, 0.6, -0.1)$

(c) $(a, b, c) = (0.5, 0.8, -0.3)$ (d) $(a, b, c) = (0.8, 0.8, -0.6)$

Figure 4. One Sheet Hyperboloid

Subcase $1/4 < a \leq b < 3/4$. $\mathcal{E}^1_{a,b,c}$ is contained in the horizontal face and $\mathcal{Q}_{a,b,c}$ cuts the cube in two connected domains.

Subcase $1/4 < a < 3/4 \leq b$. $\mathcal{E}^1_{a,b,c} \cap \partial\mathcal{C}$ consists of two curves but $\mathcal{Q}_{a,b,c}$ continue to cut the cube in two domains. For joining two points of $\Omega^-_{a,b,c} \cap \partial\mathcal{C}$ one can always go to a point in $\{w = 0\}$ outside of $\mathcal{E}^0_{a,b,c}$ and use the connexity (inside the square $\mathcal{C} \cap \{w = 0\}$) of the complementary of the full ellipse.

Subcase $3/4 \leq a$. $\mathcal{E}^1_{a,b,c} \cap \mathcal{C}$ consists of four curves. $\mathcal{Q}_{a,b,c}$ continue to cut the cube in two domains.

7.7 Two sheets hyperboloid

This corresponds to $abc \neq 0$, a, b, c not of the same sign and $(abc)(a+b+c) > 0$.

We can assume $b \geq a > 0$, $c < 0$ and $a + b + c = -1$. The equation of $\mathcal{Q}_{a,b,c}$

can be written as:

$$au^2 + bv^2 = -\frac{1}{4} - cw^2.$$

The hyperplane $\{w = 0\}$ is contained in one connected component. Hence looking at the symmetry $w \mapsto -w$, we get that necessarily an odd number (≥ 3) of nodal domains and ≤ 8 by Courant's theorem. Hence we know that it cannot be Courant sharp.

More precisely, $\mathcal{Q}_{a,b,c}$ meets the hyperplane $\{w = 1\}$ along the ellipse $\mathcal{E}_{a,b,c}$ which this time contains the horizontal upper face of the cube. The analysis of the intersection along each of the vertical faces (two symmetric curves by $w \mapsto -w$) shows that we always have exactly three connected components. See Figure 5.

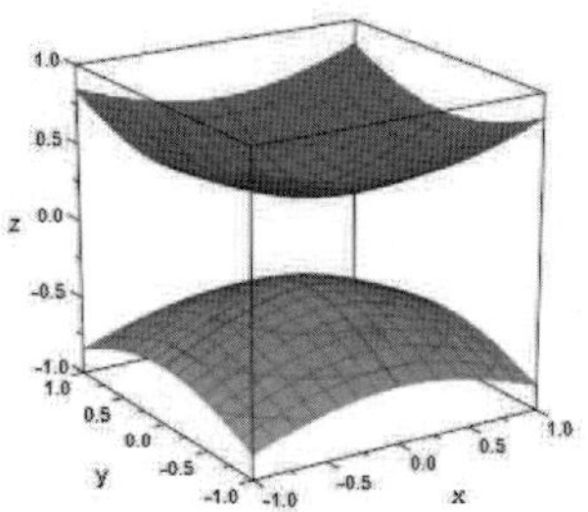

Figure 5. Two Sheets Hyperboloid: $(a, b, c) = (0.8, 0.8, -2.6)$

8 Conclusion

In this paper we have analyzed the problem in the simplest example proposed by Pleijel. One can of course ask for similar questions for other geometries starting with the parallelepipeds, the ball (this case has been solved in [17]), the flat torus (C. Léna has announced the characterization in this case). It is probably difficult to find other solvable cases. The situation for $(0, \alpha\pi) \times (0, \beta\pi) \times (0, \gamma\pi)$ is in principle easier in the "irrational" case when $\alpha\ell^2 + \beta m^2 + \gamma n^2 = \alpha\ell_1^2 + \beta m_1^2 + \gamma_1 n_1^2$ implies $(\ell, m, n) = (\ell_1, m_1, n_1)$. Each eigenvalue $\alpha\ell^2 + \beta m^2 + \gamma n^2$ is indeed of multiplicity 1 and the corresponding eigenfunction has ℓmn nodal domains.

One can also think of analyzing "thin structures" (for example γ small or $\beta + \gamma$ small, where previous results in lower dimension can probably be used) in the spirit of [11] and get partial results. Another interesting question would be to analyze the Neumann problem for the cube in the spirit of [16].

Appendix. Proof of Lemma 4.2

We follow an idea appearing in the (2D)-case in a course of R. Laugesen [21]. We start by assuming that λ is not an eigenvalue. With each triple (ℓ, m, n) with $\ell \geq 1$, $m \geq 1$, $n \geq 1$, we associate the cube

$$Q_{\ell,m,n} = (\ell - 1, \ell) \times (m - 1, m) \times (n - 1, n).$$

We observe that

$$N(\lambda) = \sum_{\substack{\ell^2+m^2+n^2<\lambda,\, \ell\geq 1,\\ m\geq 1,\, n\geq 1}} |Q_{\ell,m,n}| \leq \frac{\pi}{6}\lambda^{3/2},$$

where we recall that $|\Omega|$ denotes the volume of Ω.

We are interested in the lower bound. The claim of Laugesen is that

$$N(\lambda) > |B_\lambda|, \tag{8}$$

where

$$B_\lambda := \{(x+1)^2 + (y+1)^2 + (z+1)^2 < \lambda, x > 0, y > 0, z > 0\}.$$

The observation is that

$$B_\lambda \subset \bigcup_{\substack{\ell^2+m^2+n^2<\lambda,\, \ell\geq 1,\\ m\geq 1,\, n\geq 1}} Q_{\ell,m,n}.$$

For $t > 0$, $[t]_+$ denotes the smallest integer $\geq t$.

Let $(x, y, z) \in B_\lambda$, then it is immediate to see that $(x, y) \in Q_{[x]_+,[y]_+,[z]_+}$. It remains to verify that

$$Q_{[x]_+,[y]_+,[z]_+} \subset D(0, \sqrt{\lambda}),$$

where $D(0, \rho)$ denotes the open ball in $\mathbb{R}^3$ of radius ρ.

But we have, for $(x, y, z) \in B_\lambda$,

$$[x]_+^2 + [y]_+^2 + [z]_+^2 \leq (x+1)^2 + (y+1)^2 + (z+1)^2 < \lambda.$$

Coming back to (8), we have to find a lower bound for the area of B_λ. We note that by translation by the vector $(1, 1, 1)$:

$$|B_\lambda| = |C_\lambda|,$$

where

$$C_\lambda := D(0, \sqrt{\lambda}) \cap \{x > 1\} \cap \{y > 1\} \cap \{z > 1\}.$$

Let χ the characteristic function of the interval $(0, 1)$. We have to compute the integral

$$|C_\lambda| = \int_{D(0,\sqrt{\lambda})} (1 - \chi(x))(1 - \chi(y))(1 - \chi(z))\, \mathrm{d}x\, \mathrm{d}y\, \mathrm{d}z.$$

Developing the formula and using the symmetry by permutation of the variables, we get, if $\lambda \geq 3$,

$$\begin{aligned} |C_\lambda| = &\int_{D(0,\sqrt{\lambda})} \mathrm{d}x\, \mathrm{d}y\, \mathrm{d}z \\ &- 3\int_{D(0,\sqrt{\lambda})} \chi(x)\, \mathrm{d}x\, \mathrm{d}y\, \mathrm{d}z \\ &+ 3\int_{D(0,\sqrt{\lambda})} \chi(x)\chi(y)\, \mathrm{d}x\, \mathrm{d}y\, \mathrm{d}z \\ &- \int_{D(0,\sqrt{\lambda})} \chi(x)\chi(y)\chi(z)\, \mathrm{d}x\, \mathrm{d}y\, \mathrm{d}z. \end{aligned}$$

It is then immediate to get the lemma by observing that

$$\int_{D(0,\sqrt{\lambda})} \chi(x)\chi(y)\, \mathrm{d}x\, \mathrm{d}y\, \mathrm{d}z > \sqrt{\lambda - 2}. \tag{9}$$

We have assumed till now that λ was not an eigenvalue. But if λ is an eigenvalue > 3, we can apply the previous result for an increasing sequence $\hat{\lambda}_j$ such that $\hat{\lambda}_j \to \lambda$ (where $\hat{\lambda}_j > 3$ is not an eigenvalue). According to our definition of $N(\lambda)$ in (4), we can pass to the limit and observing that in (9) the inequality is uniformly strict when applied to the sequence λ_j, we keep the strict inequality when passing to the limit. The case $\lambda = 3$ can be verified directly.

Acknowledgements. This paper was achieved when the two authors were invited at the Isaac Newton Institute for mathematical sciences in Cambridge. Moreover Bernard Helffer was supported there as Simons foundation visiting fellow.

References

[1] V. I. Arnold, Topological properties of eigenoscillations in mathematical physics. *Tr. Mat. Inst. Steklova* **273** (2011), Sovremennye Problemy Matematiki, 30–40. In Russian. English translation, *Proc. Steklov Inst. Math.* **273** (2011), no. 1, 25–34. MR 2893541 Zbl 1229.35220

[2] P. Bérard, Inégalités isopérimétriques et applications. Domaines nodaux des fonctions propres. In *Séminaire Goulaouic-Meyer-Schwartz, 1981–1982.* Équations aux dérivées partielles. École Polytechnique, Centre de Mathématiques, Palaiseau, 1982, exp. no. XI, 10 pp. MR 0671608 Zbl 0489.35058

[3] P. Bérard and B. Helffer, Dirichlet eigenfunctions of the square membrane: Courant's property, and A. Stern's and Å. Pleijel's analyses. In A. Baklouti, A. El Kacimi, S. Kallel, and N. Mir (eds.), *Analysis and geometry.* Selected papers presented at the MIMS (Mediterranean Institute for the Mathematical Sciences)—GGTM (Geometry and Topology Grouping for the Maghreb). Conference held at the Cité des Sciences, Tunis, March 24–27, 2014. In honour of M. S. Baouendi. Springer Proceedings in Mathematics & Statistics, 127. Springer, Cham, 2015, 69–114. MR 3445517 Zbl 06499037

[4] P. Bérard and B. Helffer, A. Stern's analysis of the nodal sets of some families of spherical harmonics revisited. *Monatsh. Math.* **180** (2016), no. 3, 435–468. MR 3513215 Zbl 06608233

[5] P. Bérard and B. Helffer, Courant sharp eigenvalues for the equilateral torus, and for the equilateral triangle. Preprint 2015. arXiv:1503.00117 [math.AP] To appear in *Lett. Math. Phys.*

[6] P. Bérard and D. Meyer, Inégalités isopérimétriques et applications. *Ann. Sci. École Norm. Sup.* (4) **15** (1982), no. 3, 513–541. MR 0690651 Zbl 0527.35020

[7] R. Courant, Ein allgemeiner Satz zur Theorie der Eigenfunktionen selbstadjungierter Differentialausdrücke, *Gött. Nachr.* **1923**, 81–84. JFM 49.0342.01

[8] R. Courant and D. Hilbert, *Methods of mathematical physics.* Vol. I. Translated and revised from the German original. Interscience Publishers, New York, N.Y., 1953. MR 0065391 Zbl 0051.28802 Zbl 0053.02805

[9] C. Faber, Beweis, daß unter allen homogenen Membranen von gleicher Fläche und gleicher Spannung die kreisförmige die tiefsten Grundton gibt. *Sitzungsber. Bayer. Akad. Wiss., Math. Phys. Munich* **1923**, 169–172. JFM 49.0342.03

[10] G. M. L. Gladwell and H. Zhu, The Courant–Herrmann conjecture. *ZAMM Z. Angew. Math. Mech.* **83** (2003), no. 4, 275–281. MR 1977003 Zbl 1221.35413

[11] B . Helffer and T. Hoffmann-Ostenhof, Minimal partitions for anisotropic tori. *J. Spectr. Theory* **4** (2014), no. 2, 221–233. MR 3232810 Zbl 1321.58006

[12] B. Helffer and T. Hoffmann-Ostenhof, A review on large k minimal spectral k-partitions and Pleijel's theorem. In G. Eskin, L. Friedlander, and J. Garnett (eds.), *Spectral theory and partial differential equations.* Proceedings of the conference in honor of James Ralston's 70th birthday held at the University of California, Los Angeles, CA, June 17–21, 2013, 39–57. MR 3381015 Zbl 06622437

[13] B. Helffer, T. Hoffmann-Ostenhof, and S. Terracini, Nodal domains and spectral minimal partitions. *Ann. Inst. H. Poincaré Anal. Non Linéaire* **26** (2009), no. 1, 101–138. MR 2483815 Zbl 1171.35083

[14] B. Helffer, T. Hoffmann-Ostenhof, and S. Terracini, On spectral minimal partitions: the case of the sphere. In A. Laptev (ed), *Around the research of Vladimir Maz'ya.* III. Analysis and applications. International Mathematical Series (New York), 13. Springer, New York etc., and Tamara Rozhkovskaya Publisher, Novosibirsk, 2010, 153–178. MR 2664708 Zbl 1230.35072

[15] B. Helffer, T. Hoffmann-Ostenhof, and S. Terracini, Nodal minimal partitions in dimension 3. *Discrete Contin. Dyn. Syst.* **28** (2010), no. 2, 617–635. Special issues Part I, dedicated to Professor Louis Nirenberg on the occasion of his 85th birthday. MR 2644760 Zbl 1193.35162

[16] B. Helffer and M. Persson-Sundqvist, Nodal domains in the square – the Neumann case. *Mosc. Math. J.* **15** (2015), no. 3, 455–495, 605. MR 3427435 Zbl 1342.35198

[17] B. Helffer and M. Persson-Sundqvist, On nodal domains in Euclidean balls. *Proc. Amer. Math. Soc.* **144** (2016), no. 11, 4777–4791. MR 3544529 Zbl 06629324

[18] H. Herrmann, Beziehungen zwischen den Eigenwerten und Eigenfunktionen verschiedener Eigenwertprobleme. *Math. Z.* **40** (1936), no. 1, 221–241. MR 1545554 JFM 61.0443.01 Zbl 0011.35103

[19] E. Krahn, Über eine von Rayleigh formulierte Minimaleigenschaft des Kreises. *Math. Ann.* **94** (1925), no. 1, 97–100. MR 1512244 JFM 51.0356.05

[20] N. Kuznetsov, On delusive nodal sets of free oscillations. *Eur. Math. Soc. Newsl.* **96** (2015), 34–40. MR 3379499 Zbl 1322.35099

[21] R. S. Laugesen, *Spectral theory of partial differential equations.* Lecture Notes. University of Illinois at Urbana–Champaign, Urbana, 2011. arXiv:1203.2344 [math.AP]

[22] C. Léna, Courant-sharp eigenvalues of a two-dimensional torus. *C. R. Math. Acad. Sci. Paris* **353** (2015), no. 6, 535–539. MR 3348988 Zbl 1319.35119

[23] J. Leydold, *Knotenlinien und Knotengebiete von Eigenfunktionen.* Diplom-Arbeit. Universität Wien, Wien, 1989.

[24] J. Leydold, *On the number of nodal domains of spherical harmonics.* Ph.D. thesis. Universität Wien, Wien, 1992.

[25] J. Leydold, On the number of nodal domains of spherical harmonics. *Topology* **35** (1996), no. 2, 301–321. MR 1380499 Zbl 0853.33012

[26] J. Peetre, A generalization of Courant's nodal domain theorem. *Math. Scand.* **5** (1957), 15–20. MR 0092917 Zbl 0077.30101

[27] Å. Pleijel, Remarks on Courant's nodal line theorem. *Comm. Pure Appl. Math.* **9** (1956), 543–550. MR 0080861 Zbl 0070.32604

[28] F. Pockels, *Ueber die partielle Differentialgleichung* $-\Delta u - k^2 u = 0$ *and deren Auftreten in mathematischen Physik.* With a foreword by F. Klein. Teubner, Leipzig, 1891. Electronic version, Cornell University Library, Historical Math Monographs, `http://ebooks.library.cornell.edu/m/math/` JFM 23.0403.01

[29] I. Polterovich, Pleijel's nodal domain theorem for free membranes. *Proc. Amer. Math. Soc.* **137** (2009), no. 3, 1021–1024. MR 2457442 Zbl 1162.35005

[30] A. Stern, Bemerkungen über asymptotisches Verhalten von Eigenwerten und Eigenfunktionen. Diss. Göttingen, 1925. JFM 51.0356.01

[31] J. Toth and S. Zelditch, Counting nodal lines which touch the boundary of an analytic domain. *J. Differential Geom.* **81** (2009), no. 3, 649–686. MR 2487604 Zbl 1180.35395

[32] H. Weyl, Über die asymptotische Verteilung der Eigenwerte. *Gött. Nachr.* **1911**, 110–117. JFM 42.0432.03

A mathematical modeling of electron–phonon interaction for small wave numbers close to zero

Masao Hirokawa

This paper is dedicated to Pavel Exner
on the occasion of his 70th birthday
by introducing a part of our attempts in material science,
that is, a mathematical modeling of electron–phonon interaction
for small wave numbers close to zero.
We extrapolate the electron–phonon interaction
on the presupposition that the phonon dispersion relation
and the electron–phonon coupling function
are estimated using some experimental data.

1 Introduction

Modeling for electron–phonon interaction plays an important role in experiments to analyze the materials such as metal and crystal, see [25], [26], [38], [52], [53], and [58]: A model is proposed using the experimental data and following theory. Another experiment is performed to check the validity of the model. Such trial and error are always repeated in the circumstances of the experiment scene. The form of the interaction between an electron and a phonon-field depends on the concrete material. We have to extrapolate the interaction for the as-yet-unrecognized material based on the data obtained in the experiments. What we can estimate from the experimental data are usually the phonon dispersion relation and the electron–phonon coupling function. Several methods for estimating the phonon dispersion relation has been established basically by the so-called *Raman spectroscopy* which is an experimental method using the Raman effect (i.e., the Raman scattering), see [6], [27], and [29]. Some methods for estimating the electron–phonon coupling function have lately been developed with vivacity, see [25], [26], [37], [38], [52], [53], and [58].

This paper devotes special attention to the interaction between the electron and the massless phonon. We then meet the so-called *infrared (IR) problem for massless phonon* to argue the existence of ground state in the same way as for massless photon, see [1], [2], [3], [4], [5], [9], [15], [16], [17], [18], [19], [21], [22], [23], [24], [39], and [46]. We investigate the condition to distinguish between the existence and absence of the ground state, though in this paper we argue neither instability nor phase transition caused by the so-called *soft phonon* (i.e., phonon's softening), see [6], [29], [34], [40], and [55]. The IR problem is primarily the difficulty occurred as the wave number vanishes; it may be described by the interaction information for small wave number. Thus, as the first step of this intuition, we make the following set-ups: We assume that we can estimate the phonon dispersion relation $\omega(k)$ from the experimental data of the dispersion curve observed in the same way by the Raman spectroscopy, see [6], [8], [11], [45], [49], and [58]. In addition, we assume that we can estimate the electron–phonon coupling function $\lambda(k)$ from the experimental data obtained in a roughly similar way of, for instance, the angle-resolved photoemission spectroscopy – see [25] and [26] – or the tunneling experiment (resp. tunneling measurement), see [37], [52], [53], and [56]. The former is an experimental technique which enables us to observe a kink by electron–phonon interaction, see [7], [30], and [57]. The latter is the probe of the electron–phonon interaction using the tunneling of single particle states, see [20], [43], [44], and [51]. These experimental methods are based on the Migdal–Eliashberg theory [10] and [33]. In their theory the so-called *Fröhlich Hamiltonian* [14] is employed, and thus, it is described by the electron-field and the phonon-field. In this paper, however, we employ the Schrödinger operator for the electron instead of the electron-field. Whether the IR problem takes place is determined on the behavior of those two functions, $\omega(k)$, $\lambda(k)$, at very small wave number $|k|$. The massless phonon has $\omega(k) \sim 0$, the decay of the dispersion relation, at $|k| \approx 0$. We, in particular, are interested in the decay order because it is difficult to grasp only from the data observed in experiments (see, for example, figures of the estimated phonon dispersion relation in [6], [8], [11], [34], [45], [48], [49], [55], and [58]). Meanwhile, the electron–phonon coupling function also has the decay, $|\lambda(k)| \sim 0$, at $|k| \approx 0$ to avoid the IR catastrophe. Since the electron–phonon coupling function $\lambda(k)$ governs the coupling strength between the electron and the phonon-field, its decay means that we have only to consider the small coupling strength, and therefore, we can assume the linear coupling of the Lee–Low–Pines Hamiltonian (see [31], [32], and [36]) for the unknown interaction $H_{\rm int}$ in our modeling. The interaction $H_{\rm int}$ in the total Hamiltonian $H_{\rm QFT}$ of the material is concretely determined by several effects of the lattice structure. The lattice structure makes its own lattice vibration, and then, the lattice vibration makes its own phonon-field. The individual effect, of course, comes from some reasons such as the lattice defect as can be seen in the nitrogen-vacancy center in diamond [35] as well as the lattice's own vibration. We assume that the estimated coupling function $\lambda(k)$ almost describes the interaction between the electron and the phonon field, and thus, it includes almost all of those effects, though we actually cannot grasp all of them.

In the light of mathematics, we define a Carleman operator [54] using the estimated dispersion relation $\omega(k)$ and the estimated coupling function $\lambda(k)$, and then we consider the IR problem with the Carleman operator. As for the use of the Carleman operator for the IR problem, there is the preceding work by Hiroshima [24]. He characterizes the existence of a ground state using the Carleman operator in the case where any IR divergence does not occur. Conversely, we characterize the IR divergence by the maximal Carleman operator, and moreover, we investigate how the singularity of the maximal Carleman operator leads to the IR divergence. Then, we give a criterion for IR problem in terms of the estimated dispersion relation and the estimated coupling function for small wave numbers close to zero by making good use of Dereziński and Gérard's idea.

2 Set-ups in mathematics

2.1 Preliminaries

We here prepare some notion and tools in mathematics to define a self-adjoint operator called a *Hamiltonian* describing the total energy of a particle coupled with a Bose field.

Let $X = (X, \mathcal{A}, \mu)$ be a σ-finite measurable space. We denote by X^n n-fold Cartesian product of X. The measure, $d\mu^n(k_1, \ldots, k_n)$, for X^n is naturally defined by $d\mu(k_1) \otimes \cdots \otimes d\mu(k_n)$. We define the boson Fock space $\mathcal{F}_\mathrm{b}(L^2(X))$ over $L^2(X) := L^2(X, \mathcal{A}, \mu)$ by $\bigoplus_{n=0}^\infty \bigotimes_\mathrm{s}^n L^2(X)$. Here, $\bigotimes_\mathrm{s}^n L^2(X)$ is the n-fold symmetric tensor product of $L^2(X)$ for each $n \in \mathbb{N}$ with convention $\bigotimes_\mathrm{s}^0 L^2(X) := \mathbb{C}$. For every $\psi \in \mathcal{F}_\mathrm{b}(L^2(X))$, we use the expression $\psi = \bigoplus_{n=0}^\infty \psi^{(n)} = \psi^{(0)} \oplus \psi^{(1)} \oplus \cdots \oplus \psi^{(n)} \oplus \cdots$, where $\psi^{(n)} \in \bigotimes_\mathrm{s}^n L^2(X)$ for each $n \in \{0\} \cup \mathbb{N}$. We often abbreviate $\mathcal{F}_\mathrm{b}(L^2(X))$ to $\mathcal{F}_X$ for simplicity in this paper. We employ the standard norm $\|\cdot\|_{\mathcal{F}_X}$ in $\mathcal{F}_X$. We denote by $\|\cdot\|_{\mathcal{H}}$ the norm of a Hilbert space $\mathcal{H}$ induced by its inner product throughout this paper. We define a special vector Ω_X in the boson Fock space $\mathcal{F}_X$ by $1 \oplus 0 \oplus 0 \oplus \cdots$, and call it the *Fock vacuum*.

For each $n \in \{0\} \cup \mathbb{N}$ and every $f \in L^2(X)$, we define an operator

$$a_X(f) \colon \bigotimes_{\mathrm{s}}^n L^2(X) \ni \psi^{(n+1)} \longmapsto (a_X(f)\psi)^{(n)} \in \bigotimes_{\mathrm{s}}^n L^2(X)$$

by

$$(a_X(f)\psi)^{(n)}(k_1, \ldots, k_n) := \sqrt{n+1} \int_X \overline{f(k)} \psi^{(n+1)}(k, k_1, \ldots, k_n) \mathrm{d}\mu(k).$$

We can extend $a_X(f)$ to a closed operator acting in $\mathcal{F}_X$ by extending its domain to $\{\psi \in \mathcal{F}_X \mid \sum_{n=0}^{\infty} \|(a_X(f)\psi)^{(n)}\|^2_{\bigotimes_{\mathrm{s}}^n L^2(X)} < \infty\}$. We call $a_X(f)$ the *annihilation operator*. Since we can regard it as an operator-valued distribution, we symbolically write it as $a_X(f) = \int_X \overline{f(k)} a_X(k) \mathrm{d}\mu(k)$ with its kernel $a_X(k)$. Meanwhile, we define the *creation operator* $a_X^{\dagger}(f)$ by $a_X(f)^*$, the adjoint operator of $a_X(f)$. The kernel of $a_X^{\dagger}(f)$ is denoted as $a_X^{\dagger}(k)$ frequently.

Remark 2.1. In the case, for example, where $X = \mathbb{R}^d$, we note that it is difficult to handle $a_X(k)$ and $a_X^{\dagger}(k)$ as operators acting in $\mathcal{F}_X$ (e.g., see Remark 1 of [22]): The domain of $a_X(k)^* \equiv a_X^{\dagger}(k)$ consists of only the zero vector, and thus, the kernel $a_X(k)$ of the annihilation operator $a_X(f)$ is not closable.

From now on, we denotes by I the identity operator and $D(S)$ the domain of an operator S.

Let T be a closable operator densely defined in $L^2(X)$. For each $n \in \{0\} \cup \mathbb{N}$ we set $T^{(0)}$ as $T^{(0)} := 0$, and $T^{(n)}$ as $T^{(n)} := \overline{\sum_{j=1}^n I \otimes \cdots \otimes I \otimes T \otimes I \otimes \cdots \otimes I}$, where T sits on the j-th. We denote by $\overline{S}$ the closure of a closable operator S. Then, we define an operator $\mathrm{d}\Gamma_X(T)$ acting in $\mathcal{F}_X$ by $\bigoplus_{n=0}^{\infty} T^{(n)}$. We call $\mathrm{d}\Gamma_X(T)$ *second quantization* of T. For the second quantization the following facts are well known.

I. If $T \neq 0$, then $\mathrm{d}\Gamma_X(T)$ is unbounded.

II. If T is self-adjoint, then $\mathrm{d}\Gamma_X(T)$ is also self-adjoint.

III. Let T be non-negative, injective, and self-adjoint. Then, the inclusion relation, $D(\mathrm{d}\Gamma_X(T)^{1/2}) \subset D(a_X(f)) \cap D(a_X^{\dagger}(f))$, holds for for every $f \in D(T^{-1/2})$. In addition to this inclusion relation, the following inequalities hold for every $\psi \in D(\mathrm{d}\Gamma_X(T)^{1/2})$:

$$\|a_X^{\sharp}(f)\psi\|_{\mathcal{F}_X} \leq \|T^{-1/2} f\|_{L^2(X)} \|\mathrm{d}\Gamma_X(T)^{1/2}\psi\|_{\mathcal{F}_X} + c_{\sharp}\|f\|_{L^2(X)}\|\psi\|_{\mathcal{F}_X},$$

where $c_{\sharp} = 0$ if $a_X^{\sharp}(f) = a_X(f)$; $c_{\sharp} = 1$ if $a_X^{\sharp}(f) = a_X^{\dagger}(f)$. In particular, we define the number operator N_X by $\mathrm{d}\Gamma_X(1)$. Here 1 stands for the multiplication operator of the constant function $1(k) \equiv 1$ of $k \in X$.

Suppose now that X can be decomposed into the disjoint union of X_1 and X_2. Then, $L^2(X)$ is also decomposed into the direct sum of $L^2(X_1)$ and $L^2(X_2)$. The following is well known. There is a unique unitary operator

$$U : \mathcal{F}_X \equiv \mathcal{F}_{\mathrm{b}}(L^2(X)) \longrightarrow \mathcal{F}_{X_1} \otimes \mathcal{F}_{X_2} \equiv \mathcal{F}_{\mathrm{b}}(L^2(X_1)) \otimes \mathcal{F}_{\mathrm{b}}(L^2(X_2))$$

such that

IV.-a) the unique unitary operator U gives $U\Omega_X = \Omega_{X_1} \otimes \Omega_{X_2}$ for the individual Fock vacuums, $\Omega_X \in \mathcal{F}_X$, $\Omega_{X_1} \in \mathcal{F}_{X_1}$ and $\Omega_{X_2} \in \mathcal{F}_{X_2}$;

IV.-b) $U\,\mathrm{d}\Gamma_X(h) = \overline{\mathrm{d}\Gamma_{X_1}(h_1) \otimes I + I \otimes \mathrm{d}\Gamma_{X_2}(h_2)}$ for the decomposition $h = h_1 \oplus h_2$, where h acts in $L^2(X)$ and h_j in $L^2(X_j)$, $j = 1, 2$, respectively.

Let $\mathcal{V}$ be a separable Hilbert space. We denote its inner product and norm by $(\cdot,\cdot)_{\mathcal{V}}$ and $\|\cdot\|_{\mathcal{V}}$, respectively. Then, for each $n \in \mathbb{N}$ we define the Hilbert space $L^2_{\mathrm{sym}}(X^n;\mathcal{V})$ of all square-integrable, $\mathcal{V}$-valued, symmetric functions. We say that $f : X^n \to \mathcal{V}$ is *measurable* if $(v, f(\cdot))_{\mathcal{V}} : X^n \to \mathbb{C}$ is measurable for every $v \in \mathcal{V}$. It is known that

V. the two spaces, $\mathcal{V} \otimes \mathcal{F}_X$ and $\bigoplus_{n=0}^{\infty} L^2_{\mathrm{sym}}(X^n;\mathcal{V})$, are unitarily equivalent, that is, there is a unitary operator

$$U_{\mathcal{V}} : \mathcal{V} \otimes \mathcal{F}_X \longrightarrow \bigoplus_{n=0}^{\infty} L^2_{\mathrm{sym}}(X^n;\mathcal{V})$$

with convention $L^2_{\mathrm{sym}}(X^0;\mathcal{V}) := \mathcal{V}$.

Through this unitary transformation $U_{\mathcal{V}}$, for every $\Psi \in \mathcal{V} \otimes \mathcal{F}_X$ we denote $U_{\mathcal{V}}\Psi$ by $\Psi_{\mathcal{V}}$. Moreover, $\Psi_{\mathcal{V}}$ is often expressed by

$$\bigoplus \sum_{n=0}^{\infty} \Psi_{\mathcal{V}}^{(n)} = \Psi_{\mathcal{V}}^{(0)} \oplus \Psi_{\mathcal{V}}^{(1)} \oplus \cdots \oplus \Psi_{\mathcal{V}}^{(n)} \oplus \cdots,$$

where $\Psi_{\mathcal{V}}^{(n)} \in L^2_{\mathrm{sym}}(X^n;\mathcal{V})$ for each $n \in \{0\} \cup \mathbb{N}$. Therefore, the norm $\|\Psi_{\mathcal{V}}\|_{\mathcal{V}\otimes\mathcal{F}_X}$ is $\sqrt{\|\Psi_{\mathcal{V}}^{(0)}\|_{\mathcal{V}}^2 + \sum_{n=1}^{\infty} \|\Psi_{\mathcal{V}}^{(n)}\|_{L^2(X^n;\mathcal{V})}^2}$.

It is easy to give a small generalization of Corollary 5.1 in [22]:

Proposition 2.2. *Let $\{f_\ell\}_{\ell=1}^{\infty}$ be an arbitrary complete orthonormal system of $L^2(X)$. Then, the equation, $\|I \otimes N_X^{1/2}\Psi\|_{\mathcal{V}\otimes\mathcal{F}_X}^2 = \sum_{\ell=1}^{\infty} \|I \otimes a_X(f_\ell)\Psi\|_{\mathcal{V}\otimes\mathcal{F}_X}^2$, holds for every $\Psi \in D(I \otimes N_X^{1/2})$.*

As a special case of Proposition 2.2, we have

Corollary 2.3 (Proposition 5.1 in [22]). *Let* $\{f_\ell\}_{\ell=1}^{\infty}$ *be an arbitrary complete orthonormal system of* $L^2(X)$. *Then* $\|N_X^{1/2}\psi\|_{\mathcal{F}_X}^2 = \sum_{\ell=1}^{\infty} \|a_X(f_\ell)\psi\|_{\mathcal{F}_X}^2$, *holds for every* $\psi \in D(N_X^{1/2})$.

This equation is symbolically written as

$$\|N_X^{1/2}\psi\|_{\mathcal{F}_X}^2 = \int_X \|a_X(k)\psi\|_{\mathcal{F}_X}^2 \, \mathrm{d}\mu(k)$$

using the kernel $a_X(k)$.

2.2 Definition of general Hamiltonian

Let us suppose that the state space of a particle set in a quantum field by a separable, complex Hilbert space $\mathcal{H}$, and the quantum field's momenta are in $\mathbb{R}^d$. Only when $X = \mathbb{R}^d$, we use the abbreviation, $\mathcal{F}_\mathrm{b} := \mathcal{F}_{\mathbb{R}^d} \equiv \mathcal{F}_\mathrm{b}(L^2(\mathbb{R}^d))$. Corresponding to this abbreviation, we abbreviate $a_{\mathbb{R}^d}(f)$, $a_{\mathbb{R}^d}^\dagger(f)$, and $\mathrm{d}\Gamma_{\mathbb{R}^d}(h)$ to $a_\mathrm{b}(f)$, $a_\mathrm{b}^\dagger(f)$, and $\mathrm{d}\Gamma_\mathrm{b}(h)$, respectively. In particular, we often use the notation N_b for $\mathrm{d}\Gamma_\mathrm{b}(1)$. The total state space of the particle coupled with the Bose field is given by $\mathcal{F} := \mathcal{H} \otimes \mathcal{F}_\mathrm{b}$. We denote $I \otimes a_\mathrm{b}(f)$ and $I \otimes a_\mathrm{b}^\dagger(f)$ by $a(f)$ and $a\dagger(f)$, respectively.

Let A be a self-adjoint operator acting in $\mathcal{H}$ bounded from below. We suppose that this describes the energy of the particle. The phonon-field is among Bose fields, and thus, we are interested in the massless Bose field case. Since we focus on the IR behavior around the 0 wave length in this paper, we suppose an idealization for the dispersion relation $\omega(k)$ for the massless Bose field, namely, we assume that the dispersion relation $\omega\colon \mathbb{R}^d \longrightarrow [0\,,\,\infty)$ is a continuous function such that $0 < \omega(k) < \infty$ for every $k \in \mathbb{R}^d \setminus \{0\}$ and $\inf_{|k|>\varepsilon}\omega(k) > 0$ for every $\varepsilon > 0$. The unperturbed Hamiltonian of our model is defined by $H_0 := A \otimes I + I \otimes \mathrm{d}\Gamma_\mathrm{b}(\omega)$ with domain $D(H_0) := D(A \otimes I) \cap D(I \otimes \mathrm{d}\Gamma_\mathrm{b}(\omega)) \subset \mathcal{F}$. The operator H_0 is self-adjoint and bounded from below. We suppose that our total Hamiltonian has the form, $H_{\mathrm{QFT}} := H_0 + H_{\mathrm{int}}$, and *we always assume* H_{QFT} *to be a self-adjoint operator acting in* $\mathcal{F}$ in this paper. We then suppose that H_{int} describes the interaction between the particle and the Bose field. Our purpose is to extrapolate this unknown interaction operator H_{int}.

Let $\ker(S)$ stand for the kernel of an operator S. In addition, when S is closed, let us denote by $\sigma(S)$ the spectrum of a closed operator S.

Definition 2.4. By *ground state energy* we mean $\inf \sigma(H_{\mathrm{QFT}})$, the lowest spectrum of H_{QFT}. We denote the ground state energy by $E_0(H_{\mathrm{QFT}})$. We say H_{QFT} *has a ground state* Ψ_{QFT} if $0 \neq \Psi_{\mathrm{QFT}} \in \ker(H_{\mathrm{QFT}} - E_0(H_{\mathrm{QFT}}))$. We say Ψ_{QFT} to be normalized if $\|\Psi_{\mathrm{QFT}}\|_{\mathcal{F}} = 1$.

For simplicity, we denote $H_{\mathrm{QFT}} - E_0(H_{\mathrm{QFT}})$ by $\widehat{H}_{\mathrm{QFT}}$. We always suppose that the ground state Ψ_{QFT} has been normalized whenever it exists.

2.3 From Dereziński–Gérard's idea

When we estimate the ground-state expectation $\big(\Psi_{\mathrm{QFT}}\,,\, N\Psi_{\mathrm{QFT}}\big)_{\mathcal{F}}$ of the total number of bosons, where N is the *boson number operator* defined by $I \otimes \mathrm{d}\Gamma_{\mathrm{b}}(1)$, it is convenient to use the following symbolical equation:

$$\|N^{1/2}\Psi_{\mathrm{QFT}}\|_{\mathcal{F}}^2 = \int_{\mathbb{R}^d} \|a(k)\Psi_{\mathrm{QFT}}\|_{\mathcal{F}}^2 \,\mathrm{d}k, \tag{1}$$

where $a(k)$ denotes the kernel of the annihilation operator $a(f)$. Its established meaning in operator theory is given by Proposition 2.2. On the other hand, when the integrand $\|a(k)\Psi_{\mathrm{QFT}}\|_{\mathcal{F}}^2$ in (1) has a singularity at $k = 0$, whether RHS of (1) converges is not certain. So, in such a case, we employ the equation corresponding to the following symbolical expression instead of (1):

$$\|N_{>\varepsilon}^{1/2}\Psi_{\mathrm{QFT}}\|_{\mathcal{F}}^2 = \int_{|k|>\varepsilon} \|a(k)\Psi_{\mathrm{QFT}}\|_{\mathcal{F}}^2 \,\mathrm{d}k \tag{2}$$

for every $\varepsilon > 0$, where N_ε is the number operator defined as the second quantization of $1^{>\varepsilon}$, the constant function $1(k) = 1$ cut off within the radius of ε from the origin. Thus, by taking $\varepsilon \to 0$ in (2), we can investigate the IR problem. This is the Dereziński–Gérard's idea and an interpretation of [9] (equation (2.9) of Lemma 2.6) which we adopt in our method, though they did not clearly write it in [9].

3 Domain of the Carleman operator and IR problem

When a ground state Ψ_{QFT} of the total Hamiltonian H_{QFT} exists, we consider the $\mathcal{F}$-valued function $K_{\mathrm{PT}}\colon \mathbb{R}^d \setminus \{0\} \to \mathcal{F}$ satisfying

$$a(f)\Psi_{\mathrm{QFT}} = -\int_{\mathbb{R}^d} \overline{f(k)} K_{\mathrm{PT}}(k)\,\mathrm{d}k \tag{3}$$

for every $f \in C_0^\infty(\mathbb{R}^d \setminus \{0\})$. To guarantee the existence of such the function, we here assume the following conditions.

1) There exists an operator $B_{\mathrm{PT}}(k)$ for every $k \in \mathbb{R}^d \setminus \{0\}$ such that the $\mathcal{F}$-valued function $B_{\mathrm{PT}}(\cdot)\Psi$ is continuous for every $\Psi \in D(H_0)$. That is, $D(B_{\mathrm{PT}}(k)) \supset D(H_0)$ for every $k \in \mathbb{R}^d \setminus \{0\}$ and $(\Phi, B_{\mathrm{PT}}(\cdot)\Psi)_{\mathcal{F}} \colon \mathbb{R}^d \setminus \{0\} \longrightarrow \mathbb{C}$ is a continuous function for every $\Phi \in \mathcal{F}$.

2) The operator $(\widehat{H}_{\mathrm{QFT}} + \omega(k))^{-1} B_{\mathrm{PT}}(k)$ is bounded and

$$K_{\mathrm{PT}}(k) = (\widehat{H}_{\mathrm{QFT}} + \omega(k))^{-1} B_{\mathrm{PT}}(k) \Psi_{\mathrm{QFT}} \tag{4}$$

for every $k \in \mathbb{R}^d \setminus \{0\}$, and then, for every $\varepsilon > 0$,

$$M_\varepsilon := \left\{ \int_{|k|>\varepsilon} \|(\widehat{H}_{\mathrm{QFT}} + \omega(k))^{-1} B_{\mathrm{PT}}(k)\|^2_{\mathcal{B}(\mathcal{F})} \, \mathrm{d}k \right\}^{1/2} < \infty,$$

where $\|\cdot\|_{\mathcal{B}(\mathcal{F})}$ denotes the operator norm of $\mathcal{B}(\mathcal{F})$, the C^*-algebra of bounded operators on $\mathcal{F}$.

According to theory of physics, the operator $B_{\mathrm{PT}}(k)$ should be $[a(k), H_{\mathrm{QFT}}] = [a(k), H_0] + [a(k), H_{\mathrm{int}}]$, but the interaction H_{int} is unknown now. Thus, we estimate $B_{\mathrm{PT}}(k)$ using the estimated coupling function $\lambda(k)$ as well as making theoretical argument in physics. How to estimate it will be in the next section.

For every $\varepsilon \geq 0$, we respectively give $\mathbb{R}^d_{<\varepsilon}$ and $\mathbb{R}^d_{>\varepsilon}$ by $\{k \in \mathbb{R}^d \mid |k| < \varepsilon\}$ and $\{k \in \mathbb{R}^d \mid |k| > \varepsilon\}$, where $\mathbb{R}^d_{<0} = \emptyset$. For every $f \in L^2(\mathbb{R}^d)$ we set $f^{<\varepsilon}$ as $1^{<\varepsilon}(k) f(k)$, and $f^{>\varepsilon}$ as $1^{>\varepsilon}(k) f(k)$ in $L^2(\mathbb{R}^d)$, where $1^{<\varepsilon}$ and $1^{>\varepsilon}$ are characteristic functions:

$$1^{<\varepsilon}(k) := \begin{cases} 1 & \text{if } |k| < \varepsilon, \\ 0 & \text{otherwise,} \end{cases} \quad \text{and} \quad 1^{>\varepsilon}(k) := \begin{cases} 1 & \text{if } |k| > \varepsilon, \\ 0 & \text{otherwise.} \end{cases}$$

Since we can regard $f^{<\varepsilon}$ (resp. $f^{>\varepsilon}$) as a function in $L^2(\mathbb{R}^d_{<\varepsilon})$ (resp. $L^2(\mathbb{R}^d_{>\varepsilon})$), we often handle it as $f^{<\varepsilon} \in L^2(\mathbb{R}^d_{<\varepsilon})$ (resp. $f^{>\varepsilon} \in L^2(\mathbb{R}^d_{>\varepsilon})$) in this paper. According to this decomposition, we introduce some abbreviations:

$$\mathrm{d}\Gamma_{<\varepsilon}(h^{<\varepsilon}) := \mathrm{d}\Gamma_{\mathbb{R}^d_{<\varepsilon}}(h^{<\varepsilon}), \quad \mathrm{d}\Gamma_{>\varepsilon}(h^{>\varepsilon}) := \mathrm{d}\Gamma_{\mathbb{R}^d_{>\varepsilon}}(h^{>\varepsilon}),$$

$$a^\sharp_{<\varepsilon}(f^{<\varepsilon}) := a^\sharp_{\mathbb{R}^d_{<\varepsilon}}(f^{<\varepsilon}), \quad a^\sharp_{>\varepsilon}(f^{>\varepsilon}) := a^\sharp_{\mathbb{R}^d_{>\varepsilon}}(f^{>\varepsilon}),$$

where $a^\sharp_X$ denotes a_X or $a^\dagger_X$.

Since $\mathbb{R}^d = \mathbb{R}^d_{<\varepsilon} \oplus \{k \in \mathbb{R}^d \mid |k| = \varepsilon\} \oplus \mathbb{R}^d_{>\varepsilon}$ and the Lebesgue measure of the set $\{k \in \mathbb{R}^d \mid |k| = \varepsilon\}$ is equal to 0, we have $L^2(\mathbb{R}^d) = L^2(\mathbb{R}^d_{<\varepsilon}) \oplus L^2(\mathbb{R}^d_{>\varepsilon})$. So, by Fact IV, there exists a unitary operator U_ε for every $\varepsilon > 0$ such that

$$U_\varepsilon \mathcal{F} = \mathcal{H} \otimes \mathcal{F}_{\mathbb{R}^d_{<\varepsilon}} \otimes \mathcal{F}_{\mathbb{R}^d_{>\varepsilon}} \equiv \mathcal{H} \otimes \mathcal{F}_{\mathrm{b}}(L^2(\mathbb{R}^d_{<\varepsilon})) \otimes \mathcal{F}_{\mathrm{b}}(L^2(\mathbb{R}^d_{>\varepsilon})).$$

We denote $U_\varepsilon \mathcal{F}$ by $\mathcal{F}_\varepsilon$. Write $U_\varepsilon \Psi \in \mathcal{F}_\varepsilon$ as Ψ_ε for every $\Psi \in \mathcal{F}$. Then, Fact IV.-b) leads to the relation:

$$U_\varepsilon(I \otimes d\Gamma_{\mathrm{b}}(h))U_\varepsilon^* = \overline{I \otimes d\Gamma_{<\varepsilon}(h^{<\varepsilon}) \otimes I + I \otimes I \otimes d\Gamma_{>\varepsilon}(h^{>\varepsilon})}$$

for every real-valued Lebesgue-measurable function $h\colon \mathbb{R}^d \to \mathbb{R}$.

Symbolically we give the ground-state expectation $\langle O \rangle_{\mathrm{gs}}$ for an operator O acting in $\mathcal{F}$ by $(\Psi_{\mathrm{QFT}}, O\Psi_{\mathrm{QFT}})_{\mathcal{F}}$. Here we recall that Ψ_{QFT} is normalized. Then, we can consider $\langle O \rangle_{\mathrm{gs}}$ to be finite if $\Psi_{\mathrm{QFT}} \in D(O)$, on the other hand, to be infinite if $\Psi_{\mathrm{QFT}} \notin D(O)$. Here we also write $\Psi_{\mathrm{QFT}} \notin D(O)$ when Ψ_{QFT} does not exist in $\mathcal{F}$. That is, $\langle O \rangle_{\mathrm{gs}} < \infty$ if $\Psi_{\mathrm{QFT}} \in D(O)$; $\langle O \rangle_{\mathrm{gs}} = \infty$ if $\Psi_{\mathrm{QFT}} \notin D(O)$ or Ψ_{QFT} does not exist in $\mathcal{F}$.

Definition 3.1. We say the *soft boson (SB) divergence takes place* if $\Psi_{\mathrm{QFT}} \notin D(N^{1/2})$, and the *infrared (IR) catastrophe occurs* if Ψ_{QFT} does not exist in $\mathcal{F}$.

Remark 3.2. Since $D(N) \subset D(N^{1/2})$, the naive meaning of the soft boson divergence in Definition 3.1 is symbolically: $\langle N \rangle_{\mathrm{gs}} = \|N^{1/2}\Psi_{\mathrm{QFT}}\|^2_{\mathcal{F}} = \infty$.

For every $\varepsilon > 0$, we define $N_{>\varepsilon}$ acting in $\mathcal{F}$ by $U_\varepsilon^*(I \otimes I \otimes d\Gamma_{>\varepsilon}(1^{>\varepsilon}))U_\varepsilon$. Then, we can easily obtain the following lemma:

Lemma 3.3. $D(H_0) \subset \bigcap_{\varepsilon>0} D(N_{>\varepsilon}^{1/2})$.

To find a relation between N and $N_{>\varepsilon}$, we introduce the domain,

$$D_{\mathrm{CNB}} := \Big\{\Psi \in \bigcap_{\varepsilon>0} D(N_{>\varepsilon}^{1/2}) \;\Big|\; \sup_{\varepsilon>0} \|N_{>\varepsilon}^{1/2}\Psi\|^2_{\mathcal{F}} < \infty\Big\}.$$

The following lemma is a mathematical establishment of (2).

Lemma 3.4. *Let* $\{f_\ell^{>\varepsilon}\}_{\ell=1}^\infty$ *be an arbitrary complete orthonormal system of* $L^2(\mathbb{R}^d_{>\varepsilon})$ *for every* $\varepsilon > 0$. *Then, the equation,* $\|N_{>\varepsilon}^{1/2}\Psi\|^2_{\mathcal{F}} = \sum_{\ell=0}^\infty \|a(f_\ell^{>\varepsilon})\Psi\|^2_{\mathcal{F}}$, *holds for every* $\Psi \in D(N_{>\varepsilon}^{1/2})$.

The following lemma gives a relation between N and $N_{>\varepsilon}$. It tells us that for all vectors $\Psi \in D(H_0)$ we can check whether $(\Psi\,,\,N\Psi)_{\mathcal{F}}$ converges by making good use of Lemma 3.3 and estimating $\sup_{\varepsilon>0}\|N_{>\varepsilon}^{1/2}\Psi\|_{\mathcal{F}}^2$. Thus, the following lemma plays an important role to prove Theorems 3.7 and 3.8 below.

Lemma 3.5. *We have*

$$D_{\mathrm{CNB}} = D(N^{1/2})$$

and

$$\sup_{\varepsilon>0}\|N_{>\varepsilon}^{1/2}\Psi\|_{\mathcal{F}}^2 = \|N^{1/2}\Psi\|_{\mathcal{F}}^2$$

for $\Psi \in D_{\mathrm{CNB}}$.

We here introduce two notions of the Carleman operator (see §6.2 of [54]): Let Ω be a Lebesgue-measurable space of $\mathbb{R}^d$. A linear operator T from a Hilbert space $\mathcal{V}$ into $L^2(\Omega)$ is called the *Carleman operator* if there exists a function $k\colon\Omega\to\mathcal{V}$ such that $(Tv)(x) = (k(x), v)_{\mathcal{V}}$ (a.e. $x\in\Omega$) for all $v \in D(T)$. If

$$D(T) = \{v \in \mathcal{V} \mid (k(\cdot), v)_{\mathcal{V}} \in L^2(\Omega)\},$$

then T is called the *maximal Carleman operator.*

When a ground state Ψ_{QFT} of H_{QFT} exists, we assumed the existence of the $\mathcal{F}$-valued function $K_{\mathrm{PT}}\colon\mathbb{R}^d\setminus\{0\}\to\mathcal{F}$ given by the operator $B_{\mathrm{PT}}(k)$ as in (4). We make the following definition:

Definition 3.6. For the ground state Ψ_{QFT}, we define the *maximal Carleman operator* $T_{\mathrm{PT}}\colon\mathcal{F}\to L^2(\mathbb{R}^d)$ induced by K_{PT} by

$$D(T_{\mathrm{PT}}) := \{\Phi\in\mathcal{F} \mid (K_{\mathrm{PT}}(\cdot),\,\Phi)_{\mathcal{F}} \in L^2(\mathbb{R}^d)\},$$

$$(T_{\mathrm{PT}}\Phi)(k) := (K_{\mathrm{PT}}(k),\,\Phi)_{\mathcal{F}},\qquad \Phi\in D(T_{\mathrm{PT}}),$$

for every $k\in\mathbb{R}^d\setminus\{0\}$. Then, we call K_{PT} the *inducing function* of T_{PT}. We say that the *maximal Carleman operator* T_{PT} *has the IR singularity* at $k=0$ if $\lim_{k\to0}|\,(T_{\mathrm{PT}}\Phi)\,(k)| = \infty$.

We note that T_{PT} is closed by Theorem 6.13 in [54].

We can slightly modify Theorem 2.9 in [24] to meet our interests, and we can restate it in terms of the maximal Carleman operator.

Theorem 3.7. *Assume $D(H_{\mathrm{QFT}}) = D(H_0)$ and there exists a ground state Ψ_{QFT} of the Hamiltonian H_{QFT}. Then, the following conditions are equivalent*:

(i) $\Psi_{\mathrm{QFT}} \in D(N^{1/2})$;

(ii) $\|K_{\mathrm{PT}}(\cdot)\|_{\mathcal{F}} \in L^2(\mathbb{R}^d)$;

(iii) *T_{PT} is a Hilbert–Schmidt operator.*

If one of them holds, then $\|T_{\mathrm{PT}}\| \leq M_0 := \lim_{\varepsilon\to 0} M_\varepsilon < \infty$.

The above theorem can be proved by obeying the Dereziński–Gérard idea mentioned in §2.3 and making good use of Lemma 3.5, and tells us that $D(T_{\mathrm{PT}}) = \mathcal{F}$ if Ψ_{QFT} exists in $D(N^{1/2})$. Thus, it is trivial that $D(N^{1/2}) \subset D(T_{\mathrm{PT}})$ in this case. But, more generally, we have this inclusion relation in the following theorem, even though Ψ_{QFT} exists outside $D(N^{1/2})$:

Theorem 3.8. *Suppose that $D(H_{\mathrm{QFT}}) = D(H_0)$. If a ground state Ψ_{QFT} of the Hamiltonian H_{QFT} exists, then $D(T_{\mathrm{PT}}) \supset D(N^{1/2})$.*

This theorem is also proved by making good use of Lemma 3.5, and gives a sufficient condition of the SB divergence:

Corollary 3.9. *Suppose that $D(H_{\mathrm{QFT}}) = D(H_0)$. If $\Psi_{\mathrm{QFT}} \notin D(T_{\mathrm{PT}})$, then the SB divergence takes place.*

Thus, the next problem is when Ψ_{QFT} is not in $D(T_{\mathrm{PT}})$ if Ψ_{QFT} exists. Theorem 4.6 below deals with this question in the case where T_{PT} has an IR singularity. To prove Theorem 4.6, we need the following theorem:

Theorem 3.10. *Suppose $D(H_{\mathrm{QFT}}) = D(H_0)$. Assume a function λ on $\mathbb{R}^d$ represents IR singularity of T_{PT} as the following* (1)–(3):

(1) *there is an $\varepsilon_0 > 0$ such that $\lambda(k) \neq 0$ for every $k \in \mathbb{R}^d$ with $0 < |k| < \varepsilon_0$;*

(2) *$\lambda/\omega \notin L^2(K)$ for every neighborhood K of $k = 0$;*

(3) *there is an operator $B_0(0)$ acting in $\mathcal{F}$ such that $\lambda(k)^{-1} B_{\mathrm{PT}}(k)$ converges to $B_0(0)$ on $D(H_0)$ as $k \to 0$.*

If there exists a ground state Ψ_{QFT} such that

$$\frac{1}{\omega(\cdot)}(\Phi,\, (\widehat{H}_{\mathrm{QFT}} + \omega(\cdot))^{-1} \widehat{H}_{\mathrm{QFT}} B_{\mathrm{PT}}(\cdot)\Psi_{\mathrm{QFT}})_{\mathcal{F}} \in L^2(\mathbb{R}^d)$$

for a vector $\Phi \in D(T_{\mathrm{PT}})$, then $(\Phi,\, B_0(0)\Psi_{\mathrm{QFT}})_{\mathcal{F}} = 0$.

4 IR singularity and Carleman operator

In this section we suppose that we can estimate the dispersion relation $\omega(k)$ of phonon and the coupling function $\lambda(k)$ from the experimental data. We here show how to define the operator $B_{\mathrm{PT}}(k)$ in the inducing function $K_{\mathrm{PT}}(k)$ of the Carleman operator T_{PT}. We theoretically estimate and approximate the ideal form of the interaction between an electron and a phonon-field (see [6], [29], and [28]) so that the approximated interaction $H_{\mathrm{int}}^{\mathrm{theo}}$ does not have information about the coupling strength. We then set an operator $B_0(k)$ as $B_0(k) = [a(k), H_{\mathrm{int}}^{\mathrm{theo}}]$. Here we do not mind the coupling strength because we make an estimated coupling function $\lambda(k)$ play its role as in the condition (S1) below. We know the IR difficulty may take place in a case where the dispersion relation $\omega(k)$ satisfies $\omega(k) \to 0$ as $k \to 0$. So, to avoid IR divergence, the estimated coupling function $\lambda(k)$ has to lessen the effect of the decay. Meanwhile, equations (3) and (4) together with Proposition 2.2 say that the operator $B_{\mathrm{PT}}(k)$ plays the role of lessening the decay of the dispersion relation $\omega(k)$. Thus, we define the operator $B_{\mathrm{PT}}(k)$ by using the estimated coupling function $\lambda(k)$ and the theoretically approximated operator $B_0(k)$. We assume that the coupling function $\lambda(k)$ satisfies the condition (1) of Theorem 3.10 and that the approximated operator $B_0(k)$ satisfies:

(S1) $B_{\mathrm{PT}}(k) = \lambda(k) B_0(k)$ on $D(H_0)$ for every $k \in \mathbb{R}^d \setminus \{0\}$;

(S2) $B_0(k)\Psi \to B_0(0)\Psi$ in $\mathcal{F}$ as $k \to 0$ for every $\Psi \in D(H_0)$.

We reinterpret the IR singularity condition (see [3] and [4]) as the condition in the neighborhood of $k = 0$:

Definition 4.1. We say that *ω and λ satisfy the IR singularity condition* if there are constants $\gamma_1, \gamma_2, \varepsilon_2 > 0$ with $\gamma_1 < \gamma_2$ such that $\lambda/(\omega^\gamma) \in L^2(\mathbb{R}^d)$ for every γ with $\gamma < \gamma_1$, and $\lambda/(\omega^\gamma) \notin L^2(\mathbb{R}^d_{<\varepsilon})$ for every γ with $\gamma > \gamma_2$ and every ε with $\varepsilon_2 \geq \varepsilon > 0$. We say that *$\gamma$ is in the IR-safe region* (resp. *the IR-divergent region*) if $\gamma < \gamma_1$ (resp. $\gamma > \gamma_2$). In particular, we call γ_{c} the order of the IR singularity condition when $\gamma_1 = \gamma_2 = \gamma_{\mathrm{c}}$ and $\lambda/(\omega^{\gamma_{\mathrm{c}}}) \notin L^2(\mathbb{R}^d_{<\varepsilon})$ for every ε with $\varepsilon_2 \geq \varepsilon > 0$. In this case, we also say $\gamma = \gamma_{\mathrm{c}}$ is in the IR-divergent region.

We say *a symmetric operator S strongly commutes with H_{QFT}* if

$$e^{itH_{\mathrm{QFT}}} S \subset S e^{itH_{\mathrm{QFT}}} \quad \text{for all } t \in \mathbb{R}.$$

Then, we can derive the following theorem from Theorem 3.8. This is a generalization of Dereziński and Gérard's Lemma 2.6 in [9] and ours Theorem 3.4 in [4].

Theorem 4.2. *Suppose $D(H_{\mathrm{QFT}}) = D(H_0)$. Assume ω and λ satisfy the IR singularity condition with the order γ_{c} less than or equal to 1 (i.e., $\gamma_{\mathrm{c}} \le 1$). Assume that a ground state Ψ_{QFT} satisfies:*

(1) *$B_0(0)$ is symmetric and strongly commutes with H_{QFT};*

(2) *there is a γ in the IR-safe region such that*

$$\sup_{k \in \operatorname{supp} \widetilde{B}_0} \omega(k)^{\gamma-1} \|(B_0(k) - B_0(0))\Psi_{\mathrm{QFT}}\|_{\mathcal{F}} < \infty,$$

where $\widetilde{B}_0(k) := B_0(k) - B_0(0)$.

Then, the ground state satisfies $B_0(0)\Psi_{\mathrm{QFT}} \neq 0$.

Proof. We suppose that the ground state Ψ_{QFT} satisfies (1) and (2). For all $\Phi \in D(N^{1/2})$ and every $k \in \mathbb{R}^d \setminus \{0\}$ we have

$$\begin{aligned}(K_{\mathrm{PT}}(k), \Phi)_{\mathcal{F}} &= \lambda(k)((\hat{H}_{\mathrm{QFT}} + \omega(k))^{-1} B_0(0)\Psi_{\mathrm{QFT}}, \Phi)_{\mathcal{F}} \\ &\quad + \lambda(k)((\widehat{H}_{\mathrm{QFT}} + \omega(k))^{-1}(B_0(k) - B_0(0))\Psi_{\mathrm{QFT}}, \Phi)_{\mathcal{F}} \\ &= \frac{\lambda(k)}{\omega(k)}(B_0(0)\Psi_{\mathrm{QFT}}, \Phi)_{\mathcal{F}} \\ &\quad + \lambda(k)((\widehat{H}_{\mathrm{QFT}} + \omega(k))^{-1}(B_0(k) - B_0(0))\Psi_{\mathrm{QFT}}, \Phi)_{\mathcal{F}}\end{aligned}$$

by (1). This equation implies

$$\begin{aligned}0 &\le |(B_0(0)\Psi_{\mathrm{QFT}}, \Phi)_{\mathcal{F}}|^2 \int_{|k|>\varepsilon} \frac{|\lambda(k)|^2}{\omega(k)^2}\,\mathrm{d}k \\ &\le 2\int_{\mathbb{R}^d} |(K_{\mathrm{PT}}(k), \Phi)_{\mathcal{F}}|^2\,\mathrm{d}k \\ &\quad + 2\|\Phi\|_{\mathcal{F}}^2 \Big(\sup_{k \in \operatorname{supp} \widetilde{B}_0} \omega(k)^{\gamma-1}\|(B_0(k) - B_0(0))\Psi_{\mathrm{QFT}}\|_{\mathcal{F}}\Big)^2 \int_{\mathbb{R}^d} \frac{|\lambda(k)|^2}{\omega(k)^{2\gamma}}\,\mathrm{d}k.\end{aligned}$$

Here we note that the first integral of RHS is finite by Theorem 3.8, and the second one is also finite for the γ in (2). In addition, they are independent of every $\varepsilon \in (0, \varepsilon_2)$. Taking $\varepsilon \to 0$, Lebesgue's monotone convergence theorem tells us that $(B_0(0)\Psi_{\mathrm{QFT}}, \Phi)$ is bound to be 0 (i.e., $(B_0(0)\Psi_{\mathrm{QFT}}, \Phi) = 0$) for all $\Phi \in D(N^{1/2})$ since $\lambda/\omega \notin L^2(\mathbb{R}^d)$. Since $D(N^{1/2})$ is dense in $\mathcal{F}$, we reach $B_0(0)\Psi_{\mathrm{QFT}} = 0$ finally. □

We can obtain Dereziński and Gérard's result (Lemma 2.6 in [9]) as a corollary of Theorem 4.2:

Corollary 4.3. *Let* $D(H_{\mathrm{QFT}}) = D(H_0)$. *Suppose that there are functions* $g(k)$ *and* $J_{\mathrm{err}}(k)$ *with* $\operatorname{supp} g \subset \operatorname{supp} J_{\mathrm{err}}$ *such that* $B_{\mathrm{PT}}(k)$ *can be decomposed into*

$$B_{\mathrm{PT}}(k) = g(k) I \otimes I + J_{\mathrm{err}}(k)$$

on $D(H_0)$ *for every* $k \in \mathbb{R}^d \setminus \{0\}$. *Assume the following* (1)–(3):

(1) $g/\omega \notin L^2(\mathbb{R}^d)$;

(2) $g/(\omega^{\gamma_0}) \in L^2(\mathbb{R}^d)$ *for a* γ_0 *with* $0 < \gamma_0 < 1$;

(3) $g(k)^{-1} J_{\mathrm{err}}(k)\Psi \to 0$ *as* $k \to 0$ *for every* $\Psi \in D(H_0)$.

Then, there is no ground state Ψ_{QFT} *satisfying*

$$\sup_{k \in \operatorname{supp} J_{\mathrm{err}}} \omega(k)^{\gamma_0 - 1} g(k)^{-1} \| J_{\mathrm{err}}(k)\Psi_{\mathrm{QFT}} \|_{\mathcal{F}} < \infty.$$

Proof. Set $\lambda(k) := g(k)$ and $B_0(0) := I \otimes I$. We $B_0(k)$ by

$$B_0(k) := g(k)^{-1} J_{\mathrm{err}}(k) + I \otimes I$$

for every $k \in \operatorname{supp} J_{\mathrm{err}}$ with $k \neq 0$; by $B_0(k) := I \otimes I$ for other $k \neq 0$. The only thing we have to do is applying Theorem 4.2. □

As a corollary of Theorem 4.2 we also obtain Theorem 3.4 in [4] of which statement can be applied to several models.

Corollary 4.4. *Suppose* $D(H_{\mathrm{QFT}}) = D(H_0)$ *and there is an operator* C_{PT} *with* $D(C_{\mathrm{PT}}) \supset D(H_0)$ *such that* $B_0(k) = C_{\mathrm{PT}}$ *on* $D(H_0)$ *for all* $k \in \mathbb{R}^d$. *Assume the following* (1)–(3):

(1) $\lambda/\omega \notin L^2(\mathbb{R}^d)$;

(2) $\lambda/(\omega^{\gamma_0}) \in L^2(\mathbb{R}^d)$ *for a* γ_0 *with* $0 < \gamma_0 < 1$;

(3) C_{PT} *is symmetric and strongly commutes with* H_{QFT}.

Then, there is no ground state Ψ_{QFT} *satisfying* $C_{\mathrm{PT}}\Psi_{\mathrm{QFT}} \neq 0$.

We make another statement for our concrete case in (7) below.

Theorem 4.5. *Suppose* $D(H_{\mathrm{QFT}}) = D(H_0)$ *and* $B_0(0)$ *is symmetric and strongly commutes with* H_{QFT}. *Assume there is an* $\varepsilon_0 > 0$ *and operators* $B_j(k)$, $j = 1, \dots, d$, *acting in* $\mathcal{F}$ *for every* $k \in \mathbb{R}^d \setminus \{0\}$ *such that* $B_0(k)\Psi_{\mathrm{QFT}}$ *is decomposed into*

$$B_0(k)\Psi_{\mathrm{QFT}} = B_0(0)\Psi_{\mathrm{QFT}} + \sum_{j=1}^{d} k_j B_j(k)\Psi_{\mathrm{QFT}}$$

for $|k| < \varepsilon_0$. *If* ω *and* λ *satisfy the IR singularity condition with the order* γ_{c}, *and moreover,*

$$\int_{|k|<\varepsilon_0} \frac{|k_j|\,|\lambda(k)|^2}{\omega(k)^{1+\gamma}} \mathrm{d}k < \infty$$

for a $\gamma > 0$ *with* $\gamma < \gamma_{\mathrm{c}} < (1+\gamma)/2$ *and* $j = 1, \dots, d$, *then there is no ground state* Ψ_{QFT} *satisfying* $B_0(0)\Psi_{\mathrm{QFT}} \neq 0$ *and* $\sup_{|k|<\varepsilon_0} \|B_j(k)\Psi_{\mathrm{QFT}}\|_{\mathcal{F}} < \infty$ *for all* $j = 1, \dots, d$.

Proof. We use the reduction to absurdity. So, we suppose that there is such a ground state Ψ_{QFT}. Let us fix $\Phi \in D(N^{1/2})$ arbitrarily and define a function $F_\Phi(k)$ by

$$F_\Phi(k) := (K_{\mathrm{PT}}(k)\,,\ \Phi)_{\mathcal{F}}.$$

We define another function $F_\gamma(k)$ by

$$F_\gamma(k) := \lambda(k)\omega(k)^{-\gamma}.$$

Then, we have $F_\Phi \in L^2(\mathbb{R}^d)$ by Theorem 3.8 and $F_\gamma \in L^2(\mathbb{R}^d)$ by our assumption. For every ε with $\varepsilon < \min\{\varepsilon_0, \varepsilon_2\} =: \varepsilon_0 \wedge \varepsilon_2$, where ε_2 is in Definition 4.1, we have

$$\begin{aligned} &\int_{\varepsilon<|k|<\varepsilon_0\wedge\varepsilon_2} F_\Phi(k) F_\gamma(k)\,\mathrm{d}k \\ &= (B_0(0)\Psi_{\mathrm{QFT}},\ \Phi)_{\mathcal{F}} \int_{\varepsilon<|k|<\varepsilon_0\wedge\varepsilon_2} \frac{|\lambda(k)|^2}{\omega(k)^{1+\gamma}} \mathrm{d}k \\ &\quad + \sum_{j=1}^{d} \int_{\varepsilon<|k|<\varepsilon_0\wedge\varepsilon_2} \frac{k_j|\lambda(k)|^2}{\omega(k)^{\gamma}} ((\widehat{H}_{\mathrm{QFT}} + \omega(k))^{-1} B_j(k)\Psi_{\mathrm{QFT}},\ \Phi)_{\mathcal{F}}\,\mathrm{d}k. \end{aligned} \tag{5}$$

In the first term of RHS of the above, we used the assumption that $B_0(0)$ commutes with H_{QFT}. We can estimate the last integrals as

$$\begin{aligned}
&\left| \int_{\varepsilon<|k|<\varepsilon_0\wedge\varepsilon_2} \frac{k_j|\lambda(k)|^2}{\omega(k)^\gamma}((\widehat{H}_{\mathrm{QFT}}+\omega(k))^{-1}B_j(k)\Psi_{\mathrm{QFT}},\, \Phi)_{\mathcal{F}}\,\mathrm{d}k\right| \\
&\le \|\Phi\|_{\mathcal{F}} \sup_{|k|<\varepsilon_0} \|B_j(k)\Psi_{\mathrm{QFT}}\|_{\mathcal{F}} \int_{|k|<\varepsilon_0} \frac{|k_j|\,|\lambda(k)|^2}{\omega(k)^{1+\gamma}}\,\mathrm{d}k \\
&< \infty.
\end{aligned} \tag{6}$$

Combining (5) and the inequality (6) gives us the inequality

$$\begin{aligned}
0 &\le |(B_0(0)\Psi_{\mathrm{QFT}},\, \Phi)_{\mathcal{F}}| \int_{\varepsilon<|k|<\min\{\varepsilon_0,\varepsilon_2\}} \frac{|\lambda(k)|^2}{\omega(k)^{1+\gamma}}\,\mathrm{d}k \\
&\le \|F_\Phi\|_{L^2(\mathbb{R}^d)}\|F_\gamma\|_{L^2(\mathbb{R}^d)} \\
&\quad + \|\Phi\|_{\mathcal{F}} \sum_{j=1}^d \sup_{|k|<\varepsilon_0} \|B_j(k)\Psi_{\mathrm{QFT}}\|_{\mathcal{F}} \int_{|k|<\varepsilon_0} \frac{|k_j|\,|\lambda(k)|^2}{\omega(k)^{1+\gamma}}\,\mathrm{d}k \\
&< \infty.
\end{aligned}$$

Taking $\varepsilon \to 0$, Lebesgue's monotone convergence theorem tells us that the inner product $(B_0(0)\Psi_{\mathrm{QFT}},\, \Phi)_{\mathcal{F}}$ is bound to be 0 (i.e., $(B_0(0)\Psi_{\mathrm{QFT}},\, \Phi)_{\mathcal{F}} = 0$) for all $\Phi \in D(N^{1/2})$ since $\lambda/(\omega^{(1+\gamma)/2}) \notin L^2(\mathbb{R}^d)$. Since $D(N^{1/2})$ is dense in $\mathcal{F}$, we reach $B_0(0)\Psi_{\mathrm{QFT}} = 0$ finally. This is a contradiction. □

The following theorem follows from Theorem 3.10:

Theorem 4.6. *Assume $D(H_{\mathrm{QFT}}) = D(H_0)$ and $\lambda/\omega \notin L^2(K)$ for every neighborhood K of $k = 0$. Then, there is no ground state Ψ_{QFT} in $D(T_{\mathrm{PT}})$ satisfying $\langle B_0(0)\rangle_{\mathrm{gs}} \neq 0$. Thus, in particular, if $B_0(0)\Psi \neq 0$ for every $\Psi \in D(H_0)$ with $\Psi \neq 0$, then no ground state exists in $D(T_{\mathrm{PT}})$, and thus, the SB divergence takes place.*

Proof. Let us suppose there is a ground state Ψ_{QFT} in $D(T_{\mathrm{PT}})$ now. We easily have

$$\frac{1}{\omega(\cdot)}(\Psi_{\mathrm{QFT}},\, (\widehat{H}_{\mathrm{QFT}}+\omega(\cdot))^{-1}\widehat{H}_{\mathrm{QFT}}B_{\mathrm{PT}}(\cdot)\Psi_{\mathrm{QFT}})_{\mathcal{F}} = 0.$$

Thus, it follows immediately from Theorem 3.10 that $\langle B_0(0)\rangle_{\mathrm{gs}} = 0$, which means our theorem holds. □

5 Extrapolation of electron–phonon interaction

In this section, we consider the concrete approximated interaction $H_{\rm int}^{\rm theo}$, and then, we investigate the conditions between the estimated dispersion relation $\omega(k)$ and the estimated coupling function $\lambda(k)$ for the IR problem.

As the operator A in the free Hamiltonian H_0, we employ a Hamiltonian $H_{\rm at}$ given by the Schrödinger operator, $H_{\rm at} \equiv 1/2\, p^2 + V$, acting in $\mathcal{H} = L^2(\mathbb{R}^d)$, where $p := -i\nabla_x$ is the momentum of the electron, and V a potential. We use the natural units here. As in [22] we consider potentials V in the class either (N1-1) or (N1-2) below. Here we say that V *is in class* (N1-1) – resp. (N1-2) – if the following conditions, (N1-1-1) and (N1-1-2) (see [1]) – resp. (N1-2-1) and (N1-2-2) (see [50]) – hold. These conditions are set so that if V is in class (N1-1) or (N1-2), then $H_{\rm at}$ becomes a self-adjoint operator bounded from below with $D(H_{\rm at}) \subset D(p^2)$, and moreover, $H_{\rm at}$ has a ground state $\psi_{\rm at}$. When we say that *we assume* (N1), we mean that either (N1-1) or (N1-2) is assumed.

(N1-1-1) $H_{\rm at}$ is sclf-adjoint on $D(H_{\rm at}) = D(p^2) \cap D(V)$ and bounded from below.

(N1-1-2) There exist positive constants c_1 and c_2 such that $|x|^2 \le c_1 V(x) + c_2$ for almost every (a.e.) $x \in \mathbb{R}^d$, and $\int_{|x|\le R} |V(x)|^2 \, dx < \infty$ for all $R > 0$.

(N1-2-1) $V \in L^2(\mathbb{R}^d) + L^\infty(\mathbb{R}^d)$ and $\lim_{|x|\to\infty} |V(x)| = 0$.

Following [41] (Theorem X15) and [42] (§XIII.4, Example 6) the condition (N1-2-1) implies that $H_{\rm at}$ is self-adjoint on $D(p^2)$; V is infinitesimally p^2-bounded; and the essential spectrum $\sigma_{\rm ess}(H_{\rm at})$ of $H_{\rm at}$ is equal to $[0\,,\,\infty)$. So, we assume the following in addition:

(N1-2-2) $H_{\rm at}$ has a ground state $\psi_{\rm at}$ satisfying $\psi_{\rm at}(x) > 0$ for a.e. $x \in \mathbb{R}^d$ and $E_{\rm at} := \inf \sigma(H_{\rm at}) < 0$.

In order to define the interaction Hamiltonian $H_{\rm int}$ of the models, we use the fact that $\mathcal{F}$ is unitarily equivalent to the constant fiber direct integral $L^2(\mathbb{R}^d; \mathcal{F}_{\rm b})$, i.e.,

$$\mathcal{F} \equiv L^2(\mathbb{R}^d) \otimes \mathcal{F}_{\rm b} \cong L^2(\mathbb{R}^d; \mathcal{F}_{\rm b}) \equiv \int_{\mathbb{R}^d}^{\oplus} \mathcal{F}_{\rm b}\, dx$$

(see [42] and [47]). Throughout this section, we identify $\mathcal{F}$ to the constant fiber direct integral, i.e., $\mathcal{F} = \int_{\mathbb{R}^d}^{\oplus} \mathcal{F}_{\rm b}\, dx$.

To lessen the decay of the estimated dispersion relation, $\omega(k) \to 0$ as $k \to 0$, in the IR problem, the estimated coupling function $\lambda(k)$ also has to decay, $\lambda(k) \to 0$ as $k \to 0$. Namely, the coupling strength $|\lambda(k)|$ is very small in the neighborhood of $k = 0$. This sufficiently small coupling strength usually allows us to consider the first-order approximation, i.e., the linear coupling between the particle and the phonon-field. We assume that the true coupling function $\rho(k)$ can be well approximated by a function $\rho_{\mathrm{app}}(k)$ and a coupling constant q for small wave number $|k|$, and then, the estimated coupling function $\lambda(k)$ can be obtained as $\lambda(k) = q1^{<\Lambda}\rho_{\mathrm{app}}(k)$.

For a measurable function $\rho_{\mathrm{app}}(k)$ satisfying $1^{<\Lambda}\rho_{\mathrm{app}} \in L^2(\mathbb{R}^d)$, we give our extrapolated interaction H_{int} by the so-called *Fröhlich interaction* [14]. Thus, symbolically using the kernels of the annihilation and creation operators, the extrapolated interaction is

$$H_{\mathrm{int}} = \int_{\mathbb{R}^d} \lambda(k)(e^{-ikx}a(k) + e^{-ikx}a^\dagger(k))\,\mathrm{d}k,$$

where

$$H_{\mathrm{int}}^{\mathrm{theo}} = \int_{\mathbb{R}^d} (e^{ikx}a(k) + e^{-ikx}a^\dagger(k))\,\mathrm{d}k.$$

We mathematically assume the following:

(N2) $1^{<\Lambda}\rho_{\mathrm{app}},\ 1^{<\Lambda}\,\rho_{\mathrm{app}}/\sqrt{\omega} \in L^2(\mathbb{R}^d)$.

The total Hamiltonian H_{QFT} in our mathematical modeling is

$$H_{\mathrm{QFT}} = H_{\mathrm{at}} \otimes I + I \otimes H_{\mathrm{b}} + q\int_{\mathbb{R}^d}^{\oplus} \{a(1^{<\Lambda}\rho_{\mathrm{app}}e^{-ikx}) + a^\dagger(1^{<\Lambda}\rho_{\mathrm{app}}e^{-ikx})\}\,\mathrm{d}x$$

acting in $\mathcal{F}$, see [31] and [32]. Then, we call this H_{QFT} the *Lee–Low–Pines* (*LLP*) *Hamiltonian* in this paper, though it is called the *Pauli–Fierz Hamiltonian* for non-relativistic quantum electro dynamics (NQED) in [9], [16], and [17]. That is, we are interested in the model describing polaron [36] in the solid state physics.

Remark 5.1. Roughly speaking, phonon in a material is the quantization of the lattice vibration of the material. The vibration makes a wave, and then, the wave has the relation between its frequency ω and its wave number vector k. This is the so-called *dispersion relation*, $\omega = \omega(k)$. After the quantization of this vibration, through the Einstein–planck formula, the dispersion relation $\omega(k)$ and the phonon number $n(k)$ gives the energy $E = \hbar n(k)\omega(k)$ of the phonon with the momentum $\hbar k$ given by the de Broglie formula. Thus, each dispersion relation $\omega(k)$ is determined for the individual property of the material. The true coupling function $\rho(k)$ also depends on the material property, and we estimate it at our estimated coupling function $\lambda(k)$. Consider an electron put in a material now. For instance, suppose that the material is a crystal. The electron in the material is negatively charged and thus is attracted by a plus-charged source which is caused by the positively charged ion cores caused by, for instance, the crystal lattice deformation (§10.3 of [27]), also called the *crystal lattice distortion* [12]. Thus, we can suppose that the electron's speed decreases. Namely, we regard the electron as non-relativistic, and employ such H_{at} with V as satisfies our assumptions. This observation leads to our total Hamiltonian H_{QFT}.

As in [22], we have the following assertion:

Proposition 5.2. *Assume* $(N1)$ *and* $(N2)$. *Then,* H_{QFT} *is self-adjoint with* $D(H_{\mathrm{QFT}}) = D(H_0) \equiv D(H_{\mathrm{at}} \otimes I) \cap D(I \otimes \mathrm{d}\Gamma(1))$. H *is bounded from below for arbitrary values of* q.

Once we assume the existence of a ground state, it has to have the property of the spatial localization as stated in Propositions 5.3 and 5.4 below.

In the same way as in Proposition 6.1 in [22] we can prove the following:

Proposition 5.3. *Assume* (N1-1) *and* (N2). *If* H_{QFT} *has a ground state* Ψ_{QFT}, *then* $\Psi_{\mathrm{QFT}} \in D(x^2 \otimes I)$.

In the same way as in Proposition 6.3 in [22], obeying the idea in [21] with a little modification to meet our models, we have the following:

Proposition 5.4. *Assume* (N1-2) *and* (N2). *If* H_{QFT} *has a ground state* Ψ_{QFT}, *then there is* $C_0 > 0$ *such that* $\Psi_{\mathrm{QFT}} \in D(e^{C_0|x|})$.

The following operator-theoretical pull-through formula can be proved in the same way as in Proposition 3.1 in [22]:

Proposition 5.5. *Assume* (N1) *and* (N2). *Then, for all* $f \in C_0^\infty(\mathbb{R}^d \setminus \{0\})$,

$$a(f)\Psi_{\mathrm{QFT}} = -q \int_{\mathbb{R}^d} \overline{f(k)} \rho_{\mathrm{app}}(k) (\hat{H}_{\mathrm{QFT}} + \omega(k))^{-1} e^{-ikx} \Psi_{\mathrm{QFT}} \, \mathrm{d}k,$$

$\operatorname{supp} f \subset \operatorname{supp} 1^{<\Lambda}$ *and* $\Psi_{\mathrm{QFT}} \in D(x^2 \otimes I)$. *Therefore,* $B_{\mathrm{PT}}(k)$ *is the product of* $\lambda(k) = q 1^{<\Lambda}(k) \rho_{\mathrm{app}}(k)$ *and* $B_0(k) = e^{-ikx} \otimes I$.

To consider the problem of the mathematical modeling mentioned in Section 1, we give a concrete situation of $\omega(k)$ and $\lambda(k)$ here. Foreseeing from the experimental data that the phonon has the decay of the dispersion relation as the wave number goes to zero, it is difficult to estimate the decay order only from the data (see, for example, figures of the estimated phonon dispersion relation in [8], [11], [34], [45], [48], [49], [55], and [58]). Thus, we here consider a mathematical method to do that in the light of the IR problem. Through a simplification and an idealization, let us set $\omega(k)$ and $\lambda(k)$ as continuous functions on $(0, \Lambda]$, and assume that there exist constants, $\mu, \nu > 0$, such that

$$\omega(k) \sim |k|^{\mu} \text{ and } \lambda(k) \sim |k|^{\nu} \quad \text{as } |k| \to 0 \tag{7}$$

because we are interested in IR situation around $k = 0$. Then, we have

$$\gamma_{\mathrm{c}} = \frac{d + 2\nu}{2\mu}. \tag{8}$$

The condition, $d \leq 2(\mu - \nu)$, implies $\lambda/\omega \notin L^2(\mathbb{R}^d)$. A sufficient condition so that we can obtain γ_0 in the assumption (2) of Corollary 4.3 is $d > 2(\mu - \nu - 1)$ as shown in the proof of (iii) of Theorem 5.6 below. So, since H_{QFT} should be defined to be self-adjoint, a sufficient condition so that Corollary 4.3 (Dereziński and Gérard's result [9]) works together with the Kato–Rellich theorem [41] is

$$\max\left\{\frac{\mu}{2} - \nu,\, \mu - \nu - 1\right\} < \frac{d}{2} \leq \mu - \nu. \tag{9}$$

The dimension d has haven a restriction from below if we use Corollary 4.3. However, since $\mu/2 - \nu < \mu - \nu - 1$ if and only if $\mu > 2$, there is a possibility that $\mu/2 - \nu < d/2 < \mu - \nu - 1$ when $\mu > 2$. Thus, Corollary 4.3 does not work in this case. We try to remove this restriction in the case $\mu > 2$ by using Theorem 4.5 from now on.

Let us take μ with $2 < \mu$ now. If γ and ν satisfy

$$\begin{cases} 0 < \gamma < 1 - \dfrac{2}{\mu}, \\ \dfrac{d}{2} - \dfrac{1+\gamma}{2}\mu \leq -\nu < \dfrac{d+1}{2} - \dfrac{1+\gamma}{2}\mu, \end{cases} \tag{10}$$

then we have

$$\gamma\mu - \nu < \frac{1+\gamma}{2}\mu - \nu - \frac{1}{2} < \frac{d}{2} \leq \frac{1+\gamma}{2}\mu - \nu < \mu - \nu - 1. \tag{11}$$

Namely, d, μ and ν do not satisfy the condition (9) under (10). But we have the following criterion.

Theorem 5.6 (criterion for SB Divergence and IR catastrophe). *Suppose that H_{QFT} is self-adjoint. Set the estimated dispersion relation $\omega(k)$ and the estimated coupling function $\lambda(k)$ as* (7). *Assume that $\mu - 2\nu < d$. Let V is in class* (N1) *and* (N2). *Then, the following* (i)–(iv) *hold.*

(i) *If $\mu - \nu < d/2$, then the SB divergence does not take place, and moreover, there is a constant $q_0 \in \mathbb{R} \cup \{\infty\}$ such that ground state exists in $\mathcal{F}$ for every q with $|q| < q_0$.*

(ii) *If $\mu - \nu \geq d/2$, then the SB divergence takes place.*

(iii) *If d, μ, ν satisfy* (9), *then the IR catastrophe occurs.*

(iv) *Set $\gamma := 2\gamma_{\mathrm{c}} - 1$, where γ_{c} is in* (8). *If $\mu > 2$ and $1 - \mu < -\nu < (d-\mu)/2$, then* (10) *holds and the IR catastrophe occurs.*

Proof. We note that the condition $\mu - \nu < d/2$ implies $\lambda/\omega \in L^2(\mathbb{R}^d)$. Hence $\|K_{\mathrm{PT}}(\cdot)\|_{\mathcal{F}} \in L^2(\mathbb{R}^d)$ follows from this condition. Thus, Theorem 3.7 tells us that $\Psi_{\mathrm{QFT}} \in D(N^{1/2})$ if Ψ_{QFT} exists. Namely, IR catastrophe does not occur. The existence of a ground state Ψ_{QFT} is due to Spohn's result [50]. Thus, part (i) is completed. Part (ii) follows from Theorems 3.8 and 4.6.

To prove part (iii) we use the reduction of absurdity. Suppose that there is a ground state Ψ_{QFT}. The inequality $d/2 \leq \mu - \nu$ in (9) implies that $\gamma_{\mathrm{c}} \equiv (d+2\nu)/2\,\mu \leq 1$. Thus, 1 is in the IR-divergent region. Moreover, we have $1 - \mu^{-1} < \gamma_{\mathrm{c}}$ by $\mu - \nu - 1 < d/2$ in (9). Thus, every γ_0 with $1 - \mu^{-1} \leq \gamma_0 < \gamma_{\mathrm{c}}$ is in the IR-safe region. Proposition 5.5 says that $B_{\mathrm{PT}}(k) = \lambda(k) B_0(k)$ with $\lambda(k) = q 1^{<\Lambda}(k) \rho_{\mathrm{app}}(k)$ and $B_0(k) = e^{-ikx} \otimes I$. Taking $g(k) = \lambda(k)$ and $J_{\mathrm{err}}(k) = \lambda(k)(e^{-ikx} - 1) \otimes I$ in Corollary 4.3 shows that assumptions (1)–(3) of Corollary 4.3 are satisfied. Moreover, since $1 - \mu^{-1} \leq \gamma_0$ implies $(\gamma_0 - 1)\mu + 1 \geq 0$, there are positive constants K_Λ and C_Λ such that

$$
\begin{aligned}
&\sup_{k \in \mathrm{supp}\, J_{\mathrm{err}}} \omega(k)^{\gamma_0 - 1} \lambda(k)^{-1} \| J_{\mathrm{err}}(k) \Psi_{\mathrm{QFT}} \|_{\mathcal{F}} \\
&\quad \leq |K_\Lambda|^{(\gamma_0 - 1)\mu + 1} \| (|x| \otimes I)\, \Psi_{\mathrm{QFT}} \|_{\mathcal{F}} + C_\Lambda \\
&\quad < \infty
\end{aligned}
$$

by Propositions 5.3–5.5. This contradicts the assertion of Corollary 4.3.

Using the reduction of absurdity, we prove part (iv). Thus, we suppose there is a ground state Ψ_{QFT}. Our assumption of (iv) yields (10) immediately. It is clear that γ is in the IR-safe region due to (11) and $(1+\gamma)/2 = \gamma_{\mathrm{c}}$ in the IR-divergent region. Recall Proposition 5.5 here again. It is easy to check

$$
\int_{\mathbb{R}^d} \frac{|k_j|\,|\lambda(k)|^2}{\omega(k)^{1+\gamma}}\, dk = \int_{\mathbb{R}^d} \frac{|k_j|\,|\lambda(k)|^2}{\omega(k)^{2\gamma_{\mathrm{c}}}}\, dk \leq \int_{\mathbb{R}^d} \frac{|\lambda(k)|^2}{\omega(k)^{2(\gamma_{\mathrm{c}} - 1/2)}}\, dk < \infty
$$

since $\gamma_c - 1/2$ is in the IR-safe region. We note $B_0(0)\Psi_{\mathrm{QFT}} = \Psi_{\mathrm{QFT}} \neq 0$. Apply Maclaurin's theorem to $f(t) := e^{-itkx}$ $(t \in [0,1])$ and insert 1 into t. Then, we realize that we should define $B_0(0)$ and $B_j(k)$ in Theorem 4.5 by $B_0(0) = I \otimes I$ and $B_j(k) = -ix_j e^{-i\theta kx} \otimes I$ for a θ with $0 < \theta < 1$, respectively. Thus, Propositions 5.3 and 5.4 lead to the conclusion that $\sup_{k\in\mathbb{R}^d} \|B_j(k)\Psi_{\mathrm{QFT}}\|_{\mathcal{F}} \leq \||x|\Psi_{\mathrm{QFT}}\|_{\mathcal{F}} < \infty$. However, the last two facts contradict the statement of Theorem 4.5. □

Acknowledgements. The author is very grateful to C. Gérard for the explanation about Dereziński and Gérard's idea and an interpretation of [9] (equation (2.9) of Lemma 2.6), and to J. Dereziński, L. Bruneau, F. Hiroshima, and Y. Sekine for their useful comments.

The author acknowledges the support from JSPS, Grant-in-Aid for Scientific Research (B) 26310210 and Grant-in-Aid for Scientific Research (C) 26400117.

References

[1] A. Arai, Ground state of the massless Nelson model without infrared cutoff in a non-Fock representation. *Rev. Math. Phys.* **13** (2001), no. 9, 1075–1094. MR 1853826 Zbl 1029.81021

[2] A. Arai and M. Hirokawa, On the existence and uniqueness of ground states of a generalized spin-boson model. *J. Funct. Anal.* **151** (1997), no. 2, 455–503. MR 1491549 Zbl 0898.47048

[3] A. Arai and M. Hirokawa, Ground states of a general class of quantum field Hamiltonians. *Rev. Math. Phys.* **12** (2000), no. 8, 1085–1135. MR 1791436 Zbl 0983.81016

[4] A. Arai, M. Hirokawa, and F. Hiroshima, On the absence of eigenvectors of Hamiltonians in a class of massless quantum field models without infrared cutoffs. *J. Funct. Anal.* **168** (1999), no. 2, 470–497. MR 1719225 Zbl 0997.47061

[5] A. Arai, M. Hirokawa, and F. Hiroshima, Regularities of ground states of quantum field models. *Kyushu J. Math.* **61** (2007), no. 2, 321–372. MR 2362890 Zbl 1210.81070

[6] N. W. Aschcroft and N. D. Mermin, *Solid state physics.* Thomson Learning, South Melbourne, Australia, 1976.

[7] A. Damascelli, Z. Hussain, and Z.-X. Shen, Angle-resolved photoemission studies of the cuprate superconductors. *Rev. Mod. Phys.* **75** (2003), no. 2, 473–541.

[8] V. Yu. Davydov, Yu. E. Kitaev, I. N. Goncharuk, A. N. Smirnov, J. Graul, O. Semchinova, D. Uffmann, M. B. Smirnov, A. P. Mirgorodsky, and R. A. Evarestov, Phonon dispersion and Raman scattering in hexagonal GaN and AlN. *Phys. Rev. B* **58** (1998), no. 19, 12899–12907.

[9] J. Dereziński and C. Gérard, Scattering theory of infrared divergent Pauli–Fierz Hamiltonians. *Ann. Henri Poincaré* **5** (2004), no. 3, 523–577. MR 2067788 Zbl 1060.81060

[10] G. M. Èliašberg, Interaction between electrons and lattice vibrations in a superconductor. *Ž. Èksper. Teoret. Fiz.* **38** (1960), no. 3, 966–976. In Russian. English translation, *Soviet Physics. JETP* **11** (1960), 696–702. MR 0114618

[11] D. W. Feldman, H. Parker Jr. James, W. J. Choyke, and L. Patrick, Phonon dispersion curves by Raman scattering in SiC, polytypes $3C$, $4H$, $6H$, $15R$, and $21R$. *Phys. Rev.* **173** (1968), no. 3, 787–790.

[12] R. P. Feynman, Slow electrons in a polar crystal. *Phys. Rev.* (2) **97** (1955), 660–665. Zbl 0065.23903

[13] E. Fraval, M. J. Sellars, and J. J. Longdell, Dynamic decoherence control of a solid-state nuclear-quadrupole qubit. *Phys. Rev. Lett.* **95** (2005), no. 3, 030506, 4 pp.

[14] H. Fröhlich, Electrons in lattice fields. *Adv. in Phys.* **3** (1954), 325–362. Zbl 0056.23703

[15] J. Fröhlich, On the infrared problem in a model of scalar electrons and massless, scalar bosons. *Ann. Inst. H. Poincaré Sect. A (N.S.)* **19** (1973), 1–103. MR 0368649 Zbl 1216.81151

[16] V. Georgescu, C. Gérard, and J. S. Møller, Spectral theory of massless Pauli–Fierz models. *Comm. Math. Phys.* **249** (2004), no. 1, 29–78. MR 2077252 Zbl 1091.81059

[17] C. Gérard, On the existence of ground states for massless Pauli–Fierz Hamiltonians. *Ann. Henri Poincaré* **1** (2000), no. 3, 443–459. MR 1777307 Zbl 1004.81012

[18] C. Gérard, F. Hiroshima, A. Panati, and A. Suzuki, Infrared problem for the Nelson model on static space-times. *Comm. Math. Phys.* **308** (2011), no. 2, 543–566. MR 2851152 Zbl 1269.83036

[19] C. Gérard, F. Hiroshima, A. Panati, and A. Suzuki, Absence of ground state for the Nelson model on static space-times. *J. Funct. Anal.* **262** (2012), no. 1, 273–299. MR 2852262 Zbl 1238.81161

[20] I. Giaever, H. R. Hart, Jr., and K. Megerle, Tunneling into superconductors at temperatures below 1°K. *Phys. Rev.* **126** (1962), no. 3, 941–948.

[21] M. Griesemer, E. H. Lieb, and M. Loss, Ground states in non-relativistic quantum electrodynamics. *Invent. Math.* **145** (2001), no. 3, 557–595. MR 1856401 Zbl 1044.81133

[22] M. Hirokawa, Infrared catastrophe for Nelson's model. – Non-existence of ground state and soft-boson divergence. *Publ. Res. Inst. Math. Sci.* **42** (2006), no. 4, 897–922. MR 2289153 Zbl 1117.81097

[23] M. Hirokawa, F. Hiroshima, and H. Spohn, Ground state for point particles interacting through a massless scalar bose field. *Adv. Math.* **191** (2005), no. 2, 339–392. MR 2103217 Zbl 1093.81046

[24] F. Hiroshima, Multiplicity of ground states in quantum field models: applications of asymptotic fields. *J. Funct. Anal.* **224** (2005), no. 2, 431–470. MR 2146048 Zbl 1082.81030

[25] D. Haberer, L. Petaccia, A. V. Fedorov, C. S. Praveen, S. Fabris, S. Piccinin, O. Vilkov, D. V. Vyalikh, A. Preobrajenski, N. I. Verbitskiy, H. Shiozawa, J. Fink, M. Knupfer, B. Buchner, and A. Gruneis, Anisotropic Eliashberg function and electron–phonon coupling in doped graphene. *Phys. Rev. B* **88** (2013), no. 8, 081401, 5 pp.

[26] Ph. Hofmann, I. Yu. Sklyadneva, E. D. L. Rienks, and E. V. Chulkov, Electron–phonon coupling at surfaces and interfaces. *New J. Phys.* **11** (2009), 125005, 29 pp.

[27] H. Ibach and H. Lüth, *Solid-state physics.* An Introduction to principles of materials science. Springer, Berlin, Germany, 1990. ISBN 978-3-540-93803-3.

[28] A. Ishihara, *Statistical physics.* Academic Press, New York and London, 1971. MR 0446243

[29] C. Kittel, *Introduction to solid state physics.* John Wiley & Sons: New York, 1990.

[30] J. D. Koralek, J. F. Douglas, N. C. Plumb, Z. Sun, A. V. Fedorov, M. M. Murnane, H. C. Kapteyn, S. T. Cundiff, Y. Aiura, K. Oka, H. Eisaki, and D. S. Dessau, Laser based angle-resolved photoemission, the sudden approximation, and quasiparticle-like spectral peaks in $Bi_2Sr_2CaCu_2O_{8+\delta}$. *Phys. Rev. Lett.* **96** (2006), no. 1, 017005, 4 pp.

[31] T. D. Lee, F. E. Low, and D. Pines, The motion of slow electrons in a polar crystal. *Phys. Rev.* (2) **90** (1953), 297–302. MR 0103072 Zbl 0053.18205

[32] T. D. Lee and D. Pines, Interaction of a nonrelativistic particle with a scalar field with application to slow electrons in polar crystals. *Phys. Rev.* (2) **92** (1953), 883–889. MR 0103073 Zbl 0053.18206

[33] A. B. Migdal, Interaction between electrons and lattice vibrations in a normal metal. *Ž. Èksper. Teoret. Fiz.* **34** (1958), no. 6, 1438–1446. In Russian. English translation, *Soviet Physics. JETP* **34** (1958), 996–1001.

[34] V. J. Minkiewicz, and G. Shirane, Soft phonon modes in $KMnF_3$. *J. Phys. Soc. Jpn.* **26** (1969), 674–680.

[35] P. Neumann, N. Mizuochi, F. Rempp, P. Hemmer, H. Watanabe, S. Yamasaki, V. Jacques, T. Gaebel, F. Jelezko, and J. Wrachtrup, Multipartite entanglement among single spins in diamond. *Science* **320** (2008), no. 5881, 1326–1329.

[36] J. S. Møller, The polaron revisited. *Rev. Math. Phys.* **18** (2006), no. 5, 485–517. MR 2252041 Zbl 1109.81032

[37] M. Monni, I. Pallecchi, C. Ferdeghini, V. Ferrando, A. Floris, E. Galleani d'Agliano, E. Lehmann, I. Sheikin, C. Tarantini, X. X. Xi, S. Massidda, and M Putti, Probing the electron–phonon coupling in MgB_2 through magnetoresistance measurements in neutron irradiated thin films. *Europhysics Lett.* **81** (2008), no. 6, 67006, 6 pp.

[38] J. Noffsingera, F. Giustinoc, B. D. Malonea, C.-H. Parka, S. G. Louiea, and M. L. Cohena, EPW: a program for calculating the electron–phonon coupling using maximally localized Wannier functions. *Computer Phys. Commun.* **181** (2010), no. 12, 2140–2148. Zbl 1225.82004

[39] A. Panati, Existence and non existence of a ground state for the massless Nelson model under binding condition. *Rep. Math. Phys.* **63** (2009), no. 3, 305–330.

[40] J. V. Pulé, A. Verbeure, and V. A. Zagrebnov, Peierls–Fröhlich instability and Kohn anomaly. *J. Stat. Phys.* **76** (1994), no. 1-2, 159–182. MR 1297875 Zbl 0839.60096

[41] M. Reed and B. Simon, *Methods of modern mathematical physics.* II. Fourier analysis, self-adjointness. Academic Press [Harcourt Brace Jovanovich, Publishers], New York and London, 1975. MR 0493420 Zbl 0308.47002

[42] M. Reed and B. Simon, *Methods of modern mathematical physics.* IV. Analysis of operators. Academic Press [Harcourt Brace Jovanovich, Publishers], New York and London, 1978. ISBN 0-12-585004-2 MR 0493421 Zbl 0401.47001

[43] J. M. Rowell, A. G. Chynoweth, and J. C. Phillips, Multiphonon effects in tunnelling between metals and superconductors. *Phys. Rev. Lett.* **9** (1962), 59–61.

[44] J. M. Rowell, P. W. Anderson, and D. E. Thomas, Image of the phonon spectrum in the tunneling characteristic between superconductors. *Phys. Rev. Lett.* **10** (1963), 334–336.

[45] T. Ruf, J. Serrano, M. Cardona, P. Pavone, M. Pabst, M. Krisch, M. D'Astuto, T. Suski, I. Grzegory, and M. Leszczynski, Phonon dispersion curves in Wurtzite-structure GaN determined by inelastic x-ray scattering. *Phys. Rev. Lett.* **86** (2001), 906–909.

[46] I. Sasaki, Ground state of the massless Nelson model in a non-Fock representation. *J. Math. Phys.* **46** (2005), 102107, 12 pp. MR 2178579 Zbl 1111.81117

[47] K. Schmüdgen, *Unbounded operator algebras and representation theory.* Operator Theory: Advances and Applications, 37. Birkhäuser, Basel, 1990. ISBN 3-7643-2321-3 MR 1056697 Zbl 0697.47048

[48] J. Serrano, F. J. Manjon, A. H. Romero, A. Ivanov, M. Cardona, R. Lauck, A. Bosak, and M. Krisch, Phonon dispersion relations of zinc oxide: Inelastic neutron scattering and ab initio calculations. *Phys. Rev. B* **81** (2010), no. 17, 174304, 6 pp.

[49] H. Siegle, G. Kaczmarczyk, L. Filippidis, A. P. Litvinchuk, A. Hoffmann, and C. Thomsen, Zone-boundary phonons in hexagonal and cubic GaN. *Phys. Rev. B* **55** (1997), no. 11, 7000–7004.

[50] H. Spohn, Ground state of quantum particle coupled to a scalar boson field. *Lett. Math. Phys.* **44** (1998), no. 1, 9–16.

[51] J. R. Schrieffer, D. J. Scalapino, and J. W. Wilkins, Effective tunneling density of states in superconductors. *Phys. Rev. Lett.* **10** (1963), no. 8, 336–339.

[52] J. Teyssier, R. Lortz, A. Petrovic, and D. van der Marel, Effect of electron–phonon coupling on the superconducting transition temperature in dodecaboride superconductors: A comparison of LuB_{12} with ZrB_{12}. *Phys. Rev. B* **78** (2008), no. 13, 134504, 7 pp.

[53] A. Wälte, G. Fuchs, K.-H. Muller, A. Handstein, K. Nenkov, V. N. Narozhnyi, S.-L. Drechsler, S. Shulga, and L. Schultz, Evidence for strong electron–phonon coupling in $MgCNi_3$. *Phys. Rev. B* **70** (2004), no. 18, 174503, 18 pp.

[54] J. Weidmann, *Linear operators in Hilbert spaces.* Translated from the German by J. Szücs. Graduate Texts in Mathematics, 68. Springer, Berlin etc., 1980. ISBN: 0-387-90427-1 MR 0566954 Zbl 0434.47001

[55] U. D. Wdowik, K. Parlinski, S. Rols, and T. Chatterji, Soft-phonon mediated structural phase transition in GeTe. *Phys. Rev. B* **89** (2014), no. 22, 224306, 7 pp.

[56] E. L. Wolf, *Principles of electron tunneling spectroscopy.* Second edition. Oxford University Press, London, 2012. ISBN 0199589496.

[57] W. Zhang, *Photoemission spectroscopy on high temperature superconductor. A study of* $Bi_2Sr_2CaCu_2O_8$ *by laser-based angle-resolved photoemission.* Springer Theses. Springer, New York, 2013. ISBN 978-3-642-32471-0

[58] X.-L. Zhang and W.-M. Liu, Electron–phonon coupling and its implication for the superconducting topological insulators. *Sci. Rep.* **5** (2014), 8964, 6 pp.

[59] M. Zhong, M. P. Hedges, R. L. Ahlefeldt, J. G. Bartholomew, S. E. Beavan, S. M. Wittig, J. J. Longdell, and M. J. Sellars, Optically addressable nuclear spins in a solid with a six-hour coherence time. *Nature* **517** (2015), 177–180.

The modified unitary Trotter–Kato and Zeno product formulas revisited

Takashi Ichinose

Dedicated to my friend Pavel Exner on the occasion of his 70th birthday

1 Introduction

The aim of this note is to give some modified versions of the *unitary Trotter–Kato product formula for the form sum* C of nonnnegative selfadjoint operators A and B as well as the *Zeno product formula* for a nonnnegative selfadjoint operator H in a Hilbert space $\mathcal{H}$. The original problems are the product formulas which have not yet been established, still remaining as *conjectures*, but if they hold valid, to be expected to have the following forms:

$$\operatorname*{s-lim}_{n\to\infty} [\, e^{-i(t/n)A} e^{-i(t/n)B}\,]^n = e^{-itC}, \tag{1}$$

$$\operatorname*{s-lim}_{n\to\infty} [\, Pe^{-i(t/n)H} P\,]^n = e^{-itH_P}, \tag{2}$$

the convergence being uniform on each bounded t-interval in $\mathbb{R}$.

Some additional matters on notations and assumptions follow. First in formula (1), A and B further are for simplicity assumed such that $D[A^{1/2}] \cap D[B^{1/2}]$ is dense, where $D[T]$ stands for the domain of an operator T. Then the form sum

$$C := A \dotplus B \tag{3}$$

is defined as the nonnegative selfadjoint operator with domain $D[C]$ associated with the quadratic form

$$u \longmapsto \|A^{1/2}u\|^2 + \|B^{1/2}u\|^2$$

defined on $D[C^{1/2}] = D[A^{1/2}] \cap D[B^{1/2}]$. Next in formula (2), P is an orthogonal projection onto a closed subspace of $\mathcal{H}$, and H_P a nonnegative selfadjoint operator defined by

$$H_P := (H^{1/2}P)^*(H^{1/2}P), \tag{4}$$

which is assumed densely defined in $\mathcal{H}$.

Now we are going to state our results on these two formulas.

Unitary Trotter and Trotter–Kato product formulas

Let us begin with a brief background of the issue. In 1959, Trotter [20] proved among others, given two selfadjoint operators A and B, which need not be semibounded, the *unitary exponential* product formula, when their *operator sum* $A + B$ is *essentially selfadjoint* on the common domain $D[A] \cap D[B]$ with C being the closure of $A + B$, and also the *selfadjoint exponential* product formula together, as far as both A and B are bounded from below. In 1978, Kato [13] exploited an ingenious technique to extend the latter formula to the case for the *form sum* of two selfadjoint operators bounded from below.

The problem which we are now interested in is the very case for the *unitary* exponential product formula with the *form sum*, which we should like to call "unitary Trotter–Kato" product formula but not "unitary Trotter" one, because it concerns the *form sum*. The first attempt to this problem were a humble work of ours [8] with a technique of high-spectrum cutoff of both operators A and B, and Lapidus' ([14], [15], and [16], cf. [11]) with the short-time unitary groups $e^{-i(t/n)A}$ and $e^{-i(t/n)B}$ appearing as factors in the product on the left-hand side of (1) replaced by the resolvents. A further study with this aspect is pursued in the work [5]. These works will turn out to have treated "modified" unitary Trotter–Kato product formulas.

However, the genuine original problem is still an open question as mentioned at the beginning of this section. So we content ourselves to revisiting a modified version of *unitary Trotter–Kato product formula* for the form sum C in (3) of A and B with the short-time unitary groups in the product on the left of (1) replaced by the resolvents. So the result will not be new. The proof appeals, as is usual, to the Chernoff theorem ([1] and [2], cf. Supplement XIII.8 in [19]), which in fact entails showing the approximate generators defined through the product of the resolvents of A and B converge to C in the strong resolvent sense. We show this by solely using a now well-understood, simple method which was originally elaborated by Kato [13] to prove his celebrated selfadjoint exponential product formula [13], thus differing from and somewhat simpler than Lapidus' and that in [5].

The identity operator on $\mathcal{H}$ is denoted by I. The inner product $\langle \cdot, \cdot \rangle$ of our Hilbert space $\mathcal{H}$ is anti-linear in the first argument and linear in the second.

Theorem 1.1. *Let A and B be nonnegative selfadjoint operators in $\mathcal{H}$ and assume that $D[A^{1/2}] \cap D[B^{1/2}]$ is dense in $\mathcal{H}$. Then it holds with $\varepsilon = \pm 1$ that*

$$\operatorname*{s-lim}_{n\to\infty} \Big[\Big(I + i\varepsilon\frac{t}{n}A\Big)^{-1}\Big(I + i\varepsilon\frac{t}{n}B\Big)^{-1}\Big]^n = e^{-i\varepsilon t C}, \tag{5}$$

uniformly on each bounded t-interval in $[0, \infty)$ and therefore in $\mathbb{R}$.

Theorem 1.1 is valid for A, B and C being selfadjoint operators bounded from below, as well as from above.

By the argument in Chernoff (Section 7, p. 82, in [2]), Theorem 1.1 in particular implies the following selfadjoint version of it:

Corollary 1.2. *Under the same hypothesis as in Theorem* 1.1, *it holds that*

$$\operatorname*{s-lim}_{n\to\infty}\Big[\Big(I+\frac{t}{n}A\Big)^{-1}\Big(I+\frac{t}{n}B\Big)^{-1}\Big]^{n}=e^{-tC},$$

uniformly on each bounded t-interval in $[0,\infty)$.

Some irrelevant but notable two remarks follow: It was Nelson [18] who first mentioned that the unitary Trotter product formula is a good device which can impart a meaning to the Feynman path integral [7] which represents the solution of the Schrödinger equation as a time-sliced approximation. It was shown in [10] (cf. [9]) that there are some nontrivial special cases where even the *norm convergence* of *unitary Trotter product formulas* holds.

Zeno product formula

An attractive problem in quantum mechanical measurement is connected with this formula. However, we will not touch it here but only refer to, e.g., [17] and [6]. Some mathematical treatment has been done, e.g., in [3] and [4].

In this note, we revisit Zeno problem to give a modified version, where for the nonnegative selfadjoint operator H, its short-time unitary group $e^{-i(t/n)H}$ on the left of (2) are replaced by the resolvent $(I+i(t/n)H)^{-1}$. However, the result is not new but included in [4] (Theorem 2, p. 70), though ours is treating a slightly more general case with *t-dependent* orthogonal projections $P(t)$. However, the proof in [4] uses the quadratic form technique, while ours here does again the same simple method as used in the proof of Theorem 1.1, which is based on Kato's original idea [13] having established his selfadjoint exponential product formula for the form sum.

Theorem 1.3. *With H as above, let $P(t)$ be a strongly continuous function with values orthogonal projections on $\mathcal{H}$ defined in some neighborhood of* 0 *with $P(0)=:P$. Then it holds for H_P in* (4) *that*

$$\operatorname*{s-lim}_{n\to\infty}\Big[P\Big(\frac{t}{n}\Big)\Big(I+i\Big(\frac{t}{n}\Big)H\Big)^{-1}P\Big(\frac{t}{n}\Big)\Big]^{n}=e^{-itH_P}P, \tag{6}$$

uniformly on each bounded t-interval in $[0, \infty)$, and so in particular, for a t-independent projection P,

$$\text{s-}\lim_{n\to\infty} \Big[P\Big(I + i\Big(\frac{t}{n}\Big)H\Big)^{-1} P\Big]^n = e^{-itH_P} P,$$

uniformly on each bounded t-interval in $[0, \infty)$, and also in $\mathbb{R}$.

From this theorem, it is easy to see that it holds also for the following other products on the left-hand side, namely,

$$\begin{aligned}\text{s-}\lim_{n\to\infty} \Big[\Big(I + i\Big(\frac{t}{n}\Big)H\Big)^{-1} P\Big(\frac{t}{n}\Big)\Big]^n &= \text{s-}\lim_{n\to\infty} \Big[P\Big(\frac{t}{n}\Big)\Big(I + i\Big(\frac{t}{n}\Big)H\Big)^{-1}\Big]^n \\ &= e^{-itH_P} P.\end{aligned}$$

In Section 2 we prove Theorem 1.1 and Theorem 1.3 in Section 3.

2 Proof of Theorem 1.1

We have only to prove the theorem for $\varepsilon = 1$. The case $\varepsilon = -1$ is similarly shown.

Since A and B are nonnegative, there exist spectral measures $E_A(d\lambda)$ and $E_B(d\lambda)$ on the real line such that $A = \int_{0-}^{\infty} \lambda E_A(d\lambda)$ and $B = \int_{0-}^{\infty} \lambda E_B(d\lambda)$.

For $\tau > 0$, put

$$F(\tau) = (I + itA)^{-1}(I + itB)^{-1},$$

which is a contraction, and

$$S(\tau) = \tau^{-1}[I - F(\tau)] = \tau^{-1}[I - (I + i\tau A)^{-1}(I + i\tau B)^{-1}]. \tag{7}$$

Since $S(\tau)$ satisfies

$$\begin{aligned}\operatorname{Re}\langle f, S(\tau)f\rangle &= \tau^{-1}\operatorname{Re}[\langle f, f\rangle - \langle (I - i\tau A)^{-1} f, (I + i\tau B)^{-1} f\rangle] \\ &\geq \tau^{-1}[\|f\|^2 - \|(I - i\tau A)^{-1} f\| \, \|(I + i\tau B)^{-1} f\|] \\ &= \tau^{-1}[\|f\|^2 - \|f\|^2] \\ &= 0, \quad f \in \mathcal{H},\end{aligned}$$

$S(\tau)$ is an m-accretive operator (e.g., [12]). Therefore $I + S(\tau)$ has a bounded inverse $(I + S(\tau))^{-1}$, which is also a contraction.

In order to show (5), by the Chernoff theorem ([1] and [2], cf. Supplement XIII.8, p. 386, in [19]), we have only to prove that

$$(I + S(\tau))^{-1} \xrightarrow{s} (I + iC)^{-1}, \quad \tau \to 0_+. \tag{8}$$

We will employ the method with analogous arguments, but with slightly more deliberation, used by Kato [13] to prove the selfadjoint exponential product formula for the form sum of nonnegative selfadjoint operators (cf. [8]).

To rewrite $S(\tau)$ in (7), define $A(\tau)$ and $B(\tau)$ for $\tau > 0$ by

$$\begin{aligned} iA(\tau) &:= \frac{I-(I+i\tau A)^{-1}}{\tau} = \frac{iA}{I+i\tau A} = \frac{\tau A^2}{I+\tau^2A^2} + i\frac{A}{I+\tau^2A^2} \\ &=: G_A(\tau) + iH_A(\tau), \\ iB(\tau) &:= \frac{I-(I+i\tau B)^{-1}}{\tau} = \frac{iB}{I+i\tau B} = \frac{\tau B^2}{I+\tau^2B^2} + i\frac{B}{I+\tau^2B^2} \\ &=: G_B(\tau) + iH_B(\tau). \end{aligned}$$

Here the four operators $G_A(\tau), G_B(\tau)$ and $H_A(\tau)$, $H_A(\tau)$ are all bounded nonnegative selfadjoint for $\tau > 0$, so that $iA(\tau)$ and $iB(\tau)$ are m-accretive normal operators. We have

$$|A(\tau)| = (G_A(\tau)^2 + H_A(\tau)^2)^{1/2} = \frac{A}{(I+\tau^2A^2)^{1/2}}, \tag{9a}$$

$$|B(\tau)| = (G_B(\tau)^2 + H_B(\tau)^2)^{1/2} = \frac{B}{(I+\tau^2B^2)^{1/2}}. \tag{9b}$$

Noting that

$$\begin{aligned} S(\tau) &= \frac{I-(I+i\tau A)^{-1}}{\tau} + (I+i\tau A)^{-1}\frac{I-(I+i\tau B)^{-1}}{\tau} \\ &= iA(\tau) + i(I+i\tau A)^{-1}B(\tau) \\ &= (G_A(\tau)+iH_A(\tau)) + (G_B(\tau)+iH_B(\tau)) + \tau A(\tau)B(\tau), \end{aligned} \tag{10}$$

we consider the following modification of $S(\tau)$:

$$\widehat{S}(\tau) = (G_A(\tau)+iH_A(\tau)) + (G_B(\tau)+iH_B(\tau)). \tag{11}$$

We see $I + \widehat{S}(\tau)$ also have bounded inverse with norm not exceeding 1, since $\widehat{S}(\tau)$ is also m-accretive. Then for the proof of (8), we need to show

$$(I+\widehat{S}(\tau))^{-1} \xrightarrow{s} (I+iC)^{-1}, \tag{12}$$

$$(I+S(\tau))^{-1} - (I+\widehat{S}(\tau))^{-1} \xrightarrow{s} 0. \tag{13}$$

First we show (12).

Proof of (12). For $f \in \mathcal{H}$, put $u_\tau = (I + \widehat{S}(\tau))^{-1} f$, so that

$$\begin{aligned} f &= (I + \widehat{S}(\tau))u_\tau \\ &= u_\tau + (G_A(\tau) + G_B(\tau))u_\tau + i(H_A(\tau) + H_B(\tau))u_\tau. \end{aligned}$$

We have $\|u_\tau\| \le \|f\|$ since $(I + \widehat{S}(\tau))^{-1}$ is a contraction. Taking the inner product with u_τ,

$$\begin{aligned} \langle u_\tau, f\rangle &= \|u_\tau\|^2 + \langle u_\tau, (G_A(\tau) + G_B(\tau))u_\tau\rangle + i\langle u_\tau, (H_A(\tau) + H_B(\tau))u_\tau\rangle \\ &= \|u_\tau\|^2 + \|G_A^{1/2}(\tau)u_\tau\|^2 + \|G_B(\tau)^{1/2}u_\tau\|^2 \\ &\quad + i[\,\|H_A(\tau)^{1/2}u_\tau\|^2 + \|H_B(\tau)^{1/2}u_\tau\|^2]. \end{aligned} \tag{14}$$

Then we see from (14) that the five τ-families $\{u_\tau\}$, $\{G_A(\tau)^{1/2}u_\tau\}$, $\{G_B(\tau)^{1/2}u_\tau\}$, $\{H_A(\tau)^{1/2}u_\tau\}$ and $\{H_B(\tau)^{1/2}u_\tau\}$ are all uniformly bounded by $\|f\|$ for $\tau > 0$. Therefore there exists a (sub)sequence $\{\tau'\}$ with $\tau' \to 0$ along which these sequences are weakly convergent:

$$u_\tau \xrightarrow{w} u, \tag{15a}$$

$$G_A(\tau)^{1/2}u_\tau \xrightarrow{w} g_A, \tag{15b}$$

$$G_B(\tau)^{1/2}u_\tau \xrightarrow{w} g_B, \tag{15c}$$

$$H_A(\tau)^{1/2}u_\tau \xrightarrow{w} h_A, \tag{15d}$$

$$H_B(\tau)^{1/2}u_\tau \xrightarrow{w} h_B, \tag{15e}$$

for some vectors u, g_A, g_B, h_A, h_B in $\mathcal{H}$. We first claim that

$$u \in D[C^{1/2}], \quad g_A = g_B = 0, \quad h_A = A^{1/2}u, \quad h_B = B^{1/2}u, \tag{16}$$

and next

$$u \in D[C], \quad f = (I + iC)u \quad \text{or} \quad u = (I + iC)^{-1} f. \tag{17}$$

To show (16), note that by the spectral theorem we have for $v \in D[A^{1/2}]$,

$$G_A(\tau)^{1/2}v = \int_{0-}^{\infty} \Big[\frac{\tau\lambda^2}{1 + \tau^2\lambda^2}\Big]^{1/2} E(d\lambda)v \xrightarrow{s} 0,$$

$$H_A(\tau)^{1/2}v = \int_{0-}^{\infty} \Big[\frac{\lambda}{1 + \tau^2\lambda^2}\Big]^{1/2} E(d\lambda)v \xrightarrow{s} A^{1/2}v,$$

so that

$$\langle g_A, v\rangle = \lim\langle G_A(\tau)^{1/2}u_\tau, v\rangle = \lim\langle u_\tau, G_A(\tau)^{1/2}v\rangle = \langle u, 0\rangle = 0,$$

$$\langle h_A, v\rangle = \lim\langle H_A(\tau)^{1/2}u_\tau, v\rangle = \lim\langle u_\tau, H_A(\tau)^{1/2}v\rangle = \langle u, A^{1/2}v\rangle,$$

the limits taken along $\tau' \to 0_+$. Since $D[A^{1/2}]$ is dense in $\mathcal{H}$, we have $u \in D[A^{1/2}]$ and $g_A = 0, h_A = A^{1/2}u$. The same is true for B. This yields (16).

To show (17), note that, for all $v \in D[C^{1/2}]$,

$$\begin{aligned}\langle v, f\rangle = \langle v, u_\tau\rangle &+ \langle G_A(\tau)^{1/2}v, G_A(\tau)^{1/2}u_\tau\rangle + \langle G_B(\tau)^{1/2}v, G_B(\tau)^{1/2}u_\tau\rangle \\ &+ i[\langle H_A(\tau)^{1/2}v, H_A(\tau)^{1/2}u_\tau\rangle + \langle H_B(\tau)^{1/2}v, H_B(\tau)^{1/2}u_\tau\rangle],\end{aligned}$$

the right-hand side of which converges to

$$\langle v, u\rangle + i[\langle A^{1/2}v, A^{1/2}u\rangle + \langle B^{1/2}v, B^{1/2}u\rangle] = \langle v, u\rangle + i\langle C^{1/2}v, C^{1/2}u\rangle,$$

along $\tau' \to 0_+$. Hence we have

$$\langle v, f\rangle = \langle v, u\rangle + i\langle C^{1/2}v, C^{1/2}u\rangle,$$

so that $C^{1/2}u \in D[C^{1/2}]$ and hence $u \in D[C]$, and further $f = (I + iC)u$. This yields (17), showing what we claimed in (16)/(17). We have thus seen the weak limits of (15) do not depend on the (sub)sequence chosen. A standard argument concludes that (15) with (16) holds as $\tau \to 0_+$ without taking any (sub)sequence.

As the last step to conclude (12), we need to show the strong convergence of the five τ-families $\{u_\tau\}$, $\{G_A(\tau)^{1/2}u_\tau\}$, $\{G_B(\tau)^{1/2}u_\tau\}$, $\{H_A(\tau)^{1/2}u_\tau\}$ and $\{H_B(\tau)^{1/2}u_\tau\}$. To do so, it suffices to show the norms of these vectors converge. To this end, observe the real and imaginary parts of (14):

$$\mathrm{Re}\,\langle u_\tau, f\rangle = \|u_\tau\|^2 + \|G_A(\tau)^{1/2}u_\tau\|^2 + \|G_B(\tau)^{1/2}u_\tau\|^2,$$

$$\mathrm{Im}\,\langle u_\tau, f\rangle = \|H_A(\tau)^{1/2}u_\tau\|^2 + \|H_B(\tau)^{1/2}u_\tau\|^2.$$

Then by the fact of weak convergence of $\{u_\tau\}$ and (17), as the left-hand sides converge if $\tau \to 0_+$, i.e.,

$$\mathrm{Re}\,\langle u_\tau, f\rangle \longrightarrow \mathrm{Re}\,\langle u, f\rangle = \|u\|^2,$$

$$\mathrm{Im}\,\langle u_\tau, f\rangle \longrightarrow \langle u, Cu\rangle = \|C^{1/2}u\|^2 = \|A^{1/2}u\|^2 + \|B^{1/2}u\|^2,$$

so do the right-hand sides. Therefore we have for the real part

$$\begin{aligned}\|u\|^2 &= \lim [\, \|u_\tau\|^2 + \|G_A(\tau)^{1/2}u_\tau\|^2 + \|G_B(\tau)^{1/2}u_\tau\|^2 \,] \\ &= \liminf [\, \|u_\tau\|^2 + \|G_A(\tau)^{1/2}u_\tau\|^2 + \|G_B(\tau)^{1/2}u_\tau\|^2 \,] \\ &\geq \|u\|^2 + \|0\|^2 + \|0\|^2 \\ &= \|u\|^2,\end{aligned}$$

and for the imaginary part

$$\begin{aligned}\|C^{1/2}u\|^2 &= \lim [\, \|H_A(\tau)^{1/2}u_\tau\|^2 + \|H_B(\tau)^{1/2}u_\tau\|^2 \,] \\ &= \liminf [\, \|H_A(\tau)^{1/2}u_\tau\|^2 + \|H_B(\tau)^{1/2}u_\tau\|^2 \,] \\ &\geq \|A^{1/2}u\|^2 + \|B^{1/2}u\|^2 \\ &= \|C^{1/2}u\|^2.\end{aligned}$$

Here in the above argument we have used the fact that weak convergence of $\{w_\tau\}$ to w implies $\liminf \|w_\tau\| \geq \|w\|$. This shows the convergence of the norms of these five families and consequently their strong convergence. □

Proof of (13). We need to estimate the left-hand side of (13). Rewrite it with (10) and (11) as

$$(I + S(\tau))^{-1} - (I + \widehat{S}(\tau))^{-1} = -(I + S(\tau))^{-1}[\tau A(\tau)B(\tau)](I + \widehat{S}(\tau))^{-1}. \quad (18)$$

Four lemmas are provided. However, the one which we in fact need for the proof of (13) is only Lemma 2.4. The others, though giving some supplementary inequalities among the operators concerned, may be skipped.

Lemma 2.1. (i) *Both* $(\tau|A(\tau)|)^{1/2}$ *and* $(\tau|B(\tau)|)^{1/2}$ *are contractions, and converge strongly to* 0 *as* $\tau \to 0_+$.

(ii) *For* $v \in \mathcal{H}$,

$$\begin{aligned}&|\langle v, \tau A(\tau)B(\tau)v\rangle| \\ &\quad\leq \langle |A(\tau)|^{1/2}v, \tau|A(\tau)|\,|A(\tau)|^{1/2}v\rangle^{1/2}\langle |B(\tau)|^{1/2}v, \tau|B(\tau)|\,|B(\tau)|^{1/2}v\rangle^{1/2} \\ &\quad\leq \frac{1}{2}\langle v, (|A(\tau)| + |B(\tau)|)v\rangle.\end{aligned} \quad (19)$$

Proof. (i) It is clear from (9) that $(\tau|A(\tau)|)^{1/2}$ is a contraction which converges strongly to 0 as $\tau \to 0_+$. The same is also valid for $B(\tau)$.

(ii) First note that the normal operators $A(\tau)$ and $B(\tau)$ have the polar decompositions (see, e.g. [12]):

$$A(\tau) = U(\tau)|A(\tau)| = |A(\tau)|U(\tau), \quad B(\tau) = V(\tau)|B(\tau)| = |B(\tau)|V(\tau),$$

where $U(\tau)$ and $V(\tau)$ are partial isometries on $\mathcal{H}$. So we have

$$\tau A(\tau)B(\tau) = (\tau|A(\tau)|)^{1/2}|A(\tau)|^{1/2}(U(\tau)V(\tau))(\tau|B(\tau)|)^{1/2}|B(\tau)|^{1/2}.$$

Since $U(\tau)$ and $V(\tau)$ are contractions, it follows that for $v \in \mathcal{H}$,

$$\begin{aligned}
&|\langle v, \tau A(\tau)B(\tau)v\rangle| \\
&\quad \le \langle |A(\tau)|^{1/2}v, \tau|A(\tau)|\,|A(\tau)|^{1/2}v\rangle^{1/2}\langle |B(\tau)|^{1/2}v, \tau|B(\tau)|\,|B(\tau)|^{1/2}v\rangle^{1/2} \\
&\quad \le \frac{1}{2}\langle v, (|A(\tau)| + |B(\tau)|)v\rangle.
\end{aligned}$$

This shows the lemma. □

Lemma 2.2. *We have*

$$\|[I + G_A(\tau) + G_B(\tau) + (H_A(\tau) + H_B(\tau))]^{-1/2}[I + |A(\tau)| + |B(\tau)|]^{1/2}\| \le 1, \tag{20}$$

and

$$\begin{aligned}
&\|[I + G_A(\tau) + G_B(\tau) + i(H_A(\tau) + H_B(\tau))]^{-1/2} \\
&\qquad [I + G_A(\tau) + G_B(\tau) + (H_A(\tau) + H_B(\tau))]^{1/2}\| \\
&= \|[I + G_A(\tau) + G_B(\tau) + (H_A(\tau) + H_B(\tau))]^{1/2} \\
&\qquad [I + G_A(\tau) + G_B(\tau) + i(H_A(\tau) + H_B(\tau))]^{-1/2}\| \\
&\le \sqrt{2}.
\end{aligned}$$

Proof. The first inequality (20) is equivalent to the inequality

$$|A(\tau)| + |B(\tau)| \le G_A(\tau) + G_B(\tau) + (H_A(\tau) + H_B(\tau)),$$

which holds because $|A(\tau)| \ge G_A(\tau) + H_A(\tau)$ by the expression (9) of $|A(\tau)|$ and the same for $|B(\tau)|$.

The second inequality is a consequence of the following lemma. □

Lemma 2.3. *Let A and B be nonnegative bounded selfadjoint operators on $\mathcal{H}$. Then for $v \in \mathcal{H}$*

$$|\langle v, (I + B + iA)v\rangle| \geq \frac{1}{\sqrt{2}}\langle v, (I + B + A)v\rangle = \frac{1}{\sqrt{2}}\|(I + B + A)^{1/2}v\|^2, \quad (21)$$

so that

$$\|(I + B + A)^{1/2}(I + B + iA)^{-1}(I + B + A)^{1/2}\| \leq \sqrt{2}, \quad (22)$$

in particular,

$$\|(I+B+A)^{1/2}(I+B+iA)^{-1}\| = \|(I+B+iA)^{-1}(I+B+A)^{1/2}\| \leq \sqrt{2}. \quad (23)$$

Proof. Taking the absolute value and using $(a^2 + b^2)^{1/2} \geq 1/\sqrt{2}\,(a + b)$, we have

$$\begin{aligned}|\langle v, (I + B + iA)v\rangle| &= [\langle v, (I + B)v\rangle^2 + \langle v, Av\rangle^2]^{1/2} \\ &\geq \frac{1}{\sqrt{2}}[\langle v, (I + B)v\rangle + \langle v, Av\rangle] \\ &= \frac{1}{\sqrt{2}}\langle v, (I + B + A)v\rangle.\end{aligned}$$

This shows (21). The other assertions (22) and (23) are easy to see. □

Lemma 2.4. *For all $v \in \mathcal{H}$,*

$$|\langle v, (I + S(\tau))v\rangle| \geq \frac{\sqrt{2}-1}{2}\langle v, (I + |A(\tau)| + |B(\tau)|)v\rangle,$$

so that

$$\|(I + K(\tau))^{1/2}(I + S(\tau))^{-1}(I + K(\tau))^{1/2}\| \leq \frac{2}{\sqrt{2}-1}, \quad (24)$$

where

$$K(\tau) := |A(\tau)| + |B(\tau)|.$$

Proof. Noting the expression (10) of $S(\tau)$ and the fact that $G_A(\tau) + H_A(\tau) \geq |A(\tau)|$ and the same for $H_B(\tau)$, we have

$$
\begin{aligned}
&|(v, (I + S(\tau))v)| \\
&\quad \geq |(v, [I + (G_A(\tau) + G_B(\tau)) + i(H_A(\tau) + H_B(\tau))]v)| - |(v, \tau A(\tau)B(\tau)v)| \\
&\quad \geq \frac{1}{\sqrt{2}}(v, [I + (G_A(\tau) + G_B(\tau)) + (H_A(\tau) + H_B(\tau))]v) - |(v, \tau A(\tau)B(\tau)v)| \\
&\quad \geq \frac{1}{\sqrt{2}}[\,\|v\|^2 + (v, (|A(\tau)| + |B(\tau)|)v)] - \frac{1}{2}(v, (|A(\tau)| + |B(\tau)|))v) \\
&\quad = \frac{1}{\sqrt{2}}\Big[\,\|v\|^2 + \Big(1 - \frac{1}{\sqrt{2}}\Big)(v, (|A(\tau)| + |B(\tau)|)v)\Big] \\
&\quad \geq \frac{1}{\sqrt{2}}\Big(1 - \frac{1}{\sqrt{2}}\Big)[\,\|v\|^2 + (v, (|A(\tau)| + |B(\tau)|)v)] \\
&\quad = \frac{\sqrt{2}-1}{2}(v, (I + K(\iota))v).
\end{aligned}
$$

The transition among the above inequalities will be easy to follow; if not, with a slight help of the previous three lemmas, the second inequality by Lemma 2.2 or 2.3 and the third by Lemma 2.1 (ii). This shows Lemma 2.4. □

Now we are going to finalize the proof of (13). By (19) we have

$$
\begin{aligned}
&(I + S(\tau))^{-1} - (I + \widehat{S}(\tau))^{-1} \\
&\quad = -[(I + S(\tau))^{-1}(I + K(\tau))^{1/2}]\,[(I + K(\tau))^{-1/2}|A(\tau)|^{1/2}] \\
&\qquad\quad [(\tau|A(\tau)|)^{1/2}U(\tau)V(\tau)(\tau|B(\tau)|)^{1/2}] \\
&\qquad\qquad [|B(\tau)|^{1/2}(I + K(\tau))^{-1/2}]\,[(I + K(\tau))^{1/2}(I + \widehat{S}(\tau))^{-1}].
\end{aligned}
$$

Here the five factor in the product on the right-hand side are all bounded operators uniformly for $\tau > 0$, which are strongly continuous in τ. Indeed, by (24) in Lemma 2.4 the first and the last factor are bounded by $2/(\sqrt{2}-1)$, and further the second and the last second by 1. The crucial is that the third factor in the middle

$$
[(\tau|A(\tau)|)^{1/2}U(\tau)V(\tau)(\tau|B(\tau)|)^{1/2}]
$$

converges strongly to 0 as $\tau \to 0_+$. Thus we see (18) converge strongly to 0, showing (13), and completing the proof of Theorem 1.1. □

3 Proof of Theorem 1.3

The proof will proceed similarly as in Section 2. For $\tau > 0$, put

$$F(\tau) = P(\tau)(I + i\tau H)^{-1}P(\tau),$$

which is a contraction, and

$$S(\tau) = \tau^{-1}[I - F(\tau)] = \tau^{-1}[I - P(\tau)(I + i\tau H)^{-1}P(\tau)].$$

Since $S(\tau)$ satisfies

$$\begin{aligned}\mathrm{Re}(f, S(\tau)f) &= \tau^{-1}\mathrm{Re}[(f, f) - (P(\tau)f, (I + i\tau H)^{-1}P(\tau)f)]\\ &\geq \tau^{-1}[\,\|f\|^2 - \|P(\tau)f\|\,\|(I + i\tau H)^{-1}P(\tau)f\|\,]\\ &= \tau^{-1}[\,\|f\|^2 - \|f\|^2]\\ &= 0,\end{aligned}$$

for all $f \in \mathcal{H}$, $S(\tau)$ is an m-accretive operator again. Therefore $I + S(\tau)$ has a bounded inverse $(I + S(\tau))^{-1}$, which is also a contraction.

In order to show (6), by the Chernoff theorem we have only to show that

$$(I + S(\tau))^{-1} \xrightarrow{s} (I + iH_P)^{-1}P, \quad \tau \to 0_+. \tag{25}$$

Since

$$\begin{aligned}\frac{1}{\tau}[I - (I + i\tau H)^{-1}] &= \frac{iH}{I + \tau H}\\ &= \frac{\tau^2 H}{I + \tau H^2} + i\frac{iH}{I + \tau H^2}\\ &=: G(\tau) + iH(\tau),\end{aligned}$$

where both $G(\tau)$ and $H(\tau)$ are nonnegative selfadjoint operators, we have

$$\begin{aligned}S(\tau) &= \tau^{-1}[I - P(\tau)(I + i\tau H)^{-1}P(\tau)]\\ &= \tau^{-1}(I - P(\tau)) + P(\tau)G(\tau)P(\tau) + i\,P(\tau)H(\tau)P(\tau),\end{aligned}$$

so that

$$\begin{aligned}I + S(\tau) &= I + \tau^{-1}(I - P(\tau)) + P(\tau)G(\tau)P(\tau) + iP(\tau)H(\tau)P(\tau)\\ &= (1 + \tau^{-1})(I - P(\tau)) + P(\tau)(I + G(\tau) + iH(\tau))P(\tau).\end{aligned}$$

Now we going to show assertion (25). For $f \in \mathcal{H}$ put $u_\tau = (I + S(\tau))^{-1} f$, so that

$$\begin{aligned} f &= (I + S(\tau))u_\tau \\ &= u_\tau + \tau^{-1}(I - P(\tau))u_\tau + P(\tau)G(\tau)P(\tau)u_\tau + iP(\tau)H(\tau)P(\tau)u_\tau. \end{aligned} \tag{26}$$

Then taking the inner product with u_τ, we have

$$\begin{aligned} \langle u_\tau, f\rangle &= \langle u_\tau, u_\tau\rangle + \tau^{-1}\langle u_\tau, (I - P(\tau))u_\tau\rangle + \langle G(\tau)^{1/2} P(\tau)u_\tau, G(\tau)^{1/2} P(\tau)u_\tau\rangle \\ &\quad + i\langle H(\tau)^{1/2} P(\tau)u_\tau, H(\tau)^{1/2} P(\tau)u_\tau\rangle \\ &= \|u_\tau\|^2 + \tau^{-1}\|(I - P(\tau))u_\tau\|^2 + \|G(\tau)^{1/2} P(\tau)u_\tau\|^2 \\ &\quad + i\|H(\tau)^{1/2} P(\tau)u_\tau\|^2. \end{aligned} \tag{27}$$

From equation (27), we see that for $\tau > 0$, the four families

$$\{u_\tau\}, \quad \{\tau^{-1/2}(I - P(\tau))u_\tau\}, \quad \{G(\tau)^{1/2} P(\tau)u_\tau\}, \quad \{H(\tau)^{1/2} P(\tau)u_\tau\}$$

are bounded by $\|f\|$, so that there exist a (sub)sequence $\{\tau'\}$, $\tau' \to 0_+$, along which

$$u_\tau \xrightarrow{w} u, \tag{28a}$$

$$\tau^{-1/2}(I - P(\tau))u_\tau \xrightarrow{w} u_0, \tag{28b}$$

$$G(\tau)^{1/2} P(\tau)u_\tau \xrightarrow{w} g, \tag{28c}$$

$$H(\tau)^{1/2} P(\tau)u_\tau \xrightarrow{w} h, \tag{28d}$$

for some $u, u_0, g, h \in \mathcal{H}$. We claim first that

$$u = Pu \in D[H^{1/2}], \quad u_0 = g = 0, \quad h = H^{1/2}Pu, \quad Pf = P(I + iH_P)u, \tag{29}$$

and second

$$u = Pu \in D[H_P], \quad Pf = P(I + iH_P)u \tag{30a}$$

or

$$u = Pu = (I + iH_P)^{-1}Pf. \tag{30b}$$

In fact, first from the second constituent of (28), we see that

$$(I - P(\tau))u_\tau \xrightarrow{s} 0,$$

so that we have $(I - P)u = 0$ or $u = Pu$. Applying $(I - P(\tau))$ to (26), we have

$$(I - P(\tau))f = (1 + \tau^{-1})(I - P(\tau))u_\tau.$$

and taking the inner product with u_τ,

$$\begin{aligned}\langle u_\tau, (I - P(\tau))f\rangle &= \langle (I - P(\tau))u_\tau, f\rangle \\ &= \|(I - P(\tau))u_\tau\|^2 + \|\tau^{-1/2}(I - P(\tau))u_\tau\|^2.\end{aligned}$$

By letting $\tau' \to 0_+$, we have

$$(u, (I - P)f) = (0, f) = 0 + \liminf \|u_0\|^2 \geq \|u_0\|^2,$$

whence $u_0 = 0$.

Next, for every $w \in D[H^{1/2}]$, we have, with the limit taken along $\tau' \to 0_+$,

$$\langle w, g\rangle = \lim\langle w, G(\tau)^{1/2}P(\tau)u_\tau\rangle = \lim\langle G(\tau)^{1/2}w, P(\tau)u_\tau\rangle = \langle 0, Pu\rangle,$$

$$\langle w, h\rangle = \lim\langle w, H(\tau)^{1/2}P(\tau)u_\tau\rangle = \lim\langle H(\tau)^{1/2}w, P(\tau)u_\tau\rangle = \langle H^{1/2}w, Pu\rangle,$$

because $G(\tau)^{1/2}w \xrightarrow{s} 0$ and $H(\tau)^{1/2}w \xrightarrow{s} H^{1/2}w$ as $\tau \to 0_+$. Hence $g = 0$, and $u = Pu$ belongs to $D[H^{1/2}]$ and $h = H^{1/2}Pu$ because $D[H^{1/2}]$ is dense by assumption. Applying the projection $P(\tau)$ to (26) and taking the inner product with $w \in D[H^{1/2}]$,

$$\begin{aligned}\langle w, P(\tau)f\rangle = \langle w, P(\tau)u_\tau\rangle &+ \langle G(\tau)^{1/2}P(\tau)w, G(\tau)^{1/2}P(\tau)u_\tau\rangle \\ &+ i\langle H(\tau)^{1/2}P(\tau)w, H(\tau)^{1/2}P(\tau)u_\tau\rangle.\end{aligned}$$

Taking the limit along $\tau' \to 0_+$, we get

$$\langle w, Pf\rangle = \langle w, Pu\rangle + \langle 0, 0\rangle + i\langle H^{1/2}Pw, h\rangle.$$

By density of $D[H^{1/2}]$ again, we see $h \in D[(H^{1/2}P)^*]$ and

$$Pf = Pu + i(H^{1/2}P)^*h = Pu + i(H^{1/2}P)^*(H^{1/2}P)u = Pu + iH_P u.$$

This shows our claim (29)/(30). Thus we have seen the weak limits of (28) do not depend on the (sub)sequence chosen. A standard argument concludes that (28) with (30) holds as $\tau \to 0_+$ without taking any (sub)sequence.

Finally, as the last step to conclude (25), we need to show the strong convergence of the three families $\{u_\tau\}$, $\{G(\tau)^{1/2}P(\tau)u_\tau\}$ and $\{H(\tau)^{1/2}P(\tau)u_\tau\}$. To do so, it suffices to show the norms of these vectors converge. To this end, as in Section 2, observe the real and imaginary parts of (27), however, with $P(\tau)f$ in place of f:

$$\operatorname{Re}\langle u_\tau, P(\tau)f\rangle = \|P(\tau)u_\tau\|^2 + \|G(\tau)^{1/2}P(\tau)u_\tau\|^2,$$

$$\operatorname{Im}\langle u_\tau, P(\tau)f\rangle = \|H(\tau)^{1/2}P(\tau)u_\tau\|^2.$$

By the fact of weak convergence of $\{u_\tau\}$ for $\tau \to 0_+$, as the left-hand sides converge, i.e.,

$$\operatorname{Re}\langle u_\tau, P(\tau)f\rangle \longrightarrow \langle u, Pu\rangle = \|Pu\|^2,$$

$$\operatorname{Im}\langle u_\tau, P(\tau)f\rangle \longrightarrow \langle u, H_P u\rangle = \|H^{1/2}Pu\|^2,$$

so do the right-hand sides. Therefore we have for the real part

$$\begin{aligned}\|Pu\|^2 &- \lim\,[\|P(\tau)u_\tau\|^2 + \|G(\tau)^{1/2}P(\tau)u_\tau\|^2]\\ &= \liminf\,[\|u_\tau\|^2 + \|G(\tau)^{1/2}P(\tau)u_\tau\|^2]\\ &\geq \|Pu\|^2 + \|0\|^2\\ &= \|Pu\|^2,\end{aligned}$$

and for the imaginary part

$$\begin{aligned}\|H^{1/2}Pu\|^2 &= \lim\ \|H(\tau)^{1/2}P(\tau)u_\tau\|^2\\ &= \liminf\ \|H(\tau)^{1/2}P(\tau)u_\tau\|^2\\ &\geq \|H^{1/2}Pu\|^2.\end{aligned}$$

This shows the convergence of the norm of these three vectors and as a result their strong convergence. Thus we have shown (25), completing the proof of Theorem 1.3.

Acknowledgement. Thanks are due to the referee for valuable comments.

References

[1] P. R. Chernoff, Note on product formulas for operator semigroups. *J. Functional Analysis* **2** (1968), 238–242. MR 0231238 Zbl 0157.21501

[2] P. R. Chernoff, *Product formulas, nonlinear semigroups, and addition of unbounded operators.* Memoirs of the American Mathematical Society, No. 140. American Mathematical Society, Providence, R.I., 1974. MR 0417851 Zbl 0283.47041

[3] P. Exner and T. Ichinose, A product formula related to quantum Zeno dynamics. *Ann. Henri Poincaré* **6** (2005), no. 2, 195–215. MR 2136188 Zbl 1092.81004

[4] P. Exner, T. Ichinose, H. Neidhardt, and V. A. Zagrebnov, Zeno product formula revisited. *Integral Equations Operator Theory* **57** (2007), no. 1, 67–81. MR 2294275 Zbl 1119.47016

[5] P. Exner, H. Neidhardt, and V. A. Zagrebnov, Remarks on the Trotter–Kato product formula for unitary groups. *Integral Equations Operator Theory* **69** (2011), no. 4, 451–478. MR 2784574 Zbl 1228.47042

[6] P. Facchi, S. Pascazio, A. Scardicchio and L. S. Schulman, Zeno dynamics yields ordinary constraints. *Phys. Rev. A* **65** (2002), 012108, 5 pp.

[7] R. P. Feynman, Space-time approach to non-relativistic quantum mechanics. *Rev. Modern Physics* **20** (1948), 367–387. Zbl 0026940

[8] T. Ichinose, A product formula and its application to the Schrödinger equation. *Publ. Res. Inst. Math. Sci.* **16** (1980), no. 2, 585–600. MR 0594917 Zbl 0452.47050

[9] T. Ichinose, Time-sliced approximation to path integral and Lie–Trotter–Kato product formula. In J. Arafune, A. Arai, M. Kobayashi, K. Nakamura, T. Nakamura, I. Ojima, N. Sakai, A. Tonomura and K. Watanabe (eds.), *A garden of quanta.* Essays in honor of H. Ezawa. World Scientific, River Edge, N.J., 2003, 77–93. MR 2045959 Zbl 1053.81062

[10] T. Ichinose and H. Tamura, Note on the norm convergence of the unitary Trotter product formula. *Lett. Math. Phys.* **70** (2004), no. 1, 65–81. MR 2107706 Zbl 1075.47025

[11] G. W. Johnson and M. L. Lapidus, *The Feynman integral and Feynman's operational calculus.* Oxford Mathematical Monographs. The Clarendon Press, Oxford, 2000. MR 1771173 Zbl 0952.46044

[12] T. Kato, *Perturbation theory for linear operators.* Second edition. Grundlehren der Mathematischen Wissenschaften, 132. Springer, Berlin etc., 1976. MR 0407617 Zbl 0342.47009

[13] T. Kato, Trotter's product formula for an arbitrary pair of self-adjoint contraction semigroups. In I. Gohberg and M. Kac (eds.), *Topics in functional analysis.* Essays dedicated to M. G. Kreĭn on the occasion of his 70th birthday. Advances in Mathematics, Supplementary Studies, 3. Academic Press [Harcourt Brace Jovanovich, Publishers], New York and London, 1978, 185–195. MR 0538020 Zbl 0461.47018

[14] M. L. Lapidus, Formules de moyenne et de produit pour les résolvantes imaginaires d'opérateurs auto-adjoints. *C. R. Acad. Sci. Paris Sér. A-B* **291** (1980), no. 7, A451–A454. MR 0595913 Zbl 0446.47010

[15] M. L. Lapidus, Generalization of the Trotter–Lie formula. *Integral Equations Operator Theory* **4** (1981), no. 3, 366–415. MR 0623544 Zbl 0463.47024

[16] M. L. Lapidus, Product formula for imaginary resolvents with application to a modified Feynman integral. *J. Funct. Anal.* **63** (1985), no. 3, 261–275. MR 0808263 Zbl 0601.47025

[17] B. Misra and E. C. G. Sudarshan, The Zeno's paradox in quantum theory. *J. Mathematical Phys.* **18** (1977), no. 4, 756–763. MR 0432085

[18] E. Nelson, Feynman integrals and the Schrödinger equation. *J. Mathematical Phys.* **5** (1964), 332–343. MR 0161189 Zbl 0133.22905

[19] M. Reed and B. Simon, *Methods of modern mathematical physics.* I. Functional Analysis. Revised, and enlarged edition. Academic Press [Harcourt Brace Jovanovich, Publishers], New York, 1980. ISBN 0-12-585050-6 MR 0751959 Zbl 0459.46001

[20] H. F. Trotter, On the product of semi-groups of operators. *Proc. Amer. Math. Soc.* **10** (1959), 545–551. MR 0108732 Zbl 0099.10401

Spectral asymptotics induced by approaching and diverging planar circles

Sylwia Kondej

Dedicated to Pavel Exner on the occasion of his 70th birthday

1 Introduction

This paper discusses research into the so-called Schrödinger operators with delta potentials. We study a special model: a two-dimensional quantum system with delta potential supported by two concentric circles: C_R and C_{R_d}, where $C_R := \{(x, y) \in \mathbb{R}^2 \colon (x^2 + y^2)^{1/2} = R\}$ and C_{R_d} is defined analogously for $R_d := R + d$, $d > 0$. The Hamiltonian of such a system can be symbolically written as

$$- \Delta - \beta \delta_{C_R} - \alpha \delta_{C_{R_d}}, \quad \text{where } \alpha, \beta \in \mathbb{R}, \tag{1}$$

where δ_{C_r} stands for the Dirac delta supported on C_r. To define a self-adjoint operator $H_{\alpha,\beta,d}$ coresponding formally to (1) we employ the form sum method.

The main results. We investigate the behaviour of the discrete eigenvalues for $d \to 0$ and $d \to \infty$. In fact, in both asymptotics one can observe certain "spectral memory" on a single circle system. Therefore, it is convenient to introduce a special notation $H_{\gamma,R}$ for the Hamiltonian corresponding to the formal expression

$$-\Delta - \gamma \delta_{C_R}, \quad \gamma \in \mathbb{R}.$$

In the following γ will be expressed by means of the coupling constants α and β, however, this dependence will be different in two considered cases. If $\gamma > 0$ then operator $H_{\gamma,R}$ has $2M_{\gamma,R} + 1$ eigenvalues (counting multiplicity), where $M_{\gamma,R} := \max\{m \in \mathbb{Z} \colon 2|m| < R\gamma\}$.

The first result concerns the eigenvalue asymptotics in the approaching circles system and the statement can be formulated as follows.

• Let E_m denote an eigenvalue of $H_{\alpha+\beta,R}$. Then the eigenvalues of $H_{\alpha,\beta,d}$ admit the following asymptotics

$$E_m + t_m d + o(d),$$

for $d \to 0$. The explicit form for the first correction term t_m is derived in Theorem 3.1. The analysed system enables separation of variables and, consequently, relying on the implicit function theorem we can reproduce t_m in the terms of the Bessel functions and their derivatives. In Section 3.1 we also study certain properties of t_m; for example, we show that the sign of t_m is not defined generally.

The second result is addressed to the system with circles separated by a large distance.

• Assume that $d \to \infty$. Then the system has a "tendency for the decoupling." This is manifested as the localization of eigenvalues of $H_{\alpha,\beta,d}$ near the eigenvalues of $H_{\beta,R}$ as well as H_{α,R_d}. Precisely, the eigenvalues of $H_{\alpha,\beta,d}$ behave as

$$\begin{cases} -\dfrac{\alpha^2}{4} + \dfrac{m^2 - 1/4}{d^2} + o(d^{-2}), & |m| \leq M_{\alpha,R_d}, \\ E_{m,\beta} + w_m\, \varepsilon + o(\varepsilon), & |m| \leq M_{\beta,R}, \end{cases} \tag{2}$$

where $\varepsilon := \exp(-2d\kappa_{m,\beta})$ and $E_{m,\beta}$ stand for the eigenvalues of $H_{\beta,R}$. Note that the expression in the first line of (2) reflects the asymptotics of eigenvalues of H_{α,R_d}.

The models of delta interactions supported by circles or spheres has been already studied in the various contexts and dimensions, see for example [3], [4], [6], [7], [8], [9], [12], and [17].

2 Preliminaries and the main result

Single ring: spectral properties of the system. Spectral properties of the single circle Hamiltonian will be essential for both asymptotics considered in this paper. Therefore, we start our analysis by recalling some useful known facts, cf. [9]. Consider the Hamiltonian $H_{\gamma,R}$ associated to the sesquilinear form

$$h_{\gamma,R}(f,g) = (\nabla f, \nabla g)_{L^2(\mathbb{R}^2)} - \gamma \int_{C_R} \bar{f} g \, \mathrm{d}s, \quad f, g \in W^{1,2}(\mathbb{R}^2),\ \gamma \in \mathbb{R},$$

where the functions in the second component are understood in the sense of the trace embedding $W^{1,2}(\mathbb{R}^2) \hookrightarrow L^2(C_R)$ and the arc length parameter s ranges $s \in [0, 2\pi R)$. In fact, $L^2(C_R)$ can be identified with $L^2((0, 2\pi R))$. We define $H_{\Gamma,R}$ as the operator associated to $h_{\gamma,R}$ via the first representation theorem, cf. [11]. Applying the results of [5] we conclude that $H_{\gamma,R}$ is self-adjoint and it gives a mathematical meaning to the formal expression (1).

To be specific we introduce the polar system of coordinates (r,ϕ) where[1] $r>0$, $\phi\in[0,2\pi)$. The delta potential support C_R decomposes $\mathbb{R}^2$ onto two disjoint open sets Ω^{i}, Ω^{e}; denote by $\bar{\Omega}^{\mathrm{i}}$, $\bar{\Omega}^{\mathrm{e}}$ their closures and $\mathcal{C}^1_R := C^1(\bar{\Omega}^{\mathrm{i}})\cup C^1(\bar{\Omega}^{\mathrm{e}})$. Assume that $f\in\mathcal{C}^1_R$ satisfies

$$\lim_{r\to R+} f(r,\phi) = \lim_{r\to R-} f(r,\phi) =: f_R(\phi), \tag{3a}$$

$$\lim_{r\to R+} \partial_r f(r,\phi) - \lim_{r\to R-} \partial_r f(r,\phi) = -\gamma\, f_R(\phi). \tag{3b}$$

Then the operator which acts as

$$\breve{H}_{\gamma,R} f = -\Delta f \quad \text{a.e. in } \mathbb{R}^2,$$

on the domain

$$D(\breve{H}_{\gamma,R}) = \{ f\in\mathcal{C}^1_R \cap W^{2,2}(\mathbb{R}^2\setminus C_R) \colon f \text{ satisfies (3)}\}$$

is essentially self-adjoint and its closure coincides with $H_{\gamma,R}$, cf. [5].

Since the delta potential is compactly supported the essential spectrum of $H_{\gamma,R}$ is stable under such "perturbant," i.e.,

$$\sigma_{\mathrm{ess}}(H_{\gamma,R}) = [0,\infty),$$

cf. [5].

Henceforth we will consider negative eigenvalues. In view of the rotational symmetry we postulate that the eigenfunctions of $H_{\gamma,R}$ take the form $(1/\sqrt{2\pi})\varrho_m(r)\mathrm{e}^{im\phi}$, where $m\in\mathbb{Z}$. Let $\kappa>0$. The behaviour of eigenfunctions at the infinity and origin imposes

$$\varrho_m(r) = c_1 K_m(\kappa r), \quad \text{for } r>R, \tag{4a}$$

$$\varrho_m(r) = c_2 I_m(\kappa r), \quad \text{for } r<R, \tag{4b}$$

where $K_m(\cdot)$ and $I_m(\cdot)$ denote the modified Bessel functions, cf. [1]. Using the boundary conditions (3) we get the spectral condition

$$K_m(\kappa R) I_m(\kappa R) = \frac{1}{\gamma R}, \quad m\in\mathbb{Z}, \tag{5}$$

cf. [9].

[1] In view of the above trace embedding operator it seems natural to consider s parameter instead of ϕ; however, since we going to implement the second circle it is more convenient to stay with standard polar coordinates.

It follows from (4) that κ determines the spectral parameter and the solutions of (5) reproduce negative eigenvalues E of $H_{\gamma,R}$ by means of the relation $E = -\kappa^2$.

Remarks 2.1. A. Relying on asymptotics formulae (25)–(28) and using the fact that $(K_m I_m)(\cdot)$ is monotonously decreasing we state that the equation (5) has exactly one solution for $\kappa > 0$ provided $2|m| < R\gamma$ or equivalently $|m| \le M_{\gamma,R}$ and no solution otherwise; recall that the notation $M_{\gamma,R}$ was introduced in introduction.

B. It is also useful to recall that for γR large the solution of (5) behaves as

$$-\kappa_m^2 = -\frac{\gamma^2}{4} + \frac{m^2 - 1/4}{R^2} + O(\gamma^{-2} R^{-4}).$$

Hamiltonian with the delta potential supported by two concentric rings. Let $s \in [0, 2\pi R)$ and $s_d \in [0, 2\pi R_d)$ stand for the arc length parameters associated to C_R and C_{R_d} respectively. For $\alpha, \beta \in \mathbb{R}$ let us define the sesquilinear form

$$h_{\alpha,\beta,d}(f, g) = (\nabla f, \nabla g)_{L^2(\mathbb{R}^2)} - \beta \int_{C_R} \bar{f} g \, \mathrm{d}s - \alpha \int_{C_{R_d}} \bar{f} g \, \mathrm{d}\mathrm{d}s_d, \quad f, g \in W^{1,2}(\mathbb{R}^2),$$

where we employ the trace embedding of $W^{1,2}(\mathbb{R}^2)$ to $L^2(C_r) \simeq L^2((0, 2\pi r))$, $r = R, R_d$. Similarly as for the single circle case we define the operator $H_{\alpha,\beta,d}$ associated to $h_{\alpha,\beta,d}$ via the first representation theorem.

Analogously for the single circle we can characterize $H_{\alpha,\beta,d}$ by means of boundary conditions. Note that circles C_R and C_{R_d} decompose $\mathbb{R}^2$ onto three open sets Ω_1^{i}, Ω_2^{i} and Ω^{e}. Denote $\mathcal{C}^1_{R,R_d} := C^1(\bar{\Omega}_1^{\mathrm{i}}) \cup C^1(\bar{\Omega}_2^{\mathrm{i}}) \cup C^1(\bar{\Omega}^{\mathrm{e}})$ and assume that $f \in \mathcal{C}^1_{R,R_d}$ satisfies

$$\lim_{r \to R+} f(r, \phi) = \lim_{r \to R^-} f(r, \phi) =: f_R(\phi), \tag{6a}$$

$$\lim_{r \to R+} \partial_r f(r, \phi) - \lim_{r \to R^-} \partial_r f(r, \phi) = -\beta f_R(\phi), \tag{6b}$$

$$\lim_{r \to R_d^+} f(r, \phi) = \lim_{r \to R_d^-} f(r, \phi) =: f_{R_d}(\phi), \tag{6c}$$

$$\lim_{r \to R_d^+} \partial_r f(r, \phi) - \lim_{r \to R_d^-} \partial_r f(r, \phi) = -\alpha f_{R_d}(\phi). \tag{6d}$$

In fact, $H_{\alpha,\beta,d}$ stands for the closure

$$\check{H}_{\alpha,\beta,d} f = -\Delta f \quad \text{a.e. in } \mathbb{R}^2,$$

$$D(\check{H}_{\alpha,\beta,d}) = \{f \in \mathcal{C}^1_{R,R_d} \cap W^{2,2}(\mathbb{R}^2 \setminus (C_R \cup C_{R_d})) \colon f \text{ satisfies } (6)\}.$$

2.1 Spectral equation for the double ring system

To derive the spectral equation for the double ring system we proceed analogously as in the previous case. The system again admits separation variables. Consequently, the eigenfunctions of $H_{\alpha,\beta,d}$ can be written as $1/\sqrt{2\pi}\,\rho_m(r)\mathrm{e}^{im\phi}$, where $m \in \mathbb{Z}$ and

$$\rho_m(r) = C_1 K_m(\kappa r), \qquad \text{for } r > R_d,$$

$$\rho_m(r) = C_2 K_m(\kappa r) + C_3 I_m(\kappa r), \quad \text{for } R < r < R_d,$$

and

$$\rho_m(r) = C_4 I_m(\kappa r), \quad \text{for } r < R.$$

Inserting the above formulae to (6) we obtain four equations:

$$C_1 K_m(\kappa R_d) - C_2 K_m(\kappa R_d) - C_3 I_m(\kappa R_d) = 0,$$

$$C_1(\kappa K'_m(\kappa R_d) + \alpha K_m(\kappa R_d)) - C_2\kappa K'_m(\kappa R_d) - C_3\kappa I'_m(\kappa R_d) = 0,$$

$$C_2 K_m(\kappa R) + C_3 I_m(\kappa R) - C_4 I_m(\kappa R) = 0,$$

$$C_2\kappa K'_m(\kappa R) + C_3\kappa I'_m(\kappa R) + \ C_2(\beta I_m(\kappa R) - \kappa I'_m(\kappa R)) = 0.$$

Spectral equation. The above system of equations admits a solution if and only if the determinant of the corresponding matrix vanishes. This condition can be written by means of the equation

$$\eta_m(\kappa, d) = 0, \quad m \in \mathbb{Z},\ \kappa > 0,\ d \geq 0, \tag{7}$$

where

$$\eta_m(\kappa, d) = \nu_m(\kappa, d) - \xi_{m,\alpha}(\kappa)\xi_{m,\beta}(\kappa), \tag{8}$$

with

$$\xi_{m,\alpha}(\kappa) \equiv \xi_{m,\alpha,d}(\kappa) := \alpha R_d (K_m I_m)(\kappa R_d) - 1, \tag{9a}$$

$$\xi_{m,\beta}(\kappa) := \beta R (K_m I_m)(\kappa R) - 1, \tag{9b}$$

and

$$\nu_m(\kappa, d) := \alpha\beta R_d R\, K_m^2(\kappa R_d) I_m^2(\kappa R). \tag{10}$$

Formulae (7) constitute the spectral equations for $H_{\alpha,\beta,d}$.

Remark 2.2. Note that the functions $\xi_{m,\tau}$, where $\tau = \alpha, \beta$ are related to the single circle systems. More precisely, the relations

$$\xi_{m,\tau}(\kappa) = 0,$$

determine spectral equations of H_{α,R_d} and $H_{\beta,R}$.

3 Approaching rings

In this section we consider the eigenvalue asymptotics for $d \to 0$ and $\alpha + \beta > 0$. Note that for $d = 0$ equation (7) reads

$$K_m(\kappa R) I_m(\kappa R) = \frac{1}{(\alpha+\beta)R}; \tag{11}$$

the latter corresponds to the single ring Hamiltonian $H_{\gamma,R}$ with coupling constant $\gamma = \alpha+\beta$, cf. (5). As follows from the previous discussion, see Remark 2.1, equation (11) has exactly one solution κ_m provided $|m| \le M_{\alpha+\beta,R}$.

The following theorem provides the spectral asymptotics for approaching rings.

Theorem 3.1. *Assume $\gamma = \alpha + \beta > 0$. Let E_m, where $|m| \le M_{\gamma,R}$, stand for an eigenvalue of $H_{\gamma,R}$. Then the eigenvalues of $H_{\alpha,\beta,d}$ admit the following asymptotics for $d \to 0$:*

$$E_m(d) = E_m + t_m\, d + o(d),$$

where t_m is given by

$$t_m := \frac{2\kappa_m I_m K_m \left(-\alpha\beta R I_m K_m + \alpha\kappa_m R (I_m K_m)' + \alpha I_m K_m\right)}{R(I_m K_m)'}; \tag{12}$$

moreover, functions $K_m(\cdot)$ and $I_m(\cdot)$ as well as their derivatives contributing to (12) *are defined for the value $R\kappa_m$.*

Proof. Suppose $|m| \le M_{\gamma,R}$. Eigenvalues of $H_{\alpha,\beta,d}$ are determined by the solutions of (7). Note that

$$\eta_m(\kappa_m, 0) = 0.$$

Using the regularity of K_m and I_m we state for $d \in \mathbb{R}$ and $\kappa > 0$ the functions $\partial\eta_m/\partial\kappa$ and $\partial\eta_m/\partial d$ are C^∞. Furthermore, using (11) we get

$$\frac{\partial\eta_m}{\partial\kappa} = (\alpha+\beta)R^2 (I_m K_m)' = R\frac{(I_m K_m)'}{(I_m K_m)},$$

where the derivative at the left-hand side is defined at $(\kappa_m, 0)$. Moreover, $Z_m = Z_m(R\kappa_m)$, $Z_m = K_m, I_m$ and the analogous notation is applied for the derivatives contributing to the right-hand side of the above expression. Since the function $(I_m K_m)(\cdot)$ is monotonously decreasing we have $\partial\eta_m/\partial\kappa < 0$. Consequently, we can employ the implicit function theorem which states that there exists a neighbourhood $U \in \mathbb{R}$ of 0 and the unique function $U \ni d \mapsto \kappa_m(d) \in \mathbb{R}$ such that $\eta_m(\kappa_m(d), d) = 0$ and

$$\kappa_m(d) = \kappa_m - \left(\frac{\partial\eta_m}{\partial d}\right)\left(\frac{\partial\eta_m}{\partial\kappa}\right)^{-1} d + o(d), \tag{13}$$

where all derivatives in the second component are determined for $d = 0$, $\kappa = \kappa_m$.

Using (8) and the Wroskian equation

$$(I'_m K_m)(z) - (K'_m I_m)(z) = \frac{1}{z} \tag{14}$$

we get by a straightforward calculation

$$\frac{\partial \eta_m}{\partial d} = -\alpha\beta R(K_m I_m) + \alpha \kappa_m R(I_m K_m)' + \alpha(I_m K_m).$$

Combining the above derivatives together with (13) we arrive at

$$\begin{aligned} E_m(d) &= -\kappa_m(d)^2 \\ &= -\kappa_m^2 + 2\kappa_m \frac{\partial \eta_m}{\partial d}\Big(\frac{\partial \eta_m}{\partial \kappa}\Big)^{-1} d + o(d) \\ &= E_m + t_m d + o(d), \end{aligned}$$

with t_m given by (12). □

3.1 Discussion on the first order correction

In this section we discuss some properties of the first order correction for converging rings.

The first order correction in the terms of unperturbed eigenfunctions. The following analysis will be conducted for $m = 0$. Let f_0 stand for the ground state of $H_{\alpha+\beta,R}$. Using (3) and (4) we conclude that $c_1 = (I_0/K_0)c_2$; recall $K_0 = K_0(R\kappa_0)$, $I_0 = I_0(R\kappa_0)$ and the analogous notation is employed for derivatives. Applying the relation

$$\int x Z_0^2(x)\,\mathrm{d}x = \frac{x^2}{2}(Z_0^2(x) - (Z'_0(x))^2, \quad Z_0 = I_0, K_0,$$

one can show that the norm of eigenfunction f_0 is given by

$$\|f_0\|^2_{L^2(\mathbb{R}^2)} = |c_2|^2 \frac{R^2}{2K_0^2}((IK')^2 - (KI')^2).$$

Using again the Wronskian equation (14) one gets

$$\|f_0\|^2_{L^2(\mathbb{R}^2)} = -|c_2|^2 \frac{R}{2\kappa_0 K_0^2}(I_0 K_0)'.$$

Applying the above formula together with boundary conditions (3) and comparing this with (12) we obtain

$$t_0 = \frac{-\alpha\Big(\int_{C_R} \partial_r^+ |f_0|^2 \,\mathrm{d}s + \alpha \int_{C_R} |f_0|^2 \,\mathrm{d}s\Big) - \frac{\alpha}{R}\int_{C_R} |f_0|^2 \,\mathrm{d}s}{\|f_0\|^2_{L^2(\mathbb{R}^2)}}, \tag{15}$$

where we abbreviate $\partial_r^+ f(r,\phi) = \lim_{r\to R^+} \partial_r f(r,\phi)$; recall that the equation $s = R\phi$ states the relation between $\phi \in [0, 2\pi)$ and $s \in [0, 2\pi R)$.

Let us mention that the above formula describes a very particular case of the class considered in the forthcoming paper [13]. In this paper the spectral asymptotics for approaching hypersurfaces in $\mathbb{R}^d$ is analyzed. The method developed in [13] allows to reconstruct asymptotics of eigenvalues by means of the "unperturbed" eigenfunctions. The technics enables generalization for complex coupling constants.

The last component of (15) reflects contribution of the curvature to the first correction term. More general situation shows a presence of the first mean curvature in eigenvalue asymptotics, cf. [13]. Furthermore, let us note that a contribution of the first mean curvature in spectral asymptotics has been recently shown in related problems, see [14], [15], and [16].

The second component of (15) is a consequence of singular character of delta potential. Suppose f_0 and f_d denote the normalized ground states of $H_{\alpha+\beta,R}$ and $H_{\alpha,\beta,d}$, respectively. The second component of (15) comes directly from the fact that $\partial_r(f_0 - f_d)(r,\phi)$ do not tend to 0 if $d \to 0$ and $r \in (R, R_d)$.

The sign of t_m. For the one-dimensional system with two converging points of interaction the first order correction is always *positive*, cf. [2]. This means that the splitting of the singular potential from one point to two points leads to pushing up the eigenvalue. The situation is slightly different in the case of converging circles. As formula (12) shows the sign of t_0 depends on

$$\varsigma := \alpha(1-\beta R)I_0K_0 + \alpha\kappa_0 R(I_0K_0)',$$

i.e., $\operatorname{sign}\varsigma = -\operatorname{sign} t_0$.

• First, let us consider the situation when $R \to 0$; then $\kappa_0 \to 0$ as well, cf. [9]. Employing asymptotics formulae for Z_0, where $Z_0 = I_0, K_0$, see (27) and (28) together with (29)-(31) one gets

$$\varsigma \sim -\alpha(1-\beta R)\ln(\kappa_0 R),$$

which implies $\varsigma > 0$ for R small enough and, consequently, leads to $t_0 < 0$.

• Now we assume that $R \to \infty$. Then $\kappa_0 \sim \alpha/2$ and using again formulae (27) and (28) we obtain

$$\varsigma = -\alpha\beta R\Big(\frac{1}{2\kappa_0 R} + O((\kappa_0 R)^{-3})\Big).$$

This shows that for R large enough we have $\varsigma < 0$ and $t_0 > 0$.

The above discussion establishes that the sign of the first order correction term is generally undefined.

4 Diverging rings

In this section we consider the asymptotics for circles separated by a large distance, i.e., for $d \to \infty$.

Following the convention introduced in the previous discussion we denote by H_{α,R_d} and $H_{\beta,R}$ the corresponding single circle Hamiltonians. Operator H_{α,R_d} has $2M_{\alpha,R_d} + 1$ (counting multiplicities) eigenvalues $\{E_{m,\alpha}\}_{|m|\le M_{\alpha,R_d}}$ and $H_{\beta,R}$ has $2M_{\beta,R} + 1$ eigenvalues $\{E_{m,\beta}\}_{|m|\le M_{\beta,R}}$. Suppose $\tau = \alpha, \beta$. In fact, $E_{m,\tau}$ can be recovered from the spectral equations, i.e., $E_{m,\tau} = -\kappa^2_{m,\tau}$ where $\kappa_{m,\tau}$ stand for the solutions of

$$\xi_{m,\tau}(\kappa) = 0, \qquad \text{for } \kappa > 0.$$

Recall that $\xi_{m,\tau}$ are defined by (9). Moreover, using the statement of Remark 2.1 we conclude that

$$E_{m,\alpha} = -\frac{\alpha^2}{4} + \frac{m^2 - 1/4}{R_d^2} + O(d^{-4}). \tag{16}$$

Theorem 4.1. *Assume that α and β are positive and $E_{m,\beta} \ne -\alpha^2/4$ for all $|m| \le M_{\beta,R}$. Then the eigenvalues of $H_{\alpha,\beta,d}$ admit the following asymptotics for $d \to \infty$:*

$$\epsilon_d = \begin{cases} -\dfrac{\alpha^2}{4} + \dfrac{m^2 - 1/4}{d^2} + o(d^{-2}), & |m| \le M_{\alpha,R_d}, \\ E_{m,\beta} + w_m\,\varepsilon + o(\varepsilon), & |m| \le M_{\beta,R}, \end{cases} \tag{17}$$

where

$$\varepsilon := \exp(-2d\kappa_{m,\beta}), \qquad w_m := \frac{\pi\alpha\beta R e^{-2\kappa_{m,\beta}R} I_m(R\kappa_{m,\beta})^2}{\Big(1 - \dfrac{\alpha}{2\kappa_{m,\beta}}\Big)\xi'_{m,\beta}(\kappa_{m,\beta})}. \tag{18}$$

Remark 4.2. In fact, ϵ_d reflects the asymptotics of $2(M_\alpha + M_\beta) + 2$ eigenvalues of H_d. However, since ϵ_d converge to $E_{m,\alpha}$ and $E_{m,\beta}$ we leave the labelling inherited from the discrete eigenvalues of the single circle Hamiltonians.

Proof. The analysis is based on investigating spectral equation (7) which reads

$$\eta_m(\kappa, d) = \nu_m(\kappa, d) - \xi_m(\kappa, d) = 0, \tag{19}$$

where

$$\xi_m(\kappa, d) := \xi_{m,\alpha,d}(\kappa)\xi_{m,\beta}(\kappa).$$

First, assume that $|m| \leq M_{\beta,R}$. Then for d large enough we have $|m| \leq M_{\alpha,R_d}$. Combining equations (9) and (10) with the formulae (25) and (26) we get the following asymptotics for $\kappa \to \infty$ and any $m \in \mathbb{Z}$:

$$\begin{cases} \xi_m(\kappa, d) = 1 - \dfrac{\alpha + \beta}{2\kappa} + O(\kappa^{-2}), \\ \nu_m(\kappa, d) = \dfrac{\alpha\beta}{4\kappa^2} \mathrm{e}^{-2d\kappa}(1 + o_\kappa(1)); \end{cases} \tag{20}$$

the error terms in the above expressions are uniform with respect to $d > C$ where C is a positive number [2]. The symbol $o_\kappa(1)$ donotes that the asymptotics understood with respect to κ. On the other hand, for $\kappa \to 0$ we have

$$\xi_m(\kappa, d) = \begin{cases} \Big(\dfrac{\alpha R_d}{2m} - 1\Big)\Big(\dfrac{\beta R}{2m} - 1\Big)(1 + o_\kappa(1)), & m \neq 0, \\ \alpha\beta \log(\kappa R) \log(\kappa R_d) R R_d (1 + o_\kappa(1)), & m = 0, \end{cases} \tag{21}$$

and

$$\nu_m(\kappa, d) = \begin{cases} \dfrac{\alpha}{4m^2} \dfrac{R^{2m+1}}{R_d^{2m+1}}(1 + o_\kappa(1)), & m \neq 0, \\ \alpha\beta \log^2(\kappa R_d) R R_d (1 + o_\kappa(1)), & m = 0, \end{cases} \tag{22}$$

where all error terms are uniform with respect to $d > C$. We have $\nu_m(\kappa, d) > 0$. Moreover, $\nu_m(\kappa, d) \to 0$ as $d \to \infty$ and the limit is uniform with respect to $\kappa > C$. It follows from (21) and (22) that if $m \neq 0$ then $\xi_m(0, d) > \nu_m(0, d)$ for d large enough. If $m = 0$ then the corresponding limits for $\kappa \to 0$ do not exist, however, $\xi_m(\kappa, d) > \nu_m(\kappa, d)$ holds for κ from a neighbourhood of 0 and d large enough. The function $\xi_m(\kappa, d)$ has two roots: $\kappa_{m,\alpha}$ and $\kappa_{m,\beta}$. Moreover $\xi_m(\kappa, d) \to 1$ for $\kappa \to \infty$ and the limit is uniform with respect to $d > C$. It follows from the discussed

[2] Note that in this proof C denotes a positive constant which can change from line to line

properties of ξ_m and ν_m and their continuity that equation (19) has at least two solutions. Furthermore, for $d \to \infty$ we have $\xi_m(\kappa, d) = (\alpha/(2\kappa) - 1 + o_d(1))\xi_{m,\alpha}(\kappa)$ and the error is uniform with respect to $\kappa > C$. This implies that (19) has exactly two solutions for $\kappa > C$ which we denote as $\kappa_{m,\alpha}(d)$ and $\kappa_{m,\beta}(d)$ and they approach to $\kappa_{m,\alpha}$ and $\kappa_{m,\beta}$ as $d \to \infty$. Note that, since C can be chosen arbitrary small we can conclude, in view of the behaviour of ν_m and ξ_m in a neighbourhood of 0, that (19) has exactly two solutions for $\kappa \in (0, \infty)$. Let us consider $\kappa_{m,\beta}(d)$. We have

$$\kappa_{m,\beta}(d) = \kappa_{m,\beta} + \delta_{m,\beta}(d),$$

where $\delta_{m,\beta}(d)$ converges to 0 as $d \to \infty$. Inserting $\kappa_{m,\beta}(d)$ to $\eta(\kappa, d)$ one gets

$$\eta(\kappa_{m,\beta}(d), d) = \nu_m(\kappa_{m,\beta}(d), d) - \xi_{m,\alpha}(\kappa_{m,\beta})\xi'_{m,\beta}(\kappa_{m,\beta})\delta_{m,\beta} + o(\delta_{m,\beta}). \quad (23)$$

The equation

$$\eta(\kappa_{m,\beta}(d), d) = 0 \quad (24)$$

and the behaviour of $\nu_m(\kappa_{m,\beta}(d), d)$ imposes $d\delta_{m,\beta}(d) \to 0$ as $d \to \infty$. Therefore, we get

$$\nu_m(\kappa_{m,\beta}(d), d) = \frac{\pi\alpha\beta R}{2\kappa_{m,\beta}} \mathrm{e}^{-2\kappa_{m,\beta}R} I_m(R\kappa_{m,\beta})^2 \varepsilon + o(\varepsilon).$$

Implementing the above expression and (23) to (24) and comparing appropriated terms leads to (18). To derive the second eigenvalue we employ (16) together with the asymptotics of $\xi_{m,\alpha}$ which depends on d as well. The analogous analysis as above establishes the asympotics of eigenvalues localized near $-\alpha^2/4$, see the first line of (17).

If $|m| > M_{\beta,R}$ and $|m| \le M_{\alpha,R_d}$ then $\xi_m(\kappa, d) = 0$ has only one solution $\kappa_{m,\alpha}$. Then repeating the above steps one shows the existence of one solution of the spectral equation; this solution admits the asymptotics specified in the first line of (17). □

The result of the above theorem corresponds to the phenomena known for regular potentials. It was shown in [10] that the introducing a second well to the single well system leads to the splitting of original eigenvalues and the corresponding spectral gaps can be expressed by the current. Consequently, the asymptotics of the gaps, if the wells are separated by a large distance, is determined by the exponential decay of eigenvectors. Theorem 4.1 shows that the introducing interaction supported by circle C_{R_d} to the system governed by $H_{\beta,R}$ leads to the shifting of original energies $E_{m,\beta}$ and this spectral shifting is determined by exponential decay of corresponding eigenfunctions.

One the other hand, the system governed by H_{α,R_d} admits eigenvalues $E_{m,\alpha}$ which depend on d, see (16). Formula (17) shows that the leading terms of this eigenvalues asymptotics are preserved if we introduce also interaction supported by C_R.

5 Appendix

We complete here the asymptotics of functions contributing to the spectral equations, see [1]. Namely, for $z \to \infty$ and $m \in \mathbb{Z}$ we have

$$I_m(z) = \frac{\mathrm{e}^z}{\sqrt{2\pi}}\Big(1 - \frac{4m^2-1}{8z} + O(z^{-2})\Big), \tag{25}$$

and

$$K_m(z) = \sqrt{\frac{\pi}{2z}}\,\mathrm{e}^{-z}\Big(1 + \frac{4m^2-1}{8z} + O(z^{-2})\Big). \tag{26}$$

Furthermore, for $z \to 0$ we have

$$I_m(z) \sim \frac{1}{\Gamma(m+1)}\Big(\frac{z}{2}\Big)^m, \quad m \in \mathbb{Z} \tag{27}$$

and

$$\begin{cases} K_m(z) \sim \dfrac{\Gamma(m)}{2}\Big(\dfrac{2}{z}\Big)^m, & m \neq 0, \\ K_0(z) \sim -\ln\Big(\dfrac{z}{2}\Big). \end{cases} \tag{28}$$

Recall $\Gamma(m) = (m-1)!$.

For the asymptotics of derivatives the following formulae will be useful:

$$I'_m(z) = \frac{I_{m-1}(z) + I_{m+1}(z)}{2} \tag{29}$$

and

$$K'_m(z) = -\frac{K_{m-1}(z) + K_{m+1}(z)}{2}. \tag{30}$$

Furthermore, since $Z_m = Z_{-m}$, where $Z_m = I_m, K_m$ we, for example, have

$$I'_0(z) = I_1(z), \qquad K'_0(z) = -K_1(z). \tag{31}$$

Acknowledgements. The author is grateful to the referee for reading carefully the first version of the paper and for the useful remarks and suggestions of changes.

The work was supported by the project No. DEC-2013/11/B/ST1/03067 of the Polish National Science Centre.

References

[1] M. Abramovitz and I. A. Stegun (ed.), *Handbook of mathematical functions with formulas, graphs, and mathematical tables.* Reprint of the 1972 edition. Dover Publications, New York, 1992. ISBN 0-486-61272-4 MR 1225604 MR MR0167642 (original) Zbl 0643.33001 Zbl 0171.38503 (original)

[2] S. Albeverio, F. Gesztesy, R. Høegh-Krohn, and H. Holden, *Solvable models in quantum mechanics.* Second edition. AMS Chelsea Publishing, Providence, R.I., 2005. With an Appendix by P. Exner. ISBN 0-8218-3624-2/hbk MR 2105735 Zbl 1078.81003

[3] J. P. Antoine, F. Gesztesy, and J. Shabani, Exactly solvable models of sphere interactions in quantum mechanics. *J. Phys. A* **20** (1987), no. 12, 3687–3712. MR 0913638

[4] J. F. Brasche and A. Teta, Spectral analysis and scattering theory for Schrödinger operators with an interaction supported by a regular curve. In S. Albeverio, J. E. Fenstad, H. Holden, and T. Lindstrøm (eds.), *Ideas and methods in quantum and statistical physics.* In memory of R. Høegh-Krohn (1938–1988). Vol. 2. Papers from the Symposium on Ideas and Methods in Mathematics and Physics held at the University of Oslo, Oslo, September 1988. Cambridge University Press, Cambridge, 1992, 197–211. MR 1190526 Zbl 0787.35057

[5] J. F. Brasche, P. Exner, Yu. A. Kuperin, and P. Šeba, Schrödinger operators with singular interactions. *J. Math. Anal. Appl.* **184** (1994), no. 1, 112–139. MR 1275948 Zbl 0820.47005

[6] P. Exner, An isoperimetric problem for leaky loops and related meanchord inequalities. *J. Math. Phys.* **46** (2005), no. 6, 062105, 10 pp.

[7] P. Exner and F. Fraas, On the dense point and absolutely continuous spectrum for Hamiltonians with concentric δ shells. *Lett. Math. Phys.* **82** (2007), no. 1, 25–37. MR 2367872 Zbl 1136.35019

[8] P. Exner and M. Fraas, Interlaced dense point and absolutely continuous spectra for Hamiltonians with concentric-shell singular interactions. In I. Beltita, Gh. Nenciu and R. Purice (eds.), *Mathematical results in quantum mechanics.* Selected papers from the Quantum Mathematics International Conference (QMATH 10) held in Moieciu, September 10–15, 2007. World Scientific, Hackensack, N.J., 2008, 48–65. MR 2466678 Zbl 1156.81383

[9] P. Exner and M. Tater, Spectra of soft ring graphs. *Waves Random Media* **14** (2004), no. 1, S47–S60. Special section on quantum graphs. MR 2042544 Zbl 1063.81534

[10] E. M. Harrel, Double wells. *Comm. Math. Phys.* **75** (1980), no. 3, 239–261. MR 0581948 Zbl 0445.35036

[11] T. Kato, *Perturbation theory for linear operators.* Reprint of the 1980 edition. Classics in Mathematics, Springer, Berlin, 1995. ISBN 3-540-58661-X MR 1335452 Zbl 0836.47009

[12] S. Kondej, Resonances induced by broken symmetry in a system with a singular potential. *Ann. Henri Poincaré* **13** (2012), no. 6, 1451–1467. MR 2966468 Zbl 1252.81076

[13] S. Kondej and D. Krejčiřík, Asymptotic spectral analysis in colliding leaky quantum layers. Preprint 2016. arXiv:1606.07273 [math-ph]

[14] D. Krejčiřík, Spectrum of the Laplacian in a narrow curved strip with combined Dirichlet and Neumann boundary conditions. *ESAIM Control Optim. Calc. Var.* **15** (2009), no. 3, 555–568. MR 2542572 Zbl 1173.35618

[15] D. Krejčiřík, Spectrum of the Laplacian in narrow tubular neighbourhoods of hypersurfaces with combined Dirichlet and Neumann boundary conditions. *Math. Bohem.* **139** (2014), no. 2, 185–193. MR 3238833 Zbl 1340.35239

[16] K. Pankrashkin and N. Popoff, An effective Hamiltonian for the eigenvalue asymptotics of a Robin Laplacian with a large parameter. *J. Math. Pures Appl.* **106**, no. 4, 615–650. Zbl 1345.35068

[17] J. Shabani, Finitely many δ interactions with supports on concentric spheres. *J. Math. Phys.* **48** (2002), no. 3, 6064–6084. MR 0931469 Zbl 0658.47026

Spectral estimates for the Heisenberg Laplacian on cylinders

Hynek Kovařík, Bartosch Ruszkowski, and Timo Weidl

Dedicated to Pavel Exner on the occasion of his 70th birthday

1 Introduction

Let $\Omega \subset \mathbb{R}^3$ be an open bounded domain. We consider the Heisenberg Laplacian on Ω with Dirichlet boundary condition, formally given by

$$\mathrm{A}(\Omega) := -X_1^2 - X_2^2 \quad \text{with } X_1 := \partial_{x_1} + \frac{x_2}{2}\partial_{x_3},\ X_2 := \partial_{x_2} - \frac{x_1}{2}\partial_{x_3}.$$

This operator is associated with the closure of the quadratic form

$$\mathrm{a}[u] := \int_\Omega |X_1 u(x)|^2 + |X_2 u(x)|^2 \,\mathrm{d}x, \tag{1}$$

initially defined on $C_0^\infty(\Omega)$. The study of the Heisenberg Laplacian, also called Kohn Laplacian, appears in different fields of mathematics like for example quantum mechanics, harmonic analysis, representation theory, control theory and sub-Riemannian geometry.

In the literature this operator appears with different multiplication factor in front of ∂_{x_3}. In general we do not distinguish between these operators since they can be transformed into each other by a simple substitution depending on the x_3-variable. In this work we use the Heisenberg Laplacian with normalized commutation relation, which is from mathematical point of view more comfortable.

It is known, see, e.g., [6], [7], and [9], that $\mathrm{A}(\Omega)$ has purely discrete spectrum. We denote by $(\lambda_k(\Omega))_{k\in\mathbb{N}}$ the non-decreasing unbounded sequence of the eigenvalues of $\mathrm{A}(\Omega)$, where we repeat entries according to their finite multiplicities. We are interested in uniform upper bounds on the quantity

$$\mathrm{Tr}\,(\mathrm{A}(\Omega) - \lambda)_- = \sum_{k=1}^{\infty} (\lambda_k(\Omega) - \lambda)_-.$$

In [7] Hansson and Laptev proved the following Berezin-type inequality for $\mathrm{A}(\Omega)$:

$$\operatorname{Tr}(\mathrm{A}(\Omega)-\lambda)_- \leq \frac{|\Omega|}{96}\lambda^3 \quad \text{for all } \lambda > 0. \tag{2}$$

It is also shown in [7] that

$$\sum_{k=1}^{\infty}(\lambda-\lambda_k(\Omega))_+ = \frac{|\Omega|}{96}\lambda^3 + o(\lambda^3) \qquad \text{as } \lambda \to +\infty, \tag{3}$$

which implies that the constant $\frac{1}{96}$ on the right-hand side of (2) is sharp.

Nevertheless, the authors of the present paper proved in [9] that inequality (2) can be improved in the following sense; for a any bounded domain $\Omega \subset \mathbb{R}^3$, there exists a constant $C(\Omega) > 0$ such that for any $\lambda \geq 0$ it holds that

$$\operatorname{Tr}(\mathrm{A}(\Omega)-\lambda)_- \leq \max\Big\{0, \frac{|\Omega|}{96}\lambda^3 - C(\Omega)\,\lambda^2\Big\}. \tag{4}$$

In other words, a negative remainder term of a lower order can be added to the right-hand side of (2) without violating the inequality.

In this paper we will prove that the order of the remainder term in (4) can be further improved if we consider cylindrical domains of the type $\Omega = \omega \times (a,b)$, where $\omega \subset \mathbb{R}^2$ is open and bounded, and $a, b \in \mathbb{R}$ are such that $a < b$. In particular for cylinders with convex cross-section ω our main result, Theorem 2.3, implies that

$$\operatorname{Tr}(\mathrm{A}(\Omega)-\lambda)_- \leq \max\Big\{0, \frac{|\Omega|}{96}\lambda^3 - \frac{\lambda^{2+1/4}}{2^7\cdot 3^{5/2}}\,\frac{|\Omega|}{\mathrm{R}(\omega)^{3/2}}\Big\}, \tag{5}$$

where $\mathrm{R}(\omega)$ is the in-radius of ω, see Corollary 2.7. The proof of (5) is based on the unitary equivalence of $\mathrm{A}(\Omega)$ to the two-dimensional Laplacian with constant magnetic field. To estimate the remainder term we use a boundary estimate for the magnetic Laplacian based on an application of a Hardy inequality in the spirit of [5], see Proposition 3.1.

2 Notation and main results

As for the cross-section ω, throughout the paper we will suppose that the following condition is satisfied.

Assumption 2.1. $\omega \subset \mathbb{R}^2$ is open bounded and simply connected.

In the sequel we will decompose the vector $x = (x', x_3) \in \mathbb{R}^3$. Let us denote by

$$\delta(x') := \operatorname{dist}(x', \partial\omega),$$

the distance function between a given $x' \in \omega$ and $\partial\omega$. The in-radius of ω is then given by

$$\mathrm{R}(\omega) := \sup_{x' \in \omega} \delta(x').$$

Hardy inequality. Let $c = c(\omega)$ be defined by

$$c^{-2} := \inf_{u \in C_0^\infty(\omega)} \frac{\int_\omega |\nabla_{x'} u(x')|^2 \, \mathrm{d}x'}{\int_\omega \left| \frac{u(x')}{\delta(x')} \right|^2 \mathrm{d}x'}, \tag{6}$$

where $\nabla'_x := (\partial_{x_1}, \partial_{x_2})$. Clearly, c is the best constant in Hardy's inequality

$$\int_\omega \frac{u(x')^2}{\delta(x')^2} \, \mathrm{d}x' \le c^2 \int_\omega |\nabla_x u(x')|^2 \, \mathrm{d}x', \quad u \in C_0^\infty(\omega). \tag{7}$$

Remark 2.2. Under assumption 2.1 it follows from [1] that

$$2 \le c \le 4.$$

The best possible value of c is $c = 2$. For a survey on Hardy inequalities we refer to [14] and [3].

To continue we define, for any $\beta > 0$, the set ω^β by

$$\omega^\beta := \{x' \in \omega \mid \delta(x') < \beta\},$$

and introduce the quantity

$$l(\omega) := (b - a) \inf_{0 < \beta \le \mathrm{R}(\omega)} \frac{|\omega^\beta|}{\beta}.$$

Now we can state the main result of this paper.

Theorem 2.3. *Let $\Omega := \omega \times (a, b)$ and let c is given by* (6). *Then*

$$\operatorname{Tr}(\mathrm{A}(\Omega) - \lambda)_- \le \max \left\{ 0, \frac{|\Omega|}{96} \lambda^3 - \Lambda \right\}, \tag{8}$$

where

$$\Lambda := \lambda^{(2c+5)/(c+2)} \frac{(1 + 2/c)}{96} \, l(\omega)^{(2c+2)/(c+2)} \, |\Omega|^{-c/(c+2)} \, (4c + 4)^{-(2c+2)/(c+2)},$$

holds for all $\lambda \ge 0$.

Remark 2.4. Note that the order of the remainder term is larger than λ^2 whenever c is finite. So far the order of the second term in the asymptotic expansion (3) is not known.

Remark 2.5. For analogous improvements of the classical spectral estimates obtained in [2, 11] for Dirichlet Laplacian on bounded domains we refer to [13], [10], [15], [16], and references therein.

Remark 2.6. Following [9] it can be shown that $l(\omega)$ is strictly positive. In particular, it holds $l(\omega) \geq (b-a)\mathrm{R}(\omega)\pi$.

Corollary 2.7. *Let $\Omega := \omega \times (a, b)$. If ω is convex, then*

$$\operatorname{Tr}(\mathrm{A}(\Omega) - \lambda)_{-} \leq \max\left\{0, \frac{1}{96}|\Omega|\lambda^3 - \lambda^{2+1/4}\frac{1}{2^7 \cdot 3^2\sqrt{3}}\frac{|\Omega|}{\mathrm{R}(\omega)^{3/2}}\right\}$$

holds for all $\lambda \geq 0$.

Proof. In case that ω is convex we have $c = 2$ in (6), see, e.g., [3]. In addition, $|\omega^{\beta}|/\beta$ is a decreasing function of β on $(0, \mathrm{R}(\omega)]$, see Lemma 4.2 in [8]. Hence we compute

$$l(\omega) = \frac{|\Omega|}{\mathrm{R}(\omega)}$$

and simplify the constant in Theorem 2.3. □

3 Preliminaries

3.1 Magnetic Dirichlet Laplacian

Let $\mathrm{P}_{\mathrm{k,B}}$ be the orthogonal projection onto the k-th Landau level $\mathrm{B}(2\mathrm{k}-1)$ of the Landau Hamiltonian with constant magnetic field for $\mathrm{B} > 0$ in $L^2(\mathbb{R}^2)$ and $\mathrm{k} \in \mathbb{N}$. Denote by $\mathrm{P}_{\mathrm{k,B}}(x, y)$ the integral kernel of $\mathrm{P}_{\mathrm{k,B}}$. We will need these well-known characteristics

$$\mathrm{P}_{\mathrm{k,B}}(y,y) = \frac{1}{2\pi}\mathrm{B}, \quad \text{where } y \in \mathbb{R}^2, \tag{9a}$$

$$\begin{aligned}\int_{\mathbb{R}^2}\Big(\int_{\Omega} |\mathrm{P}_{\mathrm{k,B}}(x,y)|^2 \,\mathrm{d}y\Big)\,\mathrm{d}x &= \int_{\Omega}\Big(\int_{\mathbb{R}^2} \mathrm{P}_{\mathrm{k,B}}(x,y)\mathrm{P}_{\mathrm{k,B}}(y,x)\,\mathrm{d}x\Big)\,\mathrm{d}y \\ &= \int_{\Omega} \mathrm{P}_{\mathrm{k,B}}(y,y)\,\mathrm{d}y \\ &= \frac{\mathrm{B}}{2\pi}|\Omega|.\end{aligned} \tag{9b}$$

3.2 A boundary estimate for the Heisenberg Laplacian

In this subsection we will derive a boundary estimate for the operator $\mathrm{A}(\Omega)$ which will be crucial in estimating the size of the remainder term in Theorem 2.3.

Proposition 3.1. *Let $\Omega := \omega \times (a,b) \subset \mathbb{R}^3$ and let c be given by* (6). *Then*

$$\int_a^b \int_{\omega^\beta} |u(x',x_3)|^2 \,\mathrm{d}x'\,\mathrm{d}x_3 \le c^{2+2/c}\beta^{2+2/c}\, \|\mathrm{A}(\Omega)u\|_{L^2(\Omega)} \|\mathrm{A}(\Omega)^{1/c}u\|_{L^2(\Omega)}$$

holds for all $u \in \mathrm{Dom}(\mathrm{A}(\Omega))$ and any $\beta > 0$

For the proof we need the following Lemma.

Lemma 3.2. *Let $\Omega := \omega \times (a,b) \subset \mathbb{R}^3$. Then for all $u \in d[a]$, the form domain of the closure of* (1), *and any $\beta > 0$ we have*

$$\int_a^b \int_{\omega} \frac{|u(x',x_3)|^2}{\delta(x')^2} \,\mathrm{d}x'\,\mathrm{d}x_3 \le c^2\, a[u].$$

Proof. Let u be in $C_0^\infty(\Omega)$. In addition let us denote by $\mathfrak{F}_3$ the Fourier transform in x_3-direction, which is a unitary map in $L^2(\mathbb{R})$. Because Ω is a cylinder, the function $|\mathfrak{F}_3 u(x',\xi_3)|$ lies in $\mathrm{H}_0^1(\omega)$ for fixed $\xi_3 \in \mathbb{R}$. Therefore we can apply inequality (7) to get

$$\begin{aligned}\int_a^b \int_{\omega} \frac{|u(x',x_3)|^2}{\delta(x')^2} \,\mathrm{d}x'\,\mathrm{d}x_3 &= \int_{\mathbb{R}}\int_{\omega} \Big(\frac{|\mathfrak{F}_3 u(x',\xi_3)|}{\delta(x')}\Big)^2 \mathrm{d}x'\,\mathrm{d}\xi_3 \\ &\le c^2 \int_{\mathbb{R}}\int_{\omega} (\nabla_{x'}|\mathfrak{F}_3 u(x',\xi_3)|)^2 \,\mathrm{d}x'\,\mathrm{d}\xi_3.\end{aligned}$$

Now we set

$$\mathbf{A}(x') := \frac{1}{2}(-x_2, x_1), \tag{10}$$

and apply the diamagnetic inequality which states that

$$|\nabla|\psi|| \leq |(\mathrm{i}\nabla + \mathbf{A})\psi| \quad \text{a.e.}$$

holds for all $\psi \in H^1(\omega)$, see, e.g., [12]. This gives

$$\begin{aligned}\int_{\mathbb{R}}\int_{\omega} &(\nabla_{x'}|\mathcal{F}_3 u(x',\xi_3)|)^2 \,\mathrm{d}x'\,\mathrm{d}\xi_3 \\ &\leq \int_{\mathbb{R}}\int_{\omega} |(\mathrm{i}\nabla_{x'} + \xi_3\mathbf{A}(x'))\mathcal{F}_3 u(x',\xi_3)|^2 \,\mathrm{d}x'\,\mathrm{d}\xi_3.\end{aligned}$$

Integration by parts in the x_3-direction yields the inequality for $u \in C_0^\infty(\Omega)$. A density argument completes the proof. □

Proof of Proposition 3.1. We follow the proof of Theorem 4 in [5]. Let us fix $u \in \mathrm{Dom}(\mathrm{A}(\Omega))$ and set

$$\varphi(x) := (\max\{\delta(x'), \beta\})^{-1/c}.$$

for $x := (x', x_3) \in \Omega$ and $\beta > 0$. In what follows we will use the notation

$$\nabla_{\mathbb{H}} = (X_1, X_2)$$

to denote the Heisenberg gradient. First we check that $\varphi u \in d[a]$. Since $u \in \mathrm{Dom}(\mathrm{A}(\Omega)) \subseteq d[a]$, $\varphi \in H_0^1(\omega)$ and get

$$\int_{\Omega} |\nabla_{\mathbb{H}}(\varphi(x)u(x))|^2 \,\mathrm{d}x \leq 2\int_{\Omega} |\varphi(x)\nabla_{\mathbb{H}} u(x)|^2 \,\mathrm{d}x + 2\int_{\Omega} |\nabla_{x'}\varphi(x)|^2 |u(x)|^2 \,\mathrm{d}x. \tag{11}$$

Note that we used here $\nabla_{\mathbb{H}}\varphi(x) = \nabla_{x'}\varphi(x)$ for all $x \in \Omega$. The Eikonal equation

$$|\nabla_{x'}\varphi(x)|^2 = 1 \quad \text{a.e. } x \in \Omega, \tag{12}$$

and the boundedness of Ω yield the finiteness of (11). Hence $\varphi u \in d[a]$ and we may use Lemma 3.2 to get

$$\begin{aligned}c^{-2}\int_{\Omega} \frac{|\varphi(x)u(x)|^2}{\delta(x')^2}\,\mathrm{d}x &\leq \int_{\Omega} |\varphi(x)\nabla_{\mathbb{H}} u(x) + u(x)\nabla_{\mathbb{H}}\varphi(x)|^2 \,\mathrm{d}x \\ &= \langle \varphi^2 \nabla_{\mathbb{H}} u, \nabla_{\mathbb{H}} u\rangle + \langle u, |\nabla_{\mathbb{H}}\varphi|^2 u\rangle \\ &\quad + \frac{1}{2}\langle \nabla_{\mathbb{H}} u, u\nabla_{\mathbb{H}}(\varphi^2)\rangle + \frac{1}{2}\langle u\nabla_{\mathbb{H}}(\varphi^2), \nabla_{\mathbb{H}} u\rangle,\end{aligned}$$

where we denote by $\langle\cdot,\cdot\rangle$ the scalar product in $L^2(\Omega)$. An integration by parts in the last two terms yields

$$c^{-2}\int_\Omega \frac{|\varphi(x)u(x)|^2}{\delta(x')^2}\,dx \le \operatorname{Re}\langle\varphi^2 u,\, \mathrm{A}(\Omega)u\rangle + \langle u, |\nabla_{\mathbb{H}}\varphi|^2 u\rangle.$$

Next we will estimate the first term on the right-hand side. To this end we use Lemma 3.2, which gives

$$\delta^{-2} \le c^2\Lambda(\Omega)$$

in the operator sense. Then, by the Heinz inequality, see Lemma 4.20 in [4],

$$\varphi^4 \le (\delta^{-2})^{2/c} \le (c^2\mathrm{A}(\Omega))^{2/c}.$$

Since $\mathrm{A}(\Omega)^{-1/c}$ is bounded in $L^2(\Omega)$ we obtain

$$\|\varphi^2\mathrm{A}(\Omega)^{-1/c}\| < c^{2/c},$$

where $\|\cdot\|$ stands for the operator norm in $L^2(\Omega)$. Hence

$$\begin{aligned}|\langle\mathrm{A}(\Omega)u,\, \varphi^2 u\rangle| &= |\langle\mathrm{A}(\Omega)u,\, \varphi^2\mathrm{A}(\Omega)^{-1/c}\mathrm{A}(\Omega)^{1/c}u\rangle| \\ &\le \|\mathrm{A}(\Omega)u\|_{L^2(\Omega)}c^{2/c}\|\mathrm{A}(\Omega)^{1/c}u\|_{L^2(\Omega)}.\end{aligned}$$

So we arrive at

$$c^{-2}\int_\Omega \frac{|\varphi(x)u(x)|^2}{\delta(x')^2}\,dx \le \|\mathrm{A}(\Omega)u\|_{L^2(\Omega)}c^{2/c}\|\mathrm{A}(\Omega)^{1/c}u\|_{L^2(\Omega)} + \langle u,\, |\nabla_{\mathbb{H}}\varphi|^2 u\rangle. \tag{13}$$

On the other hand, the Eikonal equation (12) implies that

$$|\nabla_{\mathbb{H}}\varphi(x)|^2 = |\nabla_{x'}\varphi(x)|^2 = c^{-2}\delta(x')^{-(2/c)-2}\chi_{\{\delta(x')\ge\beta\}}(x'),$$

where $\chi_{\{\delta(x')>\beta\}}$ is the characteristic function of the set $\{x\in\Omega \mid \delta(x')\ge\beta\}$. Inserting the above identity into (13) yields

$$\int_{\{x\in\Omega \mid \delta(x')<\beta\}} \frac{|u(x)|^2}{\delta(x')^2}\,dx \le \beta^{2/c}\|\mathrm{A}(\Omega)u\|_{L^2(\Omega)}c^{2+2/c}\|\mathrm{A}(\Omega)^{1/c}u\|_{L^2(\Omega)}.$$

The result now follows from the estimate

$$\int_{\{x\in\Omega|\delta(x')<\beta\}} |u(x)|^2\,dx \le \beta^2\int_{\{x\in\Omega \mid \delta(x')<\beta\}} \frac{|u(x)|^2}{\delta(x')^2}\,dx. \qquad \square$$

4 Proof of Theorem 2.3

Here and below we write a vector $x \in \mathbb{R}^3$ as $x = (x', x_3)$. Let v_j denote the orthonormal basis of the eigenfunctions of $\mathrm{A}(\Omega)$ for $j \in \mathbb{N}$;

$$\mathrm{A}(\Omega) v_j = \lambda_j v_j, \quad \|v_j\|_{L^2(\Omega)} = 1.$$

Let $\mathcal{F}_3$ be the partial Fourier transform in the x_3 variable. Then

$$\mathcal{F}_3 \mathrm{A}(\mathbb{R}^3) \mathcal{F}_3^* = \Big(\mathrm{i}\partial_{x_1} - \frac{1}{2} x_2 \xi_3\Big)^2 + \Big(\mathrm{i}\partial_{x_2} + \frac{1}{2} x_1 \xi_3\Big)^2 = (\mathrm{i}\nabla_{x'} + \xi_3 \mathbf{A}(x'))^2,$$

where $\mathbf{A}(x')$ is given by (10). At this point we use the properties of the magnetic Laplacian, see section 3.1, to obtain

$$\mathcal{F}_3\, \mathrm{A}(\mathbb{R}^3)\, u(x', \xi_3) = \sum_{\mathrm{k}=1}^{\infty} |\xi_3|(2\mathrm{k} - 1) \int_{\mathbb{R}^2} \mathrm{P}_{\mathrm{k},\xi_3}(x', y') \mathcal{F}_3 u(y', \xi_3)\, \mathrm{d}y' \tag{14}$$

for $\mathcal{F}_3\, u(\cdot, \xi_3)$ in the domain of the magnetic Laplacian.

4.1 The sharp leading term

First of all we extend for every $j \in \mathbb{N}$ the eigenfunctions by $v_j(x) := 0$ for all $x \in \Omega^c$. Now we consider

$$\begin{aligned}
&\operatorname{Tr}(\mathrm{A}(\Omega) - \lambda)_- \\
&= \sum_{j:\, \lambda_j < \lambda} \lambda \|v_j\|^2_{L^2(\mathbb{R}^3)} - \|X_1 v_j\|^2_{L^2(\mathbb{R}^3)} - \|X_2 v_j\|^2_{L^2(\mathbb{R}^3)} \\
&= \int_{\mathbb{R}} \sum_{j:\, \lambda_j < \lambda} \lambda \|\mathcal{F}_3 v_j(\cdot, \xi_3)\|^2_{L^2(\mathbb{R}^2)} - \Big\| \Big(\mathrm{i}\partial_{x_1} - \frac{1}{2} x_2 \xi_3\Big) \mathcal{F}_3 v_j(\cdot, \xi_3) \Big\|^2_{L^2(\mathbb{R}^2)} \mathrm{d}\xi_3 \\
&- \int_{\mathbb{R}} \sum_{j:\, \lambda_j < \lambda} \Big\| \Big(\mathrm{i}\partial_{x_2} + \frac{1}{2} x_1 \xi_3\Big) \mathcal{F}_3 v_j(\cdot, \xi_3) \Big\|^2_{L^2(\mathbb{R}^2)} \mathrm{d}\xi_3.
\end{aligned}$$

At this point we apply the spectral decomposition (14) of the free Heisenberg Laplacian. An application of Fatou's Lemma then yields

$$\operatorname{Tr}(\mathrm{A}(\Omega)-\lambda)_- \le \int_{\mathbb{R}} \sum_{j:\,\lambda_j<\lambda} \sum_{\mathrm{k}=1}^{\infty} (\lambda - |\xi_3|(2\mathrm{k}-1))_+ \, \| f_{j,\mathrm{k},\xi_3}\|^2_{L^2(\mathbb{R}^2)} \, \mathrm{d}\xi_3,$$

where

$$\begin{aligned} f_{j,\mathrm{k},\xi_3}(x') &:= \int_{\mathbb{R}^2} \mathrm{P}_{\mathrm{k},,_3}(x',y')\mathcal{F}_3 v_j(y',\xi_3)\,\mathrm{d}y' \\ &= \frac{1}{\sqrt{2\pi}} \int_{\Omega} \mathrm{P}_{\mathrm{k},,_3}(x',y')\mathrm{e}^{-\mathrm{i}y_3\xi_3} v_j(y',y_3)\,\mathrm{d}y'\,\mathrm{d}y_3 \\ &= \frac{1}{\sqrt{2\pi}} \langle \mathrm{P}_{\mathrm{k},,_3}(x',\cdot)\mathrm{e}^{-\mathrm{i}\cdot\xi_3},\ v_j(\cdot)\rangle_{L^2(\Omega)}. \end{aligned}$$

We split the sum as follows:

$$\begin{aligned} \operatorname{Tr}(\mathrm{A}(\Omega)-\lambda)_- \le{}& \int_{\mathbb{R}} \sum_{\mathrm{k}:\,\lambda>|\xi_3|(2\mathrm{k}-1)} (\lambda-|\xi_3|(2\mathrm{k}-1)) \sum_{j=1}^{\infty} \| f_{j,\mathrm{k},\xi_3}\|^2_{L^2(\mathbb{R}^2)}\,\mathrm{d}\xi_3 \\ &- \int_{\mathbb{R}} \sum_{\mathrm{k}:\,\lambda>|\xi_3|(2k-1)} (\lambda-|\xi_3|(2\mathrm{k}-1)) \sum_{j:\,\lambda_j\ge\lambda} \| f_{j,\mathrm{k},\xi_3}\|^2_{L^2(\mathbb{R}^2)}\,\mathrm{d}\xi_3, \end{aligned} \tag{15}$$

noting that the first term on the right-hand side is positive and the other one is negative. The completeness of v_j and equation (9) yield

$$\begin{aligned} \sum_{j=1}^{\infty} \| f_{j,\mathrm{k},\xi_3}\|^2_{L^2(\mathbb{R}^2)} &= \frac{1}{2\pi} \int_{\mathbb{R}^2} \sum_{j=1}^{\infty} |\langle \mathrm{P}_{\mathrm{k},,_3}(x',\cdot)\mathrm{e}^{-\mathrm{i}\cdot\xi_3},\ v_j(\cdot)\rangle_{L^2(\Omega)}|^2\,\mathrm{d}x' \\ &= \frac{|\xi_3|}{4\pi^2}|\Omega|. \end{aligned}$$

To obtain the sharp leading term in (8) we apply this identity in the first integral on the right-hand side of (15). Using the fact that

$$\sum_{j=1}^{\infty} \frac{1}{(2j-1)^2} = \frac{\pi^2}{8},$$

we get

$$\begin{aligned}
&\int_{\mathbb{R}} \sum_{\mathrm{k}:\, \lambda>|\xi_3|(2\mathrm{k}-1)} (\lambda - |\xi_3|(2\mathrm{k}-1)) \sum_{j=1}^{\infty} \| f_{j,\mathrm{k},\xi_3}\|^2_{L^2(\mathbb{R}^2)}\, \mathrm{d}\xi_3 \\
&\quad = \frac{|\Omega|}{4\pi^2} \int_{\mathbb{R}} \sum_{\mathrm{k}:\, \lambda>|\xi_3|(2\mathrm{k}-1)} (\lambda - |\xi_3|(2\mathrm{k}-1))|\xi_3|\, \mathrm{d}\xi_3 \\
&\quad = \frac{|\Omega|}{2\pi^2} \sum_{\mathrm{k}=1}^{\infty} \frac{1}{(2\mathrm{k}-1)^2} \int_0^{\infty} s(\lambda - s)_+\, \mathrm{d}s \\
&\quad = \frac{|\Omega|}{96} \lambda^3.
\end{aligned}$$

Inserting this back into (15) gives

$$\operatorname{Tr}(\mathrm{A}(\Omega) - \lambda)_- \le \frac{|\Omega|}{96}\lambda^3 - \int_{\mathbb{R}} \sum_{k=1}^{\infty} (\lambda - |\xi_3|(2\mathrm{k}-1))_+ \sum_{j:\lambda_j \ge \lambda}^{\infty} \| f_{j,\mathrm{k},\xi_3}\|^2_{L^2(\mathbb{R}^2)}\, \mathrm{d}\xi_3. \tag{16}$$

4.2 The lower order term

In order to establish a suitable lower bound on the second term in (16) we use the same technique as in [8]. The key point of this approach is to estimate the quantity

$$\mathcal{R}_\lambda := \sum_{j:\, \lambda_j \ge \lambda} \| f_{j,\mathrm{k},\xi_3}\|^2_{L^2(\mathbb{R}^2)}$$

from below by a power function of λ. Note that

$$\begin{aligned}
\mathcal{R}_\lambda &= \frac{|\xi_3|}{4\pi^2}|\Omega| - \sum_{j:\, \lambda_j<\lambda} \| f_{j,\mathrm{k},\xi_3}\|^2_{L^2(\mathbb{R}^2)} \\
&= \frac{1}{2\pi} \int_{\mathbb{R}^2} \int_{\Omega} \Big| \mathrm{P}_{\mathrm{k},,3}(x', y')\mathrm{e}^{-\mathrm{i}y_3\xi_3} - \sum_{j:\, \lambda_j<\lambda} \sqrt{2\pi}\, f_{j,\mathrm{k},\xi_3}(x') v_j(y', y_3) \Big|^2 \mathrm{d}y'\, \mathrm{d}y_3\, \mathrm{d}x'.
\end{aligned}$$

The inclusion $\omega \supseteq \omega^\beta$ and an application of the elementary inequality $|z-w|^2 \ge 1/2\,|z|^2 - |w|^2$, which holds for all $z, w \in \mathbb{C}$, imply that

$$\begin{aligned}
\mathcal{R}_\lambda \ge{}& \frac{|\xi_3|}{8\pi^2}(b-a)\, |\omega^\beta| \\
&- \frac{1}{2\pi} \int_{\mathbb{R}^2} \int_a^b \int_{\omega^\beta} \Big| \sum_{j:\, \lambda_j<\lambda} \langle \mathrm{P}_{\mathrm{k},,3}(x', \cdot)\mathrm{e}^{-\mathrm{i}\cdot\xi_3},\, v_j(\cdot)\rangle_{L^2(\Omega)} v_j(y', y_3) \Big|^2 \mathrm{d}y'\, \mathrm{d}y_3\, \mathrm{d}x'.
\end{aligned}$$

Next we estimate the negative integral from above. Note that the linear combinations of v_j lie in Dom(A(Ω)). Therefore we may apply Proposition 3.1 and obtain

$$\begin{aligned}
&\frac{1}{2\pi}\int_{\mathbb{R}^2}\Big(\int_a^b\int_{\omega^\beta}\Big|\sum_{j:\,\lambda_j<\lambda}\langle \mathrm{P}_{\mathrm{k},,3}(x',\cdot)\mathrm{e}^{-\mathrm{i}\cdot\xi_3},\, v_j(\cdot)\rangle_{L^2(\Omega)} v_j(y',y_3)\Big|^2 \mathrm{d}y'\,\mathrm{d}y_3\Big)\mathrm{d}x' \\
&\leq c^{2+2/c}\beta^{2+2/c}\lambda^{1+1/c}\frac{1}{2\pi}\int_{\mathbb{R}^2}\Big(\sum_{j:\,\lambda_j<\lambda}|\langle P_{k,\xi_3}(x',\cdot)\mathrm{e}^{-\mathrm{i}\cdot\xi_3},\, v_j(\cdot)\rangle_{L^2(\Omega)}|^2\Big)\mathrm{d}x' \\
&\leq c^{2+2/c}\beta^{2+2/c}\lambda^{1+1/c}\frac{1}{2\pi}\int_{\mathbb{R}^2}\Big(\int_\Omega |P_{k,\xi_3}(x',y')|^2\,\mathrm{d}y'\,\mathrm{d}x_3\Big)\mathrm{d}x' \\
&= c^{2+2/c}\beta^{2+2/c}\lambda^{1+1/c}\frac{|\Omega|}{4\pi^2}|\xi_3|,
\end{aligned}$$

which yields the following lower bound on $\mathcal{R}_\lambda$:

$$\begin{aligned}
\mathcal{R}_\lambda &\geq \frac{|\xi_3|}{8\pi^2}(b-a)|\omega^\beta| - c^{2+2/c}\beta^{2+2/c}\lambda^{1+1/c}\frac{|\Omega|}{4\pi^2}|\xi_3| \\
&\geq \frac{|\xi_3|}{8\pi^2}\beta\big(l(\omega) - 2c^{2+2/c}\beta^{1+2/c}\lambda^{1+1/c}|\Omega|\big).
\end{aligned}$$

Now we set

$$\beta^{1+2/c} = \frac{l(\omega)}{c^{2+2/c}\lambda^{1+1/c}(4+4/c)|\Omega|},$$

which is possible for $\lambda \geq \lambda_1(\Omega)$, because of

$$\beta^{1+2/c} \leq \frac{1}{c^{2+2/c}\lambda_1(\Omega)^{1+1/c}(4+4/c)\mathrm{R}(\omega)} \leq \frac{\mathrm{R}(\omega)^{1+2/c}}{4}.$$

The last inequality was obtained by applying Proposition 3.1 to $u = v_1$ and $\beta = \mathrm{R}(\omega)$. Summing up we thus arrive at

$$\begin{aligned}
\mathcal{R}_\lambda &\geq \frac{|\xi_3|}{8\pi^2}\lambda^{-(c+1)/(c+2)} l(\omega)^{(2c+2)/(c+2)}|\Omega|^{-c/(c+2)}(2+4/c)(4c+4)^{-(2c+2)/(c+2)} \\
&= \lambda^{-(c+1)/(c+2)} G(\Omega)|\xi_3|,
\end{aligned}$$

where

$$G(\Omega) := \frac{l(\omega)^{(2c+2)/(c+2)}}{8\pi^2}|\Omega|^{-c/(c+2)}(2+4/c)(4c+4)^{-(2c+2)/(c+2)}.$$

This in combination with (16) gives

$$\begin{aligned}&\operatorname{Tr}(\mathrm{A}(\Omega)-\lambda)_- \\ &\quad\le \frac{|\Omega|}{96}\lambda^3 - G(\Omega)\lambda^{-(c+1)/(c+2)}\int_{\mathbb{R}} \sum_{k:\,\lambda>|\xi_3|(2k-1)} (\lambda-|\xi_3|(2\mathrm{k}-1))|\xi_3|\,\mathrm{d}\xi_3.\end{aligned}$$

To finish the proof we note that

$$\sum_{\mathrm{k}=1}^{\infty}\int_0^{\infty}(\lambda-\xi_3(2\mathrm{k}-1))_+\xi_3\,\mathrm{d}\xi_3 = \sum_{\mathrm{k}=1}^{\infty}\frac{1}{(2\mathrm{k}-1)^2}\int_0^{\infty}s(\lambda-s)_+\,\mathrm{d}s = \frac{\pi^2\lambda^3}{48}.$$

This gives the estimate stated in Theorem 2.3.

Acknowledgements. Hynek Kovařík was supported by the Gruppo Nazionale per Analisi Matematica, la Probabilità e le loro Applicazioni (GNAMPA) of the Istituto Nazionale di Alta Matematica (INdAM). The support of MIUR-PRIN2010-11 grant for the project "Calcolo delle variazioni" (Hynek Kovařík) is also gratefully acknowledged. Bartosch Ruszkowski was supported by the German Science Foundation through the Research Training Group 1838: *Spectral Theory and Dynamics of Quantum Systems.*

References

[1] A. Ancona, On strong barriers and an inequality of Hardy for domains in $\mathbb{R}^n$. *J. London Math. Soc.* (2) **34** (1986), no. 2, 274–290. MR 0856511 Zbl 0629.31002

[2] F. A. Berezin, Covariant and contravariant symbols of operators. *Izv. Akad. Nauk SSSR Ser. Mat.* **36** (1972), 1134–1167. In Russian. English translation, *Math. USSR-Izv.* **6** (1972), 1117–1151. MR 0350504 Zbl 0259.47004

[3] E. B. Davies, A review of Hardy inequalities. In J. Rossmann, P. Takáč, and G. Wildenhain (eds.), *The Maz'ya anniversary collection.* Vol. 2. Rostock Conference on Functional Analysis, Partial Differential Equations and Applications. Papers from the conference held at the University of Rostock, Rostock, August 31–September 4, 1998. Operator Theory: Advances and Applications, 110. Birkhäuser Verlag, Basel, 1999, 55–67. MR 1747888 Zbl 0936.35121

[4] E. B. Davies, *One-parameter semigroups.* London Mathematical Society Monographs, 15. Academic Press [Harcourt Brace Jovanovich, Publishers], New York and London, 1980. ISBN 0-12-206280-9 MR 0591851 Zbl 0457.47030

[5] E. B. Davies, Sharp boundary estimates for elliptic operators. *Math. Proc. Cambridge Philos. Soc.* **129** (2000), no. 1, 165–178. MR 1757786 Zbl 0963.35146

[6] G. B. Folland, A fundamental solution for a subelliptic operator. *Bull. Amer. Math. Soc.* **79** (1973), 373–376. MR 0315267 Zbl 0256.35020

[7] A. M. Hansson and A. Laptev, Sharp spectral inequalities for the Heisenberg Laplacian. In K. Tent (ed.), *Groups and analysis.* The legacy of H. Weyl. London Mathematical Society Lecture Note Series, 354. Cambridge University Press, Cambridge, 2008, 100–115. MR 2528463 Zbl 1157.58008

[8] H. Kovařík and T. Weidl, Improved Berezin–Li–Yau inequalities with magnetic field. *Proc. Roy. Soc. Edinburgh Sect. A* **145** (2015), no. 1, 145–160. MR 3304579 Zbl 1323.35116

[9] H. Kovařík, B. Ruszkowski, and T. Weidl, Melas-type bounds for the Heisenberg Laplacian on bounded domains. Preprint 2015. arXiv:1511.04223 [math.SP]

[10] H. Kovařík, S. Vugalter, and T. Weidl, Two dimensional Berezin–Li–Yau inequalities with a correction term. *Comm. Math. Phys.* **287** (2009), no. 3, 959–981. MR 2486669 Zbl 1197.35182

[11] P. Li and S. T. Yau, On the Schrödinger equation and the eigenvalue problem. *Comm. Math. Phys.* **88** (1983), no. 3, 309–318. MR 0701919 Zbl 0554.35029

[12] E. H. Lieb and M. Loss, *Analysis.* Graduate Studies in Mathematics, 14. American Mathematical Society, Providence, R.I., 1997. ISBN 0-8218-0632-7 MR 1415616 Zbl 0873.26002

[13] A. D. Melas, A lower bound for sums of eigenvalues of the Laplacian. *Proc. Amer. Math. Soc.* **131** (2003), no. 2, 631–636. MR 1933356 Zbl 1015.58011

[14] B. Opic and A. Kufner, *Hardy-type inequalities.* Pitman Research Notes in Mathematics Series, 219. Longman Scientific & Technical, Harlow, 1990. ISBN 0-582-05198-3 MR 1069756 Zbl 0698.26007

[15] S. Y. Yolcu, An improvement to a Berezin–Li–Yau type inequality. *Proc. Amer. Math. Soc.* **138** (2010), no. 11, 4059–4066. MR 2679626 Zbl 1202.35366

[16] S. Y. Yolcu and T. Yolcu, Estimates for the sums of eigenvalues of the fractional Laplacian on a bounded domain. *Commun. Contemp. Math.* **15** (2013), no. 3, 1250048, 15 pp. MR 3063552 Zbl 1273.35197

Variational proof of the existence of eigenvalues for star graphs

Konstantin Pankrashkin

Dedicated to Pavel Exner on the occasion of his 70th birthday

1 Introduction

The mathematically rigorous study of multidimensional Schrödinger operators with potentials supported by hypersurfaces was initiated in 1994 by Brasche, Exner, Kuperin and Šeba [2]. The two-dimensional Hamiltonians with interactions supported by curves have become a prominent class of solvable models of quantum mechanics [8] and are usually referred to as leaky quantum graphs. A summary of various questions and results in the spectral theory of such operators can be found in the review by Exner [7], and for the most recent developments we refer to the papers [1], [5], [10], [14], [15], [17], and [18] and to Chapter 10 in the recent monograph by Exner and Kovařík [11].

In the present contribution, we are interested in some properties of Schrödinger operators with δ-interactions supported by the so-called star graphs. By a star graph Γ we mean a subset of $\mathbb{R}^2$ obtained as the union of finitely many rays emanating from the origin. If (r,θ) is the standard polar coordinate system, then Γ is naturally identified with a family $(\theta_1,\dots,\theta_N)$ in which $0 \le \theta_1 < \dots < \theta_N < 2\pi$ by

$$\Gamma := \bigcup_{j=1}^{N} \Big\{(r,\theta) \colon \theta = \theta_j,\ r \ge 0\Big\}.$$

The associated Schrödinger operator $H_{\Gamma,\alpha} = -\Delta - \alpha\delta_\Gamma$, where δ_Γ is the Dirac δ-distribution supported by Γ and $\alpha > 0$ is a coupling constant, is defined as the unique self-adjoint operator in $L^2(\mathbb{R}^2)$ associated with the closed lower semibounded quadratic form

$$Q_{\Gamma,\alpha}(u) = \iint_{\mathbb{R}^2} |\nabla u|^2 \,\mathrm{d}x - \alpha \int_\Gamma |u|^2 \,\mathrm{d}s, \quad u \in H^1(\mathbb{R}^2).$$

where ds is the one-dimensional Hausdorff measure, cf. [2]. Such configurations appear naturally as a mathematical model for a junction of quantum wires, and they were first analyzed by Exner and Němcová [12] and [13]. The basic spectral properties of the operator are well known: the essential spectrum coincides with the semi-axis $[-\alpha^{2}/4, +\infty)$, and the discrete spectrum is non-empty except in the degenerate cases when Γ is a single ray ($N = 1$) or a straight line ($N = 2$ and $|\theta_1 - \theta_2| = \pi$). Despite the simple geometrical picture, the only available proof of the existence of eigenvalues is based on a rather involved analysis of integral operators carried out by Exner and Ichinose [9]. On the other hand, by the min-max principle, the non-emptiness of the discrete spectrum would follow from the existence of a trial function $v \in H^1(\mathbb{R}^2)$ satisfying the strict inequality

$$Q_{\Gamma,\alpha}(v) < -\frac{\alpha^2}{4}\|v\|^2_{L^2(\mathbb{R}^2)}. \tag{1}$$

Surprisingly, the construction of such a function appeared to be a difficult task. The construction of Exner and Němcová [13] works only if there is a pair of rays with $|\theta_j - \theta_k| \mod 2\pi < 0.092$. Brown, Eastham, and Wood [3], [4], and [6] managed to find a trial function for all possible configurations with $N \geq 3$ as well as for the configurations with $N = 2$ and $|\theta_1 - \theta_2| < 0.9271$. In the present note we show how to construct such a function for all possible cases (Theorem 2.1), and our approach uses a likeliness between the star graphs and a spectral problem of the surface superconductivity with a similar geometry discussed by Lu and Pan [19] and Helffer and Morame [16]. We remark again that Theorem 2.1 itself does not provide any new spectral information, but suggests a new method to show the presence of a non-empty discrete spectrum as an alternative to the analytical proof by Exner and Ichinose [9]. On the other hand, the presence of explicitly given trial functions allows one to obtain a universal upper bound for the lowest eigenvalue (Theorem 3.1), which is a new result.

2 Construction of a trial function

By the min-max principle, it is sufficient to consider the case $N = 2$ (a broken line), then, up to isometries, all possible configurations can be described by a single parameter $\theta \in (0, \pi/2)$ through $\Gamma = \overline{\Gamma_+ \cup \Gamma_-}$ with $\Gamma_\pm := \{(t, \pm t \tan\theta) : t > 0\}$, and the associated operator $H_{\Gamma,\alpha}$ will be denoted by $H(\theta, \alpha)$.

We remark first that in order to show that the discrete spectrum is non-empty it is sufficient to consider the problem in the half-plane $\mathbb{R} \times \mathbb{R}_+$, i.e., to find a function $u \in H^1(\mathbb{R} \times \mathbb{R}_+)$ satisfying

$$\iint_{\mathbb{R}\times\mathbb{R}_+} |\nabla u|^2 \mathrm{d}x - \alpha \int_{\Gamma_+} |u|^2 \mathrm{d}x < -\frac{\alpha^2}{4}\|u\|^2_{L^2(\mathbb{R}\times\mathbb{R}_+)},$$

as its extension v to the whole of $\mathbb{R}^2$ by parity automatically satisfies (1). For subsequent constructions, it is handy to perform an additional rotation to put the support of the interaction onto the positive semi-axis of ordinates. In other words, we will work with the domain $\Omega := \{(x_1, x_2)\colon x_1 < x_2 \tan\theta\}$ and the quadratic form

$$Q(u) = \iint_\Omega |\nabla u|^2 \,\mathrm{d}x - \alpha \int_{\mathbb{R}_+} |u(0, x_2)|^2 \,\mathrm{d}x_2, \quad u \in H^1(\Omega).$$

Theorem 2.1. *Pick any $\rho \in (0, \cot^2\theta)$ and any Lipschitz function $\chi\colon \mathbb{R} \to [0,1]$ with $\chi(t) = 1$ for $|t| \le 1$ and $\chi(t) = 0$ for $|t| \ge 2$, then for sufficiently large $n > 0$ the function u defined by*

$$u(x_1, x_2) = e^{-\alpha|x_1|/2}\Big(\frac{2}{\alpha}\mathbf{1}_{\mathbb{R}_+}(x_2) - \frac{1}{\alpha}e^{-\alpha|x_2|\tan\theta}\operatorname{sgn} x_2\Big)^\rho \chi\Big(\frac{x_2}{n}\Big) \tag{2}$$

satisfies the strict inequality

$$Q(u) < -\frac{\alpha^2}{4}\|u\|^2_{L^2(\Omega)}.$$

Proof. For further use, denote

$$F(t) := \int_{-\infty}^t e^{-\alpha|x_1|}\,\mathrm{d}x_1 = \frac{2}{\alpha}\mathbf{1}_{\mathbb{R}_+}(t) - \frac{1}{\alpha}e^{-\alpha|t|}\operatorname{sgn} t.$$

For the functions u of the form $u(x_1, x_2) = e^{-\alpha|x_1|/2} g(x_2)$ with real-valued g we have

$$\|u\|^2_{L^2(\Omega)} = \int_{\mathbb{R}}\int_{-\infty}^{x_2\tan\theta} e^{-\alpha|x_1|} g(x_2)^2 \,\mathrm{d}x_1\,\mathrm{d}x_2 = \int_{\mathbb{R}} g(x_2)^2 F(x_2\tan\theta)\,\mathrm{d}x_2. \tag{3}$$

Furthermore,

$$\begin{aligned} Q(u) = {} & \frac{\alpha^2}{4}\int_{\mathbb{R}} g(x_2)^2 \int_{-\infty}^{x_2\tan\theta} e^{-\alpha|x_1|}\,\mathrm{d}x_1\,\mathrm{d}x_2 \\ & + \int_{\mathbb{R}} g'(x_2)^2 \int_{-\infty}^{x_2\tan\theta} e^{-\alpha|x_1|}\,\mathrm{d}x_1\,\mathrm{d}x_2 - \alpha\int_{\mathbb{R}_+} g(x_2)^2\,\mathrm{d}x_2. \end{aligned}$$

Due to

$$\begin{aligned} \frac{\alpha^2}{4}\int_{-\infty}^{x_2\tan\theta} e^{-\alpha|x_1|}\,\mathrm{d}x_1 &= \frac{\alpha^2}{4}F(x_2\tan\theta) \\ &= -\frac{\alpha^2}{4}F(x_2\tan\theta) + \frac{\alpha^2}{2}F(x_2\tan\theta) \\ &= -\frac{\alpha^2}{4}F(x_2\tan\theta) + \alpha\,\mathbf{1}_{\mathbb{R}_+}(x_2) - \frac{\alpha}{2}e^{-\alpha|x_2|\tan\theta}\operatorname{sgn} x_2 \end{aligned}$$

we have

$$
\begin{aligned}
Q(u) = -\frac{\alpha^2}{4}\int_{\mathbb{R}} g(x_2)^2 F(x_2\tan\theta)\,\mathrm{d}x_2 + \int_{\mathbb{R}} g'(x_2)^2 F(x_2\tan\theta)\,\mathrm{d}x_2 \\
-\frac{\alpha}{2}\int_{\mathbb{R}} g(x_2)^2 e^{-\alpha|x_2|\tan\theta}\operatorname{sgn} x_2\,\mathrm{d}x_2.
\end{aligned}
\tag{4}
$$

Using the integration by parts we obtain

$$
\begin{aligned}
\int_{\mathbb{R}} g(x_2)^2 e^{-\alpha|x_2|\tan\theta}\operatorname{sgn} x_2\,\mathrm{d}x_2 &= \frac{2}{\alpha}\cot\theta\int_{\mathbb{R}} g(x_2)g'(x_2)e^{-\alpha|x_2|\tan\theta}\,\mathrm{d}x_2 \\
&= \frac{2}{\alpha}\cot\theta\int_{\mathbb{R}} g(x_2)g'(x_2)F'(x_2\tan\theta)\,\mathrm{d}x_2,
\end{aligned}
\tag{5}
$$

and the substitution of (3) and (5) into (4) gives the representation

$$
\begin{aligned}
Q(u) &= -\frac{\alpha^2}{4}\|u\|^2_{L^2(\Omega)} + R(g),\\
R(g) &:= \int_{\mathbb{R}} g'(x_2)\big(g'(x_2)F(x_2\tan\theta) - g(x_2)F'(x_2\tan\theta)\cot\theta\big)\,\mathrm{d}x_2.
\end{aligned}
$$

Hence, we need to find a function g with $R(g) < 0$.

Pick $\rho \in (0, \cot^2\theta)$ and introduce a function g_ρ by $g_\rho(x_2) = F(x_2\tan\theta)^\rho$, then

$$
R(g_\rho) = \rho\tan^2\theta(\rho - \cot^2\theta)\int_{\mathbb{R}} e^{-2\alpha|x_2|\tan\theta}F(x_2\tan\theta)^{2\rho-1}\,\mathrm{d}x_2 < 0.
$$

Remark that the integral is finite, but the function g_ρ has a non-zero finite limit at $+\infty$, and the associated function u does not belong to $H^1(\Omega)$ due to (3).

Choose a Lipschitz function $\chi\colon \mathbb{R} \to [0,1]$ with $\chi(t) = 1$ for $|t| \le 1$ and $\chi(t) = 0$ for $|t| \ge 2$, and for $n > 0$ denote $h_n := g_\rho\chi(\cdot/n)$. By construction, the associated functions u_n given by

$$
u_n(x_1, x_2) = e^{-\alpha|x_1|/2}h_n(x_2), \tag{6}
$$

belong to $H^1(\Omega)$ and coincide with (2). In addition,

$$\begin{aligned} R(h_n) - R(g_\rho) &= \int_{\mathbb{R}} \Big(\chi\Big(\frac{x_2}{n}\Big)^2 - 1\Big) g_\rho'(x_2)(g_\rho'(x_2)F(x_2\tan\theta) \\ &\qquad\qquad - g_\rho(x_2)F'(x_2\tan\theta)\cot\theta)\,\mathrm{d}x_2 \\ &\quad + \frac{1}{n}\int_{\mathbb{R}} \chi\Big(\frac{x_2}{n}\Big)\chi'\Big(\frac{x_2}{n}\Big)(2g_\rho(x_2)g_\rho'(x_2)F(x_2\tan\theta) \\ &\qquad\qquad - g_\rho(x_2)^2F'(x_2\tan\theta)\cot\theta)\,\mathrm{d}x_2 \\ &\quad + \frac{1}{n^2}\int_{\mathbb{R}} \chi'\Big(\frac{x_2}{n}\Big)^2 g_\rho(x_2)^2 F(x_2\tan\theta)\,\mathrm{d}x_2 \\ &=: I_1 + I_2 + I_3. \end{aligned}$$

Due to the finiteness of $R(g_\rho)$, for large n we have

$$\begin{aligned} |I_1| &\le \int_{\mathbb{R}\setminus(-n,n)} |g_\rho'(x_2)(g_\rho'(x_2)F(x_2\tan\theta) - g_\rho(x_2)F'(x_2\tan\theta)\cot\theta)|\,\mathrm{d}x_2 \\ &= \rho\tan^2\theta\cdot|\rho-\cot^2\theta|\int_{\mathbb{R}\setminus(-n,n)} e^{-2\alpha|x_2|\tan\theta}F(x_2\tan\theta)^{2\rho-1}\,\mathrm{d}x_2 \\ &= o(1). \end{aligned}$$

Furthermore,

$$\begin{aligned} |I_2| &= \left|\frac{1}{n}\int_{\mathbb{R}}(2\rho\tan\theta-\cot\theta)\chi\Big(\frac{x_2}{n}\Big)\chi'\Big(\frac{x_2}{n}\Big)e^{-\alpha|x_2|\tan\theta}F(x_2\tan\theta)^{2\rho}\,\mathrm{d}x_2\right| \\ &\le \frac{|2\rho-\cot^2\theta|\cdot\tan\theta\cdot\|\chi'\|_\infty}{n}\int_{\mathbb{R}} e^{-\alpha|x_2|\tan\theta}F(x_2\tan\theta)^{2\rho}\,\mathrm{d}x_2 \\ &= \mathcal{O}\Big(\frac{1}{n}\Big) \end{aligned}$$

due to the convergence of the integral. Finally, as the integrand is bounded, we have

$$|I_3| \le \frac{1}{n^2}\Big(\int_{-2n}^{-n} + \int_n^{2n} \|\chi'\|_\infty^2\, g_\rho(x_2)^2F(x_2\tan\theta)\,\mathrm{d}x_2\Big) = \frac{1}{n^2}\cdot\mathcal{O}(n) = \mathcal{O}\Big(\frac{1}{n}\Big),$$

and we arrive at $R(h_n) = R(g_\rho) + o(1)$ as n tends to $+\infty$. As $R(g_\rho) < 0$, we have $R(h_n) < 0$ for large n, which shows that the functions (6) have the sought property. □

3 Upper bound for the lowest eigenvalue

We remark first that various estimates for the lowest eigenvalue $\lambda(\theta,\alpha)$ of $H(\theta,\alpha)$ were obtained in earlier works. In particular, Duchêne and Raymond [5] showed that

$$\lambda(\theta,\alpha) = -\alpha^2[1 - c_1\theta^{2/3} + \mathcal{O}(\theta)], \quad \theta \to 0_+, \tag{7}$$

and Exner and Kondej [10] proved that

$$\lambda(\theta,\alpha) = -\alpha^2\Big[\frac{1}{4} + c_2\Big(\frac{\pi}{2} - \theta\Big)^4 + o\Big(\Big(\frac{\pi}{2} - \theta\Big)^4\Big)\Big], \quad \theta \to (\pi/2)_-, \tag{8}$$

where c_1 and c_2 are some explicit positive constants.

Recall that by the min-max principle there holds $\lambda(\theta,\alpha) \le \mathcal{Q}(v)/\|v\|^2_{L^2(\Omega)}$ for any non-zero $v \in H^1(\Omega)$. We would like to use the trial functions u from Theorem 2.1 to obtain an explicit upper estimate for the eigenvalue valid for all values of θ. As the limit $\lim_{n\to+\infty} \mathcal{Q}(u)/\|u\|^2_{L^2(\Omega)} = -\alpha^2/4$ coincides with the bottom of the essential spectrum, we cannot hope for the best possible result. Nevertheless, the estimate and the method can be of some interest as, to our best knowledge, no analogous bound has been available so far.

Theorem 3.1. *For any $\theta \in (0, \pi/2)$ there holds*

$$\lambda(\theta,\alpha) \le -\alpha^2\Big(\frac{1}{4} + \Lambda(\theta)\Big),$$

where

$$\Lambda(\theta) := \frac{3\cos^6\theta(2^{2\cos^2\theta} - 1)^2}{2(1 + 2\cos^2\theta)^3(108 + 180\cos^2\theta - 132\cos^4\theta + 45\cos^6\theta - 5\cos^8\theta)} \tag{9}$$

is strictly positive.

A comparison with (7) and (8) shows that the upper estimate is away of an optimal one. For θ close to 0 our estimate gives $\lambda(\theta,\alpha) \le -99\alpha^2/392 + \mathcal{O}(\theta)$ which is very weak when compared with the true behavior given by (7). At $\theta = \pi/2$, the value of $\Lambda(\theta)$ vanishes at the tenth order, which is also very far from the true fourth order given in (8). Our bound resulted from various experiments with the parameters and used a number of very rough inequalities, and the interested reader should feel free to improve the estimate using an alternative choice of parameters.

Proof. The result is based on a more accurate estimate of the quantities appearing in the proof of Theorem 2.1 for an explicit choice of the function χ and of the parameter ρ. Namely, we set

$$\chi(t) := \begin{cases} 1, & |t| \leq 1, \\ 2 - |t|, & |t| \in (1,2), \\ 0, & |t| \geq 2, \end{cases} \qquad \rho := \cos^2\theta,$$

then $\|\chi'\|_\infty = 1$. We have

$$R(g_\rho) = \frac{\rho \tan^2\theta(\rho - \cot^2\theta)}{\alpha^{2\rho-1}} \Big(\int_{-\infty}^{0} e^{(2\rho+1)x_2 \tan\theta} \, \mathrm{d}x_2 + \int_0^{+\infty} e^{-2\alpha x_2 \tan\theta} (2 - e^{-\alpha x_2 \tan\theta})^{2\rho-1} \, \mathrm{d}x_2 \Big).$$

We calculate

$$\int_{-\infty}^{0} e^{(2\rho+1)x_2 \tan\theta} \, \mathrm{d}x_2 = \frac{1}{(2\rho+1)\alpha \tan\theta}$$

and, using the change of variables $s = e^{-\alpha x_2 \tan\theta}$,

$$\begin{aligned} \int_0^{+\infty} & e^{-2\alpha x_2 \tan\theta} (2 - e^{-\alpha x_2 \tan\theta})^{2\rho-1} \, \mathrm{d}x_2 \\ &= \frac{1}{\alpha \tan\theta} \int_0^1 s(2-s)^{2\rho-1} \, \mathrm{d}s \\ &= \frac{1}{\alpha \tan\theta} \int_0^1 (2(2-s)^{2\rho-1} - (2-s)^{2\rho}) \, \mathrm{d}s \\ &= \frac{1}{\alpha \tan\theta} \Big(\frac{2^{2\rho}-1}{\rho} - \frac{2^{2\rho+1}-1}{2\rho+1} \Big), \end{aligned}$$

which gives

$$R(g_\rho) = \frac{1}{\alpha^{2\rho}} \frac{\tan\theta\,(\rho - \cot^2\theta)(2^{2\rho}-1)}{2\rho+1} = -\frac{\cos^3\theta(2^{2\cos^2\theta}-1)}{\sin\theta(1+2\cos^2\theta)}.$$

In what follows we will use the following estimates valid for $s \in [0,1]$ due to the convexity argument:

$$\frac{1}{1+2s} \le 1 - \frac{2}{3}s, \quad \frac{1}{(1+2s)^2} \le 1 - \frac{8}{9}s, \quad 2^{2s} \le 1 + 3s.$$

We estimate

$$\begin{aligned}
|I_1| \le{}& \frac{\cos^4\theta}{\alpha^{2\rho-1}}\Big(\int_{-\infty}^{-n} e^{(2\rho+1)\alpha x_2 \tan\theta}\,\mathrm{d}x_2 \\
&\qquad + \int_n^{+\infty} e^{-2\alpha x_2 \tan\theta}(2 - e^{-\alpha x_2\tan\theta})^{2\rho-1}\,\mathrm{d}x_2\Big) \\
\le{}& \frac{\cos^4\theta}{\alpha^{2\rho-1}}\Big(\int_{-\infty}^{-n} e^{(2\rho+1)\alpha x_2 \tan\theta}\,\mathrm{d}x_2 + 2^{2\rho}\int_n^{+\infty} e^{-2\alpha x_2\tan\theta}\,\mathrm{d}x_2\Big) \\
={}& \frac{\cos^4\theta}{\alpha^{2\rho-1}}\Big(\frac{1}{(2\rho+1)\alpha\tan\theta}e^{-(2\rho+1)\alpha n\tan\theta} + \frac{2^{2\rho}}{2\alpha\tan\theta}e^{-2\alpha n\tan\theta}\Big) \\
\le{}& \frac{\cos^4\theta}{\alpha^{2\rho-1}}\Big(\frac{1}{((2\rho+1)\alpha\tan\theta)^2 n} + \frac{2^{2\rho}}{(2\alpha\tan\theta)^2 n}\Big) \\
\le{}& \frac{1}{\alpha^{2\rho+1}}\cdot\frac{\cos^6\theta}{\sin^2\theta}\Big(\frac{1}{(2\rho+1)^2} + \frac{1}{4}\cdot 2^{2\rho}\Big)\cdot\frac{1}{n} \\
\le{}& \frac{1}{\alpha^{2\rho+1}}\cdot\frac{\cos^6\theta}{\sin^2\theta}\Big(1 - \frac{8}{9}\rho + \frac{1}{4}(1+3\rho)\Big)\cdot\frac{1}{n} \\
={}& \frac{1}{\alpha^{2\rho+1}}\cdot\frac{45\cos^6\theta - 5\cos^8\theta}{36\sin^2\theta}\cdot\frac{1}{n}
\end{aligned}$$

and

$$\begin{aligned}
|I_2| \le{}& \frac{|2\cos^2\theta - \cot^2\theta|\cdot\tan\theta}{n\alpha^{2\rho}}\Big(\int_{-\infty}^{0} e^{(2\rho+1)\alpha x_2\tan\theta}\,\mathrm{d}x_2 \\
&\qquad + 2^{2\rho}\int_0^{\infty} e^{-\alpha x_2\tan\theta}\,\mathrm{d}x_2\Big) \\
={}& \frac{|2\sin^2 - 1|\cdot\cos\theta}{n\alpha^{2\rho}\sin\theta}\Big(\frac{1}{(2\rho+1)\alpha\tan\theta} + \frac{2^{2\rho}}{\alpha\tan\theta}\Big) \\
={}& \frac{1}{\alpha^{2p+1}}\frac{\cos^2\theta\cdot|2\sin^2\theta - 1|}{\sin^2\theta}\Big(\frac{1}{2\cos^2\theta + 1} + 2^{2\cos^2\theta}\Big) \\
\le{}& \frac{1}{\alpha^{2p+1}}\frac{\cos^2\theta}{\sin^2\theta}\Big(1 - \frac{2}{3}\cos^2\theta + 1 + 3\cos^2\theta\Big) \\
={}& \frac{1}{\alpha^{2p+1}}\cdot\frac{72\cos^2\theta + 84\cos^4\theta}{36\sin^2\theta}\cdot\frac{1}{n}.
\end{aligned}$$

Finally, the bounds $|F| \le 1/\alpha$ on $\mathbb{R}_-$ and $|F| \le 2/\alpha$ on $\mathbb{R}_+$ give

$$\begin{aligned}
|I_3| &\le \frac{1}{n^2}\Big(\int_{-2n}^{-n} + \int_n^{2n} g_\rho(x_2)^2 F(x_2\tan\theta)\,\mathrm{d}x_2\Big)\\
&\le \frac{1}{n^2}\Big(\Big(\frac{1}{\alpha}\Big)^{2\rho+1} n + \Big(\frac{2}{\alpha}\Big)^{2\rho+1} n\Big)\\
&= \frac{1}{\alpha^{2\rho+1}}(1+2^{2\rho+1})\cdot\frac{1}{n}\\
&\le \frac{1}{\alpha^{2\rho+1}}(1+2(1+3\rho))\cdot\frac{1}{n}\\
&= \frac{1}{\alpha^{2\rho+1}}\cdot(3+6\cos^2\theta)\cdot\frac{1}{n}\\
&= \frac{1}{\alpha^{2\rho+1}}\cdot\frac{108\sin^2\theta+216\sin^2\theta\cos^2\theta}{36\sin^2\theta}\cdot\frac{1}{n}.
\end{aligned}$$

As a result, we obtain

$$R(h_n) \le R(g_\rho) + |I_1| + |I_2| + |I_3| \le -\Big(a - \frac{b}{n}\Big),$$

$$a := -R(g_\rho), \quad b := \frac{1}{\alpha^{2\rho+1}}\cdot\frac{B}{36\sin^2\theta},$$

$$\begin{aligned}
B &:= 108\sin^2\theta + 72\cos^2\theta + (84\cos^2\theta + 216\sin^2\theta)\cos^2\theta\\
&\quad + 45\cos^6\theta - 5\cos^8\theta\\
&= 108 - 36\cos^2\theta + (216 - 132\cos^2\theta)\cos^2\theta + 45\cos^6\theta - 5\cos^8\theta\\
&= 108 + 180\cos^2\theta - 132\cos^4\theta + 45\cos^6\theta - 5\cos^8\theta,
\end{aligned}$$

implying

$$Q(u) + \frac{\alpha^2}{4}\,\|u\|^2_{L^2(\Omega)} \le R(h_n) \le -\Big(a - \frac{b}{n}\Big).$$

On the other hand,

$$\begin{aligned}
\|u\|^2_{L^2(\Omega)} &\le \int_{-2n}^{2n} g_\rho(x_2)^2 F(x_2\tan\theta)\,\mathrm{d}x_2\\
&= \int_{-2n}^{0} F(x_2\tan\theta)^{2\rho+1}\,\mathrm{d}x_2 + \int_0^{2n} F(x_2\tan\theta)^{2\rho+1}\,\mathrm{d}x_2\\
&\le 2n\Big(\frac{1}{\alpha}\Big)^{2\rho+1} + 2n\Big(\frac{2}{\alpha}\Big)^{2\rho+1}\\
&\le \frac{1}{\alpha^{2\rho+1}}\cdot(2+4\cdot 2^{2\rho})\cdot n\\
&\le cn,
\end{aligned}$$

with

$$c := \frac{6(1 + 2\cos^2\theta)}{\alpha^{2p+1}},$$

and we have

$$\mu(\theta,\alpha) := -\frac{\alpha^2}{4} - \lambda(\theta,\alpha) \geq \frac{an - b}{cn^2} \quad \text{provided } an > b.$$

The right-hand side is optimized by $n = 2\, b/a$ resulting in

$$\mu(\theta,\alpha) \geq \frac{a^2}{4bc} = \alpha^2 \Lambda(\theta)$$

with $\Lambda(\theta)$ given in (9). □

References

[1] J. Behrndt, P. Exner, and V. Lotoreichik, Schrödinger operators with δ- and δ'-interactions on Lipschitz surfaces and chromatic numbers of associated partitions. *Rev. Math. Phys.* **26** (2014), no. 8, 1450015, 43 pp. MR 3256859 Zbl 1326.47050

[2] J. F. Brasche, P. Exner, Yu. A. Kuperin, and P. Šeba, Schrödinger operators with singular interactions. *J. Math. Anal. Appl.* **184** (1994), no. 1, 112–139. MR 1275948 Zbl 0820.47005

[3] B. M. Brown, M. S. P. Eastham, and I. G. Wood, An example on the discrete spectrum of a star graph. In P. Exner, J. P. Keating, P. Kuchment, T. Sunada, and A. Teplyaev (eds.), *Analysis on graphs and its applications.* Papers from the program held in Cambridge, January 8–June 29, 2007. Proceedings of Symposia in Pure Mathematics, 77. American Mathematical Society, Providence, R.I., 2008, 331–335. MR 2459878 Zbl 1165.35033

[4] B. M. Brown, M. S. P. Eastham, and I. G. Wood, Estimates for the lowest eigenvalue of a star graph. *J. Math. Anal. Appl.* **354** (2009), no. 1, 24–30. MR 2510414 Zbl 1194.05082

[5] V. Duchêne and N. Raymond, Spectral asymptotics of a broken δ-interaction. *J. Phys. A* **47** (2014), no. 15, 155203, 19 pp. MR 3191674 Zbl 1288.81044

[6] M. S. P. Eastham, Geometrically induced spectrum of the Schrödinger operator. *J. Assoc. Arab Univ. Basic Appl. Sci.* **7** (2009), 1–10.

[7] P. Exner, Leaky quantum graphs: a review. In P. Exner, J. P. Keating, P. Kuchment, T. Sunada, and A. Teplyaev (eds.), *Analysis on graphs and its applications.* Papers from the program held in Cambridge, January 8–June 29, 2007. Proceedings of Symposia in Pure Mathematics, 77. American Mathematical Society, Providence, R.I., 2008, 523–564. MR 2459890 Zbl 1153.81487

[8] P. Exner, Seize ans après. Appendix K in S. Albeverio, F. Gesztesy, R. Høegh-Krohn, and H. Holden, *Solvable models in quantum mechanics.* Second edition. AMS Chelsea Publishing, Providence, R.I., 2005, 453–484. MR 2105735 Zbl 1078.81003

[9] P. Exner and T. Ichinose, Geometrically induced spectrum in curved leaky wires. *J. Phys. A* **34** (2001), no. 7, 1439–1450. MR 1819942 Zbl 1002.81024

[10] P. Exner and S. Kondej, Gap asymptotics in a weakly bent leaky quantum wire. *J. Phys. A* **48** (2015), no. 49, 495301, 19 pp. MR 3434828 Zbl 1330.81101

[11] P. Exner and H. Kovařík, *Quantum waveguides.* Theoretical and Mathematical Physics. Springer, Cham, 2015. ISBN 978-3-319-18575-0 MR 3362506 Zbl 1314.81001

[12] P. Exner and K. Němcová, Bound states in point interaction star graphs. *J. Phys. A* **34** (2001), no. 38, 7783–7794. MR 1863337 Zbl 0994.81038

[13] P. Exner, K. Němcová, Leaky quantum graphs: approximations by point interaction Hamiltonians. *J. Phys. A* **36** (2003), no. 40, 10173–10193. MR 2023303 Zbl 1116.81312

[14] P. Exner and K. Pankrashkin, Strong coupling asymptotics for a singular Schrödinger operator with an interaction supported by an open arc. *Comm. Partial Differential Equations* **39** (2014), no. 2, 193–212. MR 3169783 Zbl 1291.35247

[15] P. Exner and S. Vugalter, On the existence of bound states in asymmetric leaky wires. *J. Math. Phys.* **57** (2016), no. 2, 022104, 15 pp. MR 3456722 Zbl 1332.81070

[16] B. Helffer and A. Morame, Magnetic bottles for the Neumann problem: the case of dimension 3. *Proc. Indian Acad. Sci. Math. Sci.* **112** (2002), no. 1, 71–84. Spectral and inverse spectral theory (Goa, 2000). MR 1894543 Zbl 1199.81016

[17] S. Kondej and V. Lotoreichik, Weakly coupled bound state of 2-D Schrödinger operator with potential-measure. *J. Math. Anal. Appl.* **420** (2014), no. 2, 1416–1438. MR 3240086 Zbl 1298.35128

[18] V. Lotoreichik, Note on 2D Schrödinger operators with δ-interactions on angles and crossing lines. *Nanosyst. Phys. Chem. Math.* **4** (2013), no. 2, 166–172. Zbl 1343.81093

[19] K. Lu, X.-B. Pan, Surface nucleation of superconductivity in 3-dimensions. *J. Differential Equations* **168** (2000), no. 2, 386–452. Special issue in celebration of J. K. Hale's 70th birthday, Part 2 (Atlanta, GA/Lisbon, 1998). MR 1808455 Zbl 0972.35152

On the boundedness and compactness of weighted Green operators of second-order elliptic operators

Yehuda Pinchover

Dedicated to Pavel Exner on the occasion of his 70th birthday

1 Introduction

Positive (Dirichlet) Green functions of second-order linear elliptic operators with real coefficients and their induced integral operators are among the most important building blocks of the elliptic theory for such operators, and in particular, for the qualitative theory of positive solutions of the corresponding homogeneous equations. In many problems, and in particular in the study of positive solutions, the underling topology is the open compact topology, i.e., the topology of locally compact convergence (e.g., in the Martin boundary theory, in the study of the heat kernel, and in criticality theory). On the other hand, when dealing with spectral theory of such operators, in the study of semigroups generated by such operators, or in the study of well-posedness of boundary value problems, one should usually specify a relevant Banach space. Frequently, one takes one of the classical Lebesgue spaces as the underlying space, but a priori, it is not clear why these spaces are the appropriate functional spaces to study various elliptic problems for a specific operator in a given domain.

In the present paper, we introduce for a given second-order linear elliptic operator L which is defined on a noncompact manifold Ω of dimension d and admits a positive Green function, and for a given positive weight function W, a family of weighted Lebesgue spaces $L^p(\phi_p)$, where $1 \leq p \leq \infty$. Fix $\mu \leq \lambda_0$, where $\lambda_0 = \lambda_0(L, W, \Omega)$ is the generalized principal eigenvalue (see (3)). The weight ϕ_p is given by

$$\phi_p := \phi^{-1}(\phi W \tilde{\phi})^{1/p} = \begin{cases} \phi^{-1}(\phi W \tilde{\phi})^{1/p} & 1 \leq p < \infty, \\ \phi^{-1} & p = \infty, \end{cases}$$

where ϕ (resp. $\tilde{\phi}$) is a fixed positive solution of the equation $(L - \mu W)u = 0$ (resp. $(L^\star - \mu W)u = 0$) in Ω, and $L^\star$ is the formal adjoint of L (see Section 2 for a detailed discussion on these spaces, and also for the needed terminology and some preliminary results).

Remark 1.1. Clearly, if the positive Liouville theorem holds for $L_\mu := L - \mu W$ and $L^\star - \mu W$ in Ω, then $L^p(\phi_p)$ is uniquely defined. In addition, $L^1(\phi_1)$ is always independent of ϕ while $L^\infty(\phi_\infty)$ is independent of $\tilde{\phi}$ and W. In [36], $L^\infty(\phi_\infty)$ was introduced, and some properties of weighted Green operators on $L^\infty(\phi_\infty)$ were studied. Moreover, if L is symmetric, then one may choose $\tilde{\phi} = \phi$, and then $L^2(\phi_2) = L^2(\Omega, W)$, and hence, this space is ϕ independent. We also note that $\{L^p(\phi_p)\}_{p\geq 1}$ is a family of real interpolation spaces. The latter observation is used to prove several results of the paper.

The aim of the present paper is to study some fundamental properties of the induced weighted Green operators $\mathcal{G}_\lambda$ on the weighted Lebesgue spaces $L^p(\phi_p)$. Here

$$\mathcal{G}_\lambda f(x) := \int_\Omega G^\Omega_{L_\lambda}(x, y) W(y) f(y) \mathrm{d}\nu(y),$$

where $G^\Omega_{L_\lambda}$ is the positive minimal Green function of the elliptic operator L_λ in Ω, and $\lambda < \mu \leq \lambda_0$.

We prove in Section 3 that $\mathcal{G}_\lambda$ is bounded on $L^p(\phi_p)$ for any $1 \leq p \leq \infty$ with a bound independent of p, ϕ, and $\tilde{\phi}$ (see Theorem 3.1). In Section 4, we study the existence and uniqueness of a principal eigenfunction for these weighted Green operators on $L^p(\phi_p)$, and the simplicity of the corresponding principal eigenvalue for $1 \leq p \leq \infty$. In particular, Theorem 4.4 gives sufficient conditions under which the positive solution ϕ is an eigenfunction of $\mathcal{G}_\lambda$ in $L^p(\phi_p)$. Theorem 4.5 asserts that in the positive critical case ϕ is the unique (up to a multiplicative constant) nonnegative eigenfunction of the operator $\mathcal{G}_\lambda \restriction_{L^p(\phi_p)}$. In the special cases $p = 1, 2, \infty$, we obtain *simplicity* results of the corresponding principal eigenvalue (see theorems 4.5, 4.6 and 4.7, respectively).

Next, we show in Section 5 that for $1 \leq p < \infty$, the weighted Green operator is a resolvent of a densely defined closed operator A_p such that $A_p = -W^{-1}L$ on $C_0^\infty(\Omega)$. It turns out that under some further assumptions, A_p generates a strongly continuous contraction semigroup (see Theorem 5.2). Finally, in Section 6 we prove that if W is a *(semi)small perturbation* of L in Ω, then for any $1 \leq p \leq \infty$, the associated weighted Green operator is compact on $L^p(\phi_p)$ (Theorem 6.1), and the corresponding spectrum is p-independent (Theorem 6.4).

We note that if in addition L is symmetric and W is strictly positive, then it follows from [36] that for any $k \geq 1$ the quotient ϕ_k/ϕ has a continuous extension up to the Martin boundary of the pair (Ω, L), where ϕ is the ground state of L with a principal eigenvalue $\lambda_0 = \lambda_0(L, W, \Omega)$, and ϕ_k is the k-th (weighted) eigenfunction in $L^2(\Omega, W\mathrm{d}\nu)$. It follows from the p-independence of the spectrum that in fact, $\phi, \phi_k \in L^p(\phi_p)$, for all $1 \leq p \leq \infty$.

Norm estimates for the Green operator has been studied by many authors. For example, for second-order elliptic operators with up to the boundary regular coefficients defined on smooth bounded domains, one may use the well known behavior of the Green function [19] to prove the boundedness of the Green operators in L^p spaces. Weighted L^∞-norm estimates for $\mathcal{G}_\lambda$ has been established by Hansen in [20] under the assumption that the Green function satisfies a pointwise generalized triangle property. In [2], H. Aikawa proved for $L = -\Delta$ and $W = \mathbf{1}$, that under certain conditions on Ω, the Green operator $\mathcal{G}$ is a bounded linear operator from L^p to L^∞, where $d/2 < p \leq \infty$. We also mention that A. Grigor′yan and J. Hu [18] have shown that for a regular Dirichlet form, the Green operator is bounded on L^2, while for diffusion operators with bounded drifts on $\mathbb{R}^d$, A. T. Hill has established in [21] bounds on $\|\mathcal{G}_\lambda\|_{L^1(\mathbb{R}^d)}$.

Let us mention some results concerning the existence, uniqueness and simplicity of the principal eigenvalue in nonsmooth domains. In the celebrated paper [8], H. Berestycki, L. Nirenberg and S. R. S. Varadhan considered a general *bounded* domain and a *uniformly elliptic* operator L with bounded coefficients, and proved (in the framework of strong solutions) the existence of a unique and simple principal eigenvalue (for related results in unbounded domains see [9]). Moreover, in [10] I. Birindelli proved that for bounded domains such Green operators are compact on $L^d(\Omega)$. Various sufficient conditions that guarantee that L has a pure point-spectrum in certain spaces are given for example in [7], [24], [28], and references therein.

Finally, the problem of the p-independence of the spectrum of generators of semigroups on L^p was raised by B. Simon in [41] for Schrödinger semigroups in $\mathbb{R}^d$ and has been studied in many papers, see for example [5], [14], [15], [23], [26], [27], [39], [40], [41], [42], and references therein.

2 Preliminaries

2.1 Elliptic operators and positive solutions

Let Ω be a domain in $\mathbb{R}^d$, or more generally, a noncompact connected C^2-smooth Riemannian manifold of dimension d. By a positive function we mean a strictly positive function. We assume that ν is a positive measure on Ω, satisfying $\mathrm{d}\nu = f\,\mathrm{vol}$ with f a positive measurable function; vol being the volume form of Ω (which is just the Lebesgue measure in the case of a domain Ω in $\mathbb{R}^d$). We write $\Omega_1 \Subset \Omega_2$ if Ω_2 is open, $\overline{\Omega_1}$ is compact and $\overline{\Omega_1} \subset \Omega_2$. Denote by $B(x_0, \delta)$ the open ball of radius $\delta > 0$ centered at x_0. Let $\mathbf{1}$ be the constant function on Ω taking at any point $x \in \Omega$ the value 1. For a matrix $A(x) = \left[a^{ij}(x)\right]$ and a vector field $b(x) = (b^j(x))$

we denote

$$(A(x)\,\xi)^i := \sum_{j=1}^{d} a^{ij}(x)\,\xi_j,\ \ b(x)\cdot\xi := \sum_{j=1}^{d} b^j(x)\,\xi_j,\ \ \text{where } \xi=(\xi_1,\dots,\xi_d)\in\mathbb{R}^d.$$

We associate to Ω a fixed *exhaustion* $\{\Omega_k\}_{k=1}^{\infty}$, i.e., a sequence of smooth, relatively compact subdomains such that $\Omega_1 \neq \emptyset$, $\Omega_k \Subset \Omega_{k+1}$ and $\bigcup_{k=1}^{\infty} \Omega_k = \Omega$. For every $k \geq 1$, we denote $\Omega_k^\star = \Omega \setminus \overline{\Omega_k}$.

Let L be a linear, second-order, elliptic operator defined on Ω. We assume that the coefficients of L are real, and that in any coordinate system $(U; x_1, \dots, x_d)$, the operator L is of the divergence form

$$Lu := -\operatorname{div}(A(x)\nabla u + u\tilde{b}(x)) + b(x)\cdot\nabla u + c(x)u. \tag{1}$$

Here, the minus divergence is the formal adjoint of the gradient with respect to the measure ν.

We assume that for every $x \in \Omega$ the matrix $A(x) := [a^{ij}(x)]$ is symmetric, and the associated real quadratic form

$$\xi\cdot A(x)\,\xi := \sum_{i,j=1}^{d} \xi_i a^{ij}(x)\,\xi_j \quad \xi\in\mathbb{R}^d$$

is positive definite. Moreover, throughout the paper it is assumed that L is locally uniformly elliptic, and the coefficients of L are real valued and locally sufficiently regular in Ω. All our results hold for example when L is of the form (1), and A and f are locally Hölder continuous, b, $\tilde{b} \in L^p_{\mathrm{loc}}(\Omega;\mathbb{R}^n, \mathrm{d}x)$, and $c \in L^{p/2}_{\mathrm{loc}}(\Omega;\mathbb{R}, \mathrm{d}x)$ for some $p > d$. However it would be apparent from the proofs that any conditions that guarantee standard elliptic regularity theory are sufficient. By a *potential* defined in Ω, we mean a function $V \in L^{p/2}_{\mathrm{loc}}(\Omega;\mathbb{R},\mathrm{d}x)$ for some $p > d$.

The formal adjoint $L^\star$ of the operator L is defined on its natural space $L^2(\Omega,\mathrm{d}\nu)$. When L is in divergence form (1) and $b = \tilde{b}$, the operator

$$Lu = -\operatorname{div}(A\nabla u + ub) + b\cdot\nabla u + cu,$$

is *symmetric* in the space $L^2(\Omega, \mathrm{d}\nu)$. Throughout the paper, we call this setting the *symmetric case*. We note that if L is symmetric and b is smooth enough, then L is in fact a Schrödinger-type operator of the form

$$Lu = -\operatorname{div}\big(A\nabla u\big) + Vu, \quad \text{where } V := (c - \operatorname{div} b).$$

By a solution v of the equation $Lu = 0$ in Ω, we mean $v \in W^{1,2}_{\mathrm{loc}}(\Omega)$ that satisfies the equation $Lu = 0$ in Ω in the weak sense. Subsolutions and supersolutions are defined similarly. We denote the cone of all positive solutions of the equation $Lu = 0$ in Ω by $\mathcal{C}_L(\Omega)$.

Remark 2.1. We would like to point out that the theory of positive solutions of the equation $Lu = 0$ in Ω (the so-called *criticality theory*), and in particular the results of this paper, are also valid for the class of classical solutions of locally uniformly elliptic operators of the form

$$Lu = -\sum_{i,j=1}^{d} a^{ij}(x)\partial_i\partial_j u + b(x)\cdot\nabla u + c(x)u, \qquad (2)$$

with real and locally Hölder continuous coefficients, and for the class of strong solutions of locally uniformly elliptic operators of the form (2) with locally bounded coefficients (provided that the formal adjoint operator also satisfies the same assumptions). Nevertheless, for the sake of clarity, we prefer to present our results only for the class of weak solutions.

Fix a nonzero nonnegative potential W defined in Ω, and for $\lambda \in \mathbb{R}$ denote by L_λ the elliptic operator $L - \lambda W$. Consider the (weighted) *generalized principal eigenvalue* of the operator L

$$\lambda_0 = \lambda_0(L, W, \Omega) := \sup\{\lambda \in \mathbb{R} \mid \mathcal{C}_{L_\lambda}(\Omega) \neq \emptyset\}. \qquad (3)$$

Note that if $\lambda_0 \neq -\infty$ (as assumed throughout the present paper), then in fact, $\lambda_0 = \max\{\lambda \in \mathbb{R} \mid \mathcal{C}_{L_\lambda}(\Omega) \neq \emptyset\}$.

If $\lambda_0 \geq 0$, then for every $k \geq 1$ the Dirichlet Green function $G_L^{\Omega_k}(x,y)$ of the operator L in Ω_k exists and is positive. By the generalized maximum principle, $\{G_L^{\Omega_k}(x,y)\}_{k=1}^{\infty}$ is an increasing sequence converging as $k \to \infty$ either to $G_L^{\Omega}(x,y)$, the positive *minimal Green function* of L in Ω, and then L is said to be a *subcritical operator* in Ω , or to infinity and in this case L is *critical* in Ω. If $\mathcal{C}_L(\Omega) = \emptyset$, then L is *supercritical* in Ω, see [29] and [32] (cf. [41]).

It follows that L is critical (resp. subcritical) in Ω, if and only if $L^\star$ is critical (resp. subcritical) in Ω. Clearly, L_λ is subcritical in Ω for every $\lambda \in (-\infty, \lambda_0)$, and supercritical for $\lambda > \lambda_0$. Furthermore, if L is critical in Ω, then $\lambda_0 = 0$, $\mathcal{C}_L(\Omega)$ is a one-dimensional cone, and any positive supersolution of the equation $Lu = 0$ in Ω is a solution. So, in the critical case, $\phi \in \mathcal{C}_L(\Omega)$ is uniquely defined (up to a multiplicative positive constant), and such ϕ is called the *Agmon ground state of L in* Ω, see [1], [29], and [32].

Subcriticality is a stable property in the following sense. If L is subcritical in Ω and V is a potential with a compact support in Ω, then there exists $\varepsilon > 0$ such that $L - \mu V$ is subcritical, for all $|\mu| < \varepsilon$, see [29] and [32]. On the other hand, if L is critical in Ω and V is a nonzero, *nonnegative* potential, then for any $\varepsilon > 0$ the operator $L + \varepsilon V$ is subcritical and $L - \varepsilon V$ is supercritical in Ω.

Definition 2.2 (Agmon [1]). Let L be an elliptic operator of the form (1) defined in Ω. A function $u \in \mathcal{C}_L(\Omega_n^\star)$ is said to be a *positive solution of the equation* $Lu = 0$ *of minimal growth in a neighborhood of infinity in* Ω, if for any $k > n$ and any $v \in C(\overline{\Omega_k^\star})$ which is a positive supersolution of the equation $Lw = 0$ in $\Omega_k^\star$, the inequality $u \le v$ on $\partial\Omega_k$ implies that $u \le v$ in $\Omega_k^\star$.

In the sequel, in order to simplify our terminology, we will call a positive minimal Green function – a *Green function*, an Agmon ground state – a *ground state*, and a positive solution of minimal growth in a neighborhood of infinity in Ω, a *positive solution of minimal growth* in Ω.

It turns out that if L is subcritical in Ω, then for any fixed $y \in \Omega$ the Green function $G_L^\Omega(\cdot, y)$ is a positive solution of the equation $Lu = 0$ in $\Omega \setminus \{y\}$ of minimal growth in Ω. On the other hand, if L is critical in Ω, then the ground state ϕ is a (global) positive solution of the equation $Lu = 0$ in Ω of minimal growth in Ω.

Fix a nonzero nonnegative potential W defined in Ω. Let v and $\tilde{v}$ be positive solutions of the equations $L_\mu u = 0$ and $L_\mu^\star u = 0$ in Ω, respectively, where $\mu \le \lambda_0$. Then [34] for every $\lambda < \mu$ we have

$$
\begin{aligned}
&\int_\Omega G_{L_\lambda}^\Omega(x, y) W(y) v(y)\, \mathrm{d}\nu(y) \le \frac{v(x)}{\mu - \lambda} \quad \text{for all } x \in \Omega,\\
&\int_\Omega G_{L_\lambda}^\Omega(x, y) W(x) \tilde{v}(x)\, \mathrm{d}\nu(x) \le \frac{\tilde{v}(y)}{\mu - \lambda} \quad \text{for all } y \in \Omega.
\end{aligned}
\tag{4}
$$

Moreover, in each of the inequalities in (4) either equality or strict inequality holds for all points in Ω and $\lambda < \mu$. If equality holds for v (resp. $\tilde{v}$), then v (resp. $\tilde{v}$) is said to be a positive *invariant solution* of the equation $L_\mu u = 0$ (resp. $L_\mu^\star u = 0$) in Ω, or μ*-invariant solution* of the operator L (resp. $L^\star$) in Ω.

Assume now that L_{λ_0} is critical in Ω with $\lambda_0 \in \mathbb{R}$, and let ϕ and $\tilde{\phi}$ be the ground states of L_{λ_0} and $L_{\lambda_0}^\star$, respectively. Then (see Theorem 2.1 in [34]) ϕ and $\tilde{\phi}$ are positive invariant solutions of the equations $L_{\lambda_0} u = 0$ and $L_{\lambda_0}^\star u = 0$, respectively,

Hence, for every $\lambda < \lambda_0$ we have

$$\int_\Omega G^\Omega_{L_\lambda}(x,y)W(y)\phi(y)\,\mathrm{d}\nu(y) = \frac{\phi(x)}{\lambda_0 - \lambda} \quad \text{for all } x \in \Omega,$$
$$\int_\Omega G^\Omega_{L_\lambda}(x,y)W(x)\tilde{\phi}(x)\,\mathrm{d}\nu(x) = \frac{\tilde{\phi}(y)}{\lambda_0 - \lambda} \quad \text{for all } y \in \Omega. \tag{5}$$

Assume further that W is a positive function. If $\phi\,\tilde{\phi} \in L^1(\Omega, W\,\mathrm{d}\nu)$, then L_{λ_0} is called *positive-critical in* Ω *with respect to* W. Otherwise, L_{λ_0} is called *null-critical in* Ω *with respect to* W.

Remark 2.3. Let $\mu \leq \lambda_0$, and suppose that L_μ is a subcritical operator in Ω. Let v and $\tilde{v}$ be positive solutions of the equations $L_\mu u = 0$ and $L^\star_\mu u = 0$ in Ω, respectively, such that $v\,\tilde{v} \in L^1(\Omega, W\,\mathrm{d}\nu)$. Since for every fixed x (resp. y) the function $G^\Omega_{L_\mu}(x,\cdot)$ (resp. $G^\Omega_{L_\mu}(\cdot,y)$) is a positive solution of the operator $L^\star_\mu$ (resp. L_μ) of minimal growth in Ω, the integrability condition $v\,\tilde{v} \in L^1(\Omega, W\,\mathrm{d}\nu)$ implies that

$$\int_\Omega G^\Omega_{L_\mu}(x,y)W(y)v(y)\,\mathrm{d}\nu(y) < \infty \quad \text{and} \quad \int_\Omega G^\Omega_{L_\mu}(x,y)W(x)\tilde{v}(x)\,\mathrm{d}\nu(x) < \infty. \tag{6}$$

In light of Lemma 2.1 in [35], (6) implies that v and $\tilde{v}$ are not μ-invariant positive solutions of the equations $L_\mu u = 0$ and $L^\star_\mu u = 0$ in Ω, respectively. In other words, if $v \in \mathcal{C}_{L_\mu}(\Omega)$ (resp. $\tilde{v} \in \mathcal{C}_{L^\star_\mu}(\Omega)$) is a positive μ-invariant solution of the operator L (resp. $L^\star$), and $v\,\tilde{v} \in L^1(\Omega, W\,\mathrm{d}\nu)$, then $\mu = \lambda_0$ and L_{λ_0} is positive-critical with respect to W in Ω.

The following example, a modification of the counterexamples to Stroock's conjecture given in [35], demonstrates that for $\mu = \lambda_0$ there exists a subcritical operator L_{λ_0} and a potential $W > 0$ satisfying all the properties of Remark 2.3.

Example 2.4. Consider the operator $L := -\rho\Delta$ on $\mathbb{R}^d$, where $d \geq 3$, and ρ is a strictly positive smooth function. Let $W := \mathbf{1}$. Then L is a subcritical operator in $\mathbb{R}^d$, and it follows from Liouville's theorem that the functions $v = \mathbf{1}$ and $\tilde{v} = {}^1\!/\!_\rho$ are (up to a multiplicative constant) the unique positive solutions of the equations $Lu = 0$ and $L^\star u = 0$ in $\mathbb{R}^d$, respectively. We claim that there exists a smooth positive function ρ so that $v\tilde{v} = {}^1\!/\!_\rho \in L^1(\mathbb{R}^d, \mathrm{d}x)$, and $\lambda_0(L, \mathbf{1}, \mathbb{R}^d) = 0$.

Indeed, let $0 < \beta < 1$, and $x_k := (k, 0, \dots, 0)$, where, $k = 1, 2, \dots$. Finally let $\{\varepsilon_k\} \subset (0,1)$ be a sequence satisfying $\sum_{k=1}^\infty \varepsilon_k^{d-(2+\beta)} < \infty$. Take a smooth positive function $\tilde{v} \in L^1(\mathbb{R}^d, \mathrm{d}x)$ satisfying $\tilde{v}(x)\restriction_{B(x_k,\varepsilon_k)} = (\varepsilon_k)^{-(2+\beta)}$. In particular, $v\tilde{v} \in L^1(\mathbb{R}^d, \mathbf{1}\mathrm{d}x)$.

On the other hand, we clearly have $\lambda_0(L, \mathbf{1}, B(x_k, \varepsilon_k)) < C\varepsilon_k^{\beta}$, and therefore, $\lambda_0(L, \mathbf{1}, \mathbb{R}^d) = 0$. Moreover, by Remark 2.3, the unique positive solution v (resp. $\tilde{v}$) of the equation $L_{\lambda_0}u = 0$ (resp. $L^{\star}_{\lambda_0}u = 0$) is not λ_0-invariant.

Remark 2.5. We note that Example 2.4 is in fact a strengthening of the counterexamples to Stroock's conjecture given in [35]. It gives an example of a subcritical operator L on $\mathbb{R}^d$, $d \geq 3$, with $\lambda_0 = 0$, such that the operators L and $L^\star$ do not admit λ_0-invariant positive solutions, and in addition, the product of positive entire solutions of the equations $Lu = 0$ and $L^\star u = 0$ is in $L^1(\mathbb{R}^d)$. Recall that if a Schrödinger-type operator admits a 'small' positive solution ψ in Ω (and in particular an L^2-positive solution), then the operator is critical in Ω, and in particular, ψ is an invariant positive solution, see [37].

The following notions of small and semismall perturbations play a fundamental role in criticality theory, see [29], [30], [32], and [33]. Semismall perturbations revisit in the present paper. It turns out that they guarantee the compactness of the weighted Green operators in $L^p(\phi_p)$ for all $1 \leq p \leq \infty$ (see Section 6).

Definition 2.6. Let L be a subcritical operator in Ω, and let V be a potential.

(i) We say that V is a *small perturbation* of L in Ω if

$$\lim_{k\to\infty}\left\{\sup_{x,y\in\Omega_k^\star}\int_{\Omega_k^\star}\frac{G_L^\Omega(x,z)|V(z)|G_L^\Omega(z,y)}{G_L^\Omega(x,y)}\,\mathrm{d}\nu(z)\right\} = 0.$$

(ii) We say that V is a *semismall perturbation* of L in Ω if for some (all) fixed $x_0 \in \Omega$ we have

$$\lim_{k\to\infty}\left\{\sup_{y\in\Omega_k^\star}\int_{\Omega_k^\star}\frac{G_L^\Omega(x_0,z)|V(z)|G_L^\Omega(z,y)}{G_L^\Omega(x_0,y)}\,\mathrm{d}\nu(z)\right\} = 0.$$

Remark 2.7. (i) A small perturbation of L in Ω is a semismall perturbation of L and $L^\star$ in Ω, see [30].

(ii) We note that λ_0 is well defined by (3) even if the potential W does not have a definite sign. It turns out (see [30] and [33]) that if L is subcritical and $W \not\leq 0$ is a semismall (resp. small) perturbation of $L^\star$ in Ω, then $\lambda_0 > 0$, and L_{λ_0} is critical in Ω with a ground state ϕ. Moreover, for each $\lambda < \lambda_0$ such that the positive Green function $G_{L_\lambda}^\Omega$ exists there exists a positive constant $C_{\lambda,x_0,\varepsilon}$ (resp. C_λ) such that

$$\begin{aligned}(C_{\lambda,x_0,\varepsilon})^{-1}G_{L_\lambda}^\Omega(x,x_0) &\leq \phi(x)\\ &\leq C_{\lambda,x_0,\varepsilon}G_{L_\lambda}^\Omega(x,x_0) \quad \text{for all } x\in\Omega,\ \mathrm{dist}(x,x_0)>\varepsilon,\end{aligned} \tag{7a}$$

resp.

$$(C_\lambda)^{-1}G^\Omega_{L_\lambda}(x,y) \le G^\Omega_L(x,y) \le C_\lambda G^\Omega_{L_\lambda}(x,y) \quad \text{for all } x,y \in \Omega, x \ne y. \tag{7b}$$

Since L_{λ_0} is critical if and only if $L^\star_{\lambda_0}$ is critical, (5) and (7) imply that if $W > 0$ is a semismall (resp. small) perturbation of $L^\star$ in Ω, then L_{λ_0} is positive-critical. In particular, ϕ satisfies (5).

(iii) Murata [31] proved that if L is symmetric and the corresponding (Dirichlet) semigroup generated by L is *intrinsically ultracontractive* on $L^2(\Omega)$ (see [16]), then **1** is a small perturbation of L in Ω. On the other hand, an example of Bañuelos and Davis in [6] gives us a finite area domain $\Omega \subset \mathbb{R}^2$ such that **1** is a small perturbation of the Laplacian in Ω, but the corresponding semigroup is not intrinsically ultracontractive.

2.2 Functional spaces

Let B be a Banach space and $B^\star$ its dual. If $T\colon B \to B$ is a (bounded) operator on B, we denote by $T^\star$ its dual, and the operator norm of T by $\|T\|_B$. The range and the kernel of T are denoted by $R(T)$ and $N(T)$, respectively. We denote by $\sigma(T)$, $\sigma_{\mathrm{point}}(T)$ and $\rho(T)$ the spectrum, the point-spectrum, and the resolvent set of the operator T. If $\lambda \in \rho(T)$, then we denote by $R(\lambda, T) := (\lambda I - T)^{-1}$ the resolvent of T, where I is the identity map on B. For every $f \in B$ and $g^\star \in B^\star$ we use the notation $\langle g^\star, f\rangle := g^\star(f)$. If T acts on two Banach spaces X and Y, we distinguish the operators by using the notation $T\restriction_X$, $T\restriction_Y$, respectively.

Let $1 \le p < \infty$, and let w be a fixed (strictly) positive measurable weight function defined on Ω. Denote the real ordered Banach space

$$L^p(w) := L^p(\Omega, w^p \,\mathrm{d}\nu) = \{u \mid uw \in L^p(\Omega, \mathrm{d}\nu)\}$$

equipped with the norm

$$\|u\|_{p,w} := \|uw\|_p = \left[\int_\Omega |u(x)\,w(x)|^p \,\mathrm{d}\nu(x)\right]^{1/p}.$$

For $p = \infty$, let

$$L^\infty(w) := \{u \mid \; uw \in L^\infty(\Omega, \mathrm{d}\nu)\}$$

equipped with the norm

$$\|u\|_{\infty,w} := \|uw\|_\infty = \operatorname*{ess\,sup}_\Omega(|u|w).$$

The ordering on $L^p(w)$ is the natural pointwise ordering of functions. For the purpose of spectral theory, we consider also the canonical complexification of $L^p(w)$ without changing our notation.

For $1 \le p \le \infty$, let p' be the usual conjugate exponent of p, so, $1/p + 1/p' = 1$. It is well-known that for $1 \le p < \infty$, $(L^p(w))^\star = L^{p'}(w^{-1})$, and in particular, the space $L^p(w)$ is reflexive for all $1 < p < \infty$.

Let $W, \phi, \tilde{\phi}$ be positive continuous functions in Ω. For $1 \le p \le \infty$, denote

$$\phi_p := \phi^{-1}(\phi W \tilde{\phi})^{1/p}, \quad \tilde{\phi}_p := \tilde{\phi}^{-1}(\phi W \tilde{\phi})^{1/p}, \tag{8}$$

and consider the corresponding family of weighted Lebesgue spaces $L^p(\phi_p)$, and $L^p(\tilde{\phi}_p)$.

We note that $L^1(\phi_1)$ is independent of ϕ while $L^\infty(\phi_\infty)$ is independent of $\tilde{\phi}$ and W. Moreover, if $\tilde{\phi} = \phi$ (which is often the case when L is symmetric), then $L^2(\phi_2) = L^2(\Omega, W\,\mathrm{d}\nu)$ and this space is ϕ independent.

It can be easily checked that for $1 \le p < \infty$ we have

$$(L^p(\phi_p))^\star = L^{p'}(\tilde{\phi}_{p'}), \tag{9}$$

where the pairing between $L^p(\phi_p)$ and $L^{p'}(\tilde{\phi}_{p'})$ is given by

$$\langle g^\star, f\rangle = \int_\Omega g^\star(x) W(x) f(x)\,\mathrm{d}\nu(x) \quad \text{for all } g^\star \in L^{p'}(\tilde{\phi}_{p'}),\ f \in L^p(\phi_p).$$

Here the duality is provided by the bilinear rather than the sesquilinear form.[1]

Suppose now that

$$\phi W \tilde{\phi} \in L^1(\Omega, \mathrm{d}\nu), \quad \int_\Omega \phi(x) W(x) \tilde{\phi}(x)\,\mathrm{d}\nu(x) = 1. \tag{10}$$

Then by the Hölder inequality we have the continuous embeddings

$$L^\infty(\phi_\infty) \subset L^q(\phi_q) \subset L^p(\phi_p) \subset L^1(\phi_1), \tag{11}$$

for all $1 \le p \le q \le \infty$, and for $f \in L^\infty(\phi_\infty)$ we have

$$\|f\|_{1,W\tilde{\phi}} = \|f\|_{1,\phi_1} \le \|f\|_{p,\phi_p} \le \|f\|_{q,\phi_q} \le \|f\|_{\infty,\phi_\infty} = \|f\|_{\infty,\phi^{-1}}.$$

[1] This is not essential, but simplifies somewhat the calculations.

Moreover, $\|f\|_{1,\phi_1} = \|f\|_{\infty,\phi_\infty}$ if and only if $|f| = \phi = 1$ almost everywhere. In particular, $\phi \in L^p(\phi_p)$ for every $1 \le p \le \infty$, and (10) implies that

$$\|\phi\|_{1,\phi_1} = \|\phi\|_{p,\phi_p} = \|\phi\|_{\infty,\phi_\infty} = 1 \quad \text{for all } 1 \le p \le \infty,$$

so, the norms of the embeddings in (11) equal 1. Moreover, these embeddings are dense. We also note that if $\tilde{\phi} = \phi$ (as in the symmetric case), then

$$(L^1(W\phi))^\star = L^\infty(\phi^{-1}) \subset L^2(\Omega, W\,\mathrm{d}\nu) \subset L^1(W\phi).$$

Remark 2.8. Throughout the paper we fix an operator L of the form (1), a positive potential W, $\mu \le \lambda_0$, and ϕ, $\tilde{\phi}$ two positive solutions of the equations $L_\mu u = 0$ and $L_\mu^\star u = 0$ in Ω, respectively. We study properties of a family of the corresponding weighted Green operators on $L^p(\phi_p)$ and $L^p(\tilde{\phi}_p)$. We note that if $\mu = \lambda_0$ and L_{λ_0} is critical in Ω, then the spaces $L^p(\phi_p)$ and $L^p(\tilde{\phi}_p)$ are uniquely defined.

Remark 2.9. Let ϕ and $\tilde{\phi}$ be two fixed positive solutions of the equations $Lu = 0$ and $L^\star u = 0$ in Ω, respectively. For $\mu \le \lambda_0$, define the operator

$$L_\mu^\phi := \frac{1}{\phi} L_\mu \phi = \frac{1}{\phi} L\phi - \mu W = L^\phi - \mu W,$$

which is called *Doob's ϕ-transform* (or the *ground state transform with respect to ϕ*) of the operator L_μ. Note that for $\mu \le \lambda_0$ the operator L_μ is subcritical in Ω if and only if L_μ^ϕ is subcritical in Ω, and we have

$$G_{L_\mu^\phi}^\Omega(x, y) = \frac{1}{\phi(x)} G_{L_\mu}^\Omega(x, y)\phi(y).$$

Clearly, $L^\phi \mathbf{1} = 0$ and $(L^\phi)^\star(\phi\tilde{\phi}) = 0$. In particular, L^ϕ is a *diffusion operator.* We note that for $1 \le p \le \infty$, the weighted L^p-spaces associated with the positive solutions $\mathbf{1}$ and $\phi\tilde{\phi}$ of the equations $L^\phi u = 0$ and $(L^\phi)^\star u = 0$, respectively, are just $L^p(\Omega, \phi W \tilde{\phi}\,\mathrm{d}\nu)$. So, in this case (which corresponds to the class of diffusion operators) the corresponding one-parameter weights are p-independent.

3 Boundedness of the Green operators

Fix a positive potential W and $\mu \le \lambda_0$. Let ϕ and $\tilde{\phi}$ be two fixed positive solutions of the equations $L_\mu u = 0$ and $L_\mu^\star u = 0$ in Ω, respectively. For $1 \le p \le \infty$ let ϕ_p and $\tilde{\phi}_p$ be the functions defined in (8). Note that we do not assume below neither that ϕ and $\tilde{\phi}$ are invariant solutions nor that the integrability condition (10) is satisfied.

For $\lambda < \mu$, we introduce the integral operators

$$\mathcal{G}_\lambda f(x) := \int_\Omega G^\Omega_{L_\lambda}(x,y)W(y)f(y)\mathrm{d}\nu(y),$$

$$\mathcal{G}^\odot_\lambda f(y) := \int_\Omega G^\Omega_{L_\lambda}(x,y)W(x)f(x)\mathrm{d}\nu(x).$$

In the present section we study for $1 \le p \le \infty$ the boundedness of the *weighted Green operators* $\mathcal{G}_\lambda$ and $\mathcal{G}^\odot_\lambda$ on $L^p(\phi_p)$ and $L^p(\tilde{\phi}_p)$, respectively.

Theorem 3.1. *Let L be an elliptic operator on Ω of the form* (1), *and let W be a positive potential. Fix $\mu \le \lambda_0$, and let ϕ and $\tilde{\phi}$ be two fixed positive solutions of the equations $L_\mu u = 0$ and $L^\star_\mu u = 0$ in Ω, respectively. Then*

(1) *for $1 \le p \le \infty$, the operator $\mathcal{G}_\lambda \restriction_{L^p(\phi_p)}$ (resp. $\mathcal{G}^\odot_\lambda \restriction_{L^p(\tilde{\phi}_p)}$) is a well defined bounded and positive improving operator on $L^p(\phi_p)$ (resp. $L^p(\tilde{\phi}_p)$); moreover, we have*

$$\|\mathcal{G}_\lambda\|_{L^p(\phi_p)} \le (\mu-\lambda)^{-1}, \quad (\textit{resp. } \|\mathcal{G}^\odot_\lambda\|_{L^p(\tilde{\phi}_p)} \le (\mu-\lambda)^{-1}); \tag{12}$$

(2) *for $1 \le p < \infty$, the operator $\mathcal{G}^\odot_\lambda \restriction_{L^{p'}(\tilde{\phi}_{p'})}$ is the dual operator of $\mathcal{G}_\lambda \restriction_{L^p(\phi_p)}$, and $\mathcal{G}_\lambda \restriction_{L^{p'}(\phi_{p'})}$ is the dual of $\mathcal{G}^\odot_\lambda \restriction_{L^p(\tilde{\phi}_p)}$;*

(3) *suppose that ϕ is a μ-invariant positive solution of the operator L, and $p = \infty$, then $\|\mathcal{G}_\lambda\|_{L^\infty(\phi_\infty)} = (\mu-\lambda)^{-1}$;*

(4) *suppose that ϕ is a μ-invariant positive solution of the operator L satisfying* (10), *then, for any $1 \le p \le \infty$, $\|\mathcal{G}_\lambda\|_{L^p(\phi_p)} = (\mu-\lambda)^{-1}$.*

Proof. (1) Let $f \in L^\infty(\phi_\infty)$, then by (4)

$$\begin{aligned}
|\mathcal{G}_\lambda f(x)| &\le \int_\Omega G^\Omega_{L_\lambda}(x,y)W(y)|f(y)|\mathrm{d}\nu(y)\\
&\le \|f\|_{\infty,\phi_\infty} \int_\Omega G^\Omega_{L_\lambda}(x,y)W(y)\phi(y)\mathrm{d}\nu(y)\\
&\le \frac{\|f\|_{\infty,\phi_\infty}}{\mu-\lambda}\phi(x),
\end{aligned}$$

so, $\|\mathcal{G}_\lambda\|_{L^\infty(\phi_\infty)} \le (\mu-\lambda)^{-1}$. Similarly, $\|\mathcal{G}^\odot_\lambda\|_{L^\infty(\tilde{\phi}_\infty)} \le (\mu-\lambda)^{-1}$.

Assume now that $f \in L^1(\phi_1)$, then by the Tonelli–Fubini theorem and (4) we obtain

$$\begin{aligned}
\|\mathcal{G}_\lambda f(x)\|_{1,\phi_1} &= \int_\Omega W(x)\tilde{\phi}(x) \Big| \int_\Omega G^\Omega_{L_\lambda}(x,y) W(y) f(y) \,\mathrm{d}\nu(y) \Big| \,\mathrm{d}\nu(x) \\
&\le \int_\Omega W(x)\tilde{\phi}(x) \int_\Omega G^\Omega_{L_\lambda}(x,y) W(y) |f(y)| \,\mathrm{d}\nu(y) \,\mathrm{d}\nu(x) \\
&= \int_\Omega \Big(\int_\Omega W(x)\tilde{\phi}(x) G^\Omega_{L_\lambda}(x,y) \,\mathrm{d}\nu(x) \Big) W(y) |f(y)| \,\mathrm{d}\nu(y) \\
&\le \frac{1}{\mu-\lambda} \int_\Omega \tilde{\phi}(y) W(y) |f(y)| \,\mathrm{d}\nu(y) \\
&= \frac{\|f\|_{1,\phi_1}}{\mu-\lambda}.
\end{aligned}$$

Hence, $\|\mathcal{G}_\lambda\|_{L^1(\phi_1)} \le (\mu-\lambda)^{-1}$. Similarly, $\|\mathcal{G}^{\odot}_\lambda\|_{L^1(\tilde{\phi}_1)} \le (\mu-\lambda)^{-1}$.

For $1 < p < \infty$, the boundedness of $\mathcal{G}_\lambda \restriction_{L^p(\phi_p)}$ with norm estimate

$$\|\mathcal{G}_\lambda\|_{L^p(\phi_p)} \le (\mu-\lambda)^{-1}$$

follows now directly from a Riesz–Thorin-type interpolation theorem with weights proved by Stein, see Theorem 2 in [43].

(2) The duality claim follows now directly from (9).

(3) and (4) follow from part (1) and (5). □

Remark 3.2. Theorem 3.1 (and (13)) for $1 < p < \infty$ follows also from the Schur test with weights, see Lemma 5.1 in [22]. Indeed, set

$$K(x,y) := \frac{G(x,y)}{\phi^{1-p}(y)\tilde{\phi}(y)}, \quad w(x,y) := \frac{\phi^p(y)}{\phi^p(x)}, \quad \mathrm{d}\rho(y) := \big(\phi_p(y)\big)^p \,\mathrm{d}\nu(y).$$

Then (4) implies that

$$\int_\Omega w(x,y)^{1/p} K(x,y) \,\mathrm{d}\rho(y) = \frac{\int_\Omega G^\Omega_{L_\lambda}(x,y) W(y)\phi(y) \,\mathrm{d}\nu(y)}{\phi(x)} \le \frac{1}{\mu-\lambda},$$

for all $x \in \Omega$, and

$$\int_\Omega w(x,y)^{-1/p'} K(x,y) \,\mathrm{d}\rho(x) = \frac{\int_\Omega G^\Omega_{L_\lambda}(x,y) W(x)\tilde{\phi}(x) \,\mathrm{d}\nu(x)}{\tilde{\phi}(y)} \le \frac{1}{\mu-\lambda},$$

for all $y \in \Omega$. Applying the aforementioned Schur test we get

$$\|\mathcal{G}_\lambda\|_{L^p(\phi_p)} \le (\mu - \lambda)^{-1}.$$

The Schur test with weights is essentially a theorem of Aronszajn, and in fact follows from Stein's Riesz–Thorin-type interpolation theorem with weights, see Theorem 2 in [43].

Remark 3.3. It follows from part (ii) of Theorem 4.4 that the assumptions of part (4) of Theorem 3.1 imply that in fact $\mu = \lambda_0$ and L_{λ_0} is positive-critical in Ω with respect to W.

Remark 3.4. The norm estimate (12) does not depend on ϕ, $\tilde{\phi}$ and W and p.

Remark 3.5. The requirement that W is strictly positive can be weakened, and Theorem 3.1 holds in a slightly weaker sense if W is a nonzero nonnegative function. Indeed, let $1 \le p \le \infty$. Since Stein's Riesz–Thorin-type interpolation theorem with weights (see Theorem 2 in [43]) holds for nonnegative weights, we have for $\lambda < \mu$

$$\|(\mathcal{G}_\lambda f)\phi_p\|_{L^p(\Omega,\mathrm{d}\nu)} \le \frac{1}{\mu - \lambda}\|f\phi_p\|_{L^p(\Omega,\mathrm{d}\nu)}$$

for all f such that $f\phi_p \in L^p(\Omega, \mathrm{d}\nu)$.

4 Principal eigenfunction

The Krein–Rutman theorem roughly asserts that if T is a compact operator defined on a Banach space X with a total cone P such that T is positive improving and its spectral radius $r(T)$ is strictly positive, then T admits a positive eigenfunction with an eigenvalue $r(T)$. Moreover, under an irreducibility assumption, $r(T)$ is simple. The weighted Green operator $\mathcal{G}_\lambda$ in the weighted Lebesgue spaces $L^p(\phi_p)$ is positive improving but in general, $\mathcal{G}_\lambda$ is not compact. Nevertheless, under some further conditions it admits a positive eigenfunction with an eigenvalue $(\mu - \lambda)^{-1}$.

Throughout the present section, as in Section 3, W is a fixed positive potential, $\mu \le \lambda_0$, and $\phi, \tilde{\phi}$ are fixed positive solutions of the equations $L_\mu u = 0$ and $L_\mu^\star u = 0$ in Ω, respectively. We study eigenvalues and eigenfunctions of the weighted Green operators $\mathcal{G}_\lambda$.

Remark 4.1. Eigenfunctions of $\mathcal{G}_\lambda \restriction_{L^1(\phi_1)}$ might be not smooth enough to solve weakly the corresponding partial differential equation. Therefore, if $p = 1$, we always *assume* that such eigenfunctions are also in $L^q_{\mathrm{loc}}(\Omega)$ for some $q > 1$.

Theorem 4.2. *Let W, μ, ϕ, and $\tilde{\phi}$ be as above, and let $\lambda < \mu$. Then for any $1 \leq p \leq \infty$, zero is not an eigenvalue of the operators $\mathcal{G}_\lambda \restriction_{L^p(\phi_p)}$ and $\mathcal{G}_\lambda^{\odot} \restriction_{L^p(\tilde{\phi}_p)}$.*

Moreover, any eigenfunction φ (resp. $\tilde{\varphi}$) of $\mathcal{G}_\lambda \restriction_{L^p(\phi_p)}$ (resp. $\mathcal{G}_\lambda^{\odot} \restriction_{L^p(\tilde{\phi}_p)}$) with an eigenvalue ν solves the equation

$$(L - [\lambda + (\nu)^{-1}]W)\varphi = 0 \quad (\text{resp. } (L^\star - [\lambda + (\nu)^{-1}]W)\tilde{\varphi} = 0) \quad \text{in } \Omega.$$

Proof. Let $1 \leq p \leq \infty$ and let $\varphi \in L^p(\phi_p)$, $\|\varphi\|_{p,\phi_p} = 1$ be an eigenfunction of the operator $\mathcal{G}_\lambda \restriction_{L^p(\phi_p)}$ with an eigenvalue ν, and define

$$\varphi_k(x) := \int_{\Omega_k} G_{L_\lambda}^{\Omega_k}(x, y) W(y) \varphi(y) \, d\nu(y) \quad x \in \Omega_k, \ k \geq 1.$$

Clearly,

$$G_{L_\lambda}^{\Omega_k}(x, y) W(y) |\varphi(y)| \leq G_{L_\lambda}^{\Omega}(x, y) W(y) |\varphi(y)| \quad \text{in } \Omega_k.$$

On the other hand, by Theorem 3.1 we have $G_{L_\lambda}^{\Omega}(x, \cdot) W\varphi \in L^1(\Omega, d\nu)$ for almost every $x \in \Omega$. Therefore, $\varphi_k(x)$ is well-defined almost everywhere in Ω_k, and Lebesgue's dominated convergence theorem implies that

$$\lim_{k \to \infty} \varphi_k(x) = \int_{\Omega} G_{L_\lambda}^{\Omega}(x, y) W(y) \varphi(y) \, d\nu(y) = \nu \varphi(x)$$

almost everywhere in Ω.

Since $|\varphi| \in L^p(\phi_p)$, Theorem 3.1 implies that $\mathcal{G}_\lambda |\varphi| \in L^p(\phi_p)$. Obviously, $|\varphi_k| \leq \mathcal{G}_\lambda |\varphi| \in L^p(\phi_p)$. Consequently, $\{\varphi_k\}$ is bounded in $L^p(\phi_p)$. Note that for $1 \leq p < \infty$, Lebesgue's dominated convergence theorem implies that

$$\|\varphi_k\|_{L^p(\phi_p)} \longrightarrow |\nu| \|\varphi\|_{L^p(\phi_p)},$$

and this holds true also for $p = \infty$.

On the other hand, taking into account Remark 4.1 in case $p = 1$, it follows that each φ_k solves the equation

$$(L - \lambda W)\varphi_k = W\varphi \quad \text{in } \Omega_k.$$

A standard elliptic regularity argument implies that $\nu\varphi$ solves the equation

$$(L - \lambda W)\nu\varphi = W\varphi \neq 0 \quad \text{in } \Omega.$$

In particular, $\nu \neq 0$. Thus, φ solves the equation

$$(L - [\lambda + (\nu)^{-1}]W)\varphi = 0 \quad \text{in } \Omega.$$

□

Remark 4.3. It was proved in [36] that zero is not an eigenvalue of $\mathcal{G}_\lambda \restriction_{C(\phi_\infty)}$.

The next result concerns conditions under which the positive solution ϕ is an eigenfunction of $\mathcal{G}_\lambda$ in $L^p(\phi_p)$.

Theorem 4.4. *Let W, μ, ϕ, and $\tilde{\phi}$ be as above, and let $\lambda < \mu$.*

(i) *The function ϕ (resp. $\tilde{\phi}$) is a nonnegative eigenfunction of the operator $\mathcal{G}_\lambda \restriction_{L^\infty(\phi_\infty)}$ (resp. $\mathcal{G}_\lambda^\odot \restriction_{L^\infty(\tilde{\phi}_\infty)}$) with an eigenvalue $(\mu - \lambda)^{-1}$ if and only if ϕ (resp. $\tilde{\phi}$) is a μ-invariant positive solution with respect to L (resp. $L^\star$). In this case*

$$\|\mathcal{G}_\lambda\|_{L^\infty(\phi_\infty)} = \|\mathcal{G}_\lambda^\odot\|_{L^1(\tilde{\phi}_1)} = (\mu - \lambda)^{-1},$$

resp.

$$\|\mathcal{G}_\lambda^\odot\|_{L^\infty(\tilde{\phi}_\infty)} = \|\mathcal{G}_\lambda\|_{L^1(\phi_1)} = (\mu - \lambda)^{-1}.$$

Furthermore, if ϕ and $\tilde{\phi}$ are both μ-invariant positive solutions, then for $1 \le p \le \infty$

$$\|\mathcal{G}_\lambda\|_{L^p(\phi_p)} = \|\mathcal{G}_\lambda^\odot\|_{L^p(\tilde{\phi}_p)} = (\mu - \lambda)^{-1}. \tag{13}$$

(ii) *Let $1 \le p < \infty$. Then ϕ (resp. $\tilde{\phi}$) is a nonnegative eigenfunction of the operator $\mathcal{G}_\lambda \restriction_{L^p(\phi_p)}$ (resp. $\mathcal{G}_\lambda^\odot \restriction_{L^p(\tilde{\phi}_p)}$) with an eigenvalue $\nu = (\mu - \lambda)^{-1}$ if and only if $\mu = \lambda_0$, and the operator L_{λ_0} is positive-critical with respect to W. In this case, ϕ and $\tilde{\phi}$ are the ground states of L_{λ_0} and $L^\star_{\lambda_0}$, respectively, and*

$$\|\mathcal{G}_\lambda\|_{L^p(\phi_p)} = \|\mathcal{G}_\lambda^\odot\|_{L^p(\tilde{\phi}_p)} = \frac{1}{\lambda_0 - \lambda} \quad \text{for all } 1 \le p \le \infty.$$

Proof. (i) All the claims of this part can be checked easily and left to the reader. In particular, use Theorem 3.1 and Stein's Riesz–Thorin-type interpolation theorem to prove (13).

(ii) Let $1 \le p < \infty$. The positive solution ϕ is an eigenfunction of the operator $\mathcal{G}_\lambda \restriction_{L^p(\phi_p)}$ with an eigenvalue $(\mu - \lambda)^{-1}$ if and only if $\phi W \tilde{\phi} \in L^1(\Omega)$ and ϕ is μ-invariant positive solution.

In particular, if $\mu = \lambda_0$, and L_{λ_0} is positive-critical with respect to W, then ϕ is an eigenfunction of the operator $\mathcal{G}_\lambda \restriction_{L^p(\phi_p)}$ with an eigenvalue $(\lambda_0 - \lambda)^{-1}$.

On the other hand, if ϕ is an eigenfunction of the operator $\mathcal{G}_\lambda \restriction_{L^p(\phi_p)}$ with an eigenvalue $(\mu - \lambda)^{-1}$, then $\phi W \tilde{\phi} \in L^1(\Omega)$.

Assume that L_μ is subcritical in Ω, then

$$\int_\Omega G^\Omega_{L_\mu}(x,y)W(y)\phi(y)\,\mathrm{d}\nu(y) < \infty.$$

By Remark 2.3, ϕ is not a μ-invariant solution, and we get a contradiction. Therefore, L_μ is critical in Ω and hence, $\mu = \lambda_0$. Since $\phi W \tilde{\phi} \in L^1(\Omega)$, it follows that L_{λ_0} is positive-critical with respect to W. □

In the critical case we have the following result.

Theorem 4.5. *Let L be an elliptic operator on Ω of the form* (1), *and let W be a positive potential. Assume that the operator L_{λ_0} is critical, and let ϕ and $\tilde{\phi}$ be the ground states of L_{λ_0} and $L^\star_{\lambda_0}$, respectively. Fix $\lambda < \lambda_0$.*

(i) *For $1 \le p \le \infty$, we have*

$$\|\mathcal{G}_\lambda\|_{L^p(\phi_p)} = \|\mathcal{G}^{\odot}_\lambda\|_{L^p(\tilde{\phi}_p)} = \frac{1}{\lambda_0 - \lambda}.$$

(ii) *The operator $\mathcal{G}_\lambda \restriction_{L^\infty(\phi_\infty)}$ (resp. $\mathcal{G}^{\odot}_\lambda \restriction_{L^\infty(\tilde{\phi}_\infty)}$) admits a unique eigenvalue $\nu = (\lambda_0 - \lambda)^{-1}$ with a nonnegative eigenfunction. Moreover, $(\lambda_0 - \lambda)^{-1}$ is a simple eigenvalue of $\mathcal{G}_\lambda \restriction_{L^\infty(\phi_\infty)}$ (resp. $\mathcal{G}^{\odot}_\lambda \restriction_{L^\infty(\tilde{\phi}_\infty)}$). The corresponding eigenfunction is ϕ (resp. $\tilde{\phi}$), and ϕ (resp. $\tilde{\phi}$) is the unique $L^\infty(\phi_\infty)$ (resp. $L^\infty(\tilde{\phi}_\infty)$) solution of the equation $L_{\lambda_0}u = 0$ in Ω.*

(iii) *Suppose further that the operator L_{λ_0} is positive-critical with respect to W. Then for all $1 \le p < \infty$ the function ϕ (resp. $\tilde{\phi}$) is the unique (up to a multiplicative constant) nonnegative eigenfunction of the operator $\mathcal{G}_\lambda \restriction_{L^p(\phi_p)}$ (resp. $\mathcal{G}^{\odot}_\lambda \restriction_{L^p(\tilde{\phi}_p)}$).*

Proof. (i)–(ii) Since ϕ is a ground state, it is a λ_0-invariant positive solution with respect to the operator L and the weight W, part (*i*) and the existence assertion of (ii) follow from part (i) of Theorem 4.4.

It remains to prove the uniqueness and simplicity of $(\lambda_0 - \lambda)^{-1}$ for the operator $\mathcal{G}_\lambda \restriction_{L^\infty(\phi_\infty)}$. Let φ be a nonnegative eigenfunction of the operator $\mathcal{G}_\lambda \restriction_{L^\infty(\phi_\infty)}$ with an eigenvalue $\nu = (\kappa - \lambda)^{-1}$. Without loss of generality we may assume that $\|\varphi\|_{\infty,\phi_\infty} = 1$, thus $\phi - \varphi \ge 0$. By Theorem 4.2, φ is a positive solution of the equation $(L - \kappa W)u = 0$ in Ω. Hence, $\kappa \le \lambda_0$. Therefore, $v := \phi - \varphi$ is a nonnegative supersolution of the equation $(L - \lambda_0 W)u =$ in Ω. On the other hand, L_{λ_0} is critical in Ω if and only if ϕ is the unique (up to a multiplicative constant) nonzero nonnegative supersolution. Thus, $\kappa = \lambda_0$ and $\varphi = \phi$. Hence, for all $\lambda < \lambda_0$, ϕ is the unique nonnegative eigenfunction of $\mathcal{G}_\lambda \restriction_{L^\infty(\phi_\infty)}$.

Moreover, if $(\lambda_0 - \lambda)^{-1}$ is an eigenvalue of $\mathcal{G}_\lambda \restriction_{L^\infty(\phi_\infty)}$ with a normalized eigenfunction φ, then $u := \phi - \varphi \geq 0$, and by Theorem 4.2 u is a nonnegative solution of the equation $L_{\lambda_0} u = 0$ in Ω. Since L_{λ_0} is critical, it follows that $u = c\phi$ for some $c \geq 0$. So, $\varphi = (1 - c)\phi$, and either $c = 0$ or $c = 2$. Hence, $(\lambda_0 - \lambda)^{-1}$ is a simple eigenvalue of $\mathcal{G}_\lambda \restriction_{L^\infty(\phi_\infty)}$.

(iii) For $1 \leq p < \infty$, part (ii) of Theorem 4.4 implies that ϕ is a positive eigenfunction of $\mathcal{G}_\lambda \restriction_{L^p(\phi_p)}$ with an eigenvalue $(\lambda_0 - \lambda)^{-1}$.

Let $\varphi \in L^p(\phi_p)$ be a nonnegative normalized eigenfunction of the operator $\mathcal{G}_\lambda \restriction_{L^p(\phi_p)}$ with an eigenvalue ν. Note that φ is strictly positive since $\mathcal{G}_\lambda \restriction_{L^p(\phi_p)}$ is positivity improving. Clearly, $\nu \geq 0$, and by Theorem 4.2 $\nu \neq 0$. Therefore, ν can be written as $\nu = (\kappa - \lambda)^{-1}$, where $\lambda < \kappa$.

On the other hand, one has $\nu \leq \|\mathcal{G}_\lambda\|_{L^p(\phi_p)} = (\lambda_0 - \lambda)^{-1}$, and hence $\lambda_0 \leq \kappa$. By Theorem 4.2, φ is a nonnegative solution of the equation $L_\kappa u = 0$ in Ω, therefore $\kappa \leq \lambda_0$, Thus, $\kappa = \lambda_0$, and $\varphi = \phi$. □

The next result deals with the case $p = 2$. In this case, any L^2-eigenfunction of $\mathcal{G}_\lambda \restriction_{L^2(\phi_2)}$ with the 'maximal' eigenvalue has a definite sign and this eigenvalue is simple.

Theorem 4.6. *Let L be an elliptic operator on Ω of the form* (1)*, and let W be a positive potential. Fix $\lambda < \mu \leq \lambda_0$, and let ϕ and $\tilde{\phi}$ be two positive solutions of the equation $L_\mu u = 0$ and $L^\star_\mu u = 0$ in Ω, respectively.*

If φ is an eigenfunction of $\mathcal{G}_\lambda \restriction_{L^2(\phi_2)}$ with an eigenvalue $(\mu - \lambda)^{-1}$, then φ has a definite sign, and $(\mu - \lambda)^{-1}$ is a simple eigenvalue of $\mathcal{G}_\lambda \restriction_{L^2(\phi_2)}$.

Assume further that $\phi W \tilde{\phi} \in L^1(\Omega)$. Then $\varphi = c\phi$ for some constant c, $\mu = \lambda_0$, and the operator L_{λ_0} is positive-critical. Moreover, $(\lambda_0 - \lambda)^{-1}$ is the unique eigenvalue of $\mathcal{G}_\lambda \restriction_{L^p(\phi_p)}$ with a nonnegative eigenfunction for all $1 \leq p \leq \infty$ and all $\lambda < \lambda_0$. Furthermore, $(\lambda_0 - \lambda)^{-1}$ is a simple eigenvalue of $\mathcal{G}_\lambda \restriction_{L^p(\phi_p)}$ for all $2 \leq p \leq \infty$, and all $\lambda < \lambda_0$.

Proof. Recall that

$$\|\mathcal{G}_\lambda\|_{L^2(\phi_2)} \leq \frac{1}{\mu - \lambda}.$$

Let $\varphi \in L^2(\phi_2)$ be an eigenfunction of the operator $\mathcal{G}_\lambda \restriction_{L^2(\phi_2)}$ with an eigenvalue $(\mu - \lambda)^{-1}$. Thus, $\|\mathcal{G}_\lambda\|_{L^2(\phi_2)} = (\mu - \lambda)^{-1}$. Without loss of generality, we may assume that φ is a real function. Therefore, due to the positivity improving of

$\mathcal{G}_\lambda\restriction_{L^2(\phi_2)}$, and the Cauchy–Schwarz inequality, we obtain

$$\begin{aligned}\frac{1}{\mu-\lambda}\|\varphi\|^2_{L^2(\phi_2)} &= (\varphi, \mathcal{G}_\lambda\restriction_{L^2(\phi_2)}\varphi)_{L^2(\phi_2)} \\ &\le (|\varphi|, |\mathcal{G}_\lambda\restriction_{L^2(\phi_2)}\varphi|)_{L^2(\phi_2)} \\ &\le (|\varphi|, \mathcal{G}_\lambda\restriction_{L^2(\phi_2)}|\varphi|)_{L^2(\phi_2)} \\ &\le \|\varphi\|_{L^2(\phi_2)}\|\mathcal{G}_\lambda\restriction_{L^2(\phi_2)}|\varphi|\|_{L^2(\phi_2)} \\ &\le \frac{1}{\mu-\lambda}\|\varphi\|^2_{L^2(\phi_2)}.\end{aligned} \tag{14}$$

As a result we have equality signs in all the inequalities of (14). The equality in the Cauchy–Schwarz inequality implies that

$$\mathcal{G}_\lambda\restriction_{L^2(\phi_2)}|\varphi| = \frac{1}{\mu-\lambda}|\varphi|,$$

and therefore, $|\varphi|$ is a nonnegative eigenfunction of the operator $\mathcal{G}_\lambda\restriction_{L^2(\phi_2)}$. Since $\mathcal{G}_\lambda\restriction_{L^2(\phi_2)}$ is positivity improving, we have $|\varphi| > 0$. It follows that any such eigenfunction has a definite sign. Consequently, a standard orthogonality argument shows that $(\mu-\lambda)^{-1}$ is simple (cf. Theorem XIII.43 in [38]).

Assume further that $\phi W\tilde{\phi} \in L^1(\Omega)$, and denote $\hat{\phi}_p := |\varphi|^{-1}(|\varphi|W\tilde{\phi})^{1/p}$. Then by the Cauchy–Schwarz inequality $\varphi W\tilde{\phi} \in L^1(\Omega)$, and hence $|\varphi|$ is a positive eigenfunction of the operator $\mathcal{G}_\lambda\restriction_{L^2(\hat{\phi}_2)}$. Therefore, by Theorem 4.4 *(ii)*, $\mu = \lambda_0$, the operator L_{λ_0} is positive-critical, and for some constant c we have $\varphi = c\phi$. In addition, part *(iii)* of Theorem 4.5 implies that $(\lambda_0-\lambda)^{-1}$ is the unique eigenvalue of $\mathcal{G}_\lambda\restriction_{L^p(\phi_p)}$ with a nonnegative eigenfunction ϕ for all $1 \le p < \infty$ and all $\lambda < \lambda_0$.

The simplicity of $(\lambda_0-\lambda)^{-1}$ as an eigenvalue of $\mathcal{G}_\lambda\restriction_{L^p(\phi_p)}$ for $2 \le p \le \infty$ follows now from the simplicity for $p = 2$ and the embedding (11). □

Next we study the case $p = 1$, and obtain the simplicity of the eigenvalue $(\lambda_0-\lambda)^{-1}$ for all p under the assumption that $(\mu-\lambda)^{-1}$ is an eigenvalue of $\mathcal{G}_\lambda\restriction_{L^1(W\tilde{\phi})}$. Note that for $p = 1$ the result does not depend on a particular positive solution ϕ of the equation $L_\mu u = 0$ in Ω since $L^1(\phi_1) = L^1(W\tilde{\phi})$. As a result, we obtain L^1- and L^p-Liouville theorems for solutions of the equation $L_\mu u = 0$ which are eigenfunctions of the operator $\mathcal{G}_\lambda$ with an eigenvalue $(\mu-\lambda)^{-1}$.

Theorem 4.7. *Let L be an elliptic operator on Ω of the form* (1), *and let W be a positive potential. Fix $\mu \le \lambda_0$. Let $\tilde{\phi}$ be a positive solution of the equation $L^\star_\mu u = 0$ in Ω, For $\lambda < \mu$ consider the operator $\mathcal{G}_\lambda\restriction_{L^1(W\tilde{\phi})}$. Suppose further that $(\mu-\lambda)^{-1}$ is an eigenvalue of $\mathcal{G}_\lambda\restriction_{L^1(W\tilde{\phi})}$ with an eigenfunction φ.*

(i) *The eigenfunction φ has a definite sign, $\mu = \lambda_0$, and L_{λ_0} is positive-critical with respect to W with a ground state φ. In particular, $(\lambda_0 - \lambda)^{-1}$ is a simple eigenvalue of the operator $\mathcal{G}_\lambda \restriction_{L^1(W\tilde{\phi})}$.*

(ii) *Set $\varphi_p := |\varphi|^{-1}(|\varphi| W\tilde{\phi})^{1/p}$ $(1 \le p \le \infty.)$ Then for all $\lambda < \lambda_0$ and all $1 \le p \le \infty$, φ is an eigenfunction of $\mathcal{G}_\lambda \restriction_{L^p(\varphi_p)}$ with an eigenvalue $(\lambda_0 - \lambda)^{-1}$. Moreover, $(\lambda_0 - \lambda)^{-1}$ is a simple eigenvalue, and it is the unique eigenvalue with a nonnegative eigenfunction.*

Proof. (i) Fix any positive solution ϕ of the equation $L_\mu u = 0$ in Ω, and let $\phi_p := \phi^{-1}(\phi W \tilde{\phi})^{1/p}$. Clearly, $L^1(\phi_1) = L^1(W\tilde{\phi})$. By Theorem 3.1, for any $\lambda < \mu$ the operator $\mathcal{G}_\lambda \restriction_{L^1(W\tilde{\phi})}$ is bounded with a norm $\|\mathcal{G}_\lambda \restriction_{L^1(W\tilde{\phi})}\|_{L^1(W\tilde{\phi})} \le (\mu - \lambda)^{-1}$.

Let φ be an eigenfunction of $\mathcal{G}_\lambda \restriction_{L^1(W\tilde{\phi})}$ with an eigenvalue $(\mu - \lambda)^{-1}$. By our assumption,

$$\frac{|\varphi(x)|}{\mu - \lambda} = \left| \int_\Omega G^\Omega_{L_\lambda}(x, y) W(y) \varphi(y) \,\mathrm{d}\nu(y) \right| \quad \text{for all } x \in \Omega.$$

Therefore, using (4) we obtain

$$\begin{aligned}
&\frac{1}{\mu - \lambda} \int_\Omega |\varphi(x)|\, W(x) \tilde{\phi}(x) \,\mathrm{d}\nu \\
&\quad = \int_\Omega \left| \int_\Omega G^\Omega_{L_\lambda}(x, y) W(y) \varphi(y) \,\mathrm{d}\nu(y) \right| W(x) \tilde{\phi}(x) \,\mathrm{d}\nu(x) \\
&\quad \le \int_\Omega \left(\int_\Omega G^\Omega_{L_\lambda}(x, y) W(y) |\varphi(y)| \,\mathrm{d}\nu(y) \right) W(x) \tilde{\phi}(x) \,\mathrm{d}\nu(x) \qquad (15) \\
&\quad = \int_\Omega \left(\int_\Omega G^\Omega_{L_\lambda}(x, y) W(x) \tilde{\phi}(x) \,\mathrm{d}\nu(x) \right) W(y) |\varphi(y)| \,\mathrm{d}\nu(y) \\
&\quad \le \frac{1}{\mu - \lambda} \int_\Omega \tilde{\phi}(y) W(y) |\varphi(y)| \,\mathrm{d}\nu(y).
\end{aligned}$$

Thus, the two inequalities in (15) are equalities, and in particular, for almost all $x \in \Omega$ we have

$$\left| \int_\Omega G^\Omega_{L_\lambda}(x, y) W(y) \varphi(y) \,\mathrm{d}\nu(y) \right| = \int_\Omega G^\Omega_{L_\lambda}(x, y) W(y) |\varphi(y)| \,\mathrm{d}\nu(y),$$

and hence φ does not change its sign in Ω. Moreover, the equality in (15) implies also that $\tilde{\phi}$ is an invariant solution of the equation $L^\star_\mu u = 0$ in Ω. Since $|\varphi| W \tilde{\phi} \in L^1(\Omega)$, Remark 2.3 implies that $\mu = \lambda_0$, L_{λ_0} is positive-critical, and $|\varphi|$ is its ground state. Hence, the simplicity of the 'maximal' eigenvalue follows. Consequently, part (ii) follows using the embedding (11), part (i) of the present theorem, and part (iii) of Theorem 4.5. □

Corollary 4.8. *Let L be an elliptic operator on Ω of the form* (1)*, and let W be a positive potential. Assume further that $L^\star \mathbf{1} = 0$. For $\lambda < 0$ consider the operator $\mathcal{G}_\lambda \restriction_{L^1(W)}$.*

Suppose that $|\lambda|^{-1}$ is an eigenvalue of $\mathcal{G}_\lambda \restriction_{L^1(W)}$ with an eigenfunction φ, and set $\varphi_p := |\varphi|^{-1}(|\varphi| W)^{1/p}$, where $1 \le p \le \infty$. Then φ has a definite sign, $\lambda_0 = 0$, L is positive-critical, and for all $\lambda < 0$ and all $1 \le p \le \infty$, $|\lambda|^{-1}$ is the unique eigenvalue of $\mathcal{G}_\lambda \restriction_{L^p(\varphi_p)}$ with a nonnegative eigenfunction. Moreover, $|\lambda|^{-1}$ is a simple eigenvalue of $\mathcal{G}_\lambda \restriction_{L^p(\varphi_p)}$

Example 4.9. Let $\Omega = \mathbb{R}^d$, and consider a uniformly elliptic operator L with bounded smooth coefficients on $\mathbb{R}^d$ such that $L^\star \mathbf{1} = 0$ in $\mathbb{R}^d$ (these conditions can be relaxed). For example, assume that L is of the form

$$Lu := -\operatorname{div}(A(x)\nabla u + u\tilde{b}(x)) \quad x \in \mathbb{R}^d.$$

Suppose that the equation $Lu = 0$ in $\mathbb{R}^d$ admits a solution φ satisfying $\varphi \in L^1(\mathbb{R}^d)$. Let $k_L^{\mathbb{R}^d}(x, y, t)$ the heat kernel associated with the operator L on $\mathbb{R}^d$. Then

$$v(x,t) := \int_{\mathbb{R}^d} k_L^{\mathbb{R}^d}(x, y, t)\varphi(y)\,\mathrm{d}y$$

is a well defined L^1-solution of the Cauchy problem with the initial condition φ. Since the uniqueness of the Cauchy problem for L^1-initial conditions holds true (see [3] and [4]), it follows that $v = \varphi$. Fix $\lambda < 0$. It follows that

$$\begin{aligned}
\frac{\varphi(x)}{|\lambda|} &= \int_0^\infty e^{\lambda t} v(x,t)\,\mathrm{d}t \\
&= \int_0^\infty e^{\lambda t}\Big(\int_{\mathbb{R}^d} k_L^{\mathbb{R}^d}(x, y, t)\varphi(y)\,\mathrm{d}y\Big)\,\mathrm{d}t \\
&= \int_{\mathbb{R}^d}\Big(\int_0^\infty e^{\lambda t} k_L^{\mathbb{R}^d}(x, y, t)\,\mathrm{d}t\Big)\varphi(y)\,\mathrm{d}y \\
&= \int_{\mathbb{R}^d} G_{L_\lambda}^{\mathbb{R}^d}(x, y)\varphi(y)\,\mathrm{d}y.
\end{aligned}$$

So, φ is an eigenfunction of $\mathcal{G}_\lambda \restriction_{L^1(\mathbb{R}^d)}$ with an eigenvalue $|\lambda|^{-1}$. Corollary 4.8 implies that L is positive-critical with respect to $W = \mathbf{1}$, and $|\varphi| > 0$ is the corresponding ground state. In particular, $\mathbf{1}$ is an invariant positive solution of the operator $L^\star$. Moreover, $|\lambda|^{-1}$ is a simple eigenvalue of $\mathcal{G}_\lambda \restriction_{L^1(\mathbb{R}^d)}$.

Let $1 \le p \le \infty$, and $\varphi_p := |\varphi|^{1/p - 1}$. Then, by Corollary 4.8 for all $\lambda < 0$ and all $1 \le p \le \infty$, $|\lambda|^{-1}$ is the unique eigenvalue of $\mathcal{G}_\lambda \restriction_{L^p(\varphi_p)}$ with a nonnegative eigenfunction $|\varphi|$. Moreover, $|\lambda|^{-1}$ is a simple eigenvalue of $\mathcal{G}_\lambda \restriction_{L^p(\varphi_p)}$ for all $1 < p < \infty$.

Remark 4.10. For L^p-Liouville theorems for symmetric diffusion operators on complete Riemannian manifolds, see [25], [44], and references therein.

5 Semigroups and generators

Definition 5.1. Let B be a Banach space and $\Lambda \subset \mathbb{C}$, and consider a one-parameter family of operators $\mathcal{J}(\lambda) \in \mathcal{L}(B)$ defined for each $\lambda \in \Lambda$. The family $\{\mathcal{J}(\lambda) \mid \lambda \in \Lambda\}$ is called a *pseudoresolvent* if

$$\mathcal{J}(\lambda) - \mathcal{J}(\nu) = (\nu - \lambda)\mathcal{J}(\lambda)\mathcal{J}(\nu)$$

holds for all $\lambda, \nu \in \Lambda$ (see Definition 4.3 in [17]).

Let L be an elliptic operator on Ω of the form (1), and W a positive potential. Fix $\mu \leq \lambda_0 = \lambda_0(L, W, \Omega)$, and let

$$\Lambda := \begin{cases} \{\lambda \in \mathbb{R} \mid \lambda \leq \mu\} & \text{if } L_\mu \text{ is subcritical,} \\ \{\lambda \in \mathbb{R} \mid \lambda < \mu\} & \text{if } \mu = \lambda_0 \text{ and } L_{\lambda_0} \text{ is critical.} \end{cases} \tag{16}$$

Recall that by (2.10) of [34], for all $\lambda, \nu \in \Lambda$ the corresponding Green functions satisfy the (pointwise) resolvent equation

$$G_{L_\lambda}^{\Omega}(x, y) = G_{L_\nu}^{\Omega}(x, y) + (\lambda - \nu)\int_\Omega G_{L_\lambda}^{\Omega}(x, z)W(z)G_{L_\nu}^{\Omega}(z, y)\,\mathrm{d}\nu(z) \tag{17}$$

for all $x, y \in \Omega$, $x \neq y$.

Let ϕ and $\tilde{\phi}$ be two fixed positive solutions of the equations $L_\mu u = 0$ and $L_\mu^\star u = 0$ in Ω, respectively. It follows from Theorem 3.1 and (17) that for any $1 \leq p \leq \infty$, the family

$$\{\mathcal{G}_{(-\lambda)}\restriction_{L^p(\phi_p)} \mid -\lambda \in \Lambda\} = \{\mathcal{G}_{L+\lambda W}\restriction_{L^p(\phi_p)} \mid -\lambda \in \Lambda\}$$

is a pseudoresolvent on $L^p(\phi_p)$.

We claim that for $1 \leq p < \infty$ and $\lambda \in \Lambda$, the range of $\mathcal{G}_\lambda\restriction_{L^p(\phi_p)}$ is dense in $L^p(\phi_p)$. Indeed, take $u \in C_0^\infty(\Omega)$, and let $N_u \in \mathbb{N}$ be such that $\operatorname{supp} u \Subset \Omega_{N_u}$. Set $f(x) := (W(x))^{-1}L_\lambda u(x)$. For $n \geq N_u$ denote $u_n := \mathcal{G}_\lambda^{\Omega_n}\restriction_{L^p(\phi_p)} f$. Clearly, $\operatorname{supp} f \subset \Omega_n$ for all $n \geq N_u$. Therefore, by uniqueness, for any such n we have $u_n = u$ in Ω_n, and consequently, $u = \mathcal{G}_\lambda^{\Omega}\restriction_{L^p(\phi_p)} f$, and u belongs to the range of $\mathcal{G}_\lambda\restriction_{L^p(\phi_p)}$. Since for $1 \leq p < \infty$ the space $C_0^\infty(\Omega)$ is dense in $L^p(\phi_p)$, it follows that the range of $\mathcal{G}_\lambda\restriction_{L^p(\phi_p)}$ is dense in $L^p(\phi_p)$.

On the other hand, by Theorem 4.2, zero is not an eigenvalue of the operators $\mathcal{G}_\lambda \restriction_{L^p(\phi_p)}$ for $\lambda < \mu$ and all $1 < p \le \infty$. Moreover, for $1 \le p \le \infty$ we have

$$\|\mathcal{G}_\lambda\|_{L^p(\phi_p)} \le \frac{1}{\mu - \lambda}, \tag{18}$$

and in particular,

$$\limsup_{\lambda \to -\infty} \|\lambda \mathcal{G}_\lambda\|_{L^p(\phi_p)} \le \lim_{\lambda \to -\infty} \frac{-\lambda}{\mu - \lambda} = 1.$$

Moreover, (18) implies also that if $\mu \ge 0$ and $\lambda < 0$, then

$$\|\lambda \mathcal{G}_\lambda\|_{L^p(\phi_p)} \le 1.$$

Therefore, by Proposition III.4.6 and Corollary III.4.7 of [17], and the Hille–Yosida theorem (Theorem II.3.5 in [17]), we have

Theorem 5.2. *Let L be an elliptic operator on Ω of the form* (1), *and let W be a positive potential. Fix $1 < p < \infty$, $\mu \le \lambda_0$, and let Λ be as in* (16). *Let ϕ and $\tilde{\phi}$ be two fixed positive solutions of the equations $L_\mu u = 0$ and $L_\mu^\star u = 0$ in Ω, respectively.*

(i) *The pseudoresolvent family*

$$\{\mathcal{G}_{(-\lambda)} \restriction_{L^p(\phi_p)} \mid -\lambda \in \Lambda\}$$

is a resolvent of a densely defined closed operator

$$A_p := -(\mathcal{G}_{\lambda_1} \restriction_{L^p(\phi_p)})^{-1} - \lambda_1 \quad \textit{where } \lambda_1 \in \Lambda$$

on $L^p(\phi_p)$ with a domain $D(A_p) = R(\mathcal{G}_{\lambda_1} \restriction_{L^p(\phi_p)})$. In particular, $A_p = -(1/W)L$ on $C_0^\infty(\Omega)$. Moreover, $(-\mu, \infty) \subset \rho(A_p)$, and for $\lambda \in (-\mu, \infty)$ we have

$$R(\lambda, A_p) := (\lambda - A_p)^{-1} = \mathcal{G}_{(-\lambda)} \restriction_{L^p(\phi_p)}.$$

(ii) *Zero is not an eigenvalue of $\mathcal{G}_\lambda \restriction_{L^p(\phi_p)}$.*

(iii) *If $\mu \ge 0$, then $(A_p, D(A_p))$ generates a strongly continuous contraction semigroup. Moreover, for every $\lambda \in \mathbb{C}$ with $\operatorname{Re} \lambda > 0$ one has $\lambda \in \rho(A_p)$, and*

$$\|R(\lambda, A_p)\|_{L^p(\phi_p)} \le \frac{1}{\operatorname{Re} \lambda}.$$

6 Compactness and semismall perturbations

Throughout this section, we assume that L is subcritical in Ω and $W > 0$ is a *semismall perturbation* of L and $L^\star$ in Ω. By Remark 2.7, $\lambda_0 > 0$, and L_{λ_0} is positive-critical. Denote by ϕ and $\tilde{\phi}$ the ground states of L_{λ_0} and $L^\star_{\lambda_0}$, respectively. We may assume that $\phi(x_0) = 1$. So,

$$(L - \lambda_0 W)\phi = (L^\star - \lambda_0 W)\tilde{\phi} = 0 \quad \text{in } \Omega$$

and

$$\tilde{\phi} W \phi \in L^1(\Omega).$$

Without loss of generality, we may assume that

$$\int_\Omega \tilde{\phi}(x) W(x) \phi(x) \,\mathrm{d}\nu(x) = 1.$$

It follows that $\tilde{\phi} \in (L^1(\tilde{\phi}_1)) \subset (L^\infty(\phi_\infty))^\star$, and for every $f \in L^\infty(\phi_\infty)$ we have $\langle \tilde{\phi}, f \rangle = \int_\Omega \tilde{\phi}(x) W(x) f(x) \,\mathrm{d}\nu$.

The aim of the present section is to prove that under the above assumptions, the integral operator $\mathcal{G}_\lambda$ is compact on $L^p(\phi_p)$ for any $1 \le p \le \infty$ and $\lambda < \lambda_0$, and its spectrum is p-independent. We first prove the compactness of $\mathcal{G}_\lambda$.

Theorem 6.1. *Let L be a subcritical operator in Ω. Assume that $W > 0$ is a semismall perturbation of $L^\star$ and L in Ω. Then for any $1 \le p \le \infty$ and $\lambda < \lambda_0$, the integral operators*

$$\mathcal{G}_\lambda f(x) = \int_\Omega G^\Omega_{L-\lambda W}(x, y) W(y) f(y) \,\mathrm{d}\nu(y),$$

$$\mathcal{G}^\odot_\lambda f(y) = \int_\Omega G^\Omega_{L-\lambda W}(x, y) W(x) f(x) \,\mathrm{d}\nu(x)$$

are compact on $L^p(\phi_p)$ and $L^p(\tilde{\phi}_p)$, respectively.

Proof. By Theorem 5.1 in [36], the operators $\mathcal{G}_\lambda \restriction_{L^\infty(\phi_\infty)}$ and $\mathcal{G}^\odot_\lambda \restriction_{L^\infty(\tilde{\phi}_\infty)}$ are compact on $L^\infty(\phi_\infty)$ and $L^\infty(\tilde{\phi}_\infty)$, respectively. For the sake of completeness, we prove the compactness of $\mathcal{G}_\lambda \restriction_{L^\infty(\phi_\infty)}$; the proof of the compactness of $\mathcal{G}^\odot_\lambda \restriction_{L^\infty(\tilde{\phi}_\infty)}$ is identical.

Let $\{f_n\}$ be a bounded sequence in $L^\infty(\phi_\infty)$. By Theorem 3.1, the sequence $u_n := \mathcal{G}_\lambda \restriction_{L^\infty(\phi_\infty)} f_n$ is bounded in $L^\infty(\phi_\infty)$, and satisfies

$$|u_n(x)| \le \int_\Omega G^\Omega_{L_\lambda}(x, y) W(y) |f_n(y)| \,\mathrm{d}\nu(y) \le C\phi_0(x),$$

where $C := (\lambda_0 - \lambda)^{-1} \sup_n \|f_n\|_{\infty,\phi_\infty}$ is independent of n. Moreover, it follows that u_n is the unique function in $L^\infty(\phi_\infty)$ which is a (weak) solution of the equation $L_\lambda u = f_n$ in Ω (cf. Theorem 4.6 in [36]). Consequently, a standard elliptic argument implies that the sequence $\{u_n\}$ admits a subsequence which converges in the compact open topology to a function u. Clearly, $\|u\|_{\infty,\phi_\infty} \le C$, so, $u \in L^\infty(\phi_\infty)$

Since W is a semismall perturbation, it follows that for any given $\varepsilon > 0$ there exists K such that for any $k \ge K$ and $n, m \in \mathbb{N}$

$$\begin{aligned}\int_{\Omega_k^*} & G_{L_\lambda}^{\Omega}(x,y)W(y)|f_n(y) - f_m(y)|\,\mathrm{d}\nu(y) \\ &\le 2C\int_{\Omega_k^*} G_{L_\lambda}^{\Omega}(x,y)W(y)\phi_0(y)\,\mathrm{d}\nu(y) < \varepsilon\phi_0(x) \quad \text{for all } x \in \overline{\Omega_k^*},\end{aligned} \tag{19}$$

and by the generalized maximum principle in Ω_k, (19) holds for any $x \in \Omega$.

The local uniform convergence of $\{u_n\}$ implies that there exists $N_\varepsilon \in \mathbb{N}$ such that $|u_n - u_m| < \varepsilon\phi_0$ in Ω_K for all $n, m \ge N_\varepsilon$.

Fix $n, m \ge N_\varepsilon$. It follows from Lemma 4.3 in [36] and the linearity that on Ω_K^* we have

$$u_n(x) - u_m(x) = h_{n,m}(x) + \int_{\Omega_K^*} G_{L_\lambda}^{\Omega_K^*}(x,y)W(y)(f_n(y) - f_m(y))\,\mathrm{d}\nu(y),$$

where $h_{n,m} \in L^\infty(\phi_\infty)$ satisfies

$$L_\lambda h_{n,m} = 0 \quad \text{in } \Omega_K^*,$$

and

$$h_{n,m}(x) = u_n(x) - u_m(x) \quad x \in \partial\Omega_K.$$

Since $|h_{n,m}| \le \varepsilon\phi_0$ on $\partial\Omega_K$, and $h_{n,m}$ has minimal growth in Ω, it follows that $|h_{n,m}| \le 2\varepsilon\phi_0$ in Ω_K^*. On the other hand, by (19) we have

$$\begin{aligned}\Big|\int_{\Omega_K^*} & G_{L_\lambda}^{\Omega_K^*}(x,y)W(y)(f_n(y) - f_m(y))\,\mathrm{d}\nu(y)\Big| \\ &\le \int_{\Omega_K^*} G_{L_\lambda}^{\Omega}(x,y)W(y)|f_n(y) - f_m(y)|\,\mathrm{d}\nu(y) < \varepsilon\phi_0(x) \quad \text{for all } x \in \Omega_K^*.\end{aligned}$$

Consequently, we infer that $|u_n - u_m| \le 3\,\varepsilon\phi_0$ in Ω_K^* for all $n, m \ge N_\varepsilon$. Thus, $u_n \to u$ in $L^\infty(\phi_\infty)$.

Since for each $1 \le p \le \infty$ the operator $\mathcal{G}_\lambda \restriction_{L^p(\phi_p)}$ is bounded on $L^p(\phi_p)$, and $\mathcal{G}_\lambda \restriction_{L^\infty(\phi_\infty)}$ is compact on $L^\infty(\phi_\infty)$, it follows from a variant of the Riesz–Thorin interpolation theorem with respect to compact operators (see Theorem 1.1 in [12]) that $\mathcal{G}_\lambda \restriction_{L^p(\phi_p)}$ are compact for all $1 \le p \le \infty$. The same is true for $\mathcal{G}_\lambda^{\odot} \restriction_{L^p(\tilde{\phi}_p)}$. $\square$

Remark 6.2. For $\lambda < \lambda_0$ the operators $\mathcal{G}_\lambda \restriction_{L^\infty(\phi_\infty)}$ and $\mathcal{G}_\lambda^\odot \restriction_{L^\infty(\tilde\phi_\infty)}$ are the dual operators of $\mathcal{G}_\lambda^\odot \restriction_{L^1(\tilde\phi_1)}$ and $\mathcal{G}_\lambda \restriction_{L^1(\phi_1)}$, respectively. Therefore, the well-known Schauder theorem directly implies that $\mathcal{G}_\lambda^\odot \restriction_{L^1(\tilde\phi_1)}$ and $\mathcal{G}_\lambda \restriction_{L^1(\phi_1)}$ are compact on $L^1(\tilde\phi_1)$ and $L^1(\phi_1)$, respectively.

Remark 6.3. In the proof of Theorem 6.1 we used the fact that *real* interpolation preserves the compactness of an operator. We recall that in his remarkable paper [11] A. Calderón implicitly asked a question which is apparently still open today: does *complex* interpolation preserve the compactness of an operator? For a recent survey on this question see [13].

The next theorem discusses the spectral properties of $\mathcal{G}_\lambda \restriction_{L^p(\phi_p)}$.

Theorem 6.4. *Under the assumptions of Theorem* 6.1 *we have*

(1) *for* $1 \le p \le \infty$, *the spectrum of* $\mathcal{G}_\lambda \restriction_{L^p(\phi_p)}$ *contains* 0, *and besides, consists of at most a sequence of eigenvalues of finite multiplicity which has no point of accumulation except* 0;

(2) *for any* $1 \le p \le \infty$, ϕ *(resp.* $\tilde\phi$*) is the unique nonnegative eigenfunction of the operator* $\mathcal{G}_\lambda \restriction_{L^p(\phi_p)}$ *(resp.* $\mathcal{G}_\lambda^\odot \restriction_{L^p(\tilde\phi_p)}$*); the corresponding eigenvalue* $\nu = (\lambda_0 - \lambda)^{-1}$ *is simple*;

(3) *the spectrum of* $\mathcal{G}_\lambda \restriction_{L^p}(\phi_p)$ *is* p*-independent for all* $1 \le p \le \infty$, *and we have*

$$0 \in \sigma(\mathcal{G}_\lambda \restriction_{L^p(\phi_p)}) = \sigma(\mathcal{G}_\lambda^\odot \restriction_{L^p(\tilde\phi_p)}) \subset \overline{B(0, (\lambda_0 - \lambda)^{-1})}.$$

Proof. (1) The characterization of the spectrum of $\mathcal{G}_\lambda \restriction_{L^p(\phi_p)}$ for each p follows from the Riesz–Schauder theory for compact operators.

(2) Follows from Theorem 4.7.

(3) The compactness of all the operators $\mathcal{G}_\lambda \restriction_{L^p(\phi_p)}$ implies that it is enough to show that $\sigma_{\mathrm{point}}(\mathcal{G}_\lambda \restriction_{L^p(\phi_p)})$ is p-independent.

By (11) we have for any $1 \le p \le \infty$ that

$$\sigma_{\mathrm{point}}(\mathcal{G}_\lambda \restriction_{L^\infty(\phi_\infty)}) \subset \sigma_{\mathrm{point}}(\mathcal{G}_\lambda \restriction_{L^p(\phi_p)}) \subset \sigma_{\mathrm{point}}(\mathcal{G}_\lambda \restriction_{L^1(\phi_1)}), \tag{20}$$

and

$$\sigma_{\mathrm{point}}(\mathcal{G}_\lambda^\odot \restriction_{L^\infty(\tilde\phi_\infty)}) \subset \sigma_{\mathrm{point}}(\mathcal{G}_\lambda^\odot \restriction_{L^p(\tilde\phi_p)}) \subset \sigma_{\mathrm{point}}(\mathcal{G}_\lambda^\odot \restriction_{L^1(\tilde\phi_1)}). \tag{21}$$

On the other hand, ϕ and $\tilde\phi$ are λ_0-invariant positive solutions of the operator L and $L^\star$, respectively. Therefore, Theorem 4.4 implies that $\|\mathcal{G}_\lambda\|_{L^p(\phi_p)} = (\mu - \lambda)^{-1}$.

Recall that by Theorem 3.1, for any $1 \leq p < \infty$, the operator $\mathcal{G}_\lambda^{\odot} \restriction_{L^{p'}(\tilde{\phi}_{p'})}$ is the dual operator of $\mathcal{G}_\lambda \restriction_{L^p(\phi_p)}$, and $\mathcal{G}_\lambda \restriction_{L^{p'}(\phi_{p'})}$ is the dual of $\mathcal{G}_\lambda^{\odot} \restriction_{L^p(\tilde{\phi}_p)}$. Since the spectra of a bounded operator and its dual are equal, we have

$$\sigma(\mathcal{G}_\lambda \restriction_{L^1(\phi_1)}) = \sigma(\mathcal{G}_\lambda^{\odot} \restriction_{L^\infty(\tilde{\phi}_\infty)}), \quad \sigma(\mathcal{G}_\lambda^{\odot} \restriction_{L^1(\tilde{\phi}_1)}) = \sigma(\mathcal{G}_\lambda \restriction_{L^\infty(\phi_\infty)}).$$

Thus all the point-spectra in (20) and (21) are equal. □

Acknowledgments. The author is grateful to Professor M. Cwikel and the late professor V. Liskevich for valuable discussions. The author thanks the anonymous referee for his careful reading and valuable comments.

This research was supported by the Israel Science Foundation (grants No. 963/11) founded by the Israel Academy of Sciences and Humanities.

References

[1] S. Agmon, Bounds on exponential decay of eigenfunctions of Schrödinger operators. In S. Graffi (ed.), *Schrödinger operators.* Lectures given at the 2nd 1984 session of the Centro Internationale Matematico Estivo (CIME) held in Como, August 26–September 4, 1984. Lecture Notes in Mathematics, 1159. Springer-Verlag, Berlin, 1985, 1–38. MR 0824986 Zbl 0583.35027

[2] H. Aikawa, Norm estimate of Green operator, perturbation of Green function and integrability of superharmonic functions. *Math. Ann.* **312** (1998), no. 2, 289–318. MR 1671780 Zbl 0917.31001

[3] D. G. Aronson and P. Besala, Uniqueness of solutions of the Cauchy problem for parabolic equations. *J. Math. Anal. Appl.* **13** (1966), 516–526. MR 0192197 Zbl 0137.29501

[4] D. G. Aronson and P. Besala, Corrections to "Uniqueness of solutions of the Cauchy problem for parabolic equations." *J. Math. Anal. Appl.* **17** (1967), 194–196. Zbl 0137.29501

[5] W. Arendt, Gaussian estimates and interpolation of the spectrum in L^p. *Differential Integral Equations* **7** (1994), no. 5-6, 1153–1168. MR 1269649 Zbl 0827.35081

[6] R. Bauñelos and B. Davis, Heat kernel, eigenfunctions, and conditioned Brownian motion in planar domains. *J. Funct. Anal.* **84** (1989), no. 1, 188–200. MR 0999496 Zbl 0676.60073

[7] A. Ben Amor and W. Hansen, Continuity of eigenvalues for Schrödinger operators, L^p-properties of Kato type integral operators. *Math. Ann.* **321** (2001), no. 4, 925–953. MR 1872535 Zbl 1067.35053

[8] H. Berestycki, L. Nirenberg and S. R. S. Varadhan, The principal eigenvalue and maximum principle for second-order elliptic operators in general domains. *Comm. Pure Appl. Math.* **47** (1994), no. 1, 47–92. MR 1258192 Zbl 0806.35129

[9] H. Berestycki and L. Rossi, Generalizations and properties of the principal eigenvalue of elliptic operators in unbounded domains. *Comm. Pure Appl. Math.* **68** (2015), no. 6, 1014–1065. MR 3340379 Zbl 1332.35243

[10] I. Birindelli, Hopf's lemma and anti-maximum principle in general domains. *J. Differential Equations* **119** (1995), no. 2, 450–472. MR 1340547 Zbl 0831.35114

[11] A. P. Calderón, Intermediate spaces and interpolation, the complex method. *Studia Math.* **24** (1964), 113–190. MR 0167830 Zbl 0204.13703

[12] M. Cwikel, Real and complex interpolation and extrapolation of compact operators. *Duke Math. J.* **65** (1992), no. 2, 333–343. MR 1150590 Zbl 0787.46062

[13] M. Cwikel, Nigel Kalton and complex interpolation of compact operators. Preprint 2014. arXiv:1410.4527 [math.FA]

[14] E. B. Davies, L^p spectral independence and L^1 analyticity. *J. London Math. Soc.* (2) **52** (1995), no. 1, 177–184. MR 1345724 Zbl 0913.47032

[15] E. B. Davies, L^p spectral independence for certain uniformly elliptic operators. In L. Hörmander and A. Melin (eds.), *Partial differential equations and mathematical physics.* Papers from the Danish–Swedish Analysis Seminar held at the University of Copenhagen, Copenhagen, March 17–19, 1995, and the University of Lund, Lund, May 19–21, 1995. Progress in Nonlinear Differential Equations and their Applications, 21. Birkhäuser Boston, Inc., Boston, MA, 1996, 122–125. MR 1380986 Zbl 0864.35077

[16] E. B. Davies and B. Simon, Ultracontractivity and the heat kernel for Schrödinger operators and Dirichlet Laplacians. *J. Funct. Anal.* **59** (1984), no. 2, 335–395. MR 0766493 Zbl 0568.47034

[17] K-J. Engel and R. Nagel, *One-parameter semigroups for linear evolution equations.* With contributions by S. Brendle, M. Campiti, T. Hahn, G. Metafune, G. Nickel, D. Pallara, C. Perazzoli, A. Rhandi, S. Romanelli, and R. Schnaubelt. Graduate Texts in Mathematics, 194. Springer, Berlin etc., 2000. ISBN 0-387-98463-1 MR 1721989 Zbl 0952.47036

[18] A. Grigor′yan and J. Hu, Heat kernels and Green functions on metric measure spaces. *Canad. J. Math.* **66** (2014), no. 3, 641–699. MR 3194164 Zbl 1293.35128

[19] M. Grüter and K.-O. Widman, The Green function for uniformly elliptic equations. *Manuscripta Math.* **37** (1982), no. 3, 303–342. MR 0657523 Zbl 0485.35031

[20] W. Hansen, Global comparison of perturbed Green functions. *Math. Ann.* **334** (2006), no. 3, 643–678. MR 2207878 Zbl 1123.31001

[21] A. T. Hill, Estimates on the Green's function of second-order elliptic operators in $\mathbb{R}^N$. *Proc. Roy. Soc. Edinburgh Sect. A* **128** (1998), no. 5, 1033–1051. MR 1642132

[22] R. Killip, M. Visan, and X. Zhang, Riesz transforms outside a convex obstacle. *Int. Math. Res. Not. IMRN* **2016** (2016), no. 19, 5875–5921. MR 3567262

[23] P. C. Kunstmann and H. Vogt, Weighted norm estimates and L_p-spectral independence of linear operators. *Colloq. Math.* **109** (2007), no. 1, 129–146. MR 2308831 Zbl 1126.35034

[24] D. Lenz, P. Stollmann, and D. Wingert, Compactness of Schrödinger semigroups. *Math. Nachr.* **283** (2010), no. 1, 94–103. MR 2598595 Zbl 1243.47070

[25] X. D. Li, Liouville theorems for symmetric diffusion operators on complete Riemannian manifolds. *J. Math. Pures Appl.* (9) **84** (2005), no. 10, 1295–1361. MR 2170766 Zbl 1082.58036

[26] V. Liskevich and H. Vogt, On L^p-spectra and essential spectra of second-order elliptic operators. *Proc. London Math. Soc.* (3) **80** (2000), no. 3, 590–610. MR 1744777 Zbl 1053.35114

[27] V. Liskevich, Z. Sobol and H. Vogt, On the L_p-theory of C_0-semigroups associated with second-order elliptic operators. II. *J. Funct. Anal.* **193** (2002), no. 1, 55–76. MR 1923628 Zbl 1020.47029

[28] G. Metafune, D. Pallara and M. Wacker, Compactness properties of Feller semigroups. *Studia Math.* **153** (2002), no. 2, 179–206. MR 1948923 Zbl 1033.47030

[29] M. Murata, Structure of positive solutions to $(-\Delta + V)u = 0$ in $\mathbb{R}^n$. *Duke Math. J.* **53** (1986), no. 4, 869–943. MR 0874676 Zbl 0624.35023

[30] M. Murata, Semismall perturbations in the Martin theory for elliptic equations. *Israel J. Math.* **102** (1997), 29–60. MR 1489100 Zbl 0891.35013

[31] M. Murata, Integral representations of nonnegative solutions for parabolic equations and elliptic Martin boundaries. *J. Funct. Anal.* **245** (2007), no. 1, 177–212. MR 2310806 Zbl 1116.58026

[32] Y. Pinchover, On positive solutions of second-order elliptic equations, stability results and classification. *Duke Math. J.* **57** (1988), no. 3, 955–980. MR 0975130 Zbl 0685.35035

[33] Y. Pinchover, Criticality and ground states for second-order elliptic equations. *J. Differential Equations* **80** (1989), no. 2, 237–250. MR 1011149 Zbl 0697.35036

[34] Y. Pinchover, On criticality and ground states of second order elliptic equations. II. *J. Differential Equations* **87** (1990), no. 2, 353–364. MR 1072906 Zbl 0714.35055

[35] Y. Pinchover, On nonexistence of any λ_0-invariant positive harmonic function, a counterexample to Stroock's conjecture. *Comm. Partial Differential Equations* **20** (1995), no. 9-10, 1831–1846. MR 1349233 Zbl 0839.58053

[36] Y. Pinchover, Maximum and anti-maximum principles and eigenfunctions estimates via perturbation theory of positive solutions of elliptic equations. *Math. Ann.* **314** (1999), no. 3, 555–590. MR 1704549 Zbl 0928.35010

[37] Y. Pinchover, Liouville-type theorem for Schrödinger operators. *Comm. Math. Phys.* **272** (2007), no. 1, 75–84. MR 2291802 Zbl 1135.35021

[38] M. Reed and B. Simon, *Methods of modern mathematical physics.* IV. Analysis of operators. Academic Press [Harcourt Brace Jovanovich, Publishers], New York and London, 1978. ISBN 0-12-585004-2 MR 0493421 Zbl 0401.47001

[39] Yu. A. Semenov, Stability of L^p-spectrum of generalized Schrödinger operators and equivalence of Green's functions. *Internat. Math. Res. Notices* **1997**, no. 12, 573–593. MR 1456565 Zbl 0905.47031

[40] H. Shindoh, Kernel estimates and L^p-spectral independence of generators of C_0-semigroups. *J. Funct. Anal.* **255** (2008), no. 5, 1273–1295. MR 2455498 Zbl 1225.47043

[41] B. Simon, Brownian motion, L^p properties of Schrödinger operators and the localization of binding. *J. Funct. Anal.* **35** (1980), no. 2, 215–229. MR 0561987 Zbl 0446.47041

[42] Z. Sobol and H. Vogt, On the L_p-theory of C_0-semigroups associated with second-order elliptic operators. I. *J. Funct. Anal.* **193** (2002), no. 1, 24–54. MR 1923627 Zbl 1020.47028

[43] E. M. Stein, Interpolation of linear operators. *Trans. Amer. Math. Soc.* **83** (1956), 482–492. MR 0082586 Zbl 0072.32402

[44] J.-Y. Wu, L^p-Liouville theorems on complete smooth metric measure spaces. *Bull. Sci. Math.* **138** (2014), no. 4, 510–539. MR 3215940 Zbl 1295.53030

Abstract graph-like spaces and vector-valued metric graphs

Olaf Post

Dear Pavel, thank you for having supported me over all the time;
I hope you will find this new piece of "heavy German machinery"
useful for our future collaboration,
and that we can continue working together for a long time.

Všechno nejlepší k narozeninám, Pavle!

Prologue

I became interested in graph-like spaces by a question of Vadim Kostrykin, asking whether a Laplacian on a family of open sets $(X_\varepsilon)_{\varepsilon>0}$, converging to a metric graph X_0 converges to some suitable Laplacian on X_0. At that time, I was not aware of the work of Kuchment and Zeng [10] and wrote down some ideas. Somehow Pavel must have heard about this; he invited me to visit him in Řež in October 2002, just two months after the big flood, which covered even the high-lying tracks with water, resulting in a very reduced schedule. At that time one had to buy the local ticket at *Praha Masarykovo nádraží* at a counter where one was forced to pronounce the most complicated letter in Czech language, the „*Ř*" in „*Řež*". At least I got the ticket I wanted, and enjoyed staying in this little pension *Hudec*. Řež at night has something very special and rare nowadays in our noise-polluted world – *Silence*! Only the dogs bark and from time to time, trains pass by on the other side of the *Vltava*... Also Řež was a good opportunity to pick up some Czech words, as people in that little village only spoke Czech (and sometimes a little bit German) „*Máte smažený sýr*?" – „*Dobrou chuť*!" – „*Pivo, prosím*"... This invitation was the start of a very fruitful collaboration with Pavel over many years, resulting in several publications, see [5], [6], [7], and [8]. Pavel inspired my research on graph-like spaces, resulting even in an entire book [15]. Pavel sometimes cites it with the words "... *and then we apply the heavy German machinery*..."

1 Introduction

This present note shall serve as a unified approach on how to work with spaces that can be decomposed into building blocks (the *analytic* viewpoint) or that can be built up from building blocks (the *synthetic* viewpoint) according to a graph. We will call such spaces *(abstract) graph-like spaces*. They can be obtained in basically two different ways, depending on whether the graph-like space is decomposed into pieces indexed by *vertices* or *edges*, respectively. We call them *vertex-coupled* or *edge-coupled*, respectively. There is also a mixed case, when one has a decomposition into parts indexed by vertices and edges (like for thin ε-neighbourhoods of embedded graphs or graph-like manifolds in the spirit of [15]). This case can be reduced to the vertex-coupled case by considering the *subdivision graph* as an underlying graph (see Definition 2.1 in Section 2.1 for details).

In the edge-coupled case, one can also choose a suitable subspace at each vertex determining the vertex conditions, very much in the spirit of a quantum graph. Indeed, one can consider edge-coupled spaces as *general* or *vector-valued quantum graphs* (see [12] and also [1] for a another point of view). Explaining the concept of metric and quantum graphs in an article dedicated to Pavel would be (in his own words…) *to bring owls to Athens* or *coal to Newcastle* or *firewood to the forest*… instead we refer to the book of Berkolaiko and Kuchment [2] or to Section 2.2 of [15]). We define the coupling via the language of abstract boundary value problems. Such a theory has been developed mostly for operators, in order to describe (all) self-adjoint extensions of a given minimal operator. As we are interested only in "geometric" non-negative operators such as Laplacians we find it more suitable to start with the corresponding quadratic or energy forms. A theory of abstract boundary value problems expressed entirely in terms of quadratic forms has been developed recently under the name *boundary pairs* in [16], and under the name *boundary maps* in [15] (see also [16] and references therein for related concepts, as well as [9], especially Chapter 3 by Arlinskiĭ). In particular, one has an abstract Dirichlet and Neumann operator, a solution operator for the Dirichlet problem and a Dirichlet–to–Neumann operator, see Section 2.

The coupling of abstract boundary value problems in Section 3 is – of course – not new (see, e.g., Chapter 7 in [9] and references therein). For our graph-like spaces, the new point is the interpretation of the coupled operators such as the Neumann or Dirichlet–to–Neumann operator as a discrete vector-valued graph Laplacian.

In Section 4 of this note, we explain the concept of a distance of two abstract graph-like spaces based on their building blocks (such as the vertex or edge part of a graph-like space). This concept can be used to show convergence of a family of abstract boundary value problems to a limit one. The motivation is to give a unified approach for the convergence of many types of (concrete) graph-like spaces such as thick graphs, ε-neighbourhoods of embedded graphs or graph-like manifolds, including different types of boundary conditions (Neumann, Dirichlet).

I'd like to thank the anonymous referee for very carefully reading this manuscript, valuable suggestions and pointing out quite a lot of typos. I'm afraid there are still some left…

2 Preliminaries

In this section we fix the notation and collect briefly some facts on discrete graphs, as well as on abstract boundary value problems (boundary pairs) and convergence of operators acting in different Hilbert spaces.

2.1 Discrete Graphs

Let $G = (V, E, \partial)$ be a countable graph, i.e., V and E are disjoint and at most countable sets and $\partial\colon E \to V \times V$ is a map defining the incidence between edges and vertices, namely, $\partial e = (\partial_- e, \partial_+ e)$ is the pair of the *initial* resp. *terminal vertex* of a given edge $e \in E$. Let

$$E(V_1, V_2) := \{\, e \in E \mid \partial_- e \in V_1, \partial_+ e \in V_2 \text{ or } \partial_+ e \in V_1, \partial_- e \in V_2 \,\}$$

for $V_1, V_2 \subset V$. We denote by $E_v = E(\{v\}, V) \subset E$ the set of edges adjacent with the vertex $v \in V$ and call the number $\deg v := |E_v|$ the *degree* of a vertex $v \in V$. We always assume that the graph is locally finite, i.e., that $\deg v < \infty$ for all $v \in V$ (but not necessarily uniformly bounded). For ease of notation, we also assume that the graph has no loops, i.e., edges e with $\partial_- e = \partial_+ e$.

We use the convention that we have chosen already an *orientation* of each edge via $\partial e = (\partial_- e, \partial_+ e)$, i.e., for each edge e there is not automatically an edge in E with the opposite direction. In particular, we assume that

$$\sum_{v \in V} \sum_{e \in E_v} a_e(v) = \sum_{e \in E} \sum_{v = \partial_\pm e} a_e(v) \tag{1}$$

holds for any numbers $a_e(v) \in \mathbb{C}$, and this also implies that $\sum_{v \in V} \deg v = 2|E|$ by setting $a_e(v) = 1$. We make constant use of this reordering in the sequel.

Given a graph $G = (V, E, \partial)$, we construct another graph by introducing a new vertex on each edge:

Definition 2.1. Let $G = (V, E, \partial)$ a graph. The *subdivision graph* $\mathrm{S}G = (A, B, \hat{\partial})$ is the graph with vertex set

$$A = V \mathbin{\dot\cup} E$$

(disjoint union) and edge set

$$B = \bigcup_{v \in V} \{v\} \times E_v.$$

Moreover,

$$\hat{\partial}\colon B \longrightarrow A \times A, \quad b = (v, e) \longmapsto \begin{cases} (\hat{\partial}_- b, \hat{\partial}_+ b) = (v, e), & v = \partial_- e, \\ (\hat{\partial}_- b, \hat{\partial}_+ b) = (e, v), & v = \partial_+ e. \end{cases}$$

2.2 Boundary pairs and abstract boundary value problems

Following a good tradition („*Was interessiert mich mein Geschwätz von gestern, nichts hindert mich, weiser zu werden…*“), we use a slightly different terminology than in [15] and [16]; basically, we collect *all* data involved in a boundary pair and put it into a quintuple:

Definition 2.2. (1) We say that the quintuple $\Pi := (\Gamma, \mathcal{G}, \mathfrak{h}, \mathcal{H}^1, \mathcal{H})$ is an *abstract boundary value problem* if

- $\mathfrak{h}$ is a closed, non-negative quadratic form densely defined in a Hilbert space $\mathcal{H}$; such a form is also called an *energy form*; we endow its domain $\operatorname{dom} \mathfrak{h} = \mathcal{H}^1$ with a norm given by $\|f\|^2_{\mathcal{H}^1} = \mathfrak{h}(f) + \|f\|^2_{\mathcal{H}}$; we also say that the energy form is given by $(\mathfrak{h}, \mathcal{H}^1, \mathcal{H})$;
- $\mathcal{G}$ is another Hilbert space and $\Gamma\colon \mathcal{H}^1 \to \mathcal{G}$ is a bounded operator, called a *boundary map*, such that $\mathcal{G}^{1/2} := \operatorname{ran} \Gamma (= \Gamma(\mathcal{H}^1))$ is dense in $\mathcal{G}$.

(2) If, in addition, $\mathcal{H}^{1,\mathrm{D}} := \ker \Gamma$ is dense in $\mathcal{H}$, we say that the abstract boundary value problem Π *has a dense Dirichlet domain.*[1]

(3) We say that the abstract boundary value problem Π is *bounded* if Γ is surjective, i.e., if $\operatorname{ran} \Gamma = \mathcal{G}$.

(4) We say that the abstract boundary value problem Π is *trivial* if $\mathcal{G} = \mathcal{H}$ and $\Gamma = \mathrm{id}$.

[1] A pair $(\Gamma, \mathcal{G})$ is called *boundary pair associated with the quadratic form* $\mathfrak{h}$ in [16] if $\operatorname{ran} \Gamma$ is dense in $\mathcal{G}$ and $\ker \Gamma$ is dense in $\mathcal{H}$. If only $\operatorname{ran} \Gamma$ is dense in $\mathcal{G}$, then $(\Gamma, \mathcal{G})$ is called a *generalised boundary pair* in [16].

A typical situation is $\mathcal{H} = \mathsf{L}_2(X, \mu)$ and $\mathcal{G} = \mathsf{L}_2(Y, \nu)$, where (X, μ) and (Y, ν) are measured spaces such that $Y \subset X$ is measurable. The abstract boundary value problem has a dense Dirichlet domain iff $\mu(Y) = 0$. The abstract boundary value problem is trivial if and only if $(X, \mu) = (Y, \nu)$ and $\Gamma = \mathrm{id}$.

Given an abstract boundary value problem, we can define the following objects (details can be found in [16]):

- the *Neumann operator* H as the operator associated with $\mathfrak{h}$;
- the *Dirichlet operator* H^{D} as the operator associated with the closed (!) form $\mathfrak{h}\restriction_{\ker \Gamma}$ with domain $\mathcal{H}^{1,\mathrm{D}} := \ker \Gamma$;
- the *space of weak solutions*

$$\mathcal{N}^1(z) = \{h \in \mathcal{H}^1 \mid \mathfrak{h}(h, f) = z\langle h, f\rangle \text{ for all } f \in \mathcal{H}^{1,\mathrm{D}}\};$$

- for $z \notin \sigma(H^{\mathrm{D}})$, $\mathcal{H}^1 = \mathcal{H}^{1,\mathrm{D}} \dot{+} \mathcal{N}^1(z)$ (direct sum with closed subspaces); in particular, the *Dirichlet solution operator*

$$S(z) = (\Gamma\restriction_{\mathcal{N}^1(z)})^{-1}\colon \operatorname{ran}\Gamma = \mathcal{G}^{1/2} \longrightarrow \mathcal{N}^1(z) \subset \mathcal{H}^1$$

  is defined; we also set $S := S(-1)$, i.e., the default value of z is -1;
- for $z \notin \sigma(H^{\mathrm{D}})$, the *Dirichlet–to–Neumann (sesquilinear) form* $\mathfrak{l}_z$ is defined via $\mathfrak{l}_z(\varphi, \psi) = (\mathfrak{h} - z\mathbb{1})(S(z)\varphi, S(-1)\psi)$, $\varphi, \psi \in \mathcal{G}^{1/2}$;
- we endow $\mathcal{H}^1$ with its natural norm given by $\|f\|^2_{\mathcal{H}^1} = \mathfrak{h}(f) + \|f\|^2_{\mathcal{H}}$;
- we endow $\mathcal{G}^{1/2}$ with the norm given by $\|\varphi\|^2_{\mathcal{G}^{1/2}} = \mathfrak{l}_{-1}(\varphi) = \|S\varphi\|^2_{\mathcal{H}^1}$;
- if the abstract boundary value problem is bounded, then $\mathcal{G}^{1/2} = \mathcal{G}$, and the two norms are equivalent; moreover, $\mathfrak{l}_z$ is a bounded sesquilinear form on $\mathcal{G} \times \mathcal{G}$.

For an abstract boundary value problem, one can always construct another boundary map $\Gamma'\colon \mathcal{W} \to \mathcal{G}$ which is defined on a subspace $\mathcal{W}$ of $\mathcal{H}^1 \cap \operatorname{dom} H^{\max}$, where $H^{\max} := (H^{\min})^*$ and $H^{\min} := H^{\mathrm{D}} \cap H$ denote the maximal resp. minimal operator, and on which Γ' is bounded. Moreover, one has the following *abstract Green's (first) formula*

$$\mathfrak{h}(f, g) = \langle H^{\max} f, g\rangle_{\mathcal{H}} + \langle \Gamma' f, \Gamma g\rangle_{\mathcal{G}} \tag{2}$$

for all $f \in \mathcal{W}$ and $g \in \mathcal{H}^1$.

Another property is also important (see [16] for details):

Definition 2.3. We say that an abstract boundary value problem Π (or the boundary pair $(\Gamma, \mathcal{G})$) is *elliptically regular* if the associated Dirichlet solution operator

$$S := S(-1) \colon \mathcal{G}^{1/2} \longrightarrow \mathcal{H}^1$$

extends to a bounded operator $\overline{S} \colon \mathcal{G} \to \mathcal{H}$, or equivalently, if there exists a constant $c > 0$ such that $\|S\varphi\|_{\mathcal{H}} \le c\|\varphi\|_{\mathcal{G}}$ for all $\varphi \in \mathcal{G}^{1/2}$.

All our abstract boundary value problems treated in this note will be elliptically regular. They have the important property that the Dirichlet–to–Neumann form $\mathfrak{l}_z$ is *closed* as form in $\mathcal{G}$ with domain $\operatorname{dom} \mathfrak{l}_z = \mathcal{G}^{1/2} = \operatorname{ran} \Gamma$, and hence is associated with a closed operator $\Lambda(z)$, called *Dirichlet–to–Neumann operator*; moreover, the domain $\mathcal{G}^1 := \operatorname{dom} \Lambda(z)$ of $\Lambda(z)$ is independent of $z \in \mathbb{C} \setminus \sigma(H^{\mathrm{D}})$. Another important consequence is the following formula on the difference of resolvents: Let $z \in \mathbb{C} \setminus (\sigma(H) \cup \sigma(H^{\mathrm{D}}))$, then

$$(H - z)^{-1} = (H^{\mathrm{D}} - z)^{-1} + \overline{S}(z)\Lambda(z)^{-1}\overline{S}(\bar{z})^{*}. \tag{3}$$

As a consequence of (3), one has, e.g., the spectral characterisation

$$\lambda \in \sigma(H) \iff 0 \in \sigma(\Lambda(\lambda)) \tag{4}$$

for all $\lambda \in \mathbb{R} \setminus \sigma(H^{\mathrm{D}})$.

Examples 2.4. Important examples of elliptically regular abstract boundary value problems are the following ones.

(1) Let (X, g) be a Riemannian manifold with compact smooth boundary (Y, h), then

$$\Pi = (\Gamma, \mathrm{L}_2(Y, h), \mathfrak{h}, \mathrm{H}^1(X, g), \mathrm{L}_2(X, g))$$

is an elliptically regular abstract boundary value problem with dense Dirichlet domain. Here, $\Gamma f = f \restriction_Y$ is the Sobolev trace, and the energy form is

$$\mathfrak{h}(f) = \int_X |df|_g^2 \,\mathrm{d}\operatorname{vol}_g .$$

This example is actually the godfather of the above-mentioned names for the derived objects: e.g., the Dirichlet resp. Neumann operators are actually the Dirichlet and Neumann Laplacians, the Dirichlet solution operator is the operator solving the Dirichlet problem (also called *Poisson operator*), the abstract Green's formula (2) is the usual one with $\Gamma' f$ being the normal outwards derivative and $\mathcal{W} = \mathrm{H}^2(X)$, e.g., and the Dirichlet–to–Neumann operator has its standard interpretation.

(2) Bounded abstract boundary value problems (i.e., abstract boundary value problems, where $\operatorname{ran}\Gamma = \mathcal{G}$, or equivalently, where the Dirichlet–to–Neumann operator is bounded), and in particular abstract boundary value problems with finite dimensional boundary space $\mathcal{G}$, are elliptically regular.

(3) Let $G = (V, E, \partial)$ be a graph. For simplicity, we consider only the normalised Laplacian here. We define an energy form via

$$\mathfrak{h}(f) = \sum_{e \in E} |f(\partial_+ e) - f(\partial_- e)|^2$$

for $f \in \mathcal{H}^1 = \mathcal{H} = \ell_2(V, \deg)$, where $\|f\|^2_{\ell_2(V,\deg)} = \sum_{v \in V} |f(v)|^2 \deg v$. Using (1) it is not hard to see that $0 \le \mathfrak{h}(f) \le 2\|f\|^2_{\ell_2(V,\deg)}$. The *boundary* of G is just an arbitrary non-empty subset ∂V of V (in particular, the degree of a "boundary vertex" can be arbitrary). Set $\mathcal{G} = \ell_2(\partial V, \deg)$ and $\Gamma f = f \restriction_{\partial V}$. Then $\Pi = (\Gamma, \ell_2(\partial V, \deg), \mathfrak{h}, \ell_2(V, \deg), \ell_2(V, \deg))$ is an elliptically regular abstract boundary value problem without dense Dirichlet domain (see Section 6.7 of [16]).

The Neumann operator acts as

$$(Hf)(v) = (\Delta_G f)(v) := \frac{1}{\deg v} \sum_{e \in E_v} (f(v) - f(v_e)) \tag{5}$$

for $v \in V$, where v_e denotes the vertex adjacent with e and opposite to v. The Dirichlet operator acts in the same way on $\ell_2(\mathring{V}, \deg)$ where $\mathring{V} := V \setminus \partial V$ are the *interior* vertices (note that the Dirichlet Laplacian is not the Laplacian on the subgraph $\mathring{G} := (\mathring{V}, \mathring{E}, \mathring{\partial})$ with $\mathring{E} := E(\mathring{V}, \mathring{V})$ and $\mathring{\partial} := \partial \restriction_{\mathring{E}}$, as the degree is still calculated in the entire graph G and not in $\mathring{G}$).

Moreover, the decomposition

$$\mathcal{H} = \ell_2(V, \deg) = \ell_2(\partial V, \deg) \oplus \ell_2(\mathring{V}, \deg) = \mathcal{G} \oplus \ker \Gamma$$

yields a block structure for H, namely,

$$H = \begin{pmatrix} A & B \\ B^* & D \end{pmatrix}$$

with $A\colon \mathcal{G} \to \mathcal{G}$, $B\colon \ker\Gamma \to \mathcal{G}$ and Dirichlet operator

$$D = H^{\mathrm{D}}\colon \ker\Gamma \longrightarrow \ker\Gamma.$$

The Dirichlet–to–Neumann operator is

$$\Lambda(z) = (A - z) - B(D - z)^{-1} B^*$$

provided $z \notin \sigma(H^{\mathrm{D}}) = \sigma(D)$. Moreover, the second boundary map

$$\Gamma' \colon \mathcal{W} = \ell_2(V, \deg) \longrightarrow \mathcal{G} = \ell_2(\partial V, \deg)$$

in Green's formula (2) is here

$$(\Gamma' f)(v) = \frac{1}{\deg v} \sum_{e \in E_v} (f(v) - f(v_e)), \quad v \in \partial V,$$

for $f \in \mathcal{W} = \ell_2(V, \deg)$, or in block structure, $\Gamma' = (A, B)$.

Note that we have not excluded the extreme (or trivial) case $\partial V = V$ leading to a trivial abstract boundary value problem with $\Gamma = \mathrm{id}_{\ell_2(V)}$. In this case, $\ker \Gamma = \{0\}$, hence $A = H$, $B = 0$, $H^{\mathrm{D}} = D = 0$ and $\sigma(H^{\mathrm{D}}) = \emptyset$. Moreover, $\Lambda(z) = H - z$.

(4) Let X be a metric graph (with underlying discrete graph $G = (V, E, \partial)$ and edge length function

$$\ell \colon E \longrightarrow (0, \infty), \quad e \longmapsto \ell_e,$$

(see, e.g., [2] or Section 2.2 of [15]) such that

$$\ell_0 = \inf_{e \in E} \ell_e > 0.$$

A bounded (hence elliptically regular) abstract boundary value problem is given by $\Pi = (\Gamma, \ell_2(V, \deg), \mathfrak{h}, \mathrm{H}^1(X), \mathrm{L}_2(X))$, where

$$\Gamma f = f \restriction_V$$

is the restriction of functions on X to the set of vertices,

$$\mathfrak{h}(f) = \int_X |f'(x)|^2 \,\mathrm{d}x = \sum_{e \in E} \int_0^{\ell_e} |f_e'(x_e)|^2 \,\mathrm{d}x_e,$$

and

$$f \in \mathrm{H}^1(X) = \bigoplus_{e \in E} \mathrm{H}^1([0, \ell_e]) \cap \mathrm{C}(X).$$

In this case, the Neumann operator H is the Laplacian with *standard* or (*generalised*) *Neumann* or *Kirchhoff*[2] *vertex conditions* and the Dirichlet operator H^{D} is the direct sum of the Dirichlet Laplacians on the intervals $[0, \ell_e]$, hence decoupled (see [13] and [15] for details).

[2] When one calls these vertex conditions "*Kirchhoff*" as a coauthor of Pavel, one always ends up with at least a footnote (as in my first collaboration with Pavel [5]). For Pavel, the current conservation usually associated with this name, refers to the *probability* current, which is preserved for any self-adjoint vertex condition. Many other authors think of a more naive current, defined by a derivative considered as vector field.

2.3 Convergence of abstract boundary value problems acting in different spaces

We now define a concept of a "distance" δ for objects of abstract boundary value problems Π and $\widehat{\Pi}$ acting in different spaces. One can think of $\widehat{\Pi}$ as being a perturbation of Π, and δ measures quantitatively, how far away $\widehat{\Pi}$ is from being isomorphic with Π (see Example 2.10 below for the case $\delta = 0$). The term "convergence" refers to the situation where we consider a family $(\Pi_\varepsilon)_{\varepsilon \geq 0}$ of abstract boundary value problems; one can think of $\widehat{\Pi} = \Pi_\varepsilon$ and $\Pi = \Pi_0$ with "distance" δ_ε. If $\delta_\varepsilon \to 0$ as $\varepsilon \to 0$ then we say that Π_ε *converges to* Π_0. Details of this concept of a "distance" between operators acting in different spaces can also be found in Chapter 4 in [15].

To be more precise, let $\Pi = (\Gamma, \mathcal{G}, \mathfrak{h}, \mathcal{H}^1, \mathcal{H})$ and $\widehat{\Pi} = (\widehat{\Gamma}, \widehat{\mathcal{G}}, \widehat{\mathfrak{h}}, \widehat{\mathcal{H}}^1, \widehat{\mathcal{H}})$ be two abstract boundary value problems. Recall that $\mathcal{H}^1$ is the domain of a closed non-negative form $\mathfrak{h}$ in the Hilbert space $\mathcal{H}$, and that $\Gamma \colon \mathcal{H}^1 \to \mathcal{G}$ is bounded with dense range, and similarly for the tilded objects. We need bounded operators

$$J \colon \mathcal{H} \longrightarrow \widehat{\mathcal{H}}, \quad J' \colon \widehat{\mathcal{H}} \longrightarrow \mathcal{H}, \quad I \colon \mathcal{G} \longrightarrow \widehat{\mathcal{G}}, \quad I' \colon \widehat{\mathcal{G}} \longrightarrow \mathcal{G}, \tag{6a}$$

called *identification operators* which replace unitary or isomorphic operators. The quantity $\delta > 0$ used later on measures how far these operators differ from isomorphisms. We also need *identification operators* on the level of the energy form domains, namely

$$J^1 \colon \mathcal{H}^1 \longrightarrow \widehat{\mathcal{H}}^1 \quad \text{and} \quad J'^1 \colon \widehat{\mathcal{H}}^1 \longrightarrow \mathcal{H}^1. \tag{6b}$$

In contrast to [3] we will not assume in this note that the identification operators I and I' on the boundary spaces $\mathcal{G}$ and $\widehat{\mathcal{G}}$ also respect the form domains $\mathcal{G}^{1/2}$ and $\widehat{\mathcal{G}}^{1/2}$ of the Dirichlet–to–Neumann operators.

We start with the energy forms and boundary maps:

Definition 2.5. Let $\delta > 0$. We say that the energy forms $\mathfrak{h}$ and $\widehat{\mathfrak{h}}$ are *δ-close* if there are identification operators J^1 and J'^1 as in (6) such that

$$|\widehat{\mathfrak{h}}(J^1 f, u) - \mathfrak{h}(f, J'^1 u)| \leq \delta \|u\|_{\widehat{\mathcal{H}}^1} \|f\|_{\mathcal{H}^1}$$

holds for all $f \in \mathcal{H}^1$ and $u \in \widehat{\mathcal{H}}^1$.

Definition 2.6. Let $\delta > 0$. We say that the boundary maps Γ and $\widehat{\Gamma}$ are *δ-close* if there exist identification operators J^1, J'^1, I and I' as in (6) such that

$$\|(I\Gamma - \widehat{\Gamma}J^1)f\|_{\widehat{\mathcal{G}}} \le \delta \|f\|_{\mathcal{H}^1}$$

and

$$\|(I'\widehat{\Gamma} - \Gamma J'^1)u\|_{\mathcal{G}} \le \delta \|u\|_{\widehat{\mathcal{H}}^1}$$

hold for all $f \in \mathcal{H}^1$ and $u \in \widehat{\mathcal{H}}^1$.

So far, we have only dealt with forms and their domains. Let us now define the following compatibility between the identification operators on the Hilbert space and the energy form level:

Definition 2.7. We say that the identification operators J, J', J^1 and J'^1 are *δ-quasi-unitarily equivalent* with respect to the energy forms $\mathfrak{h}$ and $\widehat{\mathfrak{h}}$ if

$$|\langle Jf, u\rangle_{\widehat{\mathcal{H}}} - \langle f, J'u\rangle_{\mathcal{H}}| \le \delta \|f\|_{\mathcal{H}} \|u\|_{\widehat{\mathcal{H}}},$$

$$\|f - J'Jf\|_{\mathcal{H}} \le \delta \|f\|_{\mathcal{H}^1},$$

$$\|u - JJ'u\|_{\widehat{\mathcal{H}}} \le \delta \|u\|_{\widehat{\mathcal{H}}^1},$$

$$\|J^1 f - Jf\|_{\widehat{\mathcal{H}}} \le \delta \|f\|_{\mathcal{H}^1},$$

$$\|J'^1 u - J'u\|_{\mathcal{H}} \le \delta \|u\|_{\widehat{\mathcal{H}}^1}$$

hold for f and u in the respective spaces. We say that the forms $\mathfrak{h}$ and $\widehat{\mathfrak{h}}$ are *δ-quasi-unitarily equivalent*, if they are δ-close with δ-quasi-unitarily equivalent identification operators.

For the boundary identification operators I and I' we define:

Definition 2.8. We say that the identification operators I and I' are *δ-quasi-isomorphic* with respect to the abstract boundary value problems Π and $\widehat{\Pi}$ if

$$\|\varphi - I'I\varphi\|_{\mathcal{H}} \le \delta \|\varphi\|_{\mathcal{G}^{1/2}}, \quad \|\psi - II'\psi\|_{\widehat{\mathcal{H}}} \le \delta \|\psi\|_{\widehat{\mathcal{G}}^{1/2}}$$

hold for $\varphi \in \mathcal{G}^{1/2}$ and $\psi \in \widehat{\mathcal{G}}^{1/2}$. We say that the boundary maps Γ and $\widehat{\Gamma}$ are δ-quasi-isomorphic if they are δ-close with δ-quasi-unitarily equivalent J, J', J^1 and J'^1 resp. δ-quasi-isomorphic I and I'.

The δ-quasi-isomorphy only refers to the Dirichlet–to–Neumann form $\mathfrak{l}_{-1}$ in $z = -1$ as $\|\varphi\|^2_{\mathcal{G}^{1/2}} = \mathfrak{l}(\varphi) = \|S(-1)\varphi\|^2_{\mathcal{H}^1}$ and no other structure of Π; a similar note holds for $\widehat{\Pi}$. We do not assume that I^* is closed to I', as this is too restrictive for Definition 2.9 (see, e.g., the proof of Proposition 2.11: $I^* = I'$ would mean $\gamma = 1$).

Finally, we define what it means for abstract boundary value problems to be "close" to each other, by combining the last four definitions:

Definition 2.9. Let $\delta > 0$. We say that the abstract boundary value problems Π and $\widehat{\Pi}$ are *δ-quasi-isomorphic* if there exist δ-quasi-unitarily equivalent identification operators J, J', J^1 and J'^1 and δ-quasi-isomorphic identification operators I and I' for which $\mathfrak{h}$ and $\hat{\mathfrak{h}}$, respectively, Γ and $\widehat{\Gamma}$ are δ-close.

Let us illustrate this concept in two examples.

Example 2.10. A good test for a reasonable definition of a "distance" is the case $\delta = 0$: if Π and $\widehat{\Pi}$ are 0-quasi-isomorphic then J is unitary with adjoint J'; J^1 and J'^1 are restrictions of J and J^*, respectively. Moreover, J intertwines H and $\widehat{H}$ in the sense that $J(H+1)^{-1} = (\widehat{H}+1)^{-1}J$; and I is a bi-continuous isomorphism with inverse I', and Γ and $\widehat{\Gamma}$ are equivalent in the sense that $\widehat{\Gamma} = I\Gamma J'^1$. We call such abstract boundary value problems *isomorphic*.

Another rather trivial case is the following: it nevertheless plays an important role in the study of shrinking domains like an ε-homothetic vertex neighbourhood shrinking to a point in the limit $\varepsilon \to 0$ (i.e., we use the abstract boundary value problem $\widehat{\Pi} = \Pi_\varepsilon$ associated with a compact and connected manifold X of dimension $d \geq 2$ with boundary $Y = \partial X$ and metric $\varepsilon^2 g$ as in Example 2.4 (1); in this case, $\delta = \mathrm{O}(\sqrt{\varepsilon})$, see Section 5.1.4 of [15] for details, also for the validity of (7)):

Proposition 2.11. *Assume that $\widehat{\Pi} = (\widehat{\Gamma}, \widehat{\mathcal{G}}, \hat{\mathfrak{h}}, \widehat{\mathcal{H}}^1, \widehat{\mathcal{H}})$ is an abstract boundary value problem such that the corresponding Neumann operator $\widehat{H}$ has* 0 *as simple and isolated eigenvalue in its spectrum. Assume also that there is $a \in (0, 1]$ such that*

$$\|\widehat{\Gamma}u\|^2_{\widehat{\mathcal{G}}} \leq a\hat{\mathfrak{h}}(u) + \frac{2}{a}\|u\|^2_{\widehat{\mathcal{H}}} \tag{7}$$

holds for all $u \in \widehat{\mathcal{H}}^1$.

Moreover, let $\Pi = (\mathrm{id}, \mathbb{C}, 0, \mathbb{C}, \mathbb{C})$ be a trivial abstract boundary value problem. Then $\widehat{\Pi}$ and Π are δ-quasi-isomorphic with δ depending only on parameters of $\widehat{\Pi}$ and a, see (8) *for a precise definition of δ.*

Proof. Let Φ_0 be a normalised eigenvector associated with the eigenvalue 0 of $\widehat{H}$. As $0 \in \sigma(\widehat{H})$, we also have $0 \in \sigma(\widehat{\Lambda}(0))$ with eigenvector $\Psi_0 = \widehat{\Gamma}\Phi_0$ (see Theorem 4.7 (i) in [16]). In particular, $\gamma := \|\widehat{\Gamma}\Phi_0\|_{\widehat{\mathcal{G}}}^{-2}$ is defined. For the identification operators, we set

$$\begin{aligned} Jf &= f\Phi_0, \quad J^1 f = Jf, \\ J'u &= J^*u = \langle u, \Phi_0\rangle_{\widehat{\mathcal{H}}}, \\ J'^1 u &= J^*u, \quad I\varphi = \varphi\Psi_0 \end{aligned}$$

and $I' = \gamma I^*$, where $I^*\psi = \langle \psi, \Psi_0\rangle_{\widehat{\mathcal{G}}}$. The choice of γ implies that $I'I\varphi = \varphi$, and

$$\|\psi - II'\psi\|_{\widehat{\mathcal{G}}}^2 = \|\psi - \gamma\langle\psi, \Psi_0\rangle_{\widehat{\mathcal{G}}}\Psi_0\|_{\widehat{\mathcal{G}}}^2 \le \frac{1}{\mu_1}\hat{\mathfrak{l}}_0(\psi)$$

as $\sqrt{\gamma}\Psi_0$ is a normalised eigenfunction of $\widehat{\Lambda}(0)$ corresponding to the eigenvalue 0, where $\mu_1 := d(\sigma(\widehat{\Lambda}(0)) \setminus \{0\}, 0)$ and $\hat{\mathfrak{l}}_0$ is the associated quadratic form. As $\lambda \mapsto \hat{\mathfrak{l}}_\lambda$ is monotonously decreasing (see Theorem 2.12 (v) in [16]), we have the estimate $\hat{\mathfrak{l}}_0(\psi) \le \hat{\mathfrak{l}}_{-1}(\psi) =: \|\psi\|_{\widehat{\mathcal{G}}^{1/2}}^2$. In particular, I and I' are $(1/\sqrt{\mu_1})$-quasi-isomorphic, see Definition 2.8.

For the δ-closeness of the forms resp. the boundary maps we have

$$\begin{aligned} \hat{\mathfrak{h}}(J^1 f, u) - \mathfrak{h}(f, J'^1 u) &= 0, \\ (I\Gamma - \widehat{\Gamma}J^1)f = If - \widehat{\Gamma}f\Phi_0 &= f \cdot (\Psi_0 - \widehat{\Gamma}\Phi_0) = 0, \end{aligned}$$

and

$$\begin{aligned} (I'\widehat{\Gamma} - \Gamma J'^1)u &= \langle\gamma\widehat{\Gamma}u, \Psi_0\rangle_{\widehat{\mathcal{G}}} - \langle u, \Phi_0\rangle_{\widehat{\mathcal{H}}} \\ &= \langle\gamma\widehat{\Gamma}u, \Psi_0\rangle_{\widehat{\mathcal{G}}} - \gamma\langle\Psi_0, \Psi_0\rangle_{\widehat{\mathcal{G}}}\langle u, \Phi_0\rangle_{\widehat{\mathcal{H}}} \\ &= \gamma\langle\widehat{\Gamma}(u - \langle u, \Phi_0\rangle_{\widehat{\mathcal{H}}}\Phi_0), \Psi_0\rangle_{\widehat{\mathcal{G}}}. \end{aligned}$$

The latter inner product can be estimated in squared absolute value by

$$\begin{aligned} |(I'\widehat{\Gamma} - \Gamma J'^1)u|^2 &\le \gamma\|\widehat{\Gamma}(u - \langle u, \Phi_0\rangle_{\widehat{\mathcal{H}}}\Phi_0)\|_{\widehat{\mathcal{G}}}^2 \\ &\le \gamma\Big(a\hat{\mathfrak{h}}(u) + \frac{2}{a}\|u - \langle u, \Phi_0\rangle_{\widehat{\mathcal{H}}}\Phi_0\|_{\widehat{\mathcal{H}}}^2\Big) \\ &\le \gamma\Big(a + \frac{2}{a\lambda_1}\Big)\hat{\mathfrak{h}}(u), \end{aligned}$$

using (7), where $\lambda_1 := d(\sigma(\widehat{H}) \setminus \{0\}, 0)$. Note that $\hat{\mathfrak{h}}(u - \langle u, \Phi_0\rangle_{\widehat{\mathcal{H}}}\Phi_0) = \hat{\mathfrak{h}}(u)$ as $\widehat{H}\Phi_0 = 0$ and hence $\hat{\mathfrak{h}}(w, \Phi_0) = \langle w, \widehat{H}\Phi_0\rangle = 0$ for any $w \in \widehat{\mathcal{H}}^1$.

Finally, $J^*Jf = f$ and

$$\|u - JJ^*u\|^2_{\widehat{\mathcal{H}}} = \|u - \langle u, \Phi_0\rangle_{\widehat{\mathcal{H}}} \Phi_0\|^2_{\widehat{\mathcal{H}}} \le \frac{1}{\lambda_1} \hat{\mathfrak{h}}(u).$$

Therefore we can choose

$$\delta = \max\Big\{\frac{1}{\sqrt{\mu_1}}, \frac{1}{\|\widehat{\Gamma}\Phi_0\|_{\widehat{\mathcal{G}}}}\sqrt{a + \frac{2}{a\lambda_1}}, \frac{1}{\sqrt{\lambda_1}}\Big\}. \tag{8}$$

This concludes the proof. □

3 Abstract graph-like spaces

Let us first explain the philosophy briefly. In the below-mentioned different couplings of abstract boundary value problems according to a graph, we show that the Neumann operator is coupled, while the Dirichlet operator is always a direct sum of the building blocks, i.e., decoupled. Moreover, we give formulas for how the coupled operators can be calculated from the building blocks. We also analyse how the coupled operators such as the Dirichlet–to–Neumann operator resemble discrete Laplacians on the underlying or related graphs, allowing a deeper understanding of the problem and relating it to problems of graph Laplacians.

In particular, the resolvent formula (3) gives an expression of a globally defined object, namely the coupled Neumann operator in terms of objects from the building blocks (see, e.g., the formulas for H^{D}, $S(z)$ and $\Lambda(z)$ in Theorems 3.3 and 3.7). Hence the understanding of the nature how $\Lambda(z)$ is obtained from the building blocks is essential in understanding the global operator H.

3.1 Direct sum of abstract boundary value problems

Given a family $(\Pi_\alpha)_{\alpha\in\mathrm{A}}$ of abstract boundary value problems, we define the *direct sum* via[3]

$$\bigoplus_{\alpha\in\mathrm{A}} \Pi_\alpha := \Big(\bigoplus_{\alpha\in\mathrm{A}} \Gamma_\alpha, \bigoplus_{\alpha\in\mathrm{A}} \mathcal{G}_\alpha, \bigoplus_{\alpha\in\mathrm{A}} \mathfrak{h}_\alpha, \bigoplus_{\alpha\in\mathrm{A}} \mathcal{H}^1_\alpha, \bigoplus_{\alpha\in\mathrm{A}} \mathcal{H}_\alpha\Big).$$

[3] The direct sum of Hilbert spaces always refers to the Hilbert space closure of the algebraic direct sum in this note.

The direct sum is an abstract boundary value problem provided $\sup_{\alpha \in A} \|\Gamma_\alpha\| < \infty$. As the direct sum is not coupled, we also call them *decoupled* and write

$$\Pi^{\mathrm{dec}} = (\Gamma^{\mathrm{dec}}, \mathcal{G}^{\mathrm{dec}}, \mathfrak{h}^{\mathrm{dec}}, \mathcal{H}^{1,\mathrm{dec}}, \mathcal{H}^{\mathrm{dec}}) := \bigoplus_{\alpha \in A} \Pi_\alpha.$$

All derived objects such as the Dirichlet solution operator or the Dirichlet–to–Neumann operator are also direct sums of the correspondent objects.

3.2 Vertex coupling

We now construct a new space from building blocks associated with each vertex. Let $G = (V, E, \partial)$ be a graph. For each *vertex* $v \in V$ we assume that there is an abstract boundary value problem $\Pi_v = (\Gamma_v, \mathcal{G}_v, \mathfrak{h}_v, \mathcal{H}^1_v, \mathcal{H}_v)$.

Definition 3.1. We say that the family of abstract boundary value problems $(\Pi_v)_{v \in V}$ allows a *vertex coupling*, if the following holds:

1. $\sup_{v \in V} \|\Gamma_v\| < \infty$;
2. there is a Hilbert space $\mathcal{G}_e$ and a bounded operator $\pi_{v,e} \colon \mathcal{G}_v \to \mathcal{G}_e$ for each edge $e \in E$. Let
$$\mathcal{G}_v^{\max} := \bigoplus_{e \in E_v} \mathcal{G}_e$$
and
$$\iota_v \colon \mathcal{G}_v \longrightarrow \mathcal{G}_v^{\max}, \quad \iota_v \varphi_v = (\pi_{v,e}\varphi_v)_{e \in E_v}.$$
We assume that ι_v is an isometric embedding.
3. Assume that
$$\operatorname{ran} \Gamma_{\partial_- e, e} = \operatorname{ran} \Gamma_{\partial_+ e, e} =: \mathcal{G}_e^{1/2}$$
for all edges $e \in E$, where
$$\Gamma_{v,e} := \pi_{v,e} \Gamma_v \colon \mathcal{H}^1_v \longrightarrow \mathcal{G}_e.$$
We set
$$\pi_v := \iota_v^* \colon \mathcal{G}_v^{\max} \longrightarrow \mathcal{G}_v,$$
then
$$\pi_v \psi = \sum_{e \in E_v} \pi_{v,e}^* \psi_e.$$
We say that the vertex coupling is *maximal* if ι_v is surjective (hence unitary). In this case, we often identify $\mathcal{G}_v$ with $\mathcal{G}_v^{\max}$.

We start with an example with maximal vertex coupling spaces $\mathcal{G}_v \cong \mathcal{G}_v^{\max}$ (an example with non-maximal vertex coupling spaces will be given in Section 3.5):

Example 3.2. Assume that we have a graph-like manifold X (without edge contributions), i.e., $X = \bigcup_{v \in V} X_v$ such that X_v is closed in X and $Y_e := X_{\partial_- e} \cap X_{\partial_+ e}$ is a smooth submanifold. Then $\mathcal{H}_v = \mathrm{L}_2(X_v)$, $\mathfrak{h}_v(f) = \int_{X_v} |df|^2$, $\operatorname{dom} \mathfrak{h}_v = \mathcal{H}_v^1 = \mathrm{H}^1(X_v)$ and $\mathcal{G}_v = \mathrm{L}_2(\partial X_v)$, $\Gamma_v f = f \restriction_{\partial X_v}$; and $\Pi_v = (\Gamma_v, \mathcal{G}_v, \mathfrak{h}_v, \mathcal{H}_v^1, \mathcal{H}_v)$ is an abstract boundary value problem. Moreover, $(\Pi_v)_v$ allows a vertex coupling with $\mathcal{G}_e = \mathrm{L}_2(Y_e)$ with maps $\pi_{v,e} \colon \mathrm{L}_2(\partial X_v) \to \mathrm{L}_2(Y_e)$ being the restriction of a function on ∂X_v onto one of its components $Y_e \subset \partial X_v$. Note that $\operatorname{ran} \Gamma_{\partial_\pm e, e} = \mathrm{H}^{1/2}(Y_e)$. As

$$\mathcal{G}_v^{\max} := \bigoplus_{e \in E_v} \mathcal{G}_e = \bigoplus_{e \in E_v} \mathrm{L}_2(Y_e) = \mathrm{L}_2(\partial X_v) = \mathcal{G}_v,$$

the vertex coupling is *maximal*. Condition (1) is typically fulfilled, if the length of each end of a building block X_v is bounded from below by some constant $\ell_0/2 > 0$.

We construct an abstract boundary value problem $\Pi = (\Gamma, \mathcal{G}, \mathfrak{h}, \mathcal{H}^1, \mathcal{H})$ from $(\Pi_v)_v$ as follows:

$$\mathcal{H} := \bigoplus_{v \in V} \mathcal{H}_v, \quad \mathcal{H}^{1,\mathrm{dec}} := \bigoplus_{v \in V} \mathcal{H}_v^1, \quad \mathfrak{h}^{\mathrm{dec}} := \bigoplus_{v \in V} \mathfrak{h}_v;$$

$$\mathcal{H}^1 := \{ f = (f_v)_{v \in V} \in \mathcal{H}^{1,\mathrm{dec}} \mid \text{for all } e \in E,\ \partial e = (v, w)\colon \\ \Gamma_{v,e} f_v = \Gamma_{w,e} f_w =: \Gamma_e f \},$$

$$\mathfrak{h} := \mathfrak{h}^{\mathrm{dec}} \restriction_{\mathcal{H}^1}, \quad \mathcal{G} := \bigoplus_{e \in E} \mathcal{G}_e, \quad \Gamma \colon \mathcal{H}^1 \longrightarrow \mathcal{G}, \quad \Gamma f = (\Gamma_{\partial_\pm e, e} f_{\partial_\pm e})_{e \in E}.$$

Denote by ι the map

$$\iota \colon \mathcal{G} = \bigoplus_{e \in E} \mathcal{G}_e \longrightarrow \mathcal{G}^{\mathrm{dec}} = \bigoplus_{v \in V} \mathcal{G}_v, \quad (\iota \varphi)_v = \pi_v \underline{\varphi}(v) = \sum_{e \in E_v} \pi_{v,e}^* \varphi_e,$$

where $\underline{\varphi}(v) = (\varphi_e)_{e \in E_v} \in \mathcal{G}_v^{\max}$ (see Definition 3.1 for the notation). It is easy to see that $\iota^* \colon \mathcal{G}^{\mathrm{dec}} \to \mathcal{G}$ acts as

$$(\iota^* \psi)_e = \sum_{v = \partial_\pm e} \pi_{v,e} \psi(v)$$

Theorem 3.3. *Assume that $(\Pi_v)_{v\in V}$ is a family of abstract boundary value problems allowing a* **vertex coupling,** *then the following holds.*

(1) *The quintuple $\Pi = (\Gamma, \mathcal{G}, \mathfrak{h}, \mathcal{H}^1, \mathcal{H})$ as constructed above is an abstract boundary value problem.*

(2) *We have*

$$\ker\Gamma = \bigoplus_{v\in V}\ker\Gamma_v, \quad H^{\mathrm{D}} = \bigoplus_{v\in V} H_v^{\mathrm{D}}$$

(i.e., the Dirichlet operator is decoupled) and

$$\sigma(H^{\mathrm{D}}) = \overline{\bigcup_{v\in V}\sigma(H_v^{\mathrm{D}})}.$$

(3) *In particular, if all abstract boundary value problems Π_v have dense Dirichlet domain then Π also has dense Dirichlet domain.*

(4) *The Neumann operator is coupled and $f \in \operatorname{dom} H$ if and only if $f \in \bigoplus_{v\in V}\mathcal{W}_v$ and*

$$\Gamma_{v,e} f_v = \Gamma_{w,e} f_w, \quad \Gamma'_{v,e} f_v + \Gamma'_{w,e} f_w = 0 \quad \text{for all } e \in E,\ \partial e = (v, w)$$

with $Hf = (H^{\max} f)_{v\in V}$ (see (2) for the notation).

(5) *We have*

$$S(z)\varphi = S^{\mathrm{dec}}(z)\iota\varphi = \bigoplus_{v\in V} S_v(z)\underline{\varphi}(v) \quad \text{for } z \notin \sigma(H^{\mathrm{D}}).$$

(6) *Moreover, if all Π_v are elliptically regular such that*

$$\sup_{v\in V}\|S_v\|_{\mathcal{G}_v\to\mathcal{H}_v} < \infty,$$

then Π is also elliptically regular.

(7) *We have $\Lambda(z) = \iota^*\Lambda^{\mathrm{dec}}(z)\iota$ for $z \in \mathbb{C}\setminus\sigma(H^{\mathrm{D}})$, i.e.,*

$$(\Lambda(z)\varphi))_e = \sum_{v=\partial_\pm e}\pi_{v,e}(\Lambda_v(z)\underline{\varphi}(v)).$$

Proof. (1) The space $\mathcal{H}^1$ is closed in $\mathcal{H}^{1,\mathrm{dec}}$ as intersection of the closed spaces

$$\{\, f \in \mathcal{H}^{1,\mathrm{dec}} \mid \Gamma_{\partial_- e,e} f_{\partial_- e} = \Gamma_{\partial_+ e,e} f_{\partial_+ e} \,\}$$

(note that since $\Gamma_{v,e}$ are bounded operators, the latter sets are closed). Hence $\mathcal{H}^1$ is closed and $\mathfrak{h}$ is a closed form. Moreover, the operator Γ is bounded, as

$$\|\Gamma f\|^2_{\mathcal{G}} = \sum_{e\in E} \|\Gamma_e f\|^2_{\mathcal{G}_e} = \frac{1}{2}\sum_{v\in V}\sum_{e\in E_v} \|\Gamma_{v,e} f_v\|^2_{\mathcal{G}_e} = \frac{1}{2}\sum_{v\in V} \|\iota_v \Gamma_v f_v\|^2_{\mathcal{G}_v^{\max}}$$

using (1) and this can be estimated by $(1/2)\sup_v \|\Gamma_v\|^2 \|f\|^2_{\mathcal{H}^1}$ as ι_v is isometric.

Finally, as $\operatorname{ran}\Gamma_v = \mathcal{G}_v^{1/2}$ is dense in $\mathcal{G}_v$ we also have that $\operatorname{ran}\Gamma_{v,e}$ is dense in $\mathcal{G}_{v,e}$ (applying the bounded operator $\pi_{v,e}\colon \mathcal{G}_v \to \mathcal{G}_e$ to a dense set). As $\operatorname{ran}\Gamma = \bigoplus_{e\in E} \operatorname{ran}\Gamma_e$ (algebraic direct sum) with $\operatorname{ran}\Gamma_e = \operatorname{ran}\Gamma_{\partial_\pm e,e}$, the density of $\operatorname{ran}\Gamma$ in $\mathcal{G}$ follows.

(2) We have $f \in \bigoplus_{v\in V} \ker\Gamma_v$ if and only if $f_v \in \ker\Gamma_v$ for all $v \in V$. Moreover, $\ker\Gamma_v = \bigcap_{e\in E_v} \ker\Gamma_{v,e}$ as $\bigcap_{e\in E_v} \ker\pi_{v,e} = \{0\} \subset \mathcal{G}_v$ (using the injectivity of ι_v, see Definition 3.1 (2)). By definition, $\Gamma_e := \Gamma_{v,e}$ for $v = \partial_\pm e$, hence we have

$$\ker\Gamma = \bigoplus_{e\in E} \ker\Gamma_e = \bigoplus_{v\in V} \ker\Gamma_v.$$

(3) In particular, if all spaces $\ker\Gamma_v$ are dense in $\mathcal{H}_v$, then $\ker\Gamma$ is dense in $\mathcal{H}$.

(4) follows from a simple calculation using (2) on each abstract boundary value problem Π_v.

(5) is obvious, as well as (6)

(7) The formula follows from $\mathfrak{l}_z(\varphi,\psi) = (\mathfrak{h} - z\mathbb{1})(S(z)\varphi, S(-1)\psi)$ and part (5). ☐

Examples 3.4. (1) In Example 3.2, the entire space is $\mathcal{H} = \mathsf{L}_2(X)$ and the Neumann operator is the usual Laplacian on the graph-like manifold X.

(2) Assume that $G = (V, E, \partial)$ is a discrete graph. We decompose G into its *star components* $G_v = (\{v\} \mathbin{\dot\cup} E_v, E_v, \hat\partial)$, i.e., each edge in G adjacent to v becomes also a vertex in G_v. As boundary of G_v we set $\partial G_v = E_v$ If we identify the new vertices $e \in V(G_{\partial_- e})$ and $e \in V(G_{\partial_+ e})$ of the star components $G_{\partial_- e}$ and $G_{\partial_+ e}$ for all edges $e \in E$ we just obtain the subdivision graph SG (see Definition 2.1).

Let Π_v be the abstract boundary value problem associated with the graph G_v and boundary $\partial G_v = E_v$, i.e.,

$$\Pi_v = (\mathbb{C}^{E_v}, \Gamma_v, \mathfrak{h}_v, \mathbb{C}^{E_v \,\dot\cup \{v\}}, \mathbb{C}^{E_v \,\dot\cup \{v\}}),$$

with $\Gamma_v f = f \restriction_{E_v}$ (see Example 2.4 (3)), where $\mathbb{C}^{E_v} := \{\varphi \colon E_v \to \mathbb{C} \,|\, \varphi \text{ map}\}$ denotes the set of maps or families with coordinates indexed by $e \in E_v$. Denote for short $d_v = \deg v$. Of course, $\mathbb{C}^{E_v}$ is isomorphic to $\mathbb{C}^{d_v}$, but this isomorphism needs a numbering of the edges which is unimportant for our purposes. The Neumann operator, written as a matrix with respect to the orthonormal basis $\varphi_v := d_v^{-1/2}\delta_v$ $(v \in V)$, has block structure $A_v = \mathrm{id}_{E_v}$ (identity matrix of dimension d_v), $B_v = -d_v^{-1/2}(1, \dots, 1)^{\mathrm{T}}$ and $D_v = 1$. In particular, the Dirichlet–to–Neumann operator is $\Lambda_v(z) = (1-z)\,\mathrm{id}_{E_v} - (d_v(1-z))^{-1}\mathbb{1}_{E_v \times E_v}$, where $\mathbb{1}_{E_v \times E_v}$ is the $(d_v \times d_v)$-matrix with all entries 1.

The vertex-coupled abstract boundary value problem Π of the family $(\Pi_v)_{v \in V}$ (it is clear that this family allows a vertex coupling) is now the abstract boundary value problem of the subdivision graph SG of G with boundary $\partial SG = E$, the edges of G. More precisely, we have already identified the subspace

$$\Big\{\varphi \in \bigoplus_{v \in V} \ell_2(G_v) \;\Big|\; f_{v,e} = f_{w,e} \text{ for all } e \in E, \partial e = (v, w)\Big\},$$

with $\ell_2(SG)$. The coupled Neumann operator is the Laplacian of the subdivision graph SG, i.e., $H = \Delta_{SG}$. Note that we can embed $\ell_2(G)$ into $\ell_2(SG)$, $f \mapsto \hat{f}$, with $\hat{f}(v) = f(v)$ and $\hat{f}(e) = (f(\partial_+ e) + f(\partial_- e))/2$; moreover, $2\mathfrak{h}_G(f) = \mathfrak{h}_{SG}(\hat{f})$ for the corresponding energy forms. The coupled Dirichlet–to–Neumann operator is

$$(\Lambda(z)\varphi))_e = 2(1-z)\varphi_e - \frac{1}{1-z}\sum_{v = \partial_\pm e} \frac{1}{\deg v} \sum_{e' \in E_v} \varphi_{e'}.$$

For $z = 0$, this is just the formula for a Laplacian on the line graph LG of G (the line graph has as vertices the edges of G, and two such edges are adjacent, if they meet in a common vertex, see, e.g., Example 3.14 (iv) in [14]). In particular, if G is r-regular, then LG is $(2r-2)$-regular and

$$\Lambda(z) = 2(1-z) - \frac{2r-2}{(1-z)r}(1 - \Delta_{LG}).$$

In particular, applying the spectral relation (4) to the last example (the Dirichlet spectrum of all star components is $\{1\}$ as $H_v^{\mathrm{D}} = D_v = 1$) we rediscover the following result (see [17]):

Corollary 3.5. *The spectra of the subdivision and line graph of an r-regular graph are related by*

$$\lambda \in \sigma(\Delta_{SG}) \iff 1 - \frac{r}{r-1}(1-\lambda)^2 \in \sigma(\Delta_{LG})$$

provided $\lambda \neq 1$.

3.3 Edge coupling

Let us now couple abstract boundary value problems indexed by the edges of a given graph $G = (V, E, \partial)$: For each *edge* $e \in E$ we assume that there is an abstract boundary value problem $\Pi_e = (\Gamma_e, \mathcal{G}_e, \mathfrak{h}_e, \mathcal{H}^1_e, \mathcal{H}_e)$.

Definition 3.6. We say that the family of abstract boundary value problems $(\Pi_e)_{e \in E}$ allows an *edge coupling*, if the following holds:

1. $\sup_{e \in E} \|\Gamma_e\| < \infty$;
2. for each vertex $e \in E$ there is a decomposition $\mathcal{G}_e = \mathcal{G}_{e,\partial_- e} \oplus \mathcal{G}_{e,\partial_+ e}$ and $\Gamma_e f = \Gamma_{e,\partial_- e} f \oplus \Gamma_{e,\partial_+ e} f$, where $\Gamma_{e,v} \colon \mathcal{H}^1_e \to \mathcal{G}_{e,v}$.

We set $\mathcal{G}^{\max}_v := \bigoplus_{e \in E_v} \mathcal{G}_{e,v}$.

Note that the sum over all maximal spaces $\mathcal{G}^{\max}_v$ is the decoupled space, as

$$\mathcal{G}^{\max} := \bigoplus_{v \in V} \mathcal{G}^{\max}_v = \bigoplus_{v \in V} \bigoplus_{e \in E_v} \mathcal{G}_{e,v} = \bigoplus_{e \in E} \bigoplus_{v = \partial_\pm e} \mathcal{G}_{e,v} = \bigoplus_{e \in E} \mathcal{G}_e = \mathcal{G}^{\mathrm{dec}}.$$

Denote $\underline{\varphi}(v) := (\varphi_e(v))_{e \in E_v} \in \mathcal{G}^{\max}_v$ the collection of all edge contributions at the vertex $v \in V$, where $\varphi_e = (\varphi_e(\partial_- e), \varphi_e(\partial_+ e)) \in \mathcal{G}_{e,\partial_- e} \oplus \mathcal{G}_{e,\partial_+ e}$.

Let $\mathcal{G}_v \subset \mathcal{G}^{\max}_v$ be a closed subspace for each $v \in V$. We construct an abstract boundary value problem $\Pi = (\Gamma, \mathcal{G}, \mathfrak{h}, \mathcal{H}^1, \mathcal{H})$ from the family $(\Pi_e)_e$ and the subspace $\mathcal{G} := \bigoplus_v \mathcal{G}_v$ as a restriction of the decoupled abstract boundary value problem $\bigoplus_{e \in E} \Pi_e$ (see Section 3.1):

$$\mathcal{H} := \bigoplus_{e \in E} \mathcal{H}_e, \quad \mathcal{H}^{1,\mathrm{dec}} := \bigoplus_{e \in E} \mathcal{H}^1_e, \quad \mathfrak{h}^{\mathrm{dec}} := \bigoplus \mathfrak{h}_e;$$

$$\mathcal{H}^1 := \{\, f = (f_e)_{e \in E} \in \mathcal{H}^{1,\mathrm{dec}} \mid \Gamma^{\mathrm{dec}} f \in \mathcal{G} \,\}, \quad \mathfrak{h} := \mathfrak{h}^{\mathrm{dec}} \restriction_{\mathcal{H}^1}, \quad \Gamma := \Gamma^{\mathrm{dec}} \restriction_{\mathcal{H}^1}$$

where $\Gamma^{\mathrm{dec}} \colon \mathcal{H}^{1,\mathrm{dec}} \to \mathcal{G}^{\mathrm{dec}} = \mathcal{G}^{\max}$. Denote by ι the embedding $\iota \colon \mathcal{G} \to \mathcal{G}^{\mathrm{dec}}$.

Theorem 3.7. *Assume that $(\Pi_e)_{e\in E}$ is a family of abstract boundary value problems allowing an edge coupling and let $\mathcal{G}_v \subset \mathcal{G}_v^{\max}$ be a closed subspace for each $v \in V$, then the following holds.*

(1) *The quintuple $\Pi = (\Gamma, \mathcal{G}, \mathfrak{h}, \mathcal{H}^1, \mathcal{H})$ as constructed above is an abstract boundary value problem.*

(2) *We have*

$$\ker \Gamma = \bigoplus_{e\in E} \ker \Gamma_e, \quad H^{\mathrm{D}} = \bigoplus_{e\in E} H_e^{\mathrm{D}}$$

(the Dirichlet operator is decoupled) and

$$\sigma(H^{\mathrm{D}}) = \overline{\bigcup_{e\in E} \sigma(H_e^{\mathrm{D}})}.$$

(3) *In particular, if all abstract boundary value problems Π_e have dense Dirichlet domain then Π also has dense Dirichlet domain.*

(4) *The Neumann operator is coupled and is given by*

$$\operatorname{dom} H = \Big\{ f \in \bigoplus_{e\in E} \mathcal{W}_e \;\Big|\; \Gamma f \in \mathcal{G},\ \Gamma' f \in \mathcal{G}^{\max} \ominus \mathcal{G} \Big\}$$

with $Hf = (H^{\max} f)_{v\in V}$ (see (2) for the notation).

(5) *We have*

$$S(z)\varphi = S^{\mathrm{dec}}(z)\varphi = \bigoplus_{e\in E} S_e(z)\varphi_e,$$

where $\varphi \in \mathcal{G} \subset \mathcal{G}^{\mathrm{dec}}$.

(6) *Moreover, if all Π_e are elliptically regular with uniformly bounded elliptic regularity constants (i.e., $\sup_{e\in E} \|S_e\|_{\mathcal{G}_e\to\mathcal{H}_e} < \infty$), then Π is also elliptically regular.*

(7) *We have $\Lambda(z) = \iota^* \Lambda^{\mathrm{dec}}(z)\iota$, i.e., if $\psi = \Lambda(z)\varphi$, then*

$$\underline{\psi}(v) := (\psi_e)_{e\in E_v} = \pi_v((\Lambda_e(z)\varphi_e)(v))_{e\in E_v}$$

for $z \notin \sigma(H^{\mathrm{D}})$, where $\pi_v \colon \mathcal{G}_v^{\max} \to \mathcal{G}_v$ is the adjoint of $\iota_v \colon \mathcal{G}_v \to \mathcal{G}_v^{\max}$.

Proof. The proof is very much as the proof of Theorem 3.3: (1) The operator Γ^{dec} is bounded, and $\mathcal{H}^1$ is closed in $\mathcal{H}^{1,\mathrm{dec}}$ as preimage of the closed subspace $\mathcal{G}$ under Γ^{dec}; in particular, Γ is bounded. Moreover, $\operatorname{ran} \Gamma = \Gamma(\mathcal{H}^1) = \Gamma^{\mathrm{dec}}(\mathcal{H}^1) = \mathcal{G} \cap \Gamma^{\mathrm{dec}}(\mathcal{H}^1)$; since $\Gamma^{\mathrm{dec}}(\mathcal{H}^1)$ is dense (as all components $\Gamma_e(\mathcal{H}_e^1)$ are dense in $\mathcal{G}_e$, the space $\operatorname{ran} \Gamma$ is also dense in $\mathcal{G}$.

(2)–(7) can be seen similarly as in the proof of Theorem 3.3. □

Examples 3.8. Let us give some important examples of subspaces $\mathcal{G}$ of $\mathcal{G}^{\max}$:

1. **Edge coupling of two-dimensional abstract boundary value problems**

   Assume that all vertex components $\mathcal{G}_{e,v}$ equal $\mathbb{C}$. Then $\mathcal{G}_v^{\max} = \mathbb{C}^{E_v}$ and we choose for example $\mathcal{G}_v := \mathbb{C}(1,\dots,1)$ (*standard* or *Kirchhoff vertex conditions*), where $(1,\dots,1) \in \mathbb{C}^{E_v}$ has all $(\deg v)$-many components 1. It is convenient to choose $|w|_{\deg v} = |w|\sqrt{\deg v}$ as norm on $\mathbb{C}$ (then $\mathbb{C}(1,\dots,1) \subset \mathbb{C}^{E_v}$ is isometrically embedded in $(\mathbb{C}, |\cdot|_{\deg v})$ via $(\eta,\dots,\eta) \mapsto \eta$). A vector $\underline{\eta}(v)$ is of course characterised by the common scalar value $\eta(v) \in \mathbb{C}$ and the projection $\pi_v \underline{\psi}(v) = (\psi_e(v))_{e\in E_v}$ is characterised by the sum $(\deg v)^{-1} \sum_{e\in E_v} \psi_e(v)$. Hence we can write the Dirichlet–to–Neumann operator as

$$(\Lambda(z)\varphi)(v) = \frac{1}{\deg v} \sum_{e\in E_v} (\Lambda_e(z)\varphi_e)(v) \tag{9}$$

   for $\varphi \in \mathcal{G} = \ell_2(V, \deg)$. This formula is a generalisation of the formula for the discrete (normalised) Laplacian. One can also choose a more general subspace $\mathcal{G}_v \subset \mathcal{G}_v^{\max} = \mathbb{C}^{E_v}$, the resulting Dirichlet–to–Neumann operators look like generalised discrete Laplacians described, e.g., in Section 3 of [13] or Section 3 of [14]. A similar approach has been used in [12].

2. **Edge coupling of two-dimensional *trivial* abstract boundary value problems gives back the original discrete graph**

   Let us now treat a special case of (1): Let $\Pi_e = (\mathrm{id}, \mathbb{C}^2, \mathfrak{h}_e, \mathbb{C}^2, \mathbb{C}^2)$ be a trivial abstract boundary value problem for each $e \in E$. The abstract boundary value problem Π_v can be understood as coming from a graph consisting only of two vertices $\partial_\pm e$ and one edge e, and both vertices belong to the boundary. The energy form is $\mathfrak{h}_e(f) = |f_2 - f_1|^2$ for $f = (f_1, f_2) \in \mathbb{C}^2$, see Example 2.4 (3)). In this case,

$$\Lambda_e(z) = \begin{pmatrix} 1-z & -1 \\ -1 & 1-z \end{pmatrix}$$

   and the Dirichlet–to–Neumann operator becomes

$$(\Lambda(z)\varphi)(v) = \frac{1}{\deg v} \sum_{e\in E_v} (\varphi(v) - \varphi(v_e)) - z\varphi(v),$$

   i.e., $\Lambda(z) = \Delta_G - z$, i.e., the Dirichlet–to–Neumann operator is just the shifted normalised Laplacian on G. Note that in this case, the Neumann Laplacian is also the Dirichlet–to–Neumann operator at $z = 0$, i.e., $H = \Lambda(0) = \Delta_G$, as the global form $\mathfrak{h}$ is $\mathfrak{h}(f) = \sum_e \mathfrak{h}_e(f_e)$.

3. Standard vertex conditions

Here, we describe an edge coupling with a special choice of vertex spaces $\mathcal{G}_v \subset \mathcal{G}_v^{\max}$, similar to the standard or Kirchhoff vertex conditions of a quantum graph.

Assume that for given $v \in V$, the vertex component $\mathcal{G}_{e,v}$ of $\mathcal{G}_e$ equals a given Hilbert space $\mathcal{G}_{v,0}$ for all $e \in E_v$. Then

$$\mathcal{G}_v := \{\, \underline{\eta}(v) = (\eta, \ldots, \eta) \in \mathcal{G}_v^{\max} \mid \eta \in \mathcal{G}_{v,0} \,\}$$

is a closed subspace (i.e., $\mathcal{G}_v$ consists of the $(\deg v)$-fold diagonal of the model space $\mathcal{G}_{v,0}$). In the above two examples, we treated the case $\mathcal{G}_{e,v} = \mathbb{C}$. A vector $\underline{\eta}(v)$ is characterised by $\eta(v)$ and the projection $\pi_v \underline{\psi}(v) = (\psi_e(v))_{e \in E_v}$ is characterised by the sum $(\deg v)^{-1} \sum_{e \in E_v} \psi_e(v)$. Hence we can write the Dirichlet–to–Neumann operator exactly as in (9). This formula is a vector-valued version of a normalised discrete Laplacian (see (13) for a more concrete formula).

3.4 Vector-valued quantum graphs

Let $I = [a, b]$ be a compact interval of length $\ell = b - a > 0$, and let $\mathcal{K}$ be a Hilbert space with non-negative closed quadratic form $\mathfrak{k} \geq 0$ such that its corresponding operator has purely discrete spectrum. We set $\mathcal{H} := \mathsf{L}_2(I, \mathcal{K}) \cong \mathsf{L}_2(I) \otimes \mathcal{K}$ and define an energy form via

$$\mathfrak{h}(f) := \int_I (\|f'(t)\|^2_{\mathcal{K}} + \mathfrak{k}(f(t)))\, \mathrm{d}t$$

for $f \in \mathsf{L}_2(I, \mathcal{K})$ such that $f \in \mathsf{C}^1(I, \mathcal{K})$ and $f(t) \in \operatorname{dom} \mathfrak{k}$ for almost all $t \in I$. Denote by $\mathfrak{h}$ also the closure of this form. As boundary space set $\mathcal{G} := \mathcal{K} \oplus \mathcal{K} \cong \mathcal{K} \otimes \mathbb{C}^2$ and define $\Gamma f := (f(a), f(b))$. It is not hard to see that $\Pi = (\Gamma, \mathcal{G}, \mathfrak{h}, \operatorname{dom} \mathfrak{h}, \mathcal{H})$ is an elliptically regular abstract boundary value problem with dense Dirichlet domain; moreover, the norm of Γ is bounded by $\sqrt{\coth(\ell/2)}$ (see, e.g., Sections 6.1 and 6.4 of [16]). We call Π the *abstract boundary value problem associated with* $(\mathfrak{k}, \mathcal{K}, I)$. Moreover, as the underlying space of Π has a product structure, we can calculate all derived objects explicitly. For example, the Dirichlet–to–Neumann operator $\Lambda(z)$ is an operator function of a matrix with respect to the decomposition $\mathcal{G} = \mathcal{K} \oplus \mathcal{K}$. In particular, we have $\Lambda(z) = \Lambda_0(z - K)$, where K is the operator associated with $\mathfrak{k}$ and where

$$\Lambda_0(z) = \frac{\sqrt{z}}{\sin(\ell\sqrt{z})} \begin{pmatrix} \cos(\ell\sqrt{z}) & -1 \\ -1 & \cos(\ell\sqrt{z}) \end{pmatrix} \tag{10}$$

is the Dirichlet–to–Neumann operator for the scalar problem ($\mathcal{K} = \mathbb{C}$)

$$\Pi_0 = (\Gamma_0, \mathbb{C}^2, \mathfrak{h}_0, \mathrm{H}^1(I), \mathrm{L}_2(I))$$

with $\Gamma_0 f = (f(a), f(b))$ and $\mathfrak{h}_0(f) = \int_I |f'(t)|^2 \,\mathrm{d}t$ (see, e.g., [3] for details). The complex square root is cut along the positive real axis. The same argument also works for abstract warped products.

Let us now consider vector-valued quantum graphs: Assume that $G = (V, E, \partial)$ is a discrete graph and that I_e is a closed interval of length ℓ_e for each $e \in E$. Assume that there is a Hilbert space $\mathcal{K}_e$ and energy form $\mathfrak{k}_e$ for each edge $e \in E$. Then we can edge-couple the family of abstract boundary value problem Π_e associated with $(\mathfrak{k}_e, \mathcal{K}_e, I_e)$. In order to formulate the next result, we define the *unoriented* and *oriented evaluation* of f at a vertex v and edge e by

$$f_e(v) = \begin{cases} f_e(\min I_e), & v = \partial_- e, \\ f_e(\max I_e), & v = \partial_+ e, \end{cases} \quad \text{and} \quad \overset{\curvearrowright}{f_e}(v) = \begin{cases} -f_e(\min I_e), & v = \partial_- e, \\ +f_e(\max I_e), & v = \partial_+ e, \end{cases}$$

for $f \in \bigoplus_{e \in E} \mathrm{H}^1(I_e, \mathcal{K}_e)$.

Theorem 3.9. *Let $\Pi_e = (\Gamma_e, \mathcal{K}_e \oplus \mathcal{K}_e, \mathfrak{h}_e, \operatorname{dom} \mathfrak{h}_e, \mathrm{L}_2(I_e, \mathcal{K}_e))$ be an abstract boundary value problems associated with $(\mathfrak{k}_e, \mathcal{K}_e, I_e)$. Assume that the length ℓ_e of I_e fulfils*

$$0 < \ell_0 := \inf_{e \in E} \ell_e. \tag{11}$$

(1) *Then the family $(\Pi_e)_{e \in E}$ allows an edge coupling. As boundary space we choose $\mathcal{G} = \bigoplus_{v \in V} \mathcal{G}_v$, where $\mathcal{G}_v$ is a closed subspace of $\mathcal{G}_v^{\max} := \bigoplus_{e \in E_v} \mathcal{K}_e$ for each $v \in V$. Then H acts as*

$$(Hf)_e(t) = -f_e''(t) + K_e f(t) \tag{12a}$$

on each edge, where K_e is the operator associated with $\mathfrak{k}_e$. Moreover, a function f in the domain of the Neumann operator H of the edge-coupled abstract boundary value problem fulfils

$$\underline{f}(v) = \big(f_e(v)\big)_{e \in E_v} \in \mathcal{G}_v \quad \textit{and} \quad \underline{\overset{\curvearrowright}{f}}'(v) = \big(\overset{\curvearrowright}{f_e}(v)\big)_{e \in E_v} \in \mathcal{G}_v^{\max} \ominus \mathcal{G}_v. \tag{12b}$$

(2) *If H is a self-adjoint operator in $\mathcal{H} = \bigoplus_{e \in E} \mathrm{L}_2(I_e, \mathcal{K}_e)$ such that* (12a) *holds for all functions $f = (f_e)_e \in \operatorname{dom} H$ such that $f_e \in \mathrm{C}^2(I_e, \operatorname{dom} K_e)$ vanishing near ∂I_e, and such that the values $\underline{f}(v)$ and $\underline{\overset{\curvearrowright}{f}}'(v)$ are not coupled in* dom H, *then there exist closed subspaces $\mathcal{G}_v$ of $\mathcal{G}_v^{\max}$ for each $v \in V$, such that* dom H *is of the form* (12b).

We call H the *vector-valued quantum graph Laplacian with vertex spaces* $\mathcal{G}_v$ *and fibre operators* K_e.

Proof. Part (1) follows already from the discussion above, the fact that we have $\Gamma f = (\underline{f}(v))_{v \in V}$ and $\Gamma' f = (\underline{\overset{\curvearrowright}{f}}'(v))_{v \in V}$, and Theorem 3.7. Note that $\|\Gamma_e\|^2$ is bounded by $2/\min\{\ell_0,1\}$, see Corollary A.2.12 in [15].

For part (2), partial integration shows that

$$\begin{aligned}
\langle Hf, g\rangle_{\mathcal{H}} &= \sum_{e \in E} \int_{I_e} \langle -f_e''(t) + K_e f_e(t), g_e(t)\rangle_{\mathcal{K}_e} \,\mathrm{d}t \\
&= \sum_{e \in E} \Big(\int_{I_e} (\langle f_e'(t), g_e'(t)\rangle_{\mathcal{K}_e} + \mathfrak{k}_e(f_e(t), g_e(t)))\,\mathrm{d}t \\
&\qquad - [\langle f_e'(t), g_e(t)\rangle_{\mathcal{K}_e}]_{\partial I_e} \Big) \\
&= \langle f, Hg\rangle_{\mathcal{H}} + \sum_{e \in E} [\langle f_e(t), g_e'(t)\rangle_{\mathcal{K}_e} - \langle f_e'(t), g_e(t)\rangle_{\mathcal{K}_e}]_{\partial I_e}
\end{aligned}$$

for $f, g \in \operatorname{dom} H$. Reordering the boundary contributions (the last sum over $e \in E$) gives

$$\begin{aligned}
&\sum_{v \in V} \sum_{e \in E_v} (\langle \overset{\curvearrowright}{f}_e'(v), g_e(v)\rangle_{\mathcal{K}_e} - \langle f_e(v), \overset{\curvearrowright}{g}_e'(v)\rangle_{\mathcal{K}_e}) \\
&\qquad = \sum_{v \in V} (\langle \underline{\overset{\curvearrowright}{f}}'(v), \underline{g}(v)\rangle_{\mathcal{G}_v^{\max}} - \langle \underline{f}(v), \underline{\overset{\curvearrowright}{g}}'(v)\rangle_{\mathcal{G}_v^{\max}}),
\end{aligned}$$

As the values $\underline{\overset{\curvearrowright}{f}}'(v)$ and $\underline{f}(v)$ resp. $\underline{\overset{\curvearrowright}{g}}'(v)$ and $\underline{g}(v)$ are not coupled, each contribution $\langle \underline{\overset{\curvearrowright}{f}}'(v), \underline{g}(v)\rangle_{\mathcal{G}_v^{\max}}$ and $\langle \underline{f}(v), \underline{\overset{\curvearrowright}{g}}'(v)\rangle_{\mathcal{G}_v^{\max}}$ has to vanish separately. We let $\mathcal{G}_v$ be the closure of the linear span of all boundary values $\underline{f}(v)$, $f \in \operatorname{dom} H$. In particular, we then have $\underline{\overset{\curvearrowright}{f}}'(v), \underline{\overset{\curvearrowright}{g}}'(v) \in \mathcal{G}_v^{\perp}$. □

Remarks 3.10. (1) For simplicity, we describe only the *energy independent* vertex conditions, not involving any condition between the values $\underline{f}(v)$ and $\underline{\overset{\curvearrowright}{f}}'(v)$. One can, of course, also consider Robin-type conditions, but one needs additional finiteness or boundedness conditions in this case.

(2) If $\mathcal{K}_e = \mathbb{C}$ for all edges $e \in E$, then we have defined an ordinary quantum graph. For the case $(\mathfrak{k}_e, \mathcal{K}_e, I_e) = (0, \mathcal{K}_0, [0,1])$ for all $e \in E$, where $\mathcal{K}_0$ is a given Hilbert space, see [1].

We have not used the whole power of abstract boundary value problems here, namely the resolvent formula (3). In this setting, the left hand side, the resolvent of H in $z \notin \sigma(H) \cup \sigma(H^{\mathrm{D}})$, equals the right hand side, which can be expressed completely in terms of the building blocks Π_e. Moreover, the Dirichlet–to–Neumann operator has the nature of a discrete Laplacian.

If we use standard vertex conditions (see Example 3.8 (3)), we have to assume that all boundary spaces $\mathcal{K}_e$ are the same and equal (or can at least be naturally identified with) a Hilbert space $\mathcal{K}_0$. Then we set $\mathcal{G}_v = \mathcal{K}_0(1, \dots, 1)$, i.e., all $\deg v$ components of $\varphi(v) \in \mathcal{G}_v$ are the same. In this case, the Dirichlet–to–Neumann operator is (see (9) and (10))

$$(\Lambda(z)\varphi)(v) = \frac{1}{\deg v} \sum_{e \in E_v} (C_e(z)\varphi(v) - S_e(z)\varphi(v_e)), \tag{13}$$

where

$$C_e(z) = \sqrt{z - K_e} \cot(\ell_e \sqrt{z - K_e}) \quad \text{and} \quad S_e(z) = \frac{\sqrt{z - K_e}}{\sin(\ell_e \sqrt{z - K_e})}.$$

The equilateral and standard (or *Kirchhoff*) case. Let us now characterise the spectrum of the vector-valued quantum graph Laplacian H in a special case:

Example 3.11. If all abstract boundary value problems Π_e are the same (or isomorphic) and all lengths ℓ_e are the same (say, $\ell_e = 1$), then we call the vector-valued quantum graph *equilateral* and $K_e = K_0$ on $\mathcal{K}_e = \mathcal{K}_0$ for all $e \in E$). A related case ($K_0 = 0$) has also been treated in [1]. In the equilateral case, we have

$$\Lambda(z) = (1 \otimes 1/\sin\sqrt{z-K_0})(1 \otimes \cos\sqrt{z - K_0} - 1 + \Delta_G \otimes 1), \tag{14}$$

where Δ_G denotes the (discrete) normalised Laplacian (see (5)) and where we have identified $\mathcal{G} \cong \ell_2(V, \deg) \otimes \mathcal{K}_0$. Since $1 \otimes (1/\sin\sqrt{z-K_0})$ is a bijective operator and $\sigma(A \otimes 1 - 1 \otimes B) = \sigma(A) - \sigma(B) = \{ a - b \mid a \in \sigma(A), b \in \sigma(B) \}$, we have in particular (using (4) for the first equivalence)

$$\begin{aligned}
\lambda \in \sigma(H) &\iff 0 \in \sigma(\Lambda(z)) \\
&\iff 0 \in \sigma\big(1 - \cos\sqrt{z - K_0}\big) - \sigma(\Delta_G) \\
&\iff \sigma\big(1 - \cos\sqrt{z - K_0}\big) \cap \sigma(\Delta_G) \neq \emptyset \\
&\iff \text{there exists } \kappa \in \sigma(K_0)\text{: } 1 - \cos\sqrt{z - \kappa} \in \sigma(\Delta_G)
\end{aligned}$$

provided

$$z \notin \sigma(H^{\mathrm{D}}) = \sigma(H^{\mathrm{D}}_{[0,1]} \otimes 1 + 1 \otimes K_e) = \{ (n\pi)^2 + \kappa \mid n = 1, 2, \dots, \kappa \in \sigma(K_0) \}.$$

We have therefore shown the following:

Corollary 3.12. *Assume that all edge abstract boundary value problems Π_e are the same, i.e., associated with* $(\mathfrak{k}_0, \mathcal{K}_0, [0,1])$ *(see the beginning of Section* 3.4*), and that all vertex spaces are standard* ($\mathcal{G}_v = \{ (\varphi, \dots, \varphi) \in \mathcal{K}_0^{\deg v} \mid \varphi \in \mathcal{K}_0 \}$)*. Then the Dirichlet–to–Neumann operator is given by* (14)*. Moreover, the spectrum of the vector-valued quantum graph Laplacian H is characterised by*

$$\lambda \in \sigma(H) \iff \text{there exists } \kappa \in \sigma(K_0)\colon\ 1 - \cos\sqrt{\lambda - \kappa} \in \sigma(\Delta_G)$$

provided $\lambda \notin \sigma(H^{\mathrm{D}}) = \{ (n\pi)^2 + \kappa \mid n = 1, 2, \dots; \kappa \in \sigma(K_0) \}$.

Molchanov and Vainberg [11] treated the asymptotic behaviour $\varepsilon \to 0$ of a Dirichlet Laplacian on the product $X \times (M, g_\varepsilon)$, where X is a metric graph, (M, g) is a compact Riemannian manifold with boundary and $g_\varepsilon = \varepsilon^2 g$. In our notation, it means that $\mathcal{K}_0 = \mathsf{L}_2(M, g_\varepsilon)$ and $K_0 = \Delta_{(M,g_\varepsilon)} = \varepsilon^{-2}\Delta_{(M,g)}$ with Dirichlet boundary conditions. It would be interesting to compare this model with the usual ε-tubular neighbourhood model with Dirichlet boundary conditions. Our methods allow such an analysis, see Section 4.

3.5 Vertex-edge coupling

Here, we treat the coupling when there are building blocks for each vertex-edge of a given graph $G = (V, E, \partial)$. Formally this coupling is just a *vertex-based coupling* for the corresponding *subdivision graph* $\mathrm{S}G = (A, B, \hat{\partial})$ of G (see Definition 2.1). Assume that Π_a is an abstract boundary value problem for each vertex $a \in A = V \uplus E$ of the subdivision graph. The family $(\Pi_a)_{a \in A}$ allows a vertex-edge coupling if the following holds:

We say that the family of abstract boundary value problems $(\Pi_v)_{v \in V}$ allows a *vertex-edge coupling*, if the following holds.

1. $\sup_{a \in A} \|\Gamma_a\| < \infty$.

2. There is a Hilbert space $\mathcal{G}_{e,v}$ for each edge $v \in V$ and $e \in E_v$ (i.e., each edge $b = (v, e)$ of the subdivision graph).

   For the vertex vertices of SG assume that there is a bounded operator

$$\pi_{v,e}\colon \mathcal{G}_v \longrightarrow \mathcal{G}_{e,v}.$$

   We set

$$\Gamma_{v,e} := \pi_{v,e}\Gamma_v\colon \mathcal{H}_v^1 \longrightarrow \mathcal{G}_{e,v}.$$

Moreover, let $\mathcal{G}_v^{\max} := \bigoplus_{e \in E_v} \mathcal{G}_e$ and

$$\iota_v : \mathcal{G}_v \longrightarrow \mathcal{G}_v^{\max}, \quad \iota_v \varphi_v = (\pi_{v,e} \varphi_v)_{e \in E_v}.$$

We assume that ι_v is an isometric embedding.

For the edge vertices of SG we assume that $\mathcal{G}_e = \bigoplus_{v=\partial_\pm e} \mathcal{G}_{e,v}$ (i.e., the vertex coupling is maximal at $e \in A$). We set

$$\Gamma_{e,v} : \mathcal{H}_e^1 \longrightarrow \mathcal{G}_{e,v}, \quad \Gamma_{e,v} f_e := (\Gamma_e \varphi_e)_v.$$

3. For each edge $(v, e) \in B$, assume that $\operatorname{ran} \Gamma_{v,e} = \operatorname{ran} \Gamma_{e,v} =: \mathcal{G}_{e,v}^{1/2}$.

The formulas for the (subdivision) vertex-coupled abstract boundary value problem can be taken from Section 3.2 verbatim. Let us give two typical examples of such couplings.

Vertex-edge coupling with maximal coupling space: shrinking graph-like manifolds. Consider a thin neighbourhood $X = X_\varepsilon$ of an embedded metric graph X_0 or a thin graph-like manifold of dimension $d \geq 2$ and decompose $X = X_\varepsilon$ into its closed vertex and edge neighbourhoods $X_v = X_{\varepsilon,v}$ and $X_e = X_{\varepsilon,e}$, respectively (see, e.g., [5], [7], or [15]). We omit the shrinking parameter $\varepsilon > 0$ in the sequel, as it does not affect the coupling.

The abstract boundary value problems Π_v and Π_e are the ones for X_v and X_e with (internal) boundary ∂X_v and ∂X_e given as follows: Let $Y_{e,v} := X_e \cap X_v$ and assume that $Y_{e,v}$ is isometric with a smooth $(d-1)$-dimensional manifold Y_e for $v = \partial_\pm e$. The (internal) boundary of X_v and X_e is now $\partial X_v = \bigcup_{e \in E_v} (X_v \cap X_e)$ and $\partial X_e = \bigcup_{v=\partial_\pm e} (X_v \cap X_e)$, respectively. We also assume that X_e is a product $I_e \times Y_e$ with $I_e = [0, \ell_e]$.

The boundary spaces for each edge $b = (v, e) \in B$ are $\mathcal{G}_{e,v} = \mathsf{L}_2(Y_{e,v}) \cong \mathsf{L}_2(Y_e)$. Note that $\mathcal{G}_v = \bigoplus_{e \in E_v} \mathcal{G}_{e,v}$, i.e., that the vertex coupling at the (original) vertices $v \in A$ is also maximal.

Under the typical uniform lower positive bound (11) and a suitable decomposition of X into X_v and X_e, one can show that $\|\Gamma_a\|$ is uniformly bounded. The other conditions above can also be seen easily.

If we consider again the shrinking parameter and if we assume that $X_{\varepsilon,v}$ is ε-homothetic (see example before Proposition 2.11), then (depending on the energy form and boundary conditions chosen on X if the external boundary ∂X is non-trivial), we can apply Proposition 2.11. *Note that the boundary values of the eigenfunction on X_v corresponding to the eigenvalue 0 determine the limit boundary space, i.e., the vertex coupling appearing below as a proper subspace of $\mathcal{G}_v^{\max}$.*

Vertex-edge coupling with non-maximal coupling space: trivial vertex abstract boundary value problems. Let us now construct another vertex-edge coupled abstract boundary value problem appearing, e.g., in the limit situation of a shrinking graph-like space.

Assume that $(\Pi_e)_{e\in E}$ allows an edge coupling and that each abstract boundary value problem Π_e is bounded. For each $v \in V$ choose a subspace $\mathcal{G}_v$ of $\mathcal{G}_v^{\max}$ (e.g., given as the boundary values of the eigenfunction corresponding to 0 of a vertex neighbourhood). For the vertex abstract boundary value problems assume that $\Pi_v = (\mathrm{id}, \mathcal{G}_v, 0, \mathcal{G}_v, \mathcal{G}_v)$, i.e., all Π_v's are trivial (see Definition 2.2).

The corresponding vertex-edge-coupled abstract boundary value problem $\widehat{\Pi} = (\widehat{\Gamma}, \mathcal{G}, \hat{\mathfrak{h}}, \widehat{\mathcal{H}}^1, \widehat{\mathcal{H}})$ is given as follows. The coupling condition in $\widehat{\mathcal{H}}^1$ here reads as $\Gamma_{e,v} f_e = \Gamma_{v,e} f_v$ for all $e \in V_v$ and $v \in V$. We define $\Gamma_v^{\mathrm{int}} f := (\Gamma_{v,e} f_v)_{e\in E_v}$ and $\Gamma_v^{\mathrm{ext}} f := (\Gamma_{e,v} f_e)_{e\in E_v}$, hence the coupling condition becomes $\Gamma_v^{\mathrm{int}} f = \Gamma_v^{\mathrm{ext}} f$. As $\Gamma_v^{\mathrm{int}} f_v = (\Gamma_{v,e} f_v)_{e\in E_v} = \Gamma_v f_v = f_v \in \mathcal{G}_v$ the coupling condition is equivalent with $f_v = \Gamma_v^{\mathrm{ext}} f \in \mathcal{G}_v$. Hence we have

$$\widehat{\mathcal{H}} = \bigoplus_{e\in E} \mathcal{H}_e \oplus \bigoplus_{v\in V} \mathcal{G}_v, \quad \mathcal{G} = \bigoplus_{v\in V} \mathcal{G}_v,$$

$$\widehat{\mathcal{H}}^1 = \{\, f = (f_a)_{a\in E\,\cup\, V} \in \widehat{\mathcal{H}}^{1,\mathrm{dec}} \mid \Gamma_v^{\mathrm{ext}} f = f_v \in \mathcal{G}_v \text{ for all } v \in V \,\}$$

$$\hat{\mathfrak{h}} = \hat{\mathfrak{h}}^{\mathrm{dec}} \restriction_{\mathcal{H}^1}, \text{ i.e., } \hat{\mathfrak{h}}(f) = \sum_{e\in E} \mathfrak{h}_e(f_e), \quad \widehat{\Gamma} f = (\Gamma_v^{\mathrm{ext}} f)_{v\in V}$$

(the $(\cdot)^{\mathrm{dec}}$-labelled objects are defined as in Section 3.1). Such operators have been treated in Section 3.4.4 of [15] under the name "*extended operator*".

We have the following result, showing that the vertex-edge coupling with trivial vertex abstract boundary value problems leads just to the edge coupling:

Theorem 3.13. *Let G be a discrete graph. Assume that Π is an abstract boundary value problem obtained from a family $(\Pi_e)_{e\in E}$ of abstract boundary value problems allowing an edge coupling and choose a closed subspace $\mathcal{G}_v$ of $\mathcal{G}_v^{\max} = \bigoplus_{e\in E_v} \mathcal{G}_{e,v}$ for each vertex. As vertex family $(\Pi_v)_{v\in V}$ we choose the trivial abstract boundary value problem $\Pi_v = (\mathrm{id}, \mathcal{G}_v, 0, \mathcal{G}_v, \mathcal{G}_v)$ for each vertex.*

Then the vertex-edge-coupled abstract boundary value problem $\widehat{\Pi}$ (i.e., vertex-coupled with respect to the subdivision graph SG*) is equivalent (see below) with the edge-coupled abstract boundary value problem Π according to the original graph G.*

Proof. Equivalence of two abstract boundary value problems Π and $\widehat{\Pi}$ means that there are bicontinuous isomorphisms $U^1: \mathcal{H}^1 \to \widehat{\mathcal{H}}^1$ and $T: \mathcal{G} \to \widehat{\mathcal{G}}$ such that $T\Gamma = \widehat{\Gamma}U^1$ and $\hat{\mathfrak{h}}(U^1 f) = \mathfrak{h}(f)$. Here we have

$$\mathcal{H} = \bigoplus_{e \in E} \mathcal{H}_e, \quad \mathcal{H}^1 = \Big\{ f \in \bigoplus_{e \in E} \mathcal{H}^1_e \;\Big|\; \text{for all } v \in V\colon\ \Gamma^{\text{ext}}_v f \in \mathcal{G}_v \Big\}$$

and $\widehat{\mathcal{H}}^1$ is given above. Set $U^1 f = (f, (\Gamma^{\text{ext}}_v f)_{v \in V}) =: (f, \Gamma^{\text{ext}} f)$ then $\hat{\mathfrak{h}}(U^1 f) = \mathfrak{h}(f)$ and

$$\begin{aligned} \|U^1 f\|^2_{\widehat{\mathcal{H}}^1} &= \hat{\mathfrak{h}}(U^1 f) + \|f\|^2_{\mathcal{H}} + \|\Gamma^{\text{ext}} f\|^2_{\mathcal{G}} \\ &= \|f\|^2_{\mathcal{H}^1} + \|\Gamma^{\text{ext}} f\|^2_{\mathcal{G}} \\ &\le (1 + \|\Gamma^{\text{ext}}\|)\|f\|^2_{\mathcal{H}^1}, \end{aligned}$$

while for the inverse $(U^1)^{-1}(f, f_0) = f$ (with $f \in \mathcal{H}$, $f_0 \in \mathcal{G}$) we have the estimate $\|(U^1)^{-1}(f, f_0)\|^2_{\mathcal{H}^1} = \|f\|^2_{\mathcal{H}^1} \le \|(f, f_0)\|^2_{\widehat{\mathcal{H}}^1}$. Moreover, we set $T\varphi = \varphi$, then $T\Gamma f = \Gamma f = \widehat{\Gamma}(U^1 f)$. □

The previous result allows us to consider convergence of vertex-edge coupled abstract boundary value problems component-wise, even if the limit problem is only edge-coupled. A typical example is the convergence of the Laplacian on a thin ε-neighbourhood of a metric graph, to a Laplacian on the metric graph. We discuss the general convergence scheme in the next section.

4 Convergence of abstract graph-like spaces

In this section we show how one can translate convergence of building blocks into a global convergence, expressed via the coupling of abstract graph-like spaces in Section 3 and the concept of quasi-unitary equivalence resp. quasi-isomorphy for abstract boundary value problems acting in different Hilbert spaces in Section 2.3.

Fix a graph $G = (V, E, \partial)$ and let Π and $\widehat{\Pi}$ be two vertex-coupled abstract boundary value problems arising from the building blocks Π_v and $\widehat{\Pi}_v$. As we have seen in Section 3.5, the vertex coupling comprises also the vertex-edge coupled and even some edge-coupled cases.

We want to show the following: Assume that all building blocks Π_v and $\widehat{\Pi}_v$ are quasi-isomorphic then the coupled Neumann forms $\mathfrak{h}$ and $\hat{\mathfrak{h}}$ of the vertex-coupled abstract boundary value problems Π and $\widehat{\Pi}$ are quasi-unitarily equivalent, as well as the coupled boundary operators Γ and $\widehat{\Gamma}$ are quasi-isomorphic. We will not treat the full (natural) problem of the quasi-isomorphy of Π and $\widehat{\Pi}$ here, as we will not show that the boundary identification operators I, I' are quasi-isomomorphic (this would mean to impose additional assumptions).

One problem with the coupling is that the naively defined identification operator acts as

$$J^{1,\mathrm{dec}} := \bigoplus_{v \in V} J_v^1 \colon \mathcal{H}^{1,\mathrm{dec}} = \bigoplus_{v \in V} \mathcal{H}_v^1 \longrightarrow \widehat{\mathcal{H}}^{1,\mathrm{dec}} = \bigoplus_{v \in V} \widehat{\mathcal{H}}_v^1$$

but it is a priori not true that $J^{1,\mathrm{dec}}(\mathcal{H}^1) \subset \widehat{\mathcal{H}}^1$, i.e., that $J^{1,\mathrm{dec}}$ respects the coupling condition along the different vertex building blocks as in Section 3.2. In order to correct this, we need the following definition (for the existence of such operators, see the propositions after our next theorem):

Definition 4.1. Let Π be a vertex-coupled abstract boundary value problem arising from the vertex building blocks $(\Pi_v)_{v \in V}$. We say that Π *allows a smoothing operator* if there is a bounded operator $B \colon \mathcal{H}^{1,\mathrm{dec}} \to \mathcal{H}^{1,\mathrm{dec}}$ such that $f - Bf \in \mathcal{H}^1$ for all $f \in \mathcal{H}^{1,\mathrm{dec}}$.

A simpler version of the following result can also be found in Section 4.8 of [15]:

Theorem 4.2. *Let $G = (V, E, \partial)$ be a discrete graph and $(\Pi_v)_v$, $(\widehat{\Pi}_v)_v$ two families of abstract boundary value problems allowing a vertex coupling. Assume that Π_v and $\widehat{\Pi}_v$ are δ_v-quasi-isomorphic for each $v \in V$. Moreover, assume that $\delta := \sup_{v \in V} \delta_v < \infty$ and that the vertex-coupled abstract boundary value problems Π and $\widehat{\Pi}$ allow smoothing operators B and $\widehat{B}$ such that*

$$\|\widehat{B} J^{1,\mathrm{dec}} f\|_{\widehat{\mathcal{H}}^{1,\mathrm{dec}}} \le \delta \|f\|_{\mathcal{H}^{1,\mathrm{dec}}} \quad \text{and} \quad \|B J'^{1,\mathrm{dec}} u\|_{\mathcal{H}^{1,\mathrm{dec}}} \le \delta \|u\|_{\widehat{\mathcal{H}}^{1,\mathrm{dec}}} \tag{15}$$

for $f \in \mathcal{H}^1$ and $u \in \widehat{\mathcal{H}}^1$. Then $\mathfrak{h}$ and $\widehat{\mathfrak{h}}$ are δ'-quasi-unitarily equivalent; and Γ and $\widehat{\Gamma}$ are δ'-close where $\delta' = \mathrm{O}(\delta)$.

Proof. We define $J^1 \colon \mathcal{H}^1 \to \widehat{\mathcal{H}}^1$, $J^1 := (\mathrm{id}_{\widehat{\mathcal{H}}^1} - \widehat{B}) J^{1,\mathrm{dec}}$. From the smoothing property, we have $J^1 f \in \widehat{\mathcal{H}}^1$ for any $f \in \mathcal{H}^{1,\mathrm{dec}}$, hence J^1 maps into the right space. Similarly, we define $J'^1 := (\mathrm{id}_{\mathcal{H}^1} - B) J'^{1,\mathrm{dec}}$. The identification operators on $\mathcal{H}$ and $\widehat{\mathcal{H}}$ are given as $J := \bigoplus_{v \in V} J_v$ and $J' := \bigoplus_{v \in V} J'_v$. Then we have

$$\|Jf - J^1 f\|^2_{\widehat{\mathcal{H}}} \le 2 \sum_{v \in V} \|J_v f_v - J_v^1 f_v\|^2_{\widehat{\mathcal{H}}_v} + 2 \|\widehat{B} J^{1,\mathrm{dec}} f\|^2_{\widehat{\mathcal{H}}^{1,\mathrm{dec}}} \le 4\delta^2 \|f\|^2_{\mathcal{H}^1}$$

using our assumptions. A similar property holds for J' and J'^1. The other properties of Definition 2.7 for J and J' follow directly from the ones of J_v and J'_v (as in Section 4.8 of [15]). For the δ-closeness of $\mathfrak{h}$ and $\hat{\mathfrak{h}}$ we have

$$\begin{aligned}|\hat{\mathfrak{h}}(J^1 f, u) - \mathfrak{h}(f, J'^1 u)|^2 &\le 3\Big(\sum_{v\in V} |\widehat{\mathfrak{h}_v}(J_v^1 f_v, u_v) - \mathfrak{h}_v(f_v, J_v'^1 u_v)|\Big)^2 \\ &\quad + 3\big|\hat{\mathfrak{h}}(\widehat{B} J^{1,\mathrm{dec}} f, u)\big|^2 + 3\big|\mathfrak{h}(f, BJ'^{1,\mathrm{dec}} u)\big|^2 \\ &\le 9\delta^2 \|f\|^2_{\mathcal{H}^1} \|u\|^2_{\widehat{\mathcal{H}}^1}\end{aligned}$$

using again (15). For the boundary identification operators we set $I\colon \mathcal{G} \to \widehat{\mathcal{G}}$ where $(I\varphi)_e = 1/2 \sum_{v=\partial_\pm e} \widehat{\pi}_{v,e} I_v \pi^*_{v,e}\varphi_e$ and similarly for I'. Then, we have the following estimates for the closeness of the boundary maps

$$\begin{aligned}\|(I\Gamma - \widehat{\Gamma} J^1) f\|^2_{\widehat{\mathcal{G}}} &= \sum_{e\subset E} \|((I\Gamma - \widehat{\Gamma} J^1) f)_e\|^2_{\widehat{\mathcal{G}}_e} \\ &\le \sum_{e\in E} \sum_{v=\partial_\pm e} \frac{1}{2} \|\widehat{\pi}_{v,e}(I_v \Gamma_v f_v - (\widehat{\Gamma} J^1 f)_v)\|^2_{\widehat{\mathcal{G}}_e} \\ &\le \frac{1}{2} \sum_{v\in V} \|\hat{\iota}_v(I_v \Gamma_v f_v - (\widehat{\Gamma} J^1 f)_v)\|^2_{\widehat{\mathcal{G}}_v^{\max}} \\ &\le \sum_{v\in V} \|(I_v \Gamma_v - \widehat{\Gamma} J_v^1) f_v\|^2_{\widehat{\mathcal{G}}_v} + \sup_v \|\widehat{\Gamma}_v\|^2 \|\widehat{B} J^{1,\mathrm{dec}} f\|^2_{\widehat{\mathcal{H}}^1} \\ &\le \delta^2 (1 + \sup_v \|\widehat{\Gamma}_v\|^2) \|f\|^2_{\mathcal{H}^1}.\end{aligned}$$

by our assumptions. Similarly, we show the related property for $I'\widehat{\Gamma} - \Gamma J'^1$.

$\delta' = \delta \max\{3, \sup_v \|\Gamma_v\| + 1, \sup_v \|\widehat{\Gamma}_v\| + 1\}$ will do the job. □

Let us now prove the existence of smoothing operators:

Proposition 4.3. *Assume that there are operators* $\chi_{e,v}\colon \mathcal{G}_e^{1/2} \to \mathcal{H}_v^1$ *such that*

$$\Gamma_{v,e}\chi_{e,v}\varphi_e = \varphi_e, \quad \Gamma_{v,e}\chi_{e',v}\varphi_{e'} = 0, \quad e, e' \in E_v, e \ne e',\ v \in V.$$

Assume in addition that $C^2 = \sup_v \sum_{e\in E_v} \|\chi_{v,e}\|^2_{\mathcal{G}_e^{1/2}\to\mathcal{H}_v^1} < \infty$ *then*

$$(Bf)_v := \frac{1}{2} \sum_{e\in E_v} \chi_{e,v}(\Gamma_{v,e} f_v - \Gamma_{v_e,e} f_{v_e})$$

defines a smoothing operator.

Proof. We have to show that $f - Bf \in \mathcal{H}^1$ whenever $f \in \mathcal{H}^{1,\mathrm{dec}}$, but this follows immediately from the fact that $\Gamma_{v,e}(f - Bf) = 1/2 \sum_{w=\partial_\pm e} \Gamma_{w,e} f_w$ is independent of $v = \partial_\pm e$. The boundedness of $B : \mathcal{H}^{1,\mathrm{dec}} \to \mathcal{H}^{1,\mathrm{dec}}$ follows easily. □

One can, e.g., choose $\chi_{e,v} = S_v \pi^*_{v,e}$ under suitable assumptions on the maps $\pi_{v,e}$ (e.g., $\pi^*_{v,e}(\mathcal{G}_e^{1/2}) \subset \mathcal{G}_v^{1/2}$), where S_v is the Dirichlet solution operator of Π_v. The necessary assumptions are typically fulfilled in our graph-like manifold example; in particular, if the boundary components Y_e, $e \in E_v$, do not touch in ∂X_v, see Example 3.2.

Finally, we show the norm bound (15) of the smoothing operators under a slightly stronger assumption than the δ-closeness of Γ and $\widehat{\Gamma}$ (see Definition 2.6):

Proposition 4.4. *Assume that a smoothing operator $\widehat{B} : \widehat{\mathcal{H}}^{1,\mathrm{dec}} \to \widehat{\mathcal{H}}^{1,\mathrm{dec}}$ as in Proposition 4.3 exists and that there is $\delta > 0$ such that*

$$\sum_{e \in E_v} \|\widehat{\pi}_{v,e}(\widehat{\Gamma}_v J^1_v - I_v \Gamma_v) f_v\|^2_{\widehat{\mathcal{G}}_e^{1/2}} \le \delta^2 \|f_v\|^2_{\mathcal{H}^1_v}$$

holds for all $v \in V$ and $f_v \in \mathcal{H}^1_v$. Then $\|\widehat{B} J^{1,\mathrm{dec}} f\|_{\widehat{\mathcal{H}}^{1,\mathrm{dec}}} \le \delta C \|f\|_{\mathcal{H}^1}$ holds for all $f \in \mathcal{H}^1$.

Proof. We have

$$\begin{aligned}
\|\widehat{B} J^{1,\mathrm{dec}} f\|^2_{\widehat{\mathcal{H}}^{1,\mathrm{dec}}} &= \sum_{v \in V} \Big\| \frac{1}{2} \sum_{e \in E_v} \widehat{\chi}_{e,v} (\widehat{\Gamma}_{v,e} J^1_v f_v - \widehat{\Gamma}_{v_e,e} J^1_{v_e} f_{v_e}) \Big\|^2_{\widehat{\mathcal{H}}^1_v} \\
&= \frac{1}{4} \sum_{v \in V} \Big\| \sum_{e \in E_v} \widehat{\chi}_{e,v} \big((\widehat{\Gamma}_{v,e} J^1_v - \widehat{\pi}_{v,e} I_v \Gamma_{v,e}) f_v \\
&\qquad\qquad + (\widehat{\pi}_{v_e,e} I_{v_e} \Gamma_{v_e,e} - \widehat{\Gamma}_{v_e,e} J^1_{v_e}) f_{v_e} \big) \Big\|^2_{\widehat{\mathcal{H}}^1_v} \\
&\le C^2 \sum_{v \in V} \sum_{e \in E_v} \|\widehat{\pi}_{v,e} l (\widehat{\Gamma}_v J^1_v - I_v \Gamma_{v,e}) f_v\|^2_{\widehat{\mathcal{G}}_e^{1/2}} \le C^2 \delta^2 \|f\|^2_{\mathcal{H}^1}
\end{aligned}$$

where we used that $\Gamma_{v,e} f_v = \Gamma_{v_e,e} f_{v_e}$ for the second equality. □

A careful observer might know that the following quote is not a rude reminder of the discomfort of aging, but just a quote from Pavel's web page…

Epilogue. *„Hlídejte si ty vzácné okamžiky, kdy vám to ještě myslí. Mohou být poslední…“*

References

[1] J. von Below and D. Mugnolo, The spectrum of the Hilbert space valued second derivative with general self-adjoint boundary conditions. *Linear Algebra Appl.* **439** (2013), no. 7, 1792–1814. MR 3090437 Zbl 1322.34048

[2] G. Berkolaiko and P. Kuchment, *Introduction to quantum graphs.* Mathematical Surveys and Monographs, 186. American Mathematical Society, Providence, R.I., 2013. ISBN 978-0-8218-9211-4 MR 3013208 Zbl 1301.00028

[3] J. Behrndt and O. Post, Convergence of the Dirichlet–to–Neumann operators on thin branched manifolds. In preparation.

[4] P. Exner, J. P. Keating, P. Kuchment, T. Sunada, and A. Teplyaev (eds.), *Analysis on graphs and its applications.* Papers from the program held in Cambridge, January 8–June 29, 2007. Proceedings of Symposia in Pure Mathematics, 77. American Mathematical Society, Providence, R.I., 2008. MR 2459860 Zbl 1143.05002

[5] P. Exner and O. Post, Convergence of spectra of graph-like thin manifolds. *J. Geom. Phys.* **54** (2005), no. 1, 77–115. MR 2135966 Zbl 1095.58007

[6] P. Exner and O. Post, Convergence of resonances on thin branched quantum waveguides. *J. Math. Phys.* **48** (2007), no. 9, 092104, 43 pp. MR 2355074 Zbl 1152.81426

[7] P. Exner and O. Post, Approximation of quantum graph vertex couplings by scaled Schrödinger operators on thin branched manifolds. *J. Phys. A* **42** (2009), no. 41, 415305, 22 pp. MR 2545638 Zbl 1179.81080

[8] P. Exner and O. Post, A general approximation of quantum graph vertex couplings by scaled Schrödinger operators on thin branched manifolds. *Comm. Math. Phys.* **322** (2013), no. 1, 207–227. MR 3073163 Zbl 1270.81099

[9] S. Hassi, H. S. V. de Snoo, and F. H. Szafraniec (eds.), *Operator methods for boundary value problems.* London Mathematical Society Lecture Note Series, 404. Cambridge University Press, Cambridge, 2012. ISBN 978-1-107-60611-1 MR 3075434 Zbl 1269.47001

[10] P. Kuchment and H. Zeng, Convergence of spectra of mesoscopic systems collapsing onto a graph. *J. Math. Anal. Appl.* **258** (2001), no. 2, 671–700. MR 1835566 Zbl 0982.35076

[11] S. Molchanov and B. Vainberg, Transition from a network of thin fibers to the quantum graphs: an explicitly solvable model. In G. Berkolaiko, R. Carlson, S. A. Fulling, and P. Kuchment (eds.), *Quantum graphs and their applications.* Proceedings of the AMS-IMS-SIAM Joint Summer Research Conference held in Snowbird, UT, June 19–23, 2005. Contemporary Mathematics, 415. American Mathematical Society, Providence, R.I., 2006, 227–239. MR 2277619 Zbl 1317.35046

[12] K. Pankrashkin, Spectra of Schrödinger operators on equilateral quantum graphs. *Lett. Math. Phys.* **77** (2006), no. 2, 139–154. MR 2251302 Zbl 1113.81056

[13] O. Post, Equilateral quantum graphs and boundary triples. In [4] (2008), 469–490. MR 2459887 Zbl 1153.81499

[14] O. Post, First order approach and index theorems for discrete and metric graphs. *Ann. Henri Poincaré* **10** (2009), no. 5, 823–866. MR 2533873 Zbl 1206.81044

[15] O. Post, *Spectral analysis on graph-like spaces.* Lecture Notes in Mathematics, 2039. Springer, Berlin etc., 2012. ISBN 978-3-642-23839-0 MR 2934267 Zbl 1247.58001

[16] O. Post, Boundary pairs associated with quadratic forms. *Math. Nachr.* **289** (2016), no. 8-9, 1052–1099. MR 3512049 Zbl 06607012

[17] T. Shirai, The spectrum of infinite regular line graphs. *Trans. Amer. Math. Soc.* **352** (2000), no. 1, 115–132. MR 1665338 Zbl 0931.05048

A Cayley–Hamiltonian theorem for periodic finite band matrices

Barry Simon

I hope Pavel Exner will enjoy this birthday bouquet.

1 Introduction – The magic formula

Let J be a doubly infinite, self-adjoint, tridiagonal Jacobi matrix (i.e., $J_{jk} = 0$ if $|j-k| > 1$ and $J_{jj+1} > 0$) that is periodic, i.e., if

$$(Su)_j = u_{j+1}, \tag{1}$$

then for some $n \in \mathbb{Z}_+$, $S^n J = J S^n$. There is a huge literature on this subject – see Simon [7], Chapter 5.

$(J - E)u = 0$ is a second order difference equation, so there is a linear map $T(E) : \mathbb{C}^2 \to \mathbb{C}^2$ so that if u_0, u_1 are given, then $T(E)\binom{u_0}{u_1} = \binom{u_n}{u_{n+1}}$ for the solution of $(J-E)u = 0$. $\Delta(E) = \operatorname{Tr}(M(E))$ is called the *discriminant* of J. We note that $\det(T(E)) = 1$ so $T(E)$ has eigenvalues λ and λ^{-1} and $\Delta(E) = \lambda + \lambda^{-1}$. If $\Delta(E) \in (-2, 2)$, then $\lambda = e^{i\theta}$ for some θ in $\pm(0, \pi)$ and then $Ju = Eu$ has Floquet solutions, $u^{\pm}$ obeying $u^{\pm}_{j+nk} = e^{\pm ik\theta} u^{\pm}_j$. These are bounded and these are only bounded solutions if $\Delta(E) \in [-2, 2]$. Thus $\operatorname{spec}(J) = \Delta^{-1}([-2, 2])$. One often writes this relation as

$$\Delta(E) = 2\cos(\theta). \tag{2}$$

In [2], Damanik, Killip and Simon emphasized and exploited the operator form of (2), namely

$$\Delta(J) = S^n + S^{-n}. \tag{3}$$

This follows from (2) and the view of J as a direct integral. More importantly, what they called the "magic formula", [2] shows that a two sided, not *a priori* periodic, Jacobi matrix, which obeys (3), is periodic and in the isospectral torus of J.

A Laurent matrix is a finite band doubly infinite matrix that is constant along diagonals, so a polynomial in S and S^{-1}. $S^n + S^{-n}$ is an example of such a matrix. The current paper had its genesis in a question asked me by Jonathan Breuer and Maurice Duits. They asked if K is finite band and periodic but not tridiagonal if there is a polynomial Q so that $Q(K)$ is a Laurent matrix. They guessed that Q might be connected to the trace of a transfer matrix.

While I don't have a formal example where I can prove there is no such Q, I have found a related result which strongly suggests that, in general, the answer is no. I found an object which replaces Δ for more general K which is width $2m + 1$ (i.e., $K_{jk} = 0$ if $|j - k| > m$), self-adjoint and non-degenerate in the sense that for all j, $K_{jj+m} \neq 0$. Namely we prove the existence of a polynomial, $p(x, y)$, in x and y of degree $2m$ in y, so that $p(K, S^n) = 0$. In the Jacobi case,

$$p(x, y) = y^2 - y\Delta(x) + 1$$

so that $p(J, S^n) = 0$ is equivalent to (3).

We prove this theorem and begin the exploration of this object in Section 1. That a scalar polynomial vanishes when the variable is replaced by an operator is the essence of the Cayley–Hamiltonian theorem which says that a matrix obeys its secular equation. This was proven in 1853 by Hamilton [4] for the two special cases of three-dimensional rotations and for multiplication by quaternions and stated in general by Cayley [1] in 1858 who proved it only for 2×2 matrices although he said he'd done the calculation for 3×3 matrices. In 1878, Frobenius [3] proved the general result and attributed it to Cayley. We regard our main result, Theorem 2.1, as a form of the Cayley–Hamiltonian Theorem.

The magic formula had important precursors in two interesting papers of Naĭman, namely [5] and [6]. These papers are also connected to our work here.

It is a pleasure to present this paper to Pavel Exner for his 70th birthday. I have long enjoyed his contributions to areas of common interest. I recall with fondness the visit he arranged for me in Prague. He was a model organizer of an ICMP. And he is an all around sweet guy. Happy birthday, Pavel.

2 Main result

By a width $2m+1$ matrix, $m \in \{1, 2, \dots\}$, we mean a doubly infinite matrix, K, with $K_{jk} = 0$ if $|j - k| > m$. If $\sup |K_{jk}| < \infty$, K defines a bounded operator on $\ell^2(\mathbb{Z})$ which we also denote by K. We say that K is *non-degenerate* if $K_{jj\pm m} \neq 0$ for all j. K is periodic (with period n) if $S^n K = K S^n$, where S is the unitary operator given by (1).

We consider width $2m + 1$, non-degenerate, period-n self-adjoint matrices. In that case, for any E, because K is non-degenerate, $Ku = Eu$, as a finite difference equation, has a unique solution for each choice of $\{u_j\}_{j=0}^{2m-1}$. $T(E)$ will be defined as the map from $\{u_j\}_{j=0}^{2m-1}$ to $\{u_j\}_{j=n}^{n+2m-1}$ – it is a $2m \times 2m$, degree n matrix. If $T(E)u = \lambda u$ for $\lambda \in \mathbb{C}$, $Ku = Eu$ has a Floquet solution with $u_{kn+j} = \lambda^k u_j$. If $T(E)$ is diagonalizable, the set of Floquet solutions is a basis for all solutions of $Ku = Eu$. If $T(E)$ has Jordan anomalies (see [8] for background on linear algebra), there is a basis of modified Floquet solutions with some polynomial growth on top of the exponential λ^k.

The values of λ are determined by

$$p(E, \lambda) = \det(\lambda \mathbf{1} - T(E)).$$

Since

$$\begin{aligned} \det(\lambda \mathbf{1} - T(E)) &= \lambda^{2m} \det(\mathbf{1} - \lambda^{-1} T(E)) \\ &= \lambda^{2m} \Big(\sum_{j=0}^{2m} (-\lambda)^j \operatorname{Tr} \Big(\bigwedge^{j} (T(E)) \Big) \Big) \\ &= \sum_{j=0}^{2m} \lambda^j p_j(E), \end{aligned} \tag{4}$$

where $\bigwedge^j$ is given by multilinear algebra (Section 1.3 of [8]) with $\bigwedge^0(T(E)) = \mathbf{1}$ on $\mathbb{C}$ so its trace is 1. Thus in (4),

$$p_{2m}(E) = 1, \quad p_j(E) = (-1)^j \operatorname{Tr} \Big(\bigwedge^{2m-j} (T(E)) \Big) \tag{5}$$

and p_j is of degree at most $(2m - j)n$ in E.

Since S^n and K are commuting bounded normal operators, they have a joint spectral resolution which is supported precisely on the solutions of $p(E, \lambda) = 0$ with $|\lambda| = 1$ because it is well known that the spectrum is precisely the set of energies with polynomially bounded solutions. By the spectral theorem (equivalently, a direct integral analysis), we thus have the main result of this note:

Theorem 2.1. *Let K be a self-adjoint, non-degenerate, width $2m + 1$, period n matrix. Then for p given by* (4)/(5), *we have that*

$$p(K, S^n) = 0. \tag{6}$$

We end with a number of comments.

(1) We used the self-adjointness of K to be able to exploit the spectral theorem. But just as the Cayley–Hamilton Theorem for finite matrices holds in the non-self-adjoint case, it seems likely that Theorem 2.1 is valid for general non-degenerate, periodic K, even if not self-adjoint.

(2) Since $K_{jj-m} \neq 0$, the transfer matrix, $T(E)$ is invertible and thus $\det(T(E))$ has no zeros. Since it is a polynomial, it must be constant, that is $p_0(E)$ is a constant. It is thus of much smaller degree than the bound, $2mn$, obtained by counting powers of E.

(3) In many cases of interest, $T(E)$ will be symplectic, i.e., there exists an antisymmetric Q on $\mathbb{C}^{2m}$ with $Q^2 = -\mathbf{1}$ so that $T(E)^t QT(E) = Q$. Such a $T(E)$ has $T(E)^{-1}$ and $T(E)^t$ similar, so the eigenvalues $\{\lambda_j\}_{j=1}^{2m}$ can be ordered so that $\lambda_{2m+1-j} = \lambda_j^{-1}$, $j = 1, \dots, m$. It follows that $\det(T(E)) = 1$ but even more, we have that

$$
\begin{aligned}
\operatorname{Tr}\Big(\bigwedge^{k}(T(E))\Big) &= \sum_{j_1<\dots<j_k} \lambda_{j_1}\dots\lambda_{j_k} \\
&= \sum_{j_1<\dots<j_{2m-k}} \lambda_{j_1}^{-1}\dots\lambda_{j_{2m-k}}^{-1} && (7)\\
&= \sum_{j_1<\dots<j_{2m-k}} \lambda_{j_1}\dots\lambda_{j_{2m-k}} && (8)\\
&= \operatorname{Tr}\Big(\bigwedge^{2m-k}(T(E))\Big)
\end{aligned}
$$

and $p_{2m-k}(E) = p_k(E)$. In the above, (7) follows from the fact that the product of all the λ's is 1, and we can sum over the complements of all k-sets. (8) then uses the fact that $\lambda_{2m+1-j} = \lambda_j^{-1}$, $j = 1, \dots, m$.

(4) One can ask whether there is a magic formula in this case, i.e., does $p(\widetilde{K}, S^n) = 0$ imply that $\widetilde{K}$ is periodic and isospectral to K. There is already one subtlety one faces at the start. If $\widetilde{K}$ is not supposed *a priori* n-periodic, then $S^{nj} p_j(\widetilde{K})$ may not equal $p_j(\widetilde{K})S^{nj}$ so there is a question of what $p(\widetilde{K}, S^n) = 0$ means. Even if one supposes that $\widetilde{K}S^n = S^n\widetilde{K}$, $p(\widetilde{K}, S^n) = 0$ and the spectral theorem only implies that $\operatorname{spec}(\widetilde{K}) \subset \operatorname{spec}(K)$, so there is more to be proven. Indeed, the isospectral set in this case remains to be explored.

(5) It seems unlikely that there is another independent relation besides (6) between a polynomial in K and Laurent polynomial in S. In general one cannot hope that $p(K, S^n) = 0$ yields a polynomial in one variable so that $Q(K)$ is a Laurent polynomial in S^n but it remains to find an explicit example where one can prove that the Breuer–Duits question has a negative answer.

There are lots of interesting open questions connected to our main result, Theorem 2.1.

Acknowledgement. Research supported in part by NSF grant DMS-1265592 and in part by Israeli BSF Grant No. 2014337.

References

[1] A. Cayley, A memoir on the theory of matrices. *Philos. Trans. Roy. Soc. London* **148** (1858), 17–37.

[2] D. Damanik, R. Killip, and B. Simon, Perturbations of orthogonal polynomials with periodic recursion coefficients. *Ann. of Math.* (2) **171** (2010), no. 3, 1931–2010. MR 2680401 Zbl 1194.47031

[3] F. G. Frobenius, Über lineare Substitutionen und bilineare Formen. *J. Reine Angew. Math.* **84** (1878), 1–63.

[4] W. R. Hamilton, *Lectures on quaternions.* Hodges and Smith, Dublin, 1853.

[5] P. B. Naĭman, *On the theory of periodic and limit-periodic Jacobian matrices. Dokl. Akad. Nauk SSSR* **143** (1962), 277–279. In Russian. English translation, *Soviet Math. Dokl.* **3** (1962), 383–385. MR 0142558 Zbl 0118.11003

[6] P. B. Naĭman On the spectral analysis of non-Hermitian periodic Jacobi matrices. *Dopovidi Akad. Nauk Ukraïn. RSR* **1963** (1963), 1307–1311. In Ukrainian. MR 0164980 Zbl 0125.00801

[7] B. Simon, *Szego's theorem and its descendants.* Spectral theory for L^2 perturbations of orthogonal polynomials. M. B. Porter Lectures. Princeton University Press, Princeton, N.J., 2011. ISBN 978-0-691-14704-8 MR 2743058 Zbl 1230.33001

[8] B. Simon, *A comprehensive course in analysis.* Part 4. Operator theory. American Mathematical Society, Providence, R.I., 2015. ISBN 978-1-4704-1103-9 MR3364494 Zbl 1334.00003

Path topology dependence of adiabatic time evolution

Atushi Tanaka and Taksu Cheon

To Pavel Exner, our friend and mentor

1 Introduction

The adiabatic theorem for isolated quantum systems is a basic principle of the quantum dynamics: Once a system is prepared to be in a stationary state, the system remains to be stationary as long as the parameters of the system are varied slow enough. There are many proofs for slowly driven systems (see [4], [17], and [2]), which are described by Hermitian Hamiltonian, as well as slowly modulated driven systems (see [35], [13], and [27]), which are described by unitary Floquet operators. The adiabatic theorem has diverse applications, e.g., in molecular science and solid state physics (see [6] and [5]), quantum holonomy (see [25], [3], and [11]) and adiabatic quantum computation (see [15], [14], and [32]).

In this article, we examine how the final stationary state of the adiabatic time evolution depends on the path in the adiabatic parameter space. In particular, we here focus on the eigenspaces corresponding to the initial and final stationary states, and we will ignore the phase information in the following. First, we will show that the final stationary state generally depends on the adiabatic path, although the initial adiabatic parameter and initial stationary state are kept fixed. It turns out that the topology of the adiabatic path plays the key role there. Second, we will show that the discrepancy between two final stationary states corresponding to two different adiabatic path is characterized by a permutation matrix, which is governed by a homotopy equivalence. Our idea is an application of the topological formulation for the exotic quantum holonomy (see [7] and [31]), which concerns the nontrivial change in eigenspaces induced by adiabatic cycles (see [28]).

The present argument heavily relies on topology, in particular the concept of homotopy and its application to the covering map. At the same time, our argument is formal in the sense of mathematics. Since our argument relies only on an elemental account of homotopy and covering map, we refer to textbooks of topology for more mathematical description (see [18], [16], and [22]). The covering map is also discussed in a study of phase holonomy of non-Hermitian quantum systems (see [19]).

The plan of this article is the following. In Section 2, we introduce the lifting of adiabatic paths for our problem. This is considered to be an extension of the lifting for the phase holonomy, see [26] and [1]. In Section 3, we present the main results. An example is shown in Section 4. We summarize the results of this article in Section 5.

2 Lifting adiabatic paths

The lifting of adiabatic paths is the central concept for the theory of conventional quantum holonomy, see [25], [3], and [11]. We take over this concept to examine the path-dependence of eigenspaces. In this sense, our approach is a straightforward extension of the works by Simon [26] and Aharonov and Anandan [1].

We here focus on the simplest case where the system is described by a N-level Hermitian Hamiltonian $H(\lambda)$ with an adiabatic parameter λ, whose space is denoted as $\mathcal{M}$. The energy spectrum of $H(\lambda)$ is assumed to be discrete and non-degenerate for an arbitrary λ in $\mathcal{M}$. Let $P_n(\lambda)$ be the eigenprojector corresponding to an eigenenergy $E_n(\lambda)$ ($n = 0, 1, \ldots, N-1$). Our assumption on the eigenenergies ensures that $E_n(\lambda)$ and $P_n(\lambda)$ are smooth in $\mathcal{M}$. Also, $P_n(\lambda)$ is rank-one. We remark that it is straightforward to extend the following analysis to unitary Floquet systems, if the spectrum of the Floquet operator is discrete and non-degenerate.

We examine all eigenprojectors at a time, which facilitates to compare the changes in eigenspaces induced by two adiabatic paths. From the eigenprojectors $P_n(\lambda)$ for a given point λ in $\mathcal{M}$, we introduce an ordered set of the eigenprojectors

$$p(\lambda) \equiv (P_0(\lambda), P_1(\lambda), \ldots, P_{N-1}(\lambda)). \tag{1}$$

We may introduce another ordered sets, since the order is arbitrary. Let σ denote a permutation of quantum numbers $0, 1, \ldots, N-1$. In other words, σ is an element of N-th symmetric group $\mathcal{S}_N$. Let $p_\sigma(\lambda) = \sigma\,(p(\lambda))$, i.e.,

$$p_\sigma(\lambda) \equiv (P_{\sigma(0)}(\lambda), P_{\sigma(1)}(\lambda), \ldots, P_{\sigma(N-1)}(\lambda)),$$

where $\sigma(n)$ is the permutated quantum number for a given quantum number n.

We introduce a fiber at λ in $\mathcal{M}$:

$$F_\lambda \equiv \bigcup_{\sigma \in \mathcal{S}_N} \{p_\sigma(\lambda)\}. \tag{2}$$

For an arbitrary pair of elements, say $p_{\sigma'}(\lambda)$ and $p_{\sigma''}(\lambda)$, of F_λ, there is a unique permutation $\sigma \in \mathcal{S}_N$ that satisfies $p_{\sigma''}(\lambda) = \sigma\,(p_{\sigma'}(\lambda))$, i.e., $\sigma''(n) = \sigma(\sigma'(n))$ for an arbitrary n. In this sense, we call $\mathcal{S}_N$ a structural group of the fiber F_λ.

A total space $\mathcal{P}$ consists of the fibers F_λ in $\mathcal{M}$:

$$\mathcal{P} \equiv \bigcup_{\lambda \in \mathcal{M}} F_\lambda, \tag{3}$$

which naturally accompanies a projection

$$\pi : \mathcal{P} \longrightarrow \mathcal{M} \tag{4}$$

by construction. Hence we obtain a fiber bundle consists of the total space $\mathcal{P}$, the projection π, the base manifold $\mathcal{M}$ and the fiber F_λ.

Utilizing the fiber bundle introduced above, we introduce a lifting of a path C in $\mathcal{M}$ to $\mathcal{P}$ in order to examine the adiabatic time evolution of p along C. Let λ_{i} and λ_{f} denote the initial and final points of C, respectively.

The adiabatic time evolution of $p \in \mathcal{P}$ can be determined by the time evolution of each eigenprojector P_n. The adiabatic theorem ensures that the final state of the adiabatic time evolution along C is unique for a given initial stationary state $P_n(\lambda_{\mathrm{i}})$. Accordingly the initial $p_{\mathrm{i}} \in F_{\lambda_{\mathrm{i}}}$ and the adiabatic path C uniquely determines the final point $p_{\mathrm{f}} \in F_{\lambda_{\mathrm{f}}}$. The corresponding trajectory of p is called the lifted path $\widetilde{C}$. Let ϕ_C from $F_{\lambda_{\mathrm{i}}}$ to $F_{\lambda_{\mathrm{f}}}$ denote the mapping from the initial to the final point, i.e.,

$$p_{\mathrm{f}} = \phi_C(p_{\mathrm{i}}). \tag{5}$$

Namely, the mapping ϕ_C describes the change of p induced by the adiabatic path C (Figure 1).

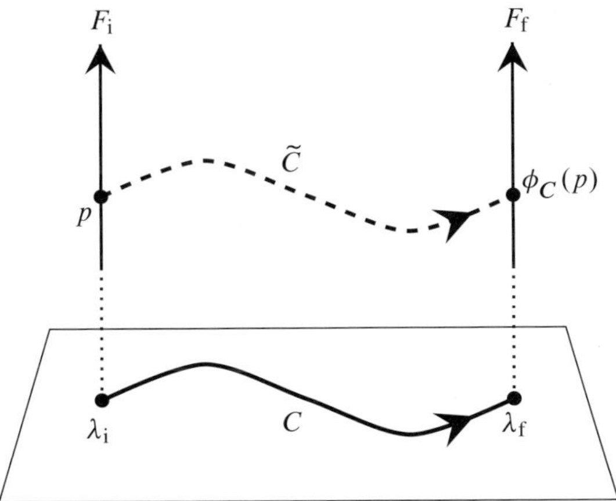

Figure 1. Lifting adiabatic path C in $\mathcal{M}$ to $\mathcal{P}$ (eq. (3)), which is made of fibers F_λ (eq. (2)). Let λ_{i} and λ_{f} denote the initial and final points of C, respectively. The lifted path $\widetilde{C}$ starts from p, which is in the initial fiber F_{i}, and satisfies the adiabatic Schrödinger equation for the ordered set of eigenprojectors. We introduce the mapping ϕ_C from F_{i} to the final fiber F_{f}, so that $\phi_C(p)$ is the final point of $\widetilde{C}$.

The projection π introduced in eq. (4) satisfies the axiom of covering projections [18]. Namely, for a given point λ in $\mathcal{M}$, there is an open subset U of $\mathcal{M}$ that satisfies the following: $\pi^{-1}(U)$ is a disjoint union of connected open subset of $\mathcal{P}$. Each of the disjoint component U_j is mapped homeomorphically onto U (Figure 2).

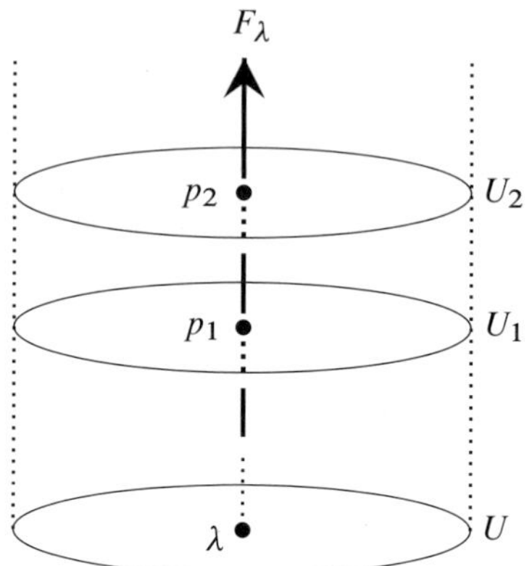

Figure 2. A schematic picture of the covering map $\pi : \mathcal{P} \to \mathcal{M}$ (eq. (4)). Let λ be a point in an open set $U \subset \mathcal{M}$. Points p_j in the fiber F_λ (eq. (2)) satisfies $\pi(p_j) = \lambda$. When π is a covering map, $\pi^{-1}(U)$ consists of disjoint union of open sets U_j, each of which is mapped homeomorphically onto U.

The covering map structure determines a various properties of ϕ_C. In particular, it will be shown below that the homotopic classification of the paths plays the central role here.

If C is a closed path with a given initial point $\lambda_{\rm i}$, the mapping ϕ_C on $F_{\lambda_{\rm i}}$ is called the monodromy action (Theorem 11.22 in [18]). We also note that ϕ_C can be regarded as a permutation of eigenprojector at $\lambda = \lambda_{\rm i}$. Since our argument in the next section much owes to the properties of the monodromy action (e.g., shown in [18]), we will quote the relevant result where appropriate.

3 Comparison of adiabatic paths

We lay out our main results in this section. We compare two adiabatic paths C_1 and C_2, which have the same initial and final points $\lambda_{\rm i}$ and $\lambda_{\rm f}$, in the adiabatic parameter space $\mathcal{M}$. For a given initial eigenprojector at $\lambda_{\rm i}$, we will elucidate how the eigenprojectors at $\lambda_{\rm f}$ depends on C_1 and C_2, by examining the adiabatic time evolutions of the ordered set of eigenprojectors (eq. (1)). In other words, we examine how ϕ_C, which is a mapping from $F_{\lambda_{\rm i}}$ to $F_{\lambda_{\rm f}}$, depends on the topology of the path.

First of all, we examine the case that C_1 is homotopic to C_2, which is denoted as $C_1 \sim C_2$. Namely, we suppose that we may smoothly deform C_1 to C_2, while keeping its initial and final points. We remark that this is the case where most conventional analyses of the periodic adiabatic time evolution have focused on.

If C_1 and C_2 are homotopic, ϕ_{C_1} and ϕ_{C_2} are identical, due to the homotopy lifting property (e.g., Theorem 11.13 in [18]). Hence, an arbitrary initial eigenprojector is adiabatically transported to the same final point through the paths C_1 and C_2.

Utilizing this result, ϕ_C may be denoted as $\phi_{[C]}$, where $[C]$ denotes the equivalence class of a path C under the homotopic classification.

Secondly, we proceed to the case where C_1 is not homotopic to C_2. We compare these paths with a closed path $C \equiv C_1 \cdot (C_2)^{-1}$, where the inverse path of C_2 follows after C_1. Hence the initial point of C is λ_{i}. If a closed path γ, whose initial point is λ_{i}, in $\mathcal{M}$ is homotopic to C, the following formula for $\phi_{[C]}$ holds:

$$\left(\phi_{[C_2]}\right)^{-1} \circ \phi_{[C_1]} = \phi_{[\gamma]}, \tag{6}$$

where $\circ$ denotes the composition of the mappings ϕ_C. Eq. (6) is shown in the following way. Because of $C \sim \gamma$, $\phi_{[C]} = \phi_{[\gamma]}$ holds. On the other hand, $\phi_{[C]} = (\phi_{[C_2]})^{-1} \circ \phi_{[C_1]}$ holds from the definition of $\phi_{[C_1 \cdot (C_2)^{-1}]}$ (eq. (5)).

Now our problem, i.e., the comparison of adiabatic time evolutions along adiabatic paths in $\mathcal{M}$, is cast into the analysis of the monodromy action $\phi_{[\gamma]}$ for an arbitrary closed path γ in $\mathcal{M}$. We remind that $\phi_{[\gamma]}$ corresponds to a permutation of eigenprojectors induced by the adiabatic time evolution along γ. For example, if γ is contractable to the point λ_{i}, which is equivalent to the case $C_1 \sim C_2$ examined above, $\phi_{[\gamma]}$ corresponds to the identical permutation, which implies $\phi_{[C_1]} = \phi_{[C_2]}$.

In order to completely solve our problem, there are two tasks. One is to enumerate all equivalence class $[\gamma]$ of closed paths in $\mathcal{M}$. Namely, we need to identify the first fundamental group $\pi_1(\mathcal{M})$ of the adiabatic parameter space $\mathcal{M}$. The other is to examine the monodromy action $\phi_{[\gamma]}$ of eigenspaces, for every $[\gamma]$ in $\pi_1(\mathcal{M})$.

There remains a question whether different equivalent classes $[\gamma]$ and $[\gamma']$ induce different permutations of eigenspaces. In other words, we need to clarify whether $\pi_1(\mathcal{M})$ completely characterizes the collection of $\phi_{[\gamma]}$. There are two cases.

1. If $\mathcal{P}$ is simply connected, i.e., $\pi_1(\mathcal{P})$ has only a single element, $\phi_{[C_1]} = \phi_{[C_2]}$ holds if and only if C_1 is homotopically equivalent to C_2. Hence $\pi_1(\mathcal{M})$ offers the complete classification of the adiabatic cycles for our problem.

2. In general, we need to modify the first case above, where the equivalence class of closed paths $\pi_1(\mathcal{M})$ is replaced with H where $H \equiv \pi_1(\mathcal{M})/\pi_*(\pi_1(\mathcal{P}))$. Namely, $\phi_{[C_1]} = \phi_{[C_2]}$ holds if and only if C_1 is equivalent to C_2 under the equivalence class H of closed paths in $\mathcal{M}$.

Here we assume that π is a normal covering map, which is equivalent to the condition that H is independent of $p_{\rm i}$ (Proposition 11.35 in [18]). As far as we see, this assumption holds in our examples.

These result concern with the group Φ consists of all possible $\phi_{[\gamma]}$ for an arbitrary closed path γ. In the theory of covering map, Φ is called a covering automorphism group, and the above result is just the one-to-one correspondence between Φ and H (Theorem 12.7 in [18]).

We examine the latter, general case, where $\mathcal{P}$ is multiply connected. There is a closed path $\widetilde{C}$ that is not contractable to a point, in $\mathcal{P}$. We assume that the initial point $p_{\rm i}$ of $\widetilde{C}$ satisfies $\pi(p_{\rm i}) = \lambda_{\rm i}$. Let C be the projection of $\widetilde{C}$ into $\mathcal{M}$, i.e., $C \equiv \pi(\widetilde{C})$. We note that an arbitrary lift of C to $\mathcal{P}$ is closed. If C is not contractable to a point, i.e., the equivalence class $[C]$ is different from $[e]$, this offers an example of $\phi_{[C]} = \phi_{[e]}$ with $[C] \neq [e]$. Accordingly such $[C]$ makes H nontrivial.

4 Example

We examine a slowly modulated periodically driven systems in this section. Here a modification of the adiabatic theorem is required for the stationary states that are described by eigenvectors of a Floquet operator, see [35], [13], and [27]. We choose the periodically driven systems instead of slowly driven Hamiltonian systems because the examples in the latter case requires either the divergence or crossing of eigenenergies, as is seen in the studies of exotic quantum holonomy, see [7], [9], and [10]. We refer to [29] to apply the present formulation for adiabatic paths that involves level crossings.

We compare an arbitrary pair (C_1, C_2) of two adiabatic paths in a two level system, where we suppose that the absence of spectral degeneracy in the adiabatic paths. After we lay out our result using a parameterization that is suitable to examine the path topology dependence, of two level systems, we will show an example of nontrivial pair of paths (C_1, C_2) using a quantum map.

First, we parameterize the adiabatic path using the set of eigenprojections P_1 and P_2

$$b \equiv \{P_1, P_2\}, \tag{7}$$

where the order of the projectors are ignored. Namely, we will specify a point in the base manifold $\mathcal{M}$ by b. This amount to the parameterization of adiabatic path by Floquet operator through the spectral decomposition

$$U = z_1 P_1 + z_2 P_2, \tag{8}$$

where z_j is j-th eigenvalue ($j = 1, 2$), since non-degenerate Floquet operator U uniquely specifies b. In contrast, there are two possible values of the ordered projector p introduced in eq. (1), i.e., (P_1, P_2) and (P_2, P_1). Note that the definition of b in eq. (7) is straightforward to extend to the systems with an arbitrary number of levels.

Second, we take up a geometric interpretation of b and p for the two level system, utilizing the following parameterization of projection operator

$$P(\boldsymbol{a}) = \frac{1}{2}(1 + \boldsymbol{a} \cdot \boldsymbol{\sigma}),$$

where $\boldsymbol{\sigma}$ is the vector consists of Pauli matrices, and $\boldsymbol{a}$ is a normalized three-dimensional real vector. The eigenprojectors in eq. (8) can be expressed as $P_1 = P(\boldsymbol{a})$ and $P_2 = P(-\boldsymbol{a})$. Now it is straightforward to see that p and $\boldsymbol{a}$ has $1:1$ correspondence, which implies that $\mathcal{P}$ can be identified with S^2. On the other hand, $\pm\boldsymbol{a}$ correspond to a single point in the b-space. Namely, the b-space can be regarded as $\mathbb{R}P^2$, the real projective plane. Hence the covering map $\pi\colon S^2 \to \mathbb{R}P^2$ for the two level system can be regarded as an identification of the antipodal points in the sphere.

Now our argument presented in the previous section is ready to apply. The fundamental class of the base space $\pi_1(\mathbb{R}P^2) = \{[e], [\gamma]\}$ has two elements, where e is the closed path that is contractable to a point, and the closed path γ is not homotopic to e. On the other hand, our total space $\mathcal{P}$ is simply connected as $\mathcal{P} = S^2$. Hence $[e]$ and $[\gamma]$, the two classes of closed paths, offers two different monodromy map $\phi_{[e]}$ and $\phi_{[\gamma]}$, which correspond to the identity and cyclic permutations of two eigenprojectors, respectively.

We summarize the analysis of two level systems. When the trails of the adiabatic paths C_1 and C_2 in $\mathbb{R}P^2$ are homotopic, the adiabatic time evolutions of an eigenprojector along C_1 and C_2 has no difference. On the other hand, when C_1 and C_2 are not homotopic, the composite closed path $C_2^{-1} \cdot C_1$ is homotopic to γ, and the corresponding discrepancy $\phi_{[\gamma]} = (\phi_{[C_2]})^{-1} \circ \phi_{[C_1]}$ (eq. (6)) is expressed by the cyclic permutation of two items.

We exemplify the above argument using a slowly modulated driven spin-1/2, where we set $\hbar = 1$. In the absence of the modulation, our example is described by the following periodically driven Hamiltonian:

$$H(t) = \frac{1}{2}\boldsymbol{B} \cdot \boldsymbol{\sigma} + \lambda \frac{1 - \sigma_z}{2} \sum_{m=-\infty}^{\infty} \delta(t - m),$$

where $\boldsymbol{B} = B(\cos\phi \boldsymbol{e}_x + \sin\phi \boldsymbol{e}_y)$ is the static magnetic field confined in xy-plane (B and ϕ are the cylindrical variables), and λ is the strength of the periodic term.

The corresponding Floquet operator is, for example

$$U = e^{-i\lambda(1-\sigma_z)/2} e^{-i\boldsymbol{B}\cdot\boldsymbol{\sigma}/2}, \tag{9}$$

which is a quantum map under a rank-1 perturbation, see [12] and [21]. In the following, we examine U under the adiabatic changes of $\boldsymbol{B}$ in (B_x, B_y)-plane except the origin $\boldsymbol{B} = 0$. Also, we set $\lambda = \phi$ along the adiabatic path. Hence U is single-valued in (B_x, B_y)-plane, since U periodically depends on λ with the period 2π. The eigenvalues of U are, as shown in [28],

$$z_{\pm} = \exp\{-i(\phi\pm\Delta)/2\},$$

where

$$\Delta = 2\arccos\Big(\cos\frac{\phi}{2}\cos\frac{B}{2}\Big).$$

The corresponding eigenprojectors are $P(\pm\boldsymbol{a})$, where

$$\boldsymbol{a} = \frac{1}{\sin(\Delta/2)}\Big[\sin\frac{B}{2}\Big(\cos\frac{\phi}{2}\boldsymbol{e}_{\rho} - \sin\frac{\phi}{2}\boldsymbol{e}_{\phi}\Big) - \sin\frac{\phi}{2}\cos\frac{B}{2}\boldsymbol{e}_z\Big],$$

$\boldsymbol{e}_{\rho} = \cos\phi\boldsymbol{e}_x + \sin\phi\boldsymbol{e}_y$ and $\boldsymbol{e}_{\phi} = -\sin\phi\boldsymbol{e}_x + \cos\phi\boldsymbol{e}_y$. Note that Δ and $\boldsymbol{a}$ are not single-valued in (B_x, B_y)-plane although U is single-valued. We depict adiabatic paths, whose initial and final points in (B_x, B_y)-plane are $\boldsymbol{B}_{\mathrm{i}} \equiv (\pi, 0)$ and $\boldsymbol{B}_{\mathrm{f}} \equiv (-\pi, 0)$, respectively, and corresponding adiabatic time evolution of eigenspaces in Figure 3.

5 Conclusion

We have shown the path topology dependence of adiabatic time evolution in closed quantum systems through a topological argument, which is based on the recent study on the exotic quantum holonomy (see [28]). We finally note that examples of systems exhibit non-trivial adiabatic path topology dependence, according to the studies of exotic quantum holonomy in, for example, quantum graphs with generalized connection conditions (see [8], [33], [23], and [24]), many-qubit systems [30], adiabatic quantum computation [32], and the Lieb–Liniger model [34].

Acknowledgement. This research was supported by the Japan Ministry of Education, Culture, Sports, Science and Technology under the Grant number 15K05216.

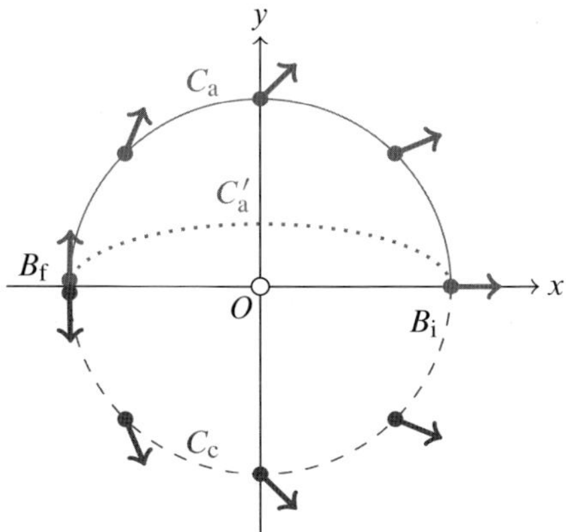

Figure 3. Adiabatic time evolution of $\boldsymbol{a}$ (thick arrow), which is equivalent to the ordered set of eigenprojectors p for the periodically driven spin-1/2 (eq. (9)). Since $\boldsymbol{a}$ is transported from $\boldsymbol{e}_x$ to $\boldsymbol{e}_y$ along the adiabatic path C_a (thick curve), the corresponding adiabatic evolution of eigenprojector is from $P(\boldsymbol{e}_x)$ to $P(\boldsymbol{e}_y)$. In other words, the adiabatic mapping of p is $\phi_{[C_a]}(p(\boldsymbol{e}_x)) = p(\boldsymbol{e}_y)$, where $p(\boldsymbol{a}) \equiv (P(\boldsymbol{a}), P(-\boldsymbol{a}))$. Other adiabatic path C'_a (dotted curve), which is homotopic to C_a, provides the same adiabatic mapping, i.e., $\phi_{[C_a]} = \phi_{[C'_a]}$. On the other hand, the adiabatic path C_c (dashed curve) is not homotopic to C_a due to an obstacle (a disclination [20] and [22]) at the origin. The corresponding adiabatic evolution is described as $\phi_{[C_c]}(p(\boldsymbol{e}_x)) = p(-\boldsymbol{e}_y)$. We may compare C_a and C_c by a closed path $C = C_c^{-1} \cdot C_a$. The discrepancy between $\phi_{[C_a]}$ and $\phi_{[C_c]}$ is given by $\phi_{[C]}$, which corresponds to the cyclic permutation of the two items in p.

References

[1] Y. Aharonov and J. Anandan, Phase change during a cyclic quantum evolution. *Phys. Rev. Lett.* **58** (1987), no. 16, 1593–1596. MR 0884314

[2] J. E. Avron, R. Seiler, and L. G. Yaffe, Adiabatic theorems and applications to the quantum Hall effect. *Comm. Math. Phys.* **110** (1987), no. 1, 33–49. MR 0885569 Zbl 0626.58033

[3] A. Bohm, A. Mostafazadeh, H. Koizumi, Q. Niu, and Z. Zwanziger, *The geometric phase in quantum systems.* Foundations, mathematical concepts, and applications in molecular and condensed matter physics. Texts and Monographs in Physics. Springer, Berlin, 2003. ISBN 3-540-00031-3 MR 1996062 Zbl 1039.81003

[4] M. Born and V. Fock, Beweis des Adiabatensatzes. *Z. Phys. A* **51** (1928), 165–180. JFM 54.0994.03

[5] M. Born and K. Huang, *Dynamical theory of crystal lattices.* International Series of Monographs on Physics. The Clarendon Press, Oxford University Press, Oxford, 1954. Reprint, Oxford Classic Texts in the Physical Sciences. The Clarendon Press, Oxford University Press, Oxford, 1988. ISBN 0-19-850369-5 (reprint) MR 1654161 (reprint) Zbl 0057.44601 Zbl 0908.01039 (reprint)

[6] M. Born and R. Oppenheimer, Zur quantentheorie der molekeln. *Annalen der Physik* **84** (1927), 457–484. JFM 53.0845.04

[7] T. Cheon, Double spiral energy surface in one-dimensional quantum mechanics of generalized pointlike potentials. *Phys. Lett. A* **248** (1998), 285–289.

[8] T. Cheon, T. Fülöp, and I. Tsutsui, Symmetry, duality and anholonomy of point interaction in one dimension. *Ann. Physics* **294** (2001), no. 1, 1–23. MR 1867783

[9] T. Cheon, A. Tanaka, and S. W. Kim, Exotic quantum holonomy in Hamiltonian systems. *Phys. Lett. A* **374** (2009), no. 2, 144–149. MR 2564075 Zbl 1235.81088

[10] T. Cheon, A. Tanaka, and O. Turek, Examples of quantum holonomy with topology changes. *Acta Polytechnica* **53** (2013), 410–415.

[11] D. Chruściński and A. Jamiołkowski, *Geometric phases in classical and quantum mechanics.* Progress in Mathematical Physics, 36. Birkhäuser Boston, Boston, MA, 2004. ISBN 0-8176-4282-X MR 2060919 Zbl 1075.81002

[12] M. Combescure, Spectral properties of a periodically kicked quantum Hamiltonian. *J. Statist. Phys.* **59** (1990), no. 3-4, 679–690. MR 1063177 Zbl 0713.58044

[13] A. Dranov, J. Kellendonk, and R. Seiler, Discrete time adiabatic theorems for quantum mechanical systems. *J. Math. Phys.* **39** (1998), no. 3, 1340–1349. MR 1608461 Zbl 1056.81505

[14] E. Farhi, J. Goldstone, S. Gutmann, J. Lapan, A. Lundgren, and D. Preda, A quantum adiabatic evolution algorithm applied to random instances of an NP-complete problem. *Science* **292** (2001), no. 5516, 472–476. MR 1838761 Zbl 1226.81046

[15] E. Farhi, J. Goldstone, S. Gutmann, and M. Sipser, Quantum computation by adiabatic evolution. Preprint 2000. arXiv:quant-ph/0001106

[16] A. Hatcher, *Algebraic topology.* Cambridge University Press, Cambridge, 2002. ISBN 0-521-79160-X, 0-521-79540-0 MR 1867354 Zbl 1044.55001

[17] T. Kato, On the adiabatic theorem of quantum mechanics. *J. Phys. Soc. Jpn.* **5** (1950), 435–439.

[18] J. M. Lee, *Introduction to topological manifolds.* second ed., Graduate Second edition. Graduate Texts in Mathematics, 202. Springer, New York, 2011. ISBN 978-1-4419-7939-1 MR 2766102 Zbl 1209.57001

[19] H. Mehri-Dehnavi and A. Mostafazadeh, Geometric phase for non-Hermitian Hamiltonians and its holonomy interpretation. *J. Math. Phys.* **49** (2008), no. 8, 082105, 17 pp. MR 2440692 Zbl 1152.81556

[20] N. D. Mermin, The topological theory of defects in ordered media. *Rev. Modern Phys.* **51** (1979), no. 3, 591–648. MR 0541885

[21] B. Milek and P. Šeba, Singular continuous quasi-energy spectrum in the kicked rotator with separable perturbation: possibility of the onset of quantum chaos. *Phys. Rev. A* (3) **42** (1990), no. 6, 3213–3220. MR 1072316 Zbl 0726.47008

[22] M. Nakahara, *Geometry, topology and physics.* Graduate Student Series in Physics. Adam Hilger, Bristol, 1990. ISBN: 0-85274-094-8, 0-85274-095-6 MR 1065614 Zbl 0764.53001

[23] S. Ohya, Parasupersymmetry in quantum graphs. *Ann. Physics* **331** (2013), 299–312. MR 3040288 Zbl 1266.81085

[24] S. Ohya, Non-Abelian monopole in the parameter space of point-like interactions. *Ann. Physics* **351** (2014), 900–913. MR 3283119

[25] A. Shapere and F. Wilczek (eds.), *Geometric phases in physics.* Advanced Series in Mathematical Physics, 5. World Scientific Publishing, Singapore etc., 1989. ISBN: 9971-50-599-1, 9971-50-621-1 MR 1084385 Zbl 0914.00014

[26] B. Simon, Holonomy, the quantum adiabatic theorem, and Berry's phase. *Phys. Rev. Lett.* **51** (1983), no. 24, 2167–2170. MR 0726866

[27] A. Tanaka, Adiabatic theorem for discrete time evolution. *J. Phys. Soc. Jpn.* **80** (2011), 125002, 2 pp.

[28] A. Tanaka and T. Cheon, Bloch vector, disclination and exotic quantum holonomy. *Phys. Lett. A* **379** (2015), no. 30-31, 1693–1698. Zbl 1343.81124

[29] A. Tanaka and T. Cheon, A topological formulation for exotic quantum holonomy. *Nanosystems: physics, chemistry, mathematics* **6** (2015), 786–792.

[30] A. Tanaka, S. W. Kim, and T. Cheon, Eigenvalue and eigenspace anholonomies in hierarchical systems. *EPL* (*Europhysics Letters*) **96** (2011), 10005, 6 pp.

[31] A. Tanaka and M. Miyamoto, Quasienergy anholonomy and its application to adiabatic quantum state manipulation. *Phys. Rev. Lett.* **98** (2007), no. 16, 160407, 4 pp.

[32] A. Tanaka and K. Nemoto, Adiabatic quantum computation along quasienergies. *Phys. Rev. A* **81** (2010), no. 2, 022320, 8 pp.

[33] I. Tsutsui, T. Fülöp, and T. Cheon, Möbius structure of the spectral space of Schrödinger operators with point interaction. *J. Math. Phys.* **42** (2001), no. 12, 5687–5697. MR 1866681 Zbl 1019.81014

[34] N. Yonezawa, A Tanaka, and T. Cheon, Quantum holonomy in the Lieb–Liniger model. *Phys. Rev. A* **87** (2013), no. 6, 062113, 6 pp.

[35] R. H. Young and W. J. Deal, Jr., Adiabatic response to an oscillatory field. *J. Math. Phys.* **11** (1970), 3298–3306. MR 0273967

On quantum graph filters with flat passbands

Ondřej Turek

Dedicated to Pavel Exner on the occasion of his 70th birthday

1 Introduction

Quantum mechanics on graphs is a useful tool for the examination of quantum motion on microscopic wires, lattices and other graph-like nanostructures. The method has been intensively developed since 1980s with regard to the technological progress achieved in microfabrication. The development of the subject led to a rich literature to date; see, e.g., monographs [1] and [7] and references therein. On the other hand, the discipline remains relatively new and is still rapidly advancing.

Quantum graph models are useful in particular for a design of quantum systems with prescribed properties. In this paper we focus on scattering problems on systems consisting of several wires connected together in one point to form a star. When a particle moving along a wire reaches the vertex, it is scattered to the other wires. The scattering characteristics depend on the energy of the particle and on the nature of the potential in the point. Such system is modelled by a star graph with a certain wave function coupling in the vertex. It is known that a vertex of degree n generally features n^2-parametric family of admissible couplings [8], and the scattering characteristics considerably vary in dependence on the coupling parameters [2]. Obviously, one can take advantage of this fact in a design of quantum devices with particular particle transmission properties. On the other hand, the role of the coupling parameters in the scattering characteristics is not well understood yet.

The problem studied in this paper concerns a star graph with n edges, some of which being subject to a constant nonzero potential V. Scattering in such a system depends i.a. on the strength of the potential. It was noticed in earlier works [11] and [12] that a certain particular choice of the vertex coupling gives rise to a "flat band" scattering behaviour. That is, the probability of transmission of a particle from an edge (we call the edge "input") to another given edge (called "output") turned out to be constant for energies E in the interval $(0, V)$ and quickly descending towards zero for $E > V$. Consequently, particles with energies exceeding the controlling potential V mostly cannot pass to the output edge. The vertex thus works as a controllable band-pass filter with a flat passband. In this paper we will deal with the problem more thoroughly. We will prove that this behaviour can occur only for certain subfamilies of vertex couplings, but at the same time we will demonstrate that there exists a multiparametric family of vertex couplings with the "flat passband" property. In other words, such a behaviour is less rare than it might seem earlier.

The paper is organized as follows. In Section 2 we bring together elementary facts and notation on vertex couplings and scattering in quantum graph vertices. Section 3 presents the concept of a controllable band-pass filter and the goals of the paper, as well as the idea of solution. Sections 4–7 are devoted to the existence of quantum graph filters featuring flat passbands. The main result is presented in Section 6, in which several designs are proposed.

2 Preliminaries

A wave function of a particle confined to a star graph having n arms consists of n components, $\Psi = (\psi_1, \psi_2, \dots, \psi_n)$. The coordinate on each arm is chosen such that 0 corresponds to the center of the star graph and the variable grows in the outgoing direction. If there are potentials $V_1, \dots, V_n$ imposed on the arms, the Hamiltonian acts as $\psi_j \mapsto -\psi_j'' + V_j \psi_j$ at each arm $j = 1, \dots, n$ (we choose the units so that $\hbar = 2m = 1$ for m being the mass of the particle).

Properties of the vertex are determined by boundary conditions that are conventionally written in the form

$$A\Psi(0) + B\Psi'(0) = 0, \tag{1}$$

where

$$\Psi(0) = \begin{pmatrix} \psi_1(0) \\ \vdots \\ \psi_n(0) \end{pmatrix} \quad \text{and} \quad \Psi'(0) = \begin{pmatrix} \psi_1'(0) \\ \vdots \\ \psi_n'(0) \end{pmatrix}$$

are the boundary vectors and A, B are complex $n \times n$ matrices satisfying

$$\operatorname{rank}(A|B) = n, \quad AB^* = BA^*, \tag{2}$$

cf. [8]. The symbol $(A|B)$ denotes the $n \times 2n$ matrix formed from columns of A and B.

In this paper we will take advantage of the so-called ST-form of boundary conditions [3], in which requirements (2) are implicitly satisfied due to a special choice of A and B. Namely, the ST-form relies on the block decomposition of A and B,

$$\begin{pmatrix} I^{(r)} & T \\ 0 & 0 \end{pmatrix} \Psi'(0) = \begin{pmatrix} S & 0 \\ -T^* & I^{(n-r)} \end{pmatrix} \Psi(0) \tag{3}$$

for a certain $r \in \{0, 1, \dots, n\}$. Matrix T is a general complex $r \times (n-r)$ matrix, S is a Hermitian matrix of order r, and $I^{(r)}$, $I^{(n-r)}$ are identity matrices of given orders. The value r corresponds to $\operatorname{rank}(B)$ in boundary conditions (1).

If a wave corresponding to a quantum particle with energy E reaches the vertex from the ℓ-th line with amplitude 1, the wave is reflected with a complex amplitude $\mathcal{S}_{jj}(E)$ and transmitted to the lines nos. $\ell \neq j$ with complex amplitudes $\mathcal{S}_{j\ell}(E)$. The scattering amplitudes form the scattering matrix of the vertex. The scattering matrix, denoted by $\mathcal{S}(E)$, is an $n \times n$ matrix function of particle energy that is given by the formula

$$\mathcal{S}(E) = -(A + \mathrm{i}\sqrt{E}B)^{-1}(A - \mathrm{i}\sqrt{E}B). \tag{4}$$

Let us emphasize that formula (4) applies only if $V_j = 0$ for all $j = 1, 2, \dots, n$. If we substitute

$$A = -\begin{pmatrix} S & 0 \\ -T^* & I^{(n-r)} \end{pmatrix} \quad \text{and} \quad B = \begin{pmatrix} I^{(r)} & T \\ 0 & 0 \end{pmatrix}$$

into equation (3), we obtain the scattering matrix expressed in terms of the ST-form of boundary conditions,

$$\mathcal{S}(E) = -I^{(n)} + 2 \begin{pmatrix} I^{(r)} \\ T^* \end{pmatrix} \Big(I^{(r)} + TT^* - \frac{1}{\mathrm{i}\sqrt{E}} S \Big)^{-1} \big(I^{(r)} \;\; T \big); \tag{5}$$

cf. [4]. It is straightforward to see from formula (5) that the scattering matrix is constant with respect to E if and only if the matrix S in the ST-form of boundary conditions (3) vanishes, i.e., when

$$\begin{pmatrix} I^{(r)} & T \\ 0 & 0 \end{pmatrix} \Psi'(0) = \begin{pmatrix} 0 & 0 \\ -T^* & I^{(n-r)} \end{pmatrix} \Psi(0).$$

Vertex couplings having energy-independent scattering matrices are called scale invariant couplings. They are widely studied; see [6], [9], [10], and [5].

3 A potential-controlled filter

Consider a quantum star graph with n edges. We will regard one of the edges as *input*, another edge as *output*. The remaining $n-2$ edges will be assumed to be of two types (see Figure 1):

- lines with constant nonzero potentials (*"controlling lines"*);
- lines without potentials (*"drains"*).

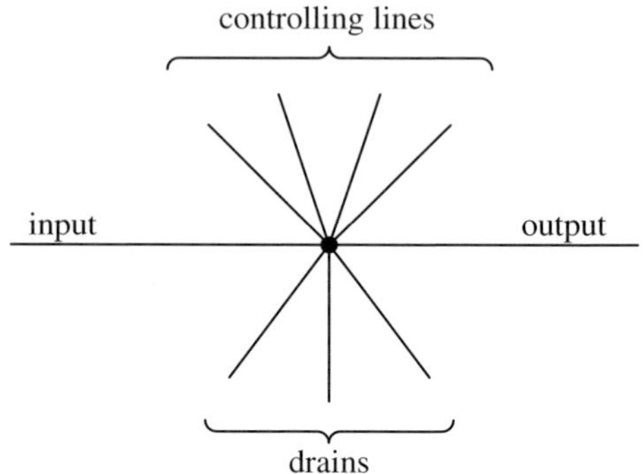

Figure 1. A controllable quantum graph filter

For a particle coming in the vertex along the input line with energy E, we denote the complex transmission amplitude to the output line by the symbol $\mathcal{T}(E)$. The corresponding transmission probability in the channel is $\mathcal{P}(E) = |\mathcal{T}(E)|^2$. This paper is concerned with the relation between the transmission probability in the input-output channel and the potentials on the controlling lines. More specifically, we will search for couplings that can serve as controllable band-pass filters with flat passbands. That is, the function $\mathcal{P}(E)$ is required to have the following three properties, cf. Figure 2:

$$\mathcal{P}(E) = \text{const} > 0 \quad \text{for } E \in (0, V) \text{ for a certain } V > 0, \tag{6}$$

$$\mathcal{P}(E) \text{ quickly decreases when } E \text{ exceeds } V, \text{ i.e., } \lim_{E \searrow V} \mathcal{P}'(E) = -\infty, \tag{7}$$

$$\lim_{E \to \infty} \mathcal{P}(E) = 0. \tag{8}$$

We assume that V is the value of the constant potential on the controlling lines.

In general, the transmission amplitude in the input-output channel is given by the term $[\mathcal{S}(E)]_{oi}$ of the scattering matrix. However, formula (4) cannot be applied straightforwardly, because the controlling lines are subject to constant potentials $V_j \neq 0$. Therefore, we will approach the problem as follows. At first we transform the original boundary conditions in the vertex of degree n to boundary conditions in a vertex of degree 2. This step is based on the idea that the controlling lines and drains

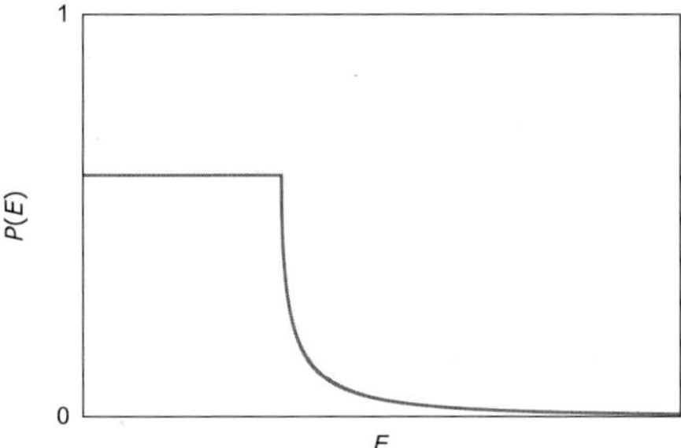

Figure 2. An example of sought transmission probability

support only outgoing waves, thus the corresponding wave function components are multiples of $e^{ik_j x}$, where

$$k_j = \begin{cases} \sqrt{E - V_j}, & \text{if } E > V_j, \\ i\sqrt{V_j - E}, & \text{if } E < V_j, \end{cases} \tag{9}$$

is the momentum on the j-th line with potential V_j. Relation $\psi_j(x) \propto e^{ik_j x}$ implies

$$\psi_j'(0) = ik_j \psi_j(0) \quad \text{for all } j \neq i, j \neq o. \tag{10}$$

Equation (10) allows us to eliminate the boundary values $\psi_j(0)$ and $\psi_j'(0)$ at all controlling edges and drains from boundary conditions (3). We obtain boundary conditions in a vertex of degree 2 that connect just the input and the output,

$$A_{\text{diss}} \Psi_{io} + B_{\text{diss}} \Psi_{io}' = 0, \tag{11}$$

where

$$\Psi_{io} = \begin{pmatrix} \psi_i(0) \\ \psi_o(0) \end{pmatrix}, \quad \Psi_{io}' = \begin{pmatrix} \psi_i'(0) \\ \psi_o'(0) \end{pmatrix}, \tag{12}$$

are boundary values at input and output line. We emphasize that $A_{\text{diss}}, B_{\text{diss}}$ are 2×2 matrices that generally do *not* obey the requirements (2), because $AB^* = BA^*$ can be broken due to the dissipation in the vertex, manifested through "hidden" drains and controllers. On the other hand, since neither input line nor the output line support a potential, formula (4) applies without reserve. Once we substitute matrices $A_{\text{diss}}, B_{\text{diss}}$ from reduced boundary conditions (11) into equation (4), we obtain the 2×2 scattering matrix that characterizes wave propagation in the input-output channel. In particular, the (2, 1)-term of the matrix is the sought transmission amplitude $\mathcal{T}(E)$.

For the derivation of matrices A_{diss}, B_{diss}, we will take advantage of the ST-form of boundary conditions. Therefore, the calculation depends on the parameter r. In the following section we begin with the case $r = 1$.

4 Case $r = 1$

The ST-form (3) of boundary conditions for $r = 1$ uses matrices $T = (t_2\ t_3\ \cdots\ t_n)$ and $S = (s)$. We may assume without loss of generality that line no. 1 is the input and line no. 2 is the output. Let us follow the steps outlined in Section 3. After eliminating $\psi_3(0), \dots, \psi_n(0)$ from the system using identities (10), we get dissipative boundary conditions (11) with

$$A_{\mathrm{diss}} = -\begin{pmatrix} s - \mathrm{i}\sum_{j=3}^{n} k_j |t_j|^2 & 0 \\ -\overline{t_2} & 1 \end{pmatrix}, \quad B_{\mathrm{diss}} = \begin{pmatrix} 1 & t_2 \\ 0 & 0 \end{pmatrix}.$$

When we substitute A_{diss}, B_{diss} into formula (4), we obtain the scattering matrix describing the input-output interface,

$$\mathcal{S}_{\mathrm{diss}}(E) = -I + \frac{2}{1 + |t_2|^2 - \frac{s}{\mathrm{i}\sqrt{E}} + \sum_{j=3}^{n} \frac{k_j}{\sqrt{E}} |t_j|^2} \begin{pmatrix} 1 & t_2 \\ \overline{t_2} & |t_2|^2 \end{pmatrix}.$$

The transmission amplitude is given as the term $[\mathcal{S}_{\mathrm{diss}}(E)]_{21}$, i.e.,

$$\mathcal{T}(E) = \frac{2\overline{t_2}}{1 + |t_2|^2 - \frac{s}{\mathrm{i}\sqrt{E}} + \sum_{j=3}^{n} \frac{k_j}{\sqrt{E}} |t_j|^2}. \tag{13}$$

Now we are ready to check whether S and T can be chosen such that the function $\mathcal{P}(E) = |\mathcal{T}(E)|^2$ satisfies conditions (6)–(8). Condition (8) is equivalent to $\lim_{E\to\infty} \mathcal{T}(E) = 0$. Equation (9) implies $\lim_{E\to\infty} k_j/\sqrt{E} = 1$ for all $j = 3, \dots, n$; hence

$$\lim_{E\to\infty} \mathcal{T}(E) = \frac{2\overline{t_2}}{1 + |t_2|^2 + \sum_{j=3}^{n} |t_j|^2}.$$

Consequently

$$\lim_{E\to\infty} \mathcal{T}(E) = 0 \iff t_2 = 0.$$

However, the choice $t_2 = 0$ implies $\mathcal{T}(E) = 0$ for all $E > 0$ (cf. (13)), which contradicts condition (6). (In physical terms, $t_2 = 0$ corresponds to a vertex with line no. 2 completely decoupled.) To sum up, conditions (8) and (6) cannot be satisfied at the same time. We conclude that a band-pass filter with flat passband cannot be constructed using a vertex coupling with $r = 1$.

5 Case $r \geq 2$ with linear dependence

Now we consider boundary conditions (1) with $r = \operatorname{rank}(B) > 2$ such that the columns of B corresponding to the input and output are linearly dependent. We can assume without loss of generality that the input corresponds to line no. 1 and the output is line no. n. When the boundary conditions are written in the ST-form, the linear dependence implies that the last column of T is a transposition of the vector $(t, 0, \dots, 0)$ for a certain $t \neq 0$. Therefore, the ST-form of boundary conditions reads as follows,

$$\begin{pmatrix} 1 & 0 & T_1 & t \\ 0 & I^{(r-1)} & T_2 & 0 \\ 0 & 0 & 0 & 0 \\ 0 & 0 & 0 & 0 \end{pmatrix} \begin{pmatrix} \psi_i' \\ \Psi_{\mathrm{cd}}' \\ \psi_o' \end{pmatrix} = \begin{pmatrix} s & S_2 & 0 & 0 \\ S_2^* & S_4 & 0 & 0 \\ -T_1^* & -T_2^* & I^{(n-r-1)} & 0 \\ -\bar{t} & 0 & 0 & 1 \end{pmatrix} \begin{pmatrix} \psi_i \\ \Psi_{\mathrm{cd}} \\ \psi_o \end{pmatrix}, \tag{14}$$

where $I^{(r-1)}, I^{(n-r-1)}$ are identity matrices of given orders and $T = \left(\begin{smallmatrix} T_1 & t \\ T_2 & 0 \end{smallmatrix}\right)$, $S = \left(\begin{smallmatrix} s & S_2 \\ S_2^* & S_4 \end{smallmatrix}\right)$. Recall that symbols ψ_i, ψ_i' and ψ_o, ψ_o' denote boundary values at the input and output line, respectively. Symbols

$$\Psi_{\mathrm{cd}} = \begin{pmatrix} \psi_2(0) \\ \vdots \\ \psi_{n-1}(0) \end{pmatrix}; \quad \Psi_{\mathrm{cd}}' = \begin{pmatrix} \psi_2'(0) \\ \vdots \\ \psi_{n-1}'(0) \end{pmatrix}$$

stand for boundary vectors at controlling edges and drains.

Values $\psi_j(0)$, $\psi_j'(0)$ obey relations (10), i.e.,

$$\Psi'_{\mathrm{cd}} = \mathrm{i}\begin{pmatrix} K_2 & 0 \\ 0 & K_3 \end{pmatrix}\Psi_{\mathrm{cd}} \tag{15}$$

for $K_2 = \mathrm{diag}(k_2, \dots, k_r)$ and $K_3 = \mathrm{diag}(k_{r+1}, \dots, k_{n-1})$. We use identity (15) to eliminate Ψ_{cd} and Ψ'_{cd} from system (14). In this way we obtain boundary conditions (11) with

$$A_{\mathrm{diss}} = -\begin{pmatrix} f & 0 \\ -\bar{t} & 1 \end{pmatrix}, \quad B_{\mathrm{diss}} = \begin{pmatrix} 1 & t \\ 0 & 0 \end{pmatrix}, \tag{16}$$

where

$$f = s - \mathrm{i}T_1K_3T_1^* + (S_2 - \mathrm{i}T_1K_3T_2^*)(\mathrm{i}K_2 + \mathrm{i}T_2K_3T_2^* - S_4)^{-1}(S_2^* - \mathrm{i}T_2K_3T_1^*).$$

The dissipative scattering matrix corresponding to matrices (16) is

$$\mathcal{S}_{\mathrm{diss}}(E) = -I + \frac{2}{1 + |t|^2 - \frac{f}{\mathrm{i}\sqrt{E}}}\begin{pmatrix} 1 & t \\ \bar{t} & |t|^2 \end{pmatrix}.$$

The transmission amplitude thus equals

$$\mathcal{T}(E) = \frac{2\bar{t}}{1 + |t|^2 - \frac{f}{\mathrm{i}\sqrt{E}}}.$$

Now we check condition (8). Since

$$\lim_{E\to\infty} \frac{1}{\sqrt{E}}K_2 = I^{(r-1)} \quad \text{and} \quad \lim_{E\to\infty} \frac{1}{\sqrt{E}}K_3 = I^{(n-r-1)},$$

we have

$$\lim_{E\to\infty} \frac{-f}{\mathrm{i}\sqrt{E}} = T_1T_1^* - T_1T_2^*(I + T_2T_2^*)^{-1}T_2T_1^* = T_1(I + T_2^*T_2)^{-1}T_1^*.$$

Hence

$$\lim_{E\to\infty} \mathcal{T}(E) = \frac{2\bar{t}}{1 + |t|^2 + T_1(I + T_2^*T_2)^{-1}T_1^*}.$$

Note that $T_1(I + T_2^*T_2)^{-1}T_1^*$ is a non-negative number for any choice of T_1, T_2. Therefore, condition (8) is equivalent to $t = 0$. However, it is easy to see from boundary conditions (14) that $t = 0$ corresponds to a completely decoupled output, which implies $\mathcal{T}(E) = 0$ for all $E > 0$. In other words, conditions (8) and (6) are contradictory. We conclude that a vertex coupling cannot serve for the construction of a band-pass filter with flat passband if the columns of B corresponding to the input and output edge are linearly dependent.

6 Case $r = 2$

This section is focused on the case $r = 2$, i.e., $\operatorname{rank}(B) = 2$. With regard to the result of Section 5, we may assume that the columns of matrix B that correspond to the input and output line are linearly independent. Without loss of generality, we associate the input and output with lines no. 1 and no. 2, respectively. The boundary conditions in the vertex are expressed in the ST-form as follows,

$$\begin{pmatrix} I^{(2)} & T \\ 0 & 0 \end{pmatrix} \begin{pmatrix} \Psi'_{io} \\ \Psi'_{\mathrm{cd}} \end{pmatrix} = \begin{pmatrix} S & 0 \\ -T^* & I^{(n-2)} \end{pmatrix} \begin{pmatrix} \Psi_{io} \\ \Psi_{\mathrm{cd}} \end{pmatrix}, \tag{17}$$

where S is a Hermitian 2×2 matrix, $T \in \mathbb{C}^{2,n-2}$, $I^{(2)}$, $I^{(n-2)}$ are identity matrices of appropriate sizes, Ψ_{io}, Ψ'_{io} are the boundary values at input and output line (cf. (12)), and

$$\Psi_{\mathrm{cd}} = \begin{pmatrix} \psi_{r+1}(0) \\ \vdots \\ \psi_n(0) \end{pmatrix}, \quad \Psi'_{\mathrm{cd}} = \begin{pmatrix} \psi'_{r+1}(0) \\ \vdots \\ \psi'_n(0) \end{pmatrix}.$$

are the boundary values at controllers and drains. Relations (9) imply

$$\Psi'_{\mathrm{cd}} = \mathrm{i}K\Psi_{\mathrm{cd}}, \tag{18}$$

where

$$K = \operatorname{diag}(k_3, \dots, k_n).$$

Identity (18) allows to eliminate Ψ_{cd} and Ψ'_{cd} from system (17). We arrive at dissipative boundary conditions connecting just the input and output,

$$\Psi'_{io} = (S - \mathrm{i}TKT^*)\Psi_{io}.$$

Formula (4) applied on $A_{\mathrm{diss}} = -(S - \mathrm{i}TKT^*)$ and $B_{\mathrm{diss}} = I$ leads to the scattering matrix

$$\mathcal{S}_{\mathrm{diss}}(E) = -I + \frac{2}{1 + \operatorname{Tr}(M(E)) + \det(M(E))}\operatorname{adj}(M(E))$$

for

$$M(E) = TDT^* - \frac{1}{\mathrm{i}\sqrt{E}}S, \tag{19}$$

where $D = \operatorname{diag}(k_3/\sqrt{E}, \cdots, k_n/\sqrt{E})$ for k_j defined in (9). The symbol $\operatorname{adj}(M(E))$ denotes the adjoint of $M(E)$. In particular, the transmission amplitude is

$$\mathcal{T}(E) = \frac{-2[M(E)]_{21}}{1 + \operatorname{Tr}(M(E)) + \det(M(E))}. \tag{20}$$

Once we have derived formula (20), our next goal is to find requirements on S and T to satisfy conditions (6)–(8). We may assume without loss of generality that the controlling lines are given numbers $3, \dots, q$ for a certain $q \in [3, \dots, n]$, and edges nos. $q+1, \dots, n$ represent drains. We write the matrix $T \in \mathbb{C}^{2,n-2}$ accordingly in the way

$$T = \begin{pmatrix} v_1 & w_1 \\ v_2 & w_2 \end{pmatrix} \tag{21}$$

with $v_1, v_2 \in \mathbb{C}^{1,q-2}$ and $w_1, w_2 \in \mathbb{C}^{1,n-q}$, where $q-2$ is the number of controllers. We start from condition (8), i.e., $\lim_{E\to\infty} \mathcal{T}(E) = 0$. Since $\lim_{E\to\infty} D = I$, equation (19) gives

$$\lim_{E\to\infty} M(E) = TT^* = \begin{pmatrix} \|v_1\|^2 + \|w_2\|^2 & v_1 v_2^* + w_1 w_2^* \\ v_2 v_1^* + w_2 w_1^* & \|v_2\|^2 + \|w_2\|^2 \end{pmatrix}.$$

Matrix TT^* is Hermitian and positive-definite; thus

$$\lim_{E\to\infty} \det(M(E)) > 0 \quad \text{and} \quad \lim_{E\to\infty} \operatorname{Tr}(M(E)) > 0.$$

Consequently, with regard to equation (20), we have $\lim_{E\to\infty} \mathcal{T}(E) = 0$ if and only if

$$v_2 v_1^* + w_2 w_1^* = 0. \tag{22}$$

Now we proceed to condition (6). If the controlling lines support a potential V, the matrix D for energies $E \in (0, V)$ equals

$$D = \begin{pmatrix} \mathrm{i}\sqrt{\frac{V}{E} - 1} \cdot I^{(q-2)} & 0 \\ 0 & I^{(n-q)} \end{pmatrix}.$$

Formula (20) gives the transmission amplitude for $E \in (0, V)$ in the form

$$\mathcal{T}(E) = \frac{-2\Big(w_2 w_1^* + \mathrm{i}\sqrt{\frac{V}{E} - 1} \cdot v_2 v_1^* + \frac{\mathrm{i}}{\sqrt{E}} s_{21}\Big)}{a + b \cdot \mathrm{i}\sqrt{\frac{V}{E} - 1} + c \cdot \Big(\frac{V}{E} - 1\Big) + d \cdot \sqrt{\frac{V}{E^2} - \frac{1}{E}} + f \cdot \frac{\mathrm{i}}{\sqrt{E}} + \frac{g}{E}}$$

with

$$
\begin{aligned}
a &= 1 + \|w_1\|^2 + \|w_2\|^2 + \|w_1\|^2 \cdot \|w_2\|^2 - |w_2 w_1^*|^2, \\
b &= \|v_1\|^2 + \|v_2\|^2 + \|w_1\|^2 \cdot \|v_2\|^2 \\
&\quad + \|v_1\|^2 \cdot \|w_2\|^2 - v_2 v_1^* w_1 w_2^* - w_2 w_1^* v_1 v_2^*, \\
c &= -\|v_1\|^2 \cdot \|v_2\|^2 + |v_2 v_1^*|^2, \\
d &= -s_{11}\|v_2\|^2 - s_{22}\|v_1\|^2 + 2\Re(s_{21} v_1 v_2^*), \\
f &= \operatorname{Tr}(S) + s_{11}\|w_2\|^2 + s_{22}\|w_1\|^2 - 2\Re(s_{21} w_1 w_2^*), \\
g &= -\det(S).
\end{aligned}
$$

Expressions for b and $\mathcal{T}(E)$ can be simplified using equation (22),

$$
b = \|v_1\|^2 + \|v_2\|^2 + \|w_1\|^2 \cdot \|v_2\|^2 + \|v_1\|^2 \cdot \|w_2\|^2 + 2|v_2 v_1^*|^2,
$$

$$
\mathcal{T}(E) = \frac{2\Big(1 - \mathrm{i}\sqrt{\frac{V}{E} - 1}\Big) v_2 v_1^* - 2\frac{\mathrm{i}}{\sqrt{E}} s_{21}}{a + b \cdot \mathrm{i}\sqrt{\frac{V}{E} - 1} + c \cdot \Big(\frac{V}{E} - 1\Big) + d \cdot \sqrt{\frac{V}{E^2} - \frac{1}{E}} + f \cdot \frac{\mathrm{i}}{\sqrt{E}} + \frac{g}{E}}. \tag{23}
$$

Lemma 6.1. *Condition* (6) *implies* $v_2 v_1^* \neq 0$.

Proof. We prove the statement by showing that $v_2 v_1^* = 0$ contradicts (6). Equation $v_2 v_1^* = 0$ implies $w_2 w_1^* = 0$ due to equation (22). Therefore, for all $E \in (0, V)$,

$$
\mathcal{T}(E) = \frac{-2\frac{\mathrm{i}}{\sqrt{E}} s_{21}}{a + b \cdot \mathrm{i}\sqrt{\frac{V}{E} - 1} + c \cdot \Big(\frac{V}{E} - 1\Big) + d \cdot \sqrt{\frac{V}{E^2} - \frac{1}{E}} + f \cdot \frac{\mathrm{i}}{\sqrt{E}} + \frac{g}{E}}.
$$

Note that $a = 1 + \|w_1\|^2 + \|w_2\|^2 + \|w_1\|^2 \cdot \|w_2\|^2 \neq 0$. Therefore, function $|\mathcal{T}(E)|^2$ is either identically zero (for $s_{21} = 0$), or non-constant. In both cases condition (6) is violated. □

With regard to Lemma 6.1, we may assume

$$v_1 \neq 0, \quad v_2 \neq 0, \quad w_1 \neq 0, \quad w_2 \neq 0. \tag{24}$$

We see from the structure of the numerator and the denominator in equation (23) that satisfying condition (6) for all $E < V$ requires

$$c = 0, \quad d = 0, \quad g = 0, \quad |a| = |b|, \quad \frac{f}{b} = \frac{s_{21}}{v_2 v_1^*} \tag{25}$$

(where $|a| = |b|$ is equivalent to $a = b$, because both a and b are obviously positive). Indeed, when (25) hold true, we have

$$\mathcal{T}(E) = \frac{2v_2 v_1^*}{a} \cdot \frac{1 - \mathrm{i}\Big(\sqrt{\frac{V}{E} - 1} + \frac{s_{21}}{v_2 v_1^*} \cdot \frac{1}{\sqrt{E}}\Big)}{1 + \mathrm{i}\Big(\sqrt{\frac{V}{E} - 1} + \frac{s_{21}}{v_2 v_1^*} \cdot \frac{1}{\sqrt{E}}\Big)} \quad \text{for all } E \in (0, V);$$

hence

$$\mathcal{P}(E) = \Big(\frac{2|v_2 v_1^*|}{1 + \|w_1\|^2 + \|w_2\|^2 + \|w_1\|^2 \cdot \|w_2\|^2 - |v_2 v_1^*|^2}\Big)^2 = \text{const.} \tag{26}$$

for $E \in (0, V)$. Now we will examine the system of conditions (25). We start from equation $c = 0$, which is equivalent to

$$\|v_1\|^2 \cdot \|v_2\|^2 = |v_2 v_1^*|^2. \tag{27}$$

Due to Cauchy–Schwarz inequality, v_1, v_2 are linearly dependent vectors. Furthermore, equation (22) together with (27) implies

$$|w_2 w_1^*| = \|v_1\| \cdot \|v_2\|. \tag{28}$$

Let us proceed to another condition from (25), $|a| = |b|$. By virtue of equation (28), we can rewrite $|a| = |b|$ in the form

$$\begin{aligned} &1 + \|w_1\|^2 + \|w_2\|^2 + \|w_1\|^2 \cdot \|w_2\|^2 - \|v_1\|^2 \|v_2\|^2 \\ &\quad = \|v_1\|^2 + \|v_2\|^2 + \|w_1\|^2 \cdot \|v_2\|^2 + \|v_1\|^2 \cdot \|w_2\|^2 + 2\|v_1\|^2 \|v_2\|^2, \end{aligned}$$

which is equivalent to

$$(1 + \|w_1\|^2 - \|v_1\|^2)(1 + \|w_2\|^2 - \|v_2\|^2) = 4\|v_1\|^2 \|v_2\|^2. \tag{29}$$

We continue to condition $g = 0$, which gives

$$|s_{21}| = \sqrt{|s_{11} s_{22}|}. \tag{30}$$

We proceed in (25) to condition $f/b = s_{21}/(v_2 v_1^*)$. This condition implies in particular that $s_{21}/(v_2 v_1^*) \in \mathbb{R}$. If we combine this fact with equations (30) and (27), we find

$$s_{21} = \pm\sqrt{|s_{11}s_{22}|}\frac{v_2 v_1^*}{\|v_2\| \cdot \|v_1\|}. \tag{31}$$

Result (31) gives $s_{21}/v_2 v_1^* = \pm\sqrt{|s_{11}s_{22}|}/(\|v_1\|\cdot\|v_2\|)$. Therefore, condition $f/b = s_{21}/(v_2 v_1^*)$ is equivalent to $\pm f \cdot \|v_1\| \cdot \|v_2| = b\sqrt{|s_{11}s_{22}|}$, i.e.,

$$\begin{aligned} &\pm (\mathrm{Tr}(S) + s_{11}\|w_2\|^2 + s_{22}\|w_1\|^2 - 2\Re(s_{21}w_1 w_2^*)) \cdot \|v_1\| \cdot \|v_2\| \\ &\quad = (\|v_1\|^2 + \|v_2\|^2 + \|w_1\|^2 \cdot \|v_2\|^2 + \|v_1\|^2 \cdot \|w_2\|^2 + 2|w_2 w_1^*|^2)\sqrt{|s_{11}s_{22}|}. \end{aligned} \tag{32}$$

We use equations (22) and (31) to rewrite $s_{21}w_1 w_2^* = -s_{21}v_1 v_2^* = \mp\sqrt{|s_{11}s_{22}|} \cdot \|v_1\| \cdot \|v_2\|$. Similarly, we rewrite term $|w_2 w_1^*|^2$ on the right hand side of (32) using equation (28). As a result of these operations certain terms in equation (32) cancel, and we get

$$\begin{aligned} &\pm (s_{11}(1 + \|w_2\|^2) + s_{22}(1 + \|w_1\|^2)) \cdot \|v_1\| \cdot \|v_2\| \\ &\quad = (\|v_1\|^2(1 + \|w_2\|^2) + \|v_2\|^2(1 + \|w_1\|^2)) \cdot \sqrt{|s_{11}s_{22}|}. \end{aligned} \tag{33}$$

The last condition among (25) to be examined is $d = 0$. We substitute for s_{21} from equation (31) into the expression for d; then $d = 0$ is equivalent to

$$- s_{11}\|v_2\|^2 - s_{22}\|v_1\|^2 \pm 2\sqrt{|s_{11}s_{22}|} \cdot \|v_1\| \cdot \|v_2\| = 0. \tag{34}$$

Now we will examine equation (34) using equation (33). We distinguish three cases.

Case $s_{11}s_{22} = 0$. In this case equation (34) together with (24) implies $s_{11} = s_{22} = 0$. Hence $S = 0$ due to equation (30). Consequently, condition (33) is always satisfied for $s_{11}s_{22} = 0$.

Case $s_{11}s_{22} > 0$. Equation (34) is equivalent to

$$(\sqrt{|s_{11}|}\|v_2\| \mp \mathrm{sgn}(s_{11})\sqrt{|s_{22}|}\|v_1\|)^2 = 0,$$

hence, due to (24),

$$\pm\,\mathrm{sgn}(s_{11}) = 1 \quad \text{and} \quad \sqrt{\frac{s_{11}}{s_{22}}} = \frac{\|v_1\|}{\|v_2\|}. \tag{35}$$

Similarly, equation (33) is equivalent to

$$\begin{aligned} &\pm \operatorname{sgn}(s_{11})(|s_{11}|(1+\|w_2\|^2)+|s_{22}|(1+\|w_1\|^2))\|v_1\|\cdot\|v_2\| \\ &\quad = (\|v_1\|^2(1+\|w_2\|^2)+\|v_2\|^2(1+\|w_1\|^2))\cdot\sqrt{|s_{11}s_{22}|}. \end{aligned} \tag{36}$$

When we substitute relations (35) into equation (36), we get an identity. In other words, condition (33) is always satisfied for $s_{11}s_{22}>0$.

Case $s_{11}s_{22}<0$. Equation (34) is equivalent to

$$\begin{aligned} &(\sqrt{|s_{11}|}\|v_2\|+(\sqrt{2}\mp\operatorname{sgn}(s_{11}))\sqrt{|s_{22}|}\|v_1\|) \\ &\quad (\sqrt{|s_{11}|}\|v_2\|-(\sqrt{2}\pm\operatorname{sgn}(s_{11}))\sqrt{|s_{22}|}\|v_1\|)=0 \end{aligned}$$

hence

$$\sqrt{\frac{|s_{11}|}{|s_{22}|}}=(\sqrt{2}\pm\operatorname{sgn}(s_{11}))\frac{\|v_1\|}{\|v_2\|}. \tag{37}$$

Equation (33) is equivalent to

$$\begin{aligned} &\pm \operatorname{sgn}(s_{11})(|s_{11}|(1+\|w_2\|^2)-|s_{22}|(1+\|w_1\|^2))\|v_1\|\|v_2\| \\ &\quad = (\|v_1\|^2(1+\|w_2\|^2)+\|v_2\|^2(1+\|w_1\|^2))\cdot\sqrt{|s_{11}s_{22}|}. \end{aligned} \tag{38}$$

When we use relation (37) in equation (38), we get the equation

$$\pm\operatorname{sgn}(s_{11})\sqrt{2}(\|v_1\|^2(1+\|w_2\|^2)-\|v_2\|^2(1+\|w_1\|^2))=0.$$

Hence

$$\frac{\|v_1\|^2}{1+\|w_1\|^2}=\frac{\|v_2\|^2}{1+\|w_2\|^2}. \tag{39}$$

We apply equivalence (39) in equation (29) and get

$$(1+\|w_1\|^2-\|v_1\|^2)^2=4\|v_1\|^4;$$

hence

$$\|w_1\|^2=3\|v_1\|^2-1,\quad \|w_2\|^2=3\|v_2\|^2-1. \tag{40}$$

At this stage we have finished the study of conditions (8) and (6). It remains to check condition (7). It is straightforward to derive the formula

$$\mathcal{P}(E)=\Big(\frac{2|v_2v_1^*|}{(1+\|w_1\|^2)(1+\|w_2\|^2)-|w_2w_1^*|^2}\Big)^2 \frac{\Big(1-\sqrt{1-\frac{V}{E}}\Big)^2+\frac{|s_{21}|^2}{\|v_1\|^2\|v_2\|^2}\cdot\frac{1}{E}}{\Big(1+\sqrt{1-\frac{V}{E}}\Big)^2+\frac{|s_{21}|^2}{\|v_1\|^2\|v_2\|^2}\cdot\frac{1}{E}}$$

for all $E>V$. It is easy to verify that $\lim_{E\searrow V}\mathcal{P}'(E)=-\infty$.

Let us summarize the results in the following theorem.

Theorem 6.2. *Consider a star graph with a vertex coupling described by boundary conditions* (17)*. The transmission probability in the input-output channel satisfies conditions* (6)–(8) *if and only if vectors v_1 and v_2 in matrix T* (21) *are linearly dependent, vectors w_1, w_2 obey requirements* (22) *and* (29)*, and one of the following three cases holds true:*

- $S = 0$;
- $s_{11}s_{22} > 0$ *and*
$$\sqrt{\frac{s_{11}}{s_{22}}} = \frac{\|v_1\|}{\|v_2\|}, \qquad s_{21} = \operatorname{sgn}(s_{11}) \cdot \sqrt{s_{11}s_{22}} \frac{v_2 v_1^*}{\|v_1\| \cdot \|v_2\|};$$
- $s_{11}s_{22} < 0$, *condition* (40) *is satisfied, and matrix S obeys conditions*
$$\sqrt{\frac{|s_{11}|}{|s_{22}|}} = (\sqrt{2} \pm \operatorname{sgn}(s_{11}) \cdot 1) \frac{\|v_1\|}{\|v_2\|}$$
and
$$s_{21} = \pm\sqrt{|s_{11}s_{22}|} \frac{v_2 v_1^*}{\|v_1\| \cdot \|v_2\|}.$$

An example of transmission probability function for S, T chosen according to Theorem 6.2 is shown in Figure 3.

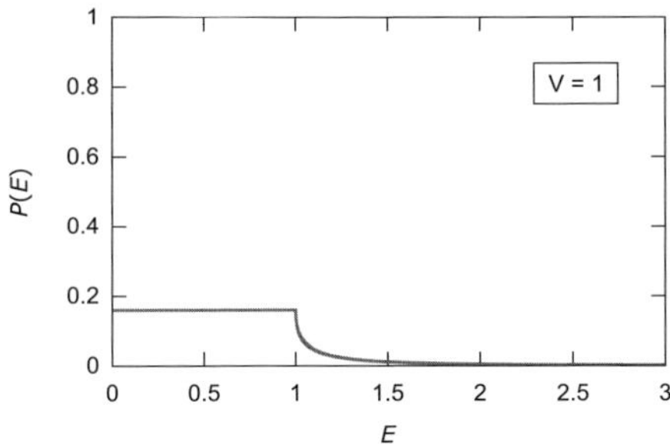

Figure 3. An example of a transmission probability featuring a flat passband. The function is obtained for the choice $\|v_1\|^2 = 5/8$, $v_2 = v_1$, $\|w_1\|^2 = \|w_2\|^2 = 7/8$, $|w_2 w_1^*|^2 = 5/8$, $S = 0$ and for the controlling potential $V = 1$.

Theorem 6.2 describes the structure of matrices S, T in boundary conditions (17) for which the star graph works as a band-pass filter with a flat passband. In the rest of the section we will find the maximal possible value of $|\mathcal{T}(E)|$ in the "flat band" interval $(0, V)$, and we will characterize the corresponding matrices S, T. With regard to equation (26), we have to find the maximum of the quantity

$$\Big(\frac{2|v_2 v_1^*|}{1 + \|w_1\|^2 + \|w_2\|^2 + \|w_1\|^2 \cdot \|w_2\|^2 - |w_2 w_1^*|^2}\Big)^2 \tag{41}$$

under conditions given in Theorem 6.2. Note that the expression (41) is independent of S, and the entries of S can be calculated after T is fixed. Therefore, we will at first find the maximum of expression (41) under conditions (22) and (29), whereas matrix S will be calculated later. We denote

$$\|w_1\|^2 = x, \quad \|w_2\|^2 = y, \quad \|v_1\|^2 = z,$$

and

$$|v_2 v_1^*|^2 = xyu \quad \text{for a certain } u \in (0, 1],$$

which is possible due to $|v_2 v_1^*|^2 = |w_2 w_1^*|^2 \le \|w_1\|^2 \cdot \|w_2\|^2$. We express $\|v_2\|^2$ using equations (22) and (27),

$$\|v_2\|^2 = \frac{\|v_1\|^2 \cdot \|v_2\|^2}{\|v_1\|^2} = \frac{|v_2 v_1^*|^2}{\|v_1\|^2} = \frac{|w_2 w_1^*|^2}{\|v_1\|^2} = \frac{xyu}{z}.$$

We shall find the maximum of the function

$$F(x, y, z, u) = \Big(\frac{2\sqrt{xyu}}{1 + x + y + xy - xyu}\Big)^2$$

(cf. (41)) under condition (29), i.e.,

$$(1 + x - z)\Big(1 + y - \frac{xyu}{z}\Big) = 4xyu.$$

We proceed in a standard way. We introduce the Langrage function

$$\begin{aligned}&\mathcal{L}(x, y, z, u, \lambda)\\ &\quad = \frac{2\sqrt{xyu}}{1 + x + y + xy - xyu} - \lambda \cdot \Big[(1 + x - z)\Big(1 + y - \frac{xyu}{z}\Big) - 4xyu\Big]\end{aligned}$$

and solve the system

$$\frac{\partial \mathcal{L}}{\partial x} = 0, \quad \frac{\partial \mathcal{L}}{\partial y} = 0, \quad \frac{\partial \mathcal{L}}{\partial z} = 0, \quad \frac{\partial \mathcal{L}}{\partial u} = 0. \tag{42}$$

It turns out that (42) has no solution. Therefore, we shall search for the maximum of F at the boundary of its domain, i.e., for $u = 1$. If we fix $u = 1$ and solve the system $\partial\mathcal{L}/\partial x = \partial\mathcal{L}/\partial y = \partial\mathcal{L}/\partial z = 0$, we obtain

$$x = y = z = \frac{1}{2}.$$

Hence we find the sought maximum of the function F,

$$F\left(\frac{1}{2}, \frac{1}{2}, \frac{1}{2}, 1\right) = \frac{1}{4}.$$

Note that $u = 1$ implies $|w_2 w_1^*| = \|w_1\| \cdot \|w_2\|$, i.e., w_1, w_2 are linearly dependent.

Theorem 6.3. *The maximal transmission probability of a band-pass filter with flat passband, constructed upon a vertex with boundary conditions* (17), *is* $1/4$. *It is obtained for*

$$T = \begin{pmatrix} v & w \\ \alpha v & -\alpha w \end{pmatrix} \quad \textit{for } \|v\| = \|w\| = \frac{1}{\sqrt{2}}, \ |\alpha| = 1, \tag{43}$$

and

$$S = s\begin{pmatrix} 1 & \bar{\alpha} \\ \alpha & 1 \end{pmatrix} \quad \textit{or} \quad S = s\begin{pmatrix} 1 \pm \sqrt{2} & \bar{\alpha} \\ \alpha & 1 \mp \sqrt{2} \end{pmatrix} \quad \textit{for } s \in \mathbb{R}. \tag{44}$$

Proof. According to calculations above, the maximal transmission probability is $1/4$, and this value is attained for $\|v_1\| = \|v_2\| = \|w_1\| = \|w_2\| = 1/\sqrt{2}$. Vectors v_1, v_2 are linearly dependent due to Theorem 6.2; hence $v_2 = \alpha \cdot v_1$ for an α satisfying $|\alpha| = 1$. Equation (22) implies $w_2 = -\alpha \cdot w_1$. Furthermore, equations listed in Theorem 6.2 imply that either $S = 0$, or the entries of S satisfy

$$s_{11} = s_{22} = s, \quad s_{21} = s \cdot \alpha$$

for a certain $s \neq 0$, or

$$s_{11} = s \cdot (\sqrt{2} \pm \operatorname{sgn}(s) \cdot 1)$$

$$s_{22} = -s \cdot (\sqrt{2} \mp \operatorname{sgn}(s) \cdot 1)$$

$$s_{21} = \pm|s| \cdot \alpha$$

for a certain $s \neq 0$. It is easy to check that all the cases above are fully covered by formulas (44). □

Figure 4 shows two examples of the transmission probability functions obtained for S, T obeying conditions from Theorem 6.3.

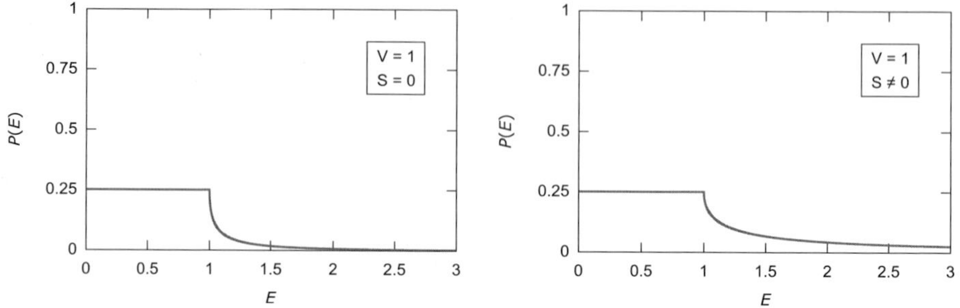

Figure 4. The maximal transmission probability in the passband for the controlling potential $V = 1$. The graphs display the function $\mathcal{P}(E)$ for T given by equation (43) and $S = \left(\begin{smallmatrix} 0 & 0 \\ 0 & 0 \end{smallmatrix}\right)$ (left) and $S = 1/2\left(\begin{smallmatrix} 1 & 1 \\ 1 & 1 \end{smallmatrix}\right)$ (right).

Remark 6.4. Matrix T given by equation (43) together with $S = 0$ generalizes an earlier result. In [11], a graph consisting of the input, output, one controlling edge and one drain, coupled in a vertex by scale invariant boundary conditions with

$$T = \frac{1}{\sqrt{2}}\begin{pmatrix} 1 & 1 \\ 1 & -1 \end{pmatrix},$$

was examined. It was demonstrated that the transmission probability is constant in the interval $(0, V)$ and quickly decreases towards zero as E exceeds the controlling potential V. Theorem 6.3 implies that the flat-band behaviour persists even if the scale invariance of the coupling is broken. This is a quite surprising fact.

7 Case $r \geq 3$

The ideas demonstrated in previous sections can be used for treating vertex couplings with $r = \operatorname{rank}(B) \geq 3$ as well. Section 5 implies that if the sought band-pass filter with flat passband exists, then the columns of B that correspond to the input and output line need to be linearly independent. This allows us to express the boundary conditions in the vertex in the ST-form as follows,

$$\begin{pmatrix} I^{(2)} & 0 & T_1 \\ 0 & I^{(r-2)} & T_2 \\ 0 & 0 & 0 \end{pmatrix}\begin{pmatrix} \Psi'_{io} \\ \Psi'_{\mathrm{cd}} \end{pmatrix} = \begin{pmatrix} S_1 & S_2 & 0 \\ S_2^* & S_4 & 0 \\ -T_1^* & -T_2^* & I^{(n-r)} \end{pmatrix}\begin{pmatrix} \Psi_{io} \\ \Psi_{\mathrm{cd}} \end{pmatrix}, \tag{45}$$

where Ψ_{io}, Ψ'_{io} are the boundary values at input and output (cf. (12)), $\Psi_{\rm cd}$, $\Psi'_{\rm cd}$ are the boundary values at controlling lines and drains (controlling edges and drains not being distinguished now), $I^{(2)}$, $I^{(r-2)}$, $I^{(n-r)}$ are identity matrices of given orders, $T = \begin{pmatrix} T_1 \\ T_2 \end{pmatrix} \in \mathbb{C}^{r,n-r}$ and $\begin{pmatrix} S_1 & S_2 \\ S_2^* & S_4 \end{pmatrix}$ is a Hermitian $r \times r$ matrix.

Relation (10) implies

$$\Psi'_{\rm cd} = \mathrm{i} \begin{pmatrix} K_2 & 0 \\ 0 & K_3 \end{pmatrix} \Psi_{\rm cd}, \tag{46}$$

where $K_2 = \mathrm{diag}(k_3, \dots, k_r)$ and $K_3 = \mathrm{diag}(k_{r+1}, \dots, k_n)$. Elimination of $\Psi_{\rm cd}$ and $\Psi'_{\rm cd}$ from system (45) using equation (46) leads to the conditions (11) with $B_{\rm diss} = I$ and

$$\begin{aligned} A_{\rm diss} &= \mathrm{i} T_1 K_3 T_1^* - S_1 \\ &\quad - (S_2 - \mathrm{i} T_1 K_3 T_2^*)(\mathrm{i} K_2 + \mathrm{i} T_2 K_3 T_2^* - S_4)^{-1}(S_2^* - \mathrm{i} T_2 K_3 T_1^*). \end{aligned}$$

Formula (4) gives the dissipative scattering matrix

$$\mathcal{S}_{\rm diss}(E) = -I + \frac{2}{1 + \mathrm{Tr}(M(E)) + \det(M(E))} \mathrm{adj}(M(E)), \tag{47}$$

where the matrix $M(E)$ is given as

$$\begin{aligned} M(E) = T_1 D_3 T_1^* + \frac{\mathrm{i}}{\sqrt{E}} S_1 - \Big(\frac{\mathrm{i}}{\sqrt{E}} S_2 + T_1 D_3 T_2^*\Big) \\ \Big(D_2 + T_2 D_3 T_2^* + \frac{\mathrm{i}}{\sqrt{E}} S_4\Big)^{-1} \\ \Big(\frac{\mathrm{i}}{\sqrt{E}} S_2^* + T_2 D_3 T_1^*\Big) \end{aligned}$$

with $D_2 = \mathrm{diag}\,(k_3/\sqrt{E}, \dots, k_r/\sqrt{E})$ and $D_3 = \mathrm{diag}\,(k_{r+1}/\sqrt{E}, \dots, k_n/\sqrt{E})$. Consequently, the transmission amplitude is

$$\mathcal{T}(E) = \frac{-2[M(E)]_{21}}{1 + \mathrm{Tr}(M(E)) + \det(M(E))}. \tag{48}$$

We require $\lim_{E\to\infty} \mathcal{T}(E) = 0$ according to (8). Since we have $\lim_{E\to\infty} D_2 = I$ and $\lim_{E\to\infty} D_3 = I$, we get

$$\lim_{E\to\infty} M(E) = T_1 T_1^* - T_1 T_2^* \left(I + T_2 T_2^*\right)^{-1} T_2 T_1^* = T_1 \left(I + T_2^* T_2\right)^{-1} T_1^*.$$

The matrix on the right hand side is Hermitian and positive-definite. The denominator of (47) thus tends to a positive number greater than 1 as $E \to \infty$. Therefore, equation (48) gives the equivalence

$$\lim_{E\to\infty} \mathcal{T}(E) = 0 \iff T_1 (I + T_2^* T_2)^{-1} T_1^* \text{ is diagonal.}$$

To sum up, a quantum star graph with the vertex coupling given by boundary conditions (45) can work as a band-pass filter with flat passband only if $T_1(I+T_2^*T_2)^{-1}T_1^*$ is a diagonal matrix.

Analyzing condition (6) needs to distinguish controllers and drains in both sets $\{3,\dots,r\}$ and $\{r+1,\dots,n\}$, which would make the problem more intricate. Therefore, the case $r \geq 3$ in general will not be addressed in this paper; nevertheless, the method presented in Sections 4–6 is in principle applicable.

Remark 7.1. Although we focused on graphs working as spectral band-pass filters with flat passbands, the same approach can be used more generally. Taking advantage of the ST-form of boundary conditions, one can explore and design quantum graphs with various other special transmission characteristics, such as filters having a sharp peak in $\mathcal{P}(E)$ at a certain energy.

Acknowledgments. The author is grateful to Prof. Taksu Cheon for inspirative discussions on the topic, and to Kochi University of Technology for hospitality during the writing of this paper.

The work was supported by the Czech Science Foundation (GAČR) within the project 14-06818S and by Youth JINR Grant No. 15-302-08.

References

[1] S. Albeverio, F. Gesztesy, R. Høegh-Krohn, and H. Holden, *Solvable models in quantum mechanics.* Second edition. AMS Chelsea Publishing, Providence, R.I., 2005. With an Appendix by P. Exner. ISBN 0-8218-3624-2/hbk MR 2105735 Zbl 1078.81003

[2] T. Cheon, P. Exner, and O. Turek, Spectral filtering in quantum Y-junction. *J. Phys. Soc. Jpn.* **78** (2009), 124004, 7 pp.

[3] T. Cheon, P. Exner, and O. Turek, Approximation of a general singular vertex coupling in quantum graphs. *Ann. Physics* **325** (2010), no. 3, 548–578. MR 2592572 Zbl 1192.81159

[4] T. Cheon, P. Exner, and O. Turek, Tripartite connection condition for a quantum graph vertex. *Phys. Lett. A* **375** (2010), no. 2, 113–118. MR 2737903 Zbl 1241.81056

[5] T. Cheon and O. Turek, Fulop–Tsutsui interactions on quantum graphs. *Phys. Lett. A* **374** (2010), no. 41, 4212–4221. MR 2684869 Zbl 1238.81124

[6] T. Fülöp and I. Tsutsui, A free particle on a circle with point interaction. *Phys. Lett. A* **264** (2000), no. 5, 366–374. MR 1739453 Zbl 0948.81537

[7] P. Exner, J. P. Keating, P. Kuchment, T. Sunada, and A. Teplyaev (eds.), *Analysis on graphs and applications.* Papers from the program held in Cambridge, January 8–June 29, 2007. Proceedings of Symposia in Pure Mathematics, 77. American Mathematical Society, Providence, R.I., 2008. ISBN 978-0-8218-4471-7 MR 2459860 Zbl 1143.05002

[8] V. Kostrykin and R. Schrader, Kirchhoff's rule for quantum wires. *J. Phys. A* **32** (1999), no. 4, 595–630. MR 1671833 Zbl 0928.34066

[9] K. Naimark and M. Solomyak, Eigenvalue estimates for the weighted Laplacian on metric trees. *Proc. London Math. Soc.* (3) **80** (2000), no. 3, 690–724. MR 1744781 Zbl 1046.34092

[10] A. V. Sobolev and M. Solomyak, Schrödinger operators on homogeneous metric trees: spectrum in gaps. *Rev. Math. Phys.* **14** (2002), 421–467. MR 1912093 Zbl 1038.81023

[11] O. Turek and T. Cheon, Threshold resonance and controlled filtering in quantum star graphs. *EPL (Europhysics Letters)* **98** (2012), no. 5, 50005, 5 pp.

[12] O. Turek and T. Cheon, Potential-controlled filtering in quantum star graphs. *Ann. Physics* **330** (2013), 104–141. MR 3011821 Zbl 1266.81098

[13] O. Turek and T. Cheon, Quantum graph as a quantum spectral filter. *J. Math. Phys.* **54** (2013), no. 3, 032104, 17 pp. MR 3059434 Zbl 1281.81041

Comments on the Chernoff $\sqrt{n}$-lemma

Valentin A. Zagrebnov

Dedicated to Pavel Exner in occasion of his 70th birthday

1 Introduction: $\sqrt{n}$-lemma

The Chernoff $\sqrt{n}$-lemma is a key point in the theory of semigroup approximations proposed in [3]. For the reader's convenience we recall this lemma below.

Lemma 1.1. *Let C be a contraction on a Banach space $\mathfrak{X}$. Then $\{e^{t(C-\mathbb{1})}\}_{t\geq 0}$ is a norm-continuous contraction semigroup on $\mathfrak{X}$ and one has the estimate*

$$\|(C^n - e^{n(C-\mathbb{1})})x\| \leq \sqrt{n}\, \|(C-\mathbb{1})x\|, \tag{1}$$

for all $x \in \mathfrak{X}$ and $n \in \mathbb{N}$.

Proof. To prove the inequality (1) we use the representation

$$C^n - e^{n(C-\mathbb{1})} = e^{-n} \sum_{m=0}^{\infty} \frac{n^m}{m!}(C^n - C^m). \tag{2}$$

To proceed we insert

$$\|(C^n - C^m)x\| \leq \|(C^{|n-m|} - \mathbb{1})x\| \leq |m-n|\|(C-\mathbb{1})x\|,$$

into (2) to obtain by the Cauchy–Schwarz inequality the estimate

$$\begin{aligned} \|(C^n - e^{n(C-\mathbb{1})})x\| &\leq \|(C-\mathbb{1})x\|\, e^{-n} \sum_{m=0}^{\infty} \frac{n^m}{m!}|m-n| \\ &\leq \Big\{\sum_{m=0}^{\infty} e^{-n}\, \frac{n^m}{m!}|m-n|^2\Big\}^{1/2} \|(C-\mathbb{1})x\|, \quad x \in \mathfrak{X}. \end{aligned} \tag{3}$$

Note that the sum in the right-hand side of (3) can be calculated explicitly. This gives the value n, which yields (1). □

The aim of the present note is to revise the Chernoff $\sqrt{n}$-lemma in two directions. First, we improve the $\sqrt{n}$-estimate (1) for contractions. Then we apply this new estimate to the proof of the Trotter product formula in the *strong* operator topology (Section 2).

Second, we use the idea of Section 2 to lift these results in Section 3 to the *operator-norm* estimates for a special class of contractions: the *quasi-sectorial* contractions.

2 Revised $\sqrt{n}$-lemma and Lie–Trotter product formula

We start by a technical lemma. It is a *revised* version of the Chernoff $\sqrt{n}$-lemma 1.1. Our estimate (4) in $\sqrt[3]{n}$-lemma 2.1 is better than (1). The scheme of the proof will be useful later (Section 3), when we use it for proving the convergence of Lie–Trotter product formula in the *operator-norm* topology.

Lemma 2.1. *Let C be a contraction on a Banach space $\mathfrak{X}$. Then $\{e^{t(C-\mathbb{1})}\}_{t\geq 0}$ is a norm-continuous contraction semigroup on $\mathfrak{X}$ and one has the estimate*

$$\|(C^n - e^{n(C-\mathbb{1})})x\| \leq \Big[\frac{1}{n^{2\delta}} + n^{\delta+1/2}\Big]\|(\mathbb{1} - C)x\|, \quad n \in \mathbb{N}. \tag{4}$$

for all $x \in \mathfrak{X}$ and $\delta \in \mathbb{R}$.

Proof. Since the operator C is bounded and $\|C\| \leq 1$, $(\mathbb{1} - C)$ is a generator of the norm-continuous semigroup, which is also a contraction:

$$\|e^{\,t(C-\mathbb{1})}\| \leq e^{-t}\Big\|\sum_{m=0}^{\infty}\frac{t^m}{m!}C^m\Big\| \leq 1, \quad t \geq 0.$$

To estimate (4) we use the representation

$$C^n - e^{n(C-\mathbb{1})} = e^{-n}\sum_{m=0}^{\infty}\frac{n^m}{m!}(C^n - C^m). \tag{5}$$

Let $\epsilon_n := n^{\delta+1/2}$, $n \in \mathbb{N}$. We split the sum (5) into two parts: the *central* part for $|m-n| \leq \epsilon_n$ and *tails* for $|m-n| > \epsilon_n$.

To estimate the *tails* we use the *Tchebychev inequality*. Let X_n be a *Poisson random variable* of the parameter n, i.e., the probability $\mathbb{P}\{X_n = m\} = {n^m e^{-n}}/{m!}$.

One obtains for the expectation $\mathbb{E}(X_n) = n$ and for the variance $\operatorname{Var}(X_n) = n$. Then by the Tchebychev inequality,

$$\mathbb{P}\{|X_n - \mathbb{E}(X_n)| > \epsilon\} \leq \frac{\operatorname{Var}(X_n)}{\epsilon^2}, \quad \text{for any } \epsilon > 0.$$

Now to estimate (5) we note that

$$\begin{aligned}\|(C^n - C^m)x\| &= \|C^{n-k}(C^k - C^{m-n+k})x\| \\ &\leq |m-n| \|C^{n-k}(\mathbb{1} - C)x\|, \quad k = 0, 1, \dots, n.\end{aligned}$$

Put in this inequality $k = [\epsilon_n]$, here $[\cdot]$ denotes the integer part. Then by $\|C\| \leq 1$ and by the Tchebychev inequality we obtain the estimate for *tails*:

$$\begin{aligned}e^{-n} \sum_{|m-n|>\epsilon_n} \frac{n^m}{m!} \|(C^n - C^m)x\| &\leq \|(\mathbb{1} - C)x\| \, e^{-n} \sum_{|m-n|>\epsilon_n} \frac{n^m}{m!} |m-n| \\ &\leq \frac{n}{\epsilon_n^2} \|(\mathbb{1} - C)x\| \\ &= \frac{1}{n^{2\delta}} \|(\mathbb{1} - C)x\|.\end{aligned} \tag{6}$$

To estimate the *central* part of the sum (5), when $|m-n| \leq \epsilon_n$, note that

$$\begin{aligned}\|(C^n - C^m)x\| &\leq |m-n| \, \|C^{n-[\epsilon_n]}(\mathbb{1} - C)x\| \\ &\leq \epsilon_n \, \|(\mathbb{1} - C)x\|.\end{aligned} \tag{7}$$

Then we obtain

$$e^{-n} \sum_{|m-n|\leq\epsilon_n} \frac{n^m}{m!} \|(C^n - C^m)x\| \leq n^{\delta+1/2} \, \|(\mathbb{1} - C)x\|,$$

for $n \in \mathbb{N}$. This last estimate together with (6) yield (4). □

Note that for $\delta = 0$ the estimate (4) gives for large n the same asymptotic as the Chernoff $\sqrt{n}$-lemma, whereas for optimal value $\delta = (-1/6)$ the asymptotic $2\sqrt[3]{n}$ is better than (1). We call this result the $\sqrt[3]{n}$-lemma.

Theorem 2.2. *Let $\Phi: t \mapsto \Phi(t)$ be a function from $\mathbb{R}^+$ to contractions on $\mathfrak{X}$ such that $\Phi(0) = \mathbb{1}$. Let $\{U_A(t)\}_{t\geq 0}$ be a contraction semigroup, and let $D \subset \operatorname{dom}(A)$ be a core of the generator A. If the function $\Phi(t)$ has a strong right-derivative $\Phi'(+0)$ at $t = 0$ and*

$$\Phi'(+0)u := \lim_{t\to+0} \frac{1}{t}(\Phi(t) - \mathbb{1})u = -Au,$$

for all $u \in D$, then

$$\lim_{n\to\infty} [\Phi(t/n)]^n\, x = U_A(t)\, x, \tag{8}$$

for all $t \in \mathbb{R}^+$ and $x \in \mathfrak{X}$

Proof. Consider the bounded approximation A_n of generator A:

$$A_n(s) := \frac{\mathbb{1} - \Phi(s/n)}{s/n}.$$

This operator is *accretive*: $(A_n(s) + \zeta\mathbb{1})^{-1} \in \mathcal{L}(\mathfrak{X})$ and $\|(A_n(s) + \zeta\mathbb{1})^{-1}\| \leq (\mathrm{Re}(\zeta))^{-1}$ for $\mathrm{Re}(\zeta) > 0$, and

$$\lim_{n\to\infty} A_n(s)\, u = A\, u,$$

for all $u \in D$ and for bounded s. Therefore, by virtue of the Trotter–Neveu–Kato generalised strong convergence theorem one gets:

$$\lim_{n\to\infty} e^{-t\, A_n(s)}\, x = U_A(t)\, x, \tag{9}$$

i.e., the strong and the uniform in t and s convergence (9) of the approximants $\{e^{-t\, A_n(s)}\}_{n\geq 1}$ for $s \in (0, s_0]$. By Lemma 2.1 for contraction $C := \Phi(t/n)$ and for $A_n(s)|_{s=t}$ we obtain

$$\begin{aligned} \|[\Phi(t/n)]^n\, x - e^{-t\, A_n(t)}\, x\| &= \|([\Phi(t/n)]^n - e^{n(\Phi(t/n)-\mathbb{1})})\, x\| \\ &\leq \frac{2}{n^{2\delta}}\|x\| + n^{\delta+1/2}\|(\mathbb{1} - \Phi(t/n))\, x\|. \end{aligned} \tag{10}$$

Since for any $u \in D$ and uniformly on $[0, t_0]$ one gets

$$\lim_{n\to\infty} n^{\delta+1/2}\|(\mathbb{1} - \Phi(t/n))\, u\| = \lim_{n\to\infty} t\, n^{\delta-1/2}\|A_n(t)\, u\| = 0, \tag{11}$$

for $\delta < 1/2$, equations (10) and (11) imply

$$\lim_{n\to\infty} \|[\Phi(t/n)]^n\, u - e^{-t\, A_n(t)}\, u\| = 0, \quad u \in D. \tag{12}$$

Then (9) and (12) together with estimate $\|[\Phi(t/n)]^n - e^{-t\, A_n(t)}\| \leq 2$ yield uniformly in $t \in [0, t_0]$:

$$\lim_{n\to\infty} [\Phi(t/n)]^n\, x = U_A(t)\, x,$$

which by density of D is extended to all $x \in \mathfrak{X}$, cf. (8). □

We call (8) the (strong) *Chernoff approximation formula* for the semigroup $\{U_A(t)\}_{t\geq 0}$.

Proposition 2.3 (Lie–Trotter product formula [3]). *Let A, B and C be generators of contraction semigroups on $\mathfrak{X}$. Suppose that algebraic sum*

$$Cu = Au + Bu, \tag{13}$$

is valid for all vectors u in a core $D \subset \operatorname{dom} C$. Then the semigroup $\{U_C(t)\}_{t\geq 0}$ can be approximated on $\mathfrak{X}$ in the strong operator topology (10) *by the Lie–Trotter product formula*:

$$e^{-tC}x = \lim_{n\to\infty}(e^{-tA/n}e^{-tB/n})^n\, x, \qquad x \in \mathfrak{X}, \tag{14}$$

for all $t \in \mathbb{R}^+$ and for $C := \overline{(A + B)}$, which is the closure of the algebraic sum (13).

Proof. Let us define the contraction $\mathbb{R}^+ \ni t \mapsto \Phi(t)$, $\Phi(0) = \mathbb{1}$, by

$$\Phi(t) := e^{-tA}e^{-tB}.$$

Note that if $u \in D$, then derivative

$$\Phi'(+0)u = \lim_{t\to+0}\frac{1}{t}(\Phi(t) - \mathbb{1})\, u = -(A + B)\, u.$$

Now we are in position to apply Theorem 2.2. This yields (14) for $C := \overline{(A + B)}$. □

Corollary 2.4. *Extension of the strong convergent Lie–Trotter product formula of Proposition 2.3 to quasi-bounded and holomorphic semigroups goes through verbatim.*

3 Quasi-sectorial contractions and $(\sqrt[3]{n})^{-1}$-theorem

Definition 3.1 ([2]). A contraction C on the Hilbert space $\mathfrak{H}$ is called *quasi-sectorial* with semi-angle $\alpha \in [0, \pi/2)$ with respect to the vertex at $z = 1$, if its numerical range $W(C) \subseteq D_\alpha$. Here

$$D_\alpha := \{z \in \mathbb{C}: |z| \leq \sin\alpha\} \cup \{z \in \mathbb{C}: |\arg(1 - z)| \leq \alpha \text{ and } |z - 1| \leq \cos\alpha\}.$$

We comment that $D_{\alpha=\pi/2} = \mathbb{D}$ (unit disc) and recall that a general contraction C verifies the weaker condition: $W(C) \subseteq \mathbb{D}$.

Note that if operator C is a quasi-sectorial contraction, then $\mathbb{1}-C$ is an m-sectorial operator with vertex $z=0$ and semi-angle α. Then for C the limits: $\alpha=0$ and $\alpha=\pi/2$, correspond respectively to self-adjoint and to standard contractions whereas for $\mathbb{1}-C$ they give a non-negative self-adjoint and an m-accretive (bounded) operators.

The resolvent of an m-sectorial operator A, with semi-angle $\alpha\in[0,\alpha_0]$, $\alpha_0<\pi/2$, and vertex at $z=0$, gives an example of the quasi-sectorial contraction.

Proposition 3.2 ([2] and [6]). *If C is a quasi-sectorial contraction on a Hilbert space $\mathfrak{H}$ with semi-angle $0\le\alpha<\pi/2$, then*

$$\|C^n(\mathbb{1}-C)\|\le\frac{K}{n+1},\qquad n\in\mathbb{N}. \tag{15}$$

The property (15) implies that the quasi-sectorial contractions belong to the class of the so-called *Ritt operators* [5]. This allows us to go beyond the $\sqrt[3]{n}$ -lemma 2.1 to the $(\sqrt[3]{n})^{-1}$-theorem.

Theorem 3.3 ($(\sqrt[3]{n})^{-1}$-theorem). *Let C be a quasi-sectorial contraction on $\mathfrak{H}$ with numerical range $W(C)\subseteq D_\alpha$, $0\le\alpha<\pi/2$. Then*

$$\|C^n-e^{n(C-\mathbb{1})}\|\le\frac{M}{n^{1/3}},\qquad n=1,2,3,\dots$$

where $M=2K+2$ and K is defined by (15).

Proof. Note that with help of inequality (15) we can improve the estimate (7) in Lemma 2.1:

$$\|C^n-C^m\|\le|m-n|\|C^{n-[\epsilon_n]}(\mathbb{1}-C)\|\le\epsilon_n\,\frac{K}{n-[\epsilon_n]+1},$$

for $\epsilon_n=n^{\delta+1/2}$. Then for $\delta<1/2$ there the above inequality together with (6) give instead of (4) (or (1)) the operator-norm estimate

$$\|C^n-e^{n(C-\mathbb{1})}\|\le\frac{2}{n^{2\delta}}+\frac{2K}{n^{1/2-\delta}},\qquad n\in\mathbb{N}. \tag{16}$$

Then the estimate $M/(n^{1/3})$ of the Theorem 3.3 results from the optimal choice of the value: $\delta=1/6$, in (16). □

Similar to $(\sqrt[3]{n})$-lemma, this $(\sqrt[3]{n})^{-1}$-theorem is only the first step in developing the *operator-norm* approximation formula à la Chernoff. To this end one needs an operator-norm analogue of Theorem 2.2. Since the last includes the Trotter–Neveu–Kato strong convergence theorem, we need the operator-norm extension of this assertion to quasi-sectorial contractions.

Proposition 3.4 ([2]). *Let $\{X(s)\}_{s>0}$ be a family of m-sectorial operators in a Hilbert space $\mathfrak{H}$ with $W(X(s)) \subseteq S_\alpha$ for some $0 < \alpha < \pi/2$ and for all $s > 0$. Let X_0 be an m-sectorial operator defined in a closed subspace $\mathfrak{H}_0 \subseteq \mathfrak{H}$, with $W(X_0) \subseteq S_\alpha$. Then the two following assertions are equivalent*:

(a) $\lim_{s\to+0} \|(\zeta\mathbb{1} + X(s))^{-1} - (\zeta\mathbb{1} + X_0)^{-1} P_0\| = 0$, *for* $\zeta \in S_{\pi-\alpha}$,

(b) $\lim_{s\to+0} \|e^{-tX(s)} - e^{-tX_0} P_0\| = 0$, *for* $t > 0$.

Here P_0 denotes the orthogonal projection from $\mathfrak{H}$ onto $\mathfrak{H}_0$.

Now $(\sqrt[3]{n})^{-1}$-theorem 3.3 and Proposition 3.4 yield a desired generalisation of the operator-norm approximation formula:

Proposition 3.5 ([2]). *Let $\{\Phi(s)\}_{s\geq 0}$ be a family of uniformly quasi-sectorial contractions on a Hilbert space $\mathfrak{H}$, i.e. such that there exists $0 \leq \alpha < \pi/2$ and $W(\Phi(s)) \subset D_\alpha$, for all $s \geq 0$. Let*

$$X(s) := \frac{\mathbb{1} - \Phi(s)}{s},$$

and let X_0 be a closed operator with non-empty resolvent set, defined in a closed subspace $\mathfrak{H}_0 \subseteq \mathfrak{H}$. Then the family $\{X(s)\}_{s>0}$ converges, when $s \to +0$, in the uniform resolvent sense to the operator X_0 if and only if

$$\lim_{n\to\infty} \|\Phi(t/n)^n - e^{-tX_0} P_0\| = 0, \quad \textit{for } t > 0.$$

Here P_0 denotes the orthogonal projection onto the subspace $\mathfrak{H}_0$.

Let A be an m-sectorial operator with semi-angle $0 < \alpha < \pi/2$ and with vertex at 0, which means that numerical range $W(A) \subseteq S_\alpha = \{z \in \mathbb{C} : |\arg(z)| \leq \alpha\}$. Then $\{\Phi(t) := (\mathbb{1} + tA)^{-1}\}_{t\geq 0}$ is the family of quasi-sectorial contractions, i.e., $W(\Phi(t)) \subseteq D_\alpha$. Let $X(s) := (\mathbb{1}-\Phi(s))/s$, $s > 0$, and $X_0 := A$. Then $X(s)$ converges, when $s \to +0$, to X_0 in the uniform resolvent sense with the asymptotic

$$\|(\zeta\mathbb{1} + X(s))^{-1} - (\zeta\mathbb{1} + X_0)^{-1}\| = s \left\| \frac{A}{\zeta\mathbb{1} + A + \zeta s A} \cdot \frac{A}{\zeta\mathbb{1} + A} \right\| = O(s),$$

for any $\zeta \in S_{\pi-\alpha}$, since we have the estimate

$$\left\| \frac{A}{\zeta\mathbb{1} + A + \zeta s A} \cdot \frac{A}{\zeta\mathbb{1} + A} \right\| \leq \left(1 + \frac{|\zeta|}{\operatorname{dist}(\zeta(1 + s\zeta)^{-1}, -S_\alpha)}\right)\left(1 + \frac{|\zeta|}{\operatorname{dist}(\zeta, -S_\alpha)}\right).$$

Therefore, the family $\{\Phi(t)\}_{t\geq 0}$ satisfies the conditions of Proposition 3.5. This implies the operator-norm approximation of the exponential function, i.e., the semi-group for m-sectorial generator, by the powers of resolvent (*the Euler approximation formula*):

Corollary 3.6. *If A is an m-sectorial operator in a Hilbert space $\mathfrak{H}$, with semi-angle $\alpha \in (0, \pi/2)$ and with vertex at 0, then*

$$\lim_{n\to\infty} \|(\mathbb{1} + tA/n)^{-n} - e^{-tA}\| = 0, \quad t \in S_{\pi/2-\alpha}. \tag{17}$$

4 Conclusion

Summarising we note that for the quasi-sectorial contractions instead of *divergent* Chernoff's estimate (1) we find the estimate (16), which converges for $n \to \infty$ to zero in the *operator-norm* topology. Note that the rate $O(1/(n^{1/3}))$ of this convergence is obtained with help of the Poisson representation and the Tchebychev inequality in the spirit of the proof of Lemma 2.1, and that it is not optimal.

The estimate $M/(n^{1/3})$ in the $(\sqrt[3]{n})^{-1}$-theorem 3.3 can be improved by a more refined lines of reasoning.

For example, by scrutinising our probabilistic arguments one can find a more precise Tchebychev-type bound for the tail probabilities. This improves the estimate (16) to the rate $O(\sqrt{\ln(n)/n})$, see [4].

On the other hand, a careful analysis of localisation the numerical range of quasi-sectorial contractions (see [6] and [1]) allows us to lift the estimate in Theorem 3.3 and in Corollary 3.6 to the ultimate optimal rate $O(1/n)$.

Note that the optimal estimate $O(1/n)$ in (16) one can easily obtain with help of the spectral representation for a particular case of the *self-adjoint* quasi-sectorial contractions, i.e., for $\alpha = 0$. This also concerns the optimal $O(1/n)$ rate of convergence of the self-adjoint Euler approximation formula (17).

Acknowledgements. The author is grateful to Mathematical Department of the University of Auckland (New Zealand) for hospitality during the writing this paper. This visit was supported by the EU Marie Curie IRSES program, project 'AOS', No. 318910.

References

[1] Yu. Arlinskiĭ and V. Zagrebnov, Numerical range and quasi-sectorial contractions. *J. Math. Anal. Appl.* **366** (2010), no. 1, 33–43. MR 2593631 Zbl 1189.47007

[2] V. Cachia and V. A. Zagrebnov, Operator-norm approximation of semigroups by quasi-sectorial contractions. *J. Funct. Anal.* **180** (2001), no. 1, 176–194. MR 1814426 Zbl 0981.47015

[3] P. R. Chernoff, Product formulas, nonlinear semigroups and addition of unbounded operators. Memoirs of the American Mathematical Society, 140. American Mathematical Society, Providence, R.I., 1974. MR 0417851 Zbl 0283.47041

[4] V. Paulauskas, On operator-norm approximation of some semigroups by quasi-sectorial operators. *J. Funct. Anal.* **207** (2004), no. 1, 58–67. MR 2027636 Zbl 1068.47047

[5] R. K. Ritt, A condition that $\lim_{n\to\infty} T^n = 0$. *Proc. Amer. Math. Soc.* **4** (1953), 898–899. MR 0059471 Zbl 0052.12501

[6] V. A. Zagrebnov, Quasi-sectorial contractions. *J. Funct. Anal.* **254** (2008), no. 9, 2503–2511. MR 2409171 Zbl 1146.47028

List of contributors

Sergio Albeverio, 1

Institut für Angewandte Mathematik, Universität Bonn, Endenicher Allee 60, 53115 Bonn, Germany;

SFB611, Bonn; HCM, Bonn; BIBOS, Bielefeld and Bonn; IZKS, Bonn;

CERFIM, Locarno; Acc. Arch., USI, Mendrisio;

Dipartimento Matematica, Università di Trento

albeverio@uni-bonn.de

Asao Arai, 31

Department of Mathematics, Hokkaido University, Kita 10–Nishi 8, Sapporo 060-0810, Japan

arai@math.sci.hokudai.ac.jp

Diana Barseghyan, 55

Department of Mathematics, Faculty of Science, University of Ostrava, 30. dubna 22, 70103 Ostrava, Czech Republic

Department of Theoretical Physics, Nuclear Physics Institute ASCR, 25068 Řež near Prague, Czech Republic

diana.barseghyan@osu.cz

Giulia Basti, 71

Dipartimento di Matematica, Sapienza Università di Roma, Piazzale A. Moro 5, 00185 Roma, Italy,

basti@mat.uniroma1.it

Jussi Behrndt, 95, 129

Institut für Numerische Mathematik, Technische Universität Graz, Steyrergasse 30, 8010 Graz, Austria

behrndt@tugraz.at

Rafael D. Benguria, 153

Instituto de Física, Pontificia Universidad Católica de Chile, Avda. Vicuña Mackenna 4860, Santiago, Chile

rbenguri@fis.puc.cl

Soledad Benguria, 153

Mathematics Department, University of Wisconsin–Madison, 480 Lincoln Dr, Madison, WI 53706, USA

benguria@math.wisc.edu

Philippe Briet, 161

Aix-Marseille Université, CNRS, CPT, UMR 7332, Case 907, 13288 Marseille, France

Université de Toulon, CNRS, CPT, UMR 7332, 83957 La Garde, France

briet@univ-tln.fr

Claudio Cacciapuoti, 1

Dipartimento di Scienza e Alta Tecnologia, Università dell'Insubria,
via Valleggio 11, 22100 Como, Italy

claudio.cacciapuoti@uninsubria.it

Giuseppe Cardone, 177

Department of Engineering, University of Sannio, Corso Garibaldi 107,
82100 Benevento, Italy

giuseppe.cardone@unisannio.it

Raffaele Carlone, 189

Università "Federico II" di Napoli,
Dipartimento di Matematica e Applicazioni "R. Caccioppoli",
MSA, via Cinthia, 80126 Napoli, Italy

raffaele.carlone@unina.it

Taksu Cheon, 531

Laboratory of Physics, Kochi University of Technology, 782-8502 Tosa Yamada, Japan

taksu.cheon@kochi-tech.ac.jp

Michele Correggi, 189

Dipartimento di Matematica e Fisica, Università degli Studi Roma Tre,
Largo San Leonardo Murialdo 1, 00146 Roma, Italy

michele.correggi@gmail.com

Ngoc T. Do, 213

Department of Mathematics, Mailstop 3368, Texas A&M University, College Station, TX 77843-3368, USA

dothanh@math.tamu.edu

Rodolfo Figari, 189

Università "Federico II" di Napoli, Dipartimento di Fisica e INFN Sezione di Napoli, MSA, via Cinthia, 80126 Napoli, Italy

rodolfo.figari@na.infn.it

Martin Fraas, 223

980 Ohlone Ave., Albany, CA 94706, USA

martin.fraas@gmail.com

Rupert L. Frank, 245

Mathematics 253-37, Caltech, Pasadena, CA 91125, USA

rlfrank@caltech.edu

Pedro Freitas, 261

Departamento de Matemática, Instituto Superior Técnico, Universidade de Lisboa, Av. Rovisco Pais, 1049-001 Lisboa, Portugal

Grupo de Física Matemática, Faculdade de Ciências, Universidade de Lisboa, Campo Grande, Edifício C6, 1749-016 Lisboa, Portugal

psfreitas@fc.ul.pt

Fritz Gesztesy, 95

Department of Mathematics, University of Missouri, Columbia, MO 65211, USA

PRESENT ADDRESS: Department of Mathematics, Baylor University, One Bear Place #97328, Waco, TX 76798-7328, USA

Fritz_Gesztesy@baylor.edu

Yaroslav Granovskyi, 271

Institute of Applied Mathematics and Mechanics, National Academy of Science of Ukraine, Dobrovol's'kogo Str. 1, 84100 Slavyansk, Ukraine

yarvodoley@mail.ru

Hiba Hammedi, 161

Aix-Marseille Université, CNRS, CPT, UMR 7332, Case 907, 13288 Marseille, France

Université de Toulon, CNRS, CPT, UMR 7332, 83957 La Garde, France

hammedi@cpt.univ-mrs.fr

Hynek Kovařík, 433

DICATAM, Sezione di Matematica, Università degli studi di Brescia, Via Branze 38, 25123 Brescia, Italy

hynek.kovarik@ing.unibs.it

David Krejčiřík, 261

Department of Mathematics, Faculty of Nuclear Sciences and Physical Engineering, Czech Technical University in Prague, Trojanova 13, 12000 Prague 2, Czech Republic

david.krejcirik@fjfi.cvut.cz

Peter Kuchment, 213

Department of Mathematics, Mailstop 3368, Texas A&M University, College Station, TX 77843-3368, USA

kuchment@math.tamu.edu

Matthias Langer, 129

Department of Mathematics and Statistics, University of Strathclyde, 26 Richmond Street, Glasgow G1 1XH, United Kingdom

m.langer@strath.ac.uk

Vladimir Lotoreichik, 129

Department of Theoretical Physics, Nuclear Physics Institute, Czech Academy of Sciences, 250 68 Řež near Prague, Czech Republic

lotoreichik@ujf.cas.cz

Mark Malamud, 271

Institute of Applied Mathematics and Mechanics, National Academy of Science of Ukraine, Dobrovol's'kogo Str. 1, 84100 Slavyansk, Ukraine

malamud3m@gmail.com

Roger Nichols, 95

Mathematics Department, The University of Tennessee at Chattanooga, 415 EMCS Building, Dept. 6956, 615 McCallie Ave, Chattanooga, TN 37403, USA

Roger-Nichols@utc.edu

Hagen Neidhardt, 271

Weierstrass Institute for Applied Analysis and Stochastics, Mohrenstr. 39,
D-10117 Berlin, Germany,

hagen.neidhardt@wias-berlin.de

Beng Ong, 213

CGG, 10300 Town Park Dr, Houston, TX 77072, USA

Beng.Ong@cgg.com

Konstantin Pankrashkin, 447

Laboratoire de Mathématiques d'Orsay,
Université Paris-Sud, CNRS, Université Paris-Saclay, 91405 Orsay, France

konstantin.pankrashkin@math.u-psud.fr

Yehuda Pinchover, 459

Department of Mathematics, Technion, Israel Institute of Technology, Haifa 32000, Israel

pincho@techunix.technion.ac.il

Andrea Posilicano, 271

Sezione di Matematica, Università dell'Insubria, via Valleggio 11, I-22100 Como, Italy

posilicano@uninsubria.it

Olaf Post, 491

Fachbereich 4 – Mathematik, Universität Trier, 54286 Trier, Germany

olaf.post@uni-trier.de

Olga Rossi, 55

Department of Mathematics, Faculty of Science, University of Ostrava,
30. dubna 22, 70103 Ostrava, Czech Republic

Department of Mathematics, Ghent University, Belgium

olga.rossi@osu.cz

Bartosch Ruszkowski, 433

Institut für Analysis, Dynamik und Modellierung, Universität Stuttgart,
Pfaffenwaldring 57, 70569 Stuttgart, Germany

Bartosch.Ruszkowski@mathematik.uni-stuttgart.de